Linear Statistical Models

THE WILEY SERIES IN PROBABILITY AND STATISTICS

Established by WALTER A. SHEWHART and SAMUEL S. WILKS

A complete list of the titles in this series appears at the end of this volume

Linear Statistical Models

JAMES H. STAPLETON
Michigan State University

A Wiley-Interscience Publication
JOHN WILEY & SONS, INC.
New York • Chichester • Brisbane • Toronto • Singapore

This text is printed on acid-free paper.

Library of Congress Cataloging in Publication Data:

Stapleton, James, 1931–
Linear statistical models / James Stapleton.
p. cm.—(Wiley series in probability and statistics. Probability and statistics section)
"A Wiley–Interscience publication."
ISBN 0-471-57150-4 (acid-free)
1. Linear models (Statistics) I. Title. II. Series. III. Series: Wiley series in probability and statistics. Probability and statistics.
QA279.S695 1995
519.5′38—dc20 94-39384

Printed in the United States of America

10 9 8 7 6 5 4 3 2 1

To Alicia, who, through all the years, never expressed
a doubt that this would someday be completed,
despite the many doubts of the author.

Contents

Preface

The first seven chapters of this book were developed over a period of about 20 years for the course Linear Statistical Models at Michigan State University. They were first distributed in longhand (those former students may still be suffering the consequences), then typed using a word processor some eight or nine years ago. The last chapter, on frequency data, is the result of a summer course, offered every three or four years since 1980.

Linear statistical models are mathematical models which are linear in the unknown parameters, and which include a random error term. It is this error term which makes the models statistical. These models lead to the methodology usually called *multiple regression* or *analysis of variance*, and have wide applicability to the physical, biological, and social sciences, to agriculture and business, and to engineering.

The linearity makes it possible to study these models from a vector space point of view. The vectors **Y** of observations are represented as arrays written in a form convenient for intuition, rather than necessarily as column or row vectors. The geometry of these vector spaces has been emphasized because the author has found that the intuition it provides is vital to the understanding of the theory. Pictures of the vectors spaces have been added for their intuitive value. In the author's opinion this geometric viewpoint has not been sufficiently exploited in current textbooks, though it is well understood by those doing research in the field. For a brief discussion of the history of these ideas see Herr (1980).

Bold print is used to denote vectors, as well as linear transformations. The author has found it useful for classroom boardwork to use an arrow notation above the symbol to distinguish vectors, and to encourage students to do the same, at least in the earlier part of the course.

Students studying these notes should have had a one-year course in probability and statistics at the post-calculus level, plus one course on linear algebra. The author has found that most such students can handle the matrix algebra used here, but need the material on inner products and orthogonal projections introduced in Chapter 1.

Chapter 1 provides examples and introduces the linear algebra necessary for later chapters. One section is devoted to a brief history of the early development of least squares theory, much of it written by Stephen Stigler (1986).

Chapter 2 is devoted to methods of study of random vectors. The multivariate normal, chi-square, t and F distributions, central and noncentral, are introduced.

Chapter 3 then discusses the linear model, and presents the basic theory necessary to regression analysis and the analysis of variance, including confidence intervals, the Gauss–Markov Theorem, power, and multiple and partial correlation coefficients. It concludes with a study of a SAS multiple regression printout.

Chapter 4 is devoted to a more detailed study of multiple regression methods, including sections on transformations, analysis of residuals, and on asymptotic theory. The last two sections are devoted to robust methods and to the bootstrap. Much of this methodology has been developed over the last 15 years and is a very active topic of research.

Chapter 5 discusses simultaneous confidence intervals: Bonferroni, Scheffé, Tukey, and Bechhofer.

Chapter 6 turns to the analysis of variance, with two- and three-way analyses of variance. The geometric point of view is emphasized.

Chapter 7 considers some miscellaneous topics, including random component models, nested designs, and partially balanced incomplete block designs.

Chapter 8, the longest, discusses the analysis of frequency, or categorical data. Though these methods differ significantly in the distributional assumptions of the models, it depends strongly on the linear representations, common to the theory of the first seven chapters.

Computations illustrating the theory were done using APL*Plus (Magnugistics, Inc.), S-Plus (Statistical Sciences, Inc.), and SAS (SAS Institute, Inc.). Graphics were done using S-Plus.). To perform simulations, and to produce graphical displays, the author recommends that the reader use a mathematical language which makes it easy to manipulate vectors and matrices.

For the linear models course the author teaches at Michigan State University only Section 2.3, Projections of Random Variables, and Section 3.9, Further Decomposition of Subspaces, are omitted from Chapters 1, 2, and 3. From Chapter 4 only Section 4.1, Linearizing Transformations, and one or two other sections are usually discussed. From Chapter 5 the Bonferroni, Tukey, and Scheffé simultaneous confidence interval methods are covered. From Chapter 6 only the material on the analysis of covariance (Section 6.6) is omitted, though relatively little time is devoted to three-way analysis of variance (Section 6.5). One or two sections of Chapter 7, Miscellaneous Other Models, are usually chosen for discussion. Students are introduced to S-Plus early in the semester, then use it for the remainder of the semester for numerical work.

A course on the analysis of frequency data could be built on Sections 1.1, 1.2, 1.3, 2.1, 2.2, 2.3, 2.4 (if students have not already studied these topics), and, of course, Chapter 8.

The author thanks Virgil Anderson, retired professor from Purdue University, now a statistical consultant, from whom he first learned of the analysis of variance and the design of experiments. He also thanks Professor James Hannan, from whom he first learned of the geometric point of view, and Professors Vaclav Fabian and Dennis Gilliland for many valuable conversations. He is grateful to Sharon Carson and to Loretta Ferguson, who showed much patience as they typed several early versions. Finally, he thanks the students who have tried to read the material, and who found (he hopes) a large percentage of the errors in those versions.

JAMES STAPLETON

CHAPTER 1

Linear Algebra, Projections

1.1 INTRODUCTION

Suppose that each element of a population possesses a numerical characteristic x, and another numerical characteristic y. It is often desirable to study the relationship between two such variables x and y in order to better understand how values of x affect y, or to predict y, given the value of x. For example, we may wish to know the effect of amount x of fertilizer per square meter on the yield y of a crop in pounds per square meter. Or we might like to know the relationship between a man's height y and that of his father x.

For each value of the independent variable x, the dependent variable Y may be supposed to have a probability distribution with mean $g(x)$. Thus, for example, $g(0.9)$ is the expected yield of a crop using fertilizer level $x = 0.9$ (kgms/m^2).

Definition 1.1.1: For each $x \in D$ suppose Y is a random variable with distribution depending on x. Then

$$g(x) = E(Y|x) \qquad \text{for} \quad x \in D$$

is the regression function for Y on x.

Often the domain D will be a subset of the real line, or even the whole real line. However, D could also be a finite set, say $\{1, 2, 3\}$, or a countably infinite set $\{1, 2, \ldots\}$. The experimenter or statistician would like to determine the function g, using sample data consisting of pairs (x_i, y_i) for $i = 1, \ldots, n$. Unfortunately, the number of possible functions $g(x)$ is so large that in order to make headway certain simplifying models for the form of $g(x)$ must be adopted. If it is supposed that $g(x)$ is of the form $g(x) = A + Bx + Cx^2$ or $g(x) = A2^x + B$ or $g(x) = A \log x + B$, etc., then the problem is reduced to one of identifying a few parameters, here labeled as A, B, C. In each of the three forms for $g(x)$ given above, g is linear in these parameters.

In one of the simplest cases we might consider a model for which $g(x) = C + Dx$, where C and D are unknown parameters. The problem of estimating

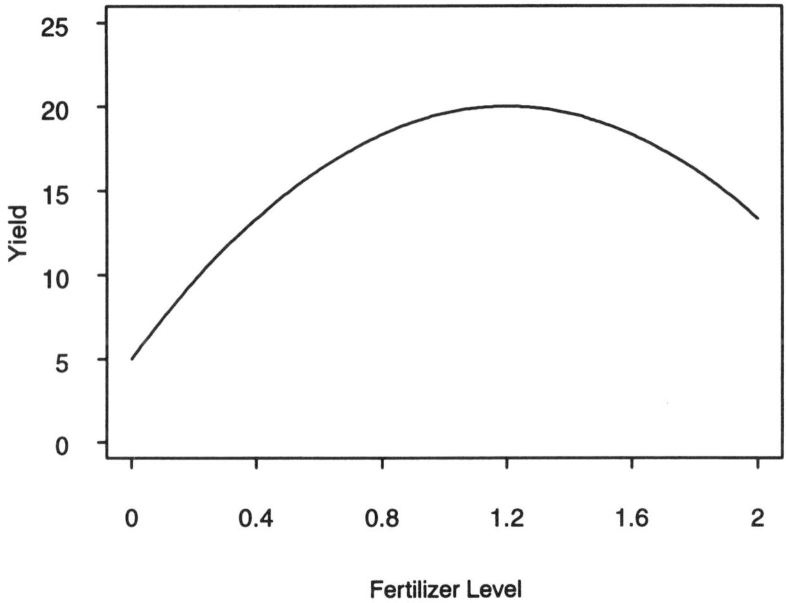

FIGURE 1.1 Regression of yield on fertilizer level.

$g(x)$ then becomes the simpler one of estimating the two parameters C and D. This model may not be a good approximation of the true regression function, and, if possible, should be checked for validity. The crop yield as a function of fertilizer level may well have the form in Figure 1.1.

The regression function g would be better approximated by a second degree polynomial $g(x) = A + Bx + Cx^2$. However, if attention is confined to the 0.7 to 1.3 range, the regression function is approximately linear, and the simplifying model $g(x) = C + Dx$, called the simple linear regression model, may be used.

In attempting to understand the relationship between a person's height Y and the heights of his/her father (x_1) and mother (x_2) and the person's sex (x_3), we might suppose

$$E(Y|x_1, x_2, x_3) = g(x_1, x_2, x_3) = \beta_0 + \beta_1 x_1 + \beta_2 x_2 + \beta_3 x_3, \quad (1.1.1)$$

where x_3 is 1 for males, 0 for females, and β_0, β_1, β_2, β_3 are unknown parameters. Thus a brother would be expected to be β_3 taller than his sister. Again, this model, called a *multiple regression model*, can only be an approximation of the true regression function, valid over a limited range of values of x_1, x_2. A more complex model might suppose

$$g(x_1, x_2, x_3) = \beta_0 + \beta_1 x_1 + \beta_2 x_2 + \beta_3 x_3 + \beta_4 x_1^2 + \beta_5 x_2^2 + \beta_6 x_1 x_2.$$

Table 1.1.1 Height Data

Indiv.	Y	x_1	x_2	x_3
1	68.5	70	62	1
2	72.5	73	66	1
3	70.0	68	67	1
4	71.0	72	64	1
5	65.0	66	60	1
6	64.5	71	63	0
7	67.5	74	68	0
8	61.5	65	65	0
9	63.5	70	64	0
10	63.5	69	65	0

This model is nonlinear in (x_1, x_2, x_3), but linear in the β's. It is the linearity in the β's which makes this model a *linear* statistical model.

Consider the model (1.1.1), and suppose we have data of Table 1.1.1 on (Y, x_1, x_2, x_3) for 10 individuals. These data were collected in a class taught by the author. Perhaps the student can collect similar data in his or her class and compare results.

The statistical problem is to determine estimates $\hat{\beta}_0, \hat{\beta}_1, \hat{\beta}_2, \hat{\beta}_3$ so that the resulting function $\hat{g}(x_1, x_2, x_3) = \hat{\beta}_0 + \hat{\beta}_1 x_1 + \hat{\beta}_2 x_2 + \hat{\beta}_3 x_3$ is in some sense a good approximation of $g(x_1, x_2, x_3)$. For this purpose it is convenient to write the model in vector form:

$$E(\mathbf{Y}) = \beta_0 \mathbf{x}_0 + \beta_1 \mathbf{x}_1 + \beta_2 \mathbf{x}_2 + \beta_3 \mathbf{x}_3,$$

where $\mathbf{x}_0$ is the vector of all ones, and $\mathbf{y}$ and $\mathbf{x}_1, \mathbf{x}_2, \mathbf{x}_3$ are the column vectors in Table 1.1.1.

This formulation of the model suggests that linear algebra may be an important tool in the analysis of linear statistical models. We will therefore review such material in the next section, emphasizing geometric aspects.

1.2 VECTORS, INNER PRODUCTS, LENGTHS

Let Ω be the collection of all n-tuples of real numbers for a positive integer n. In applications Ω will be the sample space of all possible values of the observation vector $\mathbf{y}$. Though Ω will be in one-to-one correspondence to Euclidean n-space, it will be convenient to consider elements of Ω as arrays all of the same configuration, not necessarily column or row vectors. For example, in application to what is usually called one-way analysis of variance, we might have 3, 4 and 2 observations on three different levels of some treatment effect. Then we might take

$$\mathbf{y} = \begin{bmatrix} y_{11} & y_{12} & y_{13} \\ y_{21} & y_{22} & y_{23} \\ y_{31} & y_{32} & \\ & y_{42} & \end{bmatrix}$$

and Ω the collection of all such $\mathbf{y}$. While we could easily reform $\mathbf{y}$ into a column vector, it is often convenient to preserve the form of $\mathbf{y}$. The term "n-tuple" means that the elements of a vector $\mathbf{y} \in \Omega$ are ordered. A vector $\mathbf{y}$ may be considered to be a real-valued function on $\{1, \ldots, n\}$.

Ω becomes a linear space if we define $a\mathbf{y}$ for any $\mathbf{y} \in \Omega$ and any real number a to be the element of Ω given by multiplying each component of Ω by a, and if for any two elements $\mathbf{y}_1, \mathbf{y}_2 \in \Omega$ we define $\mathbf{y}_1 + \mathbf{y}_2$ to be the vector in Ω whose ith component is the sum of the ith components of $\mathbf{y}_1$ and $\mathbf{y}_2$, for $i = 1, \ldots, n$.

Ω becomes an inner product space if for each $\mathbf{x}, \mathbf{y} \in \Omega$ we define the function

$$h(\mathbf{x}, \mathbf{y}) = \sum_1^n x_i y_i,$$

where $\mathbf{x} = (x_1, \ldots, x_n)$ and $\mathbf{y} = (y_1, \ldots, y_n)$. If Ω is the collection of n-dimensional column vectors then $h(\mathbf{x}, \mathbf{y}) = \mathbf{x}'\mathbf{y}$, in matrix notation. The inner product $h(\mathbf{x}, \mathbf{y})$ is usually written simply as $(\mathbf{x}, \mathbf{y})$, and we will use this notation. The inner product is often called the dot product, written in the form $\mathbf{x} \cdot \mathbf{y}$. Since there is a small danger of confusion with the pair $(\mathbf{x}, \mathbf{y})$, we will use bold parentheses to emphasize that we mean the inner product. Since bold symbols are not easily indicated on a chalkboard or in student notes, it is important that the meaning will almost always be clear from the context. The inner product has the properties:

$$(\mathbf{x}, \mathbf{y}) = (\mathbf{y}, \mathbf{x})$$
$$(a\mathbf{x}, \mathbf{y}) = a(\mathbf{x}, \mathbf{y})$$
$$(\mathbf{x}_1 + \mathbf{x}_2, \mathbf{y}) = (\mathbf{x}_1, \mathbf{y}) + (\mathbf{x}_2, \mathbf{y})$$

for all vectors, and real numbers a.

We define $\|\mathbf{x}\|^2 = (\mathbf{x}, \mathbf{x})$ and call $\|\mathbf{x}\|$ the (Euclidean) *length of* $\mathbf{x}$. Thus $\mathbf{x} = (3, 4, 12)$ has length 13.

The *distance* between vectors $\mathbf{x}$ and $\mathbf{y}$ is the length of $\mathbf{x} - \mathbf{y}$. Vectors $\mathbf{x}$ and $\mathbf{y}$ are said to be *orthogonal* if $(\mathbf{x}, \mathbf{y}) = 0$. We write $\mathbf{x} \perp \mathbf{y}$.

For example, if the sample space is the collection of arrays mentioned above, then

$$\mathbf{x} = \begin{bmatrix} 1 & 0 & 0 \\ 2 & 0 & 0 \\ 3 & 0 & \\ & 0 & \end{bmatrix} \quad \text{and} \quad \mathbf{y} = \begin{bmatrix} 0 & 1 & 0 \\ 0 & 3 & 0 \\ 0 & 5 & \\ & -1 & \end{bmatrix}$$

are orthogonal, with squared lengths 14 and 36. For Ω the collection of 3-tuples, $(2, 3, 1) \perp (-1, 1, -1)$.

The following theorem is perhaps the most important of the entire book. We credit it to Pythagorus (sixth century B.C.), though he would not, of course, have recognized it in this form.

Pythagorean Theorem: Let $\mathbf{v}_1, \ldots, \mathbf{v}_k$ be mutually orthogonal vectors in Ω. Then

$$\left\| \sum_1^k \mathbf{v}_i \right\|^2 = \sum_1^k \|\mathbf{v}_i\|^2$$

Proof:
$$\left\| \sum_1^k \mathbf{v}_i \right\|^2 = \left(\sum_{i=1}^k \mathbf{v}_i, \sum_{j=1}^k \mathbf{v}_j \right) = \sum_{i=1}^k \sum_{j=1}^k (\mathbf{v}_i, \mathbf{v}_j) = \sum_{i=1}^k (\mathbf{v}_i, \mathbf{v}_i)$$
$$= \sum_{i=1}^k \|\mathbf{v}_i\|^2. \qquad \square$$

Definition 1.2.1: The *projection* of a vector $\mathbf{y}$ on a vector $\mathbf{x}$ is the vector $\hat{\mathbf{y}}$ such that

1. $\hat{\mathbf{y}} = b\mathbf{x}$ for some constant b
2. $(\mathbf{y} - \hat{\mathbf{y}}) \perp \mathbf{x}$ (equivalently, $(\hat{\mathbf{y}}, \mathbf{x}) = (\mathbf{y}, \mathbf{x})$)

Equivalently, $\hat{\mathbf{y}}$ is the projection of $\mathbf{y}$ on the subspace of all vectors of the form $a\mathbf{x}$, the subspace spanned by $\mathbf{x}$ (Figure 1.2). To be more precise, these properties define othogonal projection. We will use the word projection to mean orthogonal projection. We write $p(\mathbf{y}|\mathbf{x})$ to denote this projection. Students should not confuse this will conditional probability.

Let us try to find the constant b. We need $(\hat{\mathbf{y}}, \mathbf{x}) = (b\mathbf{x}, \mathbf{x}) = b(\mathbf{x}, \mathbf{x}) = (\mathbf{y}, \mathbf{x})$. Hence, if $\mathbf{x} = \mathbf{0}$, any b will do. Otherwise, $b = (\mathbf{y}, \mathbf{x})/\|\mathbf{x}\|^2$. Thus,

$$\hat{\mathbf{y}} = \begin{cases} \mathbf{0} & \text{for} \quad \mathbf{x} = 0 \\ [(\mathbf{y}, \mathbf{x})/\|\mathbf{x}\|^2]\mathbf{x}, & \text{otherwise} \end{cases}$$

Here $\mathbf{0}$ is the vector of all zeros. Note that if $\mathbf{x}$ is replaced by a multiple $a\mathbf{x}$ of $\mathbf{x}$, for $a \neq 0$ then $\hat{\mathbf{y}}$ remains the same though the coefficient b is replaced by b/a.

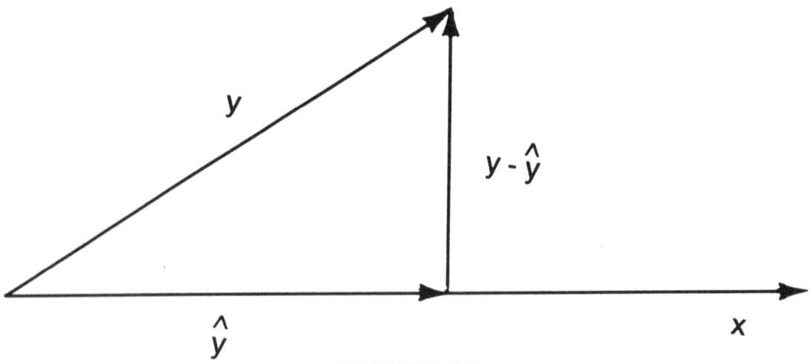

FIGURE 1.2

Example 1.2.1: Let $\mathbf{x} = \begin{pmatrix} 1 \\ -2 \\ 1 \end{pmatrix}$, $\mathbf{y} = \begin{pmatrix} 1 \\ -6 \\ 5 \end{pmatrix}$. Then $(\mathbf{x}, \mathbf{y}) = 18$, $\|\mathbf{x}\|^2 = 6$, $b = 18/6 = 3$, $\hat{\mathbf{y}} = 3\mathbf{x} = \begin{pmatrix} 3 \\ -6 \\ 3 \end{pmatrix}$, $\mathbf{y} - \hat{\mathbf{y}} = \begin{pmatrix} -2 \\ 0 \\ 2 \end{pmatrix} \perp \mathbf{x}$.

Theorem 1.2.1: Among all multiples $a\mathbf{x}$ of $\mathbf{x}$, the projection $\hat{\mathbf{y}}$ of $\mathbf{y}$ on $\mathbf{x}$ is the closest vector to $\mathbf{y}$.

***Proof*:** Since $(\mathbf{y} - \hat{\mathbf{y}}) \perp (\hat{\mathbf{y}} - a\mathbf{x})$ and $(\mathbf{y} - a\mathbf{x}) = (\mathbf{y} - \hat{\mathbf{y}}) + (\hat{\mathbf{y}} - a\mathbf{x})$, it follows that

$$\|\mathbf{y} - a\mathbf{x}\|^2 = \|\mathbf{y} - \hat{\mathbf{y}}\|^2 + \|\hat{\mathbf{y}} - a\mathbf{x}\|^2.$$

This is obviously minimum for $a\mathbf{x} = \hat{\mathbf{y}}$. □

Since $\hat{\mathbf{y}} \perp (\mathbf{y} - \hat{\mathbf{y}})$ and $\mathbf{y} = \hat{\mathbf{y}} + (\mathbf{y} - \hat{\mathbf{y}})$, the Pythagorean Theorem implies that $\|\mathbf{y}\|^2 = \|\hat{\mathbf{y}}\|^2 + \|\mathbf{y} - \hat{\mathbf{y}}\|^2$. Since $\|\hat{\mathbf{y}}\|^2 = b^2\|\mathbf{x}\|^2 = (\mathbf{y}, \mathbf{x})^2/\|\mathbf{x}\|^2$, this implies that $\|\mathbf{y}\|^2 \geq (\mathbf{y}, \mathbf{x})^2/\|\mathbf{x}\|^2$, with equality if and only if $\|\mathbf{y} - \hat{\mathbf{y}}\| = 0$, i.e., $\mathbf{y}$ is a multiple of $\mathbf{x}$. This is the famous *Cauchy–Schwarz Inequality*, usually written as $(\mathbf{y}, \mathbf{x})^2 \leq \|\mathbf{y}\|^2\|\mathbf{x}\|^2$. The inequality is best understood as the result of the equality implied by the Pythagorean Theorem.

Definition 1.2.2: Let A be a subset of the indices of the components of a vector space Ω. The indicator of A is the vector $\mathbf{I}_A \in \Omega$, with components which are 1 for indices in A, and 0 otherwise.

The *projection* $\hat{\mathbf{y}}_A$ of $\mathbf{y}$ on the vector $\mathbf{I}_A$ is therefore $b\mathbf{I}_A$ for $b = (\mathbf{y}, \mathbf{I}_A)/\|\mathbf{I}_A\|^2 = \left(\sum_{i \in A} y_i\right)/N(A)$, where $N(A)$ is the number of indices in A. Thus, $b = \bar{y}_A$, the mean of the y-values with components in A. For example, if Ω is the space of 4-component row vectors, $\mathbf{y} = (3, 7, 8, 13)$, and A is the indicator of the second and fourth components, $p(\mathbf{y}|\mathbf{I}_A) = (0, 10, 0, 10)$.

Problem 1.2.1: Let Ω be the collection of all 5-tuples of the form $\mathbf{y} = \begin{pmatrix} y_{11} & y_{21} & \\ y_{12} & y_{22} & y_{31} \end{pmatrix}$. Let $\mathbf{x} = \begin{pmatrix} 1 & 0 & \\ 2 & 1 & 3 \end{pmatrix}$, $\mathbf{y} = \begin{pmatrix} 5 & 1 & \\ 9 & 4 & 11 \end{pmatrix}$.

(a) Find $(\mathbf{x}, \mathbf{y})$, $\|\mathbf{x}\|^2$, $\|\mathbf{y}\|^2$, $\hat{\mathbf{y}} = p(\mathbf{y}|\mathbf{x})$, and $\mathbf{y} - \hat{\mathbf{y}}$. Show that $\mathbf{x} \perp (\mathbf{y} - \hat{\mathbf{y}})$, and $\|\mathbf{y}\|^2 = \|\hat{\mathbf{y}}\|^2 + \|\mathbf{y} - \hat{\mathbf{y}}\|^2$.

(b) Let $\mathbf{w} = \begin{pmatrix} -2 & 1 & \\ 0 & 2 & 0 \end{pmatrix}$ and $\mathbf{z} = 3\mathbf{x} + 2\mathbf{w}$. Show that $(\mathbf{w}, \mathbf{x}) = 0$ and that $\|\mathbf{z}\|^2 = 9\|\mathbf{x}\|^2 + 4\|\mathbf{w}\|^2$. (Why must this be true?)

(c) Let $\mathbf{x}_1$, $\mathbf{x}_2$, $\mathbf{x}_3$ be the indicators of the first, second and third columns. Find $p(\mathbf{y}|\mathbf{x}_i)$ for $i = 1, 2, 3$.

Problem 1.2.2: Is projection a linear transformation in the sense that $p(c\mathbf{y}|\mathbf{x}) = cp(\mathbf{y}|\mathbf{x})$ for any real number c? Prove or disprove. What is the relationship between $p(\mathbf{y}|\mathbf{x})$ and $p(\mathbf{y}|c\mathbf{x})$ for $c \neq 0$?

Problem 1.2.3: Let $\|\mathbf{x}\|^2 > 0$. Use calculus to prove that $\|\mathbf{y} - b\mathbf{x}\|^2$ is minimum for $b = (\mathbf{y}, \mathbf{x})/\|\mathbf{x}\|^2$.

Problem 1.2.4: Prove the converse of the Pythagorean Theorem. That is, $\|\mathbf{x} + \mathbf{y}\|^2 = \|\mathbf{x}\|^2 + \|\mathbf{y}\|^2$ implies that $\mathbf{x} \perp \mathbf{y}$.

Problem 1.2.5: Sketch a picture and prove the parallelogram law:

$$\|\mathbf{x} + \mathbf{y}\|^2 + \|\mathbf{x} - \mathbf{y}\|^2 = 2(\|\mathbf{x}\|^2 + \|\mathbf{y}\|^2)$$

1.3 SUBSPACES, PROJECTIONS

We begin the discussion of subspaces and projections with a number of definitions of great importance to our subsequent discussion of linear models. Almost all of the definitions and the theorems which follow are usually included in a first course in matrix or linear algebra. Such courses do not always include discussion of orthogonal projection, so this material may be new to the student.

Definition 1.3.1: A *subspace* of Ω is a subset of Ω which is closed under addition and scalar multiplication.

That is, $V \subset \Omega$ is a subspace if for every $\mathbf{x} \in V$ and every scalar a, $a\mathbf{x} \in V$ and if for every $\mathbf{v}_1, \mathbf{v}_2 \in V$, $\mathbf{v}_1 + \mathbf{v}_2 \in V$.

Definition 1.3.2: Let $\mathbf{x}_1, \ldots, \mathbf{x}_k$ be k vectors in an n-dimensional vector space. The subspace *spanned* by $\mathbf{x}_1, \ldots, \mathbf{x}_k$ is the collection of all vectors

$$\mathbf{y} = b_1\mathbf{x}_1 + \cdots + b_k\mathbf{x}_k$$

for all real numbers $b_1, \ldots, b_k$. We denote this subspace by $\mathscr{L}(\mathbf{x}_1, \ldots, \mathbf{x}_k)$.

Definition 1.3.3: Vectors $\mathbf{x}_1, \ldots, \mathbf{x}_k$ are *linearly independent* if $\sum_1^k b_i\mathbf{x}_i = \mathbf{0}$ implies $b_i = 0$ for $i = 1, \ldots, k$.

Definition 1.3.4: A *basis* for a subspace V of Ω is a set of linearly independent vectors which span V.

The proofs of Theorems 1.3.1 and 1.3.2 are omitted. Readers are referred to any introductory book on linear algebra.

Theorem 1.3.1: Every basis for a subspace V on Ω has the same number of elements.

Definition 1.3.5: The dimension of a subspace V of Ω is the number of elements in each basis.

Theorem 1.3.2: Let $\mathbf{v}_1, \ldots, \mathbf{v}_k$ be linearly independent vectors in a subspace V of dimension d. Then $d \geq k$.

Comment: Theorem 1.3.2 implies that if $\dim(V) = d$ then any collection of $d + 1$ or more vectors in V must be linearly dependent. In particular, any collection of $n + 1$ vectors in the n-component space Ω are linearly dependent.

Definition 1.3.6: A vector $\mathbf{y}$ is *orthogonal* to a subspace V of Ω if $\mathbf{y}$ is orthogonal to all vectors in V. We write $\mathbf{y} \perp V$.

Problem 1.3.1: Let Ω be the space of all 4-component row vectors. Let $\mathbf{x}_1 = (1, 1, 1, 1)$, $\mathbf{x}_2 = (1, 1, 0, 0)$, $\mathbf{x}_3 = (1, 0, 1, 0)$, $\mathbf{x}_4 = (7, 4, 9, 6)$. Let $V_2 = \mathscr{L}(\mathbf{x}_1, \mathbf{x}_2)$, $V_3 = \mathscr{L}(\mathbf{x}_1, \mathbf{x}_2, \mathbf{x}_3)$ and $V_4 = \mathscr{L}(\mathbf{x}_1, \mathbf{x}_2, \mathbf{x}_3, \mathbf{x}_4)$.

(a) Find the dimensions of V_2 and V_3.

(b) Find bases for V_2 and V_3 which contain vectors with as many zeros as possible.

(c) Give a vector $\mathbf{z} \neq \mathbf{0}$ which is orthogonal to all vectors in V_3.

(d) Since $\mathbf{x}_1, \mathbf{x}_2, \mathbf{x}_3, \mathbf{z}$ are linearly independent, $\mathbf{x}_4$ is expressible in the form $\sum_1^3 b_i\mathbf{x}_i + c\mathbf{z}$. Show that $c = 0$ and hence that $\mathbf{x}_4 \in V_3$, by determining $(\mathbf{x}_4, \mathbf{z})$. What is $\dim(V_4)$?

(e) Give a simple verbal description of V_3.

Problem 1.3.2: Consider the space Ω of arrays $\begin{bmatrix} y_{11} & y_{21} & y_{31} \\ y_{12} & y_{22} & \\ y_{13} & & \end{bmatrix}$ and define $\mathbf{C}_1, \mathbf{C}_2, \mathbf{C}_3$ to be the indicators of the columns. Let $V = \mathscr{L}(\mathbf{C}_1, \mathbf{C}_2, \mathbf{C}_3)$.

(a) What properties must $\mathbf{y}$ satisfy in order that $\mathbf{y} \in V$? In order that $\mathbf{y} \perp V$?

(b) Find a vector $\mathbf{y}$ which is orthogonal to V.

The following definition is perhaps the most important in the entire book. It serves as the foundation of all the least squares theory to be discussed in Chapters 1, 2, and 3.

Definition 1.3.7: The projection of a vector $\mathbf{y}$ on a subspace V of Ω is the vector $\hat{\mathbf{y}} \in V$ such that $(\mathbf{y} - \hat{\mathbf{y}}) \perp V$. The vector $\mathbf{y} - \hat{\mathbf{y}} = \mathbf{e}$ will be called the *residual* vector for $\mathbf{y}$ relative to V.

Comment: The condition $(\mathbf{y} - \hat{\mathbf{y}}) \perp V$ is equivalent to $(\mathbf{y} - \hat{\mathbf{y}}, \mathbf{x}) = 0$ for all $\mathbf{x} \in V$. Therefore, in seeking the projection $\hat{\mathbf{y}}$ of $\mathbf{y}$ on a subspace V we seek a vector $\hat{\mathbf{y}}$ in V which has the same inner products as $\mathbf{y}$ with all vectors in V (Figure 1.3).

If vectors $\mathbf{x}_1, \ldots, \mathbf{x}_k$ span a subspace V then a vector $\mathbf{z} \in V$ is the projection of $\mathbf{y}$ on V if $(\mathbf{z}, \mathbf{x}_i) = (\mathbf{y}, \mathbf{x}_i)$ for all i, since for any vector $\mathbf{x} = \sum_{j=1}^{k} b_j \mathbf{x}_j \in V$, this implies that

$$(\mathbf{z}, \mathbf{x}) = \sum b_j(\mathbf{z}, \mathbf{x}_j) = (\mathbf{y}, \sum b_j \mathbf{x}_j) = (\mathbf{y}, \mathbf{x}).$$

It is tempting to attempt to compute the projection $\hat{\mathbf{y}}$ of $\mathbf{y}$ on V by simply summing the projections $\hat{\mathbf{y}}_i = p(\mathbf{y}|\mathbf{x}_i)$. As we shall see, this is only possible in some very special cases.

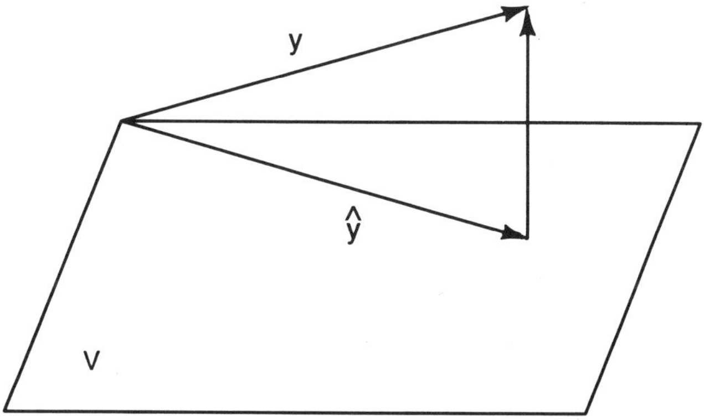

FIGURE 1.3

At this point we have not established the legitimacy of Definition 1.3.7. Does such a vector $\hat{\mathbf{y}}$ always exist and, if so, is it unique? We do know that the projection onto a one-dimensional subspace, say onto $V = \mathscr{L}(\mathbf{x})$, for $\mathbf{x} \neq \mathbf{0}$, does exist and is unique. In fact

$$\hat{\mathbf{y}} = [(\mathbf{y}, \mathbf{x})/\|\mathbf{x}\|^2]\mathbf{x} \quad \text{if} \quad \mathbf{x} \neq \mathbf{0}.$$

Example 1.3.1: Consider the 6-component space Ω of the problem above, and let $V = \mathscr{L}(\mathbf{C}_1, \mathbf{C}_2, \mathbf{C}_3)$. Let $\mathbf{y} = \begin{pmatrix} 6 & 4 & 7 \\ 10 & 8 & \\ 5 & & \end{pmatrix}$. It is easy to show that the vector $\hat{\mathbf{y}} = \sum p(\mathbf{y}|\mathbf{C}_i) = 7\mathbf{C}_1 + 6\mathbf{C}_2 + 7\mathbf{C}_3$ satisfies the conditions for a projection onto V. As will soon be shown the representation of $\hat{\mathbf{y}}$ as the sum of projections on linearly independent vectors spanning the space is possible because $\mathbf{C}_1$, $\mathbf{C}_2$, and $\mathbf{C}_3$ are mutually othogonal.

We will first show uniqueness of the projection. Existence is more difficult. Suppose $\hat{\mathbf{y}}_1$ and $\hat{\mathbf{y}}_2$ are two such projections of $\mathbf{y}$ onto V. Then $\hat{\mathbf{y}}_1 - \hat{\mathbf{y}}_2 \in V$ and $(\hat{\mathbf{y}}_1 - \hat{\mathbf{y}}_2) = (\mathbf{y} - \hat{\mathbf{y}}_2) - (\mathbf{y} - \hat{\mathbf{y}}_1)$ is orthogonal to all vectors in V, in particular to itself. Thus $\|\hat{\mathbf{y}}_1 - \hat{\mathbf{y}}_2\|^2 = (\hat{\mathbf{y}}_1 - \hat{\mathbf{y}}_2, \hat{\mathbf{y}}_1 - \hat{\mathbf{y}}_2) = \mathbf{0}$, implying $\hat{\mathbf{y}}_1 - \hat{\mathbf{y}}_2 = \mathbf{0}$, i.e., $\hat{\mathbf{y}}_1 = \hat{\mathbf{y}}_2$.

We have yet to show that $\hat{\mathbf{y}}$ always exists. In the case that it does exist (we will show that it always exists) we will write $\hat{\mathbf{y}} = p(\mathbf{y}|V)$.

If we are fortunate enough to have an *orthogonal* basis (a basis of mutually orthogonal vectors) for a given subspace V, it is easy to find the projection. Students are warned that that method applies *only* for an orthogonal basis. We will later show that all subspaces possess such orthogonal bases, so that the projection $\hat{\mathbf{y}} = p(\mathbf{y}|V)$ always exists.

Theorem 1.3.3: Let $\mathbf{v}_1, \ldots, \mathbf{v}_k$ be an orthogonal basis for V, subspace of Ω. Then

$$p(\mathbf{y}|V) = \sum_{i=1}^{k} p(\mathbf{y}|\mathbf{v}_i)$$

***Proof*:** Let $\hat{\mathbf{y}}_i = p(\mathbf{y}|\mathbf{v}_i) = b_i\mathbf{v}_i$ for $b_i = (\mathbf{y}, \mathbf{v}_i)/\|\mathbf{v}_i\|^2$. Since $\hat{\mathbf{y}}_i$ is a scalar multiple of $\mathbf{v}_i$, it is orthogonal to $\mathbf{v}_j$ for $j \neq i$. From the comment on the previous page, we need only show that $\sum \hat{\mathbf{y}}_i$ and $\mathbf{y}$, have the same inner product with each $\mathbf{v}_j$, since this implies that they have the same inner product with all $\mathbf{x} \in V$. But

$$\left(\sum_i \hat{\mathbf{y}}_i, \mathbf{v}_j\right) = \sum_i b_i(\mathbf{v}_i, \mathbf{v}_j) = b_j\|\mathbf{v}_j\|^2 = (\mathbf{y}, \mathbf{v}_j). \qquad \square$$

Example 1.3.2: Let

$$\mathbf{y} = \begin{pmatrix} 7 \\ 0 \\ 2 \end{pmatrix}, \qquad \mathbf{v}_1 = \begin{pmatrix} 1 \\ 1 \\ 1 \end{pmatrix}, \qquad \mathbf{v}_2 = \begin{pmatrix} 2 \\ -1 \\ -1 \end{pmatrix}, \qquad V = \mathscr{L}(\mathbf{v}_1, \mathbf{v}_2).$$

Then $\mathbf{v}_1 \perp \mathbf{v}_2$ and

$$p(\mathbf{y} | V) = \hat{\mathbf{y}} = p(\mathbf{y} | \mathbf{v}_1) + p(\mathbf{y} | \mathbf{v}_2) = \left(\frac{9}{3}\right)\mathbf{v}_1 + \left(\frac{12}{6}\right)\mathbf{v}_2 = \begin{pmatrix} 3 \\ 3 \\ 3 \end{pmatrix} + \begin{pmatrix} 4 \\ -2 \\ -2 \end{pmatrix} = \begin{pmatrix} 7 \\ 1 \\ 1 \end{pmatrix}.$$

Then $(\mathbf{y}, \mathbf{v}_1) = 9$, $(\mathbf{y}, \mathbf{v}_2) = 12$, $(\hat{\mathbf{y}}, \mathbf{v}_1) = 9$, and $(\hat{\mathbf{y}}, \mathbf{v}_1) = 12$. The residual vector is $\mathbf{y} - \hat{\mathbf{y}} = \begin{pmatrix} 0 \\ -1 \\ 1 \end{pmatrix}$, which is orthogonal to V.

Would this same procedure have worked if we replaced this orthogonal basis $\mathbf{v}_1, \mathbf{v}_2$ for V by a nonorthogonal basis? To experiment, let us leave $\mathbf{v}_1$ in the new basis, but replace $\mathbf{v}_2$ by $\mathbf{v}_3 = 2\mathbf{v}_1 - \mathbf{v}_2$. Note that $\mathscr{L}(\mathbf{v}_1, \mathbf{v}_3) = \mathscr{L}(\mathbf{v}_1, \mathbf{v}_2) = V$, and that $(\mathbf{v}_1, \mathbf{v}_2) \neq 0$. $\hat{\mathbf{y}}_1$ remains the same. $\mathbf{v}_3 = 2\mathbf{v}_1 - \mathbf{v}_2 = \begin{pmatrix} 0 \\ 3 \\ 3 \end{pmatrix}$, $\hat{\mathbf{y}}_3 = \frac{6}{18}\mathbf{v}_3 = \begin{pmatrix} 0 \\ 1 \\ 1 \end{pmatrix}$, and $\hat{\mathbf{y}}_1 + \hat{\mathbf{y}}_3 = \begin{pmatrix} 3 \\ 4 \\ 4 \end{pmatrix}$, which has inner products 11 and 24 with $\mathbf{v}_1$ and $\mathbf{v}_3$. $\mathbf{y} - \begin{pmatrix} 3 \\ 4 \\ 4 \end{pmatrix} = \begin{pmatrix} 4 \\ -4 \\ -2 \end{pmatrix}$, which is not orthogonal to V. Therefore, $\hat{\mathbf{y}}_1 + \hat{\mathbf{y}}_3$ is not the projection of $\mathbf{y}$ on $V = \mathscr{L}(\mathbf{v}_1, \mathbf{v}_3)$.

Since $(\mathbf{y} - \hat{\mathbf{y}}) \perp \hat{\mathbf{y}}$, we have, by the Pythagorean Theorem,

$$\|\mathbf{y}\|^2 = \|(\mathbf{y} - \hat{\mathbf{y}}) + \hat{\mathbf{y}}\|^2 = \|\mathbf{y} - \hat{\mathbf{y}}\|^2 + \|\hat{\mathbf{y}}\|^2$$

$$\|\mathbf{y}\|^2 = 53, \qquad \|\hat{\mathbf{y}}\|^2 = \frac{9^2}{3} + \frac{12^3}{6} = 51, \qquad \|\mathbf{y} - \hat{\mathbf{y}}\|^2 = \left\| \begin{pmatrix} 0 \\ -1 \\ 1 \end{pmatrix} \right\|^2 = 2.$$

Warning: We have shown that when $\mathbf{v}_1, \ldots, \mathbf{v}_k$ are mutually orthogonal the projection $\hat{\mathbf{y}}$ of $\mathbf{y}$ on the subspace spanned by $\mathbf{v}_1, \ldots, \mathbf{v}_k$ is $\sum_{j=1}^{k} p(\mathbf{y}|\mathbf{v}_j)$. This is true for all $\mathbf{y}$ *only* if $\mathbf{v}_1, \ldots, \mathbf{v}_k$ are mutually orthogonal. Students are asked to prove the "only" part in Problem 1.3.5.

Every subspace V of Ω of dimension $r > 0$ has an orthogonal basis (actually an infinity of such bases). We will show that such a basis exists by using *Gram–Schmidt orthogonalization.*

Let $\mathbf{x}_1, \ldots, \mathbf{x}_k$ be a basis for a subspace V, a k-dimensional subspace of Ω. For $1 \le i \le k$ let $V_i = \mathscr{L}(\mathbf{x}_1, \ldots, \mathbf{x}_i)$ so that $V_1 \subset V_2 \subset \cdots \subset V_k$ are properly nested subspaces. Let

$$\mathbf{v}_1 = \mathbf{x}_1, \qquad \mathbf{v}_2 = \mathbf{x}_2 - p(\mathbf{x}_2|\mathbf{v}_1).$$

Then $\mathbf{v}_1$ and $\mathbf{v}_2$ span V_2 and are othogonal. Thus $p(\mathbf{x}_3|V_2) = p(\mathbf{x}_3|\mathbf{v}_1) + p(\mathbf{x}_3|\mathbf{v}_2)$ and we can define $\mathbf{v}_3 = \mathbf{x}_3 - p(\mathbf{x}_3|V_2)$. Continuing in this way, suppose we have defined $\mathbf{v}_1, \ldots, \mathbf{v}_i$ to be mutually orthogonal vectors spanning V_i. Define $\mathbf{v}_{i+1} = \mathbf{x}_{i+1} - p(\mathbf{x}_{i+1}|V_i)$. Then $\mathbf{v}_{i+1} \perp V_i$ and hence $\mathbf{v}_1, \ldots, \mathbf{v}_{i+1}$ are mutually orthogonal and span V_{i+1}. Since we can do this for each $i \le k - 1$ we get the orthogonal basis $\mathbf{v}_1, \ldots, \mathbf{v}_k$ for V.

If $\{\mathbf{v}_1, \ldots, \mathbf{v}_k\}$ is an orthogonal basis for a subspace V then, since $\hat{\mathbf{y}} \equiv p(\mathbf{y}|V) = \sum_{j=1}^{k} p(\mathbf{y}|\mathbf{v}_j)$ and $p(\mathbf{y}|\mathbf{v}_j) = b_j\mathbf{v}_j$, with $b_j = [(\mathbf{y}, \mathbf{v}_j)/\|\mathbf{v}_j\|^2]$, it follows by the Pythagorean Theorem that

$$\|\hat{\mathbf{y}}\|^2 = \sum_{j=1}^{k} \|b_j\mathbf{v}_j\|^2 = \sum_{j=1}^{k} b_j^2\|\mathbf{v}_j\|^2 = \sum_{j=1}^{k} (\mathbf{y}, \mathbf{v}_j)^2/\|\mathbf{v}_j\|^2.$$

Of course, the basis $\{\mathbf{v}_1, \ldots, \mathbf{v}_k\}$ can be made into an *orthonormal* basis (all vectors of length one) by dividing each by its own length. If $\{\mathbf{v}_1^*, \ldots, \mathbf{v}_k^*\}$ is such an orthonormal basis then $\hat{\mathbf{y}} = p(\mathbf{y}|V) = \sum_1^k p(\mathbf{y}|\mathbf{v}_i^*) = \sum_1^k (\mathbf{y}, \mathbf{v}_i^*)\mathbf{v}_i^*$ and $\|\hat{\mathbf{y}}\|^2 = \sum_{i=1}^{k} (\mathbf{y}, \mathbf{v}_i^*)^2$.

Example 1.3.3: Consider R_4, the space of 4-component column vectors. Let us apply Gram–Schmidt orthogonalization to the columns of $\mathbf{X} = \begin{bmatrix} 1 & 1 & 4 & 8 \\ 1 & 1 & 0 & 10 \\ 1 & 5 & 12 & 0 \\ 1 & 5 & 8 & 10 \end{bmatrix}$, a matrix chosen carefully by the author to keep the arithmetic simple. Let the four columns be $\mathbf{x}_1, \ldots, \mathbf{x}_4$. Define $\mathbf{v}_1 = \mathbf{x}_1$. Let

$$\mathbf{v}_2 = \mathbf{x}_2 - \frac{12}{4}\mathbf{v}_1 = \begin{bmatrix} -2 \\ -2 \\ 2 \\ 2 \end{bmatrix}, \qquad \mathbf{v}_3 = \mathbf{x}_3 - \left[\frac{24}{4}\mathbf{v}_1 + \frac{32}{16}\mathbf{v}_2\right] = \begin{bmatrix} 2 \\ -2 \\ 2 \\ -2 \end{bmatrix},$$

and

$$\mathbf{v}_4 = \mathbf{x}_4 - \left[\frac{28}{4}\mathbf{v}_1 + \frac{(-16)}{16}\mathbf{v}_2 + \frac{(-24)}{16}\mathbf{v}_3\right] = \begin{bmatrix} 2 \\ -2 \\ -2 \\ 2 \end{bmatrix}.$$

We can multiply these $\mathbf{v}_i$ by arbitrary constants to simplify them without losing their orthogonality. For example, we can define $\mathbf{u}_i = \mathbf{v}_i/\|\mathbf{v}_i\|^2$, so that $\mathbf{u}_1$, $\mathbf{u}_2$, $\mathbf{u}_3$, $\mathbf{u}_4$ are unit length orthogonal vectors spanning Ω. Then $\mathbf{U} = (\mathbf{u}_1, \mathbf{u}_2, \mathbf{u}_3, \mathbf{u}_4)$ is an orthogonal matrix. $\mathbf{U}$ is expressible in the form $\mathbf{U} = \mathbf{XR}$, where $\mathbf{R}$ has zeros below the diagonal. Since $\mathbf{I} = \mathbf{U}'\mathbf{U} = \mathbf{U}'\mathbf{XR}$, $\mathbf{R}^{-1} = \mathbf{U}'\mathbf{X}$, and $\mathbf{X} = \mathbf{UR}^{-1}$, where $\mathbf{R}^{-1}$ has zeros below the diagonal (see Section 1.7).

As we consider linear models we will often begin with a model which supposes that $\mathbf{Y}$ has expectation $\boldsymbol{\theta}$ which lies in a subspace V_2, and will wish to decide whether this vector lies in a smaller subspace V_1. The orthogonal bases provided by the following theorem will be useful in the development of convenient formulas and in the investigation of the distributional properties of estimators.

Theorem 1.3.4: Let $V_1 \subset V_2 \subset \Omega$ be subspaces of Ω of dimensions $1 \le n_1 < n_2 < n$. Then there exist mutually orthogonal vectors $\mathbf{v}_1, \ldots, \mathbf{v}_n$ such that $\mathbf{v}_1, \ldots, \mathbf{v}_{n_i}$ span V_i, $i = 1, 2$.

Proof: Let $\{\mathbf{x}_1, \ldots, \mathbf{x}_{n_1}\}$ be a basis for V_1. Then by Gram–Schmidt orthogonalization there exists an orthogonal basis $\{\mathbf{v}_1, \ldots, \mathbf{v}_{n_1}\}$ for V_1. Let $\mathbf{x}_{n_1+1}, \ldots, \mathbf{x}_{n_2}$ be chosen consecutively from V_2 so that $\mathbf{v}_1, \ldots, \mathbf{v}_{n_1}, \mathbf{x}_{n_1+1}, \ldots, \mathbf{x}_{n_2}$ are linearly independent. (If this could not be done, V_2 would have dimension less than n_2.) Then applying Gram–Schmidt orthogonalization to $\mathbf{x}_{n_1+1}, \ldots, \mathbf{x}_{n_2}$ we have an orthogonal basis for V_2. Repeating this for V_2 replaced by Ω and $\mathbf{v}_1, \ldots, \mathbf{v}_{n_1}$ by $\mathbf{v}_1, \ldots, \mathbf{v}_{n_2}$ we get the theorem. □

For a nested sequence of subspaces we can repeat this theorem consecutively to get Theorem 1.3.5.

Theorem 1.3.5: Let $V_1 \subset V_2 \subset \cdots \subset V_k \subset \Omega = V_{k+1}$ be subspaces of Ω of dimensions $1 \le n_1 < n_2 < \cdots < n_k < n = n_{k+1}$. Then there exists an orthogonal basis $\mathbf{v}_1, \ldots, \mathbf{v}_n$ for Ω such that $\mathbf{v}_1, \ldots, \mathbf{v}_{n_i}$ is a basis for V_i for $i = 1, \ldots, k+1$.

We can therefore write for any $\mathbf{y} \in \Omega$,

$$p(\mathbf{y}|V_i) = \sum_{j=1}^{n_i} \frac{(\mathbf{y}, \mathbf{v}_j)}{\|\mathbf{v}_j\|^2} \mathbf{v}_j \qquad \text{for} \quad i = 1, \ldots, k+1,$$

and

$$\|p(\mathbf{y}|V_i)\|^2 = \sum_{j=1}^{n_i} \frac{(\mathbf{y}, \mathbf{v}_j)^2}{\|\mathbf{v}_j\|^2} \qquad \text{for} \quad i = 1, \ldots, k+1.$$

The $\mathbf{v}_j$ can be chosen to have length one, so these last formulas simplify still further.

Thus, the definition of the projection $p(\mathbf{y}|V)$ has been justified. Fortunately, it is not necessary to find an orthogonal basis in order to find the projection in the general case that the basis vectors $(\mathbf{x}_1, \ldots, \mathbf{x}_k)$ are not orthogonal. The Gram–Schmidt method is useful in the development of nonmatrix formulas for regression coefficients.

In order for $\hat{\mathbf{y}} = b_1\mathbf{x}_1 + \cdots + b_k\mathbf{x}_k$ to be the projection of $\mathbf{y}$ on $V = \mathscr{L}(\mathbf{x}_1, \ldots, \mathbf{x}_k)$ we need $(\mathbf{y}, \mathbf{x}_i) = (\hat{\mathbf{y}}, \mathbf{x}_i)$ for all i. This leads to the so-called *normal equations*:

$$(\hat{\mathbf{y}}, \mathbf{x}_i) = \sum_1^k b_j(\mathbf{x}_j, \mathbf{x}_i) = (\mathbf{y}, \mathbf{x}_i) \qquad \text{for} \quad i = 1, \ldots, k$$

It is convenient to write these k simultaneous linear equations in matrix form:

$$\underset{k \times k}{\mathbf{M}} \ \underset{k \times 1}{\mathbf{b}} = \mathbf{U},$$

where $\mathbf{M}$ is the matrix of inner products among the $\mathbf{x}_j$ vectors, $\mathbf{b}$ is the column vector of b_j's, and $\mathbf{U}$ is the $k \times 1$ column vector of inner products of $\mathbf{y}$ with the $\mathbf{x}_j$. If Ω is taken to be the space of n-component column vectors, then we can write $\mathbf{X} = (\mathbf{x}_1, \ldots, \mathbf{x}_k)$, and we get $\mathbf{M} = \mathbf{X}'\mathbf{X}$, $\mathbf{U} = \mathbf{X}'\mathbf{y}$, so the normal equations are:

$$\mathbf{M}\mathbf{b} = (\mathbf{X}'\mathbf{X})\mathbf{b} = \mathbf{X}'\mathbf{y} = \mathbf{U}$$

Of course, if $\mathbf{M} = ((\mathbf{x}_i, \mathbf{x}_j))$ has an inverse we will have an explicit solution

$$\mathbf{b} = \mathbf{M}^{-1}\mathbf{U}$$

of the normal equations. It will be shown in Section 1.6 that $\mathbf{M}$ has rank k if and only if $\mathbf{x}_1, \ldots, \mathbf{x}_k$ are linearly independent. Thus $\mathbf{b} = \mathbf{M}^{-1}\mathbf{U}$ if and only if $\mathbf{x}_1, \ldots, \mathbf{x}_k$ are linearly independent.

In the case that the elements of Ω are not column vectors, we can always rewrite its elements as column vectors, and the matrix **M** will remain unchanged. Thus, in the general case **M** possesses an inverse if and only if the vectors $\mathbf{x}_1, \ldots, \mathbf{x}_k$ are linearly independent. Of course, even in this case with $\Omega = R_n$, the space of n-component column vectors, $\mathbf{X} = (\mathbf{x}_1, \ldots, \mathbf{x}_k)$, being $n \times k$, does not have an inverse unless $n = k$. In applications we always have $n > k$.

In the computation of $\mathbf{M} = \mathbf{X}'\mathbf{X}$ it makes little sense to write **X** on its side as $\mathbf{X}'$, then **X**, and then to carry out the computation as the multiplication of two matrices, unless the computer software being used requires this. **M** is the matrix of inner products, and **U** is a vector of inner products, and this viewpoint should be emphasized.

Example 1.3.4: Let $\mathbf{y}$, $\mathbf{v}_1$ and $\mathbf{v}_2$ be as in Example 1.3.2. Let $\mathbf{x}_1 = \mathbf{v}_1$ and $\mathbf{x}_2 = 2\mathbf{v}_1 + \mathbf{v}_2$. Then

$$\mathbf{y} = \begin{bmatrix} 7 \\ 0 \\ 2 \end{bmatrix}, \qquad \mathbf{x}_1 = \begin{bmatrix} 1 \\ 1 \\ 1 \end{bmatrix}, \qquad \mathbf{x}_2 = \begin{bmatrix} 4 \\ 1 \\ 1 \end{bmatrix},$$

and $V = \mathscr{L}(\mathbf{v}_1, \mathbf{v}_2) = \mathscr{L}(\mathbf{x}_1, \mathbf{x}_2)$. We compute

$$\mathbf{M} = \begin{bmatrix} 6 & 6 \\ 6 & 18 \end{bmatrix}, \qquad \mathbf{U} = \begin{bmatrix} (\mathbf{x}_1, \mathbf{y}) \\ (\mathbf{x}_2, \mathbf{y}) \end{bmatrix} = \begin{bmatrix} 9 \\ 30 \end{bmatrix},$$

$$\mathbf{M}^{-1} = \frac{1}{18}\begin{bmatrix} 18 & -6 \\ -6 & 3 \end{bmatrix} = \frac{1}{6}\begin{bmatrix} 6 & -2 \\ -2 & 1 \end{bmatrix},$$

$$\mathbf{b} = \mathbf{M}^{-1}\mathbf{U} = \frac{1}{6}\begin{bmatrix} -6 \\ 12 \end{bmatrix} = \begin{bmatrix} -1 \\ 2 \end{bmatrix}$$

and $\hat{\mathbf{y}} = p(\mathbf{y}|V) = -\mathbf{x}_1 + 2\mathbf{x}_2 = \begin{bmatrix} 7 \\ 1 \\ 1 \end{bmatrix}$, as before.

It is easy to compute lengths of $\mathbf{y}$ and of $\mathbf{y} - \hat{\mathbf{y}}$. First,

$$\|\hat{\mathbf{y}}\|^2 = (\mathbf{y}, \hat{\mathbf{y}}) = \left(\mathbf{y}, \sum_1^k b_j\mathbf{x}_j\right) = \sum b_j(\mathbf{y}, \mathbf{x}_j) = \mathbf{b}'\mathbf{U}.$$

By the Pythagorean Theorem,

$$\|\mathbf{y} - \hat{\mathbf{y}}\|^2 = \|\mathbf{y}\|^2 - \|\hat{\mathbf{y}}\|^2.$$

For Example 1.3.2, $\|\hat{\mathbf{y}}\|^2 = b_1(\mathbf{y}, \mathbf{x}_1) + b_2(\mathbf{y}, \mathbf{x}_2) = (-1)(9) + 2(30) = 51$, as shown in Example 1.3.2.

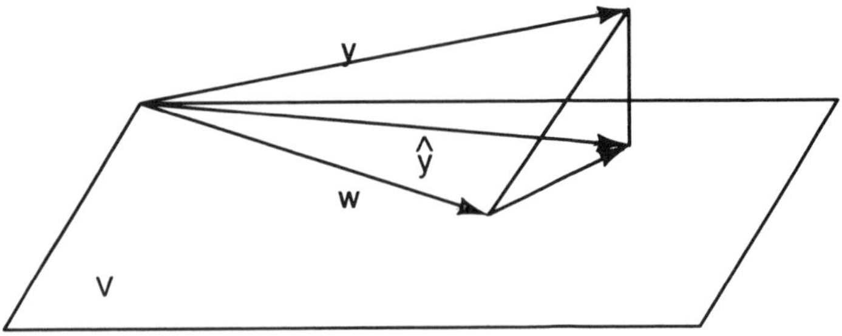

FIGURE 1.4

The projection $\hat{\mathbf{y}} = p(\mathbf{y}|V)$ is the closest vector in V to $\mathbf{y}$, since for any other vector $\mathbf{w} \in V$,

$$\|\mathbf{y} - \mathbf{w}\|^2 = \|(\mathbf{y} - \hat{\mathbf{y}}) + (\hat{\mathbf{y}} - \mathbf{w})\|^2 = \|\mathbf{y} - \hat{\mathbf{y}}\|^2 + \|\hat{\mathbf{y}} - \mathbf{w}\|^2$$

by the Pythagorean Theorem and the facts that $(\hat{\mathbf{y}} - \mathbf{w}) \in V$, and $(\mathbf{y} - \hat{\mathbf{y}}) \perp V$. Thus $\|\mathbf{y} - \mathbf{w}\|^2$ is minimized for $\mathbf{w} \in V$ by taking $\mathbf{w} = \hat{\mathbf{y}}$ (Figure 1.4).

For this reason the vectors $\mathbf{b}$ and $\hat{\mathbf{y}}$ are said to have been obtained by the principle of least squares.

Problem 1.3.3: Let Ω, $\mathbf{C}_1$, $\mathbf{C}_2$, $\mathbf{C}_3$ be defined as in problem 1.3.2. Let $V = \mathscr{L}(\mathbf{C}_1, \mathbf{C}_2, \mathbf{C}_3)$

(a) For $\mathbf{y} = \begin{bmatrix} 6 & 11 & 8 \\ 4 & 7 & \\ 2 & & \end{bmatrix}$ find $\hat{\mathbf{y}} = p(\mathbf{y}|V)$, $\mathbf{y} - \hat{\mathbf{y}}$, $\|\mathbf{y}\|^2$, $\|\hat{\mathbf{y}}\|^2$, $\|\mathbf{y} - \hat{\mathbf{y}}\|^2$.

(b) Give a general nonmatrix formula for $\hat{\mathbf{y}} = p(\mathbf{y}|V)$ for any $\mathbf{y}$.

Problem 1.3.4: Let $\mathbf{x}_1 = (1, 1, 1, 1)'$, $\mathbf{x}_2 = (4, 1, 3, 4)'$, $\mathbf{y} = (1, 9, 5, 5)'$ (so these are column vectors). Let $V = \mathscr{L}(\mathbf{x}_1, \mathbf{x}_2)$.

(a) Find $\hat{\mathbf{y}} = p(\mathbf{y}|V)$ and $\mathbf{e} = \mathbf{y} - \hat{\mathbf{y}}$.

(b) Find $\hat{\mathbf{y}}_1 = p(\mathbf{y}|\mathbf{x}_1)$ and $\hat{\mathbf{y}}_2 = p(\mathbf{y}|\mathbf{x}_2)$ and show that $\hat{\mathbf{y}} \neq \hat{\mathbf{y}}_1 + \hat{\mathbf{y}}_2$.

(c) Verify that $\mathbf{e} \perp V$.

(d) Find $\|\mathbf{y}\|^2$, $\|\hat{\mathbf{y}}\|^2$, $\|\mathbf{y} - \hat{\mathbf{y}}\|^2$, and verify that the Pythagorean Theorem holds. Compute $\|\hat{\mathbf{y}}\|^2$ directly from $\hat{\mathbf{y}}$ and also by using the formula $\|\hat{\mathbf{y}}\|^2 = \mathbf{U}'\mathbf{b}$.

(e) Use Gram–Schmidt orthogonalization to find four mutually orthogonal vectors $\mathbf{v}_1$, $\mathbf{v}_2$, $\mathbf{v}_3$, $\mathbf{v}_4$ such that $V = \mathscr{L}(\mathbf{v}_1, \mathbf{v}_2)$. *Hint*: You can choose $\mathbf{x}_3$ and $\mathbf{x}_4$ arbitrarily, as long as $\mathbf{x}_1$, $\mathbf{x}_2$, $\mathbf{x}_3$, $\mathbf{x}_4$ are linearly independent.

(f) Express $\mathbf{y}$ and $\hat{\mathbf{y}}$ in terms of the $\mathbf{v}_i$.

(g) Let $\mathbf{w} = (2, 8, 4, 2)'$. Show that $\mathbf{w} \in V$ and verify that $\|\mathbf{y} - \mathbf{w}\|^2 = \|\mathbf{y} - \hat{\mathbf{y}}\|^2 + \|\hat{\mathbf{y}} - \mathbf{w}\|^2$. (Why must this equality hold?)

(h) Does $p(\hat{\mathbf{y}}|\mathbf{x}_1) = \hat{\mathbf{y}}_1$? Is this true for any $\mathbf{y}$? That is, do we obtain the same vector by (1) first projecting $\mathbf{y}$ on V, then projecting this vector on $\mathbf{x}_1$ as by (2) projecting $\mathbf{y}$ directly on $\mathbf{x}_1$? More generally, if V is a subspace, and V_1 a subspace of V, does $p(p(\mathbf{y}|V)|V_1) = p(\mathbf{y}|V_1)$?

Problem 1.3.5: Let $\mathbf{y} = (y_1, \ldots, y_n)'$, $\mathbf{x} = (x_1, \ldots, x_n)'$, $\mathbf{J} = (1, \ldots, 1)'$, and $V = \mathscr{L}(\mathbf{J}, \mathbf{x})$.

(a) Use Gram–Schmidt orthogonalization on the vectors $\mathbf{J}$, $\mathbf{x}$ (in this order) to find orthogonal vectors $\mathbf{J}$, $\mathbf{x}^*$ spanning V. Express $\mathbf{x}^*$ in terms of $\mathbf{J}$ and $\mathbf{x}$, then find b_0, b_1 such that $\hat{\mathbf{y}} = b_0\mathbf{J} + b_1\mathbf{x}$. To simplify the notation, let $\mathbf{y}^* = \mathbf{y} - p(\mathbf{y}|\mathbf{J}) = \mathbf{y} - \bar{y}\mathbf{J}$,

$$S_{xy} = (\mathbf{x}^*, \mathbf{y}^*) = (\mathbf{x}^*, \mathbf{y}) = \sum (x_i - \bar{x})(y_i - \bar{y}) = \sum (x_i - \bar{x})y_i = \sum x_i y_i - \bar{x}\bar{y}n,$$

$$S_{xx} = (\mathbf{x}^*, \mathbf{x}^*) = \sum (x_i - \bar{x})^2 = \sum (x_i - \bar{x})^2 = \sum x_i^2 - \bar{x}^2 n,$$

$$S_{yy} = (\mathbf{y}^*, \mathbf{y}^*) = \sum (y_i - \bar{y})^2.$$

(b) Suppose $\hat{\mathbf{y}} = p(\mathbf{y}|V) = a_0\mathbf{J} + a_1\mathbf{x}^*$. Find formulas for a_1 and a_0 in terms of $\bar{y}$, S_{xy}, and S_{xx}.

(c) Express $\mathbf{x}^*$ in terms of $\mathbf{J}$ and $\mathbf{x}$, and use this to determine formulas for b_1 and b_0 so that $\hat{\mathbf{y}} = b_0\mathbf{J} + b_1\mathbf{x}$.

(d) Express $\|\hat{\mathbf{y}}\|^2$ and $\|\mathbf{y} - \hat{\mathbf{y}}\|^2$ in terms of S_{xy}, S_{xx}, and S_{yy}.

(e) Use the formula $\mathbf{b} = \mathbf{M}^{-1}\mathbf{U}$ for $\mathbf{b} = (b_0, b_1)'$ and verify that they are the same as those found in (c).

(f) For $\mathbf{y} = \begin{pmatrix} 2 \\ 6 \\ 8 \\ 8 \end{pmatrix}$, $\mathbf{x} = \begin{pmatrix} 0 \\ 1 \\ 2 \\ 3 \end{pmatrix}$ find $a_0, a_1, \hat{\mathbf{y}}, b_0, b_1, \|\mathbf{y}\|^2, \|\hat{\mathbf{y}}\|^2, \|\mathbf{y} - \hat{\mathbf{y}}\|^2$. Verify that $\|\hat{\mathbf{y}}\|^2 = b_0(\mathbf{y}, \mathbf{J}) + b_1(\mathbf{y}, \mathbf{x})$ and that $(\mathbf{y} - \hat{\mathbf{y}}) \perp V$.

Problem 1.3.6: Let Ω be the collection of 2×3 arrays of the form

$$\mathbf{y} = \begin{bmatrix} y_{11} & y_{12} & y_{13} \\ y_{21} & y_{22} & y_{23} \end{bmatrix}.$$

Let $\mathbf{R}_1, \mathbf{R}_2, \mathbf{C}_1, \mathbf{C}_2, \mathbf{C}_3$ be indicators of the 2 rows and 3 columns. For example, $\mathbf{C}_2 = \begin{bmatrix} 0 & 1 & 0 \\ 0 & 1 & 0 \end{bmatrix}$. For $V = \mathscr{L}(\mathbf{R}_1, \mathbf{R}_2, \mathbf{C}_1, \mathbf{C}_2, \mathbf{C}_3)$, $\mathbf{y} = \begin{bmatrix} 7 & 5 & 0 \\ 9 & 9 & 6 \end{bmatrix}$ find $\hat{\mathbf{y}}$, $\mathbf{e} = \mathbf{y} - \hat{\mathbf{y}}$, $\|\hat{\mathbf{y}}\|^2$, $\|\mathbf{e}\|^2$. Verify that $\mathbf{e} \perp V$. *Hint*: Find four mutually orthogonal vectors which span V. It is easier to begin with the column indicators.

Problem 1.3.7: Let $\mathbf{x}_1, \ldots, \mathbf{x}_k$ be a basis of a subspace V. Suppose that $p(\mathbf{y} | V) = \sum_{j=1}^{k} p(\mathbf{y} | \mathbf{x}_j)$ for every vector $\mathbf{y} \in \Omega$. Prove that $\mathbf{x}_1, \ldots, \mathbf{x}_k$ are mutually orthogonal. *Hint*: Consider the vector $\mathbf{y} = \mathbf{x}_i$ for each i.

Problem 1.3.8: Consider the collection $\mathscr{H}$ of all real-valued functions on the unit interval $U = [0, 1]$ having the property $\int_0^1 f^2(x)\, dx < \infty$. Define the inner product $(\mathbf{f}, \mathbf{g}) = \int_0^1 f(x)g(x)\, dx$. Such an inner product space, with the correct definition of the integral, and a more subtle property called completeness, is called a Hilbert space after the great German mathematician, David Hilbert, of the late nineteenth and early twentieth centuries. $\mathscr{H}$ is not finite dimensional, but our projection theory still applies because we will be interested in projections on finite dimensional subspaces. Consider the function $h(x) = \sqrt{x}$ for $x \in U$. For each nonnegative integer k define $p_k(x) = x^k$. The functions h, p_0, p_1, p_2 determine corresponding points $\mathbf{h}$, $\mathbf{p}_0$, $\mathbf{p}_1$, $\mathbf{p}_2$ in $\mathscr{H}$. Define $V_k = \mathscr{L}(\mathbf{p}_0, \mathbf{p}_1, \ldots, \mathbf{p}_k)$, and $\hat{\mathbf{h}}_k = p(\mathbf{h} | V_k)$. The point $\hat{\mathbf{h}}_k$ corresponds to a polynomial $\hat{h}_k$ of degree k on $[0, 1]$. Though there is a subtle difference between the point functions h, p_k, $\hat{h}_k$ and the corresponding points $\mathbf{h}$, $\mathbf{p}_k$, $\hat{\mathbf{h}}_k$ in $\mathscr{H}$, we will ignore this difference. Let $E_k = \|\mathbf{h} - \mathbf{h}_k\|^2$ be the measure of error when the function h_k is used to approximate h.

(a) Find the functions $\hat{h}_k$ for $k = 0, 1, 2$. Plot h and these three functions on the same axes. *Hint*: The inner products $(\mathbf{p}_i, \mathbf{p}_j)$ and $(\mathbf{p}_i, \mathbf{h})$ are easy to determine as functions of i and j, so that the matrices $\mathbf{M}$ and $\mathbf{U}$ are easy to determine. If possible use exact arithmetic.

(b) Evaluate E_k for $k = 0, 1, 2$.

(c) Find the Taylor approximation $\mathbf{h}^*$ of $\mathbf{h}$, using constant, linear, and quadratic terms, and expanding about $x = 1/2$. Show that the error $\|\mathbf{h} - \hat{\mathbf{h}}_2\|^2$ is smaller than the error $\|\mathbf{h} - \mathbf{h}^*\|^2$.

(d) Repeat (a) and (b) for $h(x) = 1/(1 + x)$. *Hint*: Let $c_k = (\mathbf{h}, \mathbf{p}_k) = \int_0^1 h(x)p_k(x)\, dx$. Then $c_k = \int_0^1 x^{k-1}[1 - h(x)]\, dx = (1/k) - c_{k-1}$.

1.4 EXAMPLES

In this section we discuss four real data examples, formulate them in terms of vector spaces, and carry out some of the computations. At this point we consider only ways of describing observed vectors $\mathbf{y}$ in terms of a few other vectors $\mathbf{x}_1, \ldots, \mathbf{x}_k$.

Example 1.4.1: In their classic book *Statistical Methods for Research*

Table 1.4.1 Regression of Percentage of Wormy Fruit on Size of Apple Crop

Tree Number	Size of Crop on Tree, X (Hundreds of Fruits)	Percentage of Wormy Fruits Y	Extimate of $\hat{Y}$ $E(Y\|X)$	Deviation from Regression $Y - \hat{Y} = d_{y.x}$
1	8	59	56.14	2.86
2	6	58	58.17	−0.17
3	11	56	53.10	2.90
4	22	53	41.96	11.04
5	14	50	50.06	−0.06
6	17	45	47.03	−2.03
7	18	43	46.01	−3.01
8	24	42	39.94	2.06
9	19	39	45.00	−6.00
10	23	38	40.95	−2.95
11	26	30	37.91	−7.91
12	40	27	23.73	3.27
	$\sum X = 228$	$\sum Y = 540$		
	$\bar{X} = 19$	$\bar{Y} = 45$		
	$\sum X^2 = 5{,}256$	$\sum Y^2 = 25{,}522$	$\sum XY = 9{,}324$	
	$(\sum X)^2/n = 4{,}332$	$(\sum Y)^2/n = 24{,}300$	$(\sum X)(\sum Y)/n = 10{,}260$	

Workers, Snedecor and Cochran (1980, p. 162) present the data of Table 1.4.1 accompanied by this commentary:

> 6.6—Regression of injured fruit on crop size. It is rather generally thought that the intensity of the injury by codling moth larvae is greater on apple trees bearing a small crop. Apparently the density of the flying moths is unrelated to the size of the crop on a tree so that the chance of attack for any particular fruit is augmented if there are few fruits in the tree. The data in table 6.5 are adapted from the results of an experiment (9) containing evidence about this phenomenon. The 12 trees were all given a calyx spray of lead arsenate followed by fine cover sprays made up of 3 pounds of managanese arsenate and 1 quart of fish oil per 100 gallons. There is a decided tendency for the percentage of wormy fruits to decrease as the number of apples in the tree increases.

$$x_i = X_i - \bar{X} \qquad y_i = Y_i - \bar{Y}$$

$$\sum x^2 = 924 \qquad \sum y^2 = 1222 \qquad \sum xy = -926$$

$$b = \sum xy / \sum x^2 = -936/924 = -1.013 \text{ percent per wormy apple}$$

$$\hat{Y} = \bar{Y} + b(X - \bar{X}) = 45 - 1.013(X - 19) = 64.247 - 1.103X$$

$$\sum d_{y.x}^2 = 1.222 - (-936)^2/924 = 273.88$$

$$s_{y.x}^2 = \sum d_{y.x}^2/(n-2) = 273.88/10 = 27.388$$

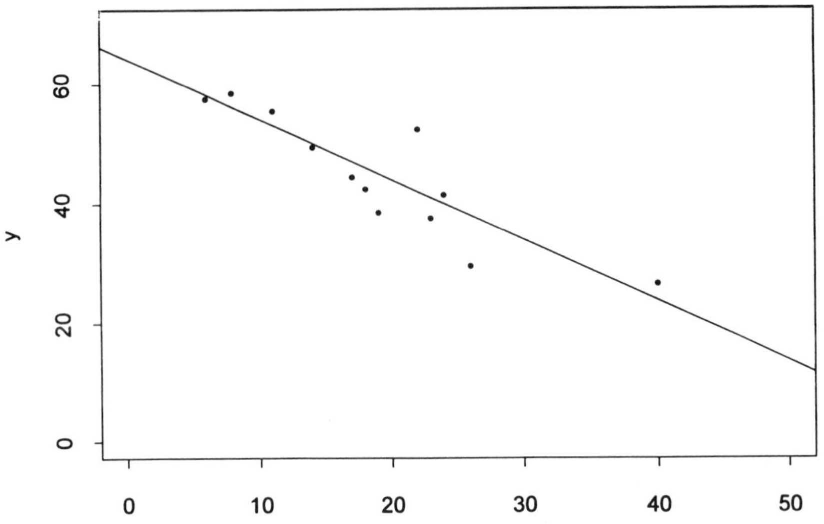

FIGURE 1.5 Regression of percentage of wormy apples on size of apple crop. From *Statistical Methods for Research Workers*, by G. W. Snedecor (1976), Iowa State Press.

The line on the scatter diagram of Figure 1.5 was obtained as follows. Suppose we try to approximate y by a linear function $g(x) = b_0 + b_1 x$. One possible criterion for the choice of the pair (b_0, b_1) is to choose that pair for which

$$Q = Q(b_0, b_1) = \sum_{i=1}^{n} [y_i - (b_0 + b_1 x_i)]^2$$

is minimum. If we define $\mathbf{y}$ and $\mathbf{x}_1$ as 12-component column vectors of y and x values, and $\mathbf{x}_0$ as the 12-component vector of all ones, then

$$Q = \|\mathbf{y} - (b_0 \mathbf{x}_0 + b_1 \mathbf{x}_1)\|^2,$$

so that Q is minimized for $b_0 \mathbf{x}_0 + b_1 \mathbf{x}_1 = \hat{\mathbf{y}}$, the projection of $\mathbf{y}$ onto $\mathscr{L}(\mathbf{x}_0, \mathbf{x}_1)$. Thus, for $\mathbf{X} = (\mathbf{x}_0, \mathbf{x}_1)$, $\mathbf{M} = \mathbf{X}'\mathbf{X}$, $\mathbf{U} = \mathbf{X}'\mathbf{y}$,

$$\mathbf{b} = \begin{pmatrix} b_0 \\ b_1 \end{pmatrix} = \mathbf{M}^{-1}\mathbf{U}.$$

$\mathbf{X}$ is the 12×2 matrix whose first column elements are all ones, and whose second column is the column labeled X in Table 1.4.1. The column vector $\mathbf{y}$ was labeled Y by Snedecor. $\hat{\mathbf{y}}$ and $\mathbf{e} = \mathbf{y} - \hat{\mathbf{y}}$ were labeled $\hat{Y}$ and $d_{y.x}$.

$$\mathbf{M} = \begin{bmatrix} 12 & 228 \\ 228 & 5{,}256 \end{bmatrix} \qquad \mathbf{M}^{-1} = \begin{bmatrix} 0.474\,030 & -0.020\,563 \\ -0.020\,563 & 0.001\,082 \end{bmatrix} \qquad \mathbf{U} = \begin{bmatrix} 540 \\ 9{,}324 \end{bmatrix}$$

$$\mathbf{b} = M^{-1}U = \begin{bmatrix} 64.247 \\ -1.013 \end{bmatrix} \qquad \|\mathbf{y}\|^2 = 25{,}522 \qquad \|\hat{\mathbf{y}}\|^2 = 25{,}248 \qquad \|\mathbf{y} - \hat{\mathbf{y}}\|^2 = 274$$

Notice that $\|\mathbf{y}\|^2 = \|\hat{\mathbf{y}}\|^2 + \|\mathbf{y} - \hat{\mathbf{y}}\|^2$, as should be the case, by the Pythagorean Theorem. Simple computations verify that $\mathbf{e} = \mathbf{y} - \hat{\mathbf{y}}$ is orthogonal to $\mathbf{x}_0$ and $\mathbf{x}_1$, that is, $\sum e_i = 0$ and $\sum e_i x_i = 0$.

We have chosen here to use the more general matrix formulas in order to determine b_0 and b_1 even though nonmatrix formulas were developed in Problem 1.2.3. A complete discussion of the simple linear regression model will be included later.

Example 1.4.2: Consider now the height data of Table 1.1.1. Let us try to approximate the 10-component vector $\mathbf{y}$ with a vector $\hat{\mathbf{y}}$ contained in $\mathscr{L}(\mathbf{x}_0, \mathbf{x}_1, \mathbf{x}_2, \mathbf{x}_3)$, where $\mathbf{x}_0$ is the 10-component column vector of ones and $\mathbf{x}_1$, $\mathbf{x}_2$, $\mathbf{x}_3$ are as given in Table 1.1.1. The approximation vectors are given in Table 1.4.2.

Table 1.4.2

X	**y**	$\hat{\mathbf{y}}$	**e**
1 70 62 1	68.5	68.66	−0.16
1 73 66 1	72.5	72.32	0.18
1 68 67 1	70.0	69.87	0.13
1 72 64 1	71.0	70.78	0.22
1 66 60 1	65.0	65.37	−0.37
1 71 63 0	64.5	63.85	0.65
1 74 68 0	67.5	67.99	−0.49
1 65 65 0	61.5	61.29	0.29
1 70 64 0	63.5	63.74	−0.25
1 69 65 0	63.5	63.63	−0.13

$$\mathbf{M} = \begin{bmatrix} 10 & 698 & 644 & 5 \\ 698 & 48{,}796 & 44{,}977 & 349 \\ 644 & 44{,}977 & 41{,}524 & 319 \\ 5 & 349 & 319 & 5 \end{bmatrix}$$

$$\mathbf{M}^{-1} = \begin{bmatrix} 10{,}927{,}530 & -55{,}341 & -108{,}380 & -150{,}056 \\ -55{,}341 & 1{,}629 & -898 & -1{,}077 \\ -108{,}380 & -898 & 2{,}631 & 3{,}158 \\ 150{,}056 & 1{,}077 & 3{,}158 & 43{,}789 \end{bmatrix} 10^{-5}$$

$$\mathbf{U} = \mathbf{X}'\mathbf{y} \begin{bmatrix} 667.5 \\ 46{,}648.0 \\ 43{,}008.5 \\ 347.0 \end{bmatrix} \qquad \mathbf{b} = \begin{bmatrix} -7.702 \\ 0.585 \\ 0.477 \\ 5.872 \end{bmatrix} \qquad \sum e_i^2 = \|\mathbf{e}\|^2 = 0.085\,75$$

The height y seems to be predicted very nicely by x_1 (father's height), x_2 (mother's height) and x_3 (sex). We must be cautious, however, in interpreting such an analysis based on 10 observations with 4 independent variables. Predictions of heights for other people, based on the coefficients determined for these data, should not be expected to be as good.

Example 1.4.3 (Snedecor, 1967, p. 278):

EXAMPLE 10.12.1—The numbers of days survived by mice inoculated with three strains of typhoid organisms are summarized in the following frequency distributions. Thus, with strains 9D, 6 mice survived for 2 days, etc. We have $n_1 = 31$, $n_2 = 60$, $n_3 = 133$, $N = 224$. The purpose of the analysis is to estimate and compare the mean numbers of days to death for the three strains.

Since the variance for strain 9D looks much smaller than for the other strains, it seems wise to calculate s_i^2 separately fro each strain, rather than use a pooled s^2 from the analysis of variance.

The calculations are given under Table 1.4.3. Again from Snedecor (1967) consider the variable days to death for three strains of typhoid organism. Let $\mathbf{y}$ be the table with three columns, having the days to death for 31 mice on 9D in column 1, for 60 mice on 11C in column 2, and 133 mice on DSC1 in column 3. Thus $\mathbf{y}$ has 224 components. Let y_{ij} be the jth component in the ith column of $\mathbf{y}$. Let $\mathbf{x}_1, \mathbf{x}_2, \mathbf{x}_3$ be the indicators of columns 1, 2, 3. The best approximation to $\mathbf{y}$ by vectors in $\mathscr{L}(\mathbf{x}_1, \mathbf{x}_2, \mathbf{x}_3) = V$ in the least squares sense is

$$\hat{\mathbf{y}} = p(\mathbf{y} \mid V) = \sum_{i=1}^{3} p(\mathbf{y} \mid \mathbf{x}_i) = \sum_{i=1}^{3} \bar{y}_i \mathbf{x}_i$$

The second equality follows by the orthogonality of $\mathbf{x}_1, \mathbf{x}_2, \mathbf{x}_3$. $\bar{y}_i$ is the mean of the values of y in the ith column. Thus $\hat{\mathbf{y}}$ is the array with 31 $\bar{y}_1$'s in column 1, 60 $\bar{y}_2$'s in column 2, $133\bar{y}_3$'s in column 3. Easy computation (remembering, for example, that 4 occurs nine times in column 1) shows that

$$\sum_j Y_{1j} = 125, \qquad \sum_j Y_{2j} = 442, \qquad \text{and} \qquad \sum_j Y_{3j} = 1{,}037.$$

We find $\bar{y}_1 = 4.032$, $\bar{y}_2 = 7.367$, $\bar{y}_3 = 7.797$, and the error sum of squares $\|\mathbf{e}\|^2 = \sum_{ij} (y_{ij} - \bar{y}_i)^2 = 1{,}278.42$, $\|\hat{\mathbf{y}}\|^2 = \sum_i n_i \bar{y}_i^2 = 11{,}845.58$, and $\|\mathbf{y}\|^2 = \sum_{ij} y_{ij}^2 = 13{,}124$.

Table 1.4.3

Days to Death	Numbers of Mice Inoculated with Indicated Strain			
	9D	11C	DSCl	Total
2	6	1	3	10
3	4	3	5	12
4	9	3	5	17
5	8	6	8	22
6	3	6	19	28
7	1	14	23	38
8		11	22	33
9		4	14	18
10		6	14	20
11		2	7	9
12		3	8	11
13		1	4	5
14			1	1
Total	31	60	133	224
$\sum X$	125	442	1,037	1,604
$\sum X^2$	561	3,602	8,961	13,124

Example 1.4.4: The following data were given in a problem in Dixon and Massey (1957, p. 185):

> The drained weight in ounces of frozen apricots was measured for various types of syrups and various concentrations of syrup. The original weights of the apricots were the same. Differences in drained weights would be attributable to differences in concentrations or type of syrups.

		Syrup Composition				
		All Sucrose	2/3 Sucrose 1/3 Corn Syrup	1/3 Sucrose 2/3 Corn Syrup	All Corn Syrup	$\bar{y}_{i\cdot}$
Conc. of Syrup	30	28.80	28.21	29.28	29.12	28.853
	40	29.12	28.64	29.12	30.24	29.280
	50	29.76	30.40	29.12	28.32	29.400
$\bar{y}_{\cdot j}$		29.227	29.083	29.173	29.227	$\bar{y}_{\cdot\cdot} = 29.178$

Let $\mathbf{y}$ be the 3×4 matrix of drained weights. Let us approximate $\mathbf{y}$ by a linear combination of indicator vectors for rows and columns. Define $\mathbf{R}_i$

to be the indicator of row i and $\mathbf{C}_j$ to be the indicator of column j. Thus, for example,

$$\mathbf{R}_2 = \begin{bmatrix} 0 & 0 & 0 & 0 \\ 1 & 1 & 1 & 1 \\ 0 & 0 & 0 & 0 \end{bmatrix} \quad \text{and} \quad \mathbf{C}_3 = \begin{bmatrix} 0 & 0 & 1 & 0 \\ 0 & 0 & 1 & 0 \\ 0 & 0 & 1 & 0 \end{bmatrix}$$

Take $V = \mathscr{L}(\mathbf{R}_1, \mathbf{R}_2, \mathbf{R}_3, \mathbf{C}_1, \ldots, \mathbf{C}_4)$. Define $\mathbf{x}_0$ to be the 3×4 vector of all ones. Then $\mathbf{x}_0 = \sum_i \mathbf{R}_i = \sum_j \mathbf{C}_j$. Let $\bar{y}_{i\cdot}$, $\bar{y}_{\cdot j}$ and $\bar{y}_{\cdot\cdot}$ be the mean of the ith row, the jth column, and the overall mean, respectively. It is not difficult to show that V has dimension $4 + 3 - 1 = 6$, and that $\hat{\mathbf{y}} = \hat{\mathbf{y}}_0 + \hat{\mathbf{y}}_R + \hat{\mathbf{y}}_C$, where

$$\hat{\mathbf{y}}_0 = p(\mathbf{y}|\mathbf{x}_0) = \bar{y}_{\cdot\cdot}\mathbf{x}_0 = \begin{bmatrix} 29.178 & 29.178 & 29.178 & 29.178 \\ 29.178 & 29.178 & 29.178 & 29.178 \\ 29.178 & 29.178 & 29.178 & 29.178 \end{bmatrix},$$

$$\hat{\mathbf{y}}_R = \sum_i (\bar{y}_{i\cdot} - \bar{y}_{\cdot\cdot})\mathbf{R}_i$$

$$= \begin{bmatrix} -0.325 & -0.325 & -0.325 & -0.325 \\ 0.102 & 0.102 & 0.102 & 0.102 \\ 0.222 & 0.222 & 0.222 & 0.222 \end{bmatrix},$$

$$\hat{\mathbf{y}}_C = \sum_i (\bar{y}_{\cdot j} - \bar{y}_{\cdot\cdot})\mathbf{C}_j$$

$$= \begin{bmatrix} 0.049 & -0.095 & -0.005 & 0.049 \\ 0.049 & -0.095 & -0.005 & 0.049 \\ 0.049 & -0.095 & -0.005 & 0.049 \end{bmatrix}.$$

Notice that $\hat{\mathbf{y}}_0$, $\hat{\mathbf{y}}_R$, and $\hat{\mathbf{y}}_C$ are orthogonal and that the ij element of $\hat{\mathbf{y}}$ is $\hat{y}_{ij} = \bar{y}_{\cdot\cdot} + (\bar{y}_{i\cdot} - \bar{y}_{\cdot\cdot}) + (\bar{y}_{\cdot j} - \bar{y}_{\cdot\cdot})$. Therefore

$$\hat{\mathbf{y}} = \begin{bmatrix} 28.902 & 28.758 & 28.848 & 28.902 \\ 29.329 & 29.186 & 29.276 & 29.329 \\ 29.449 & 29.306 & 29.396 & 29.449 \end{bmatrix},$$

$$\mathbf{e} = \begin{bmatrix} -0.102 & -0.548 & 0.432 & 0.218 \\ -0.209 & -0.546 & -0.156 & 0.911 \\ 0.311 & 1.094 & -0.276 & -1.129 \end{bmatrix}$$

Further computation gives

$$\|\mathbf{y}\|^2 = \sum_{ij} y_{ij}^2 = 10{,}221$$

$$\|\hat{\mathbf{y}}\|^2 = \|\hat{\mathbf{y}}_O\|^2 + \|\hat{\mathbf{y}}_R\|^2 + \|\hat{\mathbf{y}}_C\|^2$$

$$= \bar{y}_{..}^2(12) + 4\sum_i (\bar{y}_{i\cdot} - \bar{y}_{..})^2 + 3\sum_j (\bar{y}_{\cdot j} - \bar{y}_{..})^2$$

$$= 10{,}215.92 + 0.66 + 0.04 = 10{,}216.62$$

$$\|\mathbf{e}\|^2 = \|\mathbf{y} - \hat{\mathbf{y}}\|^2 = 4.38$$

showing again that the Pythagorean Theorem holds.

Later, after we formulate probability models, and discuss their properties, we will be able to draw further conclusions about the contributions of concentration and composition to variation in drainage weight.

1.5 SOME HISTORY

In his scholarly and fascinating history of the development of statistics before 1900, Stephen Stigler (1986) begins his first chapter, entitled "Least Squares and the Combination of Observations," with the following:

> The method of least squares was the dominant theme—the leitmotif—of nineteenth-century statistics. In several respects it was to statistics what the calculus had been to mathematics a century earlier. "Proofs" of the method gave direction to the development of statistical theory, handbooks explaining its use guided the application of the higher methods, and disputes on the priority of its discovery signaled the intellectual community's recognition of the method's value. Like the calculus of mathematics, this "calculus of observations" did not spring into existence without antecedents, and the exploration of its subtleties and potential took over a century. Throughout much of this time statistical methods were referred to as "the combination of observations." This phrase captures a key ingredient of the method of least squares and describes a concept whose evolution paced the method's development. The method itself first appeared in print in 1805.

Stigler refers to Adrien-Marie Legendre (1752–1833), who in 1805 wrote an eight-page book *Nouvelles méthodes pour le détermination des orbites des comètes* (New methods for the determination of the orbit of the planets), with a nine-page appendix, "Sur la méthode des maindres quarres" (On the method of least squares). Legendre began the appendix with a statement of his objective; here is Stigler's translation:

In most investigations where the object is to deduce the most accurate possible results from observational measurements, we are led to a system of equations of the form

$$E = a + bx + cy + fz + \cdots,$$

in which a, b, c, $f, \ldots$ are known coefficients, varying from one equation to the other, and x, y, $z, \ldots$ are known quantities, to be determined by the condition that each value of E is reduced either to zero, or to a very simple quantity.

In today's notation we might make the substitutions $E = -\varepsilon_i$, $-a = Y_i$, $b = x_{1i}$, $x = \beta_1$, $c = x_{2i}$, $y = \beta_2$, etc., and write the model as $-a = bx + cy + \cdots - E$ or $Y_i = \beta_1 x_{1i} + \cdots + \beta_k x_{ki} + \varepsilon_i$ or even as $\mathbf{Y} = \beta_1 \mathbf{x}_1 + \cdots + \beta_k \mathbf{x}_k + \boldsymbol{\varepsilon} = \mathbf{X}\beta + \boldsymbol{\varepsilon}$.

Again in Stigler's translation, Legendre wrote

> Of all the principles that can be proposed for this purpose, I think there is none more general, more exact, or more easy to apply, than that which we have used in this work; it consists of making the sum of squares of the errors a minimum. By this method, a kind of equilibrium is established among the errors which, since it prevents the extremes from dominating, is appropriate for revealing the state of the system which most nearly approaches the truth.

Legendre gave an example using data from the 1795 survey of the French meridian arc, in which there were $n = 5$ observations and $k = 3$ unknown parameters.

Though Carl Friedrich Gauss claimed in 1809 that he had used the method of least squares as early as 1795, it seems clear from published writings that Legendre should be given credit for the first development of least squares.

The statistical problem solved by Legendre had been faced earlier by astronomer Johann Tobias Mayer (1723–62), mathematician Leonhard Euler (1707–83) and scientist and mathematician Pierre-Simon Laplace (1749–1827) in considering astronomic data. We will illustrate their earlier solutions on some data concerning the motion of Saturn studied by Laplace in 1787. Table 1.5.1 is taken from Stigler's book.

Using Legendre's notation, these eighteenth century scientists considered the problem of solving the "equations"

$$E_i = a_i + w + b_i x + c_i y + d_i z \qquad (i = 1, \ldots, 24) \tag{1.5.1}$$

given by setting the E_i's all equal to zero. Observations were made on 24 occasions when Saturn, the moon, and earth were aligned over 200 years. The dependent variable a_i was the difference between the observed longitude of Saturn and that predicted by Laplace's theory. The measurements b_i, c_i, d_i were simple functions of observations made on the orbit of Saturn at those times.

They knew (or would have known) that those 24 equations in four unknowns (w, x, y, z) had no single solutions and that therefore all the E_i's could not be made zero. Mayer's idea was to reduce his collection of equations to a number equal to the number of unknowns by adding across equations. In Mayer's case he had 27 equations with three unknowns, so he grouped the 27 equations into three groups of 9 each, and simply added coefficients to get 3 equations in three unknowns. As applied to the data of the Table 1.5.1 we could add the first 6, next 6, etc. to get 4 equations in four unknowns. Mayer chose the subset of equations to add according to the sizes of the coefficients, grouping large a_i's together, etc.

Euler had available observations on Saturn and Jupiter for the years 1582–1745 ($n = 75$) and had $k = 6$ unknowns. He did not combine observations as did Mayer but instead tried to solve for his unknowns by using some periodicity of the coefficients to reduce the number of unknowns and by considering small sets of observations, trying to verify solutions on other small sets. He was largely unsuccessful, and wrote (Stigler's translation)

> Now, from these equations we can conclude nothing; and the reason, perhaps, is that I have tried to satisfy several observations exactly, whereas I should have only satisfied then approximately; and this error has then multiplied itself.

Thus, the most prolific of mathematicians, perhaps the greatest of analysts, failed even to proceed as far as Mayer.

In 1787 Laplace, eulogized by Poisson in 1827 as "the Newton of France" (Stigler 1986, p. 31), and perhaps the greatest contributor to probability and statistics before 1900, considered the Saturn data of Table 1.5.1. Laplace reduced the 24 equations in four unknowns to 4 equations. The first new equation was the sum of all equations. The second was the difference between the sum of the first 12 and the sum of the second 12. The third was the sum of equations 3, 4, 10, 11, 17, 18, 23, 24 minus the sum of equations 1, 7, 14, 20, the fourth was the sum of equations 2, 8, 9, 15, 16, 21, 22 minus the sum of equations 5, 6, 13, 19. Stigler describes some of Laplace's motivation, which now seems quite valid: Laplace obtained his jth equation by multiplying the original ith equation by a constant k_{ij} and then adding over i. His jth equation was therefore

$$0 = \sum_i k_{ij} a_i + x \sum_i k_{ij} b_i + y \sum_i k_{ij} c_i + z \sum_i k_{ij} d_i \qquad (1.5.2)$$

Laplace's k_{ij} were all 1, -1 or 0. Mayer's had all been 0 or 1. Legendre showed that the method of least squares leads to taking $k_{i1} = 1$, $k_{i2} = b_i$, $k_{i3} = c_i$, $k_{i4} = d_i$.

The column in Table 1.5.1 "Halley Residual" had been derived by Edmund Halley in 1676 using a different theory. Details are omitted.

In 1809 Gauss showed the connections among normally distributed errors, most probable parameter values (maximum likelihood estimates) and least

Table 1.5.1 Laplace's Saturn Data.*

Eq. no.	Year (i)	$-a_i$	b_i	c_i	d_i	Laplace Residual	Halley Residual	L.S. Residual
1	1591	1′11.9″	−158.0	0.22041	−0.97541	+1′33″	−0′54″	+1′36″
2	1598	3′32.7″	−151.78	0.99974	−0.02278	−0.07	+0.37	+0.05
3	1660	5′12.0″	−89.67	0.79735	0.60352	−1.36	+2.58	−1.21
4	1664	3′56.7″	−85.54	0.04241	0.99910	−0.35	+3.20	−0.29
5	1667	3′31.7″	−82.45	−0.57924	0.81516	−0.21	+3.50	−0.33
6	1672	3′32.8″	−77.28	−0.98890	−0.14858	−0.58	+3.25	−1.06
7	1679	3′9.9″	−70.01	0.12591	−0.99204	−0.14	−1.57	−0.08
8	1687	4′49.2″	−62.79	0.99476	0.10222	−1.09	−4.54	−0.52
9	1690	3′26.8″	−59.66	0.72246	0.69141	+0.25	−7.59	+0.29
10	1694	2′4.9″	−55.52	−0.07303	0.99733	+1.29	−9.00	+1.23
11	1697	2′37.4″	−52.43	−0.66945	0.74285	+0.25	−9.35	+0.22
12	1701	2′41.2″	−48.29	−0.99902	−0.04435	+0.01	−8.00	−0.07
13	1731	3′31.4″	−18.27	−0.98712	−0.15998	−0.47	−4.50	−0.53
14	1738	4′9.5″	−11.01	0.13759	−0.99049	−1.02	−7.49	−0.56
15	1746	4′58.3″	−3.75	0.99348	0.11401	−1.07	−4.21	−0.50
16	1749	4′3.8″	−0.65	0.71410	0.70004	−0.12	−8.38	+0.03
17	1753	1′58.2″	3.48	−0.08518	0.99637	+1.54	−13.39	+1.41
18	1756	1′35.2″	6.58	−0.67859	0.73452	+1.37	−17.27	+1.35
19	1760	3′14.0″	10.72	−0.99838	−0.05691	−0.23	−22.17	−0.29
20	1767	1′40.2″	17.98	0.03403	−0.99942	+1.29	−13.12	+1.34
21	1775	3′46.0″	25.23	0.99994	0.01065	+0.19	+2.12	+0.26
22	1778	4′32.9″	28.33	0.78255	0.62559	−0.34	+1.21	−0.19
23	1782	4′4.4″	32.46	0.01794	0.99984	−0.23	−5.18	−0.15
24	1785	4′17.6″	35.56	−0.59930	0.80053	−0.56	−12.07	−0.57

* Residuals are fitted values minus observed values.

Source: Laplace (1788). Reprinted with permission from *The History of Statistics: The Measurement of Uncertainty before 1900* by Stephen M. Stigler, Cambridge, MA: The Belknap Press of Harvard University Press.

squares. In 1810 Laplace published his central limit theorem and argued that this could justify the assumption of normally distributed errors, hence least squares. Laplace showed in 1811 that, at least, asymptotically, least squares estimators are normally distributed, and they are less variable than other linear estimators, i.e., solutions of (1.5.1). Normality of the errors was not needed.

In 1823 Gauss showed that the asymptotic argument was unnecessary, that the variability of the solutions to (1.5.1) could be studied algebraically, and that least squares estimators had least variability. We will make this precise in Sections 3.3 and 3.4 with a discussion of the famous Gauss–Markov Theorem. The least squares theory and applications developed by Legendre, Gauss and Laplace were widely published. Stigler cites a compilation by Mansfield Merriman in 1877 of "writings related to the method of least squares," including 70 titles between 1805 and 1834, and 179 between 1835 and 1864.

1.6 PROJECTION OPERATORS

The purpose of this section is to study the transformation $P_V: \mathbf{y} \to \hat{\mathbf{y}}$ which transforms a vector $\mathbf{y} \in \Omega$ into its projection $\hat{\mathbf{y}}$ on a subspace V.

In applications a vector $\mathbf{y}$ will be observed. The model under consideration will specify that $\mathbf{y} = \boldsymbol{\theta} + \boldsymbol{\varepsilon}$, for $\boldsymbol{\theta} \in V$, a known subspace of Ω, with $\boldsymbol{\varepsilon}$ a random vector, both $\boldsymbol{\theta}$ and $\boldsymbol{\varepsilon}$ unknown. We will usually estimate $\boldsymbol{\theta}$ by the projection of $\mathbf{y}$ onto V. We should therefore understand the properties of this projection as well as possible.

The transformation $P: \mathbf{y} \to p(\mathbf{y} \mid V)$ for a subspace V is linear, since $p(\alpha\mathbf{y} \mid V) = \alpha p(\mathbf{y} \mid V)$ and $p(\mathbf{y}_1 + \mathbf{y}_2 \mid V) = p(\mathbf{y}_1 \mid V) + p(\mathbf{y}_2 \mid V)$. (The student should check this.)

Since $\hat{\mathbf{y}} = p(\mathbf{y} \mid V)$ implies that $p(\hat{\mathbf{y}} \mid V) = \hat{\mathbf{y}}$, the projection operator P is idempotent, i.e., $P^2 = P$. In addition, P is self-adjoint, since for each $\mathbf{x}, \mathbf{y} \in \Omega$,

$$(P\mathbf{x}, \mathbf{y}) = (P\mathbf{x}, P\mathbf{y}) = (\mathbf{x}, P\mathbf{y}).$$

If Ω is the space of n-component column vectors, this means P may be represented as a symmetric matrix, a *projection matrix*. Thus, for this case the projection operator onto V is an $n \times n$ matrix $\mathbf{P}_V$ such that

$$\mathbf{P}_V' = \mathbf{P}_V \qquad \text{and} \qquad \mathbf{P}_V^2 = \mathbf{P}_V.$$

For $V = \mathscr{L}(\mathbf{x}_1, \ldots, \mathbf{x}_k)$ with $\mathbf{x}_1, \ldots, \mathbf{x}_k$ linearly independent column vectors, we have

$$p(\mathbf{y} \mid V) = \mathbf{X}\mathbf{b} = \mathbf{X}(\mathbf{X}'\mathbf{X})^{-1}\mathbf{X}'\mathbf{y},$$

where $\mathbf{X} = (\mathbf{x}_1, \ldots, \mathbf{x}_k)$, so that

$$\mathbf{P}_V = \mathbf{X}(\mathbf{X}'\mathbf{X})^{-1}\mathbf{X}'.$$

It is easy to check that $\mathbf{P}_V$ is symmetric and idempotent.

Example 1.6.1: For simplicity we will refer to a projection operator as *projection.*

1. $\mathbf{P} = \begin{bmatrix} 1 & 0 & 0 \\ 0 & 1 & 0 \\ 0 & 0 & 0 \end{bmatrix}$ = projection onto the linear subspace of vectors $\begin{bmatrix} y_1 \\ y_2 \\ 0 \end{bmatrix}$ spanned by $\begin{bmatrix} 1 \\ 0 \\ 0 \end{bmatrix}$ and $\begin{bmatrix} 0 \\ 1 \\ 0 \end{bmatrix}$.
2. $\mathbf{P} = \frac{1}{n}\mathbf{J}_n\mathbf{J}_n'$ = projection onto $\mathbf{J}_n$, the column vector fo n 1's. Then $\mathbf{P}\mathbf{x} = \bar{x}\mathbf{J}_n$, where $\bar{x} = (\mathbf{x}, \mathbf{J}_n)/\|\mathbf{J}_n\|^2 = (\sum x_i)/n$.
3. $\mathbf{P} = I_n - \frac{1}{n}\mathbf{J}_n\mathbf{J}_n'$ = projection onto the subspace of column vectors whose components add to zero, i.e., are orthogonal to $\mathbf{J}_n$. $\mathbf{P}$ adjusts $\mathbf{y}$ by subtracting $\bar{y}$ from all components. $\mathbf{Py}$ is the vector of deviations $y_i - \bar{y}$.
4. $\mathbf{P} = \mathbf{v}\mathbf{v}'/\|\mathbf{v}\|^2$ = projection onto the one-dimensional subspace $\mathscr{L}(\mathbf{v})$.
5. $\mathbf{P} = \begin{bmatrix} 1/2 & 1/2 & 0 \\ 1/2 & 1/2 & 0 \\ 0 & 0 & 1 \end{bmatrix}$ = projection onto the subspace spanned by $\begin{bmatrix} 1 \\ 1 \\ 0 \end{bmatrix}$ and $\begin{bmatrix} 0 \\ 0 \\ 1 \end{bmatrix}$. Thus, $\mathbf{P}\begin{bmatrix} y_1 \\ y_2 \\ y_3 \end{bmatrix} = \begin{bmatrix} (y_1 + y_2)/2 \\ (y_1 + y_2)/2 \\ y_3 \end{bmatrix}$.

Problem 1.6.1: Show that for $\underset{n\times k}{\mathbf{W}} = \underset{n\times k}{\mathbf{X}}\ \underset{k\times k}{\mathbf{B}}$ with $\mathbf{B}$ nonsingular, $\mathbf{X}(\mathbf{X}'\mathbf{X})^{-1}\mathbf{X}'$ remains unchanged if $\mathbf{X}$ is replaced by $\mathbf{W}$. Thus, $\mathbf{P}$ is a function of the subspace spanned by the columns of $\mathbf{X}$, not of the particular basis chosen for this subspace.

Theorem 1.6.1: Let A be a linear operator on Ω which is idempotent and self-adjoint. Then A is the projection operator onto the range of A.

***Proof*:** We must show that for all $\mathbf{y} \in \Omega$, and $\mathbf{x} \in R \equiv$ Range of A, $(A\mathbf{y}, \mathbf{x}) = (\mathbf{y}, \mathbf{x})$. If $\mathbf{x} \in R$ then $\mathbf{x} = A\mathbf{z}$ for some $\mathbf{z} \in \Omega$. But $(A\mathbf{y}, \mathbf{x}) = (\mathbf{y}, A\mathbf{x})$ by self-adjointness (symmetry) and $A\mathbf{x} = AA\mathbf{z} = A\mathbf{z} = \mathbf{x}$ because A is idempotent. □

Problem 1.6.2: Prove that the projection operator onto $V^\perp$, the collection of vectors in Ω orthogonal to V, is $I - P_V$. (I is the identity transformation.)

Subspace $V_0 \subset V$: Let V be a subspace of Ω and let V_0 be a subspace of V. Let P and P_0 be the corresponding projection operators. Then

$$(1)\quad PP_0 = P_0 \qquad \text{and} \qquad (2)\quad P_0 P = P_0.$$

Equivalently, if $\hat{\mathbf{y}} = P(\mathbf{y}|V)$ and $\hat{\mathbf{y}}_0 = P(\mathbf{y}|V_0)$ then (1) $p(\hat{\mathbf{y}}_0|V) = \hat{\mathbf{y}}_0$ and (2) $p(\hat{\mathbf{y}}|V_0) = \hat{\mathbf{y}}_0$. It is easy to check these equalities by merely noting in (1) that $\hat{\mathbf{y}}_0 \in V$ and $(\mathbf{v}, \hat{\mathbf{y}}_0) = (\mathbf{v}, \mathbf{y})$ for all $\mathbf{v} \in V_0$, and in (2) that $\hat{\mathbf{y}}_0 \in V_0$ and $(\mathbf{v}, \hat{\mathbf{y}}_0) = (\mathbf{v}, \hat{\mathbf{y}})$ for all $\mathbf{v} \in V_0$.

Direct Sums: In regression analysis and, in particular, in the analysis of variance, it will often be possible to decompose the space Ω or a subspace V into smaller subspaces, and therefore to increase understanding of the variation in the observed variable. If these smaller subspaces are mutually orthogonal, simple computational formulas and useful intepretations often result.

For any linear model it will be convenient to decompose Ω into the subspace V, and the error space $V^\perp$, so that every observation vector $\mathbf{y}$ is the sum of a vector in V and a vector in $V^\perp$.

In Example 1.4.4 V may be decomposed into the spaces $V_0 = \mathscr{L}(\mathbf{x}_0)$, $V_R = \left\{\sum_1^3 a_i\mathbf{R}_i | \sum_1^3 \mathbf{a}_i = 0\right\}$, $V_C = \left\{\sum_1^4 b_j\mathbf{C}_j | \sum b_j = 0\right\}$, so that every vector in V is the sum of its projections onto these three orthogonal subspaces. It follows that every vector $\mathbf{y}$ in Ω is the sum of four orthogonal vectors, each being the projection of $\mathbf{y}$ onto one of the four orthogonal subspaces V_0, V_R, V_C, $V^\perp$. These subspaces were chosen for their simplicity. As will be seen in later chapters, Chapter 6 in particular, the decomposition of V into orthogonal subspaces, each of a relatively simple structure, provides increased understanding of the variation in the components of $\mathbf{y}$.

Definition 1.6.1: Subspaces $V_1, \ldots, V_k$ of Ω are linearly independent if $\mathbf{x}_i \in V_i$ for $i = 1, \ldots, k$ and $\sum_{j=1}^{k} \mathbf{x}_i = \mathbf{0}$ implies that $\mathbf{x}_i = \mathbf{0}$ for $i = 1, \ldots, k$.

Let $\mathscr{M}_{ij}$ denote the property: $V_i \cap V_j = \{\mathbf{0}\}$. For $i \neq j$ linear independence of V_i and V_j is equivalent to $\mathscr{M}_{ij}$, so that linear independence of $V_1, \ldots, V_k$ implies $\mathscr{M}$: $[\mathscr{M}_{ij}$ for all $i \neq j]$. However, $\mathscr{M}$ does not imply linear independence of $V_1, \ldots, V_k$. Students are asked to prove these statements in Problem 1.6.12. Thus, linear independence of subspaces is analogous to independence of events. Pairwise independence does not imply independence of more than two events.

Definition 1.6.2: Let $V_1, \ldots, V_k$ be subspaces of Ω. Then

$$V = \left\{\mathbf{x} \mid \mathbf{x} = \sum_1^k \mathbf{x}_i, \mathbf{x}_i \in V_i, i = 1, \ldots, k\right\}$$

is called the *direct sum* of $V_1, \ldots, V_k$, and is denoted by

$$V = V_1 + V_2 + \cdots + V_k.$$

If these subspaces are linearly independent we will write

$$V = V_1 \oplus V_2 \oplus \cdots \oplus V_k.$$

The use of the $\oplus$ symbol rather than the $+$ symbol implies that the corresponding subspaces are linearly independent.

Theorem 1.6.2: The representation $\mathbf{x} = \sum_1^k \mathbf{x}_i$ for $\mathbf{x}_i \in V_i$ of elements $\mathbf{x} \in V = V_1 + V_2 + \cdots + V_k$ is unique if and only if the subspaces $V_1, \ldots, V_k$ are linearly independent.

***Proof*:** Suppose that these subspaces are linearly independent. Let $\mathbf{x} = \sum_1^k \mathbf{x}_i = \sum_1^k \mathbf{w}_i$ for $\mathbf{x}_i, \mathbf{w}_i \in V_i$, $i = 1, \ldots, k$. Then $\sum_{i=1}^k (\mathbf{x}_i - \mathbf{w}_i) = 0$ implying, by the linear independence of the V_i, that $\mathbf{x}_i - \mathbf{w}_i = \mathbf{0}$ for each i.

Suppose that the representation is unique, let $\mathbf{v}_i \in V_i$ for $i = 1, \ldots, k$, and let $\sum_{j=1}^k \mathbf{v}_i = \mathbf{0}$. Since $\mathbf{0} \in V_i$ for each i, and $\mathbf{0} = \mathbf{0} + \cdots + \mathbf{0}$, it follows that $\mathbf{v}_i = \mathbf{0}$ for each i, implying the independence of $V_1, \ldots, V_k$. □

Theorem 1.6.3: If $\{\mathbf{v}_{ij} \mid j = 1, \ldots, n_i\}$ is a basis for V_i for $i = 1, \ldots, k$ and $V_1, \ldots, V_k$ are linearly independent, then $\{v_{ij} \mid j = 1, \ldots, n_i, i = 1, \ldots, k\}$ is a basis for $V = V_1 \oplus \cdots \oplus V_k$.

***Proof*:** For any $\mathbf{x} = \sum_1^k x_i$ for $\mathbf{x}_i \in V_i$, suppose $\mathbf{x}_i = \sum_1^{n_i} b_{ij}\mathbf{v}_{ij}$. Thus, $\mathbf{x} = \sum_{ij} b_{ij}\mathbf{v}_{ij}$, so the $\mathbf{v}_{ij}$ span V. It is enough then to show that the $\mathbf{v}_{ij}$ are linearly independent. Suppose $\sum_{ij} c_{ij}\mathbf{v}_{ij} = \mathbf{0}$ for some c_{ij}'s. By the independence of $V_1, \ldots, V_k$, $\sum_j c_{ij}\mathbf{v}_{ij} = \mathbf{0}$ for each i. The independence of $\mathbf{v}_{i1}, \ldots, \mathbf{v}_{in_i}$ then implies $c_{ij} = 0$ for all j and i. □

Corollary: If $V = V_1 \oplus V_2 \oplus \cdots \oplus V_k$ then

$$\dim(V) = \dim(V_1) + \cdots + \dim(V_k).$$

Definition 1.6.3: For any subspace V of Ω, the collection of all vectors in Ω which are orthogonal to V is called the *orthogonal complement* of V. This orthogonal complement will be denoted by $V^\perp$, read "vee-perp".

It is easy to verify that $V^\perp$ is a subspace, and that $P_{V^\perp} = I - P_V$. Since $V^\perp \cap V = \{\mathbf{0}\}$, $V^\perp$ and V are linearly independent.

Theorem 1.6.4: Let V_1 and V_2 be subspaces of Ω. Then

$$(V_1 + V_2)^\perp = V_1^\perp \cap V_2^\perp \qquad \text{and} \qquad (V_1 \cap V_2)^\perp = V_1^\perp + V_2^\perp.$$

***Proof*:** We prove only the first equality. The second is proved similarly. Suppose $\mathbf{v} \in (V_1 + V_2)^\perp$. Then for each element $\mathbf{x} \in V_1 + V_2$, it follows that $\mathbf{v} \perp \mathbf{x}$. In particular, $\mathbf{v} \perp \mathbf{x}_1$, for each $\mathbf{x}_1 \in V_1$ and $\mathbf{v} \perp \mathbf{x}_2$ for each $\mathbf{x}_2 \in V_2$. Thus $\mathbf{v} \in V_1^\perp \cap V_2^\perp$ and $(V_1 + V_2)^\perp \subset V_1^\perp \cap V_2^\perp$.

If $\mathbf{v} \in V_1^\perp \cap V_2^\perp$, then $\mathbf{v} \perp \mathbf{x}_1$, $\mathbf{v} \perp \mathbf{x}_2$ for all $\mathbf{x}_1 \in V_1$, $\mathbf{x}_2 \in V_2$. It follows that $\mathbf{v} \perp (b_1\mathbf{x}_1 + b_2\mathbf{x}_2)$ for all scalars b_1, b_2, and all $\mathbf{x}_1 \in V_1$, $\mathbf{x}_2 \in V_2$, hence that $\mathbf{v} \in (V_1 + V_2)^\perp$. Thus, $(V_1 + V_2)^\perp \supset V_1^\perp \cap V_2^\perp$. □

Theorem 1.6.4 is the linear space version of De Morgan's Laws for sets:

$$(A \cup B)^c = A^c \cap B^c \qquad \text{and} \qquad (A \cap B)^c = A^c \cup B^c.$$

Theorem 1.6.5: For any subspace V and any $\mathbf{x} \in \Omega$, there exist unique elements $\mathbf{x}_1$, $\mathbf{x}_2$ such that $\mathbf{x} = \mathbf{x}_1 + \mathbf{x}_2$, $\mathbf{x}_1 = p(\mathbf{x}|V)$ and $\mathbf{x}_2 = p(\mathbf{x}|V^\perp)$.

***Proof*:** For existence take $\mathbf{x}_1 = p(\mathbf{x}|V)$, $\mathbf{x}_2 = \mathbf{x} - \mathbf{x}_1$. Uniqueness follows from the linear independence of $V^\perp$ and V. □

Example 1.6.2: Let Ω be the space of 4-component row vectors. Let $\mathbf{x}_1 = (1, 1, 1, 1)$, $\mathbf{x}_2 = (1, 1, 0, 0)$, $\mathbf{x}_3 = (1, 0, 1, 0)$, $V_1 = \mathscr{L}(\mathbf{x}_1, \mathbf{x}_2)$, $V_2 = \mathscr{L}(\mathbf{x}_3)$. Then V_1 and V_2 are linearly independent, so that $V \equiv V_1 \oplus V_2 = \{(a + b + c, a + b, a + c, a) | a, b, c \text{ real numbers}\}$ has dimension 3.

$$V_1^\perp = \{(a, -a, b, -b) | a, b \text{ real}\}$$

$$V_2^\perp = \{(a, b, -a, c) | a, b, c \text{ real}\}$$

$$V^\perp = \{(a, -a, -a, a) | a \text{ real}\}$$

so that

$$V^\perp = V_1^\perp \cap V_2^\perp$$

In general, $P_V = P_{V_1} + P_{V_2}$ only if V_1 and V_2 are orthogonal. They are *not* orthogonal in this example. Verify this by projecting $\mathbf{y} = (11, 4, 3, 8)$ onto each of V_1, V_2, and V.

Theorem 1.6.6: Let V be a subspace of Ω and let V_0 be a proper subspace of V. Let $V_1 = V_0^\perp \cap V$. Then (1) V_0 and V_1 are mutually orthogonal subspaces, (2) $V = V_0 \oplus V_1$, and (3) $P_{V_1} = P_V - P_{V_0}$.

***Proof*:** Part (1) is obvious. To prove (2) let $\mathbf{y} \in V$, and let $\hat{\mathbf{y}}_0 = p(\mathbf{y} | V_0)$. Then $\mathbf{y} = \hat{\mathbf{y}}_0 + (\mathbf{y} - \hat{\mathbf{y}}_0)$, $\hat{\mathbf{y}}_0 \in V_0$, $\mathbf{y} - \hat{\mathbf{y}}_0 \in V \cap V_0^\perp$. Thus $V \subset V_0 \oplus V_1$. Since $V \supset V_0$ and $V \supset V_1$, $V \supset V_0 \oplus V_1$, implying that $V = V_0 \oplus V_1$.

To prove (3) note that, since $V_1 \perp V_0$, $p(\mathbf{y} | V) = p(\mathbf{y} | V_0) + p(\mathbf{y} | V_1)$ for all $\mathbf{y}$. Thus $P_V = P_{V_0} + P_{V_1}$ and $P_{V_1} = P_V - P_{V_0}$. □

In fact, this theorem shows that Ω may be decomposed into three mutually orthogonal subspaces V_0, $V_0^\perp \cap V$, and $V^\perp$, whose direct sum is Ω.

Problem 1.6.3: Let Ω be Euclidean 4-space (column vectors). Let

$$\mathbf{x}_1 = \begin{bmatrix} 1 \\ 1 \\ 1 \\ 1 \end{bmatrix}, \qquad \mathbf{x}_2 = \begin{bmatrix} 1 \\ 1 \\ 0 \\ 0 \end{bmatrix}, \qquad \mathbf{x}_3 = \begin{bmatrix} 1 \\ 1 \\ 1 \\ 0 \end{bmatrix}$$

and let $V_0 = \mathscr{L}(\mathbf{x}_4)$ for $\mathbf{x}_4 = 3\mathbf{x}_3 - 2\mathbf{x}_2$, $V = \mathscr{L}(\mathbf{x}_1, \mathbf{x}_2, \mathbf{x}_3)$. Find P_{V_0}, P_V and P_{V_1} for $V_1 = V_0^\perp \cap V$. For $\mathbf{y} = \begin{bmatrix} 0 \\ 2 \\ 14 \\ 1 \end{bmatrix}$ find $p(\mathbf{y} | V_0)$, $p(\mathbf{y} | V_1)$, $p(\mathbf{y} | V)$.

Theorem 1.6.7: Let $V_1, \ldots, V_k$ be mutually orthogonal subspaces of Ω. Let $V = V_1 \oplus \cdots \oplus V_k$. Then $p(\mathbf{y} | V) = \sum_1^k p(\mathbf{y} | V_i)$ for all $\mathbf{y} \in \Omega$.

***Proof*:** Let $\hat{\mathbf{y}}_i = p(\mathbf{y} | V_i)$. We must show that for each $\mathbf{x} \in V$, $(\mathbf{y}, \mathbf{x}) = \left(\sum_1^k \hat{\mathbf{y}}_i, \mathbf{x}\right)$. Since $\mathbf{x} \in V$, $\mathbf{x} = \sum_{j=1}^k \mathbf{x}_j$ for some $\mathbf{x}_j \in V_j$ for $j = 1, \ldots, k$. Thus

$$\left(\sum_1^k \hat{\mathbf{y}}_i, \mathbf{x}\right) = \left(\sum_{i=1}^k \hat{\mathbf{y}}_i, \sum_{j=1}^k \mathbf{x}_j\right) = \sum_{i=1}^k \sum_{j=1}^k (\hat{\mathbf{y}}_i, \mathbf{x}_j)$$

$$= \sum_i (\hat{\mathbf{y}}_i, \mathbf{x}_i) = \sum_i (\mathbf{y}, \mathbf{x}_i) = \left(\mathbf{y}, \sum_{j=1}^k \mathbf{x}_i\right) = (\mathbf{y}, \mathbf{x}).$$

The third equality follows from the orthogonality of the subspaces. The fourth follows from the definition of $\hat{\mathbf{y}}_i$. □

Comment: In the case that $V = \Omega$ we see that $\mathbf{y} = p(\mathbf{y}|V) = \sum_1^k p(\mathbf{y}|V_i)$, and by the Pythagorean Theorem, $\|\mathbf{y}\|^2 = \sum_1^k \|p(\mathbf{y}|V_i)\|^2$. In applying this to the analysis of variance we will frequently make such a decomposition of the squared length of the observation vector $\mathbf{y}$. In fact, the analysis of variance may be viewed as the decomposition of the squared length of a vector into the sum of the squared lengths of several vectors, using the Pythagorean Theorem.

Example 1.6.3: Let Ω be the space of 2×3 matrices. Let $\mathbf{R}_1$, $\mathbf{R}_2$ be the row indicators and let $\mathbf{C}_1$, $\mathbf{C}_2$, $\mathbf{C}_3$, be the column indicators. Let $\mathbf{x}_0 = \sum_i \mathbf{R}_i = \sum_j \mathbf{C}_j$ be the matrix of all ones. Define $V_0 = \mathscr{L}(\mathbf{x}_0)$, $V_R = \mathscr{L}(\mathbf{R}_1, \mathbf{R}_2) \cap V_0^\perp$, $V_C = \mathscr{L}(\mathbf{C}_1, \mathbf{C}_2, \mathbf{C}_3) \cap V_0^\perp$. It is easy to show that $V_R = \{\mathbf{v} | \mathbf{v} = \sum a_i \mathbf{R}_i, a_1 + a_2 = 0\}$ and $V_C = \{\mathbf{v} = \sum b_j \mathbf{C}_j | \sum b_j = 0\}$. For example, $\begin{bmatrix} 2 & -3 & 1 \\ 2 & -3 & 1 \end{bmatrix} \in V_C$. The subspaces V_0, V_R, V_C are linearly independent and mutually orthogonal. Let $V = V_0 \oplus V_R \oplus V_C$. Then $p(\mathbf{y}|V) = \hat{\mathbf{y}}_0 + \hat{\mathbf{y}}_R + \hat{\mathbf{y}}_C$, where $\hat{\mathbf{y}}_0 = p(\mathbf{y}|V_0) = \bar{y}_{..}\mathbf{x}_0$, $\hat{\mathbf{y}}_R = p(\mathbf{y}|V_R) = \sum_i (\bar{y}_{i\cdot} - \bar{y}_{..})\mathbf{R}_i$, and $\hat{\mathbf{y}}_C = p(\mathbf{y}|V_C) = \sum_i (\bar{y}_{\cdot j} - \bar{y}_{..})\mathbf{C}_j$. Then, since $\Omega = V_0 \oplus V_R \oplus V_C \oplus V^\perp$ is the decomposition of Ω into four mutually orthogonal subspaces, $\mathbf{y} = \hat{\mathbf{y}}_0 + \hat{\mathbf{y}}_R + \hat{\mathbf{y}}_C + \mathbf{e}$, where $\mathbf{e} = \mathbf{y} - \hat{\mathbf{y}} = p(\mathbf{y}|V^\perp)$, and

$$\|\mathbf{y}\|^2 = \|\hat{\mathbf{y}}_0\|^2 + \|\hat{\mathbf{y}}_R\|^2 + \|\hat{\mathbf{y}}_C\|^2 + \|\mathbf{e}\|^2, \qquad \|\hat{\mathbf{y}}_0\|^2 = \bar{y}_{..}^2$$

$$\|\hat{\mathbf{y}}_R\|^2 = 3\sum_i (\bar{y}_{i\cdot} - \bar{y}_{..})^2, \qquad \|\hat{\mathbf{y}}_C\|^2 = 2\sum_i (\bar{y}_{\cdot j} - \bar{y}_{..})^2$$

Definition 1.6.4: The *null space* of an $m \times n$ matrix $\mathbf{A}$ is the collection of vectors $\mathbf{x} \in R_n$ such that $\mathbf{A}\mathbf{x} = \mathbf{0}$. We denote this null space by $N(\mathbf{A})$. The *column* (or range) *space* of $\mathbf{A}$ is $C(\mathbf{A}) = \{\mathbf{x} | \mathbf{x} = \mathbf{A}\mathbf{b} \text{ for some } \mathbf{b}\}$.

Theorem 1.6.8: Let $\mathbf{A}$ be an $m \times n$ matrix. Then

$$N(\mathbf{A}) = C(\mathbf{A}')^\perp \qquad \text{and} \qquad N(\mathbf{A})^\perp = C(\mathbf{A}') \tag{1.6.1}$$

***Proof*:** $\mathbf{w} \in N(\mathbf{A}) \Leftrightarrow \mathbf{w} \perp (\text{row space of } \mathbf{A}) \Leftrightarrow \mathbf{w} \perp (\text{column space of } \mathbf{A}') \Leftrightarrow \mathbf{w} \in C(\mathbf{A}')^\perp$. The second statement of (1.6.1) follows by taking complements on both sides. □

Theorem 1.6.9: Let $\mathbf{X}$ be an $n \times k$ matrix. The $C(\mathbf{X}'\mathbf{X}) = C(\mathbf{X}')$.

***Proof*:** $\mathbf{w} \in C(\mathbf{X}'\mathbf{X})$ implies the existence of $\mathbf{b}$ such that $(\mathbf{X}'\mathbf{X})\mathbf{b} = \mathbf{w} = \mathbf{X}'(\mathbf{X}\mathbf{b})$, which implies $\mathbf{w} \in C(\mathbf{X}')$. Thus $C(\mathbf{X}'\mathbf{X}) \subset C(\mathbf{X}')$.

$\mathbf{w} \in C(\mathbf{X}')$ implies that $\mathbf{w} = \mathbf{X}'\mathbf{b}$ for some $\mathbf{b} \in R_n$. Let $\hat{\mathbf{b}} = p(\mathbf{b}|C(\mathbf{X}))$. Then $\mathbf{X}'\hat{\mathbf{b}} = \mathbf{X}'\mathbf{b}$ and, since $\hat{\mathbf{b}} \in C(\mathbf{X})$, there exists $\mathbf{v}$ such that $\mathbf{X}\mathbf{v} = \mathbf{b}$. Then $\mathbf{X}'\mathbf{X}\mathbf{v} = \mathbf{X}'\hat{\mathbf{b}} = \mathbf{X}'\mathbf{b} = \mathbf{w}$, so $\mathbf{w} \in C(\mathbf{X}'\mathbf{X})$. Thus $C(\mathbf{X}'\mathbf{X}) \supset C(\mathbf{X}')$. □

It is shown in most introductory courses in linear algebra that the dimensions of the row and column spaces of any matrix $\mathbf{X}$ are equal, and this common dimension is called the *rank* of $\mathbf{X}$. We therefore conclude that $\mathbf{X}$, $\mathbf{X}'$, $\mathbf{X}'\mathbf{X}$, and $\mathbf{X}\mathbf{X}'$ all have the same rank. In particular, $\mathbf{X}'\mathbf{X} = \mathbf{M}$ has full rank (is nonsingular) if and only if $\mathbf{X}$ has full column rank, i.e., has linearly independent columns.

Problem 1.6.4: Let $\Omega = R_3$. For each subspace give the corresponding projection matrix $\mathbf{P}$. For each verify that $\mathbf{P}$ is idempotent and symmetric.
(a) $\mathscr{L}(\mathbf{x})$ for $\mathbf{x} = (1, 0, -1)'$.
(b) $\mathscr{L}(\mathbf{x}_1, \mathbf{x}_2)$ for $\mathbf{x}_1 = (1, 1, 1)'$, $\mathbf{x}_2 = (1, 0, 1)'$.

Problem 1.6.5: For the subspace $V = \mathscr{L}(\mathbf{J}, \mathbf{x})$ of Problem 1.3.5, what is P_V? (Note that $\mathscr{L}(\mathbf{J}, \mathbf{x}^*) = V$). What is $P_{V^\perp}$? Let $V_0 = \mathscr{L}(\mathbf{J})$ and $V_1 = V \cap V_0^\perp$. What is P_{V_1}?

Problem 1.6.6: Let V_1 and V_2 be subspaces of Ω and let $V_0 = V_1 \cap V_2$. Under what conditions does $P_{V_0} = P_{V_1}P_{V_2}$? Always? Never?

Problem 1.6.7: Let V_1, V_2, V_3 be subspaces. Does $V_1 \cap (V_2 + V_3) = (V_1 \cap V_2) + (V_1 \cap V_3)$ in general? If not, does this hold if V_2 and V_3 are linearly independent?

Problem 1.6.8: (a) For Example 1.6.3 find six mutually orthogonal vectors $\mathbf{v}_i$ for $i = 1, \ldots, 6$ such that

$$V_0 = \mathscr{L}(\mathbf{v}_1), \qquad V_R = \mathscr{L}(\mathbf{v}_2), \qquad V_C = \mathscr{L}(\mathbf{v}_3, \mathbf{v}_4), \qquad V^\perp = \mathscr{L}(\mathbf{v}_5, \mathbf{v}_6)$$

(b) For $\mathbf{y} = \begin{bmatrix} 12 & 7 & 11 \\ 10 & 1 & 7 \end{bmatrix}$ find $\hat{\mathbf{y}}_0$, $\hat{\mathbf{y}}_R$, $\hat{\mathbf{y}}_C$, $\hat{\mathbf{y}}$, $\mathbf{e}$, compute their lengths, and verify that the Pythagorean Theorem holds.

Problem 1.6.9: Let $\mathbf{A} = \begin{bmatrix} 2 & 3 & 7 \\ 1 & 5 & 7 \end{bmatrix}$.
(a) Find a basis for the null space of $\mathbf{A}$ (see Theorem 1.6.8).
(b) Verify Theorem 1.6.9 for $\mathbf{X} = \mathbf{A}'$.

Problem 1.6.10: Let $\mathbf{v}_1, \ldots, \mathbf{v}_n$ be an orthogonal basis for Ω.
(a) Prove Parseval's Identity: For every $\mathbf{x}, \mathbf{y} \in \Omega$

$$(\mathbf{x}, \mathbf{y}) = \sum_{i=1}^{n} (\mathbf{x}, \mathbf{v}_i)(\mathbf{y}, \mathbf{v}_i)/\|\mathbf{v}_i\|^2.$$

(b) Verify (a) for $\Omega = R^3$, $\mathbf{v}_1 = (1, 1, 1)'$, $\mathbf{v}_2 = (1, -1, 0)'$, $\mathbf{v}_3 = (1, 1, -2)'$, $\mathbf{x} = (3, 5, 8)'$, $\mathbf{y} = (2, 1, 4)'$.

Problem 1.6.11: Let V_1 and V_2 be subspaces of Ω. Let $V = V_1 \oplus V_2$. Let P_{V_1}, P_{V_2} and P_V be the corresponding projection operators. Suppose that $P_V = P_{V_1} + P_{V_2}$. (This means that $P_V\mathbf{y} = P_{V_1}\mathbf{y} + P_{V_2}\mathbf{y}$ for every $\mathbf{y} \in \Omega$.) Prove that $V_1 \perp V_2$. *Hint*: Consider $P_V\mathbf{v}_1$ for $\mathbf{v}_1 \in V_1$ and recall that $(\mathbf{v}_1 - P_{V_2}\mathbf{v}_1) \perp V_2$.

Problem 1.6.12: Prove the statements made in the paragraph following Definition 1.6.1. To prove the last statement construct an example.

Problem 1.6.13: Let $V_1, V_2, \ldots, V_k$ be mutually orthogonal subspaces, none equal to $\mathscr{L}(\mathbf{0})$. Prove that they are linearly independent.

1.7 EIGENVALUES AND EIGENVECTORS

In this section we summarize results concerning eigentheory. Though this material will not be heavily used in this course, it will be useful. Most proofs will be omitted.

(1) Let $\mathbf{A}$ be an $n \times n$ matrix. A real number λ and column vectors $\mathbf{v}$ satisfying the equation $\mathbf{A}\mathbf{v} = \lambda\mathbf{v}$ will be called an eigenpair, with λ an eigenvalue, and $\mathbf{v}$ the corresponding eigenvector. The words *characteristic* and *latent* are often used instead of *eigen*. Thus, an eigenvector $\mathbf{v}$ is transformed into a vector whose direction remains the same, but whose length is multiplied by the corresponding eigenvalue λ.

(2) A symmetric matrix $\underset{n \times n}{\mathbf{A}}$ has n real eigenvalues, though these may not all be distinct. Eigenvectors corresponding to different eigenvalues are orthogonal. If there exist k, but not more than k, independent vectors $\mathbf{v}_1, \ldots, \mathbf{v}_k$ corresponding to the same eigenvalue λ, then λ is said to have multiplicity k, and the equation $\det(\lambda\mathbf{I} - \mathbf{A}) = 0$ has root λ of multiplicity k. In this case all vectors in $\mathscr{L}(\mathbf{v}_1, \ldots, \mathbf{v}_k)$ are eigenvectors corresponding to λ, and k such vectors, say $\mathbf{w}_1, \ldots, \mathbf{w}_k$, which are mutually orthogonal, may be chosen.

If such mutually orthogonal eigenvectors are chosen for each different eigenvalue, then the entire collection $\mathbf{u}_1, \ldots, \mathbf{u}_n$ of mutually orthogonal eigenvectors corresponding to eigenvalues $\lambda_1, \ldots, \lambda_n$, where an eigenvalue is repeated k times if its multiplicity is k, span n-space.

Let $\mathbf{\Lambda} = \text{diag}(\lambda_1, \ldots, \lambda_n)$, the matrix with (ii) element λ_i, off-diagonal terms 0, and $\mathbf{U} = (\mathbf{u}_1, \ldots, \mathbf{u}_n)$. Then $\mathbf{AU} = \mathbf{U\Lambda}$, and if the $\mathbf{u}_i$ are chosen to have length one,

$$\mathbf{U}'\mathbf{U} = \mathbf{I}_n, \qquad \mathbf{U}'\mathbf{A}\mathbf{U} = \mathbf{U}'\mathbf{U}\mathbf{\Lambda} = \mathbf{\Lambda}, \qquad \mathbf{A} = \mathbf{U}\mathbf{\Lambda}\mathbf{U}'.$$

The representation $\mathbf{A} = \mathbf{U\Lambda U}'$ is called the *spectral representation* of $\mathbf{A}$.

Recall that the trace of a square matrix $\mathbf{A}$ is the sum of its diagonal elements. It is easy to show that $\text{trace}(\mathbf{BC}) = \text{trace}(\mathbf{CB})$ whenever the matrix product makes sense. It follows therefore that whenever $\mathbf{A}$ has spectral representation $\mathbf{A} = \mathbf{U\Lambda U}'$, $\text{trace}(\mathbf{A}) = \text{trace}(\mathbf{\Lambda U}'\mathbf{U}) = \text{trace}(\mathbf{\Lambda}) = \sum \lambda_i$. Similarly, $\det(\mathbf{A}) = \det(\mathbf{U}) \det(\mathbf{\Lambda}) \det(\mathbf{U}') = (\pm 1) \det(\mathbf{\Lambda}) (\pm 1) = \prod \lambda_i$.

Since, for any $r \times s$ matrix $\mathbf{C} = (\mathbf{c}_1, \ldots, \mathbf{c}_s)$ and $s \times t$ matrix

$$\mathbf{D} = \begin{pmatrix} \mathbf{d}_1 \\ \vdots \\ \mathbf{d}_s \end{pmatrix}, \qquad \mathbf{CD} = \sum_{i=1}^{s} \mathbf{c}_i \mathbf{d}_i,$$

we may express $\mathbf{A}$ in the form

$$\mathbf{A} = \mathbf{U\Lambda U}' = (\lambda_1 \mathbf{u}_1, \ldots, \lambda_n \mathbf{u}_n) \begin{pmatrix} \mathbf{u}_1' \\ \vdots \\ \mathbf{u}_n' \end{pmatrix} = \sum_1^n \lambda_i \mathbf{u}_i \mathbf{u}_i'.$$

The matrices $\mathbf{u}_i \mathbf{u}_i' = \mathbf{P}_i$ are projections onto the one-dimensional subspaces $\mathscr{L}(\mathbf{u}_i)$. If there are r different eigenvalues with multiplicities $k_1, \ldots, k_r$ then the $\mathbf{P}_i$ corresponding to the same eigenvalue may be summed to get the representation of $\mathbf{A}$,

$$\mathbf{A} = \sum_1^r \lambda_j \mathbf{P}_j^*.$$

where $\mathbf{P}_j^*$ is the projection onto the k_j-dimensional subspace spanned by the eigenvectors corresponding to λ_j.

(3) By definition a square matrix $\mathbf{A}$ is positive definite if the quadratic function $Q(\mathbf{x}) = \mathbf{x}'\mathbf{A}\mathbf{x} > 0$ for all $\mathbf{x} \neq \mathbf{0}$. It is nonnegative definite if $Q(\mathbf{x}) \geq 0$ for all $\mathbf{x}$.

Example 1.7.1: Let $\mathbf{v}_1 = (1, 1, 1, 1)'$, $\mathbf{v}_2 = (1, -1, 0, 0)'$, $\mathbf{v}_3 = (1, 1, -2, 0)'$, $\mathbf{v}_4 = (1, 1, 1, -3)'$. These $\mathbf{v}_i$ are mutually orthogonal. Let $\mathbf{P}_i$ be projection onto $\mathscr{L}(\mathbf{v}_i)$. Thus,

$$\mathbf{P}_i = \mathbf{v}_i \mathbf{v}_i' / \|\mathbf{v}_i\|^2.$$

Let

$$\mathbf{A} = 8\mathbf{P}_1 + 8\mathbf{P}_2 + 12\mathbf{P}_3 = \begin{bmatrix} 8 & 0 & -2 & 2 \\ 0 & 8 & -2 & 2 \\ -2 & -2 & 10 & 2 \\ 2 & 2 & 2 & 2 \end{bmatrix}.$$

Working backwards from $\mathbf{A}$, the roots of the fourth degree polynomial $\det(\lambda\mathbf{I}_4 - \mathbf{A}) = 0$ are $\lambda = 8, 8, 12, 0$ with corresponding eigenvectors $\mathbf{w}_1$, $\mathbf{w}_2$, $\mathbf{w}_3$, $\mathbf{w}_4$. The vectors $\mathbf{w}_1$, $\mathbf{w}_2$ may be arbitrarily chosen vectors in $\mathscr{L}(\mathbf{v}_1, \mathbf{v}_2)$, the subspace onto which $\mathbf{P}_1 + \mathbf{P}_2$ projects. They may be chosen to be orthogonal, and could be chosen to be $\mathbf{v}_1$ and $\mathbf{v}_2$; $\mathbf{w}_2$ and $\mathbf{w}_4$ are nonzero vectors in $\mathscr{L}(\mathbf{v}_3)$ and $\mathscr{L}(\mathbf{v}_4)$, respectively. The lengths of eigenvectors are arbitrary. Since one eigenvalue is 0, $\mathbf{A}$ has rank 3. The determinant of $\mathbf{A}$ is the product of its eigenvalues, 0 in this case. The trace of $\mathbf{A}$ is the sum of its eigenvalues, 28 in this example.

Let $\mathbf{u}_i = \mathbf{v}_i/\|\mathbf{v}_i\|$, so these $\mathbf{u}_i$ have length one. Let $\mathbf{U} = (\mathbf{u}_1, \mathbf{u}_2, \mathbf{u}_3, \mathbf{u}_4)$ and $\mathbf{\Lambda} = \text{diag}(8, 8, 12, 0)$. Then $\mathbf{AU} = \mathbf{U\Lambda}$, $\mathbf{U}$ is an orthogonal matrix, and $\mathbf{A} = \mathbf{U\Lambda U}'$. Here

$$\mathbf{U} = \begin{bmatrix} 0.5 & 0.707\,107 & 0.400\,248 & 0.288\,675 \\ 0.5 & -0.707\,107 & 0.400\,248 & 0.288\,675 \\ 0.5 & 0 & -0.816\,497 & 0.288\,675 \\ 0.5 & 0 & 0 & -0.866\,025 \end{bmatrix}$$

$$\mathbf{\Lambda} = \begin{bmatrix} 8 & 0 & 0 & 0 \\ 0 & 8 & 0 & 0 \\ 0 & 0 & 12 & 0 \\ 0 & 0 & 0 & 0 \end{bmatrix}$$

$$\mathbf{AU} = \mathbf{U\Lambda} = \begin{bmatrix} 4 & 5.656\,85 & 4.898\,98 & 0 \\ 4 & -5.656\,95 & 4.898\,98 & 0 \\ 4 & 0 & -9.797\,96 & 0 \\ 4 & 0 & 0 & 0 \end{bmatrix}$$

$$\mathbf{U\Lambda U}' = \begin{bmatrix} 8 & 0 & -2 & 2 \\ 0 & 8 & -2 & 2 \\ -2 & -2 & 10 & 2 \\ 2 & 2 & 2 & 2 \end{bmatrix} = \mathbf{A}$$

Consider the quadratic form

$$Q(\mathbf{x}) = \mathbf{x}'\mathbf{A}\mathbf{x} = 8x_1^2 - 4x_1x_3 + 4x_1x_4 + 8x_2^2 - 4x_2x_3 + 4x_2x_4 + 10x_3^2 + 4x_3x_4 + 2x_4^2.$$

Since

$$\mathbf{A} = \sum_{i=1}^{4} \lambda_i \mathbf{P}_i, \qquad Q(\mathbf{x}) = \sum_{i=1}^{4} \lambda_i (\mathbf{x}', \mathbf{P}_i \mathbf{x}) = \sum \lambda_i \|\hat{x}_i\|^2,$$

where $\hat{x}_i = \mathbf{P}_i\mathbf{x} = [(\mathbf{v}_i', \mathbf{x})/\|\mathbf{v}_i\|^2]\mathbf{v}_i$, and therefore $\|\hat{x}_i\|^2 = (\mathbf{v}_i', \mathbf{x})^2/\|\mathbf{v}_i\|^2$. Since one eigenvalue is zero, the others positive, **A** is nonnegative definite and $Q(\mathbf{x}) \geq 0$ for all **x**. **A** is not positive definite since $Q(\mathbf{v}_4) = 0$.

Using the representation $\mathbf{A} = \sum_1^n \lambda_i \mathbf{u}_i \mathbf{u}_i'$ above it is easy to show that a square symmetric matrix **A** is positive definite if and only if its eigenvalues are all positive, nonnegative definite if and only if its eigenvalues are all nonnegative.

If **A** is nonnegative definite we can write $\mathbf{\Lambda}^{1/2} = \text{diag}(\lambda_1^{1/2}, \ldots, \lambda_n^{1/2})$, so $\mathbf{A} = \mathbf{U}\mathbf{\Lambda}\mathbf{U}' = \mathbf{U}\mathbf{\Lambda}^{1/2}\mathbf{\Lambda}^{1/2}\mathbf{U}' = (\mathbf{U}\mathbf{\Lambda}^{1/2})(\mathbf{U}\mathbf{\Lambda}^{1/2})' = \mathbf{B}\mathbf{B}'$ for $\mathbf{B} = \mathbf{U}\mathbf{\Lambda}^{1/2}$. The decomposition $\mathbf{A} = \mathbf{B}\mathbf{B}'$ is quite useful. It is not unique, since if **C** is any orthonormal matrix (satisfying $\mathbf{C}\mathbf{C}' = \mathbf{I}$), then $(\mathbf{B}\mathbf{C})(\mathbf{B}\mathbf{C})' = \mathbf{B}\mathbf{C}\mathbf{C}'\mathbf{B}' = \mathbf{B}\mathbf{B}' = \mathbf{A}$.

Letting $\mathbf{C} = \mathbf{U}\mathbf{\Lambda}^{1/2}\mathbf{U}' = \sum \lambda_i^{1/2}\mathbf{P}_i$, we get $\mathbf{C}' = \mathbf{C}$, with $\mathbf{A} = \mathbf{C}'\mathbf{C} = \mathbf{C}^2$. The matrix **C** is the unique symmetric square root of **A**.

Letting $\mathbf{y} = \mathbf{U}'\mathbf{x}$ for **U** as defined above, we get

$$Q(\mathbf{x}) = \mathbf{x}'\mathbf{A}\mathbf{x} = (\mathbf{U}\mathbf{y})'\mathbf{A}(\mathbf{U}\mathbf{y}) = \mathbf{y}'\mathbf{U}'\mathbf{A}\mathbf{U}\mathbf{y} = \mathbf{y}'\mathbf{\Lambda}\mathbf{y} = \sum_1^k \lambda_i y_i^2$$

(4) Let $\mathbf{P}_V$ be the projection operator onto a subspace V of Ω. Then for $\mathbf{x} \in V$, $\mathbf{P}_V\mathbf{x} = \mathbf{x}$ so that all vectors in V are eigenvectors of $\mathbf{P}_V$ with eigenvalues 1. For $\mathbf{x} \in V^\perp$, $\mathbf{P}_V\mathbf{x} = \mathbf{0}$, so that all vectors in $V^\perp$ are eigenvectors of $\mathbf{P}_V$ with eigenvalue 0. The eigenvalue 1 has multiplicity equal to the dimension of V, while the eigenvalue 0 has multiplicity equal to $\dim(V^\perp) = n - \dim(V)$. Since from (2) $\text{trace}(\mathbf{A}) = \sum \lambda_i$, the trace of a projection matrix is the dimension of the subspace onto which it projects.

Partitioned Matrices (Seber, 1977):

$$\begin{bmatrix} \mathbf{A} & \mathbf{B} \\ \mathbf{B}' & \mathbf{D} \end{bmatrix}^{-1} = \begin{bmatrix} \mathbf{A}^{-1} + \mathbf{F}\mathbf{E}^{-1}\mathbf{F}' & -\mathbf{F}\mathbf{E}^{-1} \\ -\mathbf{E}^{-1}\mathbf{F}' & \mathbf{E}^{-1} \end{bmatrix}, \quad \text{where} \quad \begin{array}{l} \mathbf{E} = \mathbf{D} - \mathbf{B}'\mathbf{A}^{-1}\mathbf{B} \\ \mathbf{F} = \mathbf{A}^{-1}\mathbf{B} \end{array}$$

Singular Value Decomposition (Seber, 1977, p. 392): For $\mathbf{X}$ an $n \times k$ matrix of rank r, $n \geq k \geq r$, let the r positive eigenvalues of $\mathbf{XX}'$ be $\sigma_1^2 \geq \sigma_2^2 \geq \cdots \geq \sigma_k^2 > 0$. Let $\mathbf{D}$ be the diagonal matrix with diagonal $(\sigma_1, \ldots, \sigma_r)$. Let the length-one eigenvector of $\mathbf{XX}'$ corresponding to σ_i^2 be $\mathbf{p}_i$ for each i, $1 \leq i \leq r$, and let $\mathbf{q}_i = \mathbf{X}'\mathbf{D}^{-1}\mathbf{p}_i$. Then $\mathbf{q}_i$ is an eigenvector of $\mathbf{X}'\mathbf{X}$ corresponding to eigenvalue σ_i^2. These vectors $\mathbf{p}_i$ may be chosen to be mutually orthonormal. It follows that the $\mathbf{q}_i$ are also orthogonal. Define

$$\mathbf{P} = (\mathbf{p}_1, \ldots, \mathbf{p}_r), \qquad \mathbf{Q} = (\mathbf{q}_1, \ldots, \mathbf{q}_r)' = \mathbf{PD}^{-1}\mathbf{X}.$$

Then $\mathbf{X} = \mathbf{PDQ} = \sum \sigma_i \mathbf{p}_j \mathbf{q}_j'$. Thus, the linear transformation $\mathbf{Xx} = \mathbf{y}$, taking vectors $\mathbf{x} \in R_k$ into C = column space of $\mathbf{X}$, proceeds as follows. $\mathbf{Q}$ takes a vector $\mathbf{x} \in R_k$ with $(\mathbf{x}, \mathbf{q}_i) = c_i$ into $(c_1, \ldots, c_p)'$. $\sum$ then multiplies each c_i by σ_i. $\mathbf{P}(\mathbf{DQ})\mathbf{x} = \mathbf{Xx}$ is then $\sum_j \mathrm{c}_j \sigma_j \mathbf{p}_j$, a vector in the column space of $\mathbf{X}$.

Moore–Penrose or Pseudo-Inverse

The Moore–Penrose inverse or pseudo-inverse of the $n \times k$ matrix $\mathbf{X}$ is the $k \times n$ unique matrix $\mathbf{X}^+$ having the four properties: (1) $\mathbf{X}^+\mathbf{X}\mathbf{X}^+ = \mathbf{X}^+$, (2) $\mathbf{X}\mathbf{X}^+\mathbf{X} = \mathbf{X}$, (3) $\mathbf{X}^+\mathbf{X}$ is symmetric, (4) $\mathbf{X}\mathbf{X}^+$ is symmetric. For any vector $\mathbf{y} \in R_n$, $\mathbf{b} = \mathbf{X}^+\mathbf{y}$ is the unique vector in the row space of $\mathbf{X}$ such that $\mathbf{Xb}$ is the projection of $\mathbf{y}$ on the column space of $\mathbf{X}$. If $\mathbf{X}$ is nonsingular then $\mathbf{X}^+ = \mathbf{X}^{-1}$. The matrix $\mathbf{X}^+\mathbf{X}$ is the projection onto the row space of $\mathbf{X}$. The matrix $\mathbf{X}\mathbf{X}^+$ is the projection onto the column space of $\mathbf{X}$. If $\mathbf{X}$ has full column rank then $\mathbf{X}^+ = (\mathbf{X}'\mathbf{X})^{-1}\mathbf{X}'$. If V is the column space of $\mathbf{X}$, and $p(\mathbf{y}|V) = \mathbf{X}\hat{\boldsymbol{\beta}}$, then $\hat{\boldsymbol{\beta}} = \mathbf{X}^+\mathbf{y}$.

The Moore–Penrose inverse may be used to find solutions to the linear equation $\mathbf{Xb} = \mathbf{c}$. If this equation has a solution then $\mathbf{c}$ is in the column space of $\mathbf{X}$. That is, there exists some $\mathbf{w}$ such that $\mathbf{Xw} = \mathbf{c}$. Let $\mathbf{b} = \mathbf{X}^+\mathbf{c}$. Then $\mathbf{Xb} = \mathbf{X}\mathbf{X}^+\mathbf{Xw} = \mathbf{Xw} = \mathbf{c}$. The general solution to the equation $\mathbf{Xb} = \mathbf{c}$ is given by $\mathbf{b} = \mathbf{X}^+\mathbf{c} + (\mathbf{I}_p - \mathbf{X}^+\mathbf{X})\mathbf{d}$, for $\mathbf{d}$ any vector in R_k. Taking $\mathbf{d}$ to be any vector orthogonal to the row space of $\mathbf{X}$, we get the unique solution $\mathbf{X}^+\mathbf{c}$ in the row space of $\mathbf{X}$.

The pseudo-inverse is related to the singular value decomposition of $\mathbf{X}$ in that $\mathbf{X}^+ = \mathbf{Q}'\mathbf{D}^{-1}\mathbf{P}'$.

For a full discussion see *Regression and the Moore–Penrose Pseudoinverse* by Arthur Albert (1972).

Triangular Decomposition

Let $\mathbf{A}$ be a symmetric nonnegative definite matrix. There exist an infinite number of $n \times n$ matrices $\mathbf{B}$ such that $\mathbf{BB}' = \mathbf{A}$. Perhaps the easiest such matrix to find is one of the form (lower triangular)

$$\mathbf{B} = \begin{bmatrix} b_{11} & 0 & \cdots & 0 \\ b_{21} & b_{22} & \cdots & 0 \\ \cdot & \cdot & \cdots & 0 \\ \cdot & \cdot & \cdots & 0 \\ \cdot & \cdot & \cdots & 0 \\ b_{n1} & b_{n2} & \cdots & b_{nn} \end{bmatrix}$$

Then $b_{11}^2 = a_{11}$, so $b_{11} = \sqrt{a_{11}}$. Then, since $b_{i1}b_{11} = a_{i1}$ we have

$$b_{i1} = a_{i1}/b_{11} \qquad \text{for} \quad i = 2, \ldots, n$$

Suppose b_{ij} has already been found for $j = 1, \ldots, k-1$ and $i = 1, \ldots, n$ for $k \geq 1$. Then we can find $b_{i,k}$ inductively.

Since, $\sum_{j=1}^{k} b_{kj}b_{kj} = a_{kk}$, it follows that $b_{kk}^2 = a_{kk} - \sum_{j=1}^{k-1} b_{kj}^2$. Then

$$b_{kk} = \left(a_{kk} - \sum_{j=1}^{k-1} b_{kj}^2\right)^{1/2}.$$

Since $\sum_{j=1}^{k} b_{ij}b_{kj} = a_{ik}$ for $i > k$, it follows that

$$b_{ik} = \left(a_{ik} - \sum_{j=1}^{k-1} b_{ij}b_{kj}\right)\Big/ b_{kk} \qquad \text{for} \quad i > k.$$

Repeating for each k produces $\mathbf{B}$.

To summarize:

(1) Compute $b_{11} = (a_{11})^{1/2}$, let $b_{i1} = a_{i1}/b_{11}$, and let $k = 2$.

(2) Let $b_{kk} = \left(a_{kk} - \sum_{j=1}^{k-1} b_{kj}^2\right)^{1/2}$. ($\mathbf{A}$ is nonnegative definite if and only if the term in parentheses is nonnegative for each k.)

(3) Let $b_{ik} = \left(a_{ik} - \sum_{j=1}^{k-1} b_{ij}b_{kj}\right)\Big/ b_{kk}$ for $i > k$.

(4) Replace k by $k+1$ and repeat (2) and (3) until $k > n$.

(5) Let $b_{ij} = 0$ for $i < j$.

If any $b_{kk} = 0$ in step (3) then set $b_{ik} = 0$ for $i \geq k$.

Problem 1.7.1: Let $\mathbf{A} = \begin{pmatrix} 14 & -2 \\ -2 & 11 \end{pmatrix}$.

(a) Find the eigenvalues λ_1, λ_2 and corresponding length-one eigenvectors $\mathbf{u}_1$, $\mathbf{u}_2$ for $\mathbf{A}$.

(b) Define $\mathbf{U}$ and $\mathbf{\Lambda}$ as in Section 1.7 and show that $\mathbf{A} = \mathbf{U\Lambda U}'$ and $\mathbf{UU}' = \mathbf{I}_2$.
(c) Give the projections $\mathbf{P}_1^*$ and $\mathbf{P}_2^*$ of Section 1.7 and show that $\mathbf{A} = \lambda_1\mathbf{P}_1^* + \lambda_2\mathbf{P}_2^*$.
(d) Is $\mathbf{A}$ positive definite? Why?

Problem 1.7.2: What are the eigenvalues and eigenvectors of the projection matrices $\mathbf{P}$ of examples 1, 2, 3, 4, 5 of Example 1.6.1?

Problem 1.7.3: For $n \times k$ matrix $\mathbf{X}$ of rank k, what are the eigenvalues and vectors for $\mathbf{P} = \mathbf{X}(\mathbf{X}'\mathbf{X})^{-1}\mathbf{X}'$? What is trace($\mathbf{P}$)? What is det($\mathbf{P}$) if $n > k$? If $n = k$?

Problem 1.7.4: Let $n \times n$ matrix $\mathbf{A}$ have nonzero eigenvalue λ and corresponding eigenvector $\mathbf{v}$. Show that
(a) $\mathbf{A}^{-1}$ has an eigenvalue λ^{-1}, eigenvector $\mathbf{v}$.
(b) $\mathbf{I} - \mathbf{A}$ has an eigenvalue $1 - \lambda$, eigenvector $\mathbf{v}$.
(c) For $\mathbf{A} = \mathbf{BC}$, $\mathbf{CB}$ has eigenvalue λ, eigenvector $\mathbf{Cv}$.

Problem 1.7.5: Give 2×2 matrices which satisfy the following:
(a) Positive definite.
(b) Nonnegative definite, but not positive definite.
(c) Not nonnegative definite.

Problem 1.7.6: Let $\mathbf{A}$ be positive definite and let $\mathbf{v} \in R_n$. Prove that $(\mathbf{A} + \mathbf{vv}')^{-1} = \mathbf{A}^{-1}(\mathbf{I} - c\mathbf{vv}'\mathbf{A}^{-1})$ for $c = 1/(1 + \mathbf{v}'\mathbf{A}^{-1}\mathbf{v})$.

Problem 1.7.7: Determine whether the quadratic form $Q(x_1, x_2, x_3) = 2x_1^2 + 2x_2^2 + 11x_3^2 + 16x_1x_2 - 2x_1x_3 - 2x_2x_3$ is nonnegative definite. *Hint*: What is the matrix corresponding to Q? One of its eigenvalues is 12.

Problem 1.7.8: For $\mathbf{A} = \begin{bmatrix} 5 & -1 \\ -1 & 10 \end{bmatrix}$ find a matrix $\mathbf{B}$ such that $\mathbf{A} = \mathbf{BB}'$.

Problem 1.7.9: Let $\mathbf{G} = \begin{bmatrix} 2 & 1 & -1 \\ 1 & 3 & 0 \\ -1 & 0 & 4 \end{bmatrix}$. Find $\mathbf{G}^{-1}$ by using the formula for partitioned matrices with $\mathbf{A}$ the 1×1 matrix (2).

Problem 1.7.10: Let $\mathbf{X} = \begin{bmatrix} 5 & -1 \\ -1 & 5 \\ 2 & 2 \end{bmatrix}$. Find the singular value decomposition of $\mathbf{X}$. Also find the Moore–Penrose inverse $\mathbf{X}^+$ and verify its four defining properties.

Problem 1.7.11: Let $\mathbf{A} = \mathbf{UDV}$ be the singular value decomposition of $\mathbf{A}$. Express the following matrices in terms of $\mathbf{U}$, $\mathbf{D}$, and $\mathbf{V}$.

(a) $\mathbf{A}'\mathbf{A}$

(b) $\mathbf{A}\mathbf{A}'$

(c) $\mathbf{A}^{-1}$ (assuming $\mathbf{A}$ is nonsingular)

(d) $\mathbf{A}^n = \mathbf{A}\mathbf{A}\cdots\mathbf{A}$ (n products), assuming $\mathbf{A}$ is square. In the case that the singular values are $\sigma_1 > \sigma_2 \geq \sigma_3 \geq \cdots \geq \sigma_r > 0$, show that $\lim_{n\to\infty} \mathbf{A}^n/\sigma_1^n = \mathbf{p}_1\mathbf{p}_1'$.

(e) Projection onto the column space of $\mathbf{A}$.

(f) Projection onto the row space of $\mathbf{A}$.

(g) What are $\mathbf{U}$, $\mathbf{D}$, and $\mathbf{V}$ for the case that $\mathbf{A} = \mathbf{a}$ is an $n \times 1$ matrix? What is $\mathbf{A}^+$?

Problem 1.7.12: Let $\mathbf{A}$ be a symmetric $n \times n$ matrix of rank one.

(a) Show that $\mathbf{A}$ can be expressed in the form $\mathbf{A} = c\mathbf{v}\mathbf{v}'$, for a real number c, vector $\mathbf{v}$.

(b) Prove that either $\mathbf{A}$ or $-\mathbf{A}$ is nonnegative definite.

(c) Give the spectral decomposition for $\mathbf{A}$ in terms of c and $\mathbf{v}$.

CHAPTER 2

Random Vectors

In this chapter we discuss random vectors, n-tuples of random variables, all defined on the sample space. A random vector will usually be denoted as a capital letter from the end of the alphabet, taking values in a linear space Ω of arrays defined as in Chapter 1, e.g.,

$$\mathbf{Y} = \begin{bmatrix} Y_{11} & Y_{21} & Y_{31} \\ Y_{12} & Y_{22} & Y_{32} \\ Y_{13} & & Y_{33} \end{bmatrix}$$

We will suppose that the components of a random vector $\mathbf{Y}$ have been ordered. The particular order chosen is not important, so long as the same order is used consistently.

2.1 COVARIANCE MATRICES

Definition 2.1.1: Let $\mathbf{Y}$ be a random vector taking values in a linear space Ω. Then $E(\mathbf{Y})$ is the element in Ω whose ith component is $E(Y_i)$, where Y_i is the ith component of $\mathbf{Y}$ for each i. $E(\mathbf{Y})$ is also called the *mean vector*.

Example 2.1.1: Suppose $E(Y_1) = 3$, $E(Y_2) = 7$, $E(Y_3) = 5$. Then

$$E\begin{bmatrix} Y_1 & \\ & Y_3 \\ Y_2 & \end{bmatrix} = \begin{bmatrix} 3 & \\ & 5 \\ 7 & \end{bmatrix}.$$

Of course, the definition requires that each $E(Y_i)$ exists.

Definition 2.1.2: Let $\mathbf{Y}$ be a random vector taking values in Ω, with mean vector $\mu \in \Omega$. Then the *covariance matrix* for $\mathbf{Y}$, denoted by $D[\mathbf{Y}]$ or $\boldsymbol{\Sigma}_{\mathbf{Y}}$, is

$$(\text{cov}(Y_i, Y_j)),$$

the $n \times n$ matrix whose (ij) component is $\text{cov}(Y_i, Y_j)$.

Example 2.1.2: Let Ω be the space of arrays $\begin{pmatrix} y_1 & & y_3 \\ & y_2 & \end{pmatrix}$ with elements ordered as the subscripts are. Then the covariance matrix of a random vector **Y** taking values in Ω is

$$D[\mathbf{Y}] = \begin{bmatrix} \mathrm{Var}(Y_1) & \mathrm{cov}(Y_1, Y_2) & \mathrm{cov}(Y_1, Y_3) \\ \mathrm{cov}(Y_1, Y_2) & \mathrm{Var}(Y_2) & \mathrm{cov}(Y_2, Y_3) \\ \mathrm{cov}(Y_1, Y_3) & \mathrm{cov}(Y_2, Y_3) & \mathrm{Var}(Y_3) \end{bmatrix}$$

Of course, $D[\mathbf{Y}]$ is symmetric for any random vector **Y**. The configuration in which **Y** is written does not affect $D[\mathbf{Y}]$, but the order in which the elements are written does. Thus, if the vector **Y** of Example 2.1.2 is written instead as a column, the covariance matrix remains the same.

We will often wish to consider random vectors $\mathbf{W} = \mathbf{AY} + \mathbf{b}$, where **Y** is a random vector taking values in an n-component space Ω_n, **A** is a linear transformation from Ω_n into an m-component space Ω_m, and **b** a fixed element in Ω_m. Thus **W** takes values in Ω_m. It will be convenient to consider Ω_n and Ω_m to be the spaces of column vectors R_n and R_m, so that we can use matrix algebra. **A** may then be written as an $m \times n$ matrix. The results will generalize to the case that the elements of Ω are not written as columns simply by setting up a one-to-one correspondence between Ω_n and R_n, Ω_m and R_m.

Therefore, let

Y be a random vector taking values in R_n,
A be an $m \times n$ matrix of constants, and
b be a constant vector in R_m. Then,

Theorem 2.1.1: Let $\mathbf{W} = \mathbf{AY} + \mathbf{b}$. Then $E(\mathbf{W}) = \mathbf{A}E(\mathbf{Y}) + \mathbf{b}$.

***Proof*:** Let $(a_{i1}, \ldots, a_{in})$ be the ith row of **A** and let b_i be the ith component of **b**. Then

$$W_i = \sum_j a_{ij} Y_i + b_i$$

and $E(W_i) = \sum_j a_{ij} E(Y_j) + b_i$. This proves the theorem. □

This theorem generalizes to random matrices as follows:

Theorem 2.1.2: Let **Y** be a random $n_2 \times n_3$ matrix (an $n_2 \times n_3$ matrix of random variables). Let **A** and **B** be $n_1 \times n_2$ and $n_3 \times n_4$ matrices of constants. Then

$$E[\mathbf{AYB}] = \mathbf{A}E(\mathbf{Y})\mathbf{B}$$

***Proof*:** Let $\mathbf{A} = (a_{ij})$, $\mathbf{Y} = (Y_{jk})$, $\mathbf{B} = (B_{kl})$, $\mu_{jk} = E(Y_{jk})$. Then the $(i, 1)$th element of $\mathbf{AYB}$ is $\sum_{k=1}^{n_3} \sum_{j=1}^{n_2} a_{ij} Y_{jk} b_{kl}$, whose expectation is $\sum_{l=k}^{n_3} \sum_{k=1}^{n_2} a_{ij} \mu_{jk} b_{kl}$. But this is the (il)th element of $\mathbf{A}E(\mathbf{Y})\mathbf{B}$. □

In particular, for any linear space Ω, random vector $\mathbf{Y}$, constant vector $\mathbf{a} \in \Omega$, the linear product $(\mathbf{a}, \mathbf{Y})$ is a linear combination of the components of $\mathbf{Y}$, so that

$$E(\mathbf{a}, \mathbf{Y}) = (\mathbf{a}, E(\mathbf{Y}))$$

For a random vector $\mathbf{U}$ taking values in R_n, let $E(\mathbf{U}) = \boldsymbol{\mu}_{\mathbf{U}}$. Then $\mathbf{U} - \boldsymbol{\mu}_{\mathbf{U}}$ is the vector of *deviations*.

Definition 2.1.3: Let $\mathbf{U}$ and $\mathbf{W}$ be random vectors taking values in R_m and R_n respectively. Then the *covariance matrix* of $\mathbf{U}$ and $\mathbf{W}$ is the $m \times n$ matrix

$$C[\mathbf{U}, \mathbf{W}] = (\text{cov}(U_i, W_j)) = E[(\mathbf{U} - \boldsymbol{\mu}_{\mathbf{U}})(\mathbf{W} - \boldsymbol{\mu}_{\mathbf{W}})']$$

The covariance matrix for a single random vector $\mathbf{U}$ is $D[\mathbf{U}] = C[\mathbf{U}, \mathbf{U}]$.

Thus, we speak of the covariance matrix of a *pair* of random vectors, and also the covariance matrix of a single random vector.

For example, for $m = 2$, $n = 3$, $\sigma_{ij} = \text{cov}(U_i, W_j)$ we have

$$C[\mathbf{U}, \mathbf{W}] = \begin{bmatrix} \sigma_{11} & \sigma_{12} & \sigma_{13} \\ \sigma_{21} & \sigma_{22} & \sigma_{23} \end{bmatrix}$$

The correlation coefficient of the ith component of $\mathbf{U}$ and the jth of $\mathbf{W}$ is

$$\rho_{ij} = \frac{\text{cov}(U_i, W_j)}{\sqrt{\text{Var}(U_i)\text{Var}(W_j)}}$$

Letting

$$\sigma_{\mathbf{U}} = \text{diag}(\sqrt{\text{Var}(U_1)}, \ldots, \sqrt{\text{Var}(U_m)}),\ \sigma_{\mathbf{W}} = \text{diag}(\sqrt{\text{Var}(W_1)}, \ldots, \sqrt{\text{Var}(W_n)})$$

we define the *correlation matrix* for $\mathbf{U}$ and $\mathbf{W}$ to be

$$R[\mathbf{U}, \mathbf{W}] \equiv (\rho_{ij}) = \sigma_{\mathbf{U}}^{-1} C[\mathbf{U}, \mathbf{W}] \sigma_{\mathbf{W}}^{-1}$$

In particular, the correlation matrix for a random vector $\mathbf{Y}$ is

$$R[\mathbf{Y}] \equiv R[\mathbf{Y}, \mathbf{Y}] = \sigma_{\mathbf{Y}}^{-1} D[\mathbf{Y}] \sigma_{\mathbf{Y}}^{-1},$$

where $\sigma_{\mathbf{Y}} = \text{diag}(\sqrt{\text{Var}(Y_1)}, \ldots, \sqrt{\text{Var}(Y_n)})$.

Example 2.1.3: Suppose $D[\mathbf{Y}] = \begin{bmatrix} 4 & 3 & 6 \\ 3 & 9 & 3 \\ 6 & 3 & 16 \end{bmatrix}$. Then $\sigma_{\mathbf{Y}} = \begin{bmatrix} 2 & 0 & 0 \\ 0 & 3 & 0 \\ 0 & 0 & 4 \end{bmatrix}$ and

$$R[\mathbf{Y}, \mathbf{Y}] = \sigma_{\mathbf{Y}}^{-1} D[\mathbf{Y}] \sigma_{\mathbf{Y}}^{-1} = \begin{bmatrix} 1 & 0.5 & 0.75 \\ 0.5 & 1 & 0.25 \\ 0.75 & 0.25 & 1 \end{bmatrix}.$$

Theorem 2.1.3: Let **X** and **Y** be random vectors taking values in R_m and R_n, respectively. Then for any matrices of constants **A** and **B** of dimensions $r \times m$ and $s \times n$,

$$C[\mathbf{AX}, \mathbf{BY}] = \mathbf{A}C[\mathbf{X}, \mathbf{Y}]\mathbf{B}'.$$

***Proof*:**

$$\begin{aligned} C[\mathbf{AX}, \mathbf{BY}] &= E[(\mathbf{AX} - \mathbf{A}E(\mathbf{X}))(\mathbf{BY} - \mathbf{B}E(\mathbf{Y}))'] \\ &= \mathbf{A}E[(\mathbf{X} - E(\mathbf{X}))(\mathbf{Y} - E(\mathbf{Y}))']\mathbf{B}' \\ &= \mathbf{A}C[\mathbf{X}, \mathbf{Y}]\mathbf{B}'. \qquad \square \end{aligned}$$

Taking $\mathbf{Y} = \mathbf{X}$, we get $C[\mathbf{AX}, \mathbf{BX}] = \mathbf{A}D[\mathbf{X}]\mathbf{B}'$, and for $\mathbf{B} = \mathbf{A}$,

$$D[\mathbf{AX}] = C[\mathbf{AX}, \mathbf{AX}] = \mathbf{A}D[\mathbf{X}]\mathbf{A}'.$$

Of course, the covariance matrix is unaffected by addition of constant vectors (translations):

$$\begin{aligned} C[\mathbf{X} + \mathbf{a}, \mathbf{Y} + \mathbf{b}] &= E[(\mathbf{X} + \mathbf{a} - E(\mathbf{X}) - \mathbf{a})(\mathbf{Y} + \mathbf{b} - E(\mathbf{Y}) - \mathbf{b})'] \\ &= E[(\mathbf{X} - E(\mathbf{X}))(\mathbf{Y} - E(\mathbf{Y}))'] = C[\mathbf{X}, \mathbf{Y}]. \end{aligned}$$

The covariance "operator" on pairs of random vectors is linear in both arguments, in that Theorem 2.1.3 holds, $C[\mathbf{X}_1 + \mathbf{X}_2, \mathbf{Y}] = C[\mathbf{X}_1, \mathbf{Y}] + C[\mathbf{X}_2, \mathbf{Y})$, and $C[\mathbf{X}, \mathbf{Y}_1 + \mathbf{Y}_2] = C[\mathbf{X}, \mathbf{Y}_1] + C[\mathbf{X}, \mathbf{Y}_2]$.

Summary

(1) $C[\mathbf{X}, \mathbf{Y}]$ is linear in both arguments ('bilinear').
(2) $C[\mathbf{X}, \mathbf{Y}] = C[\mathbf{Y}, \mathbf{X}]'$
(3) $C[\mathbf{X} + \mathbf{a}, \mathbf{Y} + \mathbf{b}] = C[\mathbf{X}, \mathbf{Y}]$ for constant vectors **a**, **b**.
(4) $C[\mathbf{AX}, \mathbf{BY}] = \mathbf{A}C[\mathbf{X}, \mathbf{Y}]\mathbf{B}'$
(5) $D[\mathbf{X} + \mathbf{a}] = D[\mathbf{X}]$

(6) $D[\mathbf{X}_1 + \mathbf{X}_2] = C[\mathbf{X}_1 + \mathbf{X}_2, \mathbf{X}_1 + \mathbf{X}_2]$
$= D[\mathbf{X}_1] + C[\mathbf{X}_1, \mathbf{X}_2] + C[\mathbf{X}_2, \mathbf{X}_1] + D[\mathbf{X}_2]$.

Note that the second and third terms need not be equal.

(7) $D[\mathbf{AX}] = \mathbf{A}D[\mathbf{X}]\mathbf{A}'$

If $\mathbf{X}_1$ and $\mathbf{X}_2$ are independent then $C[\mathbf{X}_1, \mathbf{X}_2] = 0$, so that $D[\mathbf{X}_1 + \mathbf{X}_2] = D[\mathbf{X}_1] + D[\mathbf{X}_2]$. More generally, if $\mathbf{X}_1, \ldots, \mathbf{X}_n$ are independent, then $D\left[\sum_i \mathbf{X}_i\right] = \sum_i D[\mathbf{X}_i]$. If these $\mathbf{X}_i$ have the same covariance matrix $\mathbf{\Sigma}$, then $D\left[\sum_i \mathbf{X}_i\right] = n\mathbf{\Sigma}$, and the sample mean vector $\bar{\mathbf{X}} = \frac{1}{n}\sum_i \mathbf{X}_i$ has covariance matrix $D[\bar{\mathbf{X}}] = \left(\frac{1}{n}\right)^2 n\mathbf{\Sigma} = \frac{1}{n}\mathbf{\Sigma}$, a familiar formula in the univariate case.

Problem 2.1.1: Let $\mathbf{X} = \begin{bmatrix} X_1 \\ X_2 \\ X_3 \end{bmatrix}$ and $\mathbf{Y} = \begin{bmatrix} Y_1 \\ Y_2 \\ Y_3 \end{bmatrix}$ be independent random vectors with $E(\mathbf{X}) = \begin{bmatrix} 2 \\ 3 \\ 4 \end{bmatrix}$, $E(\mathbf{Y}) = \begin{bmatrix} 2 \\ 4 \\ 6 \end{bmatrix}$, $D[\mathbf{X}] \equiv \mathbf{\Sigma}_X = \begin{bmatrix} 4 & 2 & 3 \\ 2 & 9 & 1 \\ 3 & 1 & 5 \end{bmatrix}$, and $D[\mathbf{Y}] \equiv \mathbf{\Sigma}_Y = \begin{bmatrix} 4 & -2 & 3 \\ -2 & 6 & 2 \\ 3 & 2 & 8 \end{bmatrix}$. Let $\mathbf{A} = \begin{bmatrix} 1 & 0 & 2 \\ 0 & 3 & 1 \end{bmatrix}$, and $\mathbf{B} = \begin{bmatrix} 1 & 0 & -1 \\ 0 & 1 & 1 \\ 0 & -1 & 1 \end{bmatrix}$. Find

(a) $C[\mathbf{X}, \mathbf{Y}]$ (b) $D[\mathbf{X} + \mathbf{Y}]$ (c) $C[\mathbf{AX}, \mathbf{BX}]$
(d) $E[\mathbf{AX} + \mathbf{AY}]$ (e) $D[\mathbf{AX}]$ (f) $R[\mathbf{AX}, \mathbf{X}]$ (g) $R[\mathbf{AX}, \mathbf{BX}]$

Problem 2.1.2: Let $X_1, X_2, \ldots, X_n$ be independent random variables, all with variance σ^2. Define $Y_k = X_1 + \cdots + X_k$ for $k = 1, \ldots, n$. Find $D[\mathbf{Y}]$ for $\mathbf{Y}' = (Y_1, \ldots, Y_n)$. Also find $R[\mathbf{Y}]$, the correlation matrix.

Problem 2.1.3: Give an example of a joint discrete distribution of two r.v.'s X and Y such that X and Y have covariance 0, but are not independent.

Problem 2.1.4: Let X and Y be two random variables, each taking only two possible values. Show that $\text{cov}(X, Y) = 0$ implies X and Y are independent. *Hint*: Show that this is true if both X and Y take only the values 0 and 1. Then show that the general case follows from this.

Problem 2.1.5: Let $\mathbf{Y} = (Y_1, Y_2, Y_3)'$ be the vector of weights of three pigs of the same litter at 3 months of age. A reasonable model states that $Y_i = G + \varepsilon_i$, for $i = 1, 2, 3$, where G, ε_1, ε_2, ε_3 are independent random variables, with $E(G) = \mu$, $\text{Var}(G) = \sigma_G^2$, $E(\varepsilon_i) = 0$, $\text{Var}(\varepsilon_i) = \sigma_\varepsilon^2$. G is the genetic effect, while $\varepsilon_1, \varepsilon_2, \varepsilon_3$ are random deviations from G. Find $D[\mathbf{Y}]$, $R[\mathbf{Y}]$, and $\text{Var}(Y_1 + Y_2 + Y_3) = D[\mathbf{J}'\mathbf{Y}]$, where $\mathbf{J}' = (1, 1, 1)$.

2.2 EXPECTED VALUES OF QUADRATIC FORMS

In the study of regression analysis and analysis of variance we will often be interested in statistics which are quadratic functions of the observations. That is, the statistic is the sum of terms of the form $a_{ij}X_iX_j$. For example, for $\mathbf{X} = (X_1, X_2, X_3)$ we may be interested in the statistic $Q(X_1, X_2, X_3) = 2X_1^2 - X_2^2 + 3X_3^2 - 6X_1X_2 + 2X_1X_3 - 4X_2X_3$.

Definition 2.2.1: A *quadratic form* is a function $Q(\mathbf{x})$ defined on R_n for some n of the form

$$Q(\mathbf{x}) = \mathbf{x}'\mathbf{A}\mathbf{x},$$

where $\mathbf{A}$ is an $n \times n$ symmetric matrix.

Comment: The requirement that $\mathbf{A}$ be symmetric is not a restriction on Q, since otherwise we could replace $\mathbf{A}$ by $\mathbf{B} = (\mathbf{A} + \mathbf{A}')/2$, so that $\mathbf{B}$ is symmetric, and, since $\mathbf{x}'\mathbf{A}\mathbf{x} = (\mathbf{x}'\mathbf{A}\mathbf{x})' = \mathbf{x}'\mathbf{A}'\mathbf{x}$,

$$\mathbf{x}'\mathbf{B}\mathbf{x} = \mathbf{x}'(\mathbf{A} + \mathbf{A}')\mathbf{x}/2 = \mathbf{x}'\mathbf{A}\mathbf{x} = Q(\mathbf{x})$$

For the example above we could take $\mathbf{A} = \begin{bmatrix} 2 & -6 & 2 \\ 0 & -1 & -4 \\ 0 & 0 & 3 \end{bmatrix}$. However, it is more convenient to work with $\mathbf{B} = \begin{bmatrix} 2 & -3 & 1 \\ -3 & -1 & -2 \\ 1 & -2 & 3 \end{bmatrix}$, because it is symmetric and $\mathbf{x}'\mathbf{A}\mathbf{x} = \mathbf{x}'\mathbf{B}\mathbf{x}$ for all $\mathbf{x}$.

Let $\mathbf{A} = \sum_i \lambda_i \mathbf{u}_i\mathbf{u}_i'$, where $\lambda_1, \ldots, \lambda_n$ are eigenvalues and $\mathbf{u}_1, \ldots, \mathbf{u}_n$ a corresponding system of mutually orthogonal length-one eigenvectors for $\mathbf{A}$. Then $Q(\mathbf{x}) = \mathbf{x}'\mathbf{A}\mathbf{x} = \sum_{i=1}^n \lambda_i \mathbf{x}'\mathbf{u}_i\mathbf{u}_i'\mathbf{x} = \sum_{i=1}^n \lambda_i(\mathbf{u}_i'\mathbf{x})(\mathbf{u}_i'\mathbf{x}) = \sum_{i=1}^n \lambda_i w_i^2$ for $w_i = \mathbf{u}_i'\mathbf{x}$. Thus, each quadratic form may be considered to be a weighted sum of squares. It follows that if all λ_i are positive (nonnegative) $\mathbf{A}$ is positive definite (nonnegative definite).

We will be concerned with the random variable $Q(\mathbf{X}) = \mathbf{X}'\mathbf{A}\mathbf{X}$ for $\mathbf{X}$ a random vector. First we consider a more general random variable $Q(\mathbf{X}, \mathbf{Y}) = \mathbf{X}'\mathbf{A}\mathbf{Y}$ for $\mathbf{X}$ $m \times 1$, $\mathbf{Y}$ $n \times 1$, and $\mathbf{A}$ an $m \times n$ constant matrix. Let $C[\mathbf{X}, \mathbf{Y}] = (\sigma_{ij}) = C$, $E(\mathbf{X}) = \boldsymbol{\mu}_X$, $E(\mathbf{Y}) = \boldsymbol{\mu}_Y$. $Q(\mathbf{X}, \mathbf{Y})$ is a *bilinear form*, since it is linear in both arguments.

Theorem 2.2.1: Let $E(\mathbf{X}) = \boldsymbol{\mu}$, $E(\mathbf{Y}) = \mathbf{v}$, $\mathbf{C} = C[\mathbf{X}, \mathbf{Y}] = (\sigma_{ij})$. Then

$$E(\mathbf{X}'\mathbf{A}\mathbf{Y}) = \sum_{ij} a_{ij}\sigma_{ij} + \boldsymbol{\mu}'\mathbf{A}\mathbf{v} = \text{trace}(\mathbf{A}\mathbf{C}') + \boldsymbol{\mu}'\mathbf{A}\mathbf{v}.$$

Comment: $\sum_{ij} a_{ij}\sigma_{ij}$ is the sum of the products of corresponding components of $\mathbf{A}$ and $\mathbf{C}$ (*not* the sum of elements of $\mathbf{AC}$, which is not even defined unless $m = n$). The second term is $\boldsymbol{\mu}'\mathbf{A}\mathbf{v} = Q(\boldsymbol{\mu}, \mathbf{v})$, where $Q(\mathbf{X}, \mathbf{Y}) = \mathbf{X}'\mathbf{A}\mathbf{Y}$.

***Proof*:** Let the ith component of $\boldsymbol{\mu}$ be μ_i. Let the jth component of $\mathbf{v}$ be v_j. Then $\mathbf{X}'\mathbf{A}\mathbf{Y} = \sum_{i=1}^{m} \sum_{j=1}^{n} a_{ij}X_i Y_j$, and $E(X_i Y_j) = \text{cov}(X_i, Y_j) + \mu_i v_j$. Thus $E(\mathbf{X}'\mathbf{A}\mathbf{Y}) = \sum_{i=1}^{m} \sum_{j=1}^{n} a_{ij}\sigma_{ij} + \sum_{i=1}^{m} \sum_{j=1}^{n} a_{ij}\mu_i\mu_j$. The first terms may be written in the form trace($\mathbf{A}\mathbf{C}'$), since the ii term of $\mathbf{A}\mathbf{C}'$ is $\sum_{j} a_{ij}\sigma_{ij}$. □

Letting $\mathbf{Y} = \mathbf{X}$, we get

Theorem 2.2.2: Let $Q(\mathbf{X}) = \mathbf{X}'\mathbf{A}\mathbf{X}$, $D[\mathbf{X}] = \boldsymbol{\Sigma}$, and $E(\mathbf{X}) = \boldsymbol{\mu}$. Then

$$E[Q(\mathbf{X})] = \sum_{ij} a_{ij} \text{ cov}(X_i, X_j) + Q(\boldsymbol{\mu}) = \text{trace}(\mathbf{A}\boldsymbol{\Sigma}) + Q(\boldsymbol{\mu}).$$

Comment: trace($\mathbf{A}\boldsymbol{\Sigma}$) is the sum of the products of corresponding components of $\mathbf{A}$ and $\boldsymbol{\Sigma}$. Some special cases are of interest:

(1) $\boldsymbol{\Sigma} = \mathbf{I}_n\sigma^2$. Then $E[Q(\mathbf{X})] = \sigma^2$ trace($\mathbf{A}$) + $Q(\boldsymbol{\mu})$
(2) $\boldsymbol{\mu} = \mathbf{0}$. Then $E[Q(\mathbf{X})] = \text{trace}(\mathbf{A}\boldsymbol{\Sigma})$.

Example 2.2.1: Let $X_1, \ldots, X_n$ be independent r.v.'s, each with mean μ, variance σ^2. Then $E(\mathbf{X}) = \mu\mathbf{J}_n$, $D[\mathbf{X}] = \sigma^2\mathbf{I}_n$. Consider $Q(\mathbf{X}) = \sum_{i=1}^{n} (X_i - \bar{X})^2 = \|\mathbf{P}_{V^\perp}\mathbf{X}\|^2 = X'(\mathbf{I}_n - \mathbf{P}_V)\mathbf{X}$, where $V = \mathscr{L}(\mathbf{J}_n)$, and $\mathbf{P}_V = (1/n)\mathbf{J}_n\mathbf{J}_n'$. Then we may apply Theorem 2.2.2 with $\mathbf{A} = \mathbf{I}_n - \mathbf{P}_V$ and $\boldsymbol{\Sigma} = \mathbf{I}_n\sigma^2$. We get $\mathbf{A}\boldsymbol{\Sigma} = \sigma^2(\mathbf{I}_n - \mathbf{P}_V)$ and trace($\mathbf{A}\boldsymbol{\Sigma}$) $= \sigma^2(n - 1)$, since the trace of $\mathbf{P}_V$ may be determined to be one, either directly or from the general theory which states that the trace of a projection matrix is the dimension of the subspace onto which it projects.

Since $Q(\mu\mathbf{J}_n) = \sum_{i=1}^{n} (\mu_i - \mu)^2 = 0$, we find that $E[Q(\mathbf{X})] = \sigma^2(n - 1)$. For

$S^2 = \dfrac{1}{n-1}\sum_1^n (X_i - \bar{X})^2$, $E(S^2) = \dfrac{(n-1)\sigma^2}{n-1} = \sigma^2$, so that S^2 is an unbiased estimator of σ^2.

If we let $Y_i = X_i - \bar{X}$, then, from (7) of the summary to section 2.1 $\mathbf{Y} = \mathbf{P}_{V^\perp}\mathbf{X}$ has covariance matrix $\mathbf{P}_{V^\perp}(\sigma^2\mathbf{I}_n)\mathbf{P}'_{V^\perp} = \sigma^2 P_{V^\perp}$, and $\sum (X_i - \bar{X})^2 = \|\mathbf{Y}\|^2 = \mathbf{Y}'\mathbf{I}_n\mathbf{Y}$. Applying Theorem 2.2.2, we have $E(\mathbf{Y}) = \mathbf{0}$, $\mathbf{A\Sigma} = \sigma^2\mathbf{P}_{V^\perp} = \sigma^2(\mathbf{I}_n - \mathbf{P}_V)$ as before.

Problem 2.2.1: Let $\mathbf{X} = (X_1, X_2, X_3)'$, $E(\mathbf{X}) = \boldsymbol{\theta} = (2, 3, 4)'$, $\boldsymbol{\Sigma} = D[\mathbf{X}] = \begin{bmatrix} 2 & 0 & -1 \\ 0 & 1 & 1 \\ -1 & 1 & 3 \end{bmatrix}$ and $Q(\mathbf{X}) = \sum_1^3 (X_i - \bar{X})^2$. Find $E[Q(\mathbf{X})]$.

Problem 2.2.2: Let $D[\mathbf{X}] = \sigma^2\mathbf{I}_n$ and $E(\mathbf{X}) = \mu\mathbf{J}_n$. Define $Q_2(\mathbf{X}) = \sum_{i<j} (X_i - X_j)^2$.

(a) Find $E[Q_2(\mathbf{X})]$.

(b) How is $Q_2(\mathbf{X})$ related to $Q_1(\mathbf{X}) = \sum_i (X_i - \bar{X})^2$? (Compare the matrices corresponding to these quadratic forms.)

(c) Find a constant c_n such that $c_n Q_2(\mathbf{X})$ is an unbiased estimator of σ^2.

Problem 2.2.3: Let $X_1, \ldots, X_n$ be uncorrelated random variable with equal means μ and $\text{Var}(X_i) = \sigma_i^2$.

(a) For $\bar{X} = \dfrac{1}{n}\sum_1^n X_i$, what is $\text{Var}(\bar{X})$? Use Theorem 2.1.3.

(b) Find a constant K_n such that $Q = K_n \sum_1^n (X_i - \bar{X})^2$ is an unbiased estimator of $\text{Var}(\bar{X})$. Thus, even though the X_i have unequal variances, Q is still an unbiased estimator of $\text{Var}(\bar{X})$. $\bar{X}$ is not the linear unbiased estimator with the smallest variance, however. To see this, take $n = 2$ and find the unbiased linear estimator with the smallest variance. This is really not an estimator unless the σ_i^2 (or their ratios) are known.

Problem 2.2.4: Let $X_1, \ldots, X_n$ be random variables with equal means μ, variances σ^2, and covariances $\rho\sigma^2$ (correlation ρ). Assume ρ known.

(a) Find a constant $K = K(n, \rho)$ such that $K \sum_1^n (X_i - \bar{X})^2$ is an unbiased estimator of σ^2.

(b) Construct an unbiased estimator of $\text{Var}(\bar{X})$.

(c) What is the smallest possible value for ρ for each n? *Hint*: Express $\text{Var}(\bar{X})$ in terms of n and ρ. To show that ρ can take this smallest possible value, suppose that $W_1, \ldots, W_n$ have equal variances and covariances. Let $X_i = W_i - \bar{W}$, so that $\mathbf{X}$ is the vector of deviations for $\mathbf{W}$. Determine $D[\mathbf{X}]$.

Problem 2.2.5: Consider a finite population of N elements with measurements $x_1, \ldots, x_N$. A simple random sample of n of these elements is a selection of n elements taken without replacement in such a way that all permutations have the same probability. Let $X_1, \ldots, X_n$ be the corresponding measurements for the units selected. Let $\mu = \frac{1}{N}\sum_{i=1}^{N} x_i$ and $\sigma^2 = [\sum (x_i - \mu)^2]/N$. Then $E(X_j) = \mu$, $\operatorname{Var}(X_j) = \sigma^2$, and, for $j \neq k$, $\operatorname{cov}(X_j, X_k) = -\sigma^2/(N-1)$. To show this consider $y_i = x_i - \mu$. Then $\sigma^2 = \left(\sum_i y_i^2\right)\Big/N$, and, for $j \neq k$, $\operatorname{cov}(X_j, X_k) = \frac{1}{N(N-1)}\sum_{h\neq i} y_h y_i$. But $0 = \left(\sum_i y_i\right)^2 = \sum_i y_i^2 + \sum_{h\neq i} y_h y_i$, so $\sum_{h\neq i} y_h y_i = -\sum y_i^2$.

(a) Show that $D[\mathbf{X}] = [\sigma^2/(N-1)][N\mathbf{I}_n - \mathbf{J}_n\mathbf{J}_n']$ for $\mathbf{J}_n = (1, \ldots, 1)'$.

(b) For $\bar{X} = (\sum X_j)/n = (\mathbf{J}_n'\mathbf{X})/n$, show that $\operatorname{Var}(\bar{X}) = \left(\frac{\sigma^2}{n}\right)\left(\frac{N-n}{N-1}\right)$.

(c) For $S^2 = \sum (X_i - \bar{X})^2/(n-1)$, show that $E(S^2) = \sigma^2 \frac{N}{(N-1)}$.

In *Sampling Techniques*, a classic by William Cochran (1977), σ^2 is replaced by $S^2 = \left[\sum_i^N y_i^2\right]\Big/(N-1)$, and S^2, as defined above is replaced by the symbol s^2. Then, for Cochran's notation, used for much of the sampling literature, s^2 is an unbiased estimator of "the population variance" S^2.

2.3 PROJECTIONS OF RANDOM VARIABLES

This section is not required for understanding the remainder of the book, though it should be useful for those interested in multivariate analysis, the study of the joint behavior of two or more random variables.

The linear space projection theory discussed in Chapter 1 may be extended to spaces of random variables. Let Ω be the collection of all random variables defined on some probability space with mean 0, finite variance. We will refer to elements of Ω as random variables and also as vectors. For $\mathbf{X}, \mathbf{Y} \in \Omega$ define the inner product $(\mathbf{X}, \mathbf{Y}) = E(\mathbf{XY})$, $\|\mathbf{X}\|^2 = E(\mathbf{X}^2) = \operatorname{Var}(\mathbf{X})$. Ω is infinite dimensional if the probability space is infinite. However, many of the ideas developed in Chapter 1 still hold. Ω is a Hilbert Space, usually called L_2.

A (finite dimensional) subspace $\mathscr{L}(\mathbf{X}_1, \ldots, \mathbf{X}_k)$ is again the collection of linear combinations $\sum b_i\mathbf{X}_i$. The projection of an element $\mathbf{Y}$ on a subspace V is the vector $\hat{\mathbf{Y}}$ (random variable) in V such that $(\mathbf{Y} - \hat{\mathbf{Y}}, \mathbf{X}) = 0$ for all $X \in$ V. That is, $E[\mathbf{X}(\mathbf{Y} - \hat{\mathbf{Y}})] = \operatorname{cov}(\mathbf{X}, \mathbf{Y} - \hat{\mathbf{Y}}) = 0$ for all $\mathbf{X} \in V$.

Taking $\mathbf{X} = (\mathbf{X}_1, \ldots, \mathbf{X}_k)$ and using the same arguments as were used in Chapter 1, this projection is $\hat{\mathbf{Y}} = \mathbf{Xb}$, where $\mathbf{b} = \mathbf{\Sigma}_\mathbf{X}^{-1}U$, $\mathbf{\Sigma}_\mathbf{X} = D[\mathbf{X}]$, $\mathbf{U} = C[\mathbf{X}, \mathbf{Y}]$, whenever $\mathbf{\Sigma}_\mathbf{X}$ is nonsingular. Since $\operatorname{Var}(\hat{Y}) = \|\hat{Y}\|^2$, $\operatorname{Var}(\hat{Y}) = \mathbf{b}'\mathbf{\Sigma}_\mathbf{X}\mathbf{b} = \mathbf{U}'\mathbf{\Sigma}_\mathbf{X}^{-1}\mathbf{U}$. The residual vector is $\mathbf{e} = \mathbf{Y} - \hat{\mathbf{Y}}$. Since $\mathbf{Y} = \mathbf{Y} - \hat{\mathbf{Y}} + \hat{\mathbf{Y}} = \mathbf{e} + \hat{\mathbf{Y}}$, $\operatorname{cov}(\mathbf{u}, \mathbf{e}) = 0$

for all $\mathbf{u} \in V$, and $\hat{\mathbf{Y}} \in V$, it follows that $\text{cov}(\hat{\mathbf{Y}}, \mathbf{e}) = 0$. Thus, $\text{Var}(\mathbf{Y}) = \text{Var}(\mathbf{e}) + \text{Var}(\hat{\mathbf{Y}}) = (\sigma_{\mathbf{Y}}^2 - \mathbf{U}'\mathbf{\Sigma}_{\mathbf{X}}^{-1}\mathbf{U}) + \mathbf{U}'\mathbf{\Sigma}_{\mathbf{X}}^{-1}\mathbf{U}$.

$\hat{\mathbf{Y}}$ is the best linear predictor of $\mathbf{Y}$ in the sense that, among all linear combinations $\sum b_j X_j$, $\mathbf{Y} - \hat{\mathbf{Y}} = \mathbf{e}$ has the smallest possible variance. This follows from the fact that for any $\mathbf{W} \in V$, $\|\mathbf{Y} - \mathbf{W}\|^2 = \text{Var}(\mathbf{Y} - \mathbf{W}) \geq \text{Var}(\mathbf{Y} - \hat{\mathbf{Y}}) = \|\mathbf{Y} - \hat{\mathbf{Y}}\|^2$. See the proof in Chapter 1.

Now suppose $E(\mathbf{Y}) = \mu_{\mathbf{Y}}$, $E(\mathbf{X}) = \boldsymbol{\mu}_{\mathbf{X}}$. Then the best linear predictor of $\mathbf{Y} - \mu_{\mathbf{Y}}$ as a function of $\mathbf{X} - \mu_{\mathbf{X}}$ is $(\mathbf{X} - \mu_{\mathbf{X}})\mathbf{b}$, where $\mathbf{b} = \mathbf{\Sigma}_{\mathbf{X}}^{-1}\mathbf{U}$. It follows that the best (affine) linear predictor of $\mathbf{Y}$ is

$$\hat{\mathbf{Y}} = \mu_{\mathbf{Y}} + (\mathbf{X} - \mu_{\mathbf{X}})\mathbf{b}.$$

The error is $\mathbf{e} = \mathbf{Y} - \hat{\mathbf{Y}} = (\mathbf{Y} - \mu_{\mathbf{Y}}) - (\mathbf{X} - \mu_{\mathbf{X}})\mathbf{b}$, which has variance $\sigma_{\mathbf{Y}}^2 - \mathbf{U}'\mathbf{\Sigma}_{\mathbf{X}}^{-1}\mathbf{U}$. The multiple correlation coefficient of $\mathbf{Y}$ and the set $\{\mathbf{X}_1, \ldots, \mathbf{X}_k\}$ is a measure of the precision of the approximation of $\mathbf{Y}$ by $\hat{\mathbf{Y}}$, relative to that provided by $\mu_{\mathbf{Y}}$.

Definition 2.3.1: The multiple correlation coefficient of $\mathbf{Y}$ with the random vector $\mathbf{X}$ is

$$R = [\text{Var}(\hat{\mathbf{Y}})/\text{Var}(\mathbf{Y})]^{1/2} = [1 - \text{Var}(\mathbf{Y} - \hat{\mathbf{Y}})/\text{Var}(\mathbf{Y})]^{1/2}$$

All of our discussion so far has concerned the approximation of $\mathbf{Y}$ by linear combinations $\mathbf{W} = \sum b_j \mathbf{X}_j$ with the closeness of the approximation measured by $E(\mathbf{Y} - \mathbf{W})^2$. It is possible to do better if approximations are not limited to linear combinations of $\mathbf{X}_1, \ldots, \mathbf{X}_k$. Let $g(\tilde{\mathbf{x}}) = E(\mathbf{Y}|\mathbf{X} = \tilde{\mathbf{x}})$ for $\tilde{\mathbf{x}} \in R_k$. g is the regression function. Then for any real-valued function h on R_k,

$$\begin{aligned} \mathrm{E}[(\mathbf{Y} - g(\mathbf{X}))h(\mathbf{X})] &= E[E((\mathbf{Y} - g(\mathbf{X}))h(\mathbf{X})|\mathbf{X})] \\ &= E[h(\mathbf{X})E((\mathbf{Y} - g(\mathbf{X}))|\mathbf{X})] = E[h(\mathbf{X})\cdot 0] = 0 \end{aligned}$$

so that $\text{cov}(\mathbf{Y} - g(\mathbf{X}), h(\mathbf{X})) = 0$. Thus, for any predictor $h(\mathbf{X})$ of $\mathbf{Y}$,

$$E(\mathbf{Y} - h(\mathbf{X}))^2 = E[\mathbf{Y} - g(\mathbf{X})]^2 + E[g(\mathbf{X}) - h(\mathbf{X})]^2$$

It follows that $g(\mathbf{X})$ is the best predictor for $\mathbf{Y}$ in this least squares sense. If h must be chosen to be linear then, as already shown, we take

$$h(\mathbf{X}) = \mu_{\mathbf{Y}} + (\mathbf{X} - \mu_{\mathbf{X}})\mathbf{b} = \hat{\mathbf{Y}}$$

which is the *linear* least squares predictor of $\mathbf{Y}$.

In the case that $k = 1$, the linear least squares predictor of $\mathbf{Y}$ is

$$\hat{\mathbf{Y}} = \mu_{\mathbf{Y}} + \rho\sigma_{\mathbf{Y}}(\mathbf{X} - \mu_{\mathbf{X}})/\sigma_{\mathbf{X}}$$

Equivalently, writing $\mathbf{Z}_Y$ and $\mathbf{Z}_{\mathbf{X}}$ for the standardized versions of $\mathbf{X}$ and $\mathbf{Y}$, the linear predictor of $\mathbf{Z}_Y$ is $\rho\mathbf{Z}_{\mathbf{X}}$. Since $-1 \le \rho \le +1$, and $E(\mathbf{Z}_Y|\mathbf{X} = x) = \rho\mathbf{Z}_x$ it follows that $|E(\mathbf{Z}_Y|\mathbf{X} = x)| \le |Z_x|$ with equality only if $|\rho| = 1$. This inequality is often described as regression toward the mean. Note that the linear predictor $\hat{\mathbf{Y}}$ is the least squares estimator among *linear* predictors of $\mathbf{Y}$, but is not the least squares estimator in general unless $g(\mathbf{X}) = E(\mathbf{Y}|\mathbf{X})$ is linear in $\mathbf{X}$. The tendency to use $\mathbf{Z}_{\mathbf{X}}$ as the predictor of Z_Y, rather than $\rho\mathbf{Z}_{\mathbf{X}}$, is called the regression fallacy.

Consider a population of father–son pairs with heights (x, y). Suppose the fathers have mean height 69 inches, standard deviation 2.5 inches, and sons have mean height 70 inches, standard deviation 2.8 inches. Under the regression fallacy a father with height 74 inches (2 s.d.'s above average) would be predicted to produce a son with height 2 s.d.'s above average, or 75.6 inches. The best linear predictor should be just 2ρ s.d.'s above average, so that for $\rho = 1/2$, for example, the prediction is 72.8 inches. Similarly, for $\rho = 1/2$, a father below average in height should be expected to produce a son only one-half as far below average in standard units.

In this paper "Kinship and Correlation," which appeared in the *North American Review* in 1890, Francis Galton first defined the correlation coefficient as the slope of the standardized regression line for the case of "quasi-normal" data (approximate bivariate normal, so that the regression function is approximately linear). Stigler (1989) describes this paper and the circumstances under which it was written. In one of Galton's examples two clerks leave an office at the same time, and $(\mathbf{X}, \mathbf{Y})$ is the pair of times they take until they arrive at their homes. Since they ride the same omnibus together each day, but walk further, each at his own pace, $\mathbf{X}$ and $\mathbf{Y}$ are correlated so that the regression line (assuming linear regression) is described by the line

$$\mathbf{Y} = \mu_{\mathbf{Y}} + \rho\sigma_{\mathbf{Y}}(\mathbf{X} - \mu_{\mathbf{X}})/\sigma_{\mathbf{X}}.$$

In another example Galton considers a population of people with $\mathbf{X}$ = length at thigh bone, $\mathbf{Y}$ = height.

In baseball the batting averages for players in their first and second years of major league play have distributions roughly approximated by the bivariate normal distribution (so that regression is linear) with mean 0.265, standard deviations 0.30, correlation 0.40. A player who hits well the first year, say 0.310, should be expected to hit about 0.283 the second year. Instead players whose average drops this much are said to have had a sophomore slump, with all sorts of psychological reasons given. Similar deep meaning is found in the tendency for students with G.P.A.'s above 3.8 as freshmen to have lower average G.P.A.'s as sophomores.

Galton's clerk example and the baseball example may be modeled as follows. Let $\mathbf{W}$ be an r.v. denoting the common part of $\mathbf{X}$ and $\mathbf{Y}$, the bus ride time or the ability of a player, and let $\boldsymbol{\varepsilon}_{\mathbf{X}}$, $\boldsymbol{\varepsilon}_{\mathbf{Y}}$ be independent r.v.'s, the error parts of $\mathbf{X}$ and $\mathbf{Y}$, having equal variances $\sigma_{\boldsymbol{\varepsilon}}^2$. Suppose also that $\mathbf{W}$ and the pair $(\boldsymbol{\varepsilon}_{\mathbf{X}}, \boldsymbol{\varepsilon}_{\mathbf{Y}})$ are independent. Suppose $\mathbf{X} = \mathbf{W} + \boldsymbol{\varepsilon}_{\mathbf{X}}$, $\mathbf{Y} = \mathbf{W} + \boldsymbol{\varepsilon}_{\mathbf{Y}}$.

Then $\rho = \dfrac{\sigma_{\mathbf{W}}^2}{\sigma_{\mathbf{W}}^2 + \sigma_{\boldsymbol{\varepsilon}}^2}$. The best linear predictor of $\mathbf{Y}$ is $\hat{\mathbf{Y}} = \mu_{\mathbf{Y}} + \rho(\mathbf{X} - \mu_{\mathbf{X}})$, which is closer to $\mu_{\mathbf{Y}}$ than $\mathbf{X}$ is to $\mu_{\mathbf{X}}$.

Problem 2.3.1: Let $\mathbf{X} = (\mathbf{X}_1, \mathbf{X}_2, \mathbf{X}_3)$ have zero mean vector, covariance matrix $\boldsymbol{\Sigma} = \begin{bmatrix} 4 & 1 & 2 \\ 1 & 3 & -1 \\ 2 & -1 & 2 \end{bmatrix}$

(a) Find the best linear prediction $\hat{\mathbf{X}}_1$ of $\mathbf{X}_1$ as a function of $\mathbf{X}_2$ and $\mathbf{X}_3$.

(b) Find $\text{Var}(\mathbf{X}_1)$, $\text{Var}(\mathbf{X}_1 - \hat{\mathbf{X}}_1)$ and the multiple correlation coefficient of $\mathbf{X}_1$ with $\mathbf{X}_2$, $\mathbf{X}_3$.

Problem 2.3.2: Show that the multiple correlation coefficient, R, of $\mathbf{Y}$ with $(\mathbf{X}_1, \ldots, \mathbf{X}_n)$ is the ordinary correlation coefficient of $\mathbf{Y}$ with the best linear predictor $\hat{\mathbf{Y}}$.

Problem 2.3.3: Let $(\mathbf{X}_0, \ldots, \mathbf{X}_k)'$ have the equicovariance matrix $\boldsymbol{\Sigma} = \sigma^2[(1 - \rho)\mathbf{I}_{k+1} + \rho J_{k+1}\mathbf{J}'_{k+1}]$, and mean vector $\mathbf{0}$.

(a) Show that the best linear predictor of $\mathbf{X}_0$ as a function of $\mathbf{X} = (\mathbf{X}_1, \ldots, \mathbf{X}_k)$ is $\hat{\mathbf{X}}_0 = \mathbf{X}\mathbf{b}$ for $\mathbf{b} = d\mathbf{J}_k$ and $d = \rho/[1 + (k - 1)\rho]$. *Hint*: See Problem 1.6.6.

(b) Show that the multiple correlation coefficient of $\mathbf{X}_0$ with $\mathbf{X}$ is $g(\rho, k) \equiv [d\rho k]^{1/2}$. Also show that $0 \le g(\rho, k) \le 1$, g is monotone in ρ and k, and $\lim_{k \to \infty} g(\rho, k) = \rho^{1/2}$.

Problem 2.3.4: Let $k = 1$ and let $(\mathbf{X}, \mathbf{Y})$ have the joint discrete distribution with probability mass function $f(x, y)$ given by the following table:

$$y \;\begin{matrix} & x & \\ & \begin{matrix} 0 & 1 & 2 \end{matrix} & \\ \begin{matrix} 0 \\ 1 \end{matrix} & \begin{bmatrix} 0 & 0.1 & 0.2 \\ 0.2 & 0.5 & 0 \end{bmatrix} \end{matrix}$$

Find the least squares predictor $g(\mathbf{X}) = E(\mathbf{Y}|\mathbf{X})$, and the linear least squares predictor $h(\mathbf{X}) = \hat{\mathbf{Y}}$. Show that $g(\mathbf{X})$ and $h(\mathbf{X})$ are unbiased estimators of $E(Y)$. Also find $\text{Var}(g(\mathbf{X}))$, $E[\mathbf{Y} - g(\mathbf{X})]^2$, $\text{Var}(\hat{\mathbf{Y}})$ and $E[\mathbf{Y} - \hat{\mathbf{Y}}]^2$.

Problem 2.3.5: In his book *Natural Inheritance*, Galton (1889) published the data of Table 2.3.1 and Figure 2.1 on the midheights of parents (x) and that of an adult child (y). For u_i = midpoint of ith midparent interval

Table 2.3.1 Galton's Correlation Table*

Height of the Midparent (Inches)	Height of the Adult Child (Inches)														Total No. of Adult Children
	<61.7	62.2	63.2	64.2	65.2	66.2	67.2	68.2	69.2	70.2	71.2	72.2	73.2	>73.7	
>73.0	—	—	—	—	—	—	—	—	—	—	—	1	3	—	4
72.5	—	—	—	—	—	—	—	1	2	1	2	7	2	4	19
71.5	—	—	—	—	1	3	4	3	5	10	4	9	2	2	43
70.5	1	—	1	—	1	1	3	12	18	14	7	4	3	3	68
69.5	—	—	1	16	4	17	27	20	33	25	20	11	4	5	183
68.5	1	—	7	11	16	25	31	34	48	21	18	4	3	—	219
67.5	—	3	5	14	15	36	38	28	38	19	11	4	—	—	211
66.5	—	3	3	5	2	17	17	14	13	4	—	—	—	—	78
65.5	1	—	9	5	7	11	11	7	7	5	2	1	—	—	66
64.5	1	1	4	4	1	5	5	—	2	—	—	—	—	—	23
<64.0	1	—	2	4	1	2	2	1	1	—	—	—	—	—	14
Totals	5	7	32	59	48	117	138	120	167	99	64	41	17	14	928

* This cross-tabulation was compiled by Galton in 1885 and published in 1886 and again in 1889. It gives the heights of 928 adult children, classified by height of "midparents." All female heights were rescaled by multiplying by 1.08, and midparent heights were computed by averaging the height of the father and the rescaled height of the mother. For more information, see Stigler (1986, p. 286).

Source: Reprinted with permission from "Francis Galton's Account of the Invention of Correlation," *Statistical Science*, Vol. 4, No. 2, Table 1.

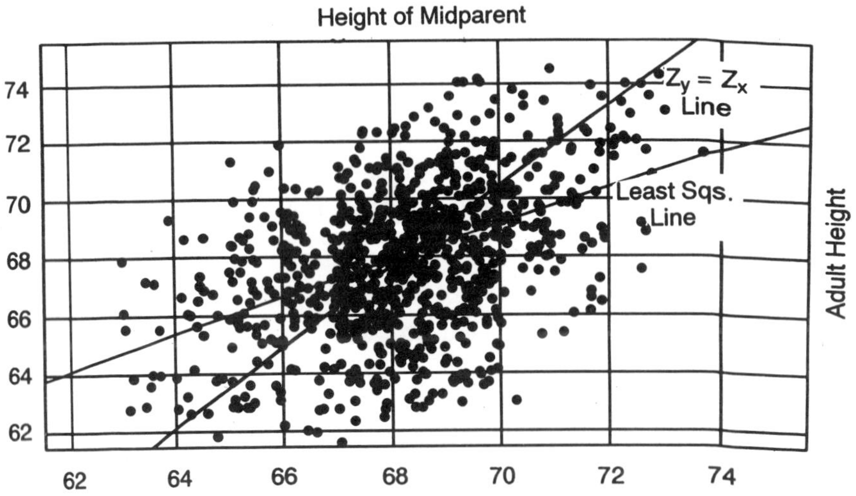

FIGURE 2.1 Galton's height data. From *Natural Inheritance* (1889), Macmillan, London.

$i = 1, \ldots, 11$, v_j = midpoint of jth adult child interval, $j = 1, \ldots, 14$, f_{ij} = frequency in ij cell, we find $\sum_{ij} f_{ij}u_i = 63{,}385$, $\sum_{ij} f_{ij}v_j = 63{,}190.6$, $\sum_{ij} f_{ij}u_i^2 = 4{,}332{,}418$ $\sum_{ij} f_{ij}v_j^2 = 4{,}308{,}850$, $\sum_{ij} f_{ij}u_iv_j = 4{,}318{,}061$, $n = 928$. Let $(\mathbf{X}, \mathbf{Y})$ take each of the values (x_i, y_i) with probability 1/928. That is, $P(\mathbf{X} = x_i, \mathbf{Y} = y_j) = f_{ij}/928$.

(a) Determine the parameters $\mu_{\mathbf{X}}, \mu_{\mathbf{Y}}, \sigma_{\mathbf{X}}, \sigma_{\mathbf{Y}}, \rho = r$, and the equation of the simple linear regression line $\hat{\mathbf{Y}} = \mu_{\mathbf{Y}} + \rho\sigma_{\mathbf{Y}}(\mathbf{X} - \mu_{\mathbf{X}})/\sigma_{\mathbf{X}}$.

(b) For tall parents (say ≥ 72) what proportion have shorter children, in the sense that $\mathbf{Z_Y} < \mathbf{Z_X}$? For short parents (say ≤ 64) what proportion have taller children, in the sense that $\mathbf{Z_Y} > \mathbf{Z_X}$?

2.4 THE MULTIVARIATE NORMAL DISTRIBUTION

Definition 2.4.1: A random vector $\mathbf{Y}$ (taking values in R_n) is said to have a multivariate normal distribution if $\mathbf{Y}$ has the same distribution as

$$\mathbf{X} \equiv \underset{n \times p}{\mathbf{A}} \; \underset{p \times 1}{\mathbf{Z}} + \underset{n \times 1}{\boldsymbol{\mu}},$$

where, for some p, $\mathbf{Z}$ is a vector of independent $N(0, 1)$ r.v.'s, $\mathbf{A}$ is an $n \times p$ matrix of constants, and $\boldsymbol{\mu}$ is an n-vector of constants. More generally, a random vector $\mathbf{X}$, taking values in an n-component linear space Ω, has a multivariate normal distribution if its column vector "version" in R_n does.

We will suppose in this section that $\Omega = R_n$, so that we can use matrix notation. Any property of the multivariate normal distribution which holds for this special case will have an obvious translation to the general case.

Since **Y** and **X** have the same distribution, we can exploit the representation of **X** to determine the density of the multivariate normal distribution. **X** does not have a density function on R_n unless **A** has rank n. In general **AZ** takes values in the subspace $C[\mathbf{A}]$. In the special case that **A** is $n \times n$ and has rank n, and therefore $\mathbf{AA}'$ has rank n, we can find the density as follows:

Let $\mathbf{X} = \mathbf{AZ} + \boldsymbol{\mu} \equiv g(\mathbf{Z})$. Then for **A** $n \times n$ of rank $ng(\mathbf{z})$ is a 1–1 function from R_n onto R_n so that

$$f_{\mathbf{X}}(\mathbf{x}) = f_{\mathbf{Z}}(g^{-1}(\mathbf{x}))\left|\det\left(\frac{\partial g^{-1}(\mathbf{x})}{\partial \mathbf{x}}\right)\right| = f_{\mathbf{Z}}(\mathbf{A}^{-1}(\mathbf{x} - \boldsymbol{\mu}))|\det \mathbf{A}^{-1}|$$

Since

$$f_{\mathbf{Z}}(\mathbf{z}) = \prod_{i=1}^{n} [(2\Pi)^{-1/2} \exp(-z_i^2/2)] = (2\Pi)^{n/2} \exp(-\mathbf{z}'\mathbf{z}/2),$$

we get

$$f_{\mathbf{X}}(\mathbf{x}) = (2\Pi)^{n/2} |\det \mathbf{A}|^{-1} \exp[-(\mathbf{x} - \boldsymbol{\mu})'(\mathbf{AA}')^{-1}(\mathbf{x} - \boldsymbol{\mu})/2]$$

Let $D[\mathbf{X}]$ be denoted by Σ. Since $E(\mathbf{X}) = \boldsymbol{\mu}$ and $\Sigma = D[\mathbf{X}] = \mathbf{AI}_n\mathbf{A}' = \mathbf{AA}'$, we have $\det(\Sigma) = \det(\mathbf{AA}') = [\det \mathbf{A}]^2$, so

$$f_{\mathbf{X}}(\mathbf{x}) = (2\Pi)^{-n/2} (\det \Sigma)^{-1/2} \exp[-(\mathbf{x} - \boldsymbol{\mu})' \Sigma^{-1}(\mathbf{x} - \boldsymbol{\mu})/2] \qquad \text{for all} \quad \mathbf{x} \in R_n.$$

Σ and its inverse are positive definite so that

$$Q(\mathbf{x}) \equiv (\mathbf{x} - \boldsymbol{\mu})' \Sigma^{-1}(\mathbf{x} - \boldsymbol{\mu}) > 0 \qquad \text{unless} \quad \mathbf{x} = \boldsymbol{\mu}.$$

The contours of $Q(\mathbf{x})$ (points **x** of equal value for Q) are ellipsoids in R_n.

The representation $\mathbf{X} = \mathbf{AZ} + \boldsymbol{\mu}$ makes it easy to compute the moment-generating function (or the characteristic function). We have

$$m_{\mathbf{X}}(\mathbf{t}) = E[e^{\mathbf{t}'\mathbf{X}}] = E[e^{\mathbf{t}'(\mathbf{AZ}+\boldsymbol{\mu})}] = e^{\mathbf{t}'\boldsymbol{\mu}}E(e^{\mathbf{t}'\mathbf{AZ}}) = e^{\mathbf{t}'\boldsymbol{\mu}}m_{\mathbf{Z}}(\mathbf{A}'\mathbf{t})$$

Recall that the moment-generating function of a standard normal r.v. is $e^{t^2/2}$ so that

$$m_{\mathbf{Z}}(\mathbf{u}) = \prod_{i=1}^{n} e^{u_i^2/2} = e^{\mathbf{u}'\mathbf{u}/2}$$

It follows that $m_{\mathbf{X}}(\mathbf{t}) = e^{\mathbf{t}'\boldsymbol{\mu}} e^{(\mathbf{A}'\mathbf{t})'(\mathbf{A}'\mathbf{t}')/2} = e^{\mathbf{t}'\boldsymbol{\mu}} e^{(\mathbf{t}'\Sigma\mathbf{t}/2)}$. Here **A** is $n \times p$ of any rank, so we have shown that the distribution of **X** depends on its mean vector $\boldsymbol{\mu}$ and covariance matrix Σ only.

Thus, if

$$\underset{n\times 1}{\mathbf{X}_1} = \underset{n\times p_1}{\mathbf{A}_1}\underset{p_1\times 1}{\mathbf{Z}_1} + \underset{n\times 1}{\boldsymbol{\mu}_1} \quad \text{and} \quad \underset{n\times 1}{\mathbf{X}_2} = \underset{n\times p_2}{\mathbf{A}_2}\underset{p_2\times 1}{\mathbf{Z}_2} + \underset{n\times 1}{\boldsymbol{\mu}_2},$$

where $\mathbf{Z}_1, \mathbf{Z}_2$ are independent, and each is a vector of independent standard normal r.v.'s, then $\mathbf{X}_1$ and $\mathbf{X}_2$ have the same distribution iff their mean vectors and covariance matrices are the same, i.e., $\boldsymbol{\mu}_1 = \boldsymbol{\mu}_2$ and $\mathbf{A}_1\mathbf{A}_1' = \mathbf{A}_2\mathbf{A}_2'$.

Theorem 2.4.1: Let $\boldsymbol{\mu}$ be an element of R_n and Σ an $n \times n$ nonnegative definite matrix. Then there exists a multivariate normal distribution with mean vector $\boldsymbol{\mu}$, covariance matrix Σ.

Comment: We will denote this distribution by $N_n(\boldsymbol{\mu}, \Sigma)$.

***Proof*:** Since Σ is symmetric, there exists a matrix $\mathbf{B}$ such that $\mathbf{BB}' = \Sigma$ (see triangular decomposition, towards the end of section 1.7). Let $\mathbf{Z}$ be an n-vector of standard normal independent r.v.'s. Let

$$\mathbf{X} = \mathbf{BZ} + \boldsymbol{\mu}.$$

Then $E(\mathbf{X}) = \boldsymbol{\mu}$ and $D[\mathbf{X}] = \mathbf{BI}_n\mathbf{B}' = \Sigma$, and by definition $\mathbf{X}$ has a multivariate normal distribution. □

The proof of this theorem suggests a method for generating multivariate normal vectors on a computer, given $\boldsymbol{\mu}$ and Σ. Find $\mathbf{B}$ such that $\mathbf{BB}' = \Sigma$. Suppose a method is available for generating independent standard normal r.v.'s, so that values of $\mathbf{Z}$ may be generated. Then $\mathbf{X} = \mathbf{BZ} + \boldsymbol{\mu}$ has the desired distribution.

Theorem 2.4.2: Let $\underset{n\times 1}{\mathbf{X}} \sim N_n(\boldsymbol{\mu}, \Sigma)$. Let $\underset{r\times 1}{\mathbf{Y}} = \underset{r\times n}{\mathbf{C}}\ \underset{n\times 1}{\mathbf{X}} + \underset{r\times 1}{\mathbf{d}}$ for $\mathbf{C}$ and $\mathbf{d}$ constant matrices. Then $\mathbf{Y} \sim N_r(\mathbf{C}\boldsymbol{\mu} + \mathbf{d}, \mathbf{C}\Sigma\mathbf{C}')$.

***Proof*:** By definition $\mathbf{X} = \mathbf{AZ} + \boldsymbol{\mu}$ for some $\mathbf{A}$ such that $\mathbf{AA}' = \Sigma$, with $\mathbf{Z} \sim N_p(\mathbf{0}, \mathbf{I}_p)$ for some p. Then $\mathbf{Y} = \mathbf{CAZ} + \mathbf{C}\boldsymbol{\mu} + \mathbf{d} = (\mathbf{CA})\mathbf{Z} + (\mathbf{C}\boldsymbol{\mu} + \mathbf{d})$, and by definition $\mathbf{Y}$ has a multivariate normal distribution with mean $\mathbf{C}\boldsymbol{\mu} + \mathbf{d}$, covariance matrix $(\mathbf{CA})(\mathbf{CA})' = \mathbf{C}\Sigma\mathbf{C}'$. □

Theorem 2.4.3: Let $\underset{n\times 1}{\mathbf{Y}}$ have a multivariate normal distribution. Let $\mathbf{Y} = \begin{bmatrix}\mathbf{Y}_1\\ \mathbf{Y}_2\end{bmatrix}$ with $\mathbf{Y}_1$ $p \times 1$ for $1 \le p < n$. Then $\mathbf{Y}_1$ and $\mathbf{Y}_2$ are independent if and only if $C[\mathbf{Y}_1, \mathbf{Y}_2] = \underset{p\times(n-p)}{\mathbf{0}}$.

***Proof*:** Suppose $\mathbf{Y}_1$, $\mathbf{Y}_2$ independent with mean vectors $\boldsymbol{\theta}_1$, $\boldsymbol{\theta}_2$. Then

$$C[\mathbf{Y}_1, \mathbf{Y}_2] = E[(\mathbf{Y}_1 - \boldsymbol{\theta}_1)(\mathbf{Y}_2 - \boldsymbol{\theta}_2)'] = E[(\mathbf{Y}_1 - \boldsymbol{\theta}_1)]E[(\mathbf{Y}_2 - \boldsymbol{\theta}_2)']$$
$$= \underset{p \times 1}{\mathbf{0}} \quad \underset{(n-p) \times 1}{(\mathbf{0})'} = \underset{p \times (n \times p)}{\mathbf{0}}.$$

Now suppose $C[\mathbf{Y}_1, \mathbf{Y}_2] = \underset{p \times (n-p)}{\mathbf{0}}$. We will use the fact that two random vectors are independent if their joint moment-generating function is the product of their marginal m.g.f.'s. Let $\underset{m \times 1}{\mathbf{t}} = \begin{bmatrix} \mathbf{t}_1 \\ \mathbf{t}_2 \end{bmatrix}$ for $\mathbf{t}$, $p \times 1$. Then $\mathbf{Y}$ has m.g.f.

$$m_{\mathbf{Y}}(\mathbf{t}) = \exp[\mathbf{t}_1'\boldsymbol{\theta}_1 + \mathbf{t}_2'\boldsymbol{\theta}_2 + \tfrac{1}{2}\mathbf{t}'\boldsymbol{\Sigma}\mathbf{t}]$$

But

$$\mathbf{t}'\boldsymbol{\Sigma}\mathbf{t} = (\mathbf{t}_1', \mathbf{t}_2')\begin{pmatrix} \boldsymbol{\Sigma}_{11} & \boldsymbol{\Sigma}_{12} \\ \boldsymbol{\Sigma}_{12} & \boldsymbol{\Sigma}_{22} \end{pmatrix}\begin{pmatrix} \mathbf{t}_1 \\ \mathbf{t}_2 \end{pmatrix} = \mathbf{t}_1'\boldsymbol{\Sigma}_{11}\mathbf{t}_1 + \mathbf{t}_2'\boldsymbol{\Sigma}_{22}\boldsymbol{\psi}_2,$$

since $\boldsymbol{\Sigma}_{12} = \underset{p \times n-p}{0}$. Thus,

$$m_{\mathbf{Y}}(\mathbf{t}) = \exp[\mathbf{t}_1'\boldsymbol{\theta}_1 + \tfrac{1}{2}\mathbf{t}_1'\boldsymbol{\Sigma}_{11}\mathbf{t}_1 + \mathbf{t}_2'\boldsymbol{\theta}_2 + \tfrac{1}{2}\mathbf{t}_2'\boldsymbol{\Sigma}_{22}\mathbf{t}_2] = m_{\mathbf{Y}_1}(\mathbf{t}_1)m_{\mathbf{Y}_2}(\mathbf{t}_2) \qquad \square$$

Theorem 2.4.4: Let $\underset{n \times 1}{\mathbf{X}}$ have a multivariate normal distribution. Let $\mathbf{X} = \begin{pmatrix} \mathbf{X}_1 \\ \mathbf{X}_2 \end{pmatrix}$ with $\mathbf{X}_1$ $p \times 1$ for $1 \le p < n$, with mean vector $\boldsymbol{\mu} = \begin{pmatrix} \boldsymbol{\mu}_1 \\ \boldsymbol{\mu}_2 \end{pmatrix}$, covariance matrix $\boldsymbol{\Sigma} = \begin{pmatrix} \boldsymbol{\Sigma}_{11} & \boldsymbol{\Sigma}_{12} \\ \boldsymbol{\Sigma}_{12}' & \boldsymbol{\Sigma}_{22} \end{pmatrix}$ for μ_1 $p \times 1$, $\boldsymbol{\Sigma}_{11}$ $p \times p$. Suppose $\boldsymbol{\Sigma}_{22}$ is nonsingular. Then

(1) $\mathbf{X}_1 \sim N_p(\boldsymbol{\mu}_1, \Sigma_{11})$

(2) The conditional distribution of $\mathbf{X}_1$, given $\mathbf{X}_2 = \mathbf{x}_2$ is $N_p(\boldsymbol{\mu}_1 + A(\mathbf{x}_2 - \boldsymbol{\mu}_2), \Sigma_{1.2})$ for $\mathbf{A} = \boldsymbol{\Sigma}_{12}\boldsymbol{\Sigma}_{22}^{-1}$, $\Sigma_{1.2} = \boldsymbol{\Sigma}_{11} - \boldsymbol{\Sigma}_{12}\boldsymbol{\Sigma}_{22}^{-1}\boldsymbol{\Sigma}_{12}'$.

***Proof*:** (1) follows directly from the representation $\mathbf{X} = \mathbf{AZ} + \boldsymbol{\mu}$.

To prove (2) let $\mathbf{Y}_1 = \mathbf{X}_1 - \boldsymbol{\mu}_1$, $\mathbf{Y}_2 = \mathbf{X}_2 - \boldsymbol{\mu}_2$. Then, from Section 2.3, $\mathbf{Y}_1 = \hat{\mathbf{Y}}_1 + (\mathbf{Y}_1 - \hat{\mathbf{Y}}_1)$, where $\hat{\mathbf{Y}}_1 = \mathbf{AY}_2$ for $\mathbf{A} = \boldsymbol{\Sigma}_{12}\boldsymbol{\Sigma}_{22}^{-1}$, with $C(\hat{\mathbf{Y}}_1, \mathbf{Y}_1 - \hat{\mathbf{Y}}_1) = \underset{p \times (n-p)}{\mathbf{0}}$. Since the pair $\hat{\mathbf{Y}}_1$, $\mathbf{Y}_1 - \hat{\mathbf{Y}}_1$ has a joint multivariate normal distribution, this implies independence of these two random vectors. Therefore, the distribution of the residual vector $\mathbf{Y}_1 - \hat{\mathbf{Y}}_1$ does not depend upon $\mathbf{Y}_2 = \mathbf{X}_2 - \boldsymbol{\mu}_2 = \mathbf{x}_2 - \boldsymbol{\mu}_2$. It follows that conditionally upon $\mathbf{X}_2 = \mathbf{x}_2$,

$$\mathbf{Y}_1 \sim N_p(\mathbf{A}(\mathbf{x}_2 - \boldsymbol{\mu}_2), \boldsymbol{\Sigma}_{11} - \boldsymbol{\Sigma}_{12}\boldsymbol{\Sigma}_{22}^{-1}\boldsymbol{\Sigma}_{12}' = \boldsymbol{\Sigma}_{1.2})$$

Since $\mathbf{X}_1 = \mathbf{Y}_1 + \boldsymbol{\mu}_1$, this implies (2). $\square$

Problem 2.4.1: Let $\mathbf{X} = (\mathbf{X}_1, \mathbf{X}_2)$ have a bivariate normal distribution with parameters $(\mu_1, \mu_2, \sigma_1^2, \sigma_2^2, \rho)$ for $-1 < \rho < +1$.

(a) Show that the density of $\mathbf{X}$ is

$$f_{\mathbf{X}}(\mathbf{x}) = [(2\pi)^2\sigma_1^2\sigma_2^2(1-\rho^2)]^{-1/2}\exp[-\tfrac{1}{2}Q(\mathbf{x})]/(1-\rho^2)]$$

for $Q(\mathbf{x}) = z_1^2 + z_2^2 - 2\rho z_1 z_2$, $z_1 = (x_1 - \mu_1)/\sigma_1$, $z_2 = (x_2 - \mu_2)/\sigma_2$. (This is the bivariate normal density.)

(b) Show that the conditional distribution of X_2, given $X_1 = x_1$, is $N\left(\mu_2 + \rho\sigma_2\left(\frac{x_1 - \mu_1}{\sigma_1}\right), \sigma_2^2(1-\rho^2)\right)$. Thus, for example, if the heights in inches of fathers (X_1) and sons (X_2) have a bivariate normal distribution with means $\mu_1 = 69$, $\mu_2 = 70$, $\sigma_1 = 2$, $\sigma_2 = 3$, $\rho = 0.4$, then the conditional distribution of a son's height for a father 73 inches tall (2 s.d.'s above average) is $N(72.4, 7.56)$ (0.8 s.d.'s above average, variance $0.84\sigma_2^2$).

Problem 2.4.2: Let $X = (X_1, X_2, X_3)$ be as in problem 2.3.1. What is the conditional distribution of X_1, given $X_2 = 1$, $X_3 = -1$?

Problem 2.4.3: Let $\mathbf{X} = (X_1, X_2, X_3)$ have covariance matrix $\boldsymbol{\Sigma} = \begin{bmatrix} 3 & 1 & 1 \\ 1 & 3 & -7 \\ 1 & -7 & 15 \end{bmatrix}$. Note that $\boldsymbol{\Sigma}\begin{bmatrix} 1 \\ -2 \\ -1 \end{bmatrix} = \mathbf{0}$ so that $\boldsymbol{\Sigma}$ has rank two. If $\boldsymbol{\mu}_{\mathbf{X}} = \begin{pmatrix} 10 \\ 20 \\ 30 \end{pmatrix}$ and $\mathbf{X}$ has a multivariate normal distribution, describe the range of $\mathbf{X}$. That is, what is the subset of R_3 in which $\mathbf{X}$ takes its values?

Problem 2.4.4: Let Z_1, Z_2 and B be independent r.v.'s, with Z_1, Z_2 standard normal and B taking the values -1, $+1$ with probabilities 1/2, 1/2. Let $Y_1 = B|Z_1|$ and $Y_2 = B|Z_2|$. (a) Argue that Y_1 and Y_2 each have the standard normal distribution, but the pair (Y_1, Y_2) does not have a bivariate normal distribution. (b) Show that $\rho(Y_1, Y_2) = 2/\pi$.

Problem 2.4.5: Suppose that a population of married couples have heights in inches, X for the wife, and Y for the husband. Suppose that (X, Y) has a bivariate normal distribution with parameters $\mu_X = 65$, $\mu_Y = 70$, $\sigma_X = 2.5$, $\sigma_Y = 2.7$, and $\rho = 0.4$. What is the probability that the husband is taller than his wife? Does the probability increase or decrease with ρ?

2.5 The χ^2, F, AND t DISTRIBUTIONS

In this section we study the distributions of certain functions of random variables which are normally distributed. Students will probably be somewhat familiar with the central χ^2, F, and t distributions, but may not have studied their noncentral counterparts. In each case these distributions are defined in terms of normally distributed random variables or, in the case of F and t, in terms of χ^2-random variables. Tests of hypotheses and confidence intervals will depend upon these central distributions. Determination of power for these tests (t and F tests) will depend upon these noncentral distributions.

Definition 2.5.1: Let $X_1, \ldots, X_n$ be independent $N(\mu_i, 1)$ random variables for $i = 1, \ldots, n$, then $Y = \sum_1^n X_i^2$ is said to have a noncentral chi-square distribution with n degrees of freedom and noncentrality parameter $\delta = \sum_1^n \mu_i^2$.

Comment: We will denote this distribution by $\chi_n^2(\delta)$. This is a legal definition only if we show that Y has a distribution which depends only on n and $\delta = \sum_1^n \mu_i^2$.

Define $\mathbf{X} = \begin{bmatrix} X_1 \\ \vdots \\ X_n \end{bmatrix}$ and $\boldsymbol{\mu} = \begin{bmatrix} \mu_1 \\ \vdots \\ \mu_n \end{bmatrix} = E(\mathbf{X})$, $\mathbf{a}_1 = \boldsymbol{\mu}/\|\boldsymbol{\mu}\|$, and let $\mathbf{a}_2, \ldots, \mathbf{a}_n$ be orthogonal vectors of length one, all orthogonal to $\mathbf{a}_1$. Thus $\mathbf{a}_1, \ldots, \mathbf{a}_n$ form an orthonormal basis for R_n. Since these $\mathbf{a}_i$ form a basis for R_n, $\mathbf{X} = \sum W_j \mathbf{a}_j$ for random variables W_j, and $(\mathbf{X}, \mathbf{a}_k) = W_k\|\mathbf{a}_k\|^2 = W_k$. Thus, $\mathbf{X} = \sum_1^n (\mathbf{X}, \mathbf{a}_i)\mathbf{a}_i$. (See immediately after Theorem 1.3.5.)

Let $\mathbf{A} = (\mathbf{a}_1, \ldots, \mathbf{a}_n)$. Then

$$\mathbf{W} = \begin{pmatrix} W_1 \\ \vdots \\ W_n \end{pmatrix} = \mathbf{A}'\mathbf{X}, \qquad \mathbf{A}'\mathbf{A} = \mathbf{I}$$

and

$$\mathbf{X} = \mathbf{A}\mathbf{W}. \quad E(\mathbf{W}) = \mathbf{A}'\boldsymbol{\mu} = \begin{pmatrix} \mathbf{a}_1'\boldsymbol{\mu} \\ 0 \\ \vdots \\ 0 \end{pmatrix} = \begin{pmatrix} \|\boldsymbol{\mu}\| \\ 0 \\ \vdots \\ 0 \end{pmatrix}, \; D[\mathbf{W}] = \mathbf{A}'\mathbf{I}_n\mathbf{A} = \mathbf{A}'\mathbf{A} = \mathbf{I}_n.$$

Thus, since $\mathbf{W}$ has a multivariate normal distribution, $W_1, \ldots, W_n$ are independent with

$$W_1 \sim N(\|\boldsymbol{\mu}\|, 1), \qquad W_i \sim N(0, 1) \qquad \text{for} \quad i = 2, \ldots, n$$

and $\sum_1^n X_i^2 = \|\mathbf{X}\|^2 = \|\mathbf{A}\mathbf{W}\|^2 = \mathbf{W}'\mathbf{A}'\mathbf{A}\mathbf{W} = \mathbf{W}'\mathbf{W} = \|\mathbf{W}\|^2 = \sum_1^n \mathbf{W}_i^2$. The distribution of $\sum_1^n \mathbf{W}_i^2$ depends only on n and $\delta = \sum_1^n \mu_i^2 = \|\boldsymbol{\mu}\|^2$.

Comment: Using the notation above and letting $Z_1 = W_1 - \delta^{1/2}$, and $Z_i = W_i$ for $i = 2, \ldots, n$ we get $\sum_1^n X_i^2 = \sum_1^n W_i^2 = (Z_1 + \delta^{1/2})^2 + \sum_2^n Z_i^2$ for $Z_1, \ldots, Z_n$ standard normal independent r.v.'s. Thus, a noncentral χ^2 r.v. is the sum of a noncentral χ^2 r.v. with one degree of freedom (d.f.) and an independent central χ^2 r.v. with $(n - 1)$ d.f.

The central χ^2 density for n degrees of freedom is

$$f(y; n) = \frac{y^{n/2-1}e^{-y/2}}{\Gamma(n/2)2^{n/2}} \qquad \text{for} \quad y > 0.$$

This is a gamma density for power parameter $n/2$, scale parameter $1/2$. The noncentral χ^2 density is a Poisson mixture of central χ^2 densities:

$$f(y; n, \delta) = \sum_{k=0}^{\infty} p(k; \delta) f(y; n + 2k) \qquad \text{for} \quad y > 0,$$

where $p(k; \delta) = [e^{-\delta/2}(\delta/2)^k]/k!$. Thus, the noncentral χ^2 distribution is a weighted average of central χ^2 distributions, with Poisson weights. We will write χ_n^2 to denote the central χ^2 distribution with n d.f.

For any δ, the representation $Y = \sum_1^n W_i^2$ above gives

$$E(Y) = \sum_1^n E(W_i^2) = [\|\boldsymbol{\mu}\|^2 + 1] + 1 + \cdots + 1 = \|\boldsymbol{\mu}\|^2 + n = \delta + n.$$

In addition,

$$\text{Var}(Y) = \sum_1^n \text{Var}(W_i^2) = (2 + 4\delta) + 2(n - 1) = 2n + 4\delta$$

As $n \to \infty$ for fixed δ with $Y_n \sim \chi_n^2(\delta)$, $\dfrac{(Y_n - (m + \delta))}{\sqrt{2n + 4\delta}}$ is asymptotically $N(0, 1)$ by the Central Limit Theorem.

For the central χ^2 distribution a better approximation is given by the cube-root transformation. If U_n has a χ_n^2 distribution then, as $n \to \infty$,

$$Z_n = [(U_n/n)^{1/3} - a_n]/b_n,$$

for $a_n = 1 - 2/(9n)$, $b_n = (2/(9n))^{1/2}$ converges in distribution to $N(0, 1)$ (Fabian and Hannan, 1985, p. 125). Thus, the 100γth percentile is given approximately by

$$u_\gamma = n[a_n + z_\gamma b_n]^3,$$

where z_γ is the 100γth percentile of the standard normal distribution. For example, for $n = 10$, from Table 3 in the back of the book we find $\chi^2_{10,0.95} = 18.3070$, while $u_{0.95} = 10[0.97778 + (1.645)(0.14907)] = 18.2928$. The approximation is even better for larger n.

Definition of a Quadratic Form: Let $\boldsymbol{\Sigma}$ be a nonsingular, positive definite $n \times n$ matrix. Suppose $\mathbf{Y} \sim N(\mathbf{0}, \Sigma)$ and let $\underset{n\times n}{\mathbf{A}}$ be a nonnegative definite matrix. Consider the random variable $Q = Y'\mathbf{A}\mathbf{Y}$. We will be primarily concerned with the special cases that $\mathbf{A}$ is a projection matrix, or that $\boldsymbol{\Sigma}$ is a multiple of the identity matrix, and for simplicity will consider these later in this section. Those looking for a respite from the heavy emphasis on matrix manipulation necessary for the general case may therefore wish to skip to the definition of the noncentral F distribution, promising, of course, not to leave forever.

Let $\boldsymbol{\Sigma}^{1/2}$ be the unique symmetric square root of $\boldsymbol{\Sigma}$, and let $\boldsymbol{\Sigma}^{-1/2}$ be its inverse. We can write $Q = (\mathbf{Y}'\boldsymbol{\Sigma}^{-1/2})(\boldsymbol{\Sigma}^{1/2}\mathbf{A}\boldsymbol{\Sigma}^{1/2})(\boldsymbol{\Sigma}^{-1/2}\mathbf{Y}) = \mathbf{Z}'\mathbf{B}\mathbf{Z}$, where $\mathbf{B}$ and $\mathbf{Z}$ are the second and third terms in parentheses. Then $D[\mathbf{Z}] = \boldsymbol{\Sigma}'^{-1/2}\boldsymbol{\Sigma}\boldsymbol{\Sigma}^{-1/2} = \mathbf{I}_n$, so that $\mathbf{Z} \sim N_n(\mathbf{0}, \mathbf{I}_n)$.

Let $\mathbf{B} = \mathbf{T}\boldsymbol{\Lambda}\mathbf{T}'$ be the spectral decomposition of $\mathbf{B}$. That is, $\boldsymbol{\Lambda}$ is the diagonal matrix of eigenvalues of $\mathbf{B}$, $\mathbf{T}$ is the $n \times n$ matrix whose columns are the corresponding eigenvectors of $\mathbf{B}$, and $\mathbf{T}$ is an orthogonal matrix, i.e., $\mathbf{T}\mathbf{T}' = \mathbf{T}'\mathbf{T} = \mathbf{I}_n$.

Thus, $Q = \mathbf{Z}'\mathbf{T}\boldsymbol{\Lambda}\mathbf{T}'\mathbf{Z} = (\mathbf{T}'\mathbf{Z})'\boldsymbol{\Lambda}(\mathbf{T}'\mathbf{Z}) = \mathbf{W}'\boldsymbol{\Lambda}\mathbf{W}$ for $\mathbf{W} = \mathbf{T}'\mathbf{Z}$. Since $D[\mathbf{W}] = \mathbf{T}'\mathbf{I}_n\mathbf{T} = \mathbf{I}_n$, $\mathbf{W} \sim N_n(\mathbf{0}, \mathbf{I}_n)$. Denoting the eigenvalues of $\mathbf{B}$ by $\lambda_1, \ldots, \lambda_n$, so $\boldsymbol{\Lambda} = \text{diag}(\lambda_1, \ldots, \lambda_n)$, we find that

$$Q = \lambda_1 W_1^2 + \lambda_2 W_2^2 + \cdots + \lambda_n W_n^2,$$

where $\mathbf{W}' = (W_1, \ldots, W_n)$. Thus, Q is a linear combination with coefficients $\lambda_1, \ldots, \lambda_n$ of independent χ_1^2 random variables. The coefficients $\lambda_1, \ldots, \lambda_n$ are the eigenvalues of $\boldsymbol{\Sigma}'^{1/2}\mathbf{A}\boldsymbol{\Sigma}^{1/2}$, and therefore also the eigenvalues of $\mathbf{A}\boldsymbol{\Sigma}$ and of $\boldsymbol{\Sigma}\mathbf{A}$.

Often, in applications, $\boldsymbol{\Sigma} = \sigma^2\mathbf{I}_n$ for some $\sigma^2 > 0$, so the λ_i are σ^2 multiples of the eigenvalues of $\mathbf{A}$. If, in addition, $\mathbf{A}$ is a projection matrix, so that the eigenvalues are all 0's and 1's, with (number of 1's) = rank($\mathbf{A}$), then Q/σ^2 has a central χ^2 distribution with d.f. = rank $\mathbf{A}$. In more generality, Q has a central χ^2 distribution if $\mathbf{A}\boldsymbol{\Sigma}$ is a projection matrix.

We can extend these results a bit by instead supposing only that $\mathbf{Y} \sim N_n(\boldsymbol{\theta}, \boldsymbol{\Sigma})$. Define matrices $\mathbf{B}$, $\mathbf{T}$, $\boldsymbol{\Lambda}$ as before and

$$\mathbf{W} = \mathbf{T}'\boldsymbol{\Sigma}^{-1/2}\mathbf{Y} = \mathbf{T}'\boldsymbol{\Sigma}^{-1/2}\boldsymbol{\theta} + \mathbf{T}'\boldsymbol{\Sigma}^{-1/2}(\mathbf{Y} - \boldsymbol{\theta})$$

Then $\mathbf{W} \sim N_n(\mathbf{T}'\boldsymbol{\Sigma}^{-1/2}\boldsymbol{\theta}, \mathbf{I}_n)$ and $Q = \mathbf{Y}'\mathbf{A}\mathbf{Y} = \mathbf{W}'\boldsymbol{\Lambda}\mathbf{W} = \sum_1^n \lambda_i W_i^2$. In this case the W_i are independent, normally distribution, with standard deviations 1. Thus, each $W_i^2 \sim \chi_1^2(\delta_i)$, where δ_i is the square of the ith component of $\mathbf{T}'\mathbf{\Sigma}^{-1/2}\mathbf{\theta}$ and we have expressed Q as a linear combination with coefficients $\lambda_1, \ldots, \lambda_n$ of independent $\chi_1^2(\delta_i)$ random variables.

In the special case that $\Sigma = \sigma^2\mathbf{I}_n$, and $Q = \mathbf{Y}'\mathbf{P}\mathbf{Y}/\sigma^2$, for $\mathbf{P}$ a projection matrix onto a d-dimensional subspace V of R_n, we have $\mathbf{A} = \mathbf{P}/\sigma^2$, $\mathbf{B} = (\sigma\mathbf{I}_n)(\mathbf{P}/\sigma^2)(\sigma\mathbf{I}_n) = \mathbf{P}$. The eigenvalues of $\mathbf{P}$ are 1, with multiplicity d, and 0, with multiplicity $(n - d)$. Without loss of generality suppose $\lambda_1 = \cdots = \lambda_d = 1$ and $\lambda_{d+1} = \cdots = \lambda_n = 0$. Then the columns $\mathbf{t}_1, \ldots, \mathbf{t}_n$ of $\mathbf{T}$ are orthogonal, of length one, and the first d span V. Thus, $\mathbf{W} \sim N_n(\mathbf{T}^{-1}\mathbf{\theta}/\sigma, \mathbf{I}_n)$, $Q = \sum_1^d W_i^2$, with the W_i^2 independent $\chi_1^2(\delta_i)$ and δ_i the square of the ith component of $\mathbf{T}^{-1}\mathbf{\theta}/\sigma^2$. Since $\mathbf{T}^{-1} = \mathbf{T}'$, the ith component of $\mathbf{T}^{-1}\mathbf{\theta}/\sigma$ is $(\mathbf{t}_i, \mathbf{\theta})/\sigma$. Since the sum of independent noncentral chi-square r.v.'s is noncentral χ^2, we conclude that

$$Q \sim \chi_d^2(\delta) \qquad \text{with} \qquad \delta = \left\| \sum_1^d (\mathbf{t}_i, \mathbf{\theta})\mathbf{t}_i/\sigma \right\|^2 = \|p(\mathbf{\theta}|V)\|^2/\sigma_2.$$

Note that $Q = \mathbf{Y}'\mathbf{P}\mathbf{Y}/\sigma^2 = \|p(\mathbf{Y}|V)\|^2/\sigma^2 \sim \chi_d^2(\|p(\mathbf{\theta})|V)\|^2/\sigma^2)$.

The result that Q has a noncentral χ^2 distribution when $\mathbf{A} = \Sigma^{-1}$ is very useful in statistics, important enough to dignify with the name Theorem, for later reference.

Theorem 2.5.1: Let $\mathbf{Y} \sim N_n(\mathbf{\theta}, \mathbf{\Sigma})$. Let $Q = (\mathbf{Y} - \mathbf{C})'\mathbf{\Sigma}^{-1}(\mathbf{Y} - \mathbf{C})$, where $\mathbf{C}$ is a vector of constants. Then $Q \sim \chi_n^2(\delta)$, for $\delta = (\mathbf{\theta} - \mathbf{C})'\Sigma^{-1}(\mathbf{\theta} - \mathbf{C})$.

Biometrika Tables for Statisticians by Pearson and Hartley (1966) gives percentile values of the $\chi_n^2(\delta)$ and χ_n^2 distributions.

Definition 2.5.2.: (Noncentral F). Let $U_1 \sim \chi_{n_1}^2(\delta)$ and $U_2 \sim \chi_{n_2}^2$ (central) be independent. Then $V = \dfrac{U_1/n_1}{U_2/n_2}$ is said to have a noncentral F distribution with noncentrality parameter δ, and n_1 and n_2 degrees of freedom. For $\delta = 0$, V is said to have a *central* F distribution (or simply, an F distribution) with n_1 and n_2 degrees of freedom.

We will denote the noncentral F by $F_{n_1,n_2}(\delta)$ and the central F by F_{n_1,n_2} or $F(n_1, n_2)$. The 100γth percentile of the central F distribution will be denoted by $F_{n_1,n_2,\gamma}$ or by $F_\gamma(n_1, n_2)$.

For completeness we give the central and noncentral F-densities. The noncentral F distribution with noncentrality parameter δ, and n_1, n_2 degrees of freedom has density

$$h(v; \delta, n_1, n_2) = \sum_{k=0}^{\infty} p(k; \delta)h(v; n_1, n_2),$$

where $p(k; \delta) = [e^{-\delta/2}(\delta/2)^k]/k!$, the Poisson probability function with mean $\delta/2$, and

$$h(v; n_1, n_2) = \frac{\Gamma\left(\frac{n_1+n_2}{2}\right)}{\Gamma\left(\frac{n_1}{2}\right)\Gamma\left(\frac{n_2}{2}\right)} \left(\frac{n_1}{n_2}\right)^{(n_1)/2} \frac{v^{n_1/2} - 1}{\left(1 + v\frac{n_1}{n_2}\right)^{(n_1+n_2)/2}} \quad \text{for} \quad v > 0.$$

$h(v; n_1, n_2) = h(v; 0, n_1, n_2)$ is the central F-density.

(Students are warned not to memorize this; the effort required has been known to debilitate them for weeks.) The mean and variance for the noncentral F are

$$\frac{n_2}{n_2 - 2}\left(1 + \frac{\delta}{n_1}\right) \quad \text{and} \quad \frac{2n_2^2}{n_1^2(n_1 - 2)}\left[\frac{(n_1 + \delta)^2}{(n_2 - 2)(n_2 - 4)} + \frac{n_1 + 2\delta}{n_2 - 2}\right],$$

for $n_2 \geq 2$ and $n_2 \geq 4$ respectively, undefined otherwise.

The Noncentral t Distribution

Definition 2.5.3: Let $W \sim N(\theta, 1)$ and $Y \sim \chi_m^2$ be independent random variables. Then

$$T = W/\sqrt{Y/m}$$

is said to have (Student's) t distribution with noncentrality parameter θ and m degrees of freedom.

We will denote this distribution by $t_m(\theta)$. The $t_m(0)$ distribution is called Student's t distribution or simply the t distribution. The 100γth percentile of the central t distribution will be denoted by $t_{m,\gamma}$. Notice that $T^2 \sim F_{1,m}(\theta^2)$. Student's t distribution was first found by William Gosset (1907) while he was on leave from his position as a brewer from the Guinness Brewery of Dublin to study with Karl Pearson at University College in London. Upon request of Guinness, Gosset published "On the probable error of the mean," in *Biometrika* in 1908 under the name "Student." A discussion is given by Ronald A. Fisher's daughter, Joan Fisher Box (1987).

Example 2.5.1: Let $X_1, \ldots, X_n$ be a random sample from a $N(\mu, \sigma^2)$ distribution. We will prove soon that $\bar{X} \sim N\left(\mu, \frac{\sigma^2}{n}\right)$, that $\frac{(n-1)S^2}{\sigma^2} \sim \chi_{n-1}^2$ for $S^2 = \frac{\sum (X_i - \bar{X})^2}{n - 1}$, and that $\bar{X}$ and S^2 are independent. Taking, $m = n - 1$, constant a,

$$W = \frac{\bar{X} - a}{\sigma/\sqrt{n}}, \qquad Y = \frac{(n-1)S^2}{\sigma^2}, \qquad \theta = \frac{\mu - a}{\sigma/\sqrt{n}},$$

we find that

$$T = W/\sqrt{Y/m} = \frac{\dfrac{\bar{X} - a}{\sigma/\sqrt{n}}}{\sqrt{S^2/\sigma^2}} = \frac{\bar{X} - a}{S/\sqrt{n}},$$

has a $t_{n-1}(\theta)$ distribution, central t if $a = \mu$.

The $t_n(\theta)$ density is

$$f(t; m, \theta) = \frac{(\pi m)^{-1/2}}{\Gamma(m/2)}\left(1 + \frac{t^2}{m}\right)^{-(m+1)/2} \sum_{k=0}^{\infty} \frac{\Gamma(m/2 + k/2 + 1/2)(2^{1/2}t\theta)^k}{k!\, m^{k/2}\left(1 + \dfrac{t^2}{m}\right)^{k/2}}$$

for all t.

This simplifies for the case $\theta = 0$ to the central t-density

$$f(t; m) = \frac{(m\pi)^{-1/2}}{\Gamma(m/2)}\Gamma(m/2 + 1/2)(1 + t^2/m)^{-(m+1)/2} \qquad \text{for all } t.$$

Problem 2.5.1: Use Stirling's Formula: $\Gamma(n + 1)/[e^{-n}n^n\sqrt{2\pi n}] \to 1$ as $n \to \infty$ to show that $f(t; m) \to (1/\sqrt{2\pi})e^{-t^2/2}$, the standard normal density, as $m \to \infty$.

Problem 2.5.2: Let X_1, X_2 be independent, each $N(\mu, \sigma^2)$. Prove that $W = \dfrac{X_1 + X_2}{|X_1 - X_2|}$ has a noncentral t distribution. What are the parameters? (It may be helpful to read the next theorem first).

Theorem 2.5.2: Let $E(\mathbf{Y}) = \boldsymbol{\theta}$, $D[\mathbf{Y}] = \sigma^2\mathbf{I}_n$. Then

(1) For any $\mathbf{a}$, $E(\mathbf{a}, \mathbf{Y}) = (\mathbf{a}, \boldsymbol{\theta})$

(2) For any $\mathbf{a}, \mathbf{b}$, $\text{cov}((\mathbf{a}, \mathbf{Y}), (\mathbf{b}, \mathbf{Y})) = \sigma^2(\mathbf{a}, \mathbf{b})$, so that $\text{Var}((\mathbf{a}, \mathbf{Y})) = \sigma^2\|\mathbf{a}\|^2$.

(3) Let $\mathbf{Y}$ have a multivariate normal distribution, let $\mathbf{a}_1, \ldots, \mathbf{a}_r$ be vectors of constants, and $W_i = (\mathbf{a}_i, \mathbf{Y})$. Then $(W_1, \ldots, W_r)$ has a multivariate normal distribution.

***Proof*:** (1) $E(\mathbf{a}, \mathbf{Y}) = (\mathbf{a}, \boldsymbol{\theta})$ follows by the linearity of expectation. For (2), compute $\text{cov}((\mathbf{a}, \mathbf{Y}), (\mathbf{b}, \mathbf{Y})) = \mathbf{a}'D[\mathbf{Y}]\mathbf{b} = \sigma^2\mathbf{a}'\mathbf{b} = \sigma^2(\mathbf{a}, \mathbf{b})$. (3) follows directly from Theorem 2.4.2 by taking the ith row of $\mathbf{C}$ to be $\mathbf{a}_i'$. □

The linear models we will consider will have the form $Y \sim N_n(\boldsymbol{\theta}, \sigma^2\mathbf{I}_n)$, where $\boldsymbol{\theta}$ is supposed to lie in a subspace V spanned by vectors $\mathbf{x}_1, \ldots, \mathbf{x}_k$ of constants, and σ^2 is an unknown parameter. For this reason we will be interested in the statistical properties of projections $\hat{\mathbf{Y}}_V = p(\mathbf{Y}|V)$ and functions of $\hat{\mathbf{Y}}_V$ and

$\mathbf{Y} - \hat{\mathbf{Y}}_V$. Expectations of linear functions of $\mathbf{Y}$ ($\hat{\mathbf{Y}}_V$ is an example) are determined by $\boldsymbol{\theta}$ alone. Expectations of quadratic functions of $\mathbf{Y}$ ($\|\hat{\mathbf{Y}}_V\|^2$ is an example) are determined by $\boldsymbol{\theta}$ and $D[\mathbf{Y}]$. The distributional form of functions of $\mathbf{Y}$ (such as $\hat{\mathbf{Y}}_V$) are determined by the distributional form of $\mathbf{Y}$. We will prove, for example, that $\mathbf{Y} \sim N_n(\boldsymbol{\theta}, \sigma^2\mathbf{I}_n)$ implies that $\|\hat{\mathbf{Y}}_V\|^2/\sigma^2$ has a noncentral χ^2 distribution.

Theorem 2.5.3: Let V be a k-dimensional subspace of R_n and let $\mathbf{Y}$ be a random vector taking values in R_n. Let $E(\mathbf{Y}) = \boldsymbol{\theta}$, $\hat{\mathbf{Y}}_V = p(\mathbf{Y}|V)$ and $\boldsymbol{\theta}_V = p(\boldsymbol{\theta}|V)$. Then

(1) $E(\hat{\mathbf{Y}}_V) = \boldsymbol{\theta}_V$,

(2) $D[\mathbf{Y}] = \sigma^2\mathbf{I}_n$ implies that $D[\hat{\mathbf{Y}}_V] = \sigma^2\mathbf{P}_V$ and $E[\|\hat{\mathbf{Y}}_V\|^2] = \sigma^2 k + \|\boldsymbol{\theta}_V\|^2$.

(3) $\mathbf{Y} \sim N(\boldsymbol{\theta}, \sigma^2\mathbf{I}_n)$ implies that $\hat{\mathbf{Y}}_V \sim N_n(\boldsymbol{\theta}_V, \sigma^2\mathbf{P}_V)$ and $\|\hat{\mathbf{Y}}_V\|^2/\sigma^2 \sim \chi_k^2(\delta)$ for $\delta = \|\boldsymbol{\theta}_V\|^2/\sigma^2$.

***Proof*:** (1) follows by the linearity of expectation and the fact that $p(\mathbf{Y}|V)$ is a linear function of $\mathbf{Y}$. To prove (2) note that $D[\hat{\mathbf{Y}}_V] = \sigma^2\mathbf{P}_V\mathbf{P}_V' = \sigma^2\mathbf{P}_V$ and $\|\hat{\mathbf{Y}}_V\|^2 = \mathbf{Y}'\mathbf{P}_V\mathbf{Y}$. By Theorem 2.2.1 $E\|\hat{\mathbf{Y}}_V\|^2 = \text{trace}(\sigma^2\mathbf{I}_n\mathbf{P}_V) + \boldsymbol{\theta}'\mathbf{P}_V\boldsymbol{\theta} = \sigma^2\,\text{trace}(\mathbf{P}_V) + \boldsymbol{\theta}'\mathbf{P}_V'\mathbf{P}_V\boldsymbol{\theta} = \sigma^2 k + \|\boldsymbol{\theta}_V\|^2$.

The first conclusion of (3) follows from (1) and (2) and the normality of $\mathbf{Y}$. To prove the second conclusion let $\mathbf{a}_1, \ldots, \mathbf{a}_k, \mathbf{a}_{k+1}, \ldots, \mathbf{a}_n$ be an orthonormal basis for R_n with $\mathbf{a}_1, \ldots, \mathbf{a}_k$ spanning V. Then $\mathbf{Y} = \sum_1^n (\mathbf{Y}, \mathbf{a}_i)\mathbf{a}_i$. Let $W_i = (\mathbf{Y}, \mathbf{a}_i)$ for each i. Then

$$\hat{\mathbf{Y}}_V = \sum_{i=1}^k W_i\mathbf{a}_i$$

and from Theorem 2.5.1, $W_1, \ldots, W_n$ are independent $N(\eta_i, \sigma^2)$ random variables, where $\eta_i = (\boldsymbol{\theta}, \mathbf{a}_i)$. Then $\|\hat{\mathbf{Y}}_V\|^2/\sigma^2 = \sum_1^k (W_i/\sigma)^2 \sim \chi_k^2(\delta)$ for $\delta = \sum_1^k (\eta_i/\sigma)^2 = \|\boldsymbol{\theta}_V\|^2/\sigma^2$. □

We will be interested in the joint distributions of projections $\hat{\mathbf{Y}}_i = p(\mathbf{Y}|V_i)$ onto different subspaces V_i. For example, for a subspace V, we shall be interested in $\hat{\mathbf{Y}}_V$ and the residual vector $\mathbf{Y} - \mathbf{Y}_V = p(\mathbf{Y}|V^\perp)$. In the case that the subspaces under consideration are mutually orthogonal, the resulting projections are, under suitable conditions, independent random vectors. Their squared lengths, the sums of squares of the analysis of variance, are therefore independent random variables.

Theorem 2.5.4: Let $V_1, \ldots, V_k$ be mutually orthogonal subspaces of R_n and let $\mathbf{Y}$ be a random vector taking values in R_n. Let $\mathbf{P}_i$ be the projection matrix onto V_i. Let $E(\mathbf{Y}) = \boldsymbol{\theta}$, $\hat{\mathbf{Y}}_i = \mathbf{P}_i\mathbf{Y}$ and $\boldsymbol{\theta}_i = \mathbf{P}_i\boldsymbol{\theta}$ for $i = 1, \ldots, k$.

Then

(1) $D[\mathbf{Y}] = \sigma^2 \mathbf{I}_n$ implies that $C(\hat{\mathbf{Y}}_i, \hat{\mathbf{Y}}_{i'}) = 0$ for $i \neq i'$.

(2) $\mathbf{Y} \sim N_n(\boldsymbol{\theta}, \sigma^2 \mathbf{I}_n)$ implies that $\hat{\mathbf{Y}}_1, \ldots, \hat{\mathbf{Y}}_k$, are independent, with $\hat{\mathbf{Y}}_i \sim N(\theta_i, \sigma^2 \mathbf{P}_i)$.

***Proof*:** Let $\mathbf{P}_i$ be the projection matrix onto V_i. $C(\hat{\mathbf{Y}}_i, \hat{\mathbf{Y}}_{i'}) = C(\mathbf{P}_i \mathbf{Y}, \mathbf{P}_{i'} \mathbf{Y}) = \mathbf{P}_i C(\mathbf{Y}, \mathbf{Y}) \mathbf{P}_{i'}' = \sigma^2 \mathbf{P}_i \mathbf{P}_{i'} = 0$.

If $\mathbf{Y}$ has a multivariate normal distribution, it follows that the linear functions $\hat{\mathbf{Y}}_i = \mathbf{P}_i \mathbf{Y}$ are jointly multivariate normal and are independent if and only if each $C(\hat{\mathbf{Y}}_i, \hat{\mathbf{Y}}_{i'}) = 0$ for $i \neq i'$. (3) of Theorem 2.5.2 then implies (2) above. □

The following important theorem, first given by William Gosset (Student) in 1908 and proved rigorously using n-space geometry by R. A. Fisher, is an easy consequence of Theorems 2.5.3 and 2.5.4. This theorem justifies the statements of Example 2.5.1.

Theorem 2.5.5: Let $Y_1, \ldots, Y_n$ be a random sample from a normal distribution with mean μ, variance σ^2. Let $\bar{Y} = \frac{1}{n}(Y_1 + \cdots + Y_n)$ and

$$S^2 = \frac{1}{n-1} \sum_1^n (Y_i - \bar{Y})^2.$$

Then

(1) $\bar{Y} \sim N(\mu, \sigma^2/n)$

(2) $\dfrac{S^2(n-1)}{\sigma^2} = \sum_1^n \dfrac{(Y_i - \bar{Y})^2}{\sigma^2} \sim \chi^2_{n-1}$

(3) $\bar{Y}$ and S^2 are independent

(4) $\dfrac{\bar{Y} - \mu_0}{S/\sqrt{n}}$ has a noncentral t distribution with $(n-1)$ d.f., noncentrality parameter $\left(\dfrac{\mu - \mu_0}{\sigma/\sqrt{n}}\right)$ for any constant μ_0.

***Proof*:** $\mathbf{Y} = (Y_1, \ldots, Y_n)' \sim N_n(\mu \mathbf{J}_n, \sigma^2 \mathbf{I}_n)$, where $\mathbf{J}_n' = (1, \ldots, 1)$. Let $V = \mathscr{L}(\mathbf{J}_n)$, a 1-dimensional subspace of R_n. Let $\hat{\mathbf{Y}} = p(\mathbf{Y} | V) = \dfrac{(\mathbf{Y}, \mathbf{J}_n)}{\|\mathbf{J}_n\|^2} \mathbf{J}_n = \bar{Y} \mathbf{J}_n$. $\bar{Y}$ has mean μ and variance $[\mathbf{J}_n' D[\mathbf{Y}] \mathbf{J}_n] / \|\mathbf{J}_n\|^4 = \sigma^2 / \|\mathbf{J}_n\|^2 = \dfrac{\sigma^2}{n}$, and of course is normally distributed. This proves (1).

$\mathbf{Y} - \hat{\mathbf{Y}} = p(\mathbf{Y} | V^\perp) = (Y_1 - \bar{Y}, \ldots, Y_n - \bar{Y})'$, the vector of deviations, is independent of $\hat{\mathbf{Y}}$, hence of $\bar{Y}$. Thus, S^2, a function of $\mathbf{Y} - \hat{\mathbf{Y}}$, and $\bar{Y}$ are independent. By (3) of Theorem 2.5.2 and the fact that $\boldsymbol{\theta} = \mu\mathbf{J}_n \perp V$, $\dfrac{S^2(n-1)}{\sigma^2} = \dfrac{\|\mathbf{Y} - \hat{\mathbf{Y}}\|^2}{\sigma^2}$ is distributed as χ^2_{n-1}. This proves (2) and (3).

Finally, for $U = \dfrac{\bar{Y} - \mu_0}{\sigma/\sqrt{n}}$, $T \equiv \dfrac{\bar{Y} - \mu_0}{S/\sqrt{n}} = \left[\dfrac{\bar{Y} - \mu_0}{\sigma/\sqrt{n}}\right]\Big/\sqrt{S^2/\sigma^2} = \dfrac{U}{\sqrt{V/(n-1)}} \sim N\left(\dfrac{\mu - \mu_0}{\sigma/\sqrt{n}}, 1\right)$ and $V \equiv \dfrac{1}{\sigma^2}\sum_1^n (Y_i - \bar{Y})^2 \sim \chi^2_{n-1}$. U and V are independent by (3). Thus, by the definition of the noncentral t distribution, $\dfrac{\bar{Y} - \mu_0}{S/\sqrt{n}} \sim t_{n-1}\left(\dfrac{\mu_1 - \mu_0}{\sigma/\sqrt{n}}\right)$, proving (4). □

Example 2.5.2: Let Y_1, Y_2, Y_3, Y_4 be independent $N(\mu, \sigma^2)$. Find a constant K such that

$$W = K\frac{\bar{Y} - \mu_0}{\sqrt{(Y_1 - Y_2)^2 + (Y_1 + Y_2 - 2Y_3)^2/3 + (Y_1 + Y_2 + Y_3 - 3Y_4)^2/6}},$$

has one of the distributions noncentral χ^2, noncentral F, or noncentral t. Identify the parameters. K must be a constant.

Solution

$$\bar{Y} - \mu_0 \sim N((\mu - \mu_0), \sigma^2/4) \qquad \text{so} \qquad \frac{\bar{Y} - \mu_0}{\sigma/\sqrt{4}} \sim N\left(\frac{\mu - \mu_0}{\sigma/\sqrt{n}}, 1\right).$$

The r.v.'s

$$U_1 = Y_1 - Y_2,\ U_2 = Y_1 + Y_2 - 2Y_3,\ U_3 = Y_1 + Y_2 + Y_3 - 3Y_4$$

are each normally distribution, with zero means, and with variances $2\sigma^2$, $6\sigma^2$, and $12\sigma^2$. Moreover, U_1, U_2, U_3 are independent of $\bar{Y}$ and of each other since their coefficient vectors are mutually orthogonal. Thus,

$$\frac{U_1^2}{2\sigma^2} + \frac{U_2^2}{6\sigma^2} + \frac{U_3^2}{12\sigma^2} = \frac{1}{2\sigma^2}\left(U_1^2 + \tfrac{1}{3}U_2^2 + \tfrac{1}{6}U_3^2\right) \sim \chi_3^2$$

and is independent of the numerator $\bar{Y} - \mu_0$. Therefore,

$$\frac{\dfrac{\bar{Y} - \mu_0}{\sigma/\sqrt{4}}}{\sqrt{\dfrac{1}{3}\dfrac{1}{2\sigma^2}\left[U_1^2 + \tfrac{1}{3}U_2^2 + \tfrac{1}{6}U_3^2\right]}} = \frac{\sqrt{6}\sigma}{\sigma/\sqrt{4}}\frac{W}{K} \sim t_3\left(\frac{\mu - \mu_0}{\sigma/\sqrt{4}}\right).$$

We therefore need $\dfrac{\sqrt{6\sigma^2}}{\sigma/\sqrt{4}}\Big/K = 1$ or $K = 2\sqrt{6}$.

Theorem 2.5.6: Let $\mathbf{Y} = \boldsymbol{\theta} + \boldsymbol{\varepsilon}$, for $\boldsymbol{\theta} \in V$, and $\boldsymbol{\varepsilon} \sim N_n(\mathbf{0}, \sigma^2\mathbf{I}_n)$. Let V_1 be a subspace of V. Suppose $\dim(V_1) \equiv k_1 < k \equiv \dim(V)$. Let $\hat{\mathbf{Y}} \equiv p(\mathbf{Y}|V)$, $\hat{\mathbf{Y}}_1 \equiv p(\mathbf{Y}|V_1)$, and $\boldsymbol{\theta}_1 \equiv p(\boldsymbol{\theta}|V_1)$. Then

$$F \equiv \frac{\|\hat{\mathbf{Y}}_1\|^2/k_1}{\|\mathbf{Y} - \hat{\mathbf{Y}}\|^2/(n-k)}$$

has a noncentral F distribution with k_1 and $n - k$ degrees of freedom and noncentrality parameter $\delta = \|\boldsymbol{\theta}_1\|^2/\sigma^2$.

***Proof*:** Since $\mathbf{Y} - \hat{\mathbf{Y}} = p(\mathbf{Y}|V^{\perp})$ it follows from (2) of Theorem 2.5.3 that the squared lengths Q_1 and Q_2 in the numerator and denominator are independent random variables. By (3) of Theorem 2.5.2 $Q_1 \sim \chi^2_{k_1}(\delta)$ and $Q_2 \sim \chi^2_{n-k}$. The definition of the noncentral χ^2 distribution then implies the theorem. □

For completeness we present the following more general theorems. Their proofs are given in Srivastava and Khatri (1979, p. 66–67). These theorems will not be needed for the statistical applications to follow.

Theorem 2.5.7: Let $\mathbf{Y} \sim N_n(\boldsymbol{\theta}, \boldsymbol{\Sigma})$, with $\boldsymbol{\Sigma}$ positive definite. Let $\mathbf{A}, \mathbf{A}_1, \ldots, \mathbf{A}_k$ be symmetric matrices of ranks $r, r_1, \ldots, r_k$, and define $Q = \mathbf{Y}'\mathbf{A}\mathbf{Y}$, and $Q_j = \mathbf{Y}'\mathbf{A}_j\mathbf{Y}$ for $j = 1, \ldots, k$. Let $\lambda = \boldsymbol{\theta}'\mathbf{A}\boldsymbol{\theta}$, and $\lambda_j = \boldsymbol{\theta}'\mathbf{A}_j\boldsymbol{\theta}$ for each j. Consider the statements:

(i) $Q_j \sim \chi^2_{r_j}(\lambda_j)$, $j = 1, \ldots, k$.
(ii) $Q_1, \ldots, Q_k$ are independent random variables.
(iii) $Q \sim \chi^2_r(\lambda)$.
(iv) $r = \sum r_j$.

Then (a) any two of (i), (ii), (iii) imply all four statements, and (b) (iii) and (iv) imply (i) and (ii).

Cochran's Theorem follows as a corollary of Theorem 2.5.6:

Cochran's Theorem: Let $\mathbf{Y}, \mathbf{A}, \mathbf{A}_1, \ldots, \mathbf{A}_k, Q, Q_1, \ldots, Q_k$ be as defined in Theorem 2.5.7. Then $Q_1, \ldots, Q_k$ are independently distributed as noncentral χ^2 random variables if and only if $\sum \text{Rank}(\mathbf{A}_j) = n$.

Problem 2.5.3: The following questions are designed to acquaint students with the central and noncentral χ^2, F and t distributions. For each question give the distribution name, and parameters.

Let $X_1, \ldots, X_{n_1}$ and let $Y_1, \ldots, Y_{n_2}$ be independent random samples from

$N(\mu_1, \sigma_1^2)$ and $N(\mu_2, \sigma_2^2)$, respectively. Find the distributions of:

(a) $\dfrac{X_1 - \mu_1}{\sigma_1}, \dfrac{X_1 - \mu_0}{\sigma_1}$ (μ_0 an arbitrary constant)

(b) $\left(\dfrac{X_1 - \mu_1}{\sigma_1}\right)^2, \left(\dfrac{X_1 - \mu_0}{\sigma_1}\right)^2$

(c) $\dfrac{1}{\sigma_1^2} \sum_1^{n_1} (X_i - \mu_1)^2, \dfrac{1}{\sigma_1^2} \sum_{i=1}^{n_1} (X_i - C_i)^2$ for arbitrary constants $C_1, \ldots, C_n$.

(d) $\dfrac{\bar{Y} - \mu_2}{\sigma_2/\sqrt{n_2}}, \dfrac{\bar{Y} - \mu_0}{\sigma_2/\sqrt{n_2}}, \dfrac{\bar{X} - \mu_2}{\sigma_1/\sqrt{n_1}}$

(e) $\dfrac{n_1}{\sigma_1^2} (\bar{X} - \mu_0)^2, \dfrac{n_1}{\sigma_1^2} (\bar{X} - \mu_1)^2$

(f) $[(\bar{X} - \bar{Y}) - \delta] \Big/ \sqrt{\dfrac{\sigma^2}{n_1} + \dfrac{\sigma^2}{n_2}}$ (δ = an arbitrary constant)

(g) $\dfrac{(n_1 - 1)S_1^2}{\sigma_1^2}$

(h) $\dfrac{\bar{X} - \mu_0}{S_1/\sqrt{n_1}} = \left[\dfrac{\bar{X} - \mu_0}{\sigma_1/\sqrt{n_1}}\right] \Big/ \sqrt{S^2/\sigma_1^2}$, for $S_1^2 = \dfrac{1}{n_1 - 1} \sum_1^{n_1} (X_i - \bar{X})^2$

(i) $\dfrac{\bar{X} - \mu_2}{S_1/\sqrt{n_1}}$

(j) $n_1(\bar{X} - \mu_1)^2/\sigma_1^2 + n_2(\bar{Y} - \mu_2)^2/\sigma_2^2$. Suppose in (k), (l) and (m) that $\sigma_1^2 = \sigma_2^2 = \sigma^2$.

(k) $[S_1^2(n_1 - 1) + S_2^2(n_2 - 1)]/\sigma^2$

(l) $\dfrac{n_2 - 1}{n_1} [\sum (X_i - \mu_0)^2]/[\sum (Y_i - \bar{Y})^2]$

(m) $[(\bar{X} - \bar{Y}) - \delta] \Big/ \left[S_1 \sqrt{\dfrac{1}{n_1} + \dfrac{1}{n_2}}\right]$

For the remainder find a constant K, so that the resulting random variable has one of the distributions: normal, χ^2, t, or F (central or noncentral). Give the degrees of freedom and noncentrality parameters.

(n) $\dfrac{X_1 + X_2}{|Y_1 - Y_2|} K$

(o) $\dfrac{\bar{X} - \mu_2}{S_2} K$

(p) $\dfrac{(X_1 - a)^2 + (X_2 - a)^2}{S_2^2} K$

Problem 2.5.4: Let $(X_1, Y_1), \ldots, (X_n, Y_n)$ be a random sample from the bivariate normal distribution with parameters μ_1, μ_2, σ_1^2, σ_2^2, and ρ. Find a constant K so that

$$T = K \frac{(\bar{X} - \bar{Y}) - \delta}{\left\{\sum_{i=1}^{n} [(X_i - Y_i) - (\bar{X} - \bar{Y})]^2\right\}^{1/2}} \sim t_m(\theta).$$

Express m and θ as a function of the parameters and the constant δ. *Hint*: Let $D_i = X_i - Y_i$. Express T as a function of the D_i.

Problem 2.5.5: Let $\mathbf{x}_1 = (1, 1, 1, 1, 1)'$, $\mathbf{x}_2 = (1, 1, 0, 0, 0)'$, $\boldsymbol{\theta} = (6, 6, 2, 2, 2)'$ and suppose that $\mathbf{Y} \sim N_5(\boldsymbol{\theta}, 9\mathbf{I}_5)$. Let $V = \mathscr{L}(\mathbf{x}_1, \mathbf{x}_2)$. Find a constant K such that $K\|\hat{\mathbf{Y}}\|^2/\|\mathbf{Y} - \hat{\mathbf{Y}}\|^2$ has a noncentral F distribution. Identify the parameters n_1, n_2, δ.

Problem 2.5.6: The gamma distribution $\Gamma(m, \theta)$ with power parameter m, scale parameter θ, has density $f(y; m, \theta) = [\Gamma(m)\theta^n]^{-1}y^{m-1}e^{-y/\theta}$ for $y > 0$. Thus, the χ_n^2 and $\Gamma(n/2, 1/2)$ distributions are identical. If $V \sim \Gamma(m, 1)$ then $V\theta \sim \Gamma(m, \theta)$.

Suppose that V_1, V_2 are independent $\Gamma(m_1, \theta)$ and $\Gamma(m_2, \theta)$. Let $W_1 = V_1 + V_2$, $W_2 = V_1/V_2$. Use a density transformation theorem to show that (1) W_1, W_2 are independent, (2) $W_1 \sim \Gamma(m_1 + m_2, \theta)$, (3) $W_2(m_2/m_1) \sim F_{m_1, m_2}$. Hence conclude that U_1, U_2 independent $\chi_{n_1}^2$, $\chi_{n_2}^2$ implies that

$$U_1 + U_2 \sim \chi_{n_1+n_2}^2, \qquad \frac{U_1/n_1}{U_2/n_2} \sim F_{n_1, n_2}$$

and that these r.v.'s are independent. (4) Let $Y_1, \ldots, Y_n$ be independent with $Y_i \sim \Gamma(m_i, \theta)$. Let $S_i = Y_1 + \cdots + Y_i$, and let $W_1 = Y_2/Y_1$, $W_2 = Y_3/S_2, \ldots,$ $W_{n-1} = Y_n/S_{n-1}$, $W_n = S_n$. Use (3) and induction to argue that $W_1, \ldots, W_n$ are independent. What are the distributions of $W_1, \ldots, W_n$?

CHAPTER 3

The Linear Model

3.1 THE LINEAR HYPOTHESIS

Suppose we observe $\mathbf{Y}$, an n-component vector, and our model states that $\mathbf{Y} = \beta_1\mathbf{x}_1 + \cdots + \beta_k\mathbf{x}_k + \boldsymbol{\varepsilon}$, where $\mathbf{x}_1, \ldots, \mathbf{x}_k$ are known vectors of constants, and $\boldsymbol{\varepsilon} \sim N_n(\mathbf{0}, \sigma^2\mathbf{I}_n)$, for $\beta_1, \ldots, \beta_k$ unknown parameters.

This model, called the *linear hypothesis*, or the *multiple linear regression model*, includes a great variety of statistical models, including simple linear regression, one-way, two-way and higher order analysis of variance, and analysis of covariance. The assumptions on ε are not all necessary in order to make useful statements, and we will point out which can be dropped under certain circumstances.

The inner product space Ω in which the vectors $\mathbf{Y}, \mathbf{x}_1, \ldots, \mathbf{x}_k$ take their values will usually be R_n, the space of column vectors. In this case we can define the $n \times k$ matrix $\mathbf{X} = (\mathbf{x}_1, \ldots, \mathbf{x}_k)$, and the column vector $\boldsymbol{\beta} = (\beta_1, \ldots, \beta_k)'$ and write the linear model in the briefer form $\mathbf{Y} = \mathbf{X}\boldsymbol{\beta} + \boldsymbol{\varepsilon}$. This briefer form makes it possible to use matrix algebra to investigate the properties of the estimators of $\boldsymbol{\beta}$ and of σ^2. However, other configurations for these vectors sometimes have an intuitive value which will make it worthwhile to give up the column form.

Example 3.1.1: We observe pairs (x_i, Y_i) for $i = 1, \ldots, n$ and suppose $Y_i = \beta x_i + \varepsilon_i$. Then in vector form $\mathbf{Y} = \beta\mathbf{x} + \boldsymbol{\varepsilon}$. This model, about the simplest of interest, is called regression through the origin. Taking $\mathbf{x} = \mathbf{J}$, the vector of all ones, and $\beta = \mu$, we get the one sample model with $Y_1 \sim N(\mu, \sigma^2)$.

Example 3.1.2: (Two regression lines with equal slopes) Suppose the yields on one-acre plots of land of varying fertility levels are recorded for two experimental conditions. Let Y_{ij} = yield of corn under condition i on jth plot, and

$$x_{ij} = \text{fertility of plot } j \text{ for condition } i, \qquad \text{for} \quad j = 1, \ldots, n_i,\ i = 1, 2.$$

Define Ω to be the collection of vectors $\begin{pmatrix} y_{11} & y_{21} \\ \vdots & \vdots \\ y_{1n_1} & y_{2n_2} \end{pmatrix}$, $\mathbf{Y}$ a random element of Ω, $\mathbf{x}$ the corresponding vector of x_{ij}'s, $\mathbf{w}_1$ the indicator of the first column, $\mathbf{w}_2$ the indicator of the second column. Then $\mathbf{Y} = \beta_1 \mathbf{w}_1 + \beta_2 \mathbf{w}_2 + \beta_c \mathbf{x} + \boldsymbol{\varepsilon}$ is a commonly used model. $\beta_1 > \beta_2$ could be interpreted to mean that experimental condition #1 provides higher average yields, given the same fertility levels, than does #2. β_c is the additional yield of corn for one unit more of fertility. Generalizing to k experimental conditions we get the *analysis of covariance* model $\mathbf{Y} = \sum_1^k \beta_j \mathbf{w}_j + \beta_c \mathbf{x} + \boldsymbol{\varepsilon}$.

Example 3.1.3: (Two-way analysis of variance) Suppose yield of corn Y_{ij} is observed on rc plots of land for seed levels $1, 2, \ldots, r$ and fertilizer levels, $1, 2, \ldots, c$. Let $\mathbf{Y} = (Y_{ij})$ be the $r \times c$ rectangular array, Ω the sample space, $\mathbf{J} \in \Omega$ the vector of all 1's $\mathbf{R}_i$ the indicator of row i, and $\mathbf{C}_j$ the indicator of column j. Then the *additive effects* model states that

$$\mathbf{Y} = \mu \mathbf{J} + \sum_1^r \alpha_i \mathbf{R}_i + \sum_1^c \beta_j \mathbf{C}_j + \boldsymbol{\varepsilon} = \boldsymbol{\theta} + \boldsymbol{\varepsilon}$$

We will be interested in deciding whether row (seed) effects are zero, i.e., that the mean vector is $\boldsymbol{\theta} = \mu \mathbf{J} + \sum_j \beta_j \mathbf{C}_j$.

Example 3.1.4: (Polynomial regression) Suppose we observe pairs (x_i, Y_i) for $i = 1, 2, \ldots, n$ and that it is reasonable to suppose that

$$Y_i = \beta_0 + \beta_1 x_i + \beta_2 x_i^2 + \beta_3 x_i^3 + \varepsilon_i$$

with $\varepsilon_1, \ldots, \varepsilon_n$ independent $N(0, \sigma^2)$. By defining $w_{0i} = 1$, $w_{1i} = x_i$, $w_{2i} = x_i^2$, $w_{3i} = x_i^3$ we get the model $Y_i = \sum_0^3 \beta_j w_{ji} + \varepsilon_i$, or in vector form, $\mathbf{Y} = \sum_0^3 \beta_j \mathbf{w}_j + \boldsymbol{\varepsilon}$. We could replace $1, x_i^1, x_i^2, x_i^3$ by any other four functions of x_i. The model remains linear in the β_j's. Or, we might have reason to expect the regression function $g(x) = E(Y|x)$ to have the form of Figure 3.1.

We might then consider the model $Y_i = \beta_0 e^{\beta_1 x_i} \varepsilon_i$. Taking logs (all logs are base e), we get $\log Y_i = \log \beta_0 + \beta_1 x_i + \log \varepsilon_i$ and, defining $Z_i = \log Y_i$, $\gamma_0 = \log \beta_0$, $\eta_i = \log \varepsilon_i$, we get $Z_i = \gamma_0 + \beta_1 x_i + \eta_i$, so that for $\mathbf{x}_0$ the vector of all ones, $\mathbf{Z} = \gamma_0 \mathbf{x}_0 + \beta_1 \mathbf{x} + \boldsymbol{\eta}$. If we suppose $\boldsymbol{\eta} \sim N_n(\mathbf{0}, \sigma^2 \mathbf{I}_n)$ then the ε_i are independent with log-normal distributions.

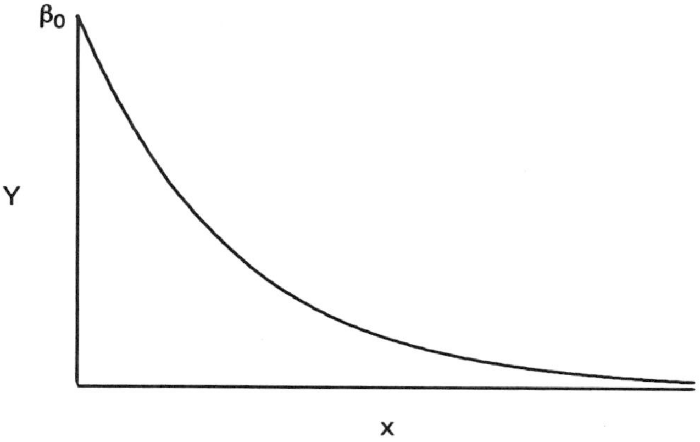

FIGURE 3.1 $Y = g(x) = \beta_0 e^{\beta_1 x}$.

We can fit the model with $g(x) = E(Y|x) = \beta_0 x^{\beta_1}$ and $Y_i = \beta_0 x_i^{\beta_1} \varepsilon_i$ by again taking logs to get the linear model $Z_i \equiv \log Y_i = \log \beta_0 + \beta_1 \log x_i + \log \varepsilon_i$.

Some Philosophy for Statisticians

The linear models we discuss make statements about (1) $\boldsymbol{\theta} = E(\mathbf{Y})$, (2) $D[\mathbf{Y}]$, and (3) the distribution of $\boldsymbol{\varepsilon} = \mathbf{Y} - \boldsymbol{\theta}$, and therefore of $\mathbf{Y}$ (almost always multivariate normal). In practice we rarely know whether these models hold. In fact, models should be viewed as idealizations, oversimplifications of a very complex real world. Models should be thought of as approximations, guidelines which enable us to understand relationships among variables better than we would without them. Since the statement of the model will hold only in approximation, the probability statements, and statements about means, variances, covariances, and correlation coefficients should be expected to hold only in reasonable approximation. In general it is difficult to make precise statements about the precision of these approximations.

In Chapter 4 we will discuss some techniques which allow us to investigate the appropriateness of the models we adopt. Fortunately, it will turn out that the procedures we use are often *robust* in the sense that they *almost* have the properties claimed for them, even though the models justifying them are satisfied in only rough approximation. Much of the reason for studying the theory and applications of linear models, rather than the applications alone, is that we must have a reasonable understanding of the effects of deviations from the models studied, and must be able to convey these effects to users of statistics, who usually do not have a strong understanding of the theory.

We end this excursion into philosophy with a quotation of John von Neumann, the great mathematician, taken from *Statistics and Truth*, the fine book of C. R. Rao (1989):

> The sciences do not try to explain, they hardly even try to interpret, they mainly make models. By a model is meant a mathematical construct which, with the addition of certain verbal interpretations, describes observed phenomena. The justification of such a mathematical construct is solely and precisely that it is expected to work.

In general, the representation of $\boldsymbol{\theta}$ as a linear combination of given vectors $\mathbf{x}_1, \ldots, \mathbf{x}_k$ is the most crucial part of the selection of a model. Second most crucial is the choice of a model for $D[\mathbf{Y}]$. Certainly, the model which supposes $D[\mathbf{Y}]$ is a multiple of the identity matrix is often inappropriate in econometrics, particularly if the the observations on $\mathbf{Y}$ are time series data (ordered by time). Fortunately, normality is often not crucial, particularly if the sample size n is large. Statisticians are always on safe grounds when recommending sample sizes of 10,000 or more. But they must be prepared for looks of horror, and perhaps early dismissal.

Estimation Theory

The linear hypothesis may be written in the equivalent form, depicted in Figure 3.2:

$$\mathbf{Y} = \boldsymbol{\theta} + \boldsymbol{\varepsilon} \qquad \text{for} \quad \boldsymbol{\theta} \in V = \mathscr{L}(\mathbf{x}_1, \ldots, \mathbf{x}_k) \qquad \text{and} \qquad \boldsymbol{\varepsilon} \sim N(\mathbf{0}, \sigma^2 \mathbf{I}_n)$$

Sometimes is is enough to estimate $\boldsymbol{\theta}$, and the representation of $\boldsymbol{\theta}$ in the form $\sum_1^k \beta_j \mathbf{x}_j$ is not important. On other occasions the coefficients $\beta_1, \ldots, \beta_k$ are themselves of interest. So that we can use matrix notation, let us write all vectors as columns, and let $\mathbf{X} = (\mathbf{x}_1, \ldots, \mathbf{x}_k)$. The matrix $\mathbf{X}$ is often called the *design matrix*, particularly when the experimenter has control over the choice of the $\mathbf{x}$-vectors. Then $\boldsymbol{\theta} = \mathbf{X}\boldsymbol{\beta}$, and if $\mathbf{x}_1, \ldots, \mathbf{x}_k$ are linearly independent, $(\mathbf{X}'\mathbf{X})^{-1}\mathbf{X}'\boldsymbol{\theta} = \boldsymbol{\beta}$.

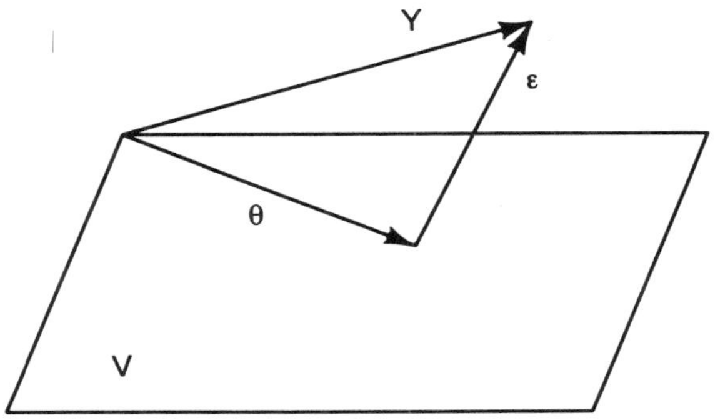

FIGURE 3.2

The principle of least squares leads to estimation of $\boldsymbol{\theta}$ by $\hat{\boldsymbol{\theta}}$, where $\boldsymbol{\theta} = \hat{\boldsymbol{\theta}}$ minimizes $\|\mathbf{Y} - \boldsymbol{\theta}\|^2 \equiv Q(\boldsymbol{\theta})$, subject to $\boldsymbol{\theta} \in V$. Thus, the principle of least squares leads to the estimator $\hat{\boldsymbol{\theta}} = p(\mathbf{Y} \mid V) = \hat{\mathbf{Y}}$.

In the case that $\mathbf{X}$ has full column rank (the $\mathbf{x}_j$ the linearly independent)

$$\hat{\boldsymbol{\theta}} = \mathbf{X}(\mathbf{X}'\mathbf{X})^{-1}\mathbf{X}'\mathbf{Y}, \qquad \text{and} \qquad \hat{\boldsymbol{\beta}} = (X'\mathbf{X})^{-1}\mathbf{X}'\mathbf{Y}.$$

Then $\hat{\boldsymbol{\beta}} = (\mathbf{X}'\mathbf{X})^{-1}\mathbf{X}'(\mathbf{X}\boldsymbol{\beta} + \boldsymbol{\varepsilon}) = \boldsymbol{\beta} + (\mathbf{X}'\mathbf{X})^{-1}\mathbf{X}'\boldsymbol{\varepsilon}$. The coefficient matrix $(\mathbf{X}'\mathbf{X})^{-1}\mathbf{X}'$ is the Moore–Penrose inverse $\mathbf{X}^+$ of $\mathbf{X}$ discussed in Section 1.7. If the column vectors of $\mathbf{X}$ are orthogonal, then $\hat{\mathbf{Y}} = \Sigma p(\mathbf{Y}|\mathbf{x}_j) = \Sigma \hat{\beta}_j \mathbf{x}_j$, where $\hat{\beta}_j = (\mathbf{Y}, \mathbf{x}_j)/\|\mathbf{x}_j\|^2 = \beta_j + (\boldsymbol{\varepsilon}, \mathbf{x}_j)/\|\mathbf{x}_j\|^2$.

Thus, each component of $\hat{\boldsymbol{\beta}}$ is equal to the corresponding component of $\boldsymbol{\beta}$ plus a linear combination of the components of $\boldsymbol{\varepsilon}$. In fact,

(1) $E(\hat{\boldsymbol{\beta}}) = \boldsymbol{\beta} + (\mathbf{X}'\mathbf{X})^{-1}\mathbf{X}'E(\boldsymbol{\varepsilon}) = \boldsymbol{\beta}$

(2) $D[\hat{\boldsymbol{\beta}}] = D[(\mathbf{X}'\mathbf{X})^{-1}\mathbf{X}'\boldsymbol{\varepsilon}] = (\mathbf{X}'\mathbf{X})^{-1}\mathbf{X}'(\sigma^2\mathbf{I}_n)[(\mathbf{X}'\mathbf{X})^{-1}\mathbf{X}']' = (\mathbf{X}'\mathbf{X})^{-1}\sigma^2$

(3) $\hat{\boldsymbol{\beta}}$ has a multivariate normal distribution (if $\boldsymbol{\varepsilon}$ does)

Of course, (1) requires only that $\mathbf{Y} = \mathbf{X}\boldsymbol{\beta} + \boldsymbol{\varepsilon}$ with $E(\boldsymbol{\varepsilon}) = \mathbf{0}$. (2) requires only that $D[\boldsymbol{\varepsilon}] = \sigma^2\mathbf{I}_n$. (3) requires only that $\boldsymbol{\varepsilon}$ has a multivariate normal distribution. In applications it is often unrealistic to suppose that $\boldsymbol{\varepsilon}$ has a multivariate normal distribution. However, a form of the Central Limit Theorem suggests that for large n, k different linear combinations of independent components of $\boldsymbol{\varepsilon}$ should have an approximate multivariate normal distribution even when $\boldsymbol{\varepsilon}$ does not.

Maximum Likelihood: The likelihood function is for each observed $\mathbf{Y} = \mathbf{y}$

$$L(\boldsymbol{\theta}, \sigma^2; \mathbf{y}) = \frac{1}{(2\pi)^{n/2}\sigma^n} e^{-(1/2)\|\mathbf{y} - \boldsymbol{\theta}\|^2/\sigma^2}$$

for $\boldsymbol{\theta} \in V = \mathscr{L}(\mathbf{x}_1, \ldots, \mathbf{x}_k)$ and $\sigma^2 > 0$. The maximum likelihood principle leads to the estimator of the pair $(\boldsymbol{\theta}, \sigma^2)$ which maximizes L for each $\mathbf{Y} = \mathbf{y}$, or, equivalently, maximizes

$$\log L = -\frac{n}{2}\log(2\pi) - \frac{n}{2}\log\sigma^2 - \frac{1}{2}\|\mathbf{y} - \boldsymbol{\theta}\|^2/\sigma^2$$

For each fixed σ^2, $\log L$ is maximized by taking $\boldsymbol{\theta} = p(\boldsymbol{\theta} \mid V) = \hat{\boldsymbol{\theta}}$. For this choice of $\boldsymbol{\theta}$ we get

$$\log L = -\frac{n}{2}\log(2\pi) - \frac{n}{2}\log\sigma^2 - \frac{1}{2}\|\mathbf{y} - \hat{\boldsymbol{\theta}}\|^2/\sigma^2$$

Replacing σ^2 by w and taking the derivative w.r.t. w we find

$$\frac{d(\log L)}{dw} = \frac{-n/2}{w} + \frac{(1/2)\|\mathbf{y} - \hat{\boldsymbol{\theta}}\|^2}{w^2}$$

which is zero for $w = \hat{\sigma}^2 = \dfrac{\|\mathbf{y} - \hat{\boldsymbol{\theta}}\|^2}{n}$. It is easy to verify that $\dfrac{\partial^2 \log L}{dw^2} < 0$, so that $\hat{\sigma}^2 = \|\mathbf{y} - \hat{\boldsymbol{\theta}}\|^2/n$ does maximize $\log L$ for each $\hat{\boldsymbol{\theta}}$. Thus, the pair $(\hat{\boldsymbol{\theta}}, \hat{\sigma}^2)$ maximizes L, i.e., this pair is the maximum likelihood estimator of $(\boldsymbol{\theta}, \sigma^2)$.

Estimation of σ^2: The maximum likelihood estimator of σ^2 is $\hat{\sigma}^2 = \|\mathbf{Y} - \hat{\boldsymbol{\theta}}\|^2/n$. We can employ Theorem 2.5.2 to determine the expectation of $\hat{\sigma}^2$. Since $\mathbf{Y} - \hat{\boldsymbol{\theta}} = \mathbf{P}_{V^\perp}\mathbf{Y}$, $D[\mathbf{Y}] = \sigma^2\mathbf{I}_n$, we get $E[\|\mathbf{P}_{V^\perp}\mathbf{Y}\|^2] = \sigma^2 \dim(V^\perp) + \|\mathbf{P}_{V^\perp}\boldsymbol{\theta}\|^2$. For $\boldsymbol{\theta} \in V$, $\mathbf{P}_{V^\perp}\boldsymbol{\theta} = 0$, so that

$$E(\hat{\sigma}^2) = \frac{\sigma^2}{n}\dim(V^\perp) = \frac{\sigma^2}{n}[n - \dim(V)].$$

Thus, unless $\dim(V) = 0$, i.e. $V = \mathscr{L}(\mathbf{0})$, $\hat{\sigma}^2$ is a biased estimator of σ^2. For this reason the most commonly used estimator of σ^2 is

$$S^2 = \|\mathbf{Y} - \hat{\boldsymbol{\theta}}\|^2/[n - \dim(V)] \tag{3.1.1}$$

In the special case that $\boldsymbol{\theta} = \mu\mathbf{J}_n$, so that $V = \mathscr{L}(\mathbf{J}_n)$, we get $\hat{\boldsymbol{\theta}} = \bar{Y}\mathbf{J}_n$, $\dim(V) = 1$, so that $S^2 = \|\mathbf{Y} - \bar{Y}\mathbf{J}_n\|^2/[n - 1] = \sum_{i=1}^{n} (Y_i - \bar{Y})^2/(n - 1)$. Students should remember, however, that this is *only* a special case, and S^2 will, in general, be as defined in (3.1.1).

If $\boldsymbol{\varepsilon}$ has a multivariate normal distribution then by Theorem 2.5.2

$$\|\mathbf{Y} - \hat{\boldsymbol{\theta}}\|^2/\sigma^2 \sim \chi^2_{n - \dim(V)}$$

Since the central χ^2 distribution with m degrees freedom has variance $2m$, we have

$$\text{Var}(S^2) = 2\sigma^4[n - \dim(V)][n - \dim(V)]^{-2} = 2\sigma^4/[n - \dim(V)]$$

for $\boldsymbol{\varepsilon}$ multivariate normal.

Properties of $\hat{\boldsymbol{\theta}}$ and S^2: Since $\hat{\boldsymbol{\theta}} = p(\mathbf{Y}|V)$ and $\mathbf{Y} - \hat{\boldsymbol{\theta}} = p(\mathbf{Y}|V^\perp)$, with V and $V^\perp$ orthogonal subspaces, $\hat{\boldsymbol{\theta}}$ and $\mathbf{Y} - \hat{\boldsymbol{\theta}}$ are uncorrelated random vectors, independent under normality. It follows that $\hat{\boldsymbol{\theta}}$ and $S^2 = \|\mathbf{Y} - \hat{\boldsymbol{\theta}}\|^2/(n - \dim(V))$ are independent, and, in the case that the columns of $\mathbf{X}$ are a basis for V, $\hat{\boldsymbol{\beta}} = (\mathbf{X}'\mathbf{X})^{-1}\mathbf{X}'\hat{\boldsymbol{\theta}} = (\mathbf{X}'\mathbf{X})^{-1}\mathbf{X}'\mathbf{Y}$ and the residual vector $\mathbf{e} = \mathbf{Y} - \hat{\boldsymbol{\theta}}$ are uncorrelated random vectors, independent if $\mathbf{Y}$ is multivariate normal.

To summarize, under the model $\mathbf{Y} = \boldsymbol{\theta} + \boldsymbol{\varepsilon}$ for $\boldsymbol{\theta} \in V$, $\boldsymbol{\varepsilon} \sim N_n(\mathbf{0}, \sigma^2\mathbf{I}_n)$ it follows that

(1) $\hat{\boldsymbol{\theta}} \sim N_n(\boldsymbol{\theta}, \mathbf{P}_V\sigma^2)$.

(2) $\mathbf{e} = \mathbf{Y} - \hat{\boldsymbol{\theta}} \sim N_n(\mathbf{0}, (\mathbf{I}_n - \mathbf{P}_V)\sigma^2)$.

(3) $\hat{\boldsymbol{\theta}}$ and $\mathbf{Y} - \hat{\boldsymbol{\theta}}$ are independent random vectors.

(4) $\|\mathbf{Y} - \hat{\boldsymbol{\theta}}\|^2/\sigma^2 \sim \chi^2_{n-\dim(V)}$, so that $S^2 = \|\mathbf{Y} - \hat{\boldsymbol{\theta}}\|^2/(n - \dim(V))$ is an unbiased estimator of σ^2.

(5) If the columns of $\mathbf{X}$ form a basis for V and $\hat{\boldsymbol{\theta}} = \mathbf{X}\hat{\boldsymbol{\beta}}$, then $\hat{\boldsymbol{\beta}} = (\mathbf{X}'\mathbf{X})^{-1}\mathbf{X}'\mathbf{Y}$ and S^2 are independent, with $\hat{\boldsymbol{\beta}} \sim N_k(\boldsymbol{\beta}, (\mathbf{X}'\mathbf{X})^{-1}\sigma^2)$. If, in addition, the columns of $\mathbf{X}$ are mutually orthogonal, the estimators $\hat{\beta}_j$ are uncorrelated, and therefore independent.

Problem 3.1.1: A scale has two pans. The measurement given by the scale is the difference between the weights in pan #1 and pan #2 plus a random error. Thus, if a weight μ_1 is put in pan #1, a weight μ_2 is put in pan #2, then the measurement is $Y = \mu_1 - \mu_2 + \varepsilon$. Suppose that $E(\varepsilon) = 0$, $\text{Var}(\varepsilon) = \sigma^2$, and that in repeated uses of the scale, observations Y_i are independent.

Suppose that two objects, #1 and #2, have weights β_1 and β_2. Measurements are then taken as follows:

(1) Object #1 is put on pan #1, nothing on pan #2.

(2) Object #2 is put on pan #2, nothing on pan #1.

(3) Object #1 is put on pan #1, object #2 on pan #2.

(4) Objects #1 and #2 are both put on pan #1.

(a) Let $\mathbf{Y} = (Y_1, Y_2, Y_3, Y_4)'$ be the vector of observations. Formulate this as a linear model.

(b) Find vectors $\mathbf{a}_1$, $\mathbf{a}_2$ such that $\hat{\beta}_1 = (\mathbf{a}_1, \mathbf{Y})$ and $\hat{\beta}_2 = (\mathbf{a}_2, \mathbf{Y})$ are the least squares estimators of β_1 and β_2.

(c) Find the covariance matrix for $\hat{\boldsymbol{\beta}} = (\hat{\beta}_1, \hat{\beta}_2)'$.

(d) Find a matrix $\mathbf{A}$ such that $S^2 = \mathbf{Y}'\mathbf{A}\mathbf{Y}$.

(e) For the observation $\mathbf{Y} = (7, 3, 1, 7)'$ find S^2, and estimate the covariance matrix of $\hat{\boldsymbol{\beta}}$.

(f) Show that four such weighings can be made in such a way that the least squares estimators of β_1 and β_2 have smaller variances than for the experiment above.

Problem 3.1.2: The following model has been used to predict the yield Y of the valuable chemical "gloxil" as a function of the temperature T. The expected yield is continuous function $g(T)$ of T. There is no yield for $T < T_0 = 20$ (degrees Celsius). For temperatures between T_0 and $T_1 = 100$, the expected yield is a linear function of T. For temperatures above 100 the expected yield is also linear, though the slope changes at $T = 100$. Suppose measurements on the yield are made for $T = 40$, 80, 120, 180 and 240.

(a) Formulate this as a linear model.

(b) Suppose that the measurements at the five temperatures were: 57, 176, 223, 161, 99. Estimate the parameters of your model and plot your estimate of the regression function of yield on T. Determine the residual vector $\mathbf{e}$ and use it to determine S^2.

Problem 3.1.3: A chemist wishes to determine the percentages of impurities β_1 and β_2 in two 100 gram containers (1 and 2) of potassium chloride (KCl). The process she uses is able to measure the weight in grams of the impurities in any 2 gram sample of KCl with mean equal to the true weight of the impurities and standard deviation 0.006 gram. She makes three measurements. Measurement #1 is on a 2 gram sample from container 1. Measurement #2 is on a 2 gram sample from container 2. Measurement #3 is on a mixture of a 1 gram sample from container 1 and a 1 gram sample from container 2.

(a) Formulate this as a linear model.

(b) Give formulas for unbiased estimators $\hat{\beta}_1$ and $\hat{\beta}_2$ of β_1 and β_2.

(c) Determine the covariance matrix of $(\hat{\beta}_1, \hat{\beta}_2)$.

(d) Estimate (β_1, β_2) for the three measurements 0.036, 0.056, 0.058. Also determine S^2, and compare it to the true variance, which we know to be $\sigma^2 = 0.006^2$.

Problem 3.1.4: Chemical processes A, B, and C have yields Y which have expectations which are each linearly affected by the amount x of a catalyst used. That is, $g(x, p) = E(Y|x, p)$, where $p =$ A, B, or C, is a linear function of x for each p, with slope which may depend on p. Suppose also that the expected yields for $x = 50$ are the same for A, B, and C. Two independent observations Y were taken for each combination of values of x and the three processes. Their values were

	$x = 20$		$x = 80$	
A	69	63	120	132
B	34	44	151	167
C	18	12	204	186

(a) Let $p =$ A, B, C index the three processes, let $j = 1, 2$ index the the two levels of x, and let $k = 1, 2$ index the two measurements made for the same process, x-level combination. Let Y_{pki} be the yield for the pki combination. Define a linear model. That is, define vectors $\mathbf{x}_1, \ldots, \mathbf{x}_k$, so that $\mathbf{Y} = \sum \beta_j \mathbf{x}_j + \boldsymbol{\varepsilon}$.

Hint: Let $\mathbf{x}_2 = \begin{bmatrix} -30 & -30 & 30 & 30 \\ 0 & 0 & 0 & 0 \\ 0 & 0 & 0 & 0 \end{bmatrix}$.

(b) Find the least squares estimate of the vector $\boldsymbol{\beta}$ of regression coefficients.

(c) Find $\hat{\mathbf{Y}}$, $\mathbf{e} = \mathbf{Y} - \hat{\mathbf{Y}}$, SSE (Sum of Squares for Error) $= \|\mathbf{Y} - \hat{\mathbf{Y}}\|^2$, and S^2. Also estimate the covariance matrix for $\hat{\boldsymbol{\beta}}$.

(d) Suppose that the assumption that the expected yield is the same for each p is dropped. What are $\hat{\mathbf{Y}}$, $\mathbf{e}$, and SSE?

Problem 3.1.5: Consider the model of Example 3.1.2 for two experimental conditions with $n_1 = 4$, $n_2 = 3$. Give explicit nonmatrix formulas for the estimators $\hat{\beta}_1$, $\hat{\beta}_2$, $\hat{\beta}_c$. *Hint*: Orthogonalize $\mathbf{x}$ with respect to $\mathbf{w}_1$ and $\mathbf{w}_2$.

3.2 CONFIDENCE INTERVALS AND TESTS ON $\eta = c_1\beta_1 + \cdots + c_k\beta_k$

We are often interested in giving confidence intervals or testing hypotheses on the β_j or on differences $\beta_j - \beta_{j'}$. More generally, we may be interested in a linear combination $\eta = (\mathbf{c}, \boldsymbol{\beta}) = c_1\beta_1 + \cdots + c_k\beta_k$, for c_j's chosen by the statistician. A natural estimator of η is $\hat{\eta} = c_1\hat{\beta}_1 + \cdots + c_k\hat{\beta}_k$. By the linearity of expectation $\hat{\eta}$ is an unbiased estimator of η. Its variance is $\mathrm{Var}(\hat{\eta}) = \mathbf{c}'\mathbf{M}^{-1}\mathbf{c}\sigma^2 \equiv d\sigma^2$, where $\mathbf{M}$ is the inner product matrix. The corresponding estimator of $\mathrm{Var}(\hat{\eta})$ is $S^2_{\hat{\eta}} = dS^2$. In the special case that $\eta = \beta_j$, c is the jth unit vector, and d is the jj term of $\mathbf{M}^{-1}$. If, for example, $\eta = \beta_2 - \beta_2$, then $\mathbf{c} = (0, 1, -1, 0, \ldots, 0)'$, and if $\mathbf{M}^{-1} = (f_{ij})$, then $d = f_{22} + f_{33} - 2f_{23}$, and $\mathrm{Var}(\hat{\eta}) = d\sigma^2 = f_{22}\sigma^2 + f_{33}\sigma^2 - 2f_{23}\sigma^2 = \mathrm{Var}(\hat{\beta}_2) + \mathrm{Var}(\hat{\beta}_3) - 2\,\mathrm{cov}(\hat{\beta}_2, \hat{\beta}_3)$.

Thus, $\hat{\eta} \sim N(\eta, d\sigma^2)$, $\dfrac{\hat{\eta} - \eta}{\sqrt{d\sigma^2}} \sim N(0, 1)$, and $\dfrac{\hat{\eta} - \eta}{\sqrt{dS^2}} \sim t_{n-k}$, so that, for $t = t_{n-k,1-\alpha}$, $1 - \alpha = P\left(-t \le \dfrac{\hat{\eta} - \eta}{\sqrt{dS^2}} \le t\right) = P(\hat{\eta} - t\sqrt{dS^2} \le \eta \le \hat{\eta} + t\sqrt{dS^2})$.

Thus, $[\hat{\eta} \pm t_{n-k,1-\alpha}\sqrt{dS^2}]$ is a $100(1 - \alpha)\%$ confidence interval on η.

Example 3.2.1: Let $\mathbf{x}_1 = (1, 1, 1, 1)'$, $\mathbf{x}_2 = (1, 1, 0, 0)'$, $\mathbf{x}_3 = (1, -1, 1, -1)'$ and suppose $\mathbf{Y} = \sum_1^3 \beta_j\mathbf{x}_j + \boldsymbol{\varepsilon}$ for $\boldsymbol{\varepsilon} \sim N_4(\mathbf{0}, \sigma^2\mathbf{I}_4)$. Then

$$\mathbf{M} = \mathbf{X}'\mathbf{X} = 2\begin{bmatrix} 2 & 1 & 0 \\ 1 & 1 & 0 \\ 0 & 0 & 2 \end{bmatrix} \quad \text{and} \quad \mathbf{M}^{-1} = \frac{1}{4}\begin{bmatrix} 2 & -2 & 0 \\ -2 & 4 & 0 \\ 0 & 0 & 1 \end{bmatrix}.$$

Thus, for $\eta = \beta_1 - \beta_2$, $\mathbf{c} = (1, -1, 0)'$ and $d = \mathbf{c}'\mathbf{M}^{-1}\mathbf{c} = [2 + 4 + 2(-1)(-2)]/4 = 10/4$, so that a 95% confidence interval on η is $\hat{\beta}_1 - \hat{\beta}_2 \pm t_{1,0.975}\sqrt{S^2(10/4)}$. The 95% confidence interval on β_2 is $[\hat{\beta}_2 \pm t_{1,0.975}S]$, since $d = 1$.

Tests of Hypothesis on $\eta = \sum c_j\beta_j$: Suppose we wish to test $H_0\colon \eta \le \eta_0$ vs. $H_1\colon \eta > \eta_0$, where η_0 is a known constant, often chosen to be 0. Since

$$t = \frac{\hat{\eta} - \eta_0}{\sqrt{dS^2}} \sim t_{n-k}\left(\frac{\eta - \eta_0}{\sqrt{d\sigma^2}}\right),$$

which becomes central t when $\eta = \eta_0$, the test which rejects H_0 for $t = \dfrac{\hat{\eta} - \eta_0}{\sqrt{S^2 d}} > t_{n-k,\,1-\alpha}$ is an α-level test. The two-sided hypotheses $H_0\colon \eta = \eta_0$ vs. $H_1\colon \eta \ne \eta_0$ is rejected for $|t| \ge t_{n-k,\,1-\alpha/2}$.

In the applications of multiple regression it is very common to take $\eta = \beta_j$ and $\eta_0 = 0$ for some j, so statistical software packages print the corresponding t-statistics $t_j = \hat{\beta}_j/S_{\hat{\beta}_j}$, where $S^2_{\hat{\beta}_j} = f_{jj}S^2$, and f_{jj} is the jj term of $\mathbf{M}^{-1}$. Usually they present $\hat{\alpha}_j$-values as well (p-values, the probabilities under the null hypothesis that $|t_j|$ would be as large or larger than the value observed).

There is a one-to-one correspondence between a family of tests of hypotheses on η and confidence intervals in the following sense. If $C(\mathbf{Y})$ is a 100 $(1-\alpha)$% confidence interval on η, then $A(\eta_0) = \{\mathbf{y} \mid \eta_0 \in C(\mathbf{y})\}$ is the acceptance region (the part of Ω for which H_0 is to be accepted) for an α-level test of $H_0\colon \eta = \eta_0$, and $C(\mathbf{y}) = \{\eta \mid \mathbf{y} \in A(\eta)\}$. For example, if [35, 47] is a 95% confidence interval on η, then we should accept $H_0\colon \eta = \eta_0$ vs. $H_a\colon \eta \ne \eta_0$ at level $\alpha = 0.05$ for any $\eta_0 \in [35, 47]$, reject otherwise. The one-sided 95% confidence interval $(\hat{\eta} - t_{0.95}\sqrt{S^2 d}, +\infty)$ on η corresponds to the 0.05-level t-test of $H_0\colon \eta \le \eta_0$ vs. $H_a\colon \eta > \eta_0$.

Example 3.2.2: (Simple linear regression) Let $Y_i = \beta_0 + \beta_1 x_i + \varepsilon_i$ for $i = 1, \ldots, n$. That is, $\mathbf{Y} = \beta_0\mathbf{J} + \beta_1\mathbf{x} + \boldsymbol{\varepsilon}$, where

$$J = \begin{bmatrix} 1 \\ \vdots \\ 1 \end{bmatrix}, \qquad \mathbf{x} = \begin{bmatrix} x_1 \\ \vdots \\ x_n \end{bmatrix}, \qquad \mathscr{L}(\mathbf{J}, \mathbf{x}) = \mathscr{L}(\mathbf{J}, \mathbf{x}^*),$$

and since $\mathbf{x}^* = \mathbf{x} - \bar{x}\mathbf{J}$, $\mathbf{J} \perp \mathbf{x}^*$. Thus

$$\hat{\mathbf{Y}} = p(\mathbf{Y} \mid V) = \frac{(\mathbf{Y}, \mathbf{J})}{\|\mathbf{J}\|^2}\mathbf{J} + \frac{(\mathbf{Y}, \mathbf{x}^*)}{\|\mathbf{x}^*\|^2}\mathbf{x}^* = \bar{Y}\mathbf{J} + \frac{S_{xy}}{S_{xx}}\mathbf{x}^*,$$

where

$$S_{xy} = (\mathbf{Y}, \mathbf{x}^*) = \sum_1^n Y_i(x_i - \bar{x}) = \sum_i (Y_i - \bar{Y})(x_i - \bar{x}) = \sum_i x_i(Y_i - \bar{Y})$$
$$= \sum x_i Y_i - n\bar{x}\bar{Y}$$

$$S_{xx} = (\mathbf{x}^*, \mathbf{x}^*) = \sum_1^n (x_i - \bar{x})^2 = \sum x_i(x_i - \bar{x}) = \sum x_i^2 - n\bar{x}^2$$

Since $\mathbf{x}^* = \mathbf{x} - \bar{x}\mathbf{J}$, we get $\hat{\mathbf{Y}} = \hat{\beta}_0\mathbf{J}_0 + \hat{\beta}_1\mathbf{x} = \bar{Y}\mathbf{J} + \dfrac{S_{xy}}{S_{xx}}(\mathbf{x} - \bar{x}\mathbf{J})$, so that $\hat{\beta}_1 = \dfrac{S_{xy}}{S_{xx}}$, $\hat{\beta}_0 = \bar{Y} - \hat{\beta}_1\bar{x}$.

The variance of the estimator $\hat{\beta}_1$ of the slope β_1 in simple linear regression is

$$\begin{aligned}\operatorname{Var}(\hat{\beta}_1) &= \operatorname{Var}\left(\frac{S_{xy}}{S_{xx}}\right) = \left(\frac{1}{S_{xx}}\right)^2 \operatorname{Var}[(\mathbf{Y}, \mathbf{x}^*)] = \left(\frac{1}{S_{xx}}\right)^2 [\sigma^2\|\mathbf{x}^*\|^2] \\ &= \left(\frac{1}{S_{xx}}\right)^2 \sigma^2 S_{xx} = \frac{\sigma^2}{S_{xx}},\end{aligned}$$

$$\operatorname{Var}(\bar{Y}) = \frac{\sigma^2}{n}, \ \operatorname{cov}(\bar{Y}, \hat{\beta}_1) = 0, \qquad \text{since} \quad \mathbf{J} \perp \mathbf{x}^*.$$

Then

$$\begin{aligned}\operatorname{cov}(\hat{\beta}_0, \hat{\beta}_1) &= \operatorname{cov}(\bar{Y} - \hat{\beta}_1\bar{x}, \hat{\beta}_1) = \operatorname{cov}(\bar{Y}, \hat{\beta}_1) - \bar{x}\operatorname{cov}(\hat{\beta}_1, \hat{\beta}_1) \\ &= 0 - \bar{x}\sigma^2/S_{xx} = -(\bar{x}/S_{xx})\sigma^2.\end{aligned}$$

Alternatively, we could have found the covariance matrix of $\hat{\boldsymbol{\beta}} = (\hat{\beta}_0, \hat{\beta}_1)'$ from

$$D[\hat{\beta}] = \sigma^2(\mathbf{X}'\mathbf{X})^{-1} \qquad \text{for} \quad \mathbf{X}'\mathbf{X} = \begin{bmatrix} n & \sum x_i \\ \sum x_i & \sum x_i^2 \end{bmatrix}$$

We now want to find $S^2 = \|\mathbf{Y} - \hat{\mathbf{Y}}\|^2/(n-2)$. Since error sum of squares $=$ $\|\mathbf{Y} - \hat{\mathbf{Y}}\|^2 = \|\mathbf{Y}\|^2 - \|\hat{\mathbf{Y}}\|^2$, we first determine $\|\hat{\mathbf{Y}}\|^2$.

$$\begin{aligned}\|\hat{\mathbf{Y}}\|^2 &= (\hat{\mathbf{Y}}, \hat{\mathbf{Y}}) = (\mathbf{Y}, \hat{\mathbf{Y}}) = (\mathbf{Y}, \bar{Y}\mathbf{J} + \hat{\beta}_1\mathbf{x}^*) = \bar{Y}^2 n + \hat{\beta}_1 S_{xy} = \bar{Y}^2 n \\ &\quad + S_{xy}^2/S_{xx} = \|p(\mathbf{Y}\,|\,\mathbf{x}_0)\|^2 + \|p(\mathbf{Y}\,|\,x_1^*)\|^2.\end{aligned}$$

Then

$$\begin{aligned}\|\mathbf{Y} - \hat{\mathbf{Y}}\|^2 &= \|\mathbf{Y}\|^2 - [\bar{Y}^2 n + \hat{\beta}_1 S_{xy}] = \Sigma(Y_i - \bar{Y})^2 - \hat{\beta}_1 S_{xy} \\ &= S_{yy} - \hat{\beta}_1 S_{xy} = S_{yy} - S_{xy}^2/S_{xx}.\end{aligned}$$

and $S^2 = \left[\dfrac{1}{n-2}\right][S_{yy} - \hat{\beta}_1 S_{xy}]$. Since $\operatorname{Var}(\hat{\beta}_1) = \sigma^2/S_{xx}$, an unbiased estimator of $\operatorname{Var}(\hat{\beta}_1)$ is $S^2(\hat{\beta}_1) = S^2/S_{xx}$. It follows that a $100(1-\alpha)\%$ confidence interval on β_1 is given by $\hat{\beta}_1 \pm t_{n-2,1-\alpha/2} S(\hat{\beta}_1)$.

We sometimes want a confidence interval on $g(x_0) = \beta_0 + \beta_1 x_0$, the mean Y for $x = x_0$, where x_0 is a specified value for x. Since $\hat{g}(x_0) = \hat{\beta}_0 + \hat{\beta}_1 x_0 = \bar{Y} + \hat{\beta}_1(x_0 - \bar{x})$ is an unbiased estimator of $g(x_0)$, with

$$\operatorname{Var}[\hat{g}(x_0)] = \sigma^2\left[\frac{1}{n} + (x_0 - \bar{x})^2/S_{xx}\right] \equiv \sigma^2 h(x_0)$$

and is normally distributed, it follows that $\frac{\hat{g}(x_0) - g(x_0)}{\sqrt{S^2 h(x_0)}} \sim t_{n-2}$, so that a $(1 - \alpha)100\%$ confidence interval on $g(x_0) = \beta_0 + \beta_1 x_0$ is given by

$$[\hat{g}(x_0) \pm t_{n-2,1-\alpha/2}\sqrt{S^2 h(x_0)}] \equiv I(x_0)$$

This means that this random interval $I(x_0)$ satisfies

$$P(g(x_0) \in I(x_0)) = 1 - \alpha \qquad \text{for all} \quad x_0. \tag{*}$$

This is not the same as $P(g(x_0) \in I(x_0)$ for *all* $x_0)$, which is smaller. (See the difference?). We will later develop methods for finding random intervals $I_s(x_0)$ (s for "simulataneous") so that $P(g(x_0) \in I_s(x_0)$ for all $x_0) = 1 - \alpha$.

Problem 3.2.1: (a) For **Y** as in Problem 3.1.1, (a) Find a 95% confidence interval on $\beta_1 - \beta_2$.

(b) Consider instead the four weighings in part (f). What is the ratio of the expected lengths of the confidence interval found in (a) to that found for these four weighings?

Problem 3.2.2: The following model is often used for the scores achieved on a standardized exam taken by students, such as a S.A.T. or A.C.T. exam required for entrance to colleges or universities. Let θ denote the student's "true score", the theoretical long-term average score that student would achieve on repetitions of the exam (assuming no learning effect). Let Y denote the student's score on the exam, and suppose that $Y = \theta + \varepsilon$ for $E(\varepsilon) = 0$, $\text{Var}(Y) = \sigma_\varepsilon^2$. Suppose that σ_ε^2 is the same for all θ (this may not be realistic). Suppose also that ε is normally distributed.

(a) Suppose σ_ε is known and Y is observed. Give a formula for a 95% confidence interval on θ.

(b) Now suppose σ_ε is unknown but n students have been given different versions of the exam twice, with scores (W_{i1}, W_{i2}) for $i = 1, \ldots, n$. Suppose that $W_{ij} = \theta_i + \varepsilon_{ij}$ for $j = 1, 2$ and $i = 1, \ldots, n$, where the θ_i are any fixed unknown parameters, and the ε_{ij} are all independent, each with variance σ_ε^2. Find an unbiased estimator S_ε^2 of σ_ε^2 and prove that $nS_\varepsilon^2/\sigma_\varepsilon^2 \sim \chi_n^2$.

(c) Now suppose another student takes the exam and achieves a score of Y_0. Let θ_0 be her (unknown) true score. Find a function of Y_0, θ_0 and S_ε which has a t distribution and use this function as a "pivot" to find a 95% confidence interval on θ_0.

Problem 3.2.3: Let $x_i > 0$ for $i = 1, \ldots, n$. Suppose that $Y_i = \beta x_i + \varepsilon_i$, for $i = 1, \ldots, n$ with the ε_i independent $N(0, \sigma^2)$

(a) For $x_0 > 0$ give a formula for a $100\gamma\%$ confidence interval on $g(x_0) \equiv \beta x_0$.

(b) Apply the formula for the observed pairs (x_i, Y_i): (2, 3), (3, 11), (4, 12), $x_0 = 5$, and $\gamma = 0.95$. Repeat for $x_0 = 10$.

Problem 3.2.4: For the wormy fruit data of Example 1.4.1 find 95% confidence intervals on $g(x_0) = \beta_0 + \beta_1 x_0$ for $x_0 = 6$, 18, and 50.

Problem 3.2.5: For the simple linear regression model H_0: $\beta_1 = 0$ or H_0: $\beta_1 \le 0$ or H_0: $\beta_1 \ge 0$ may be tested using the test statistic $t = \hat{\beta}_1/S_{\hat{\beta}_1} = \hat{\beta}_1/\sqrt{S^2/S_{xx}}$. Show that $t = [r\sqrt{n-2}]/\sqrt{1-r^2}$, where $r = S_{xy}/\sqrt{S_{xx}S_{yy}}$ is the correlation coefficient.

Problem 3.2.6: (a) Consider any collection of n pairs (x_i, y_i). Define $u_{ij} = x_i - x_j$ and $v_{ij} = y_i - y_j$. Show that $\sum_{i,j} u_{ij}v_{ij} = 2\sum_{i<j} u_{ij}v_{ij} = nS_{xy}$ and that therefore $S_{xx} = \sum_{ij} u_{ij}^2/n$.

(b) Let $b_{ij} = u_{ij}/v_{ij}$ for $i \ne j$. Then b_{ij} is the slope of the straight line from (x_i, y_i) to (x_j, y_j). Let $D_{ij} = (x_i - x_j)^2 = \det(X'_{ij}X_{ij})$, where $X_{ij} = \begin{bmatrix} 1 & x_i \\ 1 & x_j \end{bmatrix}$. Show that $\hat{\beta}_1 = \left(\sum_{i<j} D_{ij}b_{ij}\right) \Big/ \sum_{i<j} D_{ij}$. Thus, $\hat{\beta}_1$ is the weighted average of the two-at-a-time slopes. C. F. J. Wu (1986) shows that this result holds more generally. If s is a subset of the integers $1, \ldots, n$ of size $r \ge k$, and $\hat{\beta}_s$ is the least squares estimator of β based on those observations Y_i with $i \in s$ only, then $\hat{\beta} = \left[\sum_r \det(\mathbf{X}_s'\mathbf{X}_s)\hat{\beta}_s\right] \Big/ \left[\sum_s \det(\mathbf{X}_s'\mathbf{X}_s)\right]$, where $\mathbf{X}_s$ is the submatrix of $\mathbf{X}$ consisting of the rows with index in s. In the case $r = k$, as in the simple linear regression case, take $\hat{\beta}_s = (\mathbf{X}_s')^{-1}\mathbf{Y}_s$ if the inverse exists, zero otherwise.

(c) Check the formulas of (a) and (b) for the three pairs (1, 9), (3, 2), (5, 3).

3.3 THE GAUSS–MARKOV THEOREM

Each least squares estimator $\hat{\beta}_j$ of β_j and $\hat{\eta} = \sum c_j\hat{\beta}_j$ of $\eta = \sum c_j\beta_j$ is linear and unbiased, where "linear" refers to linearity in the components of $\mathbf{Y}$. In fact, for the full rank case with column vectors $\mathbf{x}_1, \ldots, \mathbf{x}_k$, $\mathbf{X} = (\mathbf{x}_1, \ldots, \mathbf{x}_k)$, $\mathbf{M} = \mathbf{X}'\mathbf{X}$, and $\mathbf{c} = (c_1, \ldots, c_k)'$, it follows that $\eta = \mathbf{c}'\boldsymbol{\beta} = \mathbf{c}'\mathbf{M}^{-1}\mathbf{X}'\theta = (\mathbf{a}, \boldsymbol{\theta})$ and $\eta = (\mathbf{a}, \mathbf{Y})$ for $\mathbf{a} = \mathbf{X}(\mathbf{X}'\mathbf{X})^{-1}\mathbf{c}$. The vector $\mathbf{a}$ is an element of V, the column space of $\mathbf{X}$, satisfying the condition

$$(\mathbf{a}, \boldsymbol{\theta}) = \mathbf{a}'\boldsymbol{\theta} = \mathbf{a}'\mathbf{X}\boldsymbol{\beta} = \mathbf{c}'\boldsymbol{\beta} \qquad \text{for all} \quad \boldsymbol{\beta}, \text{ i.e.,}$$

$\mathbf{a}'\mathbf{X} = \mathbf{c}'$ or $\mathbf{X}'\mathbf{a} = \mathbf{c}$. We have shown that $\mathbf{a}$ is the vector in V which has inner product c_j with $\mathbf{x}_j$ for $j = 1, \ldots, k$. Are there other vectors $\mathbf{d}$ such that $(\mathbf{d}, \mathbf{Y})$ is an unbiased estimator of η, and has smaller variance than $(\mathbf{a}, \mathbf{Y})$? The answer is no, as shown by the famous Gauss–Markov Theorem (Figure 3.3).

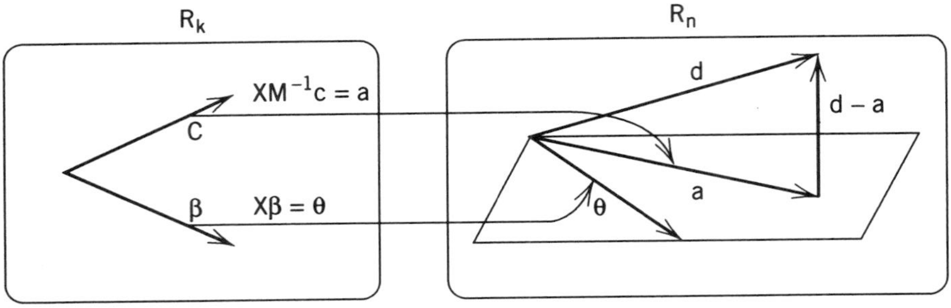

FIGURE 3.3 Illustration of the Gauss–Markov Theorem: Full-rank case.

Gauss–Markov Theorem: (Full-rank case) Suppose that $\mathbf{Y} = \sum_1^k \beta_j \mathbf{x}_j + \boldsymbol{\varepsilon}$, where $\mathbf{x}_1, \ldots, \mathbf{x}_k$ are linearly independent, $E(\boldsymbol{\varepsilon}) = \mathbf{0}$, $D[\boldsymbol{\varepsilon}] = \sigma^2 \mathbf{I}_n$. Let $\eta = \sum c_j \beta_j$, and let η^* be any linear unbiased estimator of η. Then $\text{Var}(\eta^*) \geq \text{Var}(\hat{\eta})$ with equality only if $\eta^* = \hat{\eta}$ for all $\mathbf{Y}$.

***Proof*:** First note that $\boldsymbol{\beta} = (\mathbf{X}'\mathbf{X})^{-1}\mathbf{X}'\boldsymbol{\theta}$ and $\eta = \mathbf{c}'\boldsymbol{\beta} = \mathbf{c}'(\mathbf{X}'\mathbf{X})^{-1}\mathbf{X}'\boldsymbol{\theta} = (\mathbf{a}, \boldsymbol{\theta})$ for $\mathbf{a} = \mathbf{X}(\mathbf{X}'\mathbf{X})^{-1}\mathbf{c}$. Similarly, $\hat{\eta} = \mathbf{c}'\hat{\boldsymbol{\beta}} = \mathbf{c}'(\mathbf{X}'\mathbf{X})^{-1}\mathbf{X}'\mathbf{Y} = (\mathbf{a}, \mathbf{Y})$. These representations of η and $\hat{\eta}$ as inner products facilitate the computation of variances and provide an intuitive justification for the conclusion.

Consider any linear estimator $\eta^* = (\mathbf{d}, \mathbf{Y})$ of η. Then $E(\eta^*) = (\mathbf{d}, \boldsymbol{\theta})$. η^* is unbiased for η if $(\mathbf{d}, \boldsymbol{\theta}) = (\mathbf{a}, \boldsymbol{\theta})$ for all $\boldsymbol{\theta} \in V$, i.e., if $(\mathbf{d} - \mathbf{a}, \boldsymbol{\theta}) = 0$ for all $\boldsymbol{\theta} \in V$, equivalently if $(\mathbf{d} - \mathbf{a}) \perp V$. Then

$$\eta^* = (\mathbf{d}, \mathbf{Y}) = (\mathbf{a}, \mathbf{Y}) + (\mathbf{d} - \mathbf{a}, \mathbf{Y}) = \hat{\eta} + (\mathbf{d} - \mathbf{a}, \boldsymbol{\theta} + \boldsymbol{\varepsilon}) = \hat{\eta} + (\mathbf{d} - \mathbf{a}, \boldsymbol{\varepsilon}).$$

The r.v.'s $\hat{\eta}$ and $(\mathbf{d} - \mathbf{a}, \boldsymbol{\varepsilon})$ are uncorrelated since $\mathbf{a} \perp (\mathbf{d} - \mathbf{a})$. It follows that

$$\text{Var}(\eta^*) = \text{Var}(\hat{\eta}) + \|\mathbf{d} - \mathbf{a}\|^2 \sigma^2,$$

so that $\text{Var}(\eta^*) \geq \text{Var}(\hat{\eta})$ with equality only if $\mathbf{d} = \mathbf{a}$, i.e., $\eta^* = \hat{\eta}$ for all $\mathbf{Y}$.

□

Comments

(1) The estimator $\hat{\eta}$ is often called the best linear unbiased estimator (BLUE). It is also called the least squares or the Gauss–Markov estimator.

(2) $\text{Var}(\hat{\eta}) = \|\mathbf{a}\|^2\sigma^2 = [\mathbf{c}'(\mathbf{X}'\mathbf{X})^{-1}\mathbf{c}]\sigma^2$.

(3) Figure 3.3 illustrates the proof:

Every vector $\mathbf{d}$ for which $(\mathbf{d}, \mathbf{Y})$ is an unbiased estimator of η is of the form $\mathbf{d} = \mathbf{a} + \mathbf{b}$ for $\mathbf{b} \perp V$. The vector $\mathbf{a}$ of coefficients of the Y_i in $\hat{\eta}$ lies in V. The set $\{\mathbf{d} \mid E(\mathbf{d}, \mathbf{Y}) = \eta\} = \mathbf{a} \oplus V^{\perp}$, the hyperplane of vectors of the form $\mathbf{d} = \mathbf{a} + \mathbf{b}$ for

$\mathbf{b} \in V^\perp$. The variance of η^* is $\text{Var}(\hat{\eta})$ plus $\sigma^2\|\mathbf{b}\|^2 = \sigma^2\|\mathbf{a} - \mathbf{d}\|^2$, which is minimum for $\mathbf{b} = \mathbf{d} - \mathbf{a} = \mathbf{0}$. Since $\eta^* = \hat{\eta} + (\mathbf{b}, \mathbf{Y})$, the part $(\mathbf{b}, \mathbf{Y})$ of η^* is wasteful in the sense that $E(\mathbf{b}, \mathbf{Y}) = 0$ for all $\boldsymbol{\theta} \in V$, but $(\mathbf{b}, \mathbf{Y})$ increases the variance by $\|\mathbf{b}\|^2\sigma^2$.

Let $\mathbf{u}_j$ be the jth k-component unit vector, having one in the jth component, 0 otherwise. Then $\beta_j = \mathbf{u}_j'\beta$ has least squares estimator $\mathbf{u}_j'\hat{\boldsymbol{\beta}} = \hat{\beta}_j = \mathbf{u}_j'\mathbf{AY}$ for $\mathbf{A} = (\mathbf{X}'\mathbf{X})^{-1}\mathbf{X}'$. It follows that for all linear unbiased estimators $\boldsymbol{\beta}^* = \mathbf{BY}$ of $\boldsymbol{\beta}$, $\boldsymbol{\beta}^* = \mathbf{AY} + (\mathbf{B} - \mathbf{A})\mathbf{Y}$, and $D[\boldsymbol{\beta}^*] = [\mathbf{AA}' + (\mathbf{B} - \mathbf{A})(\mathbf{B} - \mathbf{A})']\sigma^2$, which has minimum diagonal elements (minimum variances) for $\mathbf{B} = \mathbf{A}$, i.e., $\boldsymbol{\beta}^* = \hat{\boldsymbol{\beta}}$. It may not always make sense to insist that the estimators we consider be unbiased. For a discussion of this see Sections 4.2 and 4.7. For an example of a silly unbiased estimator, consider the unbiased estimator of $e^{-\lambda}$ for a single observation X from the Poisson distribution with mean λ.

Example 3.3.1: Let $\mathbf{x}_1 = (1, 0, 1, 1)'$, $\mathbf{x}_2 = (0, 1, 1, 1)'$ and suppose $\mathbf{Y} = \beta_1\mathbf{x}_1 + \beta_2\mathbf{x}_2 + \boldsymbol{\varepsilon}$. Suppose we wish to estimate $\eta = \beta_1 - \beta_2$. Then $\mathbf{c} = \begin{bmatrix} 1 \\ -1 \end{bmatrix}$. A linear unbiased estimator $\eta^* = (\mathbf{d}, \mathbf{Y})$ must satisfy $(\mathbf{d}, \mathbf{X}\boldsymbol{\beta}) = \mathbf{d}'\mathbf{X}\boldsymbol{\beta} = \mathbf{c}'\boldsymbol{\beta}$ for all $\boldsymbol{\beta}$. Thus $\mathbf{d}'\mathbf{X} = \mathbf{c}'$, equivalently, $\mathbf{X}'\mathbf{d} = \mathbf{c}$. That is, $\mathbf{d}$ must have the "correct" inner products with the $\mathbf{x}_j$'s, with $(\mathbf{d}, \mathbf{x}_j) = \mathbf{c}_j$ for $j = 1, 2$. In this case $\mathbf{d}$ must satisfy

$$(\mathbf{x}_1, \mathbf{d}) = d_1 + d_3 + d_4 = 1$$

$$(\mathbf{x}_2, \mathbf{d}) = d_2 + d_3 + d_4 = -1$$

One such vector is $\mathbf{d} = 2(3, 1, -1, -1)'$, for example. The estimator $(\mathbf{d}, \mathbf{Y})$ has variance $\|\mathbf{d}\|^2\sigma^2 = 12\sigma^2$. The BLUE for η is $\hat{\eta} = \mathbf{c}'(\mathbf{X}'\mathbf{X})^{-1}\mathbf{X}'\mathbf{Y} = (\mathbf{a}, \mathbf{Y})$ for

$$\mathbf{a} = \mathbf{X}(\mathbf{X}'\mathbf{X})^{-1}\mathbf{c} = \begin{bmatrix} 1 & 0 \\ 0 & 1 \\ 1 & 1 \\ 1 & 1 \end{bmatrix} \frac{1}{5}\begin{pmatrix} 3 & -2 \\ -2 & 3 \end{pmatrix}\begin{pmatrix} 1 \\ -1 \end{pmatrix} = (1, -1, 0, 0)',$$

so that $\hat{\beta}_1 - \hat{\beta}_2 = Y_1 - Y_2$, which has variance $\|\mathbf{a}\|^2\sigma^2 = 2\sigma^2$.

Note that $\mathbf{d} - \mathbf{a} = (2, 2, -1, -1)'$ is orthogonal to V. All unbiased estimators of η have the form $\hat{\eta} + (\mathbf{b}, \mathbf{Y})$ for $\mathbf{b} \perp V$.

The vector $\mathbf{a}$ is $p(\mathbf{d} \mid V)$ for $V = \mathscr{L}(x_1, \mathbf{x}_2)$ for any $\mathbf{d}$ such that $(\mathbf{d}, \mathbf{Y})$ is an unbiased estimator of $\eta = \beta_1 - \beta_2$. To see this note that $\mathbf{P}_V\mathbf{d} = \mathbf{X}(\mathbf{X}'\mathbf{X})^{-1}\mathbf{X}'\mathbf{d} = \mathbf{X}(\mathbf{X}'\mathbf{X})^{-1}\mathbf{c} = \mathbf{a}$.

Example 3.3.2: Consider the enrollment totals in Table 3.3.1 and Figure 3.4 for minority students at Michigan State for the years 1981 to 1990:

Table 3.3.1

	1981	1982	1983	1984	1985	1986	1987	1988	1989	1990
Men	1,357	1,393	1,493	1,477	1,528	1,539	1,661	1,793	1,919	2,012
Women	1,867	1,930	1,937	2,038	2,117	2,199	2,212	2,464	2,625	2,798

Let Y_{1i} and Y_{2i} be the logs of enrollments for men and women in year $(1980 + i)$ for $i = 1, \ldots, 10$. We have chosen to make a log transformation because the following model seems more appropriate for logs than for enrollments themselves. Let $x_{1i} = x_{2i} = (\text{year} - 1980) = i$.

Suppose $Y_{1i} = \beta_1 + \beta_3 x_{1i} + \varepsilon_{1i}$ for $i = 1, \ldots, n_1 = 10$ and $Y_{2i} = \beta_2 + \beta_3 x_{2i} + \varepsilon_{2i}$ for $i = 1, \ldots, n_2 = 10$. That is, we suppose that the Y_{1i} satisfy one

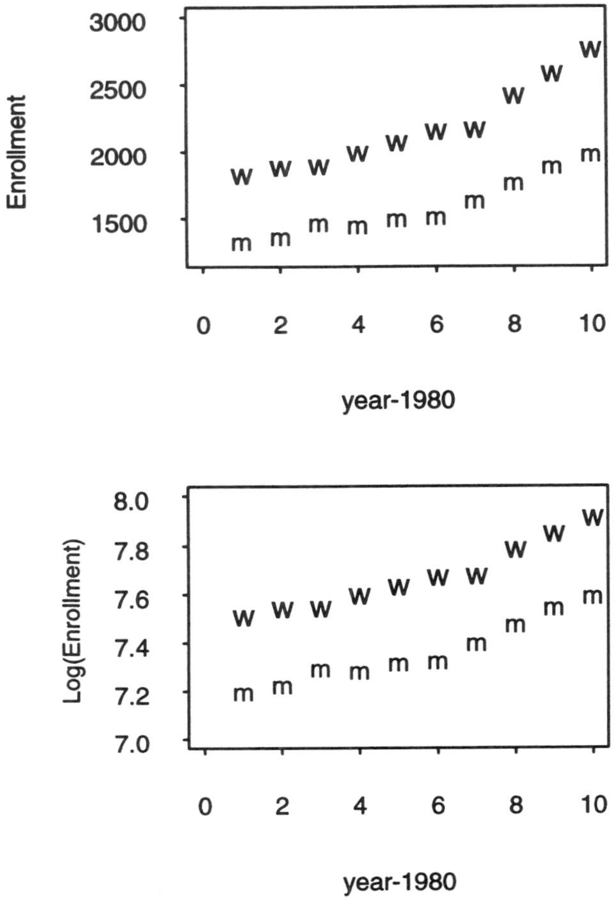

FIGURE 3.4 Minority enrollments of men and women, 1981–1990.

linear regression model and the Y_{2i} another, with the slopes being the same. By defining $\mathbf{Y}$ to be the 2-row array with Y_{ji} in the ith place in the jth row ($j = 1, 2$), $\mathbf{x}$ and $\boldsymbol{\varepsilon}$ similarly, and $\mathbf{J}_1$ and $\mathbf{J}_2$ the indicators of the rows, we can write

$$\mathbf{Y} = \beta_1 \mathbf{J}_1 + \beta_2 \mathbf{J}_2 + \beta_3 \mathbf{x} + \boldsymbol{\varepsilon}$$

We suppose that the components of $\boldsymbol{\varepsilon}$ are uncorrelated random variables with equal variances. It seems doubtful that correlations are zero, since students enrolled one year have a tendency to be enrolled the next, causing positive correlations, but let us proceed as if the model is at least a reasonable approximation.

Suppose we wish to estimate $\eta = \beta_1$, the intercept of the first regression line. It might seem that we should use the Y_{1i}'s only. However, the Gauss–Markov (G–M) Theorem states that we should use $\hat{\beta}_1 = \bar{Y}_1 - \hat{\beta}_3 \bar{x}_1$, which depends on the Y_{2i}'s as well. Similarly, the common slope β_3 of the two regression lines could be estimated unbiasedly using only the Y_{1i}'s. However, assuming the model holds, so that the regression slopes *are* the same, the G–M estimator $\hat{\beta}_3$ has smaller variance.

Example 3.3.3: For the model $\mathbf{Y} = \beta \mathbf{x} + \boldsymbol{\varepsilon}$ (regression through the origin) linear unbiased estimators of β have the form $\beta^* = (\mathbf{d}, \mathbf{Y})$ for $\mathbf{d} = \mathbf{a} + \mathbf{h}$, where $\mathbf{a} = \mathbf{x}/\|\mathbf{x}\|^2$, $\mathbf{h} \perp \mathbf{x}$.

Problem 3.3.1: Let $\Omega = R_4$, $\mathbf{x}_1 = (1, 1, 0, 0)'$, $\mathbf{x}_2 = (0, 0, 1, 1)'$, $V = \mathscr{L}(\mathbf{x}_1, \mathbf{x}_2)$, and $\eta = 2\beta_1 - \beta_2$.

(a) Find $\mathbf{a}$ so that $(\mathbf{a}, \mathbf{Y})$ is the BLUE.

(b) Find $\mathbf{d}$ so that $\eta^* = (\mathbf{d}, \mathbf{Y})$ is another unbiased linear estimator of η. Show that $p(\mathbf{d} \mid V) = \mathbf{a}$, find $\text{Var}(\eta^*)$, and show that $\text{Var}(\eta^*) - \text{Var}(\hat{\eta}) > 0$.

Problem 3.3.2: Let Ω be the space of arrays $\begin{bmatrix} y_{11} & y_{12} & y_{13} \\ y_{21} & y_{22} & y_{23} \\ & y_{32} & \end{bmatrix}$. Let $\mathbf{C}_j$ be the indicator of column j, let $\mathbf{Y} = \mu_1 \mathbf{C}_1 + \mu_2 \mathbf{C}_2 + \mu_3 \mathbf{C}_3 + \boldsymbol{\varepsilon}$, where $E(\boldsymbol{\varepsilon}) = \mathbf{0}$, $D[\boldsymbol{\varepsilon}] = \sigma^2 \mathbf{I}_7$.

(a) Find the BLUE for $\eta = 2\mu_1 - \mu_2 - \mu_3$, and determine its variance.

(b) Suggest another unbiased linear estimator of η, and show that it has larger variance.

Problem 3.3.3: In testing the "bounce factor" in baseballs, balls are dropped onto concrete from a height of x feet. The height Y in feet to which the ball bounces is then recorded by taking a picture against a linear scale. The following model seems appropriate:

Suppose that for each x, $E(Y \mid x) = \beta x$ and $\text{Var}(Y \mid x) = \sigma^2 g(x)$ for β and σ^2 unknown parameters, and $g(x)$ a known function of x. Suppose that (x_i, Y_i) are observed independently for $i = 1, 2, \ldots, n$. Consider the x_i to be constants, all nonzero. Define $\mathbf{Y} = (Y_1, \ldots, Y_n)'$.

(a) Consider estimators $(\mathbf{a}, \mathbf{Y}) = \hat{\beta}_a$ of β. What condition must $\mathbf{a}$ satisfy in order that $\hat{\beta}_a$ be an unbiased estimator of β?

(b) Show that $\hat{\beta}_a$ has minimum variance among all linear unbiased estimators when $\mathbf{a}$ is a multiple of the vector $\mathbf{G}^{-1}\mathbf{x}$, where $\mathbf{G} = \text{diag}(g(x_1), \ldots, g(x_n))$. *Hint*: Let $Z_i = Y_i/c_i$, where c_i is chosen so that the resulting vector $\mathbf{Z}$ satisfies the hypothesis of the Gauss–Markov Theorem. What is the optimum choice for $\mathbf{a}$?

(c) Let $\hat{\beta}$ be the estimator corresponding to this optimum $\mathbf{a}$. Find $\text{Var}(\hat{\beta})$.

(d) Give formulas for $\hat{\beta}$ for the cases (1) $g(x) \equiv 1$, (2) $g(x) \equiv x$, (3) $g(x) \equiv x^2$.

(e) For each of the cases in (d) find $\text{Var}(\hat{\beta})$.

(f) How would you estimate σ^2? *Hint*: Use the Z_i.

(g) Find $\hat{\beta}$ for each of the cases in (d), and estimate $\text{Var}(\hat{\beta})$ for the following (x, y) pairs (3, 2.2), 5, 3.2), (10, 7.3), (15, 10.0).

Problem 3.3.4: Let $Y_1, \ldots, Y_n$ be a random sample from the double exponential distribution, with density $f(x; \theta) = \dfrac{1}{2\eta} e^{-|x-\theta|/\eta}$, for all real x and θ, $\eta > 0$. (mean θ, variance $2\eta^2$). For the case $n = 3$ this might be a reasonable model for the distribution of the times recorded on three hand watches in the timing of swimmers or runners.

(a) Show that the vector of Y's satisfies a linear model.

(b) What is the BLUE $\hat{\theta}$ for θ?

(c) Find the maximum likelihood estimator $\hat{\theta}_M$ of θ. *Hint*: $G(c) = \sum_{j=1}^{n} |x_i - c|$ is minimized by $c = \text{median}(x_1, \ldots, x_n)$.

(d) Since $\hat{\theta}_M$ is symmetrically distributed about θ, it is an unbiased estimator of θ. Though it is not possible to write a simple expression for the variance $\hat{\theta}_M$, we can give an approximation for large n: $\text{Var}(\hat{\theta}_M) \sim 1/[4nf^2(\theta; \theta)]$. Show that this is smaller than the variance of $\hat{\theta}$. Why is $\hat{\theta}$ not a better estimator of θ in this case? Has the Gauss–Markov Theorem failed to hold?

3.4 THE GAUSS–MARKOV THEOREM FOR THE GENERAL CASE

For simplicity we have supposed that the vectors $\mathbf{x}_1, \ldots, \mathbf{x}_k$ spanning V are linearly independent. For purposes of estimation of θ this is not really a restriction in the model, since enough $\mathbf{x}_j$'s may always be dropped so that this is the case, and any $\boldsymbol{\theta} \in V$ may then be expressed as a unique linear combination of the remaining $\mathbf{x}$-vectors. There are occasions, however, when interpretations may be more easily made if the $\mathbf{x}$-vectors are linearly dependent.

For example, consider the usual one-way layout with observation Y_{ij} for $i = 1, \ldots, n_j$ and $j = 1, \ldots, k$, $Y_{ij} \sim N(\mu_j, \sigma^2)$, $n = \sum_1^k n_j$, Y_{ij}'s independent.

Define $\mu = \frac{1}{k}\sum_j^k \mu_j$, $\alpha_j = \mu_j - \mu$, $\varepsilon_{ij} = Y_{ij} - \mu_j$. Then $Y_{ij} = \mu + \alpha_j + \varepsilon_{ij}$. Writing this in vector form with k columns, we have $\mathbf{Y} = \mu\mathbf{J} + \sum_1^k \alpha_j \mathbf{C}_j + \boldsymbol{\varepsilon}$, where $\mathbf{C}_j$ is the indicator of column j, and $\mathbf{J} = \sum_j \mathbf{C}_j$. Thus, $\boldsymbol{\theta} = \sum_j \mu_j \mathbf{C}_j = \mu\mathbf{J} + \sum_j \alpha_j \mathbf{C}_i$ lies in $V = \mathscr{L}(\mathbf{C}_1, \ldots, \mathbf{C}_k)$, which has dimension k, but has been expressed as a linear combination of $(k + 1)$ linearly dependent vectors.

In general, suppose $\mathbf{X}$ has rank $r < k$, so that the columns of $\mathbf{X}$ span an r-dimensional subspace. In this case the null space of $\mathbf{X}$ (the collection of vectors b such that $\mathbf{Xb} = \mathbf{0}$) has dimension $k - r > 0$, and the set $W_{\boldsymbol{\theta}} = \{\boldsymbol{\beta} \mid \mathbf{X}\boldsymbol{\beta} = \boldsymbol{\theta}\}$ for fixed $\boldsymbol{\theta} \in W_{\boldsymbol{\theta}}$ is a hyperplane in k-space. In order to have a unique representation of $\boldsymbol{\theta}$, $\boldsymbol{\beta}$ is often required to satisfy some additional linear restriction of the form $\mathbf{H}\boldsymbol{\beta} = \mathbf{0}$, where $\mathbf{H}$ is $(k - r) \times k$ and the matrix $\mathbf{X_H} = \begin{pmatrix} \mathbf{X} \\ \mathbf{H} \end{pmatrix}$ has rank k. We could, for example, require that $\boldsymbol{\beta}$ lie in the row space of $\mathbf{X}$, in which case $\mathbf{H}$ could be any collection of $k - r$ linearly independent vectors such that $\mathbf{X_H}$ has rank k. The same linear restrictions placed on $\boldsymbol{\beta}$ may also be placed on $\hat{\boldsymbol{\beta}}$. Without these restrictions $\hat{\boldsymbol{\beta}}$ is not defined uniquely, since any $\hat{\boldsymbol{\beta}} \in W_{\hat{\mathbf{Y}}}$ implies $\mathbf{X}\hat{\boldsymbol{\beta}} = \hat{\mathbf{Y}}$. A least squares estimator of $\boldsymbol{\beta}$ is any function of $\mathbf{Y}$ satisfying $\mathbf{X}\hat{\boldsymbol{\beta}} = \hat{\mathbf{Y}}$ (any $\hat{\boldsymbol{\beta}} \in W_{\hat{\mathbf{Y}}}$).

If, in this non-full-rank case, $\boldsymbol{\beta}$ is allowed to range over all of R^k, then not all components of $\boldsymbol{\beta}$ may be estimated unbiasedly (and linearly). Consider the one-way layout example above. Can we find an unbiased linear estimator $T = \sum_{ij} a_{ij} Y_{ij}$ of α_1? $E(T) = \sum_{ij} a_{ij}\mu_j = \sum_j a_{.j}\mu_j = \mu a_{..} + \sum_j a_{.j}\alpha_j$, where the dot subscript indicates summation over the subscript replaced. This is $E(T) = \alpha_1$ for all parameter values only if $a_{.1} = 1$, $a_{.j} = 0$ for $j > 1$ and $a_{..} = 0$, which is impossible. Thus, α_1 has no unbiased linear estimator if the parameter vector is unrestricted. If the parameter vector is restricted so that $\sum_{j=1}^k \alpha_j = 0$, then α_1 does have the unbiased linear estimator $\bar{Y}_{.1} - \bar{Y}_{..}$

Definition 3.4.1: Let $\mathbf{c} = (c_1, \ldots, c_k)'$ be a vector of constants. The parameter $\eta = (\mathbf{c}, \boldsymbol{\beta}) = \sum_j c_j \beta_j$ is *estimable* if there exists a vector $\mathbf{a}$ in n-space such that $E(\mathbf{a}, \mathbf{Y}) = (\mathbf{c}, \boldsymbol{\beta})$ for all $\boldsymbol{\beta} \in R_k$.

Thus, $\eta = (\mathbf{c}, \boldsymbol{\beta})$ is estimable if there exists $\mathbf{a}$ such that $E(\mathbf{a}, \mathbf{Y}) = (\mathbf{a}, \boldsymbol{\theta}) = \mathbf{a}'\mathbf{X}\beta = \mathbf{c}'\boldsymbol{\beta}$ for all $\boldsymbol{\beta} \in R_k$. This is true if and only if there exists $\mathbf{a}$ such that $\mathbf{X}'\mathbf{a} = \mathbf{c}$, i.e., $\mathbf{c}$ lies in the row space of $\mathbf{X}$.

If $\mathbf{X}'\mathbf{a} = \mathbf{c}$ and V is the column space of $\mathbf{X}$, then for $\mathbf{a}_V = p(\mathbf{a} \mid V)$, $\mathbf{X}'\mathbf{a}_V = \mathbf{c}$, so that we can always take $\mathbf{a} \in V$ if $\eta = (\mathbf{c}, \boldsymbol{\beta})$ is estimable, and $(\mathbf{a}_V, \mathbf{Y})$ has smaller variance $\|\mathbf{a}_V\|^2\sigma^2$ than any other linear unbiased estimator.

Example 3.4.1: Let $\mathbf{x}_1 = (1, 1, 1, 1)'$, $\mathbf{x}_2 = (1, 1, 1, 1)'$, $\mathbf{x}_3 = 3\mathbf{x}_1 - 2\mathbf{x}_2 = (1, 1, 1, 3)'$. Then $\mathbf{X} = (\mathbf{x}_1, \mathbf{x}_2, \mathbf{x}_3)$ has rank 2, and for $\mathbf{c} = \mathbf{X}'\begin{bmatrix} 1 \\ 1 \\ 1 \\ 2 \end{bmatrix} = \begin{bmatrix} 5 \\ 3 \\ 9 \end{bmatrix}$, $(\mathbf{c}, \boldsymbol{\beta}) = 5\beta_1 + 3\beta_2 + 9\beta_3$ is estimable. The parameters β_1 and $\beta_1 - \beta_2$ are not estimable, since $3c_1 - 2c_2$ must equal c_3 (why?).

The Gauss–Markov Theorem (General Case): Let $\mathbf{Y} = \boldsymbol{\beta} + \boldsymbol{\varepsilon}$ for $\boldsymbol{\theta} = \sum_1^k \beta_j \mathbf{x}_j = \mathbf{X}\boldsymbol{\beta} =$ and $E(\boldsymbol{\varepsilon}) = \mathbf{0}$. $D[\boldsymbol{\varepsilon}] = \sigma^2 \mathbf{I}_n$. Let $\eta = \mathbf{c}'\boldsymbol{\beta}$ be estimable. Let $V = \mathscr{L}(\mathbf{x}_1, \ldots, \mathbf{x}_k)$ and $\hat{\mathbf{Y}} = p(\mathbf{Y} \mid V)$. Let $\hat{\boldsymbol{\beta}}$ be any least squares estimator of $\boldsymbol{\beta}$. That is, $\mathbf{X}\hat{\boldsymbol{\beta}} = \hat{\mathbf{Y}}$ for every $\mathbf{Y}$. Then

(1) $\hat{\eta} = \mathbf{c}'\hat{\boldsymbol{\beta}}$ is a linear unbiased estimator of η.
(2) For any other linear unbiased estimator η^* of η, $\mathrm{Var}(\hat{\eta}) \le \mathrm{Var}(\eta^*)$ with equality only if $\hat{\eta} = \eta^*$ for all $\mathbf{Y}$.

Comment: An estimator $\hat{\boldsymbol{\beta}}$ is called a least squares estimator if $\mathbf{X}\hat{\boldsymbol{\beta}} = \hat{\mathbf{Y}}$ for every $\mathbf{Y}$. The estimator $\hat{\boldsymbol{\beta}}$ need not be linear. In fact, $\hat{\boldsymbol{\beta}}$ is a function of $\mathbf{Y}$ which chooses one member of $W_{\hat{\mathbf{Y}}} = \{b \mid \mathbf{X}\mathbf{b} = \hat{\mathbf{Y}}\}$. This choice need not be linear. In the case of one-way analysis of variance we might, for example, choose $\hat{\mu} = \bar{Y}_{..}$, $\hat{\alpha}_i = \bar{Y}_{i.} - \bar{Y}_{..}$ whenever all components of $\mathbf{Y}$ exceed 7, but $\hat{\mu} = 0$, $\hat{\alpha}_i = \bar{Y}_{i.}$, otherwise. Clearly $\hat{\boldsymbol{\beta}} = (\hat{\mu}, \hat{\alpha}_1, \ldots, \hat{\alpha}_k)'$ is not linear in $\mathbf{Y}$, but does satisfy $\mathbf{X}\hat{\boldsymbol{\beta}} = \hat{\mathbf{Y}}$.

***Proof*:** Since $\eta = \mathbf{c}'\boldsymbol{\beta} =$ is estimable, there exists a vector $\mathbf{a}$ such that $E(\mathbf{a}, \mathbf{Y}) = (\mathbf{a}, \boldsymbol{\theta}) = \mathbf{a}'\mathbf{X}\boldsymbol{\beta} = \mathbf{c}'\boldsymbol{\beta}$ for all $\boldsymbol{\beta}$. Since $\mathbf{a}$ and $p(\mathbf{a} \mid V)$ have the same inner products with the columns of $\mathbf{X}$ we may take $\mathbf{a} \in V$. Since the equality holds for all $\boldsymbol{\beta}$, we conclude that $\mathbf{c} = \mathbf{X}'\mathbf{a}$. Thus, $\hat{\eta} = \mathbf{c}'\hat{\boldsymbol{\beta}} = \mathbf{a}'\mathbf{X}\hat{\boldsymbol{\beta}} = \mathbf{a}'\mathbf{Y} = \mathbf{a}'\mathbf{Y}$, a linear function of $\mathbf{Y}$.

Let $\eta^* = (\mathbf{d}, \mathbf{Y}) = \mathbf{d}'\mathbf{Y}$. Since $E(\eta^*) = \mathbf{d}'\boldsymbol{\theta} = \mathbf{d}'\mathbf{X}\boldsymbol{\beta} = \mathbf{c}'\boldsymbol{\beta} = \eta$ for all $\boldsymbol{\beta}$ only if $\mathbf{c} = \mathbf{X}'\mathbf{d}$, η^* is unbiased for η only if $\mathbf{c} = \mathbf{X}'\mathbf{d}$. Then $\mathbf{d}$ and $\mathbf{a}$ have the same inner products with all vectors in V. Since $\mathbf{a} \in V$, $\mathbf{a} = p(\mathbf{d} \mid V)$. Therefore $\eta^* = (\mathbf{a}, \mathbf{Y}) + (\mathbf{d} - \mathbf{a}, \mathbf{Y}) = \hat{\eta} + (\mathbf{d} - \mathbf{a}, \mathbf{Y})$ and $\hat{\eta}$ and $(\mathbf{d} - \mathbf{a}, \mathbf{Y})$ have covariance 0. Thus, $\mathrm{Var}(\eta^*) = \mathrm{Var}(\hat{\eta}) + \sigma^2 \|\mathbf{d} - \mathbf{a}\|^2$, which is minimum for $\mathbf{d} = \mathbf{a}$. □

Problem 3.4.1: Let $\mathbf{x}_1 = (1, 1, 1, 1)'$, $\mathbf{x}_2 = (1, 0, 1, 0)'$, $\mathbf{x}_3 = 3\mathbf{x}_1 - \mathbf{x}_2$.

(a) Find conditions on $\mathbf{c}' = (c_1, c_2, c_3)$ so that $\eta = c_1\beta_1 + c_2\beta_2 + c_3\beta_3$ is estimable for the linear model $\mathbf{Y} = \beta_1\mathbf{x}_1 + \beta_2\mathbf{x}_2 + \boldsymbol{\beta}_3\mathbf{x}_3 + \boldsymbol{\varepsilon}$.

(b) Show that $\eta = 3\beta_1 - \beta_2 - \beta_3$ is estimable, find $\mathbf{a}$ such that $\hat{\eta} = (\mathbf{a}, \mathbf{Y})$ is the BLUE for η, and find another unbiased estimator η^* of η. Show that $\mathrm{Var}(\hat{\eta}) < \mathrm{Var}(\eta^*)$.

Problem 3.4.2: For the one-way layout example above, find conditions on $c_0, c_1, \ldots, c_k$ such that $c_0\mu + c_1\alpha_1 + \cdots + c_k\alpha_k = \eta$ is estimable.

Problem 3.4.3: For the one-way layout with $k = 1$, $n_1 = 2$, $n_2 = 3$, $n_3 = 1$, find two vectors $\mathbf{a}_1$ and $\mathbf{a}_2$ such that $T_1 = (\mathbf{a}_1, \mathbf{Y})$ and $T_2 = (\mathbf{a}_2, \mathbf{Y})$ are both unbiased estimators of $\eta = \alpha_1 - \alpha_2$, T_1 is the Gauss–Markov estimator, and $\text{Var}(T_2) > \text{Var}(T_1)$. Also show that $\mathbf{a}_1 = p(\mathbf{a}_2 \mid V)$, where V is the subspace spanned by the indicators of the columns.

Problem 3.4.4: Let Ω be the collection of 2×3 tables with elements $\mathbf{y} = (y_{ij})$. Suppose that $\mathbf{Y} = (Y_{ij})$, with $Y_{ij} = \mu + \alpha_i + \beta_j + \varepsilon_{ij}$, for $\varepsilon_{ij} \sim N(0, \sigma^2)$.

(a) Write the model in vector form.

(b) Find conditions on c_1, c_2, c_3 such that $\eta = c_1\beta_1 + c_2\beta_2 + c_3\beta_3$ is estimable. Show that β_1 is not estimable, but $\beta_1 - \beta_2$ is. Give two unbiased estimators of $\beta_1 - \beta_2$, one of which is the Gauss–Markov estimator.

3.5 INTERPRETATION OF REGRESSION COEFFICIENTS

Let $\mathbf{Y} = \sum_{j=1}^{k} \beta_j \mathbf{x}_j + \boldsymbol{\varepsilon}$ with $\mathbf{x}_1, \ldots, \mathbf{x}_k$ linearly independent. Define $V_{k-1} = \mathscr{L}(\mathbf{x}_1, \ldots, \mathbf{x}_{k-1})$, $\hat{\mathbf{x}}_k = p(\mathbf{x}_k \mid V_{k-1})$ and $\mathbf{x}_k^\perp = \mathbf{x}_k - \hat{\mathbf{x}}_k$. Then $\mathbf{x}_k^\perp \perp V_{k-1}$, and $\|\mathbf{x}_k^\perp\|^2 = (\mathbf{x}_k^\perp, \mathbf{x}_k^\perp) = (\mathbf{x}_k^\perp, \mathbf{x}_k) - (\mathbf{x}_k^\perp, \hat{\mathbf{x}}_k) = (\mathbf{x}_k^\perp, \mathbf{x}_k)$. $\mathbf{x}_k^\perp$ is the part of $\mathbf{x}_k$ which is orthogonal to the other $\mathbf{x}_j$, or in more intuitive language, the part of $\mathbf{x}_k$ which measures something different (in a linear sense) than the other $\mathbf{x}_j$. $\mathbf{x}_k^\perp$ is sometimes called the *signal* part of $\mathbf{x}_k$. In the case of simple linear regression with $\mathbf{Y} = \beta_0\mathbf{J} + \beta_1\mathbf{x}$, $\mathbf{x}^\perp = \mathbf{x} - p(\mathbf{x} \mid \mathbf{J})$ is the vector of deviations, with ith component $x_i - \bar{x}$. We have called this $\mathbf{x}^*$ in the past.

Then $(\boldsymbol{\theta}, \mathbf{x}_k^\perp) = \sum_{j=1}^{k} \beta_j(\mathbf{x}_j, \mathbf{x}_k^\perp) = \beta_k(\mathbf{x}_k, \mathbf{x}_k^\perp) = \beta_k\|\mathbf{x}_k^\perp\|^2$, so that $\beta_k = \dfrac{(\boldsymbol{\theta}, \mathbf{x}_k^\perp)}{\|\mathbf{x}_k^\perp\|^2}$. Similarly, for $\hat{\mathbf{Y}} = p(\mathbf{Y} \mid V)$, $(\mathbf{Y}, \mathbf{x}_k^\perp) = (\hat{\mathbf{Y}}, \mathbf{x}_k^\perp) = \sum_{j=1}^{k} \hat{\beta}_j(\mathbf{x}_j, \mathbf{x}_k^\perp) = \hat{\beta}_k(\mathbf{x}_k, \mathbf{x}_k^\perp) = \hat{\beta}_k\|\mathbf{x}_k^\perp\|^2$, so that $\hat{\beta}_k = (\mathbf{Y}, \mathbf{x}_k^\perp)/\|\mathbf{x}_k^\perp\|^2$. Thus, β_k is determined solely by the relationship between $\boldsymbol{\theta}$ and $\mathbf{x}_k^\perp$. Similarly, $\hat{\beta}_k$ is determined solely by the relationship between $\mathbf{Y}$ and $\mathbf{x}_k^\perp$. Thus, for example, in any multiple regression analysis which includes the vector $\mathbf{J}$ of all ones as an $\mathbf{x}$-vector, the β's and $\hat{\beta}$'s corresponding to other vectors are not affected by adding the same constant to all elements of those vectors.

Define $\boldsymbol{\theta}_k = p(\boldsymbol{\theta} \mid V_{k-1})$. Then $\boldsymbol{\theta} = \boldsymbol{\theta}_k + \beta_k\mathbf{x}_k^\perp$ and $\|\boldsymbol{\theta}\|^2 = \|\boldsymbol{\theta}_k\|^2 + \beta_k^2\|\mathbf{x}_k^\perp\|^2 = \|\boldsymbol{\theta}_k\|^2 + (\boldsymbol{\theta}, \mathbf{x}_k^\perp)^2/\|\mathbf{x}_k^\perp\|^2$. Similarly, for $\hat{\mathbf{Y}}_k = p(\mathbf{Y} \mid V_{k-1})$, $\hat{\mathbf{Y}} = \hat{\mathbf{Y}}_k + \hat{\beta}_k\mathbf{x}_k^\perp$ and $\|\hat{\mathbf{Y}}\|^2 = \|\hat{\mathbf{Y}}_k\|^2 + \hat{\beta}_k^2\|\mathbf{x}_k^\perp\|^2 = \|\hat{\mathbf{Y}}_k\|^2 + (\mathbf{Y}, \mathbf{x}_k^\perp)^2/\|\mathbf{x}_k^\perp\|^2$.

We can express the variance of $\hat{\beta}_k$ in terms of $\mathbf{x}_k^\perp$.

$$\text{Var}(\hat{\beta}_k) = \frac{1}{\|\mathbf{x}_k^\perp\|^4}\,\text{Var}((\mathbf{x}_k^\perp, \mathbf{Y})) = \frac{\sigma^2}{\|\mathbf{x}_k^\perp\|^2}.$$

This is a useful formula in that it provides insight into the effects that "collinearity", the "near" linear relationship among the independent variables, has on the precision of the estimators of the regression coefficients. It provides a warning: independent variables which are "almost" linear combinations of other independent variables will have coefficient estimators with large variances.

More generally, if for any j, $\mathbf{x}_j^{\perp}$ is the part of $\mathbf{x}_j$ orthogonal to the other $\mathbf{x}_i$, then

$$\operatorname{cov}(\hat{\beta}_i, \hat{\beta}_j) = \frac{1}{\|\mathbf{x}_i^{\perp}\|^2 \|\mathbf{x}_j^{\perp}\|^2} \operatorname{cov}((\mathbf{Y}, \mathbf{x}_i^{\perp})), ((\mathbf{Y}, \mathbf{x}_j^{\perp})) = \sigma^2(\mathbf{x}_i^{\perp}, \mathbf{x}_j^{\perp})/\|\mathbf{x}_i^{\perp}\|^2\|\mathbf{x}_j^{\perp}\|^2.$$

Since we already knew that the covariance matrix for $\hat{\beta}$ is $\sigma^2(\mathbf{X}'\mathbf{X})^{-1}$, we have discovered that the ij element of $(\mathbf{X}'\mathbf{X})^{-1}$ is $(\mathbf{x}_i^{\perp}, \mathbf{x}_j^{\perp})/\|\mathbf{x}_i^{\perp}\|^2\|\mathbf{x}_j^{\perp}\|^2$.

The cosine of the angle ω between two vectors $\mathbf{u}$ and $\mathbf{v}$ is defined by $\cos\omega = (\mathbf{u}, \mathbf{v})/(\|\mathbf{u}\|\ \|\mathbf{v}\|)$. Thus,

$$\operatorname{cov}(\hat{\beta}_i, \hat{\beta}_j) = \sigma^2(\cos\omega_{ij})/\|\mathbf{x}_i^{\perp}\|\ \|\mathbf{x}_j^{\perp}\|$$

where ω_{ij} is the angle between $\mathbf{x}_i^{\perp}$ and $\mathbf{x}_j^{\perp}$. The correlation between $\hat{\beta}_i$ and $\hat{\beta}_j$ is therefore $\rho(\hat{\beta}_i, \hat{\beta}_j) = \cos\omega_{ij}$.

Example 3.5.1: Let $\Omega = R_5$, $\mathbf{x}_1 = (1, 1, 1, 1, 1)'$, $\mathbf{x}_2 = (1, 0, 1, 0, 1)'$, and $\mathbf{x}_3 = (1, 1, 1, 0, 0)'$. Let $V = \mathscr{L}(\mathbf{x}_1, \mathbf{x}_2, \mathbf{x}_3)$ and $V_2 = \mathscr{L}(\mathbf{x}_1, \mathbf{x}_2)$. V_2 is spanned by $\mathbf{x}_2$ and $\mathbf{w} = \mathbf{x}_1 - \mathbf{x}_2$, and $\mathbf{x}_2 \perp \mathbf{w}$. Thus, $\mathbf{x}_3 \equiv p(\mathbf{x}_3 | V_2) = (2/3)\mathbf{x}_2 + (1/2)\mathbf{w} = (1/6)(3\mathbf{x}_1 + \mathbf{x}_2) = (1/6)(4, 3, 4, 3, 4)'$, so $\mathbf{x}_3^{\perp} = \mathbf{x}_3 - \hat{\mathbf{x}}_3 = (1/6)(2, 3, 2, -3, -4)'$. Notice that $\mathbf{x}_3^{\perp} \perp V_2$ and $\|\mathbf{x}_3^{\perp}\|^2 = 7/6$.

For $\mathbf{Y} = (3, 4, 3, 3, 9)'$, $\hat{\beta}_3 = (\mathbf{Y}, \mathbf{x}_3^{\perp})/\|\mathbf{x}_3^{\perp}\|^2 = \dfrac{(-21/6)}{(7/6)} = -3$ and $\operatorname{Var}(\hat{\beta}_3) = \sigma^2/\|\mathbf{x}_3^{\perp}\|^2 = 6\sigma^2/7$.

More generally, suppose instead $\mathbf{x}_3 = \hat{\mathbf{x}}_3 + \alpha(2, 3, 2, -3, -4)'$. Then $\mathbf{x}_3^{\perp} = \alpha(2, 3, 2, -3, -4)'$, $\hat{\beta}_3 = (\mathbf{Y}, \mathbf{x}_3^{\perp})/\|\mathbf{x}_3^{\perp}\|^2 = -1/(2\alpha)$, and $\operatorname{Var}(\hat{\beta}_3) = \sigma^2/(42\alpha^2)$, so that for small α ("short $\mathbf{x}_3^{\perp}$"), $\operatorname{Var}(\hat{\beta}_3)$ is large.

Problem 3.5.1: Let $\mathbf{x}_1 = \begin{pmatrix} 1 \\ 0 \\ 0 \end{pmatrix}$, $\mathbf{u} = \begin{pmatrix} 0 \\ 1 \\ 0 \end{pmatrix}$, $\mathbf{x}_2 = \mathbf{x}_1 + \alpha\mathbf{u}$, $V = \mathscr{L}(\mathbf{x}_1, \mathbf{x}_2)$.

Find $\mathbf{x}_1^{\perp}$, $\mathbf{x}_2^{\perp}$, $\|\mathbf{x}_1^{\perp}\|^2$, $\|\mathbf{x}_2^{\perp}\|^2$, $(\mathbf{x}_1^{\perp}, \mathbf{x}_2^{\perp})$, $\operatorname{Var}(\hat{\beta}_1)$, $\operatorname{Var}(\hat{\beta}_2)$, $\rho(\hat{\beta}_1, \hat{\beta}_2)$. What happens to these variances and to the correlation as $\alpha \to 0$ or $\alpha \to +\infty$?

Problem 3.5.2: Consider Example 3.1.2, with $\mathbf{Y} = \beta_1\mathbf{w}_1 + \beta_2\mathbf{w}_2 + \beta_c\mathbf{x} + \boldsymbol{\varepsilon}$.
(a) Find $\mathbf{x}^{\perp}$, and use this to find nonmatrix expressions for $\hat{\beta}_c$ and $\operatorname{Var}(\hat{\beta}_c)$.
(b) Use $\hat{\beta}_c$ to give simple expressions for $\hat{\beta}_1$ and $\hat{\beta}_2$.

(c) Give a simple formula for the variance of the predicted yield $\hat{g}(1, x)$ on a one-acre plot with fertility level x, under experimental condition 1.

(d) For $\mathbf{Y} = \mathbf{y} = \begin{bmatrix} 91 & 107 \\ 101 & 139 \\ 124 & 115 \\ 132 & 119 \end{bmatrix}$, and $\mathbf{x} = \begin{bmatrix} 2 & 5 \\ 2 & 9 \\ 6 & 7 \\ 6 & 7 \end{bmatrix}$, find $\hat{\beta}_1$, $\hat{\beta}_2$, $\hat{\beta}_c$, $\hat{\mathbf{Y}}$, $\mathbf{e}$, S^2, and $S_{\hat{\beta}_c}$.

Problem 3.5.3: Let $V = \mathscr{L}(\mathbf{x}_1, \ldots, \mathbf{x}_k)$ have dimension k. Find simple formulae for the coefficients a_j in $p(\mathbf{y} \mid V) = \sum_j a_j \mathbf{x}_j^{\perp}$ and prove that $\mathscr{L}(\mathbf{x}_1^{\perp}, \ldots, \mathbf{x}_k^{\perp}) = V$.

Problem 3.5.4: Let $\mathbf{x}_1$, $\mathbf{x}_2$, $\mathbf{x}_3$ be linearly independent vectors in R_n, with $\mathbf{x}_1 = \mathbf{J}$, the vector of all ones. Suppose that $\mathbf{Y} = \beta_1\mathbf{x}_1 + \beta_2\mathbf{x}_2 + \beta_3\mathbf{x}_3 + \boldsymbol{\varepsilon}$, with $E(\boldsymbol{\varepsilon}) = \mathbf{0}$ and $D[\boldsymbol{\varepsilon}] = \sigma^2\mathbf{I}_n$. Define $\mathbf{x}_j^* = \mathbf{x}_j - p(\mathbf{x}_j \mid \mathbf{x}_1)$ for $j = 2, 3$. Then $r_{23} = (\mathbf{x}_2^*, \mathbf{x}_3^*)/[\|\mathbf{x}_2^*\|\,\|\mathbf{x}_3^*\|]$ is the correlation between $\mathbf{x}_2$ and $\mathbf{x}_3$. Show that

(a) $\mathbf{x}_2^{\perp} = \mathbf{x}_2^* - p(\mathbf{x}_2^* \mid \mathbf{x}_3^*)$ and $\mathbf{x}_3^{\perp} = \mathbf{x}_3^* - p(\mathbf{x}_3^* \mid \mathbf{x}_2^*)$. *Hint*: $\mathscr{L}(\mathbf{x}_1, \mathbf{x}_2^*) = \mathscr{L}(\mathbf{x}_1, \mathbf{x}_2)$.

(b) $\mathrm{Var}(\hat{\beta}_j) = \sigma^2/[\|\mathbf{x}_j^*\|^2(1 - r_{23}^2)]$ for $j = 2, 3$.

(c) $\rho(\hat{\beta}_2, \hat{\beta}_3) = -r_{23}$.

3.6 THE MULTIPLE CORRELATION COEFFICIENT

Definition 3.6.1: Let $\mathbf{y}$, $\mathbf{x}_1, \ldots, \mathbf{x}_k$ be elements in R_n. Let $V = \mathscr{L}(\mathbf{J}, \mathbf{x}_1, \ldots, \mathbf{x}_k)$ and let $\hat{\mathbf{y}} = p(\mathbf{y} \mid V)$. Let $\hat{\mathbf{y}}_0 = p(\mathbf{y} \mid \mathbf{J}) = \bar{y}\mathbf{J}$. Then the multiple correlation coefficient of $\mathbf{y}$ with $\mathbf{x}_1, \ldots, \mathbf{x}_k$ is $R = R_{\mathbf{y}.12\ldots k} = \dfrac{\|\hat{\mathbf{y}} - \hat{\mathbf{y}}_0\|}{\|\mathbf{y} - \hat{\mathbf{y}}_0\|}$.

Comments: (1) From Figure 3.5, since $(\mathbf{y} - \hat{\mathbf{y}}) \perp V$ and $(\hat{\mathbf{y}} - \hat{\mathbf{y}}_0) \in V$, $\|\mathbf{y} - \hat{\mathbf{y}}_0\|^2 = \|\hat{\mathbf{y}} - \hat{\mathbf{y}}_0\|^2 + \|\mathbf{y} - \hat{\mathbf{y}}\|^2$ by the Pythagorean Theorem.

$$\text{Total SSqs.} = \underset{\text{(about mean)}}{\text{Regression SSqs.}} + \text{Error SSqs.}$$

Thus

$$R^2 = \frac{\text{Regression SSqs.}}{\text{Total SSqs.}} = 1 - \frac{\text{Error SSqs.}}{\text{Total SSqs.}}$$

so that R^2 may be interpreted as the proportion of variation in $\mathbf{Y}$ which is explained by linear regression on $\mathbf{x}_1, \ldots, \mathbf{x}_k$.

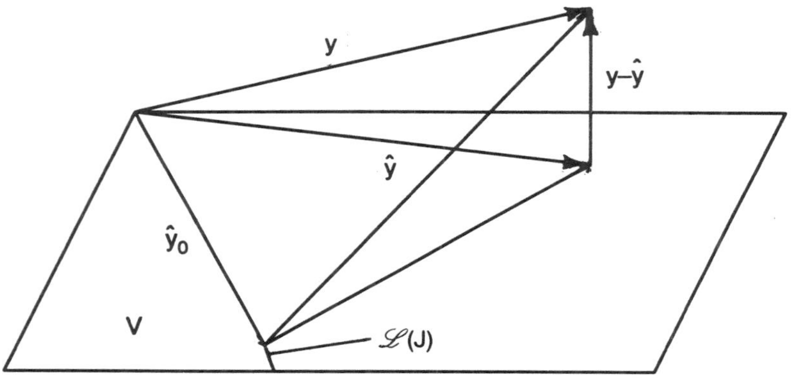

FIGURE 3.5 The multiple regression coefficient R, where $R^2 = \|\hat{\mathbf{y}} - \hat{\mathbf{y}}_0\|^2/\|\mathbf{y} - \hat{\mathbf{y}}_0\|^2$.

(2) Let $\mathbf{w} = c\mathbf{y} + d\mathbf{J}$ so that $w_i = cy_i + d$ for $i = 1, \ldots, n$, $c \neq 0$. Then $\hat{\mathbf{w}}_0 \equiv p(\mathbf{w} \mid \mathbf{J}) = c\mathbf{y}_0 + d\mathbf{J}$ and $\hat{\mathbf{w}} \equiv p(\mathbf{w} \mid V) = c\hat{\mathbf{y}} + d\mathbf{J}$. Thus,

$$\hat{\mathbf{w}} - \hat{\mathbf{w}}_0 = c(\hat{\mathbf{y}} - \hat{\mathbf{y}}_0), \qquad \mathbf{w} - \hat{\mathbf{w}} = c(\mathbf{y} - \hat{\mathbf{y}})$$

and $R_{\mathbf{w}\cdot 12\ldots k} = R_{\mathbf{y}\cdot 12\ldots k}$.

In addition, note that R is a function of $\mathbf{y}$ and the subspace V, not of the particular $\mathbf{x}_j$ vectors spanning V. For $\mathbf{J} \in V$ it follows that R remains unchanged when scale and location changes are made in the $\mathbf{x}_j$ vectors, more generally when $\mathbf{X}$ is replaced by $\mathbf{XC}$ for $\mathbf{C}$ $(k+1) \times (k+1)$ nonsingular.

(3) The ordinary correlation coefficient of $\mathbf{y}$ with $\hat{\mathbf{y}}$ is $r_{\mathbf{y}\hat{\mathbf{y}}} = \dfrac{(\mathbf{y} - \hat{y}_0, \hat{\mathbf{y}} - \mathbf{y}_0)}{\|\mathbf{y} - \hat{\mathbf{y}}_0\| \, \|\hat{\mathbf{y}} - \mathbf{y}_0\|}$. But $\mathbf{y} - \hat{\mathbf{y}}_0 = (\mathbf{y} - \hat{\mathbf{y}}) + (\hat{\mathbf{y}} - \hat{\mathbf{y}}_0)$ and $(\hat{\mathbf{y}} - \hat{\mathbf{y}}_0) \perp (\mathbf{y} - \hat{\mathbf{y}})$. Thus, $(\mathbf{y} - \hat{\mathbf{y}}_0, \hat{\mathbf{y}} - \hat{\mathbf{y}}_0) = (\hat{\mathbf{y}} - \hat{\mathbf{y}}_0, \hat{\mathbf{y}} - \hat{\mathbf{y}}_0) = \|\hat{\mathbf{y}} - \hat{\mathbf{y}}_0\|^2$ and $r_{\mathbf{y}\hat{\mathbf{y}}} = \dfrac{\|\hat{\mathbf{y}} - \hat{\mathbf{y}}_0\|^2}{\|\mathbf{y} - \hat{\mathbf{y}}_0\| \, \|\hat{\mathbf{y}} - \mathbf{y}_0\|} = \dfrac{\|\hat{\mathbf{y}} - \hat{\mathbf{y}}_0\|}{\|\mathbf{y} - \hat{\mathbf{y}}_0\|} = R$. The multiple correlation coefficient is the ordinary correlation coefficient between $\mathbf{y}$ and $\hat{\mathbf{y}}$. It must be nonnegative.

Contribution of $\mathbf{x}_k$ to the Reduction of Error Sum of Squares

Let $\mathbf{x}_1, \ldots, \mathbf{x}_k$ be linearly independent, $V_{k-1} = \mathscr{L}(\mathbf{x}_1, \ldots, \mathbf{x}_{k-1})$ and $V_k = \mathscr{L}(\mathbf{x}_1, \ldots, \mathbf{x}_k)$. Let $\hat{\mathbf{Y}}_k = p(\mathbf{Y} \mid V_k)$ and $\hat{\mathbf{Y}}_{k-1} = p(\mathbf{Y} \mid V_{k-1})$. The error sum of squares when the independent vectors are $\mathbf{x}_1, \ldots, \mathbf{x}_{k-1}$ is $\|\mathbf{Y} - \hat{\mathbf{Y}}_{k-1}\|^2 = \|\mathbf{Y}\|^2 - \|\hat{\mathbf{Y}}_{k-1}\|^2 \equiv \text{ESS}_{k-1}$. The error sum of squares when V_{k-1} is replaced by V_k is $\text{ESS}_k = \|\mathbf{Y} - \hat{\mathbf{Y}}_k\|^2 = \|\mathbf{Y}\|^2 - \|\hat{\mathbf{Y}}_k\|^2$. Since $\|\hat{\mathbf{Y}}_k\|^2 = \|\hat{\mathbf{Y}}_{k-1}\|^2 + \hat{\beta}_k^2\|\mathbf{x}_k^\perp\|^2$, the difference is $\hat{\beta}_k^2\|\mathbf{x}_k^\perp\|^2 = (\mathbf{Y}, \mathbf{x}_k^\perp)^2/\|\mathbf{x}_k^\perp\|^2$. That is, $\text{ESS}_k = \text{ESS}_{k-1} - (\mathbf{Y}, \mathbf{x}_k^\perp)^2/\|\mathbf{x}_k^\perp\|^2$.

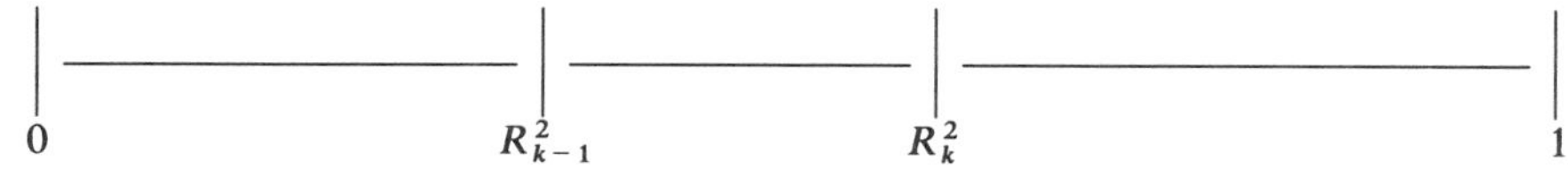

FIGURE 3.6

The t-statistic for testing H_0: $\beta_k = 0$ vs. H_1: $\beta_k \neq 0$ is $t = \hat{\beta}_k/\sqrt{S^2/\|\mathbf{x}_k^\perp\|^2} = \dfrac{\hat{\beta}_k\|\mathbf{x}_k^\perp\|}{S}$. Since $S^2 = \|\mathbf{Y} - \hat{\mathbf{Y}}_k\|^2/(n-k)$, we find that $t^2 = \dfrac{(\text{ESS}_{k-1} - \text{ESS}_k)}{\text{ESS}_k}(n-k)$ and $\text{ESS}_k = \text{ESS}_{k-1}\Big/\left(1 + \dfrac{t^2}{n-k}\right)$.

Let R_k^2 and R_{k-1}^2 be the squares of the multiple correlation coefficient for $\mathbf{Y}$ with, respectively, $\mathbf{x}_1, \ldots, \mathbf{x}_k$ and $\mathbf{x}_1, \ldots, \mathbf{x}_{k-1}$. Then

$$R_k^2 = 1 - \frac{\text{ESS}_k}{\text{TSS}}, \qquad R_{k-1}^2 = 1 - \frac{\text{ESS}_{k-1}}{\text{TSS}} \qquad \text{for} \quad \text{TSS} = \sum_1^n (Y_i - \bar{Y})^2.$$

Thus,

$$R_k^2 = 1 - \frac{\text{ESS}_{k-1}}{\text{TSS}(1 + t^2/(n-k))} = 1 - (1 - R_{k-1}^2)\Big/\left(1 + \frac{t^2}{n-k}\right) = R_{k-1}^2$$
$$+ (1 - R_{k-1}^2)\left(\frac{d}{1+d}\right) \qquad \text{for} \quad d = t^2/(n-k).$$

It follows that

$$\frac{R_k^2 - R_{k-1}^2}{1 - R_{k-1}^2} = \frac{d}{1+d}$$

is the proportion of possible improvement in the explanation of the variability of $\mathbf{Y}$ which $\mathbf{x}_k$ gives beyond that provided by $\mathbf{x}_1, \ldots, \mathbf{x}_{k-1}$. The possible improvement in the multiple correlation coefficient beyond that given by the first $k-1$ variables is $1 - R_{k-1}^2$. From Figure 3.6 the actual improvement provided by using $\mathbf{x}_k$ as well is $R_k^2 - R_{k-1}^2$. The proportion of actual improvement to possible improvement is $d/(1+d)$.

Problem 3.6.1: Show that for the case of the simple linear regression, $R_{\mathbf{y}.\mathbf{x}}$, the multiple regression coefficient with one $\mathbf{x}$-vector $\mathbf{x}$, is the absolute value of the simple correlation coefficient $r_{\mathbf{yx}}$.

Problem 3.6.2: For the data of Example 3.5.1 find $R_{y.x_2x_3}$ and $R_{y.x_2}$ and show that $R^2_{\mathbf{y}.\mathbf{x}_2\mathbf{x}_3} = R^2_{\mathbf{y}.\mathbf{x}_2} + (1 - R^2_{\mathbf{y}.\mathbf{x}_2})\dfrac{d}{1+d}$ for d as defined above.

Problem 3.6.3: Let $\mathbf{y} = \begin{bmatrix} 6 & 11 & 8 \\ 4 & 7 & \\ 2 & & \end{bmatrix}$, and let $\mathbf{C}_j$ be the indicator of column j for $j = 1, 2, 3$.

(a) Find the multiple correlation coefficient R of $\mathbf{y}$ with $\mathbf{C}_1$, $\mathbf{C}_2$, $\mathbf{C}_3$.

(b) Let ESS_2 and ESS_3 be the error SSqs. corresponding to $V_2 = \mathscr{L}(\mathbf{C}_1, \mathbf{C}_2)$ and $\mathscr{L}(\mathbf{C}_1, \mathbf{C}_2, \mathbf{C}_3)$. Verify that $\text{ESS}_3 = \text{ESS}_2 - \hat{\beta}_3^2 \|\mathbf{x}_3^\perp\|^2$ and $\text{ESS}_3 = \text{ESS}_2 \Big/ \left(1 + \dfrac{t^2}{n-3}\right)$ for these data.

(c) For the general one-way layout with three columns, n_j observations in the jth column, give a formula for R.

3.7 THE PARTIAL CORRELATION COEFFICIENT

Suppose an educational psychologist studied the relationship between the height v_1 and reading ability v_2 of children as measured by the score on a standardized test. For 200 children in the third, fourth, and fifth grades of an elementary school she measured v_1 and v_2, then found that the correlation

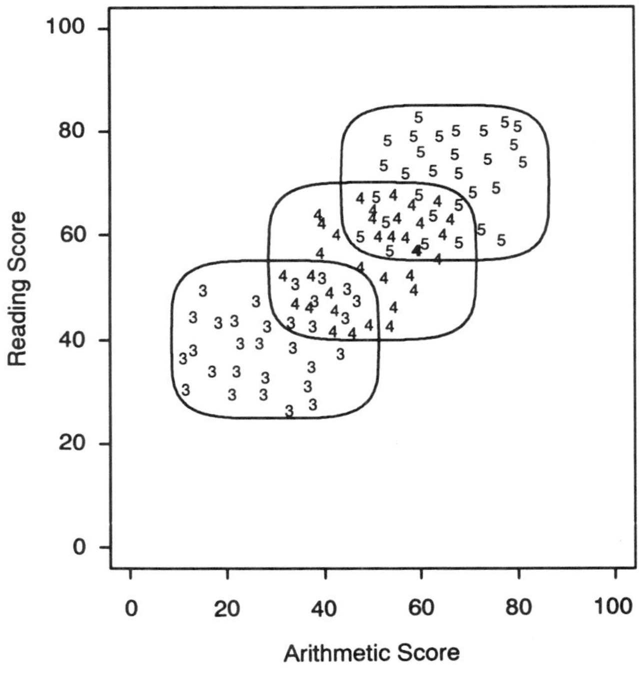

FIGURE 3.7

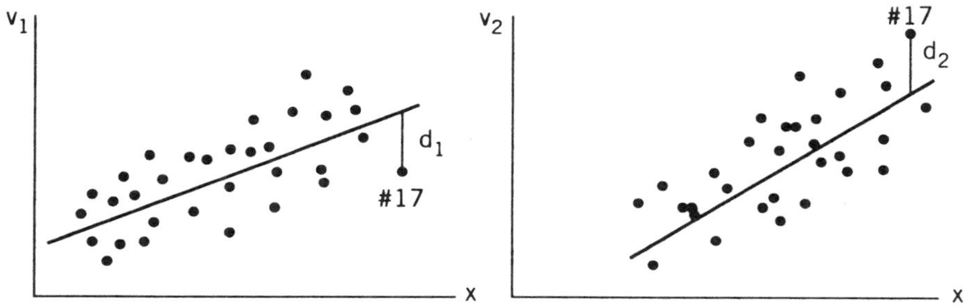

FIGURE 3.8 d_1 and d_2 for child #17.

between v_1 and v_2 was 0.56. Would she be correct in deciding that taller children read better, perhaps because they can more easily see over their classmate's heads? (The author, being fairly tall, is often tempted by such conclusions. Some of his students have disagreed vigorously.)

A little thought suggests that the data for the third, fourth, and fifth graders might be graphed as in Figure 3.7. Thus, the "spurious" correlation could be caused by differences in grades, or ages of the children, since both v_1 and v_2 would tend to increase with age. For this reason age is called a "lurking variable". Somehow the experimenter would like to confine her study to children of the same age. Even children in the same grade differ somewhat in age, however, so that confining the study to one grade might not suffice. Confining the study to children within a few months in age could result in too small a group.

The partial correlation coefficient is a measure of linear relationship between two variables, with the linear effects of one or more other variables, in this case age x, removed. In this example, we could fit the simple linear regression lines of Figure 3.8 to v_1 vs. x and to v_2 vs. x. For each child the deviations d_1 and d_2 from the fitted lines could be determined. Then the partial correlation coefficient of v_1 and v_2 with the effect of age removed is the ordinary correlation coefficient among the (d_1, d_2) pairs for all children.

More generally, the partial correlation coefficient is defined as follows:

Definition 3.7.1: Let $\mathbf{v}_1, \mathbf{v}_2, \mathbf{x}_1, \ldots, \mathbf{x}_k \in R_n$. Let $V = \mathscr{L}(\mathbf{x}_1, \ldots, \mathbf{x}_k)$, $\hat{\mathbf{v}}_1 = p(\mathbf{v}_1 \mid V)$, $\hat{\mathbf{v}}_2 = p(\mathbf{v}_2 \mid V)$. Then the *partial correlation coefficient* of $\mathbf{v}_1$ and $\mathbf{v}_2$ with the effects of $\mathbf{x}_1, \ldots, \mathbf{x}_k$ (equivalently V) removed is

$$r = r_{\mathbf{v}_1\mathbf{v}_2 \cdot \mathbf{x}_1\mathbf{x}_2 \ldots \mathbf{x}_k} = \frac{(\mathbf{v}_1 - \hat{\mathbf{v}}_1, \mathbf{v}_2 - \hat{\mathbf{v}}_2)}{\|\mathbf{v}_1 - \hat{\mathbf{v}}_1\| \, \|\mathbf{v}_2 - \hat{\mathbf{v}}_2\|}$$

(Undefined if $\mathbf{v}_1 \in V$ or $\mathbf{v}_2 \in V$).

Comments

(1) In practice it is usually the case that $\mathbf{J} \in V$, so that the additive effect of a constant on $\mathbf{v}_1$ and $\mathbf{v}_2$ is removed. We will always suppose this unless stated otherwise, so that it will be unnecessary to include $\mathbf{J}$ among the independent variables listed.

(2) The ordinary correlation coefficient is the special case $k = 1$, $\mathbf{x}_1 = \mathbf{J}$.

(3) r is unchanged by scale changes in any of the variables, or, in the usual case that $\mathbf{J} \in V$, changes in location (addition of a constant) for any of the variables. More generally, r is a function of the subspace V, not of the specific vectors spanning V, so that $\mathbf{X} = (\mathbf{x}_1, \ldots, \mathbf{x}_k)$ may be replaced by $\mathbf{XA}$ for $\mathbf{A}$ nonsingular. For example, if a vector $\mathbf{w} \in V$ is added to $\mathbf{v}_1$ then $\mathbf{v}_1 - \hat{\mathbf{v}}_1$ remains the same. r is unchanged.

(4) Consider multiple regression of $\mathbf{Y}$ on $\mathbf{x}_1, \ldots, \mathbf{x}_{k-1}, \mathbf{x}_k$. Let $V_{k-1} = \mathscr{L}(\mathbf{x}_1, \ldots, \mathbf{x}_{k-1})$ and $V_k = \mathscr{L}(\mathbf{x}_1, \ldots, \mathbf{x}_k)$. Let $\hat{\mathbf{Y}}_{k-1} = p(\mathbf{Y} \mid V_{k-1})$, $\hat{\mathbf{Y}}_k = p(\mathbf{Y} \mid V_k)$, $\mathbf{x}_k^\perp = \mathbf{x}_k - p(\mathbf{x}_k \mid V_{k-1}) = \mathbf{x}_k - \hat{\mathbf{x}}_k$, $\mathbf{e} = \mathbf{Y} - \mathbf{Y}_k$. Then

$$\mathbf{Y} = \hat{\mathbf{Y}}_{k-1} + \hat{\beta}_k \mathbf{x}_k^\perp + \mathbf{e} \qquad \text{and} \qquad \mathbf{x}_k = \hat{\mathbf{x}}_k + \mathbf{x}_k^\perp$$

are decompositions of $\mathbf{Y}$ and $\mathbf{x}_k$ into orthogonal vectors. The partial correlation coefficient of $\mathbf{Y}$ with $\mathbf{x}_k$ with the effects of $\mathbf{x}_1, \ldots, \mathbf{x}_{k-1}$ removed is

$$r = (\hat{\beta}_k \mathbf{x}_k^\perp + \mathbf{e}, \mathbf{x}_k^\perp)/[\|\hat{\beta}_k \mathbf{x}_k^\perp + \mathbf{e}\| \, \|\mathbf{x}_k^\perp\|] = \hat{\beta}_k \|\mathbf{x}_k^\perp\|^2/[\hat{\beta}_k^2 \|\mathbf{x}_k^\perp\|^2 + \|\mathbf{e}\|^2]^{1/2} \|\mathbf{x}_k^\perp\|$$

$$= \frac{\hat{\beta}_k \|\mathbf{x}_k^\perp\|}{\|\mathbf{e}\|} \Big/ \left[\frac{\hat{\beta}_k^2 \|\mathbf{x}_k^\perp\|^2}{\|\mathbf{e}\|^2} + 1\right]^{1/2} = \frac{t/(n-k)^{1/2}}{\left[1 + \dfrac{t^2}{n-k}\right]^{1/2}},$$

where $t = \hat{\beta}_k \Big/ \left[\dfrac{\|\mathbf{e}\|^2}{n-k} \Big/ \|\mathbf{x}_k^\perp\|^2\right]^{1/2}$ is the t-statistic used to test $H_0 : \beta_k = 0$ in the model $\mathbf{Y} = \sum_1^k \beta_j \mathbf{x}_i + \boldsymbol{\varepsilon}$.

(5) Let $\mathbf{x}_1, \ldots, \mathbf{x}_k$ be $k > 3$ vectors and let $I = \{3, \ldots, k\}$, $J = \{4, \ldots, k\}$. Let $r_{12 \cdot \mathbf{I}}$ and $r_{12 \cdot \mathbf{J}}$ be the partial correlation coefficients of $\mathbf{x}_1$ and $\mathbf{x}_2$ with the effects respectively of the vectors $\{\mathbf{x}_j \mid j \in I\}$ and $\{\mathbf{x}_j \mid j \in J\}$ removed. We will try to develop a formula relating $r_{12 \cdot \mathbf{I}}$ to partial correlations $r_{12 \cdot \mathbf{J}}$, $r_{13 \cdot \mathbf{J}}$, and $r_{23 \cdot \mathbf{J}}$.

Let $V_J = \mathscr{L}(\mathbf{x}_4, \ldots, \mathbf{x}_k)$ and $V_I = \mathscr{L}(\mathbf{x}_3, \ldots, \mathbf{x}_k)$. Let $\mathbf{x}_1^\perp = \mathbf{x}_1 - p(\mathbf{x}_1 \mid V_J)$, $\mathbf{x}_2^\perp = \mathbf{x}_2 - p(\mathbf{x}_2 \mid V_J)$, $\mathbf{x}_3 = \mathbf{x}_3 + \mathbf{x}_3^\perp$ for $\hat{\mathbf{x}}_3 = p(\mathbf{x}_3 \mid V_J)$. Then $p(\mathbf{x}_i \mid V_I) = p(\mathbf{x}_i \mid V_J) + p(\mathbf{x}_i \mid \mathbf{x}_3^\perp)$, so that $\mathbf{w}_i = \mathbf{x}_i - p(\mathbf{x}_i \mid V_I) = \mathbf{x}_i^\perp - p(\mathbf{x}_i \mid \mathbf{x}_3^\perp) = \mathbf{x}_i^\perp - p(\mathbf{x}_i^\perp \mid \mathbf{x}_3^\perp) = \mathbf{x}_i^\perp - [(\mathbf{x}_i^\perp, \mathbf{x}_3^\perp)/\|\mathbf{x}_3^\perp\|^2]\mathbf{x}_3^\perp$ for $i = 1, 2$. Thus, $r_{12 \cdot \mathbf{I}}$ is a function of the vectors $\mathbf{x}_1^\perp, \mathbf{x}_2^\perp, \mathbf{x}_3^\perp$ and by (3) is unaffected by scale changes in these vectors. We therefore may take their lengths each to be one. The vectors $\mathbf{w}_1, \mathbf{w}_2$ therefore have inner product $(\mathbf{x}_1^\perp, \mathbf{x}_2^\perp) - (\mathbf{x}_1^\perp, \mathbf{x}_3^\perp)(\mathbf{x}_2^\perp, \mathbf{x}_3^\perp) = r_{12 \cdot \mathbf{J}} - r_{13 \cdot \mathbf{J}} r_{23 \cdot \mathbf{J}}$ and lengths

$\|\mathbf{x}_i^{\perp}\|^2 - (\mathbf{x}_i, \mathbf{x}_3^{\perp})^2 = 1 - r_{i3.\mathbf{J}}^2$ for $i = 1, 2$. Therefore,

$$r_{12\cdot\mathbf{I}} = \frac{r_{12\cdot\mathbf{J}} - r_{13\cdot\mathbf{J}}r_{23\cdot\mathbf{J}}}{\sqrt{(1 - r_{13\cdot\mathbf{J}}^2)(1 - r_{23\cdot\mathbf{J}}^2)}}$$

In practice the vector $\mathbf{J}$ of all ones is included in $\mathscr{L}(\mathbf{x}_4, \ldots, \mathbf{x}_k)$. Of course, the choice of subscripts 1, 2, 3 here was only a notational convenience. Change of notation leads, for example, to the formula

$$r_{14.23} = \frac{r_{14.2} - r_{13.2}r_{43.2}}{\sqrt{(1 - r_{13.2}^2)(1 - r_{43.2}^2)}}$$

(6) Let R_k and R_{k-1} be the multiple correlation coefficient of $\mathbf{Y}$ with respectively $\mathbf{x}_1, \ldots, \mathbf{x}_k$ and $\mathbf{x}_1, \ldots, \mathbf{x}_{k-1}$. Suppose $\mathbf{x}_1 = \mathbf{J}$. Then we showed in Section 3.6 that $R_k^2 = R_{k-1}^2 + \dfrac{d}{1+d}(1 - R_{k-1}^2)$, where $d = t^2/(n-k)$. From (4) above the partial correlation coefficient of $\mathbf{Y}$ and $\mathbf{x}_k$ with the effects of $\mathbf{x}_1, \ldots, \mathbf{x}_{k-1}$ removed is $r = \left[\dfrac{d}{1+d}\right]^{1/2}$ (sign $\hat{\beta}_k$). Therefore,

$$R_k^2 = R_{k-1}^2 + r^2(1 - R_{k-1}^2) \qquad \text{and} \qquad r^2 = \frac{R_k^2 - R_{k-1}^2}{1 - R_{k-1}^2}.$$

We conclude that r^2 is the proportion of improvement in the explanation of the variation of $\mathbf{Y}$ caused by adding $\mathbf{x}_k$ to the collection of explanatory variables, as compared to the possible improvement $1 - R_{k-1}^2$.

(7) To see that the pair $(r_{12}, r_{12.3})$ may take arbitrary values in the square $A = (-1, +1) \times [-1, +1]$ let $\mathbf{w}_1, \mathbf{w}_2, \mathbf{x}_3$ be length one vectors with components adding to zero, $(\mathbf{w}_1, \mathbf{w}_2) = r$, $\mathbf{x}_3 \perp \mathbf{w}_1$, $\mathbf{x}_3 \perp \mathbf{w}_2$. Let $\mathbf{x}_1 = c_1\mathbf{x}_3 + \mathbf{w}_1$ and $\mathbf{x}_2 = c_2\mathbf{x}_3 + \mathbf{w}_2$. Then $r_{12.3} = r$ and $r_{12} = \dfrac{c_1c_2 + r}{\sqrt{(1 + c_1^2)(1 + c_2^2)}}$ for real numbers c_1, c_2. As (r, c_1, c_2) ranges over $(-1, 1) \times R_1 \times R_1$, $(r_{12}, r_{12.3})$ ranges over A.

(8) Here we summarize some results on the distribution of the sample correlation coefficients. Proofs are omitted. Under the bivariate normal model the sample correlation coefficient r is asymptotically normally distributed with mean ρ and asymptotic variance $(1 - \rho^2)^2/n$. However, the convergence is rather slow, particularly for ρ near -1 or $+1$. The transformed variable $g(r) = \dfrac{1}{2}\ln\dfrac{1+r}{1-r}$ converges much more rapidly in distribution to the normal, with approximate mean $g(\rho) + \rho/2(n-1)$ and approximate variance $1/(n-3)$.

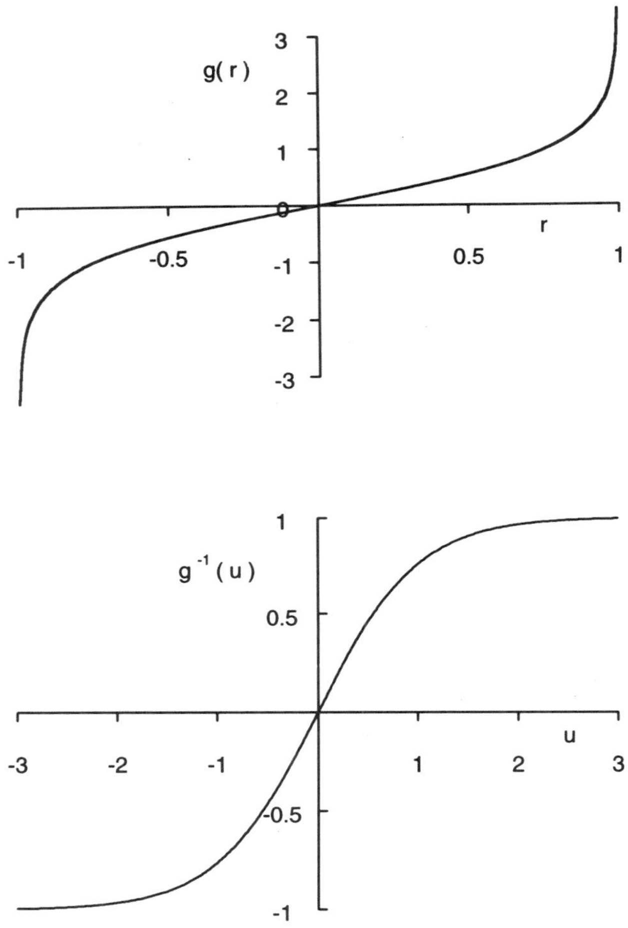

FIGURE 3.9 The functions $u = g(r)$ and $r = g^{-1}(u)$.

This leads to a confidence interval $[a, b] = [g(r) \pm z_{1-\alpha/2}/\sqrt{n-3}]$ on $g(\rho)$ and a corresponding interval $[g^{-1}(a), g^{-1}(b)]$ on ρ, where $g^{-1}(u) = \dfrac{e^u - e^{-u}}{e^u + e^{-u}} = \tanh u$ (Figure 3.9).

Under the multivariate normal model the distribution of a partial correlation coefficient is the same as that of a simple correlation coefficient, with n reduced by the number of conditioning variables (not counting $\mathbf{J}$).

Problem 3.7.1: For $\mathbf{v}_1 = (5, 1, 0, 3, 5)'$, $\mathbf{v}_2 = (5, 3, 7, 6, 10)'$ find the partial correlation coefficient of $\mathbf{v}_1$ and $\mathbf{v}_2$ with the effects of $\mathbf{J} = (1, 1, 1, 1, 1)'$ and $\mathbf{v}_3 = (0, 0, 0, 1, 1)'$ removed. Verify the formula derived under (5) among the preceding comments for this case

$$r_{12.3} = \frac{r_{12} - r_{13}r_{23}}{\sqrt{1 - r_{13}^2}\sqrt{1 - r_{23}^2}}.$$

Problem 3.7.2: Give an explicit test of $H_0: \beta_k = 0$ in the general linear model with $\mathbf{x}_1 = \mathbf{J}$, in terms of the partial correlation coefficient $R_{Yx_k.x_1 \ldots x_{k-1}} = r$.

Problem 3.7.3: Find vectors $\mathbf{x}_1$, $\mathbf{x}_2$, $\mathbf{x}_3 \in R_4$ such that $r_{12} = 1/2$, while $r_{12.3} = -1/2$. In $r_{12.3}$ suppose the effects of both $\mathbf{x}_3$ and $\mathbf{J}$ are removed. What is the multiple correlation coefficient of $\mathbf{v}_1$ with respect to $\mathbf{v}_2$ and $\mathbf{v}_3$?

Problem 3.7.4: Referring to (7) among the preceding comments, show that $|r_{12}| = 1$ implies that $r_{12.3} = r_{12}$ (whenever $r_{12.3}$ is defined).

Problem 3.7.5: The reliability of an examination is the correlation ρ of pairs (X_1, X_2) of scores obtained on repetitions of the same (or very similar) examinations given to the same individual in the population of individuals to be given the exam. Since the learning effect may preclude giving the same or even similar exams to the same individual the following technique may be useful. Split the exam into two equivalent halves and record the scores (Y_1, Y_2) on each half. Record the pair (Y_1, Y_2) for each of a number of individuals and use the sample correlation coefficient r to estimate $\rho_Y = \rho(Y_1, Y_2)$.

Suppose equivalent forms of the exam are given with scores on the two halves: (Y_{11}, Y_{12}) and (Y_{21}, Y_{22}) and total scores $X_1 = Y_{11} + Y_{12}$ and $X_2 = Y_{21} + Y_{22}$ on the two exams. Suppose $Y_{ij} = A + H_{ij}$ for $i = 1, 2$ and $j = 1, 2$, with A, H_{11}, H_{12}, H_{21}, H_{22} uncorrelated r.v.'s with $\text{Var}(A) = \sigma_A^2$, $\text{Var}(H_{ij}) = \sigma_H^2$ for all i and j. A may be considered to be the ability of the individual, while the H_{ij} are random deviations from ability.

(a) Express ρ_X as a function of ρ_Y.

(b) Suppose 100 independent observations are made on (Y_1, Y_2), with observed sample correlation coefficient $r_Y = 0.31$. Suppose these pairs have a bivariate normal distribution. Give a 95% confidence interval on ρ_Y, then on ρ_X.

3.8 TESTING $H_0:\boldsymbol{\theta} \in V_0 \subset V$

Consider the one-way analysis of variance model

$$\mathbf{Y} = \begin{pmatrix} Y_{11} & Y_{21} & & Y_{k1} \\ \vdots & \vdots & \cdots & \vdots \\ Y_{1n_1} & Y_{2n_2} & & Y_{kn_k} \end{pmatrix} = \sum_{j=1}^{k} \mu_j \mathbf{J}_j + \boldsymbol{\varepsilon},$$

where $\mathbf{J}_j$ is the array which indicates the jth column. It is often of interest to test $H_0: \mu_1 = \cdots = \mu_k$. Then H_0 is equivalent to the statement that $E(\mathbf{Y}) = \boldsymbol{\theta} = \mu_1 \sum_{j=1}^{k} \mathbf{J}_j = \mu_1 \mathbf{J}$ for some μ_1. That is, under $H_0: \boldsymbol{\theta} \in \mathscr{L}(\mathbf{J})$.

Similarly, we might fit a regression model in which $\mathbf{Y}$ is college G.P.A., x_1 is high school G.P.A., $\mathbf{x}_2$ is S.A.T. score, $\mathbf{x}_3$ is # of years of father's education, and $\mathbf{x}_4$ is # of years of mother's education. The full model might then be $\mathbf{Y} = \beta_0 \mathbf{J} + \beta_1 \mathbf{x}_1 + \cdots + \beta_4 \mathbf{x}_4 + \boldsymbol{\varepsilon}$. We might like to test $H_0: \beta_3 = \beta_4 = 0$ (Mother's and father's education are valueless, in predicting college G.P.A., as additional information beyond $\mathbf{x}_1$ and $\mathbf{x}_2$, in a linear sense). Then, under $H_0: E(\mathbf{Y}) = \boldsymbol{\theta} \in \mathscr{L}(\mathbf{x}_0, \mathbf{x}_1, \mathbf{x}_2) \equiv V_0$, a subspace of $V = \mathscr{L}(\mathbf{x}_0, \mathbf{x}_1, \ldots, \mathbf{x}_4)$.

Thus, we need a procedure which will allow us to test $H_0: \boldsymbol{\theta} \in V_0$, where V_0 is a proper subspace of V of dimension $k_0 < k = \dim(V)$. The alternative is then $H_1: \boldsymbol{\theta} \notin V_0$.

Intuitively we should select a test statistic which tends to be large for $\boldsymbol{\theta} \notin V_0$, small for $\boldsymbol{\theta} \in V_0$. We will suggest such a statistic, and show that it has desirable properties. Later we will show that the test which rejects H_0 for large values of this test statistic is the likelihood ratio test.

Refer to Figure 3.10 and let $\hat{\mathbf{Y}} = p(\mathbf{Y} \mid V)$ and $\hat{\mathbf{Y}}_0 = p(\mathbf{Y} \mid V_0)$, and

$$F = \frac{\|\hat{\mathbf{Y}} - \hat{\mathbf{Y}}_0\|^2/(k - k_0)}{\|\mathbf{Y} - \hat{\mathbf{Y}}\|^2/(n - k)} = \frac{(\|\hat{\mathbf{Y}}\|^2 - \|\hat{\mathbf{Y}}_0\|^2)/(k - k_0)}{S^2}$$

$$\hat{\mathbf{Y}} - \hat{\mathbf{Y}}_0 = p(\mathbf{Y} \mid V_0^{\perp} \cap V) \qquad \text{and} \qquad \mathbf{Y} - \hat{\mathbf{Y}} = p(\mathbf{Y} \mid V^{\perp})$$

By Theorem 2.5.3:

(1) $\|\hat{\mathbf{Y}} - \hat{\mathbf{Y}}_0\|^2/\sigma^2 \sim \chi^2_{k-k_0}(\|\boldsymbol{\theta} - \boldsymbol{\theta}_0\|^2/\sigma^2)$ for $\boldsymbol{\theta}_0 = p(\boldsymbol{\theta} \mid V_0)$

(2) $\|\mathbf{Y} - \hat{\mathbf{Y}}\|^2/\sigma^2 \sim \chi^2_{n-k}$

(3) The two r.v.'s in (1) and (2) are independent.

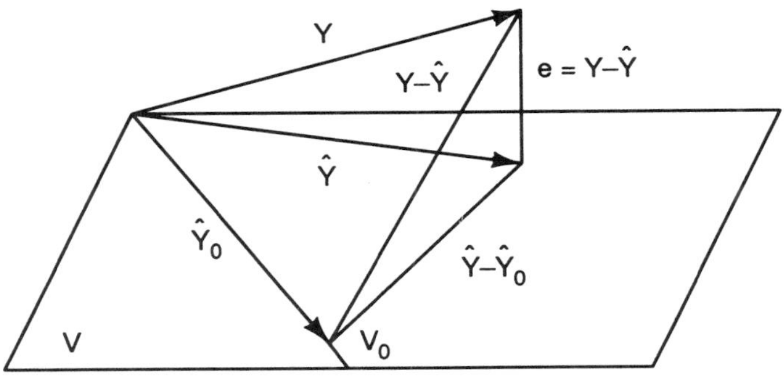

FIGURE 3.10 Illustration for the F-test of H_0: $\theta \in V_0$.

It follows that the F-statistic has a noncentral F-distribution with $(k - k_0)$ and $(n - k)$ d.f., and noncentrality parameter $\delta = \|\boldsymbol{\theta} - \boldsymbol{\theta}_0\|^2/\sigma^2$.

Since H_0 is equivalent to $\delta = 0$, F has a central F distribution under H_0. Therefore, the test which rejects H_0 for $F \geq F_{k-k_0, n-k, 1-\alpha} \equiv F_{1-\alpha}$ has level α. It has power $\gamma(\delta) = P(F \geq F_{1-\alpha})$, which depends on $k - k_0$, $n - k$, δ, and α.

Comments

(1) The numerator of F can be written in various ways by taking advantage of the Pythagorean Theorem. Under the full model the error sum of squares and regression sum of squares are $\text{ESS}_{FM} = \|\mathbf{Y} - \hat{\mathbf{Y}}\|^2$ and $\text{RSS}_{FM} = \|\hat{\mathbf{Y}}\|^2$. Under H_0 the error sum of squares and regression sum of squares are $\text{ESS}_{H_0} = \|\mathbf{Y} - \hat{\mathbf{Y}}_0\|^2$ and $\text{RSS}_{H_0} = \|\hat{\mathbf{Y}}_0\|^2$. Then, by the Pythagorean Theorem,

$$\|\hat{\mathbf{Y}} - \hat{\mathbf{Y}}_0\|^2 = \|\mathbf{Y} - \hat{\mathbf{Y}}_0\|^2 - \|\mathbf{Y} - \hat{\mathbf{Y}}\|^2 = \text{ESS}_{H_0} - \text{ESS}_{FM} = \text{RSS}_{FM} - \text{RSS}_{H_0}$$

Thus, to find the numerator of the F-statistic, we need to fit both the full model and the model under H_0, determining the error sum of squares in both cases. Of course, in order to determine the denominator S^2, we need only fit the full model.

(2) Letting $\hat{\mathbf{Y}} = \mathbf{X}\hat{\boldsymbol{\beta}}$ and $\hat{\mathbf{Y}}_0 = \mathbf{X}\hat{\boldsymbol{\beta}}_0$ we get $\hat{\mathbf{Y}} - \hat{\mathbf{Y}}_0 = \mathbf{X}(\hat{\boldsymbol{\beta}} - \hat{\boldsymbol{\beta}}_0)$ and $\|\hat{\mathbf{Y}} - \hat{\mathbf{Y}}_0\|^2 = (\hat{\boldsymbol{\beta}} - \hat{\boldsymbol{\beta}}_0)'(\mathbf{X}'\mathbf{X})(\hat{\boldsymbol{\beta}} - \hat{\boldsymbol{\beta}}_0)$, and $\|\mathbf{Y} - \hat{\mathbf{Y}}\|^2 = \|\mathbf{Y}\|^2 - (\hat{\mathbf{Y}}, \hat{\mathbf{Y}}) = \|\mathbf{Y}\|^2 - (\mathbf{Y}, \hat{\mathbf{Y}}) = \|\mathbf{Y}\|^2 - \hat{\boldsymbol{\beta}}'(\mathbf{X}'\mathbf{Y})$

(3) Once a computational method for $\|\hat{\mathbf{Y}} - \hat{\mathbf{Y}}_0\|^2$ is determined, the noncentrality parameter δ for the F-statistic may be obtained by substituting $\boldsymbol{\theta}$ for $\mathbf{Y}$ (and dividing by σ^2). Of course, δ is a function of $\boldsymbol{\theta}$ and σ^2, which are unknown.

Example 3.8.1: (One-way analysis of variance) Let $\mathbf{Y} = \sum_1^k \mu_j \mathbf{J}_j + \boldsymbol{\varepsilon}$ and let $H_0: \mu_1 = \cdots = \mu_k$. Then under the full model $\boldsymbol{\theta} \in V = \mathscr{L}(\mathbf{J}_1, \ldots, \mathbf{J}_k)$ and under $H_0: \boldsymbol{\theta} \in V_0 = \mathscr{L}(\mathbf{J})$ for $\mathbf{J} = \sum_1^k \mathbf{J}_j$. Thus, since the $\mathbf{J}_j$ are mutually orthogonal, $\hat{\mathbf{Y}} = \sum_1^k \bar{Y}_j \mathbf{J}_j$ and $\hat{\mathbf{Y}}_0 = \bar{Y}\mathbf{J}$, where $\bar{Y}_j = \frac{1}{n_i} \sum_{i=1}^{n_j} Y_{ij}$ and $\bar{Y} = \left(\sum_{ij} Y_{ij}\right)/n$ for $n = \sum_1^k n_j$. Then $\hat{\mathbf{Y}} - \hat{\mathbf{Y}}_0 = \sum_i \bar{Y}_i \mathbf{J}_i - \bar{Y}\mathbf{J} = \sum_i (\bar{Y}_i - \bar{Y})\mathbf{J}_i$ and $\|\hat{\mathbf{Y}} - \hat{\mathbf{Y}}_0\|^2 = \sum_i (\bar{Y}_i - \bar{Y})^2 n_i = \|\hat{\mathbf{Y}}\|^2 - \|\hat{\mathbf{Y}}_0\|^2 = \sum \bar{Y}_i^2 n_i - n\bar{Y}^2$

The error sum of squares under the full model is

$$\|\mathbf{Y} - \hat{\mathbf{Y}}\|^2 = \sum_{ij} (Y_{ij} - \bar{Y}_i)^2 = \|\mathbf{Y}\| - \|\hat{\mathbf{Y}}\|^2 = \sum_{ij} Y_{ij}^2 - \sum_i \bar{Y}_i^2 n_i.$$

This information is usually summarized in an analysis of variance table similar to Table 3.8.1, which gives the squared lengths of the projections of $\mathbf{Y}$

Table 3.8.1 Analysis of Variance

DF	Space	Source	Sum of Squares	Mean Squares	Expected Mean Squares
Mean	V_0	1	$\|\hat{\mathbf{Y}}_0\|^2$		$\sigma^2 + \bar{\mu}^2 n$
Difference in means	$V \cap V_0^{\perp}$	$k-1$	$\|\hat{\mathbf{Y}} - \hat{\mathbf{Y}}_0\|^2$		$\sigma^2 + \dfrac{\sum (\mu_j - \bar{\mu})^2 n_j}{k-1}$
Error	$V^{\perp}$	$n-k$	$\|\mathbf{Y} - \hat{\mathbf{Y}}\|^2$		σ^2
Total	Ω	n	$\|\mathbf{Y}\|^2$		

on the orthogonal subspace. Mean squares are obtained by dividing sums of squares by the corresponding degrees of freedom. Since each sum of squares is of the form $\|\mathbf{P}_{V*}\mathbf{Y}\|^2$ for some subspace V^*, its expected value is $\dim(V^*)\sigma^2 + \|\mathbf{P}_{V*}\boldsymbol{\theta}\|^2$, so that $E(\text{mean square}) = \sigma^2 + \|\mathbf{P}_{V*}\boldsymbol{\theta}\|^2/(\dim V^*)$. $\|\mathbf{P}_{V*}\boldsymbol{\theta}\|^2$ may be obtained by substituting $\boldsymbol{\theta}$ for $\mathbf{Y}$ in the formula for the sum of squares. Thus, for $V^* = V_0$, $\mathbf{P}_{V_0}\boldsymbol{\theta} = \bar{\mu}\mathbf{J}$ and $\|\mathbf{P}_{V_0}\boldsymbol{\theta}\|^2 = \bar{\mu}^2 n$ for $\bar{\mu} = (\sum n_j \mu_j)/n$.

The F-statistic is

$$F = \frac{\text{Mean square for differences in means}}{\text{Error mean square}},$$

which has an $F_{k-1,n-k}(\delta)$ distribution for

$$\delta = \|\boldsymbol{\theta} - \boldsymbol{\theta}_0\|^2/\sigma^2 = \left[\sum_1^k (\mu_j - \bar{\mu})^2 n_j\right]\Big/\sigma^2.$$

For a numerical example, suppose that a crop scientist wished to investigate three hybrids of corn. He had 12 acres of land available. Four (1/3) acre plots were assigned at random to each of variety 1, 2 and 3. Corn was then planted and the yield in bushels measured separately on the plots. Unfortunately one plot (variety #2) was flooded and the observation lost. From the following data we can construct an analysis of variance table, Table 3.8.2.

Table 3.8.2 Analysis of Variance

Source	DF	SSqs.	Mean Squares	Expected Mean Squares
Mean	1	35,284.45	35,284.45	$\sigma^2 + \bar{\mu}^2(11)$
Among Varieties	2	86.55	43.27	$\sigma^2 + \frac{1}{2}\sum_1^3 (\mu_i - \bar{\mu})^2 n_i$
Error	8	92.00	11.50	σ^2
Total	11	35,463.00		

	Yield (Bushels)			
	1	2	3	
	52	64	53	
	56	57	55	
	60	62	58	
	56		50	
$\bar{Y}_j$	56	61	54	Grand mean $= \bar{Y} = 56.636$
n_j	4	3	4	Mean SSqs. $= 35{,}284.45$
Total	224	183	216	Grand total $= 623$

$$\text{Among Hybrids SSqs.} = \sum_{j=1}^{3} \bar{Y}_j^2\, n_j - \bar{Y}^2 n = 35{,}371 - 35{,}284.45 = 86.55$$

$$\text{Total SSqs.} = \sum_{ij} Y_{ij}^2 = 35{,}463$$

$$\begin{aligned}\text{Error SSqs.} &= \text{Total SSqs.} - [\text{Mean SSqs.} + \text{Among Hybrids SSqs}] \\ &= 35{,}463 - 35{,}371 = 92\end{aligned}$$

To test $H_0: \mu_1 = \mu_2 = \mu_3$, i.e., no variety effect vs. $H_1: H_0$ not true, for $\alpha = 0.05$, we reject if

$$F = \frac{\text{Among hybrids MSqs.}}{\text{Error MSqs.}} > F_{2,8,0.95} = 4.46.$$

In this case we observe $F = 3.74$, so we fail to reject H_0 at 0.05 level.

In general, an analysis of variance table has the columns of Table 3.8.1 with rows corresponding to subspaces $V_1, \ldots, V_m$, where the subspaces V_i are usually mutually orthogonal and $\Omega = V_1 \oplus \cdots \oplus V_m$. In applications the "Space" column is omitted and the "Total" row is replaced by a "Corrected Total" row, usually called (somewhat confusingly) the "Total" row, corresponding to $\Omega \cap V_0^\perp$, for $V_0 = \mathscr{L}(\mathbf{J})$.

Example 3.8.2: Let $\mathbf{x}_1 = (1, 1, 1, 1, 1,)'$, $\mathbf{x}_2 = (1, 1, 1, 0, 0)'$, $\mathbf{x}_3 = (1, 0, 0, 0, 1)'$, $\mathbf{Y} = \sum_1^3 \beta_j \mathbf{x}_j + \boldsymbol{\varepsilon}$, and we wish to test $H_0: \beta_2 = \beta_3$. Since H_0 is of the form $\eta = \mathbf{c}'\boldsymbol{\beta} = 0$, it is possible to use the t-statistic $t = \hat{\eta}/S_{\hat{\eta}}$. However, we will instead compute the F-statistic, which, as will be shown, is t^2. Suppose we observe $\mathbf{Y} = \mathbf{y} = (7, 2, 3, 11, 12)'$.

For $V = \mathscr{L}(\mathbf{x}_1, \mathbf{x}_2, \mathbf{x}_3)$, we find $\mathbf{M} = \mathbf{X}'\mathbf{X} = \begin{bmatrix} 5 & 3 & 3 \\ 3 & 3 & 1 \\ 3 & 1 & 2 \end{bmatrix}$, $\mathbf{U} = \mathbf{X}'\mathbf{y} = \begin{bmatrix} 35 \\ 12 \\ 19 \end{bmatrix}$,

$$\mathbf{M}^{-1} = (1/7)\begin{bmatrix} 5 & -4 & -3 \\ -4 & 6 & 1 \\ -3 & 1 & 6 \end{bmatrix}, \quad \hat{\boldsymbol{\beta}} = \mathbf{M}^{-1}\mathbf{U} = \begin{bmatrix} 10 \\ -7 \\ 3 \end{bmatrix}, \quad \hat{\mathbf{Y}} = (6, 3, 3, 10, 13)',$$

$\mathbf{e} = (1, -1, 0, 1, -1)'$, $\text{ESS}_{FM} = \|\mathbf{e}\|^2 = 4$, $S^2 = 4/2 = 2$. Under H_0 $\boldsymbol{\theta} = \beta_1\mathbf{x}_1 + \beta_2(\mathbf{x}_2 + \mathbf{x}_3) = \beta_1\mathbf{x}_1 + \beta_2\mathbf{x}_4$ for $\mathbf{x}_4 = (2, 1, 1, 0, 1)'$, so that H_0 is equivalent to $\boldsymbol{\theta} \in V_0 = \mathscr{L}(\mathbf{x}_1, \mathbf{x}_4)$. The model $\boldsymbol{\theta} \in V_0$ is the simple linear regression model, so we can use the formulas developed for that model, or we can simply use the multiple regression approach. We find $\hat{\boldsymbol{\beta}}_0 = (9, -2)'$, $\hat{\mathbf{Y}}_0 = 9\mathbf{x}_1 - 2\mathbf{x}_4 = (5, 7, 7, 9, 7)'$, $\mathbf{e}_0 = \mathbf{Y} - \hat{\mathbf{Y}}_0 = (2, -5, -4, 2, 5)'$, $\hat{\mathbf{Y}} - \hat{\mathbf{Y}}_0 = (1, -4, -4, 1, 6)'$. Notice that $\mathbf{e}_0$ and $\hat{\mathbf{Y}} - \hat{\mathbf{Y}}_0$ are orthogonal to V_0. Then $\text{ESS}_{H_0} = \|\mathbf{e}_0\|^2 = 74$ and $\|\hat{\mathbf{Y}} - \hat{\mathbf{Y}}_0\|^2 = 70$. By the Pythagorean Theorem this is the same as $\text{ESS}_{H_0} - \text{ESS}_{FM} = 74 - 4$.

The F-statistic is therefore $F = [70/1]/2 = 35$. Since $F_{1,2,0.95} = 18.5$, $F_{1,2,0.975} = 38.5$, we reject H_0 at level $\alpha = 0.05$, but not at level 0.025. The "observed α-level" is 0.0274. Since $\hat{\beta}_3 - \hat{\beta}_2 = -7 - 3 = -10$, and $\text{Var}(\hat{\beta}_3 - \hat{\beta}_2) = \sigma^2[6 + 6 - 2]/7 = 10\sigma^2/7$, the corresponding t-statistic is $t = (-10)/\sqrt{10(2)/7} = \sqrt{35}$, so that $t^2 = F$.

The Likelihood Ratio Approach: Suppose again that $\mathbf{Y} = \boldsymbol{\theta} + \boldsymbol{\varepsilon}$ for $\boldsymbol{\theta} \in V$, a k-dimensional subspace of R_n, $\boldsymbol{\varepsilon} \sim N_n(\mathbf{0}, \sigma^2\mathbf{I}_n)$ and we wish to test $H_0: \boldsymbol{\theta} \in V_0$, a $k_0 < k$-dimensional subspace of V. Consider the likelihood function

$$L(\boldsymbol{\theta}, \boldsymbol{\sigma}^2; \mathbf{Y}) \equiv L(\boldsymbol{\theta}, \sigma^2) = (2\pi\sigma^2)^{-n/2} \exp[-\|\mathbf{Y} - \boldsymbol{\theta}\|^2/2\sigma^2]$$

as shown in Section 3.1. $L(\boldsymbol{\theta}, \sigma^2)$ is maximized under the restriction $\boldsymbol{\theta} \in V_*$, a subspace of V, by taking $\boldsymbol{\theta} = \hat{\mathbf{Y}}_* = p(\mathbf{Y} \mid V_*)$ and $\sigma^2 = \hat{\sigma}_*^2 = \|\mathbf{Y} - \hat{\mathbf{Y}}_*\|^2/n$. Define $\hat{\mathbf{Y}} = p(\mathbf{Y} \mid V)$, $\hat{\mathbf{Y}}_0 = p(\mathbf{Y} \mid V_0)$, $\hat{\sigma}^2 = \|\mathbf{Y} - \hat{\mathbf{Y}}\|^2/n$, and $\sigma_0^2 = \|\mathbf{Y} - \hat{\mathbf{Y}}_0\|^2/n$. Then the likelihood ratio statistic is

$$L = \left[\sup_{\boldsymbol{\theta} \in V_0} \mathscr{L}(\boldsymbol{\theta}, \sigma^2)\right] \Big/ \left[\sup_{\boldsymbol{\theta} \in V} L(\boldsymbol{\theta}, \sigma)\right] = L(\hat{\mathbf{Y}}_0, \hat{\sigma}_0^2)/L(\hat{\mathbf{Y}}, \hat{\sigma}^2),$$

so that $-2n \log L = \lambda = \log(\hat{\sigma}_0^2/\sigma_0^2)$. The likelihood ratio test rejects for large λ, equivalently for large $\hat{\sigma}_0^2/\hat{\sigma}_0^2 = 1 + \dfrac{\|\hat{\mathbf{Y}} - \hat{\mathbf{Y}}_0\|^2}{\|\mathbf{Y} - \hat{\mathbf{Y}}\|^2} = 1 + \left(\dfrac{k - k_0}{n - k}\right)F$, a monotone function of the F-statistic suggested earlier on heuristic grounds. Thus, the likelihood ratio test is the F-test.

Asymptotic theory for the likelihood ratio statistic states that as $n \to \infty$, λ converges in distribution under H_0 to $\chi^2_{k-k_0}$ where, as usual, $k = \dim(V)$,

$k_0 = \dim(V_0)$. Since

$$\lambda = n\log\left(1 + F\left(\frac{k-k_0}{n-k}\right)\right) \doteq nF\left(\frac{k-k_0}{n-k}\right) \doteq F(k-k_0)$$
$$\doteq \|\hat{\mathbf{Y}} - \hat{\mathbf{Y}}_0\|^2/\sigma^2 \sim \chi^2_{k-k_0}$$

for large n we have another "proof" of this same conclusion.

Testing $H_0^*: \mathbf{A}\boldsymbol{\beta} = \mathbf{0}$: The null hypothesis $\boldsymbol{\theta} \in V_0$ is a statement about $E(\mathbf{Y}) = \boldsymbol{\theta}$. It may be more natural to state a null hypothesis in terms of $\boldsymbol{\beta}$. Consider a $q \times k$ matrix $\mathbf{A}$ of known constants of rank q, and suppose we wish to test $H_0^*: \mathbf{A}\boldsymbol{\beta} = \underset{q\times 1}{\mathbf{0}}$.

Two approaches are possible. One is to devise a test directly in terms of $\mathbf{A}\hat{\boldsymbol{\beta}}$. The other is to reduce $H_0^*: \mathbf{A}\boldsymbol{\beta} = \mathbf{0}$ to an equivalent form $H_0: \boldsymbol{\theta} \in V_0$, and then to use the method already discussed to test H_0. The two approaches turn out to be equivalent.

To take the more direct approach consider the random q-dimensional vector $\mathbf{Z} = \mathbf{A}\hat{\boldsymbol{\beta}} \sim N(\mathbf{A}\boldsymbol{\beta}, \mathbf{A}(\mathbf{X}'\mathbf{X})^{-1}\mathbf{A}'\sigma^2)$. The following theorem will enable us to devise a statistic depending on $\mathbf{Z}$.

Theorem 3.8.1: Let $\underset{k\times 1}{\mathbf{Z}} \sim N_k(\boldsymbol{\eta}, \boldsymbol{\Sigma})$, with $\boldsymbol{\Sigma}$ nonsingular. Then $\mathbf{Q} = \mathbf{Z}'\boldsymbol{\Sigma}^{-1}\mathbf{Z} \sim \chi^2_k(\delta)$ for $\delta = \boldsymbol{\eta}'\boldsymbol{\Sigma}^{-1}\boldsymbol{\eta}$.

***Proof*:** This follows directly for the more general theory on quadratic forms in Chapter 2. For clarity we present a proof here. Let $\mathbf{B}$ be a matrix satisfying $\mathbf{B}\mathbf{B}' = \boldsymbol{\Sigma}$, so $\mathbf{B}^{-1\prime}\mathbf{B}^{-1} = \boldsymbol{\Sigma}^{-1}$. Let $\mathbf{W} = \mathbf{B}^{-1}\mathbf{Z}$. Then $\mathbf{W} \sim N(\mathbf{B}^{-1}\boldsymbol{\eta}, \mathbf{I}_k)$ and $Q = \mathbf{Z}'\boldsymbol{\Sigma}^{-1}\mathbf{Z} = \mathbf{W}'\mathbf{W} = \|\mathbf{W}\|^2 \sim \chi^2_k(\delta)$ for $\delta = \|\mathbf{B}^{-1}\boldsymbol{\eta}\|^2 = \boldsymbol{\eta}'\boldsymbol{\Sigma}^{-1}\boldsymbol{\eta}$. □

Taking $\mathbf{Z} = \mathbf{A}\hat{\boldsymbol{\beta}}$, we get $Q = \mathbf{Z}'[\mathbf{A}(\mathbf{X}'\mathbf{X})^{-1}\mathbf{A}']^{-1}\mathbf{Z}/\sigma^2 = H(\hat{\boldsymbol{\beta}})/\sigma^2 \sim \chi^2_q(\delta)$, for $\delta = H(\boldsymbol{\beta})/\sigma^2$, where, for each $\mathbf{b}$, $H(\mathbf{b}) \equiv (\mathbf{A}\mathbf{b})'[\mathbf{A}(\mathbf{X}'\mathbf{X})^{-1}\mathbf{A}']^{-1}(\mathbf{A}\mathbf{b})$, and, since $\hat{\boldsymbol{\beta}}$ and therefore Q are independent of $S^2 = \|\mathbf{Y} - \hat{\mathbf{Y}}\|^2/(n-k)$, the statistic

$$F^* = \frac{H(\hat{\boldsymbol{\beta}})/q\sigma^2}{S^2/\sigma^2} = \frac{H(\hat{\boldsymbol{\beta}})/q}{S^2} \sim F_{q,n-k}(\delta)$$

and we can test H_0^* at level α by rejecting H_0^* for $F^* \geq F_{q,n-k,1-\alpha}$.

Now consider a more indirect approach. Consider the subspace $C \equiv \{\boldsymbol{\beta} \mid \mathbf{A}\boldsymbol{\beta} = \mathbf{0}\} = (\text{row space of } \mathbf{A})^{\perp}$. Define $\mathbf{a}_c = \mathbf{X}\mathbf{c}$ for $\mathbf{c} \in C$, and let $V_1 = \{\mathbf{a}_c \mid \mathbf{c} \in C\}$. Since $\boldsymbol{\beta} = \mathbf{M}^{-1}\mathbf{X}'\boldsymbol{\theta}$ for each $\boldsymbol{\theta} \in V$, $\mathbf{A}\boldsymbol{\beta} = \mathbf{0} \Leftrightarrow \mathbf{A}\mathbf{M}^{-1}\mathbf{X}'\ \boldsymbol{\theta} = \mathbf{0} \Leftrightarrow \boldsymbol{\theta} \perp V_1 = (\text{row space of } \mathbf{A}\mathbf{M}^{-1}\mathbf{X}') = (\text{column space of } \mathbf{B} \equiv \mathbf{X}\mathbf{M}^{-1}\mathbf{A}') \Leftrightarrow \boldsymbol{\theta} \in (\text{column space of } \mathbf{B})^{\perp}$. Let this last space be V_0. Thus, $V_1 = V \cap V_0^{\perp}$.

The numerator sum of squares in the F-statistic used to test $H_0: \boldsymbol{\theta} \in V_0$ is $\|\hat{\mathbf{Y}} - \hat{\mathbf{Y}}_0\|^2 = \|\mathbf{P}_{V_1}\mathbf{Y}\|^2$. But the projection matrix $\mathbf{P}_{V_1}$ is

$$\mathbf{P}_{V_1} = \mathbf{B}(\mathbf{B}'\mathbf{B})^{-1}\mathbf{B}' = \mathbf{X}(\mathbf{X}'\mathbf{X})^{-1}\mathbf{A}'[\mathbf{A}(\mathbf{X}'\mathbf{X})^{-1}\mathbf{X}'\mathbf{X}(\mathbf{X}'\mathbf{X})^{-1}\mathbf{A}']^{-1}\mathbf{A}(\mathbf{X}'\mathbf{X})^{-1}\mathbf{X}'$$
$$= \mathbf{X}(\mathbf{X}'\mathbf{X})^{-1}\mathbf{A}'[\mathbf{A}'(\mathbf{X}'\mathbf{X})^{-1}\mathbf{A}]^{-1}\mathbf{A}(\mathbf{X}'\mathbf{X})^{-1}\mathbf{X}'.$$

(Isn't this a beautiful formula? It has 12 matrix products, 6 transposes, and 4 inverses.) Since $\hat{\boldsymbol{\beta}} = (\mathbf{X}'\mathbf{X})^{-1}\mathbf{X}'\mathbf{Y}$ and $\|\mathbf{P}_{V_1}\mathbf{Y}\|^2 = \mathbf{Y}'\mathbf{P}_{V_1}\mathbf{Y}$, we get $\|\mathbf{P}_{V_1}\mathbf{Y}\|^2 = (\mathbf{A}\hat{\boldsymbol{\beta}})'[\mathbf{A}(\mathbf{X}'\mathbf{X})^{-1}\mathbf{A}]^{-1}(\mathbf{A}\hat{\boldsymbol{\beta}}) = H(\hat{\boldsymbol{\beta}})$, the same numerator sum of squares obtained by the direct approach. Thus, $F^* = F$. The two approaches are equivalent.

Continuation of Example 3.8.2: For $\mathbf{Y}$, $\mathbf{x}_1$, $\mathbf{x}_2$, $\mathbf{x}_3$ as above suppose that we wish to test $H_0: \beta_1 = 0$, $\beta_3 = \beta_2$, equivalently that $\mathbf{A}\boldsymbol{\beta} = \begin{bmatrix} 0 \\ 0 \end{bmatrix}$, for $\mathbf{A} = \begin{bmatrix} 1 & 0 & 0 \\ 0 & 1 & -1 \end{bmatrix}$. The subspace C of R_3 in which $\boldsymbol{\beta}$ lies under H_0 is the orthogonal complement of the row space of $\mathbf{A}$, the collection of $\boldsymbol{\beta}$ of the form $\begin{bmatrix} 0 \\ \beta_2 \\ \beta_2 \end{bmatrix}$. The subspace V_1 is the row space of $\mathbf{B}' = \mathbf{A}\mathbf{M}^{-1}\mathbf{X}' = (1/7)\begin{bmatrix} -2 & 1 & 1 & 5 & 2 \\ -1 & 4 & 4 & -1 & -6 \end{bmatrix}$. V_0 is the image of C under the transformation $\mathbf{X}$, so that vectors $\boldsymbol{\theta} \in V_0$ are of the form $\beta_2\mathbf{x}_2 + \beta_2\mathbf{x}_3 = \beta_2(2, 1, 1, 0, 1)'$, and are orthogonal to V_1, the column space of $\mathbf{B}$. Since $V_0 = \mathscr{L}(\mathbf{x}_4 \equiv \mathbf{x}_2 + \mathbf{x}_3)$, $\hat{\mathbf{Y}}_0 = p(\mathbf{Y} \mid V_0) = [(\mathbf{Y}, \mathbf{x}_4)/\|\mathbf{x}_4\|^2]\mathbf{x}_4 = (31/7)\mathbf{x}_4$, $\hat{\mathbf{Y}}_1 \equiv p(\mathbf{Y} \mid V_1) = \hat{\mathbf{Y}} - \hat{\mathbf{Y}}_0 = (1/7)(-20, -10, -10, 70, 60)'$, $\|\hat{\mathbf{Y}}_1\|^2 = 1{,}300/7 = 185.7$. The F-statistic is $F = [185.7/2]/S^2 = 46.4$, for 2 and 1 d.f. Since $F_{2,2,0.975} = 39.0$, we reject at the $\alpha = 0.025$ level.

We could have computed $\|\hat{\mathbf{Y}}_1\|^2$ from $\mathbf{Z}'[\mathbf{A}\mathbf{M}^{-1}\mathbf{A}']^{-1}\mathbf{Z}$ for $\mathbf{Z} = \mathbf{A}\hat{\boldsymbol{\beta}} = \begin{bmatrix} 10 \\ -10 \end{bmatrix}$. Since $H \equiv \mathbf{A}\mathbf{M}^{-1}\mathbf{A}' = (1/7)\begin{bmatrix} 5 & -1 \\ -1 & 10 \end{bmatrix}$, $H^{-1} = (1/7)\begin{bmatrix} 10 & 1 \\ 1 & 5 \end{bmatrix}$, we get $1{,}300/7 = 185.7$, as before.

Still another approach may seem reasonable, and, once again turns out to be equivalent to the F-test. Again, let $\mathbf{A}$ be $q \times k$, of rank k, let C be its row space, and suppose we wish to test $H_0: \mathbf{A}\boldsymbol{\beta} = \mathbf{0}$. Let $H_c: \eta_c = \mathbf{c}'\boldsymbol{\beta} = \mathbf{0}$ for any $\mathbf{c} \in C$. Let $h(\mathbf{c}) = \mathbf{c}'\mathbf{M}^{-1}\mathbf{c}$. Then $\operatorname{Var}(\hat{\eta}_c) = h(\mathbf{c})\sigma^2$ and $S_c^2 \equiv h(\mathbf{c})S^2$ is its unbiased estimator. We can test H_c using the statistic $t_c = \hat{\eta}_c/S_c$, rejecting H_c at level α for $t_c^2 = G(\hat{\boldsymbol{\beta}}, \mathbf{c})/S^2 \ge t_{n-k,1-\alpha}^2$, where $G(\mathbf{c}, \hat{\boldsymbol{\beta}}) = (\mathbf{c}'\hat{\boldsymbol{\beta}})^2/h(\mathbf{c})$. Then $H_0: (\eta_c = 0$ for all $\mathbf{c} \in C) \Leftrightarrow (\mathbf{A}\boldsymbol{\beta} = \mathbf{0}$ for all $\boldsymbol{\beta}) \Leftrightarrow \boldsymbol{\theta} \in V_0$, where $V_0 = V \cap$ (column space of $\mathbf{B} = \mathbf{X}\mathbf{M}^{-1}\mathbf{A}')^{\perp}$. It seems reasonable to base a test on the statistic $W \equiv \sup_{\mathbf{c} \in C} t_c^2 = K(\hat{\boldsymbol{\beta}})/S^2$, where $K(\hat{\boldsymbol{\beta}}) = \sup_{\mathbf{c} \in C} G(\mathbf{c}, \hat{\boldsymbol{\beta}})$. We need to know the distribution of W, since the test which rejects for $W \geq t^2_{n-k,1-\alpha}$ will have level larger than α. We will show that $K(\hat{\boldsymbol{\beta}}) = H(\hat{\boldsymbol{\beta}})$, so that $W/q = F$, and again we arrive at the same F-test.

For each $\mathbf{c} \in C$, again let $\mathbf{a}_c = \mathbf{X}\mathbf{M}^{-1}c$. The vector $\mathbf{a}_c$ may be written in the form $\sum b_j \mathbf{x}_j$, where $\mathbf{b} = \mathbf{M}^{-1}\mathbf{c} = (b_1, \ldots, b_k)'$, and $\mathbf{c}$ is the vector of inner products of $\mathbf{a}_c$ with the $\mathbf{x}_j$. Then $\mathbf{c}'\hat{\boldsymbol{\beta}} = \mathbf{c}'\mathbf{M}^{-1}\mathbf{X}\mathbf{Y} = (\mathbf{a}_c, \mathbf{Y})$ and $h(\mathbf{c}) = \mathbf{c}'\mathbf{M}^{-1}\mathbf{c} = \|\mathbf{a}_c\|^2$, so that $G(\mathbf{c}, \hat{\boldsymbol{\beta}}) = (\mathbf{a}_c, \mathbf{Y})^2/\|\mathbf{a}_c\|^2 = \|p(\mathbf{Y} \mid \mathbf{a}_c)\|^2$. Since $\mathbf{a}_c \in V_1$, and $\hat{\mathbf{Y}}_1 \equiv p(\mathbf{Y} \mid V_1) = p(\mathbf{Y} \mid \mathbf{a}_c) + [\hat{\mathbf{Y}}_1 - p(\mathbf{Y} \mid \mathbf{a}_c)]$, the orthogonality of these two vectors implies that $\|p(\mathbf{Y} \mid \mathbf{a}_c)\|^2 = \|\hat{\mathbf{Y}}_1\|^2 - \|\hat{\mathbf{Y}}_1 - p(\mathbf{Y} \mid \mathbf{a}_c)\|^2$, so that $G(\mathbf{c}, \hat{\boldsymbol{\beta}}) \leq \|\hat{\mathbf{Y}}_1\|^2/S^2$, with equality if and only if $\mathbf{a}_c$ is a multiple of $\hat{\mathbf{Y}}_1$. We conclude that $K(\hat{\boldsymbol{\beta}}) = \|\hat{\mathbf{Y}}_1\|^2 = H(\hat{\boldsymbol{\beta}})$. Therefore, $W/q = F$. The supremum of t_c^2 for $\mathbf{c} \in C$ is taken for $\mathbf{c} = \mathbf{X}'\hat{\mathbf{Y}}_1$, or any scalar multiple.

In establishing three equivalent forms of the numerator sum of squares in the F-statistic, we have established some useful algebraic identities, which we summarize now for later use.

Theorem 3.8.2: Let $\mathbf{M}$ be a $k \times k$ positive definite matrix. Let $\mathbf{A}$ be a $q \times k$ matrix of rank q. Let C be the row space of $\mathbf{A}$, and define $Q = \mathbf{A}'[\mathbf{A}\mathbf{M}^{-1}\mathbf{A}]^{-1}\mathbf{A}$. Then, for any $\mathbf{b} \in R_k$, (1) $\sup_{\mathbf{c} \in C} \dfrac{(\mathbf{c}'\mathbf{b})^2}{\mathbf{c}'\mathbf{M}^{-1}\mathbf{c}} = \mathbf{b}'\mathbf{Q}\mathbf{b}$, with the supremum achieved for $\mathbf{c} = \mathbf{Q}\mathbf{b}$, and (2) If $\mathbf{X}$ is an $n \times k$ matrix, $\mathbf{X}'\mathbf{X} = \mathbf{M}$, $\hat{\mathbf{y}} = \mathbf{X}\mathbf{b}$ and $\hat{\mathbf{y}}_1 = \mathbf{X}\mathbf{M}^{-1}\mathbf{Q}\mathbf{b}$, then $\hat{\mathbf{y}}_1 = p(\hat{\mathbf{y}} \mid V_1)$, where V_1 is the column space of $\mathbf{X}\mathbf{M}^{-1}\mathbf{A}'$, and $\|\hat{\mathbf{y}}_1\|^2 = \mathbf{b}'\mathbf{Q}\mathbf{b}$.

***Proof*:** These identities were established above by first showing (2), then showing that $\|\hat{\mathbf{y}}_1\|^2$ was equal to the supremum in (1). In order to provide more insight, let us show (1) directly.

As before, define $G(\mathbf{c}, \mathbf{b}) = \dfrac{(\mathbf{c}'\mathbf{b})^2}{\mathbf{c}'\mathbf{M}^{-1}\mathbf{c}}$. Let $\mathbf{B}$ be any $\mathbf{k} \times \mathbf{k}$ matrix satisfying $\mathbf{B}'\mathbf{B} = \mathbf{M}^{-1}$, and let $\mathbf{d} = \mathbf{B}\mathbf{c}$. Then $G(\mathbf{c}, \mathbf{b}) = (\mathbf{B}^{-1}\mathbf{d}, \mathbf{b})^2/\|\mathbf{d}\|^2 = (\mathbf{d}, \mathbf{w})^2/\|\mathbf{d}\|^2$, where $\mathbf{w} = \mathbf{B}^{-1}\mathbf{b}$. If $\mathbf{c}$ is restricted to the row space of $\mathbf{A}$, then $\mathbf{d}$ is restricted to the column space of $\mathbf{B}'$. $G(\mathbf{c}, \mathbf{b})$ therefore remains unchanged if $\mathbf{w}$ is replace by its projection onto the column space of $\mathbf{B}\mathbf{A}'$. The projection matrix is $\mathbf{P} = \mathbf{B}\mathbf{A}'[\mathbf{A}\mathbf{B}'\mathbf{B}\mathbf{A}]^{-1}\mathbf{A}\mathbf{B}' = \mathbf{B}\mathbf{A}'[\mathbf{A}\mathbf{M}^{-1}\mathbf{A}']^{-1}\mathbf{A}\mathbf{B}'$. Thus, $G(\mathbf{c}, \mathbf{b}) = (\mathbf{d}, \mathbf{P}\mathbf{w})^2/\|\mathbf{d}\|^2$. By the Schwarz Inequality this is maximum (as a function of $\mathbf{c}$) for $\mathbf{d}$ any multiple of $\mathbf{P}\mathbf{w} = \mathbf{P}\mathbf{B}^{-1}\mathbf{Q}$, $\mathbf{c}$ any multiple of $\mathbf{B}^{-1}\mathbf{P}\mathbf{B}^{-1}\mathbf{b} = \mathbf{Q}\mathbf{b}$, with maximum value $\|\mathbf{P}\mathbf{w}\|^2 = \mathbf{b}'\mathbf{B}^{-1}\mathbf{P}\mathbf{B}^{-1}\mathbf{b} = \mathbf{b}'\mathbf{Q}\mathbf{b}$. This proves (1). (2) follows "easily" by substitution and lots of computation. The author uses "easily" when he wants the students to do the work. □

Continuation of Example 3.8.2: Since $\hat{\mathbf{Y}}_1 = (1/7)(-20, -10, -10, 70, 60)$, t_c^2 is maximized for $\mathbf{a}_c = \hat{\mathbf{Y}}_1$, $\mathbf{c} = \mathbf{X}'\hat{\mathbf{Y}}_1 = (1/7)(90, -40, -40)$, and $t_c^2 = (\mathbf{c}'\hat{\boldsymbol{\beta}})^2/\mathbf{c}'\mathbf{M}^{-1}\mathbf{c} = (1{,}300/7)^2/[1{,}300/7] = 1{,}300/7 = \|\hat{\mathbf{Y}}_1\|^2 = 185.7$.

Continuation of Example 3.8.1: Consider the corn example with $k = 3$, $n_1 = n_3 = 4$, and $n_2 = 3$. Take $\boldsymbol{\beta} = (\mu_1, \mu_2, \mu_3)'$, and $H_0: \mu_1 = \cdots = \mu_k$. Take $\mathbf{A} = \begin{bmatrix} 1 & -1 & 0 \\ 1 & 0 & -1 \end{bmatrix}$. This $\mathbf{A}$ could be replaced by $\mathbf{GA}$ for any nonsingular 2×2 matrix $\mathbf{G}$, so that the row space remains the same. The subspace C of R_3 is the row space of $\mathbf{A}$, the collection of 3-component vectors with components adding to zero. V_1 is the subspace of V of vectors of the form $\sum_{j=1}^{3} b_j \mathbf{J}_j$ which are orthogonal to $V_0 = \mathscr{L}(\mathbf{J})$. Thus $\sum b_j n_j = 0$. An example is $\mathbf{J}_1 - 4\mathbf{J}_2 + 2\mathbf{J}_3$. The ij element of $\hat{\mathbf{Y}}_1$ is $\bar{Y}_j - \bar{Y}$, where $\bar{Y} = \sum_{ij} Y_{ij}/n$. Thus, $\hat{\mathbf{Y}}_1 = \begin{bmatrix} -0.636 & 4.364 & -2.636 \\ -0.636 & 4.364 & -2.636 \\ -0.636 & 4.354 & -2.636 \\ -0.636 & & -2.636 \end{bmatrix}$, whose squared length is 86.55. t_c^2 is maximum for $\mathbf{c} = (n_1(\bar{Y}_1 - \bar{Y}), n_2(\bar{Y}_2 - \bar{Y}), n_3(\bar{Y}_3 - \bar{Y}))' = (-2.544, 13.092, -7.992)'$, and $\hat{\mathbf{Y}}_1 = \sum (c_j/n_j)\mathbf{J}_j = \sum_j (\bar{Y}_j - \bar{Y})\mathbf{J}_j$.

Problem 3.8.1: Let $\mathbf{x}_1 = (1, 1, 1, 1, 1, 1)'$, $\mathbf{x}_2 = (3, -1, 4, 6, 3, 3)'$, $\mathbf{x}_3 = (7, 3, 2, 0, 3, 3)'$, $\mathbf{x}_4 = (8, 4, 9, -5, 4, 4)'$, $\mathbf{Y} = (4, 36, 44, 12, 16, 8)'$ $V = \mathscr{L}(\mathbf{x}_1, \mathbf{x}_2, \mathbf{x}_3, \mathbf{x}_4)$. Suppose we wish to test $H_0: \beta_4 = 0, \beta_2 = \beta_3$.

(a) Find two matrices $\mathbf{A}$ so that $H_0 \Leftrightarrow \mathbf{A}\boldsymbol{\beta} = \mathbf{0}$.

(b) Find $\hat{\boldsymbol{\beta}}$, $\hat{\mathbf{Y}} = \mathbf{X}\hat{\boldsymbol{\beta}}$, and $\mathbf{Z} = \mathbf{A}\hat{\boldsymbol{\beta}}$, for one of your choices for $\mathbf{A}$.

(c) Define V_0 so that $H_0 \Leftrightarrow \boldsymbol{\theta} \in V_0$ and find $\hat{\mathbf{Y}}_0 = p(\mathbf{Y} \mid V_0)$, $\mathbf{Y} - \hat{\mathbf{Y}}_0$ and $\hat{\mathbf{Y}}_1 = \hat{\mathbf{Y}} - \hat{\mathbf{Y}}_0$.

(d) Determine $\text{ESS}_{FM} = \|\mathbf{Y} - \hat{\mathbf{Y}}\|^2$, $\text{SSE}_{H_0} = \|\mathbf{Y} - \hat{\mathbf{Y}}_0\|^2$, $\|\hat{\mathbf{Y}} - \hat{\mathbf{Y}}_0\|^2$, and the F-statistic.

(e) Verify that $\|\hat{\mathbf{Y}} - \hat{\mathbf{Y}}_0\|^2 = \mathbf{Z}'[\mathbf{AM}^{-1}\mathbf{A}']^{-1}\mathbf{Z}$.

(f) Find $\mathbf{c}$ and $\mathbf{a}_c$ so that $\|\hat{\mathbf{Y}} - \hat{\mathbf{Y}}_0\|^2/S^2 = t_c^2 = (\mathbf{a}_c, \mathbf{Y})^2/[S^2\|\mathbf{a}_c\|^2]$.

Problem 3.8.2: Let Ω be the space of arrays of the form of $\mathbf{Y}$ in one-way analysis of variance with $k = 3$, $n_1 = 3$, $n_2 = 4$, $n_3 = 3$, and let $\mathbf{J}_1, \mathbf{J}_2, \mathbf{J}_3$ be the corresponding column indicators. Let $\mathbf{x} = \begin{pmatrix} 1 & 2 & 2 \\ 2 & 1 & 3 \\ 3 & 4 & 4 \\ & 3 & \end{pmatrix}$. Suppose the model $\mathbf{Y} = \sum_1^3 \beta_j \mathbf{J}_j + \beta_4 \mathbf{x} + \boldsymbol{\varepsilon}$ holds and we observe $\mathbf{y} = \begin{pmatrix} 4 & 4 & 10 \\ 9 & 4 & 9 \\ 8 & 10 & 14 \\ & 6 & \end{pmatrix}$. Test at level

$\alpha = 0.05$ the null hypothesis that $\beta_1 = \beta_2 = \beta_3$. (The analysis justified by this model is called *analysis of covariance.* The y-values might be corn yields and the x-values fertility measurements on the corresponding plots).

Problem 3.8.3: Consider the baseball bounce example of Problem 3.3.4. Let type A, B and C baseballs be dropped from heights 5, 10 and 15 feet and let the rebound heights be as given.

	Height, x (ft)		
	5	10	15
A	2.4	6.0	9.2
B	2.4	4.9	7.6
C	3.7	7.2	10.3

State a model (assuming equal variances, a questionable hypothesis), and test the null hypotheses that the bounce coefficients are equal ($\alpha = 0.05$). If instead the standard deviation of bounce heights were proportional to height x, how could you proceed?

Problem 3.8.4: Consider the following regression model for $n = 20$ pairs (x_i, Y_i)

$$Y_i = \beta_0 + \beta_1 x_i + \beta_2 x_i^2 + \beta_3 x_1^3 + \varepsilon_i$$

for $\varepsilon_1, \ldots, \varepsilon_n$ independent $N(0, \sigma^2)$ random variables. The model above was fit, giving error sum of squares ESS(3) = 160. When the cubic term was omitted, the error sum of squares was ESS(2) = 180. When both the quadratic and cubic terms were omitted, the error sum of squares was ESS(1) = 200. Total sum of squares was ESS(0) = 1,000, after correction for the mean.

(a) Give the sample multiple correlation coefficient for the cubic model above.

(b) For $\alpha = 0.05$ test H_0: true model is the simple linear regression model $Y_i = \beta_0 + \beta_1 x_i + \varepsilon_i$.

(c) Find a matrix $\mathbf{A}$ such that H_0 of (b) is equivalent to $\mathbf{A}\boldsymbol{\beta} = 0$.

Problem 3.8.5: Show that the numerator sum of squares $H(\hat{\boldsymbol{\beta}})$ of Theorem 3.8.1 is not changed by replacing $\mathbf{A}$ by $\mathbf{GA}$ for $\mathbf{G}$ nonsingular.

Problem 3.8.6: Suppose you wished to test $H_0: \mathbf{A}\boldsymbol{\beta} = \mathbf{d}$ for $\mathbf{d} \in R_q$, a known vector of constants. How should you change $\mathbf{H}(\hat{\boldsymbol{\beta}})$ (Theorem 3.8.1)? It can be shown that $\mathbf{A}\boldsymbol{\beta} = \mathbf{d}$ is equivalent to $\boldsymbol{\theta} = \boldsymbol{\theta}_0 + \mathbf{v}$ for $\boldsymbol{\theta}_0 = \mathbf{B}(\mathbf{A}\mathbf{M}^{-1}\mathbf{A}')^{-1}\mathbf{d}$, $\mathbf{M} = \mathbf{X}'\mathbf{X}$, $\mathbf{B} = \mathbf{X}\mathbf{M}^{-1}\mathbf{A}'$, $\mathbf{v} \in V_0$ for V_0 as defined in Theorem 3.8.1. Thus, we can test the equivalent hypothesis that $\mathbf{Z} = \mathbf{Y} - \boldsymbol{\theta}_0$ has mean vector lying in V_0. The test statistic is therefore obtained by replacing $\mathbf{Y}$ by $\mathbf{Y} - \boldsymbol{\theta}_0$, equivalently by replacing $\hat{\boldsymbol{\beta}}$ by $\hat{\boldsymbol{\beta}} - \hat{\boldsymbol{\beta}}_0$ for $\hat{\boldsymbol{\beta}}_0 = \mathbf{M}^{-1}\mathbf{X}'\boldsymbol{\theta}_0$.

Problem 3.8.7: (Two-sample t-statistic) Let $(Y_{11}, \ldots, Y_{1n_1})$ and $(Y_{21}, \ldots, Y_{2n_2})$ be independent vectors, each having independent normally distributed components with variance all σ^2, common means μ_1 and μ_2. Let $\mathbf{Y}$ be the array with two columns, n_i elements in column i.

(a) Invent vectors $\mathbf{x}_1$, $\mathbf{x}_2$ so that $\mathbf{Y} = \mu_1\mathbf{x}_1 + \mu_2\mathbf{x}_2 + \boldsymbol{\varepsilon}$, with $\boldsymbol{\varepsilon}$ satisfying the usual model.

(b) Let $\bar{Y}_1$, $\bar{Y}_2$, S_1^2, S_2^2 be the sample means and variances for the two samples. Show that the least squares estimator of (μ_1, μ_2) is $(\bar{Y}_1, \bar{Y}_2,)$ and that

$$S^2 = [(n_1 - 1)S_1^2 + (n_2 - 1)S_2^2]/(n_1 + n_2 - 2).$$

(c) Show that

$$\bar{Y}_1 - \bar{Y}_2 \pm t_{\nu, 1-\alpha/2} S(1/n_1 + 1/n_2)^{1/2}, \qquad \text{for} \quad \nu = n_1 + n_2 - 2,$$

is a $100(1 - \alpha)\%$ confidence interval on $\eta = \mu_1 - \mu_2$.

(d) Consider H_0: $\eta = \mu_1 - \mu_2 = 0$. Show that H_0 is equivalent to $\boldsymbol{\theta} \in V_0$ for some V_0, and that, for this H_0, $F = t^2$ for $t = (\bar{Y}_1 - \bar{Y}_2)/[S(1/n_1 + 1/n_2)^{1/2}]$.

(e) Consider the following tire mileages (in 1,000's) of two brands of tires.

#1	41	49	45	41	
#2	51	48	46	48	47

Find a 95% confidence interval on $\mu_1 - \mu_2$ and test H_0: $\mu_1 \geq \mu_2$ vs. $H_1: \mu_1 < \mu_2$ for $\alpha = 0.05$.

(f) Suppose it is known that $\sigma_2^2 = r\sigma_1^2$ for r known but σ_1^2 and σ_2^2 unknown. Show that a $100(1 - \alpha)\%$ confidence interval on $\mu_1 - \mu_2$ is

$$\bar{Y}_1 - \bar{Y}_2 \pm t_{\nu, 1-\alpha/2} S(1/n_1 + r/n_2)^{1/2},$$

where

$$S^2 = \frac{(n_1 - 1)S_1^2 + (n_2 - 1)S_2^2/r}{n_1 + n_2 - 2}$$

Hint: Let $Z_i = Y_{2i}/\sqrt{r}$ and estimate $\mu_1 - \sqrt{r}(\mu_2/r)$. Find the interval for $r = 2$.

Problem 3.8.8: Show that the proof of (2) in Theorem 3.8.2 is really easy.

Problem 3.8.9: (Behrens–Fisher) Consider the two sample problem $Y_{i1}, \ldots, Y_{in_i} \sim N(\mu_i, \sigma_i^2)$, $i = 1, 2$, where all random variables are independent. This problem was considered in Problem 3.8.6. in the case $r\sigma_1^2 = \sigma_2^2$, and our standard linear theory applied to produce the well-known formulas for confidence intervals and tests. This theory does not apply in the case that the ratio $\sigma_2^2/\sigma_1^2 = r$ is not known. For large n_1, n_2 (say both > 20) S_i^2 with high probability will be close to σ_i^2 so that

$$Z = \frac{\bar{Y}_1 - \bar{Y}_2 - (\mu_1 - \mu_2)}{S_D} \qquad \text{for} \quad S_D^2 = \frac{S_1^2}{n_1} + \frac{S_2^2}{n_2}$$

has an approximate $N(0, 1)$ distribution, and $\bar{Y}_1 - \bar{Y}_2 \pm z_{1-\alpha/2} S_D$ is an approximate $100(1 - \alpha)\%$ confidence interval on $\mu_1 - \mu_2$.

For moderate n_1, n_2 a reasonable approximation due to Welch (1947) is obtained by replacing $z_{1-\alpha/2}$ by $t_{\nu, 1-\alpha/2}$ for ν the greatest integer less than or equal to

$$b = \frac{(n_1 - 1)(n_2 - 1)}{(n_2 - 1)a^2 + (n_1 - 1)(1 - a^2)} \qquad \text{for} \quad a = \frac{S_1^2/n_1}{(S_1^2/n_1) + (S_2^2/n_2)}$$

A study of Wang (1971) shows that the approximation is good for $\alpha \geq 0.05$ for $n_1, n_2 \geq 7$, or $\alpha \geq 0.005$ and $n_1, n_2 \geq 11$.

Apply the Welch method for the data of Problem 3.8.7(e). Even though the Wang conditions are not satisfied, the method should be reasonably good.

3.9 FURTHER DECOMPOSITION OF SUBSPACES

Every subspace W of dimension d may be decomposed into d mutually orthogonal one-dimensional subspaces $W_1, \ldots, W_d$. This is accomplished by finding an orthogonal basis $\mathbf{w}_1, \ldots, \mathbf{w}_d$ for W and taking $W_i = \mathscr{L}(\mathbf{w}_i)$. If these vectors, or, equivalently, the spaces W_i, are chosen appropriately then we may break the projection of $\mathbf{Y}$ onto W into the sum of its projections onto the subspaces W_i and, using the Pythagorean Theorem, break the squared length of this projection (sum of squares for d d.f.) into d sums of squares with one d.f. each. That is, $\|p(\mathbf{Y} \mid W)\|^2 = \sum_{i=1}^{d} (\mathbf{Y}, \mathbf{w}_i)^2/\|\mathbf{w}_i\|^2$.

For example, consider a one-way layout with n_i observations for treatment level i, $i = 1, \ldots, k$. The levels of treatment may correspond to measurements x_i (amount of a chemical, temperature, time, etc.), and it may be reasonable to suppose that $\mu_i = g(x_i)$ for some function $g(x)$. It may also be reasonable to suppose that $g(x_i) = \beta_0 + \beta_1 x_i$ for some β_0, β_1 or that $g(x_i) = \beta_0 + \beta_1 x_i + \beta_2 x_i^2$

for some β_0, β_1, β_2. In the second case the model becomes

$$\mathbf{Y} = \sum_{i=1}^{k} (\beta_0 + \beta_1 x_i + \beta_2 x_i^2)\mathbf{C}_i + \boldsymbol{\varepsilon} = \beta_0 \mathbf{J} + \beta_1 \sum_1^k x_i \mathbf{C}_i + \beta_2 \sum_1^k x_i^2 \mathbf{C}_i + \boldsymbol{\varepsilon}.$$

We can test H_0: $\{g(x)$ is linear in $x\}$, by taking $V_0 = \mathscr{L}(\mathbf{J}, \sum x_i \mathbf{C}_i) = \mathscr{L}(\mathbf{J}, \mathbf{x}^*)$ for $\mathbf{x}^* = \sum (x_i - \bar{x})\mathbf{C}_i$ in the usual F-test. Then

$$\hat{\mathbf{Y}}_0 = p(\mathbf{Y} \mid V_0) = \bar{Y}\mathbf{J} + \hat{\beta}_1 \mathbf{x}^* = \sum_{i=1}^{k} [\bar{Y} + \hat{\beta}_1(x_i - \bar{x})]\mathbf{C}_i$$

for

$$\hat{\beta}_1 = \frac{(\mathbf{Y}, \mathbf{x}^*)}{\|\mathbf{x}^*\|^2} = \frac{\sum (x_i - \bar{x})\bar{Y}_i n_i}{\sum n_i (x_i - \bar{x})^2}$$

This is the slope obtained in fitting a straight line to the points $(x_i, \bar{Y}_{i\cdot})$, each such point repeated n_i times. Then $\hat{\mathbf{Y}} - \hat{\mathbf{Y}}_0 = \sum_{i=1}^{k} [\bar{Y}_i - \bar{Y} - \hat{\beta}_1(x_i - \bar{x})]\mathbf{C}_i = \sum_{i=1}^{k} f_i \mathbf{C}_i$, so $\|\hat{\mathbf{Y}} - \hat{\mathbf{Y}}_0\|^2 = \sum_{i=1}^{k} f_i^2 n_i$ is the error sum of squares obtained in a simple linear regression on these points $(x_i, \bar{Y}_{i\cdot})$.

The noncentrality parameter is therefore

$$\|\boldsymbol{\theta} - \boldsymbol{\theta}_0\|^2/\sigma^2 = (1/\sigma^2) \sum_{i=1}^{k} [\mu_i - \bar{\mu} - \beta_1(x_i - \bar{x})]^2 n_i$$

for $\bar{\mu} = \sum n_i \mu_i / n$, and β_1 the same as $\hat{\beta}_1$ with μ_i replacing $\bar{Y}_i$. This is zero, of course, if the μ_i are linear functions of x_i.

In the notation of the first paragraph of this section we could take $W = V = \mathscr{L}(\mathbf{C}_1, \ldots, \mathbf{C}_k)$, $\mathbf{v}_0 = \mathbf{J}$,

$$\mathbf{v}_1 = \sum_1^k x_i \mathbf{C}_i, \qquad \mathbf{v}_2 = \sum x_i^2 \mathbf{C}_i, \ldots, \mathbf{v}_{k-1} = \sum_i x_i^{k-1} \mathbf{C}_i$$

and $\mathbf{w}_1, \ldots, \mathbf{w}_k$ a Gram–Schmidt orthogonalization of these $\mathbf{v}_i$'s.

$$p(\mathbf{Y} \mid V) = \sum p(\mathbf{Y} \mid \mathbf{w}_i) = \sum_i [(\mathbf{Y}, \mathbf{w}_i)/\|\mathbf{w}_i\|^2]\mathbf{w}_i$$

The treatment sum of squares is $\|\hat{\mathbf{Y}} - \hat{\mathbf{Y}}_0\|^2 = \sum_1^k (\mathbf{Y}, \mathbf{w}_i)^2/\|\mathbf{w}_i\|^2$. The ith term in this sum is the sum of squares due to the ith power of the x's. In the case that the x_i's are equally spaced, (of the form $x_i = x_0 + di$), relatively simple formulas for these sums of squares may be developed. To test

H_0: $\{g(x) = E(Y|x)$ is a quadratic function of $x\}$, assuming the one-way **A** of V model, we would use numerator sum of squares

$$\|\hat{\mathbf{Y}} - \hat{\mathbf{Y}}_0\|^2 - \sum_1^2 (\mathbf{Y}, \mathbf{w}_i)^2/\|\mathbf{w}_i\|^2 = \left[\sum_1^k \bar{Y}_{i.}^2 n_i - \bar{Y}_{..}^2 n\right] - \sum_1^2 (\mathbf{Y}, \mathbf{w})^2/\|\mathbf{w}\|^2$$

$$\text{for} \quad k - 1 - 2 = k - 3 \text{ d.f.}$$

Example 3.9.1: Take $k = 5$, $n_i = 10$ for $i = 1, \ldots, 5$, $\bar{Y}_1 = 40$, $\bar{Y}_{2.} = 45$, $\bar{Y}_{3.} = 48$, $\bar{Y}_{4.} = 46$, $\bar{Y}_{5.} = 43$, Error SSqs. = 720. Treatment SSqs. = 372, and $F = [372/4]/(720/36)] = 4.65$, which is significant at the $\alpha = 0.01$ level.

Suppose now that treatment level i corresponds to x_i for $x_1 = 3$, $x_2 = 5$, $x_3 = 7$, $x_4 = 9$, $x_5 = 11$. A plot of the $\bar{Y}_{i.}$ against these x_i (Figure 3.11) indicates a quadratic relation between the x_i and μ_i.

The treatment SSqs. may be composed into four independent sums of squares, each with one degree of freedom (Table 3.9.1). Take $\mathbf{x}_1$ to the array with 10 x_i's in the ith column. Let $\mathbf{x}_2, \mathbf{x}_3, \mathbf{x}_4$ be the arrays formed by replacing these x_i's by the second, third and fourth powers of these x_i's. Then $V = \mathscr{L}(\mathbf{J}, \mathbf{x}_1, \mathbf{x}_2, \mathbf{x}_3, \mathbf{x}_4)$. Take $W = V \cap V_0^{\perp}$. Use the Gram–Schmidt process to find vectors $\mathbf{w}_0 = \mathbf{J}$, $\mathbf{w}_1$, $\mathbf{w}_2$, $\mathbf{w}_3$, $\mathbf{w}_4$. Then $\mathscr{L}(\mathbf{w}_0, \mathbf{w}_1, \ldots, \mathbf{w}_i) = \mathscr{L}(\mathbf{J}, \mathbf{x}_1, \ldots, \mathbf{x}_i)$ for $i = 1, \ldots, 4$ and, of course, these $\mathbf{w}_i$ are mutually orthogonal. $W = \mathscr{L}(\mathbf{w}_1, \mathbf{w}_2, \mathbf{w}_3, \mathbf{w}_4)$. Since the x_i's increase linearly with i, and

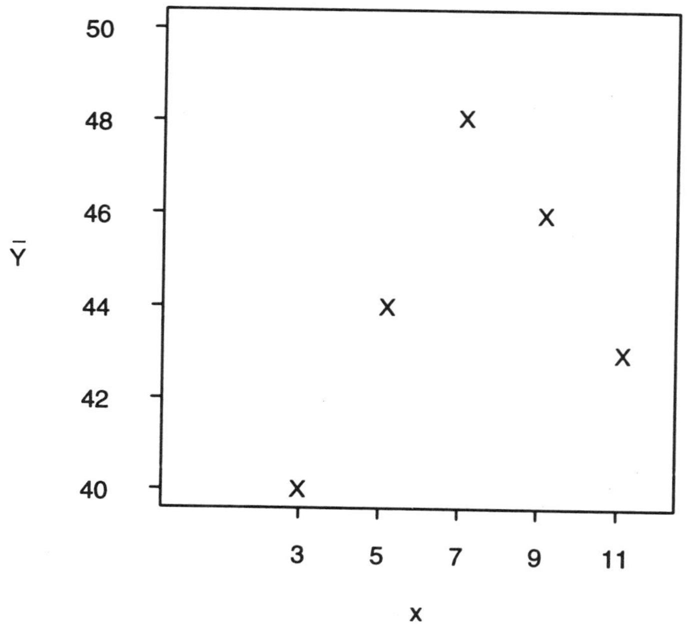

FIGURE 3.11 $\bar{Y}_i$ vs. x_i, $i = 1, 2, \ldots, 5$.

Table 3.9.1

	Column						
	1	2	3	4	5	Squared Length	Inner Product with **Y**
Linear $\mathbf{w}_1$	−2	−1	0	1	2	(10)(10)	(10)(7)
Quad. $\mathbf{w}_2$	2	−1	−2	−1	2	(10)(14)	(10)(−21)
Cubic $\mathbf{w}_3$	−1	2	0	−2	1	(10)(10)	(10)(1)
Quart. $\mathbf{w}_4$	1	−4	6	−4	1	(10)(70)	(10)(7)

the $\mathbf{w}_i$ may be multiplied by arbitrary constants, we can determine simple expressions for them. The jth column of these $\mathbf{w}_i$ have the following identical 10 values. Then

$$p(\mathbf{Y} \mid W) = \sum_{i=1}^{4} p(\mathbf{Y} \mid \mathbf{w}_i) = \sum_{i=1}^{4} \frac{(\mathbf{Y}, \mathbf{w}_i)}{\|\mathbf{w}_i\|^2} \mathbf{w}_i$$

and

$$\|p(\mathbf{Y} \mid W)\|^2 = \sum_{i=1}^{4} (\mathbf{Y}, \mathbf{w}_i)^2 / \|\mathbf{w}_i\|^2.$$

The inner products and squared lengths are given in the table above. If the elements in the jth column of $\mathbf{w}_i$ are all w_{ij} then $(\mathbf{Y}, \mathbf{w}_i) = \sum_{j=1}^{5} \bar{Y}_{j\cdot} \mathbf{w}_{ij}(10)$ and $\|\mathbf{w}_i\|^2 = \sum_{i=1}^{5} w_i^2(10)$. Thus, treatment SSqs. (372) has been decomposed into linear SSqs. ($70^2/100 = 49$), quadratic SSqs. $(-210)^2/140 = 315$), cubic SSqs. ($10^2/10 = 1$) and quartic SSqs. ($70^2/700 = 7$). Each has one degree of freedom. Obviously in this case the quadratic effect dominates. The null hypothesis H_0: $\{\mu_1$ is linear in $x_i\}$ is equivalent to $\boldsymbol{\theta} \in \mathscr{L}(\mathbf{J}, \mathbf{x}_1) = \mathscr{L}(\mathbf{J}, \mathbf{w}_1) = V_1$ (say). The numerator SSqs. in the F-statistic is therefore $\|p(\mathbf{Y} \mid V \cap V_1^{\perp})\|^2$. But $V \cap V_1^{\perp} = \mathscr{L}(\mathbf{w}_2, \mathbf{w}_3, \mathbf{w}_4)$, so that

$$\sum_{2}^{4} (\mathbf{Y}, \mathbf{w}_i)^2 / \|\mathbf{w}_i\|^2 = \text{Treatment SSqs.} - (\mathbf{Y}, \mathbf{w}_1)^2 / \|\mathbf{w}_1\|^2. \tag{3.9.1}$$

In this case, $372 - 49 = 323$, so that $F = \dfrac{323/3}{20} = 5.38$ and H_0 is rejected. Similarly, H_0: (μ_i is a quadratic function of x_i) has numerator SSqs. $= \sum_{3}^{4} (\mathbf{Y}, \mathbf{w}_i)^2 / \|\mathbf{w}_i\|^2 = 8$ so $F = \dfrac{8/2}{20} = 0.2$ and H_0 is not rejected (at reasonable α levels).

3.10 POWER OF THE F-TEST

In order to determine the power of an F-test we need a means of computing $P(F \geq F_{\nu_1, \nu_2, \alpha})$ for given values of ν_1, ν_2, α and the noncentrality parameter δ, for F. The most common means of presenting these probabilities uses graphs $p(\phi)$, for $\phi = \sqrt{\delta/(\nu_1 + 1)}$, one graph for each combination of ν_2, ν_1 and $\alpha = 0.05$ and 0.01. Pearson–Hartley charts present graphs of $p(\phi)$; see Tables 5.1–5.8 in the Appendix. For example, for $\nu_1 = 2$, $\alpha = 0.05$, $\nu_2 = 30$, $\delta = 12$ we get $\phi = \sqrt{12/3} = 2$ and power approximately 0.85. Odeh and Fox (1991) describe methods and provide charts which facilitate the finding of sample sizes necessary to achieve given power.

Recall that the noncentrality parameter is $\delta = \|\boldsymbol{\theta}_1\|^2/\sigma^2$, where $\boldsymbol{\theta}_1 = p(\boldsymbol{\theta} \mid V_1)$ and $V_1 = V \cap V_0^{\perp}$. In Section 3.9 we showed that $\|\hat{\mathbf{Y}}_1\|^2 = \mathbf{Z}'[\mathbf{A}'\mathbf{M}^{-1}\mathbf{A}]^{-1}\mathbf{Z}$, when H_0 is expressed in the form $\mathbf{A}\boldsymbol{\beta} = \mathbf{0}$, and $\mathbf{Z} = \mathbf{A}\hat{\boldsymbol{\beta}}$. We need only replace $\mathbf{Y}$ by $\boldsymbol{\theta}$ in this formula or in any other formula we have for the numerator sum of squares in the F-statistic. Thus, for $\boldsymbol{\zeta}$ (zeta) $= \mathbf{A}\boldsymbol{\beta}$, $\delta = (1/\sigma^2)\,\boldsymbol{\zeta}'[\mathbf{A}\mathbf{M}^{-1}\mathbf{A}']\boldsymbol{\zeta}$.

Power of the F-Test in One-Way A of V: Suppose the statistic F has a noncentral F distribution with ν_1 and ν_2 d.f. and noncentrality parameter δ. Consider also that under the null hypothesis H_0: $\delta = 0$ and suppose that H_0 is to be rejected for $F \geq F_{\nu_1, \nu_2, 1-\alpha}$ (which we also denote by $F_{1-\alpha}(\nu_1, \nu_2)$.) This is the situation in one-way analysis of variance when we test H_0: $\mu_1 = \mu_2 = \cdots = \mu_k$. In this case

$$F = \frac{\text{Among means mean square}}{\text{Error mean square}}$$

and

$$\delta = \frac{\|\boldsymbol{\theta} - \boldsymbol{\theta}_0\|^2}{\sigma^2} = (1/\sigma^2) \sum_{j=1}^{k} (\mu_j - \bar{\mu})^2 n_j,$$

where $\boldsymbol{\theta}_0 = p(\boldsymbol{\theta} \mid V_0)$ for $V_0 = \mathscr{L}(\mathbf{J})$ and $\bar{\mu} = \left(\sum_{1}^{k} n_j \mu_j\right)/n$.

Example 3.10.1: This example is taken from Scheffé (1959, p. 163) who in turn credits it to Cuthbert Daniel, a well-known statistical consultant. Suppose four different kinds of alloy steel are prepared by varying the method of manufacture. It is expected that the tensile strength will be of order 150,000 psi (pounds per square inch) and the standard deviation of duplicate specimens from the same batch will be about 3,000 psi. Suppose 10 specimens of each kind are tested, and that

$$\mu_1 = 150{,}000, \qquad \mu_2 = 149{,}000, \qquad \mu_3 = 148{,}000, \qquad \mu_4 = 153{,}000.$$

What is the power of the resulting $\alpha = 0.05$ level F-test?

Solution: We have

$$\delta = (1/\sigma^2) \sum_1^4 (\mu_j - \bar{\mu})^2 n_j = \frac{10}{(3{,}000)^2}\left[\sum_1^4 (\mu_j - 150{,}000)^2\right] = \frac{10(14)}{9} = 15.56,$$

$\nu_1 = 4 - 1 = 3$, $\nu_2 = 40 - 4 = 36$, $\phi = \sqrt{\dfrac{15.56}{4}} = 1.96$. Then from the Pearson–Hartley charts the power is approximately 0.88.

Similarly for $n_j = 5$ observations per alloy type we get $\delta = \dfrac{5(14)}{9} = 7.78$, $\nu_1 = 3$, $\nu_2 = 20 - 4 = 16$, $\phi = \sqrt{7.78/4} = 1.39$, power approximately 0.60.

Example 3.10.2: Continuing the alloy example, suppose we wish to design an experiment which will have power at least 0.90 for $\alpha = 0.05$ in the case that any two of the four means differ by 8,000 or more. How many observations should be taken on each alloy?

Solution: $\delta = n_0 \sum_1^4 (\mu_j - \bar{\mu})^2/\sigma^2$, where n_0 is the common sample size. If two means are to differ by 8,000 or more, then $\sum (\mu_j - \bar{\mu})^2$ is smallest when one μ_j is 4,000 larger than $\bar{\mu}$, one is 4,000 smaller than $\bar{\mu}$, and the other two μ_j are equal to $\bar{\mu}$ (see Problem 3.10.2). Since the power is an increasing function of δ we must choose n_0 so that 0.90 power is achieved for the smallest possible value of δ. Then $\delta = n_0[4{,}000^2 + 0^2 + 0^2 + (-4{,}000)^2]/\ 3{,}000^2 = \dfrac{32}{9} n_0$. Also $\nu_1 = 4 - 1 = 3$, $\nu_2 = 4(n_0 - 1)$, $\phi = \sqrt{\delta/4} = 0.943\sqrt{n_0}$. We can proceed by trial and error.

For $n_0 = 9$, $\phi = 2.828$, $\nu_2 = 32$, power $\doteq 1.00$. We should try a smaller n_0. For $n_0 = 4$, $\phi = 1{,}846$, $\nu_2 = 12$, power $= 0.76$. For $n_0 = 5$, $\phi = 2.11$, $\nu_2 = 16$, power $= 0.89$. For $n_0 = 6$, $\phi = 2.30$, $\nu_2 = 20$, power $= 0.955$. $n_0 = 5$ seems to be approximately the right sample size.

Problem 3.10.1: Consider Problem 3.8.1, let H_0: $\beta_2 = \beta_3$, $\beta_4 = 0$, let $\mathbf{A}$ be defined as before, and let $\boldsymbol{\zeta} = \mathbf{A}\boldsymbol{\theta}$.

(a) Express $\boldsymbol{\theta}_1 = \boldsymbol{\theta} - \boldsymbol{\theta}_0$ as a linear combination of two vectors $\mathbf{v}_1$ and $\mathbf{v}_2$, with coefficients $(\beta_2 - \beta_3)$ and β_4. *Hint*: Define $\mathbf{x}_5 = \mathbf{x}_2 + \mathbf{x}_3$, $\mathbf{x}_j^\perp = p(\mathbf{x}_j |\ V_1 = V \cap V_0^\perp)$, for $j = 2, 3, 4$. Show that $\boldsymbol{\theta}_1 = \beta_2 \mathbf{x}_2^\perp + \beta_3 \mathbf{x}_3^\perp + \beta_4 \mathbf{x}_4^\perp$ and that $\mathbf{x}_2^\perp + \mathbf{x}_3^\perp = \mathbf{0}$.

(b) Express the noncentrality parameter δ as a quadratic form in $(\beta_1 - \beta_2)$ and β_4 by using the result of (a) and also by using the formula $\delta = \boldsymbol{\zeta}'[\mathbf{A}\mathbf{M}^{-1}\mathbf{A}']^{-1}\boldsymbol{\zeta}/\sigma^2$.

(c) For $\boldsymbol{\beta} = (10, 3, 5, -2)'$, $\sigma^2 = 16$, and $\alpha = 0.05$, find $\boldsymbol{\theta}$, $\boldsymbol{\theta}_0$, $\boldsymbol{\theta}_1$, δ, and the power of the F-test.

Problem 3.10.2: Let $\Delta > 0$. Then $G(x_1, \ldots, x_n) = \sum (x_i - \bar{x})^2$ is minimum, subject to $\max_{ij} |x_i - x_j| \geq 2\Delta$ for $(n-2)$ x_i's equal to $\bar{x}$, one x_i equal to $\bar{x} - \Delta$, one equal to $\bar{x} + \Delta$. Prove this. *Hint*: Let the two x_i's differing by 2Δ be x_1 and x_2. Let $\bar{x}_2 = (x_1 + x_2)/2$. Show that $G = \Delta^2/2 + 2(\bar{x}_2 - \bar{x})^2 + \sum_3^n (x_i - \bar{x})^2$.

Problem 3.10.3: (a) Suppose that for the corn yield in Example 3.8.1 the true means were 70, 75, 95 and that $\sigma = 20$. Find the power of the $\alpha = 0.05$ level test for equal means.

(b) How large should n_0, the number of observations per treatment (number of plots per treatment) be in order to have power at least 0.90 for the parameters in (a)?

(c) Suppose we wish to design an experiment with the three kinds of fertilizer which will have probability at least 0.90 of rejecting H_0 for $\alpha = 0.05$ when two means differ by 10 or more, and $\sigma = 20$. How large should n_0 be?

Problem 3.10.4: Consider a one-way layout with $k = 4$, $n_1 = n_2 = n_3 = 5$, $n_4 = 6$. Let $x_1 = 2$, $x_2 = 3$, $x_3 = 5$, $x_4 = 6$.

(a) For $\mu_1 = 4$, $\mu_2 = 11$, $\mu_3 = 17$, $\mu_4 = 16$, $\sigma = 4$, find the power of the $\alpha = 0.05$ level test of the null hypotheses that μ_i is a linear function of x_i for $i = 1, 2, 3, 4$. *Hint*: See Problem 3.10.1. The noncentrality parameter is obtained by replacing $\mathbf{Y}$ by $\boldsymbol{\theta}$.

(b) For equal sample sizes n_0, how large would n_0 have to be in order for the test in (a) to have power at least 0.90?

Problem 3.10.5: Suppose $g(x) = E(\mathbf{Y} \mid x)$ and $Y_{ij} = g(x_j) + \varepsilon_{ij}$ is observed for $i = 1, \ldots, r$ and $j = 1, 2, 3, 4$ for $\varepsilon_{ij} \sim N(0, \sigma^2)$, independent. $x_1 = 1$, $x_2 = 2$, $x_3 = 4$, $x_4 = 5$.

(a) Assuming the full model for which g is an arbitrary function of x, express the noncentrality parameter for the F-test of H_0: $\{g \text{ is a linear in } x\}$ as a function of r, β_2, and σ^2 for $g(x) = \beta_0 + \beta_1 x + \beta_2 x^2$.

(b) Evaluate the power for $r = 5$, $\beta_2 = 0.05$, $\sigma^2 = 3$, $\alpha = 0.05$.

(c) Determine the minimum value of r for which the power of the test is 0.90, for $\sigma^2 = 3$, $\alpha = 0.05$, $\beta_2 = 0.05$.

3.11 CONFIDENCE AND PREDICTION INTERVALS

Let $\tilde{\mathbf{x}} = (x_1, \ldots, x_k)$, and $g(\tilde{\mathbf{x}}) = E(\mathbf{Y} \mid \tilde{\mathbf{x}})$ be the regression function for $\mathbf{Y}$ on $\tilde{\mathbf{x}}$. Suppose also that $g(\tilde{\mathbf{x}}) = \sum_1^k \beta_j x_j$, and that we observe $Y_i = g(\tilde{\mathbf{x}}_i) + \varepsilon_i$ for

$\varepsilon_i \sim N(0, \sigma^2)$ independently for $i = 1, \ldots, n$, where $\tilde{x}_1, \ldots, \tilde{x}_n$ are n values of $\tilde{x}$. That is, we independently observe pairs $(Y_i, \tilde{x}_i)$. Suppose that $\tilde{\mathbf{x}}_0 = (x_{01}, \ldots, x_{0k})$, is still another constant vector and that we wish to estimate $\eta = g(\tilde{\mathbf{x}}_0)$.

For example, for the simple linear regression model $Y_i = \beta_0 + \beta_1 x_i + \varepsilon_i$, equivalently $\mathbf{Y} = \beta_0 \mathbf{J} + \beta_1 \mathbf{x} + \boldsymbol{\varepsilon}$, suppose we want a confidence interval on $g(x_0) = \beta_0 + \beta_1 x_0$. Then $\tilde{\mathbf{x}}_0 = (1, x_0)$.

Since $\eta = g(\tilde{\mathbf{x}}_0) = \sum_1^k \beta_j x_{0j}$ is a linear function of the β_j, we can use the BLUE, $\hat{\eta} = \hat{g}(\tilde{\mathbf{x}}_0) = \tilde{\mathbf{x}}_0 \hat{\boldsymbol{\beta}}$, the best linear unbiased estimator of η. Again, from Section 3.2, $\text{Var}(\hat{\eta}) = (\tilde{\mathbf{x}}_0 \mathbf{M}^{-1} \tilde{\mathbf{x}}_0')\sigma^2 = h(x_0)\sigma^2$ (say), so $S_{\hat{\eta}}^2 = h(x_0)S^2$ and a $100(1 - \alpha)\%$ confidence interval on η is given by $g(\tilde{\mathbf{x}}_0) \pm t_{1-\alpha/2}[h(\tilde{\mathbf{x}}_0)S^2]^{1/2}$, where $t_{1-\alpha/2}$ has $(n - k)$ d.f.

For simple linear regression $\hat{\eta} = \hat{g}(\tilde{\mathbf{x}}_0) = \hat{\beta}_0 + \hat{\beta}_1 x_0 = \bar{Y} + \hat{\beta}_1(x_0 - \bar{x})$, so

$$\text{Var}(\hat{\eta}) = \sigma^2 \left[\frac{1}{n} + (x_0 - \bar{x})^2 / S_{xx} \right] = \sigma^2 h(x_0).$$

Thus, $g(x_0)$ is estimated most precisely for x_0 near $\bar{x}$. In fact the variance of $\hat{g}(x_0)$ is the sum of the variance in estimating the height of the line at $x = \bar{x}$ and the variance in estimating the slope, multiplied by the square of the distance of x from $\bar{x}$. Of course, the slope is estimated most precisely if the x-values used to estimate it are more widely spread, resulting in larger S_{xx}, subject to the suitability of the model.

Suppose that we want confidence intervals on $g(\tilde{\mathbf{x}})$ for each of the n rows of a design matrix $\mathbf{X}$, the points in k-space at which $\mathbf{Y}$ has been observed. The value $h(\tilde{\mathbf{x}})$ can be obtained for all such $\tilde{\mathbf{x}}$ very easily as the diagonal of $\mathbf{X}\mathbf{M}^{-1}\mathbf{X}'$, which is the projection matrix $\mathbf{P}_V$ onto the column space of $\mathbf{X}$. Since trace $(\mathbf{P}_V) = \dim(V) = k$, it follows that these $h(\tilde{\mathbf{x}})$ average k/n. For simple linear regression $h(\tilde{x}_i) = 1/n + (x_i - \bar{x})^2/S_{xx}$, so that $\sum h(x_i) = 1 + [\sum (x_i - \bar{x})^2]/S_{xx} = 2$. More on this in Sections 4.4 and 4.6.

Prediction Intervals: Suppose that we would like to predict the value Y_0 of a future observation, to be taken at a point $\tilde{\mathbf{x}} = \tilde{\mathbf{x}}_0$. This is a tougher problem because, while $g(\tilde{\mathbf{x}}_0)$ was a fixed target, Y_0 is random, a moving target. An analogy would have an archer shoot $n = 25$ arrows at a "bullseye" target located on a wall, after which the target is removed, with the arrows remaining. A confidence interval corresponds to a guess we make about the location of center of the target based on our observation of the location of the arrows. A prediction interval is analogous to a guess as to the location of another arrow not yet shot by the same archer. Obviously our prediction should be $\hat{g}(\tilde{\mathbf{x}}_0) \equiv \hat{Y}_0$. Let the error made be $e_0 = Y_0 - \hat{Y}_0$. Then $E(e_0) = g(\tilde{x}_0) - g(\tilde{x}_0) = 0$, and $\text{Var}(e_0) = \text{Var}(Y_0) + \text{Var}(\hat{Y}_0)$, since Y_0 is independent of the observations used to determine $\hat{Y}_0$. Thus, $\text{Var}(e_0) = \sigma^2 + \sigma^2 h(\tilde{\mathbf{x}}_0) = \sigma^2[1 + h(\tilde{\mathbf{x}}_0)]$. A $100(1 - \alpha)\%$

prediction interval on Y_0 is therefore given by $\hat{g}(\tilde{\mathbf{x}}_0) \pm t_{1-\alpha/2}[S^2 k(x_0)]^{1/2}$ for $k(\tilde{\mathbf{x}}_0) = 1 + h(\tilde{\mathbf{x}}_0)$. For simple linear regression $k(\tilde{\mathbf{x}}_0) = 1 + \frac{1}{n} + (x_0 - \bar{x})^2/S_{xx}$.

Problem 3.11.1: A Ph.D. candidate in education did a study (actually, this is fictitious data, but it could be real!) of the relationship between hours of study (x) and grade point average at a large Midwestern university, whose name shall be protected. Fifty students were chosen at random from among those who had earned at least 30 semester credits, and the number of hours each spent studying during the fall semester was carefully recorded, using personal diaries kept by the students. Interviews with the students during the semester convinced the Ph.D. candidate that the numbers of hours reported were reasonably accurate. The number of hours spent studying during the term, x_1, the previous G.P.A. (x_2), and the G.P.A. for the fall term, Y, were all recorded. Consider the data of Table 3.11.1 and the inner product matrices $\mathbf{M} = \mathbf{X}'\mathbf{X}$ and $\mathbf{U} = \mathbf{X}'\mathbf{Y}$. $\mathbf{X}$ is the 50×3 matrix with ones in the first column, x_1 and x_2 values in the second and third columns.

(a) Find the least squares simple linear regression line for Y vs. x_1, and sketch the estimated regression line on the scatter diagram. Also determine S^2 and the correlation coefficient (c.c.) r_{yx_1}.

(b) Let the 95% confidence interval for $g(x_1) = \beta_0 + \beta_1 x_1$ be $(L(x_1), U(x_1))$. Sketch the two functions $L(x_1)$ and $U(x_1)$ on the same axes.

(c) Let $(L_p(x_1), U_p(x_1))$ be the corresponding prediction intervals for a student who studies x_1 hours. Sketch these intervals.

(d) What conclusions can you reach about the relationship between studying and G.P.A. for the fall semester? Would it be better to study the partial c.c. between x_1 and Y, with the effects of x_2 removed? (Computations show that $r_{x_1x_2} = 0.3862$ and $r_{yx_1} = 0.5754$, so you should be able to find this partial c.c. without difficulty.) How much higher could you expect a student's G.P.A. to be if the student studies 100 more hours during the term? (Think about this; be careful about your conclusions.)

Problem 3.11.2: Let $x_0, x_1, \ldots, x_n$ be positive constants. Let $Y_i = \beta x_i + \varepsilon_i$ for $i = 1, \ldots, n$ for $\varepsilon_0, \ldots, \varepsilon_n$ independent $N(0, \sigma^2)$. The pairs (x_i, Y_i) have been observed for $i = 1, \ldots, n$ and the value of Y_0 is to be predicted.

(a) Give a formula for a $100(1 - \alpha)\%$ prediction interval on Y_0.

(b) Apply the formula for $n = 4$, $\alpha = 0.05$, and (x_i, Y_i) pairs (1, 2), (2, 7), (3, 10), (4, 11), $x_0 = 5$. Repeat for the same pairs, but $x_0 = 10$.

(c) Repeat (a) and (b) for confidence intervals on $g(x_0) = \beta x_0$, rather than prediction intervals.

Problem 3.11.3: Let $Y_1, \ldots, Y_n$ be a random sample from a $N(\mu, \sigma^2)$ distribution. Give a formula for a $100(1 - \alpha)\%$ prediction interval on an observation Y_0 to be taken independently from the same distribution. *Hint*: See Problem 3.11.1.

Table 3.11.1

#	x_1	x_2	Y	#	x_1	x_2	Y
1	303	2.74	2.85	26	396	3.18	3.22
2	206	2.46	2.12	27	416	2.98	3.54
3	247	3.00	2.81	28	350	2.59	2.94
4	234	2.82	2.46	29	387	2.91	3.25
5	266	2.74	2.74	30	296	3.10	3.19
6	365	2.28	3.08	31	288	2.67	2.19
7	331	2.15	2.45	32	303	2.66	2.26
8	337	2.82	2.79	33	353	2.61	3.24
9	369	3.00	3.15	34	217	3.01	2.80
10	391	3.30	3.34	35	349	3.52	3.72
11	366	2.61	2.66	36	359	2.99	2.96
12	355	3.15	3.34	37	157	2.39	2.24
13	208	3.06	2.32	38	372	3.50	3.76
14	287	3.57	2.90	39	333	3.55	2.73
15	315	2.74	3.05	40	226	2.45	2.54
16	508	3.29	3.98	41	235	2.97	3.00
17	308	2.86	2.79	42	289	2.76	2.73
18	263	2.38	2.74	43	307	2.53	2.06
19	323	2.88	2.28	44	408	2.91	3.98
20	251	2.77	2.67	45	247	2.36	2.34
21	125	2.25	1.83	46	268	3.12	3.22
22	245	3.20	2.29	47	305	3.02	3.10
23	392	2.48	2.73	48	358	3.24	2.82
24	261	2.15	2.30	49	358	3.27	2.93
25	256	2.90	2.86	50	115	2.52	1.63

$$\mathbf{M} = \begin{bmatrix} 50.00 & 15{,}204.00 & 142.41 \\ 15{,}204.00 & 4{,}910{,}158.00 & 43{,}834.26 \\ 142.41 & 43{,}834.26 & 412.18 \end{bmatrix}$$

$$\mathbf{U} = \begin{bmatrix} 140.92 \\ 44{,}300.23 \\ 406.66 \end{bmatrix}, \qquad \sum Y_i^2 = 410.778.$$

Problem 3.11.4: A chemist has two methods of determining the amount of a chemical in samples of blood. Method A is expensive, but is quite precise. Method B is inexpensive, but somewhat imprecise. Label measurements under method A by x, and under B by y. Because of the extra cost of the x measurements, it is preferable to obtain y, but not x, but to give a statement about the uncertainty of the measurement. This is the "calibration problem", discussed by Scheffé (1973) much more thoroughly.

Suppose that pairs (x_i, Y_i) are observed independently for $i = 1, \ldots, n$, where the x_i's are constants and $Y_i = g(x_i) + \varepsilon_i$, $g(x) = \beta x$, and $\varepsilon_i \sim N(0, \sigma^2)$. An additional observation Y_0 is made on another blood sample using method B, but the corresponding measurement x_0, using method A is not made.

(a) Find functions L and U, depending on the pairs (x_i, Y_i) and Y_0 so that $P(L \le x_0 \le U) = 0.95$. *Hint*: $T = (Y_0 - \hat{g}(x_0))/\sqrt{k(x_0)}\,S$ has a t distribution for the proper choice of the function $k(x_0)$. Use this as a pivotal quantity, but don't forget that both the numerator and denominator of T depend on x_0.)

(b) Apply your method to the data:

x	3.13	4.45	5.64	6.79
Y	2.6	4.1	5.1	6.0

and $Y_0 = 7.5$

(c) Let $g(x) = \beta_0 + \beta_1 x$, with β_0, β_1 unknown. For observations

x	2.21	3.54	4.89	5.96
Y	0.8	2.2	3.3	4.5

and $Y_0 = 6.7$

find a 95% confidence interval on x_0, the x-reading corresponding to Y_0.

Problem 3.11.5: Let $h(x) = 1/n + (x - \bar{x})^2/S_{xx}$, as defined earlier.

(a) Use the fact that $h(x_i)$ is the ith diagonal term of a projection matrix to prove that $0 \le h(x_i) \le 1$ for each i.

(b) Use the inequality in (a) to give an upper bound for $|x_i - \bar{x}|$ in terms of the sample standard deviation of the x_i's and n. Could the upper bound be achieved for some choice of $(x_1, \ldots, x_n)$?

3.12 AN EXAMPLE FROM SAS

The following study was carried out at North Carolina State University in order to determine the relationship between oxygen consumption (y), a measure of aerobic fitness, and several other variables related to physical fitness among 31 runners. For each individual the following measurements were made.

Y = oxygen consumption in volume per unit body weight per unit time (oxy)
x_1 = time to run $1\frac{1}{2}$ miles (runtime)
x_2 = age in years (age)
x_3 = weight in kilograms (weight)
x_4 = pulse rate at end of run (runpulse)
x_5 = maximum pulse rate (maxpulse)
x_6 = resting pulse rate (restingpulse)

The following analysis (Tables 3.12.1 through 3.12.9) was taken from the *SAS User's Guide: Statistics Version 5.*

The analysis will treat the x-variables as constants, though it may seem reasonable to consider them as random. However, we are interested in the conditional distribution of Y, given the x-variables, and not in the distribution of the x-variables. In fact the manner of selection of the participants in the study does not support conclusions about the joint distribution of the x-variables.

(1) In Table 3.12.3 under each correlation coefficient r, $P(|R| > r)$ is given for R having the distribution of a sample c.c. from a bivariate normal distribution with $\rho = 0$.

Table 3.12.1

i	x_1	x_2	x_3	x_4	x_5	x_6	Y
1	11.37	44	89.47	178	182	62	44.609
2	10.07	40	75.07	185	185	62	45.313
3	8.65	44	85.84	156	168	45	54.297
4	8.17	42	68.15	166	172	40	59.571
5	9.22	38	89.02	178	180	55	49.874
6	11.63	47	77.45	176	176	58	44.811
7	11.95	40	75.98	176	180	70	45.681
8	10.85	43	81.19	162	170	64	49.091
9	13.08	44	81.42	174	176	63	39.442
10	8.63	38	81.87	170	186	48	60.055
11	10.13	44	73.03	168	168	45	50.541
12	14.03	45	87.66	186	192	56	37.388
13	11.12	45	66.45	176	176	51	44.754
14	10.60	47	79.15	162	164	47	47.273
15	10.33	54	83.12	166	170	50	51.855
16	8.95	49	81.42	180	185	44	49.156
17	10.95	51	69.63	168	172	57	40.836
18	10.00	51	77.91	162	168	48	46.672
19	10.25	48	91.63	162	164	48	46.774
20	10.08	49	73.37	168	168	67	50.388
21	12.63	57	73.37	174	176	58	39.407
22	11.17	54	79.38	156	165	62	46.080
23	9.63	52	76.32	164	166	48	45.441
24	8.92	50	70.87	146	155	48	54.625
25	11.08	51	67.25	172	172	48	45.118
26	12.88	54	91.63	168	172	44	39.203
27	10.47	51	73.71	186	188	59	45.790
28	9.93	57	59.08	148	155	49	50.545
29	9.40	49	76.32	186	188	56	48.673
30	11.50	48	61.24	170	176	52	47.920
31	10.50	52	82.78	170	172	53	47.467

Source: Reprinted with permission from *SAS/Stat (TM) User's Guide*, Release 6.03 Edition, Cary, NC. © 1988 SAS Institute Inc.

Table 3.12.2

Variable	N	Mean	Standard Deviation	Sum	Minimum	Maximum
Oxy	31	47.375 806 45	5.327 230 50	1,468.650 000 00	37.388 000 00	60.055 000 00
Runtime	31	10.586 129 03	1.387 414 09	328.170 000 00	8.170 000 00	14.030 000 00
Age	31	47.677 419 35	5.211 443 16	1,478.000 000 00	38.000 000 00	57.000 000 00
Weight	31	77.444 516 13	8.328 567 64	2,400.780 000 00	59.080 000 00	91.630 000 00
Runpulse	31	169.645 161 29	10.251 986 43	5,259.000 000 00	146.000 000 00	186.000 000 00
Maxpulse	31	173.774 193 55	9.164 095 44	5,387.000 000 00	155.000 000 00	192.000 000 00
Restingpulse	31	53.451 612 90	7.619 443 15	1,657.000 000 00	40.000 000 00	70.000 000 00

Table 3.12.3 Pearson Correlation Coefficients, Prob > $|R|$ Under H_0: $RH_0 = 0$, $N = 31$

	Oxy	Runtime	Age	Weight	Runpulse	Maxpulse	Restingpulse
Oxy	1.000 00 0.000 0	−0.862 19 0.000 1	−0.304 59 0.095 7	−0.162 75 0.381 7	−0.397 97 0.026 6	−0.236 74 0.199 7	−0.399 36 0.026 0
Runtime	−0.862 19 0.000 1	1.000 00 0.000 0	0.188 75 0.309 2	0.143 51 0.441 2	0.313 65 0.085 8	0.226 10 0.221 3	0.450 38 0.011 0
Age	−0.304 59 0.095 7	0.1885 7 0.309 2	1.000 00 0.000 0	−0.233 54 0.0206 1	−0.337 87 0.063 0	−0.432 92 0.015 0	−0.164 10 0.377 7
Weight	−0.162 75 0.381 7	0.143 51 0.441 2	−0.233 54 0.206 1	1.000 00 0.000 0	0.181 52 0.328 4	0.249 38 0.176 1	0.043 97 0.814 3
Runpulse	−0.397 97 0.026 6	0.316 65 0.085 8	−0.337 87 0.063 0	0.181 52 0.328 4	1.000 00 0.000 0	0.929 75 0.000 1	0.352 46 0.051 8
Maxpulse	−0.236 74 0.199 7	0.226 10 0.221 3	−0.432 92 0.015 0	0.249 38 0.176 1	0.929 75 0.000 1	1.000 00 0.000 0	0.305 12 0.095 1
Restingpulse	−0.399 36 0.026 0	0.450 38 0.011 0	−0.164 10 0.377 7	0.043 97 0.814 3	0.352 46 0.051 8	0.305 12 0.095 1	1.000 00 0.000 0

Table 3.12.4

Obs.	_Type_	_Name_	Oxy	Runtime	Age	Weight	Runpulse	Maxpulse	Resting-pulse
1	Mean		47.375 8	10.586 1	47.677 4	77.444 5	169.645	173.774	53.451 6
2	Standard deviation		5.327 23	1.387 41	5.211 44	8.328 57	10.252	9.164 1	7.619 44
3	*N*		31.000 000	31.000 000	31.000 000	31.000 000	31.000 000	31.000 000	31.000 000
4	Correlation	Oxy	1.000 000	−0.862 195	−0.304 592	−0.162 753	−0.397 974	−0.236 74	−0.399 356
5	Correlation	Runtime	−0.862 195	1.000 000	0.188 745	0.143 508	0.313 648	0.226 103	0.450 383
6	Correlation	Age	−0.304 592	−0.188 745	1.000 000	−0.233 539	−0.337 87	−0.432 916	−0.164 1
7	Correlation	Weight	−0.162 753	0.143 508	−0.233 539	1.000 000	0.181 516	0.249 381	0.439 742
8	Correlation	Runpulse	−0.397 974	0.313 648	−0.337 87	0.181 516	1.000 000	0.929 754	0.352 461
9	Correlation	Maxpulse	−0.236 74	0.226 103	−0.432 916	0.249 381	0.929 754	1.000 000	0.305 124
10	Correlation	Restingpulse	−0.399 356	0.450 383	−0.164 1	0.439 742	0.352 461	0.305 124	1.000 000

Table 3.12.5

Obs.	_Name_	_Type_	Intercept	Oxy	Runtime	Age	Weight	Runpulse	Maxpulse	Restingpulse
1	Intercept	SSCP	31.00	1,469	328.2	1,478	2,401	5,259	5,387	1,657
2	Oxy	SSCP	1,468.65	70,430	15,356.1	69,768	113,522	248,497	254,867	78,015
3	Runtime	SSCP	328.17	15,356	3,531.8	15,687	25,465	55,806	57,114	17,684
4	Age	SSCP	1,478.00	69,768	15,687.2	71,282	114,159	250,194	256,218	78,806
5	Weight	SSCP	2,400.78	113,522	25,464.7	114,159	188,008	407,746	417,765	128,409
6	Runpulse	SSCP	5,259.00	248,497	55,806.3	250,194	407,746	895,317	916,499	281,928
7	Maxpulse	SSCP	5,387.00	254,867	57,113.7	256,218	417,765	916,499	938,641	288,583
8	Restingpulse	SSCP	1,657.00	78,015	17,684.0	78,806	128,409	281,928	288,583	90,311
9		*N*	31.00	31	31.0	31	31	31	31	31

Table 3.12.6a Analysis of Variance

Source	DF	Sum of Squares	Mean Square	F-Value	Prob > F
Model	6	722.543 61	120.423 93	22.433	0.000 1
Error	24	128.837 94	5.368 247 41		
C Total	30	851.381 54			
Root MSE		2.316 948	R^2	0.848 7	
Dep Mean		47.375 81	Adj. R^2	0.810 8	
C.V.		4.890 572			

(2) The matrix in Table 3.12.5 is the $\mathbf{X}'\mathbf{X}$ matrix and the $\mathbf{X}'\mathbf{Y}$ vector for $\mathbf{X}$ the x-data matrix of $\mathbf{A}$, with the attached column of ones (intercept). SSCP denotes sum of squares and cross products.

(3) Table 3.12.6 is the analysis corresponding to the full model $\mathbf{Y} = \beta_0\mathbf{x}_0 + \cdots + \beta_6\mathbf{x}_6 + \boldsymbol{\varepsilon}$. For $V = \mathscr{L}(\mathbf{x}_0, \ldots, \mathbf{x}_6)$ and $V_0 = \mathscr{L}(\mathbf{x}_0)$. The lines "Model", "Error" and "C Total" correspond to the subspaces $V \cap V_0^\perp$, $V^\perp$ and $V_0^\perp$ in 31-space.

The F-value in Table 3.12.6a (22.433) is model MS/error MS, which may be used to test H_0: All β_j for $j \geq 1$ are 0. It is easy to show that $F = (R^2/(1 - R^2))[(n - k - 1)/k]$ for k independent variables, not counting the constant term. For observed $F = f$, Prob > F is $P(F > f)$ for $F \sim F_{k,n-k-1}$ ($k = 6$ here). C.V. is the coefficient of variation $= s/\bar{y} = 2.32/47.38$. Adj. R^2 is R^2 adjusted for degrees of freedom, defined as $1 - s^2/s_y^2$. Thus, $(1 - R^2_{\text{Adj.}})(n - k - 1)/(n - 1) = 1 - R^2$, so that $R^2_{\text{Adj.}}$ is always less than R^2.

Table 3.12.6b reports for each j: $\hat{\beta}_j$, $S_{\hat{\beta}_j}$, $t_j = \hat{\beta}_j/S_{\hat{\beta}_j}$, and $P(T > |t_j|)$ for T with the t distribution for $(n - k - 1)$ d.f. Type I SS is the reduction in error sum of squares (or increase in regression SS) given by adding that variable to the model given by the variables on lines above. For example, C total SS = 851 is error SS when only the intercept is used. That error SS is reduced by 633 to 118 when runtime is also used. It is reduced still further by 18 to 100 when age is also used. The total of all type I SS's is 722.54, regression SS for the full model.

Type II SS for variable j is the reduction A_j in error SS achieved by adding variable j to the model without variable j. It is sometimes useful to compute R^2-delete, called R_j^2 for variable j, the multiple c.c. when variable j is dropped. Since $R^2 - R_j^2 = A_j^2/\text{TSS}$, R_j^2 may easily be computed. For example, for runtime $R_1^2 = 0.848\,7 - 250.8/851.4 = 0.554$. The partial c.c. r_j of variable j with y, with the effects of other variables removed is, from Section 3.7, $d_j/(1 + d_j^2)^{1/2}$ for $d_j = t_j/(n - k - 1)^{1/2}$. Some computer packages allow for printing of R_j^2 and r_j. The column "Tolerance" gives $1 - R^2$ for R^2 the m.c.c. for that variable with respect to all the other independent variables. It can be found using $\text{TOL} = S^2/[S^2(\hat{\beta}_j)S_{x_j}^2(n - 1)]$. The reciprocal of the jth tolerance, often called the *variance inflation factor*, is the jj term of $(\mathbf{X}'\mathbf{X})^{-1}$, multiplied by $\sum_i (x_{ij} - \bar{x}_j)^2$.

Table 3.12.6b Parameter Estimates

Variable	DF	Parameter Estimate	Standard Error	T for H_0: Parameter $= 0$	Prob $> \|T\|$	Type I SS	Type II SS	Standardized Estimate	Tolerance	Variance Inflation
Intercept	1	102.934 48	12.403 258 10	8.299	0.000 1	69,578.478 15	369.728 31	0.000 000 00	0.000 000 00	0.000 000 00
Runtime	1	−2.628 652 82	0.384 562 20	−6.835	0.000 1	632.900 10	250.822 10	−0.684 601 49	0.628 877 1	1.590 867 88
Age	1	−0.226 973 80	0.099 837 47	−2.273	0.032 2	17.765 632 52	27.745 771 48	−0.222 040 52	0.661 010 10	1.512 836 18
Weight	1	−0.074 177 41	0.054 593 16	−1.359	0.186 9	5.605 217 00	9.910 588 36	−0.115 968 63	0.865 554 01	1.155 329 40
Runpulse	1	−0.369 627 76	0.119 852 94	−3.084	0.005 1	38.875 741 95	51.058 058 32	−0.711 329 98	0.118 521 69	8.437 274 18
Maxpulse	1	0.303 217 13	0.136 495 19	2.221	0.036 0	26.826 402 70	26.491 424 05	0.521 605 12	0.114 366 12	8.743 848 43
Resting-pulse	1	−0.021 533 64	0.066 054 28	−0.326	0.747 3	0.570 512 99	0.570 512 99	−0.030 799 18	0.706 419 90	1.415 588 65

Table 3.12.7a Analysis of Variance

Source	DF	Sum of Squares	Mean Square	F-Value	Prob > F
Model	3	656.270 95	218.756 98	30.272	0.000 1
Error	27	195.110 60	7.226 318 35		
C Total	30	851.381 54			
Root MSE		2.680 181	R^2	0.770 8	
Dep Mean		47.375 81	Adj. R^2	0.745 4	
C.V.		5.674 165			

Table 3.12.7b Parameter Estimates

Variable	DF	Parameter Estimate	Standard Error	T for H_0: Parameter = 0	Prob > $\|T\|$
Intercept	1	93.126 150 08	7.559 156 30	12.320	0.000 1
Runtime	1	−3.140 386 57	0.367 379 84	−8.548	0.000 1
Age	1	−0.173 876 79	0.099 545 87	−1.747	0.092 1
Weight	1	−0.054 436 52	0.061 809 13	−0.881	0.386 2

Standardized estimate is $\hat{\beta}_j S_{x_j}/S_Y$, the estimated regression coefficient when both Y and x_j are scaled to have standard deviation one.

(4) Table 3.12.7 reports the analysis for the smaller model including data which might be more easily available, with independent variables runtime, age, and weight. The new error SS in 195.11, an increase of $195.11 - 128.84 = 66.27$ over error SS under the full model. Thus, the F-ratio for a test of the null hypothesis that coefficients for all other independent variables are zero is

$$F = \frac{66.27/3}{128.84/24} = 4.11$$

Since $F_{3,24,0.95} = 3.01$, we reject at the 0.05 level. The model which includes all independent variables except weight and restpulse would, based on the t_j-values, seem to be of interest.

If one variable alone is to be used as a predictor, runtime $= x_1$ would seem to be best, since its c.c. with Y is $-0.862\,19$. The equation of the simple linear regression fit of Y against x_1 has slope $r_{Yx_1}S_y/S_{x_1} = (-0.86\,219)(5.327/1.387) = -3.31$, intercept $\bar{Y} - \hat{\beta}_1\bar{x}_1 = 47.38 - (-3.31)(10.59) = 82.43$. This variable alone explains 74.3% of the variation in Y. Other variables, by themselves, explain much less.

The F-ratio for the test of the null hypotheses that only runtime and the constant term is needed is $F = 3.35$, which exceeds $F_{0.95}$. The partial correlation coefficients $r_{Yx_j \cdot x_1}$ for $j \neq 1$ would be of interest.

Table 3.12.8 Covariance of Estimates

Covariance	Intercept	Runtime	Age	Weight	Runpulse	Maxpulse	Restingpulse
Intercept	153.840 8	0.767 837 4	−0.902 049	−0.178 238	0.280 796 5	−0.832 762	−0.147 955
Runtime	0.767 837 4	0.147 888 1	−0.014 191 7	−0.004 417 67	−0.009 047 78	0.004 624 95	−0.010 915 2
Age	−0.902 849	−4.014 917	0.009 967 521	0.001 021 91	−0.001 203 91	0.003 582 384	0.001 489 753
Weight	−0.178 238	−0.004 417 67	0.001 021 91	0.002 980 413	0.000 964 468 3	−0.001 372 24	0.000 379 929 5
Runpulse	0.280 796 5	−0.009 047 78	−0.001 203 91	0.000 964 468 3	0.014 364 73	−0.014 952 5	−0.000 764 507
Maxpulse	−0.832 762	0.004 624 95	0.003 582 384	−4.001 372 24	−0.014 952 5	0.018 630 94	0.000 342 572 4
Restingpulse	−0.147 955	−0.010 915 2	0.001 489 753	0.000 379 929 5	−0.000 764 507	0.000 342 572 4	0.004 363 167

Table 3.12.9 Correlation of Estimates

Correlation	Intercept	Runtime	Age	Weight	Runpulse	Maxpulse	Restingpulse
Intercept	1.0000	0.1610	−0.7285	−0.2622	0.1889	−0.4919	−0.1806
Runtime	0.1610	1.0000	−0.3696	−0.2104	−0.1963	0.0881	−0.4297
Age	−0.7285	−0.3696	1.0000	0.1875	−0.1006	0.2629	0.2259
Weight	−0.2632	−0.2104	0.1875	1.0000	0.1474	−0.1842	0.1054
Runpulse	0.1889	−0.1963	−0.1006	0.1474	1.0000	−0.9140	−0.0966
Maxpulse	−0.4919	0.0881	0.2629	−0.1842	−0.9140	1.0000	0.0380
Restingpulse	−0.1806	−0.4297	0.2259	0.1054	−0.0966	0.0380	1.0000

Problem 3.12.1: For a simple linear regression of Y vs. age, give the equation of the estimated regression line. Estimate σ^2 for this model, and give 95% confidence and prediction intervals for a 40 year old.

Problem 3.12.2: Fit the model $\mathbf{Y} = \beta_0 \mathbf{x}_0 + \beta_1 \mathbf{x}_1 + \beta_2 \mathbf{x}_2 + \boldsymbol{\varepsilon}$ for these data. *Hint*: First fit the model with $\mathbf{x}_1$ and $\mathbf{x}_2$ replaced by the vectors of deviations from means. Then the $(\mathbf{X}'\mathbf{X})$ and $\mathbf{X}'\mathbf{Y}$ matrices may be found from the tables provided and the inverse is easy to compute by hand. Also find the error sum of squares, and test the null hypothesis that this model suffices, assuming the full model $\mathbf{Y} = \beta_0 \mathbf{x} + \beta_1 \mathbf{x}_1 + \cdots + \beta_6 \mathbf{x}_6 + \boldsymbol{\varepsilon}$.

Problem 3.12.3: Find the partial correlation coefficients $r_{Y4.1}$ and $r_{Y5.1}$.

Problem 3.12.4: Find the partial correlation coefficient of Y with age, with the effects of all other variables removed.

Problem 3.12.5: What is the coefficient of variation of weight?

Problem 3.12.6: What would R^2 be if runpulse were dropped from the analysis?

Problem 3.12.7: Assuming the full model of problem 3.12.1, test H_0: $\beta_4 = \beta_5 = \beta_6 = 0$, for $\alpha = 0.05$.

3.13 ANOTHER EXAMPLE: SALARY DATA

We consider here some faculty salary data, with the particular aim of trying to determine whether there is evidence of discrimination on the basis of gender. The College of Arts and Letters (English, History, Art, etc.) at Michigan State University was chosen for this small study because the data was readily available to the author (it is published for public use each year), and because that college had a larger number of female faculty than most. The data contains 158 salaries for full professors on nine-month appointments for 1990–1991. Also recorded were years in rank and years of experience. The gender of the faculty member was determined from the name, which shall not be given here. Thus, we let

Y = full year salary
x_1 = indicator of females (there were 26 females)
x_2 = years in rank of full professor
x_3 = total years of professional experience

Table 3.13.1a Salary Data for the College of Arts and Letters

Mean		Standard Deviation
x_1	0.1646	0.3720
x_2	10.6962	6.0336
x_3	26.4114	6.2731
Y	49,304.9304	7,313.1961

Correlations

$$\begin{array}{c} \\ Y \\ x_1 \\ x_2 \\ x_3 \end{array} \begin{array}{c} \begin{array}{cccc} Y & x_1 & x_2 & x_3 \end{array} \\ \begin{bmatrix} 1.000 & -0.072 & 0.587 & 0.359 \\ -0.072 & 1.000 & -0.009 & 0.066 \\ 0.587 & -0.009 & 1.000 & 0.718 \\ 0.359 & 0.066 & 0.718 & 1.000 \end{bmatrix} \end{array}$$

Table 3.13.1b Multiple Regression Table

Variable	$\hat{\beta}_j$	Estimate of Standard Error	t_j	R Delete	$\hat{\alpha}$
Constant	44,457.1	2,215.2	20.07		0.000
Indicator of female	−1,134.4	1,278.6	−0.88	0.594	0.376
Years in rank	815.0	112.9	7.22	0.372	0.000
Years experience	−139.4	108.9	−1.28	0.591	0.202

Total SSqs. (corrected for mean)	= 8,396,805,409	$R^2 = 0.35591$
Regression SSqs.	= 2,988,495,539	$R = 0.5966$
Error SSqs.	= 5,408,309,832	$S = 5,926.1$

We will first consider the full model: $Y_i = \beta_0 + \beta_1 x_{i1} + \beta_2 x_{i2} + \beta_3 x_{i3} + \varepsilon_i$, for $i = 1, \ldots, 158$ with the ε_i independent $N(0, \sigma^2)$. Regression analyses were performed using APL and SPSS, with the results of Tables 3.13.1 and 3.12.2. Since x_3 = years of experience seemed to contribute little beyond the other variables to the prediction of salary, it was then dropped from the model, giving the new table 3.13.2.

A reasonable conclusion, *based on these models*, is that there is no or little discrimination based on gender. However, there is danger in such analyses which must be considered carefully.

Suppose that the true regression function is $E(Y) = g(x_2)$ and that g has the form sketched in Figure 3.12 (concave).

Teachers in the public school system do in fact have regression functions of this shape, reaching an upper limit after some fixed number of years. Regression

Table 3.13.2 Multiple Regression Table

Variable	$\hat{\beta}_j$	Estimate of Standard Error	t_j	R Delete	$\hat{\alpha}$
Constant	41,916.2	988.0	42.42		0.000
Indicator of female	−1,305.3	1,274.2	−1.02	0.594	0.587
Years in rank	710.9	78.5	9.05	0.071	0.000

Total SSqs. (corrected for mean) = 8,396,805,409 $R^2 = 0.349\,05$

Regression SSqs. = 2,930,874,674 $R = 0.590\,8$

Error SSqs. = 5,465,930,698 $S = 5{,}938.4$

functions for management in some industries may instead have convex rather than concave shapes. Suppose in addition that female faculty members tend to have smaller x_2 values, perhaps due to a recent effort to increase the proportion of female faculty. In this case $\hat{\beta}_1$, the coefficient of the indicator for females would tend to be positive, suggesting that females are paid at a higher rate than males, even though the salary policy is gender neutral. If, instead, the regression function were convex, bending upward, then lower x_2 values for females and use of our linear model would suggest that males are paid more.

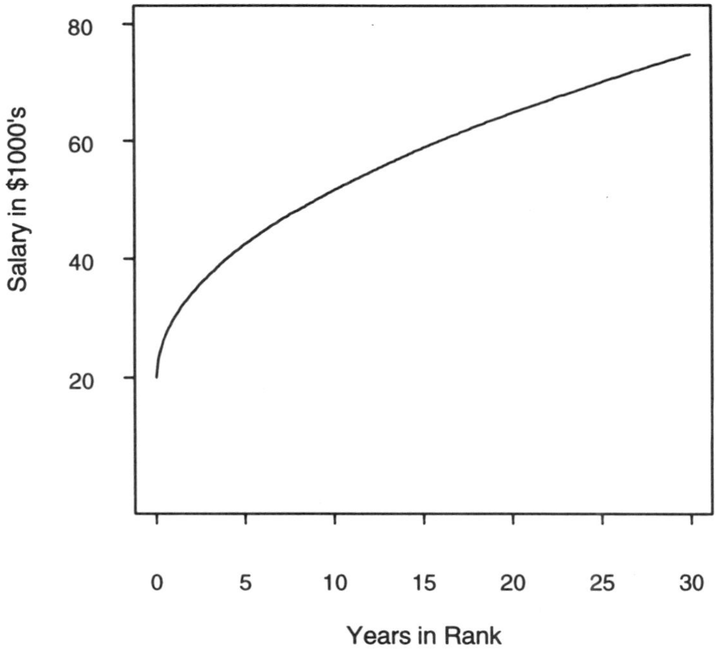

Figure 3.12 Salary in $1,000's vs. years in rank.

The fault, of course, is in the use of the model linear in x_2. Use of an additional term x_2^2 in the model for the Michigan State data did not improve the fit. The same conclusions resulted when $z = \log Y$, rather than Y, was used as the dependent variable.

Consider the model $\mathbf{Y} = \boldsymbol{\theta} + \boldsymbol{\varepsilon}$, where $\boldsymbol{\theta} = \beta_0 \mathbf{J} + \beta_1 \mathbf{f} + \beta_2 \mathbf{x}_2$, $\mathbf{J}$ is the vector of all ones, $\mathbf{f}$ is the indicator for females, and $\mathbf{x}_2$ is the vector of x_{i2} values. Let $\mathbf{m} = \mathbf{J} - \mathbf{f}$ be the indicator for males, and $\mathbf{x}_2^\perp = \mathbf{x}_2 - [p(\mathbf{x}_2 \mid \mathbf{m}) + p(\mathbf{x}_2 \mid \mathbf{f})] = \mathbf{x}_2 - \bar{x}_{2m}\mathbf{m} - \bar{x}_{2f}\mathbf{f}$. Then $\boldsymbol{\theta} \in V = \mathscr{L}(\mathbf{x}_0, \mathbf{f}, \mathbf{x}_2) = \mathscr{L}(\mathbf{f}, \mathbf{m}, \mathbf{x}_2^\perp)$, and these last three vectors are mutually orthogonal. It follows that if $\boldsymbol{\theta} = \gamma_f \mathbf{f} + \gamma_m \mathbf{m} + \gamma_2 \mathbf{x}_2^\perp$, then $\hat{\gamma}_f = \bar{Y}_f$, $\hat{\gamma}_m = \bar{Y}_m$, and $\hat{\beta}_2 = \hat{\gamma}_2 = (\mathbf{x}_2^\perp, \mathbf{Y})/\|\mathbf{x}_2^\perp\|^2$ are the least squares estimators of γ_f, γ_m, and $\gamma_2 = \beta_2$, and that $\beta_1 = \gamma_f - \gamma_m - \beta_2(\bar{x}_{f2} - \bar{x}_{m2})$. Thus, $\hat{\beta}_1 = \bar{Y}_f - \bar{Y}_m - \hat{\beta}_2(\bar{x}_{f2} - \bar{x}_{m2})$. The model states that for each sex the regression of Y on x_2 is linear with slope $\beta_2 = \gamma_2$, with intercept $\beta_0 + \beta_1$ for females, and β_0 for males. The estimates of these regression lines have slopes $\hat{\beta}_2$, and pass through the points of means $(\bar{x}_{f2}, \bar{Y}_f)$ for females, and $(\bar{x}_{m2}, \bar{Y}_m)$ for males. The coefficient $\hat{\beta}_1$ will be positive if and only if $\bar{Y}_f - \bar{Y}_m > \hat{\beta}_2(\bar{x}_{f2} - \bar{x}_{m2})$, i.e., when $\bar{x}_{2f} > \bar{x}_{2m}$, if the slope of the line from $(\bar{x}_{f2}, \bar{Y}_f)$ to $(\bar{x}_{m2}, \bar{Y}_m)$ exceeds $\hat{\beta}_2$, equivalently if the corrected female mean salary, $\bar{Y}_f - \hat{\beta}_2 \bar{x}_{f2}$ exceeds the corresponding corrected male mean score.

There are occasions when it may be more appropriate to interchange the roles of x_2 and Y, so that x_2 is the dependent variable, and Y one of the independent variables. This might be more appropriate when salary levels are fixed, but the number of years x_2 needed to reach a salary may be varied by the employer. Define $\mathbf{Y}^\perp = \mathbf{Y} - \bar{Y}_f \mathbf{f} - \bar{Y}_m \mathbf{m}$. The least squares lines for the case that the roles of Y and x_2 are reversed again pass through the points of means, but have common slopes, when the abscissa is x_2, $\hat{\beta}_2^* = \|\mathbf{Y}^\perp\|^2/(\mathbf{x}_2, \mathbf{Y}^\perp) = \|\mathbf{Y}^\perp\|^2/(\mathbf{x}_2^\perp, \mathbf{Y}) = \hat{\beta}_2/r^2$, where r is the partial correlation coefficient of Y and x_2, with the effects of $\mathbf{f}$ and $\mathbf{m}$ removed. The regression line for females will be above that for males if the slope of the line between the point of means for females to the point of means for males is greater than $\hat{\beta}_2^* = \hat{\beta}_2/r^2$. Supposing $r^2 < 1$, it follows that $\hat{\beta}_2^* > \hat{\beta}_2$. Thus, it is quite possible for the female line to be above the male line when Y is used as the dependent variable, but below when x_2 is the dependent variable, indicating that females wait longer to achieve the same salary, though they have higher average salaries for the same time in rank. Of course, the reverse conclusions are also possible. This paradoxical situation is a consequence of the use of least squares, with the sum of the squares of vertical distances being minimized in one case, and the sum of squares of horizontal distances (x_2-distance) in the other. In the courtroom, opposing lawyers in a dis-crimination suit, with their own "expert" statisticians, can each find supporting arguments.

For the College of Arts and Letters the estimates of the regression lines were: $\hat{\mathbf{Y}} = 40{,}611.0 + 710.86\ x_2$ for females, and $\hat{\mathbf{Y}} = 41{,}916.2 + 710.86\ x_2$ for males. When x_2 is treated as the dependent variable we get $\hat{x}_2 = -12.829 + 4.863\,1 \times 10^{-4} Y$ for females, and $\hat{x}_2 = -13.370 + 4.863\,1 \times 10^{-4} Y$ for males. Reversing the axes, with x_2 as the abscissa, we get the lines $Y = 263\,8.0 + 205\,6.3\ x_2$ for females, and $Y = 2749.3 + 2056.3\ x_2$ for females. The partial correlation coefficient of Y and x_2 with the effects of the gender variables removed was $r = 0.5880$. The conclusions are consistent. For the same number of years in rank, when we use Y as the dependent variable, we estimate that males tend to have a salary about \$1,300 higher. When we treat x_2 as the dependent variable, we estimate that males tend to earn about \$111 more, and to have to work about six months less to earn the same salary. Too much should not be made of this, however, because the differences are not statistically significant.

Problem 3.13.1: Consider the following salary data (fictitious) for the Department of Sociomechanics at Rich University in thousands of dollars per month (Y) and x_2 (years in rank).

Females					Males				
x_2	1	2	2	3	x_2	3	4	4	5
Y	2	3	5	6	Y	8	6	12	10

(a) Fit the model $Y = \beta_0 + \beta_1 x_1 + \beta_2 x_2 + \varepsilon$, where x_1 is the indicator for females.

(b) Estimate σ^2 and give a 95% confidence interval on β_1.

(c) Fit the model $x_2 = \beta_0 + \beta_1 x_1 + \eta Y + \varepsilon$. Sketch the regression lines for males and for females, with x_2 as the abscissa. On the same axes sketch the regression lines found in (a). Does there seem to be discrimination? Is it for or against females?

(d) Find the partial c.c. r of Y and x_2, with the effects of gender removed. How is r related to the slopes of the lines sketched in (c)?

(e) The Department of Philosophical Engineering uses the formula $Y = 40 + 10\sqrt{x_2}$. It has five female full professors, having $x_2 = 0, 1, 2, 3, 4$ and four male professors having $x_2 = 8, 12, 16, 20$. Repeat (a) for this department. Does there seem to be discrimination?

Problem 3.13.2: Suppose that for the model and x_2 values of problem 3.13.1 $\beta_1 = -2$, $\beta_2 = 2$, and $\sigma = 0.8$. What is the power of the $\alpha = 0.05$ level F-test (and two-sided t-test) of H_0: $\beta_1 = 0$?

Problem 3.13.3: For that model $\mathbf{Y} = \beta_0 \mathbf{J} + \beta_1 \mathbf{f} + \beta_2 \mathbf{x}_2 + \boldsymbol{\varepsilon}$, let $S_{22} = \|\mathbf{x}_2^{\perp}\|^2 = \sum (x_{2fi} - \bar{x}_{2f})^2 + \sum (x_{2mi} - \bar{x}_{2m})^2$. Show that

(a) $\text{Var}(\hat{\beta}_1) = [1/n_f + 1/n_m + (\bar{x}_{2f} - \bar{x}_{2m})^2/S_{22}]\sigma^2$,

(b) $S^2 = [\sum (Y_{fi} - \bar{Y}_f)^2 + \sum (Y_{mi} - \bar{Y}_m)^2 + \hat{\beta}_2^2 S_{22}]/(n-3)$.

Problem 3.13.3: The College of Natural Science (Physics, Chemistry, Mathematics, Zoology, etc., including Statistics) had 178 full professors on nine-month appointments in 1993–1994. For 15 female professors $S_Y = 7{,}167.2$, $S_{x_2} = 6.174$, with correlation coefficient $r_{Yx_2} = 0.1094$. For 163 male professors $S_Y = 13893.0$, $S_{x_2} = 7.773$, with $r_{Yx_2} = 0.1623$. The means were $\bar{Y} = 57{,}023.87$, $\bar{x}_2 = 8.867$ for females and $\bar{Y} = 65848.7$, $\bar{x}_2 = 14.436$ for males.

(a) Use the model of Problem 3.13.1 to decide whether the data indicate discrimination against females.

(b) Find a 95% confidence interval on β_1.

(c) Find the multiple correlation coefficient for this model.

Table 3.13.3

Team	Y	x_1	x_2	x_3	x_4	x_5	x_6	r
National League								
East								
New York	0.625	0.256	0.328	152	140	0.981	2.91	703
Pittsburgh	0.531	0.247	0.321	110	119	0.980	3.47	651
Montreal	0.500	0.251	0.311	107	189	0.978	3.08	628
Chicago	0.475	0.261	0.312	113	120	0.980	3.84	660
St. Louis	0.469	0.249	0.312	71	234	0.981	3.47	578
Philadelphia	0.404	0.239	0.308	106	112	0.976	4.14	597
West								
Los Angeles	0.584	0.248	0.308	99	131	0.977	2.96	628
Cincinnati	0.540	0.246	0.311	122	207	0.980	3.35	641
San Diego	0.516	0.247	0.313	94	123	0.981	3.28	594
San Francisco	0.512	0.248	0.321	113	121	0.980	3.39	670
Houston	0.506	0.244	0.308	96	198	0.978	3.41	617
Atlanta	0.338	0.242	0.301	96	95	0.976	4.09	597
American League								
East								
Boston	0.549	0.283	0.360	124	65	0.984	3.97	813
Detroit	0.543	0.250	0.326	143	87	0.982	3.71	703
Toronto	0.537	0.268	0.334	158	107	0.982	3.80	763
Milwaukee	0.537	0.257	0.316	113	159	0.981	3.45	682
New York	0.528	0.263	0.336	148	146	0.978	4.24	772
Cleveland	0.481	0.261	0.317	134	97	0.980	4.16	666
Baltimore	0.335	0.238	0.307	137	69	0.980	4.54	550
West								
Oakland	0.642	0.263	0.339	156	129	0.983	3.44	800
Minnesota	0.562	0.274	0.343	151	107	0.986	3.93	759
Kansas City	0.522	0.259	0.324	121	137	0.980	3.65	704
California	0.463	0.261	0.324	124	86	0.979	4.32	714
Chicago	0.441	0.244	0.305	132	98	0.976	4.12	631
Texas	0.435	0.252	0.323	112	130	0.979	4.05	737
Seattle	0.422	0.257	0.319	148	95	0.980	4.15	664

Problem 3.12.5: The data of Table 3.13.3 describe the performances of the 26 major league baseball teams during the 1988 season. Teams played between 160 and 163 games (rainouts caused fewer, ties caused more, than the scheduled 162). Presented are: $Y =$ percentage of games won, $x_1 =$ batting average (proportion of hits to times at bat), $x_2 =$ on base average, $x_3 = \#$ home runs, $x_4 = \#$ stolen bases, $x_5 =$ fielding average, and $x_6 =$ earned run average (mean number of runs given up per nine innings), and $r = \#$ runs scored. The data are from page 916 of the 1989 version of *Total Baseball*, Thorn and Palmer (1989). It would be of interest to determine the relationship between Y and the explanatory variables $x_1, \ldots, x_6$, (*not* including r) and to explain a reasonable proportion of the variation in Y, using as few of the explanatory variables as possible. Those who discover a good prediction formula may be hired as a team general manager. A general manager may be able to control some of the x-variables, at the expense of others, by making trades of players, drafting some players rather than others, or by spending money on the a minor league system, but could not control r directly. For this reason it is of interest to omit r as an explanatory variable.

(a) Fit the model and use appropriate tests of hypotheses to decide whether a smaller model would suffice. Prepare a one-page report which could be understood by the president of a baseball team who had a beginning statistics class 40 years ago. Particularly ambitious readers may want to analyze such data for the years 1901–1994.

(b) Since teams win when they score more runs than the opponent, it would seem that r and x_6 alone would serve as better predictors of Y. Is that true for these data? How well can r be predicted from the x-variables?

CHAPTER 4

Fitting of Regression Models

Good model-building requires knowledge of the subject matter involved, a grasp of regression techniques, access to a computer and software, and ingenuity. Rather than looking for *the* model one looks for reasonable models. Only in an idealized world is there a perfect model. The regression function is almost never exactly linear in the independent variables, the errors probably do not have equal variances, and are not normally distributed.

The purpose of this chapter is to provide some understanding of model-building techniques and of the effects of deviations from the idealizations we have made, and provide some techniques for recognizing them, and for making adjustments. The chapter should be viewed as a brief introduction rather than a complete review. Entire books have been devoted to regression techniques. See Cook and Weisberg (1982), Belsley, Kuh, and Welsch (1980), Myers (1986), Seber (1977), Searle (1971), Draper and Smith (1981), Carroll and Ruppert (1988), Hastie and Tibshirani (1990), and Koul (1992).

4.1 LINEARIZING TRANSFORMATIONS

Consider the problem of finding a regression function $g(x) = E(Y|x)$ for a single variable x. We have discussed techniques for estimating $g(x)$ for the case that $g(x)$ is a linear combination $h(x) = \sum_{j=1}^{k} \beta_j f_j(x)$ when the f_j are known functions of x; that is, $g(x)$ is expressible as a linear function of unknown parameters β_j. Presumably if g is a *smooth* function and x is not allowed to range too widely then $g(x)$ may be approximated by a function of the form of $h(x)$ over the range of interest for x. What can we do if this is not the case?

Consider a simple example, with observed (x_i, Y_i) pairs as in the scatter diagram of Figure 4.1.

It may be possible to fit a model of the form of h above, for h quadratic or cubic, say. However, the model is likely to require a number of parameters and could very well have undesirable properties at the extremes of x, particularly

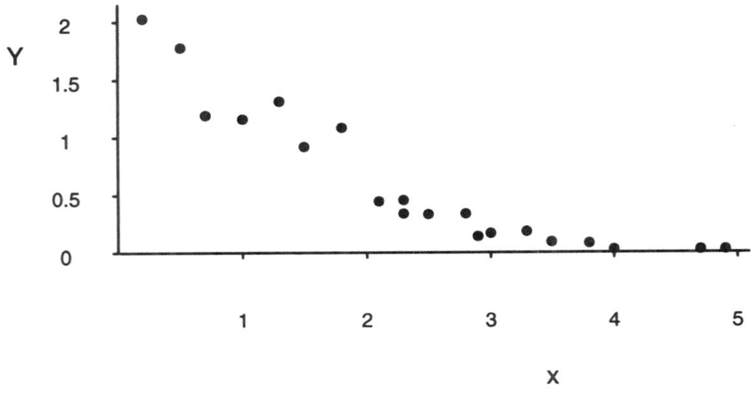

FIGURE 4.1

if extrapolation is to be used. We might instead like to consider a function $h(x) = h(x, \gamma_0, \gamma_1) = \gamma_0 e^{\gamma_1 x}$, since it is always positive (for $\gamma_0 > 0$) and approaches 0 as $x \to \infty$ (for $\gamma_1 < 0$). This function is not linear in the parameters, however. The principle of least squares would, for observations (x_i, Y_i), $i = 1, \ldots, n$, lead to the minimization of the function

$$Q(\gamma_0, \gamma_1) = \sum_{i=1}^{n} [Y_i - h(x_i, \gamma_0, \gamma_1)]^2$$

Since h is not linear in γ_0 and γ_1, the solution $(\hat{\gamma}_0, \hat{\gamma}_1)$ is not linear in the Y_i. Many statistical computer packages include routines which find the solution by iterative means.

We will instead show how the model may be *linearized* so that techniques already discussed may be employed. Ignoring an error term, consider the approximation $Y \doteq h(x, \gamma_0, \gamma_1) \doteq \gamma_0 e^{\gamma_1 x}$ or $\log Y \doteq \log \gamma_0 + \gamma_1 x_1$ (the symbol $\doteq$ means approximately equal). Setting $Z = \log Y$, $\beta_0 = \log \gamma_0$, $\beta_1 = \gamma_1$, we get $Z \doteq \beta_0 + \beta_1 x$, a function linear in β_0, β_1. If we now employ the simple linear regression model $Z_i = \beta_0 + \beta_1 x_i + \varepsilon_i$ for $Z_i = \log Y_i$, we can obtain an estimate $(\hat{\beta}_0, \hat{\beta}_1)$ for (β_0, β_1), then use these to obtain the estimate $(\hat{\gamma}_0 = e^{\hat{\beta}_0}, \hat{\gamma}_1 = \hat{\beta}_1)$ for (γ_0, γ_1). Of course, this linear model is equivalent to

$$Y_i = e^{Z_i} = \gamma_0 e^{\gamma_1 x_i} e^{\varepsilon_i}$$

so that the error term $\eta_i = e^{\varepsilon_i}$ for Y_i is now multiplicative and has a log-normal distribution if the ε_i have normal distributions. Confidence intervals on γ_0 or γ_1 may be found by first finding confidence intervals on β_0 or β_1. Lack of dependency of Y on x corresponds to $\beta_1 = 0$. Note that the function $h(x) = \gamma_0 e^{\gamma_1 x}$ is not the regression function of Y on x in general, since $E(\varepsilon_i) = 0$

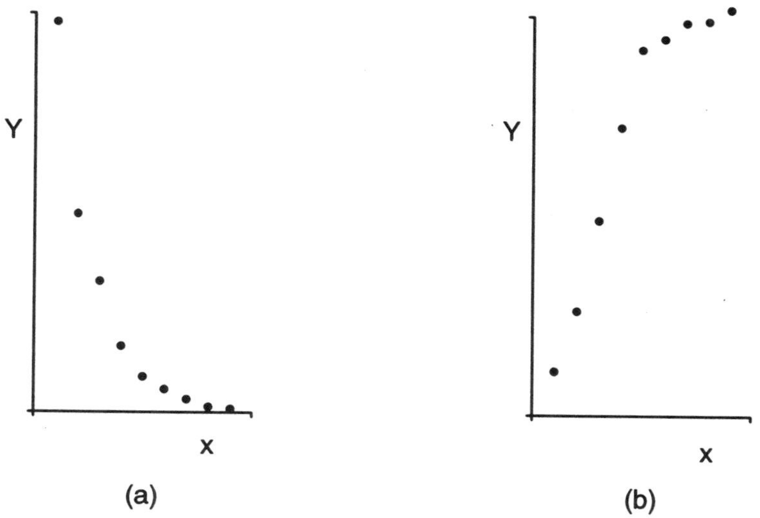

FIGURE 4.2

does not imply $E(e^{\varepsilon_i}) = 1$. However, it may be a reasonable approximation. Of course, the solution $(\hat{\gamma}_0, \hat{\gamma}_1)$ does not minimize Q.

Consider a scatter diagram of the form (a) or (b) in Figure 4.2. Data of the sort in (b) may arise in chemistry when $x = 1/v$, v is volume, and $Y = p$ is pressure at a constant temperature. For either graph we may consider the model $h(x) = \gamma_0 x^{\beta_1}$, with $\beta_1 < 0$ corresponding to (a) and $0 < \beta_1$ corresponding to (b). Linearizing again, take $Y \doteq \gamma_0 x^{\beta_1}$, $Z = \log Y = \log \gamma_0 + \beta_1 \log x = \beta_0 + \beta_1 w$. Considering the model $Z_i = \beta_0 + \beta_1 w_i + \varepsilon_i$, for $Z_i = \log Y_i$, $w_i = \log x_i$, we can again find an estimate $(\hat{\beta}_0, \hat{\beta}_1)$ for (β_0, β_1), then let $\hat{\gamma}_0 = e^{\hat{\beta}_0}$, $\hat{\gamma}_1 = \hat{\beta}_1$.

In the case that Y is necessarily between 0 and 1, a proportion for example, we might consider the model

$$h(x) = \frac{v}{1 + v} \qquad \text{for} \quad v = \gamma_0 e^{\gamma_1 x}$$

Setting $Y \doteq h(x)$ and solving for v, we get $v \doteq Y/(1 - Y)$ and $\log v \doteq \log \gamma_0 + \gamma_1 x = \log\left(\frac{Y}{1 - Y}\right) = Z$. Setting $Z_i = \log[Y_i/(1 - Y_i)]$, $\beta_0 = \log \gamma_0$, $\beta_1 = \gamma_1$, and considering the model

$$Z_i = \beta_0 + \beta_1 x_i + \varepsilon_i$$

we can now estimate (β_0, β_1), hence $\gamma_0 = e^{\beta_1}$ and $\gamma_1 = \beta_1$. Least squares is not in general the best method to be used in the estimation of (β_0, β_1), since the usual error model for the ε_i is usually not appropriate (see Chapter 8). However, least squares will usually provide quite reasonable estimates.

More generally for several independent variables $(x_1, \ldots, x_k)$ the transformation $Z = \log\left(\frac{Y}{1 - Y}\right)$ (called *log-odds* for probability Y) facilitates the fitting of the log-linear model

$$h(\tilde{\mathbf{x}}) = k(\tilde{\mathbf{x}})/[1 + k(\tilde{\mathbf{x}})]$$

for $\tilde{\mathbf{x}} = (x_1, \ldots, x_k)$ and $k(\tilde{\mathbf{x}}) = \exp\left(\sum_1^k \beta_j x_j\right)$.

The plotting of Y vs. x, $\log Y$ vs. x, $\log Y$ vs. $\log x$, etc. can suggest which model may fit reasonably well. For example, if $\log Y$ seems to be approximately linear in x, then the model $\log Y = \beta_0 + \beta_1 x$, equivalently, $Y = \gamma_0 e^{\beta_1 x}$ may be appropriate. Or, if $\log Y$ seems to be quadratic in x, then $Y = \exp(\beta_0 + \beta_1 x + \beta_2 x^2)$ may be appropriate.

Example 4.1.1: Consider 74 makes of automobiles with measurements: miles per gallon (mpg) and weight (wgt), as reported by the magazine *Consumers Report*. Figure 4.3 indicates that the regression of mpg against wgt is not linear.

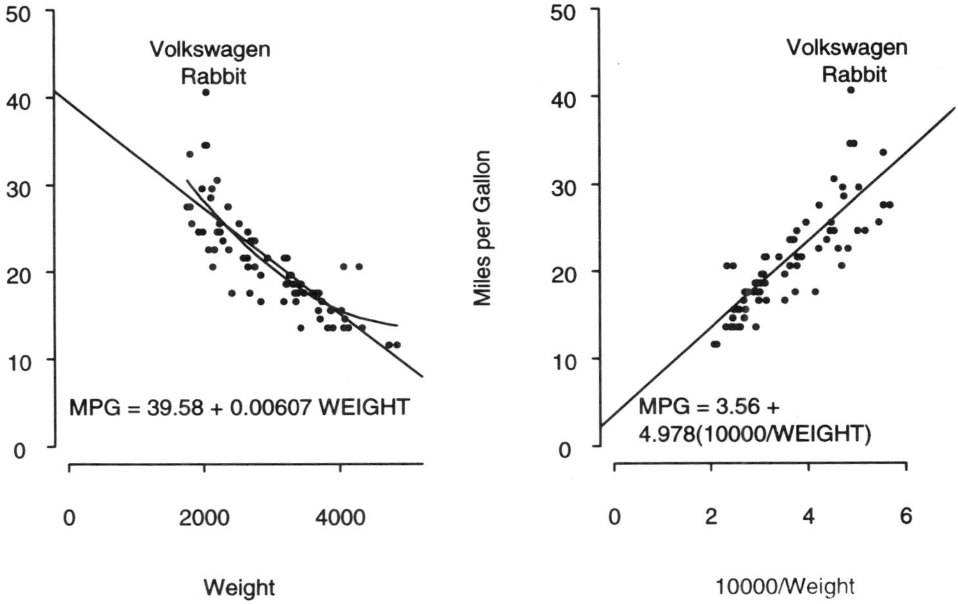

FIGURE 4.3 Miles per gallon vs. weight and 10,000/weight for 74 automobiles.

There seems to be a significant gain in mpg per pound as wgt drops below 2,000. A plot of mpg vs. 10,000/wgt is more linear and suggests the model mpg $= \beta_0 + \beta_1(10{,}000/\text{wgt}) + \varepsilon$. Least squares was used to fit this model, then the fitted model mpg $= 3.556 + 49{,}780/\text{wgt}$ plotted on the (wgt, mpg) axes for the weights observed. The resulting curve seems to be a better fit than the straight line.

Example 4.1.2: Consider ten (x, Y) pairs as follows, where we seek a model for which $0 \le Y \le 1$ for all $x \ge 0$.

x	0.5	1.0	1.5	2.0	2.5	3.0	3.5	4.0	4.5	5.0
Y	0.035	0.080	0.171	0.329	0.538	0.734	0.868	0.940	0.974	0.989

$Z = \log\left(\frac{Y}{1 - Y}\right)$ is approximately linear in x, with $\hat{Z} = -4.171 + 1.729x$ (see Figure 4.4). Thus $\hat{Y} = e^{\hat{Z}}/(1 + e^{\hat{Z}})$.

Example 4.1.3: Consider the (x, Y) pairs in the first two columns of Table 4.1.1. It is possible that a quadratic function of x may fit these data points reasonably well. However, a plot of $Z = \log Y$ vs. x indicates that Z is approximately linear in x, or, equivalently, that Y may be approximated by a function of the form $h(x) = \gamma_0 e^{\gamma_1 x}$. Taking logs, we get the approximation $Z = \log Y \doteq \log \gamma_0 + \gamma_1 x$, so we can use simple linear regression of Z on x to estimate $\log \gamma_0$ and γ_1 (Figure 4.5).

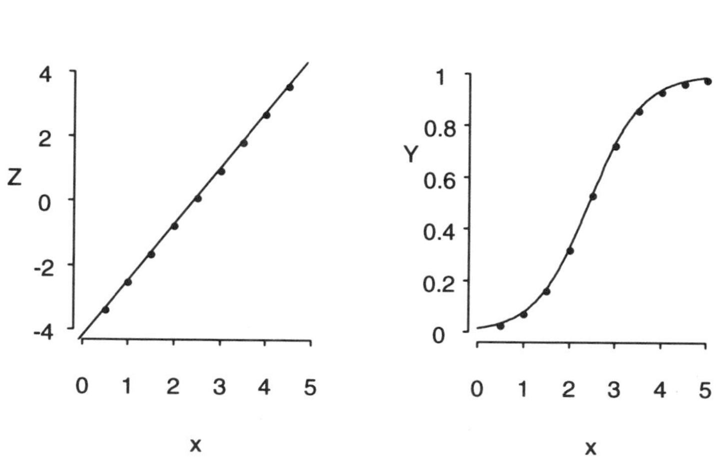

FIGURE 4.4 Least squares fit of $Z = \log(Y/(1 - Y))$ vs. x and estimate of the regression of Y on x.

Table 4.1.1

x	Y	z	$\hat{z}$	$\hat{Y}$
1	3.394	1.222	1.203	3.330
2	4.161	1.426	1.645	5.182
3	9.786	2.281	2.087	5.182
4	10.850	2.384	2.529	12.540
5	23.000	3.135	2.971	19.510
6	34.500	3.541	3.413	30.360
7	43.190	3.766	3.855	47.240
8	79.940	4.381	4.297	73.490
9	99.880	4.604	4.739	114.300

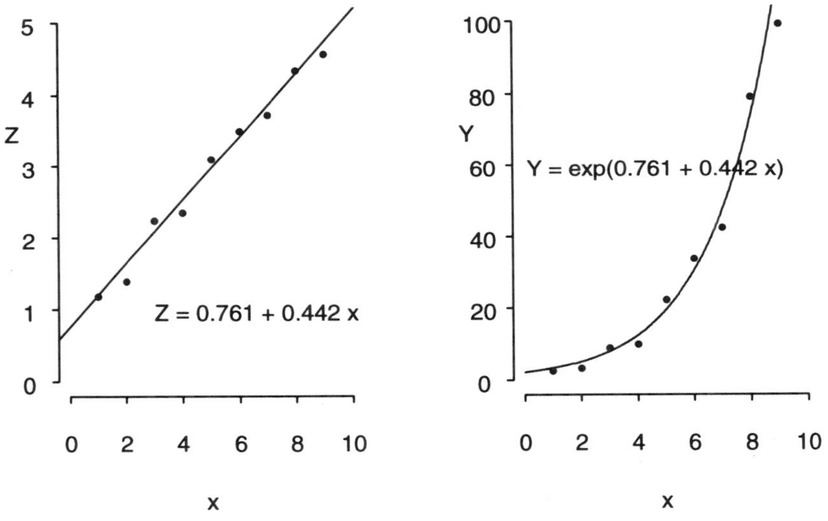

FIGURE 4.5 Fit of Y vs. x using linear fit of $Z = \log Y$ vs. x.

Measuring the Goodness of Fit: In order to compare two or more attempts at fitting models to the same data we need a measure of the closeness or *goodness* of the fit. If the approximation of y_i given by the model is $\hat{\mathbf{Y}}_i$ then one such measure is $R^2(\hat{\mathbf{Y}}, \mathbf{y}) \equiv 1 - \frac{\sum (y_i - \hat{y}_i)^2}{\sum (y_i - \bar{y})^2}$. This is the multiple correlation coefficient only if the $\hat{\mathbf{Y}}_i$ were obtained by fitting a linear model with a constant term. Consequently $R^2(\hat{\mathbf{Y}})$ can be less that zero!

Suppose that a transformation $z = g(y)$ is made. Let $\hat{z}$ be the predicted value of z by fitting a linear model for z vs. x. Let $\hat{\mathbf{Y}}$ be the predicted value of $\mathbf{y}$ for a simple linear regression of y on x, and let $\hat{\mathbf{Y}}_z = g^{-1}(\hat{\mathbf{z}})$. Scott and Wild (1991) give an example of a collection of six pairs of (x, y) values for which $R^2(\hat{\mathbf{Y}}, \mathbf{y}) = 0.88$, while $R^2(\hat{\mathbf{Y}}_{\mathbf{z}}, \mathbf{y}) = -0.316$!

$$(x, y)\colon (0, 0.1), (3, 0.4), (8, 2), (13, 10), (16, 15), (20, 16).$$

In the least squares sense $\bar{y}$ is a better predictor of y than is $\hat{\mathbf{Y}}_{\mathbf{z}}$. On the z-scale $R^2(\mathbf{z}, \hat{\mathbf{z}}) = 0.94$. Scott and Wild warn, and give examples to show, that values of R^2 based on different scales are not comparable.

Measurement of the value of a transformation should be based on the scale to be used in making judgements about the subject matter. These who use the methodology must choose the scale on which measures of the goodness of approximation are to be made. Finally, we should be reminded that the choice of R^2 as a measure was somewhat arbitrary. We could, for example, replace squared deviations by absolute or maximum deviations.

Problem 4.1.1: Consider the ten (x_i, Y_i) values of Table 4.1.2 and the corresponding plots of (x_i, Y_i) and (w_i, Y_i) for $w_i = \ln x_i$ (see Figure 4.6). State a model which will justify 95% confidence and prediction intervals on $E(Y|x)$ and on Y for $x = 10$ and find these intervals.

Table 4.1.2

x	Y	w	x	Y	w
2	20.43	0.693	6	12.45	1.792
2	20.92	0.693	8	10.35	2.079
4	15.57	1.386	8	10.18	2.079
4	14.85	1.386	10	7.78	2.303
6	13.02	1.792	10	9.09	2.303

Table 4.1.3

	x	y	w	z	u		
1	0.5	21.640	−0.693	3.075	2.000	$\sum x = 27.5$	$\sum xz = 21.50$
2	1.0	8.343	0.000	2.121	1.000	$\sum x^2 = 96.25$	$\sum wy = 4.191$
3	1.5	4.833	0.405	1.575	0.667	$\sum y = 50.57$	$\sum uy = 60.38$
4	2.0	4.300	0.693	1.459	0.500	$\sum y^2 = 603.0$	$\sum uz = 3.946$
5	2.5	2.343	0.916	0.851	0.400	$\sum w = 8.173$	
6	3.0	2.623	1.099	0.964	0.333	$\sum w^2 = 11.52$	
7	3.5	1.818	1.253	0.598	0.286	$\sum z = 11.98$	
8	4.0	1.628	1.386	0.493	0.250	$\sum z^2 = 21.25$	
9	4.5	1.909	1.504	0.647	0.222	$\sum u = 5.860$	
10	5.0	1.120	1.609	0.113	0.220	$\sum u^2 = 6.199$	

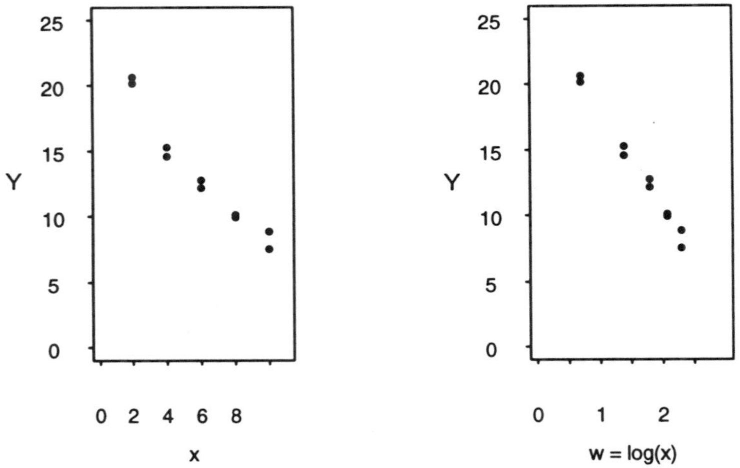

FIGURE 4.6

Problem 4.1.2: Find a 90% confidence interval on γ_1 for the data in Example 4.1.2. What model justifies this interval?

Problem 4.1.3: For the (x_i, Y_i) pairs of Table 4.1.3 and Figure 4.7 define $w_i = \log x_i$, $z_i = \log Y_i$, $u_i = 1/x_i$. Suggest a model, estimate the parameters, and sketch the resulting function $h(x)$.

Problem 4.1.4: Verify the two values of R^2 given by Scott and Wild.

4.2 SPECIFICATION ERROR

It is often difficult to determine which of many possible variables in $\tilde{\mathbf{x}} = (x_1, \ldots, x_k)$ to use in estimating the regression function $g(\tilde{\mathbf{x}}) \equiv E(\mathbf{Y}|\tilde{\mathbf{x}})$ or in predicting $\mathbf{Y}$, particularly in cases for which n is relatively small. The statistican is torn between the wish to keep the model simple and the wish for a good approximation. If a poor choice of a subset $x_{i_1}, \ldots, x_{i_r}$ of possible measurements is made, what will the penalty be?

For example, in trying to determine the regression function $g(\tilde{\mathbf{x}})$, should we use a fifth degree polynomial, or will a quadratic function suffice? Obviously we can fit the data more closely with a fifth degree polynomial, but may pay a price in increased complication, poor extrapolation, and, as we shall see, a loss of precision. On the other hand, if the true regression is cubic (it would be better to say, is *approximately* cubic for x of interest) and we fit a quadratic function some inaccuracy (bias) would seem to result.

To make the discussion precise suppose $\mathbf{Y} = \boldsymbol{\theta} + \boldsymbol{\varepsilon}$ for $\boldsymbol{\varepsilon} \sim N(0, \sigma^2\mathbf{I}_n)$ and

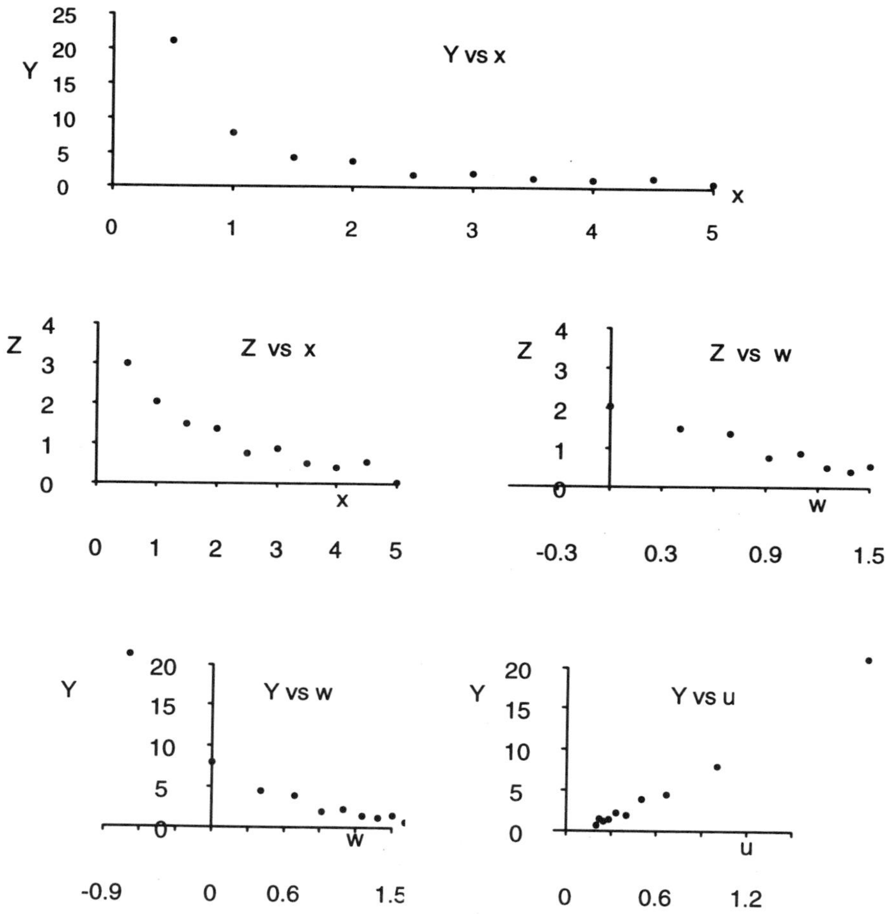

FIGURE 4.7 Scatterplots among x, $w = \log x$, $u = 1/x$, Y, and $z = \log Y$.

our postulated model is $\boldsymbol{\theta} \in V$, a known subspace of R_n of dimension k. As will be seen by the following analysis, if $\boldsymbol{\theta} \notin V$, errors may result. To see this, let $\boldsymbol{\theta} = \boldsymbol{\theta}_V + \boldsymbol{\theta}_\perp$ for $\boldsymbol{\theta}_V = p(\boldsymbol{\theta} | V)$. Let $\boldsymbol{\varepsilon}_V = p(\boldsymbol{\varepsilon} | V)$ and $\boldsymbol{\varepsilon}_\perp = \boldsymbol{\varepsilon} - \boldsymbol{\varepsilon}_V$. Then the least squares estimator of $\boldsymbol{\theta}$ is $\hat{\mathbf{Y}} = \boldsymbol{\theta}_V + \boldsymbol{\varepsilon}_V$ and the error in the estimation of $\boldsymbol{\theta}$ is $\mathbf{d} = \hat{\mathbf{Y}} - \boldsymbol{\theta} = -\boldsymbol{\theta}_\perp + \boldsymbol{\varepsilon}_V$. Thus, $\hat{\mathbf{Y}}$ has bias $-\boldsymbol{\theta}_\perp$. We can assess the expected sizes of the errors made in estimating $\boldsymbol{\theta}$ by computing

$$E(\hat{\mathbf{Y}} - \boldsymbol{\theta})(\hat{\mathbf{Y}} - \boldsymbol{\theta})' = \boldsymbol{\theta}_\perp \boldsymbol{\theta}_\perp' + E(\boldsymbol{\varepsilon}_V \boldsymbol{\varepsilon}_V') = \boldsymbol{\theta}_\perp \boldsymbol{\theta}_\perp' + E(\mathbf{P}_V \boldsymbol{\varepsilon}\boldsymbol{\varepsilon}' \mathbf{P}_V) = \boldsymbol{\theta}_\perp \boldsymbol{\theta}_\perp' + \sigma^2 \mathbf{P}_V.$$

Here we have taken advantage of the orthogonality of $\boldsymbol{\theta}_\perp$ and $\boldsymbol{\varepsilon}_V$ (Figure 4.8). To gauge the size of this we can compute the sum of the expected squared errors.

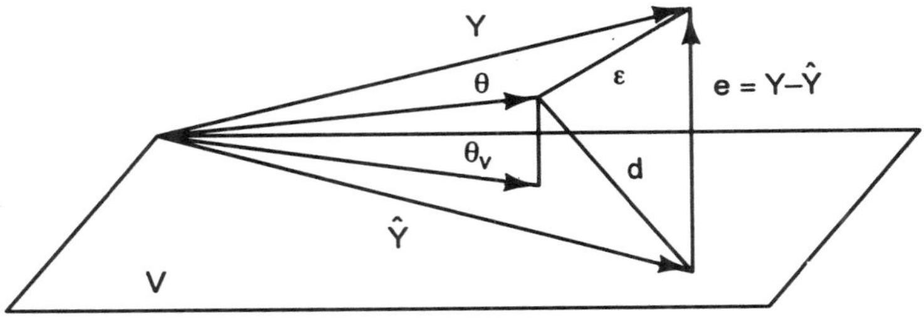

FIGURE 4.8

$$E\|\mathbf{d}\|^2 = E[\|\boldsymbol{\theta}_\perp\|^2 + \|\boldsymbol{\varepsilon}_V\|^2] = \|\boldsymbol{\theta}_\perp\|^2 + k\sigma^2,$$

since V has dimension k.

We might also study the random variable $Q = \|\mathbf{d}\|^2 = \|\boldsymbol{\theta}_\perp\|^2 + \|\boldsymbol{\varepsilon}_V\|^2$ in order to understand the sizes of these errors. Q is a constant plus σ^2 multiplied by a central χ^2 random variable (not noncentral χ^2), with expectation given above.

Error sum of squares is $\|\mathbf{Y} - \hat{\mathbf{Y}}\|^2 = \|\boldsymbol{\theta}_\perp + \boldsymbol{\varepsilon}_\perp\|^2$, so that $\|\mathbf{Y} - \hat{\mathbf{Y}}\|^2/\sigma^2 \sim \chi^2_{n-k}(\delta)$ for $\delta = \|\boldsymbol{\theta}_\perp\|^2/\sigma^2$. Thus, $E(S^2) = \sigma^2 + \|\boldsymbol{\theta}_\perp\|^2/(n-k)$.

In searching for a good model we might try to choose a subspace V so that $H_V = E\|\mathbf{d}\|^2/\sigma^2 = \|\boldsymbol{\theta}_\perp\|^2/\sigma^2 + k$ is small. Of course, H_V depends on unknown parameters. It can be estimated if we can find an estimator of *pure error variance* σ^2. We might, for example, use a particularly large subspace V_L in which we are quite sure $\boldsymbol{\theta}$ lies, and use error sum of squares for this subspace to estimate σ^2. Or we might use past data from another experiment with repeated observations on $\mathbf{Y}$ for the same $\tilde{\mathbf{x}}$ to estimate σ^2. Let σ^2 be this estimator of pure error variance. Let S^2 be the estimator for the subspace V.

Then $E(S^2 - \hat{\sigma}^2)(n-k) = \|\boldsymbol{\theta}_\perp\|^2$, so that $C_V = \dfrac{(S^2 - \hat{\sigma}^2)}{\hat{\sigma}^2}(n-k) + k$ can be used as an estimator of H_V. C_V is called *Mallows* C_p for the case that $\dim(V) = p$ (Mallows 1964). Since $H_V = \dim(V)$ for $\boldsymbol{\theta} \in V$, we should hope to find a subspace V such that C_V is close to or smaller than $\dim(V)$.

Consider, for example, a sequence of regression vectors $\mathbf{x}_1, \mathbf{x}_2, \ldots$, with order chosen by the statistician. $\mathbf{x}_j$ might, for example, be the vector of jth powers. Then for $V_k = \mathscr{L}(\mathbf{x}_1, \ldots, \mathbf{x}_k)$ and $C_k = C_{V_k}$, we can compute the sequence C_1, $C_2, \ldots$ and, as recommended by Mallows, plot the points (k, C_k), choosing the subspace V_k for the smallest k for which C_k is close to k.

One possible criterion for the choice of a subspace V_0 rather than a subspace V in which $\boldsymbol{\theta}$ is known to lie may be developed as follows. Since the bias of $\hat{\mathbf{Y}}_0 = p(\mathbf{Y}|V_0)$ is $p(\boldsymbol{\theta}|V_0) - \boldsymbol{\theta} = -\boldsymbol{\theta}_\perp$ and the sum of the squared errors is $Q = \|\boldsymbol{\theta}_\perp\|^2 + \|\boldsymbol{\varepsilon}_{V_0}\|^2$, we have $E(Q) = \|\boldsymbol{\theta}_\perp\|^2 + k_0\sigma^2$ (see Theorem 2.2.2). The sum of squared errors for $\hat{\mathbf{Y}} = p(\mathbf{Y} \mid V)$ is $\|\boldsymbol{\varepsilon}_V\|^2$, which has expectation $k\sigma^2$. Thus we should choose V_0 if

$$\|\boldsymbol{\theta}_\perp\|^2 + k_0\sigma^2 < k\sigma^2, \qquad \text{or} \qquad \|\boldsymbol{\theta}_\perp\|^2 < (k - k_0)\sigma^2, \tag{4.2.1}$$

equivalently if the noncentrality parameter $\delta = \|\boldsymbol{\theta}_\perp\|^2/\sigma^2$ in the F-test of $H_0\colon \boldsymbol{\theta} \in V_0$ is less than $k - k_0$.

Let $Q = \|\mathbf{Y} - \hat{\mathbf{Y}}\|^2$ and $Q_0 = \|\mathbf{Y} - \hat{\mathbf{Y}}_0\|^2$. Then $Q - Q_0 = \|\hat{\mathbf{Y}} - \hat{\mathbf{Y}}_0\|^2$ has expectation $\|\boldsymbol{\theta}_\perp\|^2 + (k - k_0)\sigma^2$ and $E(Q) = (n - k)\sigma^2$, so that

$$Q_0 - Q\left(\frac{k - k_0}{n - k}\right)$$

is an unbiased estimator of $\|\boldsymbol{\theta}_\perp\|^2$. Thus, if we replace the parameters in (4.2.1) by unbiased estimators we get $Q_0 < 2Q[(k - k_0)/(n - k)]$, equivalently $F = \dfrac{(Q - Q_0)/(k - k_0)}{Q/(n - k)} < 2$. This is equivalent to $C_k < k$ (see Problem 4.25).

Example 4.2.1: The data of Table 4.2.1 were generated using the regression function

$$g(\tilde{\mathbf{x}}) = \beta_0 + \beta_1 x_1 + \beta_2 x_2 + \beta_3 x_3 + \beta_4 x_1^2$$

Table 4.2.1

i	$\mathbf{x}_1$	$\mathbf{x}_2$	$\mathbf{x}_3$	$\mathbf{Y}_1$	$\mathbf{Y}_2$	$\hat{\mathbf{Y}}_1$	$\hat{\mathbf{Y}}_2$
1	1	1	1	8.04	11.46	9.81	14.33
2	1	2	2	17.57	20.17	16.93	19.20
3	2	3	4	27.43	34.60	28.20	27.11
4	2	4	3	28.96	20.35	27.78	25.23
5	3	1	1	15.78	18.72	11.48	14.69
6	3	2	5	30.91	30.04	29.92	29.70
7	4	3	4	28.95	18.27	31.72	29.55
8	4	4	3	28.79	33.89	31.30	27.67
9	5	1	5	40.74	34.71	31.94	32.72
10	5	2	4	23.04	26.81	31.52	30.84
11	6	3	3	29.60	34.78	35.18	32.77
12	6	4	1	33.98	24.67	30.98	27.50
13	7	1	1	24.65	26.36	25.93	27.88
14	7	2	1	31.42	41.04	29.28	29.37
15	8	3	1	39.82	37.68	38.56	36.76
16	8	4	5	62.77	54.81	57.00	51.76
17	9	1	5	49.23	53.78	53.79	54.21
18	9	2	3	47.81	50.30	49.60	48.95
19	10	3	3	65.14	55.03	60.73	58.41
20	10	4	1	53.56	54.30	56.53	53.15

for $\beta_0 = 10$, $\beta_1 = -1$, $\beta_2 = 2$, $\beta_3 = 3$, $\beta_4 = 0.4$, $\sigma^2 = 25$. Consider the full quadratic model

$$h(\tilde{\mathbf{x}}) = \beta_0 + \beta_1 x_1 + \beta_2 x_2 + \beta_3 x_3 + \beta_4 x_1^2 + \beta_5 x_2^2 + \beta_6 x_3^2 + \beta_7 x_1 x_2 + \beta_8 x_1 x_3 + \beta_9 x_2 x_3,$$

and the sequence of subspaces V_k spanned by the first k terms. So V_5 is the true model. For these parameters two determinations $\mathbf{Y}_1 = \boldsymbol{\theta} + \boldsymbol{\varepsilon}_1$ and $\mathbf{Y}_2 = \boldsymbol{\theta} + \boldsymbol{\varepsilon}_2$ were made for 20 triples $\tilde{\mathbf{x}}_i = (x_{1i}, x_{2i}, x_{3i})$. Table 4.2.1 presents $\mathbf{x}_1$, $\mathbf{x}_2$, $\mathbf{x}_3$, $\mathbf{Y}_1$, $\mathbf{Y}_2$ and estimates of $\boldsymbol{\theta}$ corresponding to $\mathbf{Y}_1$ and $\mathbf{Y}_2$ for model V_5.

An estimate $\hat{\sigma}^2$ of σ^2 was obtained by fitting the full quadratic model with 10 parameters. For $\mathbf{Y}_1$ and $\mathbf{Y}_2$ these were $\hat{\sigma}_1^2 = 26.04$ and $\hat{\sigma}_2^2 = 32.18$. Then consecutive models $V_2, V_3, \ldots, V_k$ were fit, and values of S_k^2 and the Mallows statistic C_k obtained for each (Table 4.2.2). If we define $\boldsymbol{\theta}_\perp^k = \boldsymbol{\theta} - p(\boldsymbol{\theta} \mid V_k)$, then C_k is an estimate of $H_{V_k} = \|\boldsymbol{\theta}_\perp^k\|^2/\sigma^2 + k$ (recall that $\dim(V_k) = k + 1$). In choosing a model we look for the smallest k for which C_k is reasonably close to k, equivalently S_k^2 is close to $\hat{\sigma}^2$. For both $\mathbf{Y}_1$ and $\mathbf{Y}_2$ this suggests the model V_5.

For these models the analyses for $\mathbf{Y}_1$ and $\mathbf{Y}_2$ are given in Table 4.2.3.

Table 4.2.2

	Y_1			Y_2		
k	S_k^2	C_k	R_k^2	S_k^2	C_k	R_k^2
2	75.740	37.255	0.684	59.952	17.395	0.696
3	63.902	28.165	0.748	60.024	17.763	0.711
4	33.827	8.080	0.875	38.127	6.140	0.828
5	21.458	1.182	0.925	22.239	−0.943	0.906
6	21.551	2.412	0.930	23.606	1.003	0.907
7	20.563	3.054	0.938	25.249	2.984	0.907
8	22.048	5.005	0.939	27.299	5.028	0.908
9	24.020	7.067	0.939	29.612	7.041	0.908
10	26.044	9.000	0.940	32.182	9.000	0.909

Table 4.2.3

	Y_1					Y_2			
j	$\hat{\beta}_j$	$S_{\hat{\beta}_j}$	t_j	R_j	j	$\hat{\beta}_j$	$S_{\hat{\beta}_j}$	t_j	R_j
0	3.235	4.579	0.707		0	10.833	4.662	2.324	
1	−1.013	1.631	−0.621	0.961	1	−1.896	1.661	−1.142	0.947
2	3.351	0.938	3.573	0.928	2	1.494	0.955	1.565	0.944
3	3.772	0.684	5.511	0.880	3	3.379	0.697	4.849	0.871
4	0.463	0.145	3.197	0.935	4	0.519	0.147	3.526	0.910

Table 4.2.4

j	$\hat{\beta}_j$	$S_{\hat{\beta}_j}$	t_j	R_j
0	8.116	6.822	1.190	
1	−1.047	1.635	−0.640	0.963
2	−1.712	5.321	−0.322	0.964
3	3.860	0.692	5.578	0.880
4	0.466	0.145	3.212	0.937
5	1.012	1.047	0.967	0.962

R_j is R-delete for the jth variable, the multiple c.c. when that variable is deleted.

Thus, for these examples the procedure worked well. It will not always work as well. Had we entered the $\beta_5 x_2^2$ term before the $\beta_4 x_1^2$ term the Mallows statistics for $\mathbf{Y}_1$ would have been C_k: 37.3, 28.2, 8.1, 9.5, 2.4, 3.1, 5.0, 7.1, 9.0 for $k = 2, \ldots, 10$, so that the model V_6 would have been chosen, producing the regression analysis of Table 4.2.4. This suggests that variable 6 or 3 be dropped. Since variable 6 is x_2^2, it seems more reasonable to drop this, reducing the model to V_5 again. The close linear relationship between x_2 and x_2^2 caused their regression coefficients to have large variance, so that both t_3 and t_6 are small.

The AIC procedure of Akaike (1973, 1978) chooses the model $\mathcal{M}_k$ in a sequence of models $\mathcal{M}_1, \ldots, \mathcal{M}_L$ of dimensions given by the subscripts for which $\text{AIC}(k) = n(\log S_k^2 + 1) + 2(k + 1)$ is minimum. Both this criterion and the Bayesian information criterion, which minimizes $\text{BIC}(k) = n \log S_k^2 + k \log n$, can be justified from the Bayesian point of view. See Schwarz (1978) for a discussion of the asymptotic properties of these procedures. Hurvich and Tsai (1993) discuss the effects of use of these criteria for model selection on the estimation of parameter vectors by confidence ellipsoids. For a full discussion of model-building methods see the book by Linhart and Zucchini (1986).

Effects of Specification Errors on $\eta = \mathbf{c}'\boldsymbol{\beta}$

In order to assess the effect of specification errors on $\boldsymbol{\beta}$, or more generally, $\eta = \mathbf{c}'\boldsymbol{\beta}$ we need to specify $\boldsymbol{\theta}$ and V in terms of specific $\mathbf{x}$-vectors. We consider two cases: overspecification and underspecification.

Overspecification

Suppose $\boldsymbol{\theta} = \sum_1^k \beta_j \mathbf{x}_j$, but our postulated model is $\boldsymbol{\theta} = \sum_1^k \beta_j \mathbf{x}_j + \sum_1^r \gamma_j \mathbf{w}_j$, where $(\mathbf{x}_1, \ldots, \mathbf{w}_r)$ are linearly independent. In matrix form we can write $\boldsymbol{\theta} = \mathbf{X}\boldsymbol{\beta} + \mathbf{W}\boldsymbol{\gamma}$.

The true model therefore corresponds to the statement that $\boldsymbol{\gamma} = \mathbf{0}$, so that the least squares estimator $(\hat{\boldsymbol{\beta}}, \hat{\boldsymbol{\gamma}})$ of $(\boldsymbol{\beta}, \boldsymbol{\gamma})$ is unbiased. To see how this overspecification affects variances, consider a parameter $\eta = c_1\beta_1 + \cdots + c_k\beta_k$. The Gauss–Markov Theorem states that $\hat{\eta}_0 = \sum_1^r c_j\hat{\beta}_{j0}$, where $\hat{\boldsymbol{\beta}}_0 = (\hat{\beta}_{10}, \ldots, \hat{\beta}_{r0})'$ is the least squares estimation of η under the true model, has the smallest variance among all linear unbiased linear estimators. Since $\hat{\eta} = \mathbf{c}'\hat{\boldsymbol{\beta}}$ is a linear unbiased estimator, we conclude that

$$\text{Var}(\hat{\eta}_0) < \text{Var}(\hat{\eta}) \qquad \text{unless} \quad \hat{\eta} \equiv \hat{\eta}_0.$$

More explicitly, from the proof of the Gauss–Markov Theorem, we get

$$\hat{\eta}_0 = (\mathbf{a}_0, \mathbf{Y}) \qquad \text{for} \quad \mathbf{a}_0 = \mathbf{X}(\mathbf{X}'\mathbf{X})^{-1}\mathbf{c},$$

the unique vector in $V_0 = \mathscr{L}(\mathbf{x}_1, \ldots, \mathbf{x}_k)$ satisfying $\mathbf{X}'\mathbf{a}_0 = \mathbf{c}$.

In order to write $\hat{\eta}$ explicitly, define $\mathbf{Z} = (\mathbf{X}, \mathbf{W}) = (\mathbf{x}_1, \ldots, \mathbf{x}_k, \mathbf{w}_1, \ldots, \mathbf{w}_T)$. Then $\hat{\eta} = (\mathbf{a}, \mathbf{Y})$ for $\mathbf{a} = \mathbf{Z}(\mathbf{Z}'\mathbf{Z})^{-1}\begin{pmatrix} \mathbf{c} \\ \mathbf{0} \\ r \times 1 \end{pmatrix} = \mathbf{Z}(\mathbf{Z}'\mathbf{Z})^{-1}\mathbf{c}$ the unique vector in $V = \mathscr{L}(\mathbf{x}_1, \ldots, \mathbf{x}_k, \mathbf{w}_1, \ldots, \mathbf{w}_r)$ satisfying $\mathbf{Z}'\mathbf{a} = \begin{pmatrix} \mathbf{c} \\ \mathbf{0} \\ r \times 1 \end{pmatrix}$. Then $p(\mathbf{a} \mid V_0) = \mathbf{a}_0$, so that

$$\text{Var}(\hat{\eta}) - \text{Var}(\hat{\eta}_0) = \|\mathbf{a} - \mathbf{a}_0\|^2\sigma^2 > 0$$

For $\hat{\mathbf{Y}} = p(\mathbf{Y} \mid V)$, $(\mathbf{Y} - \hat{\mathbf{Y}}) \perp V_0$, since V_0 is a subspace of V. Therefore, $\dfrac{\|\mathbf{Y} - \hat{\mathbf{Y}}\|^2}{\sigma^2} \sim \chi^2_{n-k-r}$ and $S^2 = \|\mathbf{Y} - \hat{\mathbf{Y}}\|^2/[n - (k + r)]$ is an unbiased estimator of σ^2 even in the case of overspecification. The degrees of freedom for error is smaller by r than it would be for the true model, so we pay a price in reduced degrees of freedom for error.

Underspecification

Suppose that the true model is $\boldsymbol{\theta} = \mathbf{X}\boldsymbol{\beta} + \mathbf{W}\boldsymbol{\gamma}$, but the postulated model is $\boldsymbol{\theta} = \mathbf{X}\boldsymbol{\beta}$. If V is the column space of $\mathbf{X}$, then $\hat{\mathbf{Y}}_V$ has bias $-\boldsymbol{\theta}_\perp = -(\mathbf{I} - \mathbf{P}_V)\mathbf{W}\boldsymbol{\gamma}$. Since $E(\hat{\boldsymbol{\beta}}) = (\mathbf{X}'\mathbf{X})^{-1}\mathbf{X}'\mathbf{Y} = (\mathbf{X}'\mathbf{X})^{-1}\mathbf{X}'(\mathbf{X}\boldsymbol{\beta} + \mathbf{W}\boldsymbol{\gamma}) = \boldsymbol{\beta} + (\mathbf{X}'\mathbf{X})^{-1}\mathbf{X}'\mathbf{W}\boldsymbol{\gamma}$, $\hat{\boldsymbol{\beta}}$ has bias $(\mathbf{X}'\mathbf{X})^{-1}\mathbf{X}'\mathbf{W}\boldsymbol{\gamma}$ as an estimator of $\boldsymbol{\beta}$. However, $E(\mathbf{X}\hat{\boldsymbol{\beta}}) = \mathbf{P}_V\mathbf{Y} = \boldsymbol{\theta}_V$, so that $\hat{\boldsymbol{\beta}}$ is an unbiased estimator of the vector $\boldsymbol{\beta}_V = \boldsymbol{\theta} + (\mathbf{X}'\mathbf{X})^{-1}\mathbf{X}'\mathbf{W}\boldsymbol{\gamma}$ of coefficients in the least squares approximation $\boldsymbol{\theta}_V$ of $\boldsymbol{\theta}$.

Suppose that we wish to estimate $\eta = E(\mathbf{Y} \mid \tilde{\mathbf{x}}, \tilde{\mathbf{w}}) = g(\tilde{\mathbf{x}}, \tilde{\mathbf{w}}) = \tilde{\mathbf{x}}\boldsymbol{\beta} + \tilde{\mathbf{w}}\boldsymbol{\gamma}$, the mean $\mathbf{Y}$ for an individual or unit with characteristics $(\tilde{\mathbf{x}}, \tilde{\mathbf{w}})$. For the postulated model the estimator is $\hat{\eta} = \tilde{\mathbf{x}}\hat{\boldsymbol{\beta}}$, where $\hat{\boldsymbol{\beta}}$ is the least squares estimator corresponding to $\mathbf{X}$. Let $(\hat{\boldsymbol{\beta}}_0, \hat{\boldsymbol{\gamma}}_0)$ and $\hat{\eta}_0 = \tilde{\mathbf{x}}\hat{\boldsymbol{\beta}}_0 + \tilde{\mathbf{w}}\hat{\boldsymbol{\gamma}}_0$ be the estimators corresponding to the true model. It is important to interpret the ith component of $\boldsymbol{\beta}$ as the change in $g(\tilde{\mathbf{x}}, \tilde{\mathbf{w}})$ given a change in one unit in the ith component of $\tilde{\mathbf{x}}$, and no changes in other components of $(\tilde{\mathbf{x}}, \tilde{\mathbf{w}})$.

Let $V = \mathscr{L}(\mathbf{X})$, $\mathbf{Z} = (\mathbf{X}, \mathbf{W})$, $V_0 = \mathscr{L}(\mathbf{Z})$, $\mathbf{d} = \mathbf{X}\mathbf{M}^{-1}\tilde{\mathbf{x}}$, $\mathbf{d}_0 = \mathbf{Z}(\mathbf{Z}'\mathbf{Z})^{-1}(\tilde{\mathbf{x}}, \tilde{\mathbf{w}})'$. Then $\hat{\eta} = \mathbf{d}'\mathbf{Y}$, $\hat{\eta}_0 = \mathbf{d}_0'\mathbf{Y}$, and

$$E(\hat{\eta}) = \mathbf{d}'\boldsymbol{\theta} = \tilde{\mathbf{x}}\mathbf{M}^{-1}\mathbf{X}'(\mathbf{X}\boldsymbol{\beta} + \mathbf{W}\boldsymbol{\gamma}) = \tilde{\mathbf{x}}\boldsymbol{\beta} + \tilde{\mathbf{x}}\mathbf{M}^{-1}\mathbf{X}'\mathbf{W}\boldsymbol{\gamma}$$

so that $\hat{\eta}$ has bias

$$b(\tilde{\mathbf{x}}, \tilde{\mathbf{w}}) = (\mathbf{d} - \mathbf{d}_0)'\boldsymbol{\theta} = (\tilde{\mathbf{x}}\mathbf{M}^{-1}\mathbf{X}'\mathbf{W} - \tilde{\mathbf{w}})\boldsymbol{\gamma}.$$

The first term within the last parentheses is the predicted value of $\tilde{\mathbf{w}}$ for given $\tilde{\mathbf{x}}$ based on the data $(\mathbf{X}, \mathbf{W})$ and a linear model $\mathbf{W} = \mathbf{X}\mathbf{B} + \boldsymbol{\omega}$, where $\boldsymbol{\omega}$ is an $n \times r$ error matrix. Thus, if $\tilde{\mathbf{w}}$ is *exactly consistent* with the relationship between $\mathbf{X}$ and $\mathbf{W}$, then $\hat{\eta}$ is unbiased for η.

For example, let Y be college grade point average, x_1 S.A.T. exam score, w_1 high school grade point average, and suppose linear regression of w_1 on x_1 predicts $w_1 = 2.90$ for $x_1 = 500$. Then $\hat{\eta}$ is an unbiased estimator of $g(500, 2.90)$, but would be biased for $g(500, 3.60)$.

In order to compare $\hat{\eta}$ and $\hat{\eta}_0$ let us compute their mean square errors. We have

$$\operatorname{Var}(\hat{\eta}) = \|\mathbf{d}\|^2\sigma^2 \qquad \text{and} \qquad \operatorname{Var}(\hat{\eta}_0) = \|\mathbf{d}_0\|^2\sigma^2.$$

Since $\mathbf{d} \in V$ and, since $\mathbf{d}_0'Z = \mathbf{d}_0'(\mathbf{X}, \mathbf{W}) = (\tilde{\mathbf{x}}, \tilde{\mathbf{w}})$, $\mathbf{X}'\mathbf{d}_0 = \tilde{\mathbf{x}}$ and $\mathbf{X}'\mathbf{d} = \tilde{\mathbf{x}}$, $\mathbf{d} = p(\mathbf{d}_0 | V)$. Let $\mathbf{d}_\perp = \mathbf{d}_0 - \mathbf{d}$. Then $\mathbf{d}_\perp \perp V$ and $\mathbf{d}_0 = \mathbf{d} + \mathbf{d}_\perp$ is a decomposition of $\mathbf{d}_0$ into orthogonal vectors. It follows that $\hat{\eta}_0 = (\mathbf{d}_0, \mathbf{Y}) = \hat{\eta} + (\mathbf{d}_\perp, \mathbf{Y})$, where the two terms on the right are uncorrelated. Thus,

$$\operatorname{MSE}(\hat{\eta}) = b^2(\tilde{\mathbf{x}}, \tilde{\mathbf{w}}) + \operatorname{Var}(\hat{\eta}) = [\mathbf{d}_\perp'\mathbf{W}\boldsymbol{\gamma}]^2 + \|\mathbf{d}\|^2\sigma^2$$

and $\operatorname{MSE}(\hat{\eta}_0) = \operatorname{Var}(\hat{\eta}_0) = \|\mathbf{d}_0\|^2\sigma^2$, so that

$$\operatorname{MSE}(\hat{\eta}) - \operatorname{MSE}(\hat{\eta}_0) = [\mathbf{d}_\perp'\mathbf{W}\boldsymbol{\gamma}]^2 - \|\mathbf{d}_\perp\|^2\sigma^2.$$

Thus $\hat{\eta}$ is a better estimator than $\hat{\eta}_0$ if

$$\delta = \frac{\|p(\mathbf{W}\boldsymbol{\gamma}|\mathbf{d}_\perp)\|^2}{\sigma^2} = \frac{[\mathbf{d}_\perp'\mathbf{W}\boldsymbol{\gamma}]^2}{\|\mathbf{d}_\perp\|^2\sigma^2} < 1.$$

δ is maximized for all $\mathbf{d}_\perp \in V^\perp \cap V_0$, equivalently all $(\tilde{\mathbf{x}}, \tilde{\mathbf{w}})$, for $\mathbf{d}_\perp$ any multiple of $p(\mathbf{W}\boldsymbol{\gamma}| V^\perp \cap V_0) = (\mathbf{I} - \mathbf{P}_V)\mathbf{W}\boldsymbol{\gamma}$. We conclude that $\hat{\eta}$ is better than $\hat{\eta}_0$ for all vectors $(\tilde{\mathbf{x}}, \tilde{\mathbf{w}})$ if

$$\Delta = [\boldsymbol{\gamma}'\mathbf{W}'(\mathbf{I} - \mathbf{P}_V)\mathbf{W}\boldsymbol{\gamma}]/\sigma^2 = \|\boldsymbol{\theta}_\perp\|^2/\sigma^2 < 1$$

Δ is the noncentrality parameter in the F-test of $H_0: \gamma = 0$. Since $E(F) = \frac{\nu_2 - 1}{\nu_2}\left(1 + \frac{\Delta}{\nu_1}\right)$ for (ν_1, ν_2) d.f., and unbiased estimator of Δ is $D = \left(\frac{\nu_2 - 2}{\nu_2}F - 1\right)\nu_1$, so we could choose to use the smaller model when $D < 1$, equivalently if $F < \left(\frac{1 + \nu_1}{\nu_1}\right)\frac{\nu_2}{\nu_2 - 2}$. This requires knowledge of $\mathbf{W}$, of course. Here $\nu_2 = n - (k - r)$, $\nu_1 = r$. For large n and $r = 1$, this suggests dropping a single $\mathbf{w}$ if the F-statistic for testing $H_0: \gamma = 0$ is less than 2, equivalently if $|t| < \sqrt{2}$. It is difficult to evaluate the properties of such a procedure.

It is possible that δ can be less than one for some $(\tilde{\mathbf{x}}, \tilde{\mathbf{w}})$, greater for others, since δ depends on the bias term $(\tilde{\mathbf{x}}\mathbf{M}^{-1}\mathbf{X}'\mathbf{W} - \tilde{\mathbf{w}})\boldsymbol{\gamma}$. In the case that $\mathbf{W} = \mathbf{w}$, a one-column matrix, the subspace $V^\perp \cap V_0$ is spanned by $\mathbf{w}_\perp = p(\mathbf{w} | V^\perp \cap V_0)$, we get

$$\delta = \gamma^2 \|\mathbf{w}_\perp\|^2/\sigma^2,$$

which does not depend upon $(\tilde{\mathbf{x}}, \tilde{\mathbf{w}})$, so that $\hat{\eta}$ is either better or worse than $\hat{\eta}_0$, uniformly for all $(\tilde{\mathbf{x}}, \tilde{\mathbf{w}})$.

It is important to interpret the ith component of $\boldsymbol{\beta}$ as the change in $E(\mathbf{Y}) = g(\tilde{\mathbf{x}}, \tilde{\mathbf{w}})$ given a change in one unit in the ith elements of $\tilde{\mathbf{x}}$, and no change in $\tilde{\mathbf{w}}$.

As noted earlier $E(S^2) = \sigma^2 + \|\boldsymbol{\theta}_\perp\|^2/(n - k)$. In the case of underspecification

$$\boldsymbol{\theta}_\perp = (\mathbf{I} - \mathbf{P}_V)\mathbf{W}\boldsymbol{\gamma}, \qquad \text{so that} \quad \|\boldsymbol{\theta}_\perp\|^2 = \Delta\sigma^2.$$

Thus $E(S^2) = \sigma^2\left(1 + \frac{\Delta}{n - k}\right)$.

Example 4.2.2: Suppose that $g(x) = E(Y|x) = \beta_0 + \beta_1 x + \beta_2 x^2 + \beta_3 x^3$, $\text{Var}(Y|x) = \sigma^2$, and we observe Y independently for $x = -2, 1, 0, 1, 2$. What penalty do we pay if our postulated model is simple linear regression: $g(x) = \beta_0 + \beta_1 x$?

Define $\mathbf{x}_0 = (1, 1, 1, 1, 1)'$, $\mathbf{x}_1 = (-2, -1, 0, 1, 2)'$, $\mathbf{w}_1 = (4, 1, 0, 1, 4)'$, $\mathbf{w}_2 = (-8, -1, 0, 1, 8)'$, so that $\mathbf{X} = (\mathbf{x}_0, \mathbf{x}_1, \mathbf{x}_2)$, $\mathbf{W} = (\mathbf{w}_1, \mathbf{w}_2)$, $\beta = (\beta_1, \beta_2)'$, $\boldsymbol{\gamma} = (\gamma_1, \gamma_2)'$, $\boldsymbol{\theta} = \boldsymbol{\theta}_1 + \boldsymbol{\theta}_2$ where $\boldsymbol{\theta}_1 = \mathbf{X}\beta$, $\boldsymbol{\theta}_2 = \mathbf{W}\boldsymbol{\gamma}$. Then $\boldsymbol{\theta}_\perp = \boldsymbol{\theta} - p(\boldsymbol{\theta}| V) = \boldsymbol{\theta}_2 - p(\boldsymbol{\theta}_2 | V) = \beta_3 \mathbf{w}_1^\perp + \beta_4 \mathbf{w}_2^\perp$, where for $i = 1$, 2, $\mathbf{w}_i^\perp$ is the part of $\mathbf{w}_i$ orthogonal to V. Thus, $\mathbf{w}_1^\perp = \mathbf{w}_1 - 2\mathbf{x}_0 = (2, -1, -2, -1, 2)'$, $\mathbf{w}_2^\perp = \mathbf{w}_2 - 3.4\mathbf{x}_1 = (-1.2, 2.4, 0, -2.4, 1.2)'$.

Thus, for example, $\hat{\mathbf{Y}}_2$ has bias $-(-\beta_3 + 2.4\beta_4)$ as an estimator of $\boldsymbol{\theta}_2$, while $\hat{\mathbf{Y}}_3$ has bias $2\beta_3$ as an estimator of $\boldsymbol{\theta}_3$. Since $\|\boldsymbol{\theta}_\perp\|^2 = \beta_3^2\|\mathbf{w}_1^\perp\|^2 + \beta_4^2\|\mathbf{w}_2^\perp\|^2 = 14\beta_3^2 + 10\beta_4^2$, $E(S^2) = \sigma^2 + (14/3)\beta_3^2 + (10/3)\beta_4^2$.

The bias in $\hat{\boldsymbol{\beta}} = (\hat{\beta}_0, \hat{\beta}_1)'$ is $(\mathbf{X}'\mathbf{X})^{-1}\mathbf{X}'\mathbf{W}\boldsymbol{\gamma} = (2\beta_3, 3.4\beta_4)'$. Thus, $\hat{\beta}_0$ is unbiased for β_0 if and only if $\beta_3 = 0$, and $\hat{\beta}_1$ is unbiased for β_1 if and only if $\beta_4 = 0$.

Table 4.2.5

Model	c_{00}	c_{11}	c_{22}	c_{33}	c_{44}
V_5	0.977 13	0.124 00	0.040 99	0.021 83	0.000 98
V_6	2.159 27	0.124 06	1.313 43	0.022 22	0.000 98
V_7	2.285 11	0.131 47	1.818 19	0.896 44	0.001 01
V_8	2.566 52	0.133 88	2.504 56	1.218 20	0.001 23
V_9	2.606 01	0.141 64	3.327 51	1.804 93	0.001 27
V_{10}	2.923 70	0.145 92	3.557 44	1.886 16	0.001 32

Example 4.2.3: For **X** as in Example 4.2.1 consider the changes in the variances of regression coefficients as the number of terms in the model grows. Table 4.2.5 gives $c_{jj} = \text{Var}(\hat{\beta}_j)/\sigma^2$ for each postulated model $V_5, V_6, \ldots, V_{10}$. c_{jj} is the jj term in the corresponding matrix $(\mathbf{X}'\mathbf{X})^{-1}$. Notice that $\text{Var}(\hat{\beta}_2)$ jumps considerably as the model is changed from V_5 to V_6, that is, the variable x_2^2, which has correlation 0.984 4 with x_2, is added. Similarly $\text{Var}(\hat{\beta}_3)$ jumps when the variable x_3^2 is added (correlation = 0.980 8).

Suppose we fit model V_3. Using the notation for underspecification $\mathbf{X} = (\mathbf{x}_0, \mathbf{x}_1, \mathbf{x}_2)$, $\mathbf{W} = (\mathbf{x}_3, \mathbf{x}_4)$. Then $\boldsymbol{\gamma} = (\beta_3, \beta_4)' = (3, 3)$ and $\hat{\boldsymbol{\beta}} = (\hat{\beta}_0, \hat{\beta}_1, \hat{\beta}_2)$, has bias $(\mathbf{X}'\mathbf{X})^{-1}\mathbf{X}'\mathbf{w}\boldsymbol{\gamma} = (-0.232, 4.36, 0.0149)'$. This bias could cause serious problems if regression coefficients are to have subject matter interpretations. $\hat{\beta}_1$ will almost certainly be positive though its true value is -1.

Example 4.2.4: Suppose **Y** satisfies the simple linear regression model $\mathbf{Y} = \beta_0\mathbf{x}_0 + \beta_1\mathbf{x}_1 + \boldsymbol{\varepsilon}$, and we ignore the $\mathbf{x}_1$ vector. That is, our postulated model is $\mathbf{Y} = \beta_0\mathbf{x}_0 + \boldsymbol{\varepsilon}$. Then $\mathbf{X} = \mathbf{x}_0$, $\mathbf{W} = \mathbf{x}_1 = \mathbf{w}$ and $\mathbf{w}_\perp$ is the vector of deviations of the x_i's from $\bar{x}$. Let $\eta = \beta_0 + \beta_1 x_0$. Under this postulated model the estimator of η is $\hat{\eta} = \bar{Y}$, which has bias $b(1, x_0) = \beta_0 + \beta_1\bar{x} - (\beta_0 + \beta_1\bar{x}_0) = -\beta_1(x_0 - \bar{x})$, and $\text{Var}(\hat{\eta}) = \sigma^2/n$.

The estimator under the true model is $\hat{\eta}_0 = \hat{\beta}_0 = \hat{\beta}_1 x_0 = \bar{Y} + \hat{\beta}_1(x_0 - \bar{x})$, which is unbiased for η, with $\text{Var}(\hat{\eta}_0) = \sigma^2\left[\frac{1}{n} + \frac{(x_0 - \bar{x})^2}{S_{xx}}\right]$. Thus, $\text{MSE}(\hat{\eta}) = \beta_1^2(x_0 - \bar{x})^2 + \frac{\sigma^2}{n}$ and $\text{MSE}(\hat{\eta}_0) = \sigma^2\left[\frac{1}{n} + \frac{(x_0 - \bar{x})^2}{S_{xx}}\right]$, so that $\text{MSE}(\hat{\eta}) - \text{MSE}(\hat{\eta}_0) = (x_0 - \bar{x})^2\left[\beta_1^2 - \frac{\sigma^2}{S_{xx}}\right]$, which is negative for $\delta = \frac{\beta_1^2}{\sigma^2/S_{xx}} < 1$. δ is the noncentrality parameter in the F-test of $H_0\colon \beta_1 = 0$.

In the case that $\eta = g(\tilde{\mathbf{x}}_1, \tilde{\mathbf{w}}) - g(\tilde{\mathbf{x}}_2, \tilde{\mathbf{w}}) = (\tilde{\mathbf{x}}_1 - \tilde{\mathbf{x}}_2)\boldsymbol{\beta}$ represents the change in $E(Y)$ as $\tilde{\mathbf{x}}$ changes from $\tilde{\mathbf{x}}_1$ to $\tilde{\mathbf{x}}_2$, while $\tilde{\mathbf{w}}$ is held constant, use of the postulated model $\boldsymbol{\theta} = \mathbf{X}\boldsymbol{\beta}$ leads to the estimator $\hat{\eta} = (\tilde{\mathbf{x}}_1 - \tilde{\mathbf{x}}_2)\hat{\boldsymbol{\beta}}$ with bias $(\tilde{\mathbf{x}}_1 - \tilde{\mathbf{x}}_2)\mathbf{M}^{-1}\mathbf{X}'\mathbf{W}\boldsymbol{\gamma}$. $\hat{\eta}$ is an unbiased estimator of η for all choices $\tilde{\mathbf{x}}_1$ and $\tilde{\mathbf{x}}_2$ from the rows of **X** if the column vectors of **X** are uncorrelated with those of **W** (see Problem 4.2.6). The comparisons between $\hat{\eta}$ and $\hat{\eta}_0$ are the same as for the more general case of $\eta = \tilde{\mathbf{x}}\boldsymbol{\beta} + \tilde{\mathbf{w}}\mathbf{Y}$ with $\tilde{\mathbf{x}}_1 - \tilde{\mathbf{x}}_2$ replacing $\tilde{\mathbf{x}}$, $\tilde{\mathbf{0}}$ replacing $\tilde{\mathbf{w}}$.

Problem 4.2.1: Suppose that $g(x) = E(Y|x) = \beta_0 + \beta_1 x + \beta_2 x^2 + \beta_3 x^3$, $\text{Var}(Y|x) = \sigma^2$, and we observe Y independently for $x = -2, -1, 0, 1, 2$. Find the biases in the estimation of (β_0, β_1), $\boldsymbol{\theta}$, and σ^2, caused by fitting the simple linear regression model.

Problem 4.2.2: Suppose that the roles of the postulated model and true model in Problem 4.2.1 are exchanged, so the simple linear regression model holds. Is $(\hat{\beta}_0, \hat{\beta}_1)$ in $(\hat{\beta}_0, \ldots, \hat{\beta}_3)$ an unbiased estimator of (β_0, β_1)? Evaluate $\text{Var}(\hat{\beta}_0 + \hat{\beta}_1 x)$ and show that it is larger for the cubic model than it is for the (correct) simple linear regression model.

Problem 4.2.3: Let $x_1, \ldots, x_{20}$ be 20 real numbers, which include at least six different values. Suppose that we observe Y corresponding to each x, and fit a polynomial of degree k in x for $k = 0, \ldots, 5$. Let $\text{SSE}(k)$ be the error sum of squares corresponding to the polynomial of degree k. Let $\text{ESS}(0) =$ 1,000, $\text{ESS}(1) = 300$, $\text{ESS}(2) = 120$, $\text{ESS}(3) = 90$, $\text{ESS}(4) = 85$, $\text{ESS}(5) = 84$. Evaluate the Mallows statistics C_k, using the "pure" estimate of error $\hat{\sigma}^2 = \text{SSE}(5)/(20 - 6)$. Which model seems to be appropriate?

Problem 4.2.4: Suppose that the relationship between weight Y in pounds and height x in inches is $Y = \beta_0 + \beta_1 x_1 + \beta_2 w + \varepsilon$, where w is the indicator for males. Observations were made on four males and three females:

Y	150	180	140	160	110	120	130
x_1	70	74	66	72	62	64	66
w	1	1	1	1	0	0	0

(a) Estimate the expected weight η of a person 68 inches tall, with $(\hat{\eta}_0)$ and without $(\hat{\eta})$ using their sex.

(b) Give an expression for the bias of $\hat{\eta}$ and evaluate it for female 68 inches tall, if $\beta_2 = 20$.

(c) For this 68 inch female for what values of pairs (β_2, σ^2) would $\hat{\eta}$ be a better estimator than $\hat{\eta}_0$?

(d) Let $S^2 = \text{SSE}/(n - 2)$ for the fit of the simple linear regression model. Find $E(S^2)$ for $\beta_2 = 20$, $\sigma^2 = 25$.

Problem 4.2.5: Prove that, for F as defined immediately before Example 4.2.1, $F < 2$ is equivalent to $C_k < k$.

Problem 4.2.6: Prove that in the case of the underspecified model zero correlation of all the column vectors of $\mathbf{X}$ with those of $\mathbf{W}$ implies that $\hat{\eta} = (\tilde{\mathbf{x}}_1 - \tilde{\mathbf{x}}_2)\hat{\boldsymbol{\beta}}$ is an unbiased estimator of $\eta = (\tilde{\mathbf{x}}_1 - \tilde{\mathbf{x}}_2)\boldsymbol{\beta}$ for all pairs $(\tilde{\mathbf{x}}_1, \tilde{\mathbf{x}}_2)$ of row vectors of $\mathbf{X}$.

4.3 "GENERALIZED" LEAST SQUARES

Under the model $\mathbf{Y} = \boldsymbol{\theta} + \boldsymbol{\varepsilon} = \sum_1^k \beta_j \mathbf{x}_j + \boldsymbol{\varepsilon}$, the assumption that $E(\boldsymbol{\varepsilon}) = \mathbf{0}$ and $D[\boldsymbol{\varepsilon}] = \sigma^2 \mathbf{I}_n$, leads, by the Gauss–Markov Theorem, to the optimality, in a certain sense, of least squares estimation. Though the condition $D[\boldsymbol{\varepsilon}] = \sigma^2 \mathbf{I}_n$ is often quite reasonable, there are certainly many occasions when it is unrealistic. If the components of $\mathbf{Y}$ correspond to observations at consecutive points in time (time series) as is often the case with economic data, there will often be correlation, usually positive but sometimes negative, between observations at consecutive points in time. Larger values for the components of $\boldsymbol{\theta}$ often lead to corresponding large values for the variance terms of $D[\boldsymbol{\varepsilon}]$.

Though lack of knowledge of $\boldsymbol{\Sigma} = D[\boldsymbol{\varepsilon}]$ can cause severe problems, there are occasions when $\boldsymbol{\Sigma}$ is known up to or nearly up to a multiplicative constant. In this case we can reduce the problem to the previous form.

Let $\mathbf{Y} = \boldsymbol{\theta} + \boldsymbol{\varepsilon} = \sum_1^k \beta_j \mathbf{x}_j + \boldsymbol{\varepsilon} = \mathbf{X}\boldsymbol{\beta} + \boldsymbol{\varepsilon}$, $D[\boldsymbol{\varepsilon}] = \boldsymbol{\Sigma} = \sigma^2 \mathbf{A}$, where $\mathbf{A}$ is a *known* $n \times n$ nonsingular matrix, σ^2 is an unknown constant (which is the common variance if the diagonal elements of $\mathbf{A}$ are all ones). Let $\mathbf{B}\mathbf{B}' = \mathbf{A}$, where $\mathbf{B}$ is $n \times n$. Since $\mathbf{A}$ must be positive definite and $(\mathbf{BF})(\mathbf{BF})' = \mathbf{A}$ for any orthogonal matrix $\mathbf{F}$, an infinity of such $\mathbf{B}$ can be chosen. Thus $\mathbf{B}$ may be chosen to have special properties. It can, for example, be chosen to be lower triangular (zeros above the diagonal). Or if $\mathbf{A}$ has eigenvalues $\lambda_1, \ldots, \lambda_n$ with corresponding eigenvectors $\mathbf{v}_1, \ldots, \mathbf{v}_n$, each of length one, then for $\mathbf{V} = (\mathbf{v}_1, \ldots, \mathbf{v}_n)$, $\mathbf{D} = \operatorname{diag}(\lambda_1, \ldots, \lambda_n)$, $\mathbf{AV} = \mathbf{VD}$, $\mathbf{A} = \mathbf{VDV}'$, so we can take $\mathbf{B} = \mathbf{VD}^{1/2}$ or $\mathbf{B} = \mathbf{VD}^{1/2}\mathbf{V}'$ (the symmetric version).

For given $\mathbf{B}$ with $\mathbf{BB}' = \mathbf{A}$, define

$$\mathbf{Z} = \mathbf{B}^{-1}\mathbf{Y} = \mathbf{B}^{-1}\boldsymbol{\theta} + \mathbf{B}^{-1}\boldsymbol{\varepsilon} = \sum_1^k \beta_j(\mathbf{B}^{-1}\mathbf{x}_j) + \mathbf{B}^{-1}\boldsymbol{\varepsilon} = \sum_1^k \beta_j \mathbf{w}_j + \boldsymbol{\eta}$$

for $\mathbf{w}_j = \mathbf{B}^{-1}\mathbf{x}_j$ and $\boldsymbol{\eta} = \mathbf{B}^{-1}\boldsymbol{\varepsilon}$. Then $E(\boldsymbol{\eta}) = \mathbf{B}^{-1}E(\boldsymbol{\varepsilon}) = \mathbf{0}$, and $D[\boldsymbol{\eta}] = \mathbf{B}^{-1}(\sigma^2\mathbf{A})\mathbf{B}'^{-1} = \sigma^2\mathbf{B}^{-1}\mathbf{A}\mathbf{B}'^{-1} = \sigma^2\mathbf{I}_n$, so that $\mathbf{Z}$ satisfies the standard linear hypothesis we have considered before. Under the model $E(\mathbf{Z}) = \mathbf{B}^{-1}\boldsymbol{\theta} \in \mathscr{L}(\mathbf{w}_1, \ldots, \mathbf{w}_k)$. The least squares estimator of $\boldsymbol{\beta}$ based on $\mathbf{Z}$ is therefore

$$(\mathbf{W}'\mathbf{W})^{-1}\mathbf{W}'\mathbf{Z} = \hat{\boldsymbol{\beta}},$$

where $\mathbf{W} = (\mathbf{w}_1, \ldots, \mathbf{w}_k) = \mathbf{B}^{-1}\mathbf{X}$. It follows that

$$\hat{\boldsymbol{\beta}} = (\mathbf{X}'\mathbf{B}'^{-1}\mathbf{B}^{-1}\mathbf{X})^{-1}\mathbf{X}'\mathbf{B}'^{-1}\mathbf{B}^{-1}\mathbf{Y},$$

so

$$\hat{\boldsymbol{\beta}} = (\mathbf{X}'\mathbf{A}^{-1}\mathbf{X})^{-1}\mathbf{X}'\mathbf{A}^{-1}\mathbf{Y} \tag{4.3.1}$$

Thus, $\hat{\boldsymbol{\beta}}$ is a function of $\mathbf{A}$ alone (as well as $\mathbf{X}$ and $\mathbf{Y}$), and does not depend on the particular decomposition $\mathbf{BB}' = \mathbf{A}$. Because $\hat{\boldsymbol{\beta}}$ is the least squares estimator of $\boldsymbol{\beta}$ as a function of $\mathbf{Z}$, and $D[\mathbf{Z}] = \sigma^2\mathbf{I}_n$, $\hat{\boldsymbol{\beta}}$ has the optimality properties of the Gauss–Markov Theorem. Any tests of hypothesis can be performed using $\mathbf{Z}$. Of course, any statistic can be rewritten as a function of $\mathbf{Y}$ by making the substitution $\mathbf{Z} = \mathbf{B}^{-1}\mathbf{Y}$.

The formula (4.3.1) above, and other formulas used in connection with this "generalized least squares" can be expressed as functions of inner products. The inner products in this case have the form

$$(\mathbf{v}_1, \mathbf{v}_2)_* = \mathbf{v}_1'\mathbf{A}^{-1}\mathbf{v}_2 \tag{4.3.2}$$

Thus, when $D[\boldsymbol{\varepsilon}] = \sigma^2\mathbf{I}_n$ the inner product should be as given in (4.3.2) and

$$\hat{\boldsymbol{\beta}} = \mathbf{M}^{-1}\mathbf{U},$$

where $\mathbf{M}$ is the inner product matrix among the $\mathbf{x}_j$'s using the * inner product, and $\mathbf{U}$ is the vector of * inner products of $\mathbf{Y}$ with the $\mathbf{x}_j$. Then $\hat{\mathbf{Y}} = \sum_1^k \hat{\beta}_j\mathbf{x}_j$ is the orthogonal projection (in the * inner product) of $\mathbf{Y}$ onto $\mathscr{L}(\mathbf{x}_1, \ldots, \mathbf{x}_k)$.

Example: Suppose $\mathbf{A} = \text{diag}(w_1^2, \ldots, w_n^2)$, $\mathbf{B} = \text{diag}(w_1, \ldots, w_n)$. Then $\mathbf{Z} = \mathbf{B}^{-1}\mathbf{Y} = \begin{pmatrix} Y_1/w_1 \\ \vdots \\ Y_n/w_n \end{pmatrix}$. This is usually called *weighted least squares*, since $\hat{\boldsymbol{\beta}}$ as defined in (4.3.1) minimizes $\sum_1^n (Y_i - \tilde{x}_i\beta)^2 w_i^2$. For $k = 1$, we get

$$\hat{\beta} = \left(\sum_1^n \frac{x_i^2}{w_i}\right)^{-1} \sum_1^n \left(\frac{x_i Y_i}{w_i}\right),$$

so that, for $w_i = Kx_i$, we get

$$\hat{\beta} = \left(\sum Y_i\right)/\sum x_i = \bar{Y}/\bar{x},$$

and, for $w_i = Kx_i^2$, we get $\hat{\beta} = \dfrac{1}{n}\sum_1^n Y_i/x_i$.

Suppose $\mathbf{A} = (a_{ij})$ for $a_{ij} = \rho^{|i-j|}$, $-1 < \rho < 1$, as is the case when the ε_i satisfy the first-order autoregressive model $\varepsilon_i = \rho\varepsilon_{i-1} + \xi_i$, where the ξ_i are independent with equal variances. Then

$$\mathbf{A}^{-1} = \frac{1}{\Delta}\begin{bmatrix} 1 & \rho & 0 & \cdot & \cdot & \cdot & 0 \\ -\rho & 1+\rho^2 & -\rho & 0 & \cdot & \cdot & 0 \\ 0 & -\rho & 1+\rho^2 & -\rho & \cdot & \cdot & \cdot \\ \cdot & \cdot & -\rho & \cdot & \cdot & \cdot & \cdot \\ \cdot & \cdot & 0 & \cdot & \cdot & \cdot & \cdot \\ \cdot & \cdot & \cdot & \cdot & \cdot & \cdot & \cdot \\ \cdot & \cdot & \cdot & \cdot & \cdot & 1-\rho^2 & -\rho \\ 0 & 0 & 0 & \cdot & \cdot & -\rho & 1 \end{bmatrix}, \qquad \Delta = 1+\rho^2$$

$$\mathbf{B} = \begin{bmatrix} 1/\Delta & 0 & 0 & 0 & \cdot & \cdot & 0 \\ \rho/\Delta & 1 & 0 & \cdot & \cdot & \cdot & 0 \\ \rho^2/\Delta & \rho & 1+\rho^2 & -\rho & \cdot & \cdot & \cdot \\ \cdot & \cdot & -\rho & \cdot & \cdot & \cdot & \cdot \\ \cdot & \cdot & 0 & \cdot & \cdot & \cdot & \cdot \\ \cdot & \cdot & \cdot & \cdot & \cdot & \cdot & \cdot \\ \cdot & \cdot & \cdot & \cdot & \cdot & 1-\rho^2 & -\rho \\ \rho^{n-1}/\Delta & \rho^{n-2} & \rho^{n-3} & \cdot & \cdot & -\rho & 1 \end{bmatrix}$$

$$\mathbf{B}^{-1} = \begin{bmatrix} \Delta & 0 & 0 & 0 & \cdot & \cdot & 0 \\ -\rho & 1 & 0 & 0 & \cdot & \cdot & 0 \\ 0 & -\rho & 1 & 0 & \cdot & \cdot & 0 \\ 0 & 0 & -\rho & 1 & \cdot & \cdot & 0 \\ \cdot & \cdot & \cdot & \cdot & \cdot & \cdot & \cdot \\ \cdot & \cdot & \cdot & \cdot & \cdot & \cdot & \cdot \\ 0 & 0 & 0 & \cdot & \cdot & -\rho & 1 \end{bmatrix}$$

Thus,

$$\mathbf{Z} = \mathbf{B}^{-1}\mathbf{Y} = \begin{bmatrix} \Delta Y_1 \\ Y_2 - \rho Y_2 \\ Y_3 - \rho Y_2 \\ \vdots \\ Y_n - \rho Y_{n-1} \end{bmatrix} \quad \text{and} \quad \mathbf{w}_j = \begin{bmatrix} \Delta x_{j1} \\ x_{j2} - \rho x_{j1} \\ \vdots \\ \vdots \\ x_{jn} - \rho x_{jn-1} \end{bmatrix} \qquad (4.3.3)$$

The first term is often discarded for simplicity. ρ can be estimated by first using least squares to estimate $\boldsymbol{\varepsilon}$, then estimating ρ by the correlation of consecutive pairs in $\mathbf{e} = \mathbf{Y} - \hat{\mathbf{Y}}$.

The Durbin–Watson statistic is often used to estimate this autocorrelation coefficient ρ or to test the null hypotheses that it is zero. Assume equal variances among the ε_i, let $\mathbf{e}$ be the residual vector in the usual least squares fit, and define

$$d = \sum_{i=1}^{n-1} (e_{i+1} - e_i)^2/\|\mathbf{e}\|^2$$

Then $E(d)$ is approximately $2(1 - \rho)$, so that $\hat{\rho} = 1 - d/2$ is approximately unbiased for ρ. The distributions of d and $\hat{\rho}$ depend on $\mathbf{X}$, so that an exact test of $H_0\colon \rho = 0$ is not available. Durbin and Watson (1950, 1951) gave an approximate test. See Theil, (1972) for details. Once $\hat{\rho}$ is obtained $\hat{\boldsymbol{\beta}}$ may be found by using $\mathbf{z}$ and $\mathbf{w}_j$ as defined in (4.3.3).

Variance Stabilizing Transformations: In some situations it may be obvious that it is unrealistic to suppose that the variances of the Y_i are equal, and it may be better to choose a transformation $Z = g(Y)$, which will cause the variance to remain relatively constant, even as the mean changes. For example, if W_i has a binomial distribution with parameters m and p_i, and $Y_i = W_i/m$ is the sample proportion, then $E(Y_i) = p_i$, and $\text{Var}(Y_i) = p_i(1 - p_i)/m$. This model will be discussed more thoroughly in Chapter 8, but let us consider how we might transform the Y_i, so that their variances are approximately equal.

Suppose that Y has mean μ, and variance $\sigma^2(\mu) = h(\mu)$. In the example $h(\mu) = \mu(1 - \mu)/m$. If g is chosen to be *smooth*, and the variation of Y around μ not too large relative to μ, then the distribution of $g(Y)$ and that of the linear function $g_L(Y) = g(\mu) + g'(\mu)(Y - \mu)$ should be approximately the same. But $g_L(Y)$ has mean $g(\mu)$ and variance $[g'(\mu)]^2 h(\mu)$. We should therefore choose g so that this function of μ is a constant, say c. Thus, $g'(\mu) = [c/h(\mu)]^{1/2}$. This implies that $g(\mu) = c^{1/2} \int h(\mu)^{-1/2}\, d\mu + C$, where C, and c, may be chosen arbitrarily. Usually C is chosen to be 0.

For the example, $h(\mu) = \mu(1 - \mu)/m$, so that we can take

$$g(\mu) = (cm)^{1/2} \int [\mu(1 - \mu)]^{-1/2}\, d\mu = 2(cm)^{1/2} \arcsin \mu^{1/2},$$

transforming the $Y_i = W_i/m$ by $g(Y_i)$. If, for convenience, we take $2(cm)^{1/2} = 1$, the resulting approximate variance is $c = 1/4m$. Especially when m is large, $g(Y_i)$ will have approximately the same variance across a wide range of $\mu = p_i$. For example, for $m = 20$, for $p_i = 0.5$, 0.2 and 0.1, simulation shows that the variances of $\arcsin Y_i^{1/2}$ are approximately 0.133, 0.154, and 0.193, as compared to $1/4m = 0.125$ given by the asymptotic theory.

Another commonly used transformation is used for count data, for which the Poisson distribution may be an appropriate model. In this case $h(\mu) = \mu$, so that an appropriate transformation is $g(Y)$, where $g(\mu) = c^{1/2} \int \mu^{-1/2}\, d\mu = \sqrt{\mu}$, for $c = 1/4$. For a table of such transformations see Kempthorne (1952, p. 156).

Though such variance stabilizing transformations often have the additional benefit of creating more nearly normal distributions of observations and estimators, they may destroy the linear relationship between the expected value of the dependent variable and the explanatory variables. They therefore should be used with care and with understanding of the subject matter of the application.

Problem 4.3.1: Let $Y_i \sim N(\mu, w_i K)$ be independent for $i = 1, \ldots, n$, for known w_i, unknown K. Find the weighted least squares estimator of μ both by using (4.3.1) and directly by writing $Z_i = Y_i/\sqrt{w_i}$. Give a formula for an unbiased estimator of K.

Problem 4.3.2: Let $\mathbf{Y} = (Y_1, Y_2, Y_3)' = \beta_1 \mathbf{x}_1 + \beta_2 \mathbf{x}_2 + \boldsymbol{\varepsilon}$ for $\mathbf{x}_1 = (1, 0, 1)'$, $\mathbf{x}_2 = (1, 1, 0)'$, $\boldsymbol{\varepsilon} \sim N(\mathbf{0}, \boldsymbol{\Sigma})$, where the components of $\boldsymbol{\varepsilon}$ have equal variance σ^2, $\rho(\varepsilon_1, \varepsilon_2) = \rho(\varepsilon_2, \varepsilon_3) = 0$, $\rho(\varepsilon_1, \varepsilon_3) = 1/2$. Find $\mathbf{a}_1, \mathbf{a}_2$ such that $((\mathbf{a}_1, \mathbf{Y}), (\mathbf{a}_2, \mathbf{Y})) = (\hat{\beta}_1, \hat{\beta}_2)$ is the generalized least squares estimator of (β_1, β_2). Give the covariance matrix for $(\hat{\beta}_1, \hat{\beta}_2)$.

Problem 4.3.3: Let $X_1, \ldots, X_n$ be a random sample (r.s.) from $N(\mu_1, \sigma^2)$ and let $Y_1, \ldots, Y_m$ be an r.s. from $N(\mu_2, \theta\sigma^2)$, with the vectors of X's and Y's independent, for unknown μ_1, μ_2, σ^2, known θ. Give the generalized least squares estimator of $\mu_1 - \mu_2$, and the corresponding confidence interval on $\mu_1 - \mu_2$ (see Problem 3.8.6).

Problem 4.3.4: Suppose that $Y_i = \beta_0 + \beta_1 x_i + \varepsilon_i$ with the ε_i's independent with variances $k_i\sigma^2$. Give an explicit nonmatrix formula for the generalized least squares estimator $\hat{\beta}_1$ of β_1 and give its variance (k_i's known). Compare the variances of $\hat{\beta}_1$ and the least squares estimator (based on the Y_i's) $\hat{\beta}_{LS}$ of β_1 for $n = 5$, $x_i = i$, $k_i = i$.

Problem 4.3.5: Find the variance stabilizing transformation g for the case that $\text{Var}(Y) = h(\mu) = \lambda\mu^2$, for a constant λ. This is the case for the gamma distribution, including the chi-square.

4.4 EFFECTS OF ADDITIONAL OR FEWER OBSERVATIONS

In this section we study the effects on $\hat{\mathbf{Y}}$ and $\hat{\boldsymbol{\beta}}$ of the addition or deletion of an observation vector $(\tilde{\mathbf{x}}_i, \mathbf{Y}_i)$ or collection of such vectors. Much work has been done over the last 15 or 20 years on this topic, and we will only give here an introduction. Readers are referred to books by Belsey, Kuh and Welsch (1980), by Cook and Weisberg (1980) and to the review paper and book by Chatterjee and Hadi (1986, 1988).

To facilitate the study let us decompose $\mathbf{X}$ and $\mathbf{Y}$ as follows:

$$\mathbf{X} = \begin{pmatrix} \mathbf{X}_1 \\ \mathbf{X}_2 \end{pmatrix}, \qquad \mathbf{Y} = \begin{pmatrix} \mathbf{Y}_1 \\ \mathbf{Y}_2 \end{pmatrix},$$

where $\mathbf{X}_i$ and $\mathbf{Y}_i$ have n_i rows for $i = 1, 2$ and $n_1 + n_2 = n$. Let $\mathbf{M}_i = \mathbf{X}_i'\mathbf{X}_i$, $\mathbf{M} = \mathbf{X}'\mathbf{X} = \mathbf{M}_1 + \mathbf{M}_2$. We suppose that $\mathbf{X}_1$ has full column rank, so that $\mathbf{M}_1$, and therefore $\mathbf{M}$, are nonsingular. Let $\hat{\boldsymbol{\beta}}$ be the least squares estimator of $\boldsymbol{\beta}$ for $(\mathbf{X}, \mathbf{Y})$ and let $\hat{\boldsymbol{\beta}}^*$ be the estimator for $(\mathbf{X}_1, \mathbf{Y}_1)$. Let $\hat{\mathbf{Y}} = \mathbf{X}\hat{\boldsymbol{\beta}} = \begin{pmatrix} \hat{\mathbf{Y}}_1 \\ \hat{\mathbf{Y}}_2 \end{pmatrix}$ and $\hat{\mathbf{Y}}^* = \mathbf{X}\hat{\boldsymbol{\beta}}^* = \begin{pmatrix} \hat{\mathbf{Y}}_1^* \\ \hat{\mathbf{Y}}_2^* \end{pmatrix}$, with $\hat{\mathbf{Y}}_i = \mathbf{X}_i\hat{\boldsymbol{\beta}}$ and $\hat{\mathbf{Y}}_i^* = \mathbf{X}_i\hat{\boldsymbol{\beta}}$. Define $= \mathbf{Y} - \hat{\mathbf{Y}} = \begin{pmatrix} \mathbf{e}_1 \\ \mathbf{e}_2 \end{pmatrix}$ and similarly define $\mathbf{e}^* = \mathbf{Y} - \hat{\mathbf{Y}}^* = \begin{pmatrix} \mathbf{e}_1^* \\ \mathbf{e}_2^* \end{pmatrix}$. Our tasks will be to develop formulas relating $\hat{\boldsymbol{\beta}}$ to $\hat{\boldsymbol{\beta}}^*$ and $\hat{\mathbf{Y}}$ to $\hat{\mathbf{Y}}^*$. We will be interested in *adding* observations (finding $\hat{\boldsymbol{\beta}}$ from $\hat{\boldsymbol{\beta}}^*$ and $\hat{\mathbf{Y}}$ from $\hat{\mathbf{Y}}^*$) and *deleting* observations (finding $\hat{\boldsymbol{\beta}}^*$ from $\hat{\boldsymbol{\beta}}$ and $\hat{\mathbf{Y}}^*$ from $\hat{\mathbf{Y}}$).

Adding Observations: The normal equation for $\hat{\boldsymbol{\beta}}$ is

$$\mathbf{M}\hat{\boldsymbol{\beta}} = (\mathbf{M}_1 + \mathbf{M}_2)\hat{\boldsymbol{\beta}} = \mathbf{X}'\mathbf{Y} = \mathbf{X}_1'\mathbf{Y}_1 + \mathbf{X}_2'\mathbf{Y}_2 = \mathbf{M}_1\hat{\boldsymbol{\beta}}^* + \mathbf{X}_2'\mathbf{Y}_2. \quad (4.4.1)$$

The last equality follows because $\hat{\boldsymbol{\beta}}^*$ satisfies the normal equation relative to $(\mathbf{X}_1, \mathbf{Y}_1)$. Thus

$$\mathbf{M}_1(\hat{\boldsymbol{\beta}} - \hat{\boldsymbol{\beta}}^*) = \mathbf{X}_2'\mathbf{Y}_2 - \mathbf{M}_2\hat{\boldsymbol{\beta}} = \mathbf{X}_2'(\mathbf{Y}_2 - \hat{\mathbf{Y}}_2) = \mathbf{X}_2'\mathbf{e}_2,$$

and

$$\hat{\boldsymbol{\beta}} - \hat{\boldsymbol{\beta}}^* = \mathbf{M}_1^{-1}\mathbf{X}_2'\mathbf{e}_2. \quad (4.4.2)$$

We want a formula depending on $\mathbf{e}_2^*$ rather than $\mathbf{e}_2$. To find one, multiply though by $\mathbf{X}_2$ to obtain

$$\hat{\mathbf{Y}}_2 - \hat{\mathbf{Y}}_2^* = \mathbf{Q}_2\mathbf{e}_2, \qquad \text{where} \quad \mathbf{Q}_2 = \mathbf{X}_2\mathbf{M}_1^{-1}\mathbf{X}_2'.$$

The left-hand side is $\mathbf{e}_2^* - \mathbf{e}_2$, so that

$$\mathbf{e}_2^* = (\mathbf{I}_{n_2} + \mathbf{Q}_2)\mathbf{e}_2$$

and (4.4.2) becomes

$$\hat{\boldsymbol{\beta}} - \hat{\boldsymbol{\beta}}^* = \mathbf{M}_1^{-1}\mathbf{X}_2'(\mathbf{I}_{n_2} + \mathbf{Q}_2)^{-1}\mathbf{e}_2^* \quad (4.4.3)$$

In the literature this *change vector* is often called DFBETA. Let us call it $\boldsymbol{\Delta}$ here. The change $\hat{\mathbf{Y}} - \hat{\mathbf{Y}}^*$ is therefore

$$\hat{\mathbf{Y}} - \hat{\mathbf{Y}}^* = \begin{pmatrix} \mathbf{X}_1 \\ \mathbf{X}_2 \end{pmatrix} \boldsymbol{\Delta} = \begin{pmatrix} \mathbf{X}_1 \\ \mathbf{X}_2 \end{pmatrix} \mathbf{M}_1^{-1} \mathbf{X}_2' (\mathbf{I}_{n_2} + \mathbf{Q}_2)^{-1} \mathbf{e}_2^* \tag{4.4.4}$$

To express SSE in terms of SSE* and $\mathbf{e}_2^*$, we write

$$\mathbf{Y} - \hat{\mathbf{Y}}^* = (\mathbf{Y} - \hat{\mathbf{Y}}) + (\hat{\mathbf{Y}} - \hat{\mathbf{Y}}^*) \tag{4.4.5}$$

The two vectors on the right are orthogonal, since $(\mathbf{Y} - \hat{\mathbf{Y}}) \perp$ (column space of $\mathbf{X}$). Then

$$\|\mathbf{Y} - \hat{\mathbf{Y}}^*\|^2 = \|\mathbf{Y}_1 - \mathbf{Y}_1^*\|^2 + \|\mathbf{e}_2^*\|^2 = \text{SSE}^* + \|\mathbf{e}_2^*\|^2, \tag{4.4.6}$$

and, using (4.4.4), we find $\|\hat{\mathbf{Y}} - \hat{\mathbf{Y}}^*\|^2 = \mathbf{e}_2^{*\prime}(\mathbf{I}_{n_2} + \mathbf{Q}_2)^{-1}\mathbf{Q}_2\mathbf{e}_2^*$. Then using (4.4.6) and the Pythagorean Theorem we find

$$\text{SSE} = \text{SSE}^* + \mathbf{e}_2^{*\prime}(\mathbf{I} + \mathbf{Q}_2)^{-1}\mathbf{e}_2^*. \tag{4.4.7}$$

The case $n_2 = 1$ is of special interest. Let $\mathbf{X}_2 = \tilde{\mathbf{x}}_0$, $\mathbf{e}_2^* = e_2^* = Y_2 - \tilde{\mathbf{x}}_0\hat{\boldsymbol{\beta}}^*$, $Q_2 = \tilde{\mathbf{x}}_0\mathbf{M}_1^{-1}\mathbf{x}_0'$. We get $\boldsymbol{\Delta} = \hat{\boldsymbol{\beta}} - \hat{\boldsymbol{\beta}}^* = \mathbf{M}_1^{-1}\mathbf{x}_0'(e_2^*/(1 + Q_2))$. The term $1 + Q_2$ is $k(\tilde{\mathbf{x}}_0)$, the multiplier of σ^2 in the variance of prediction error (see Section 3.11). Notice that the change $\boldsymbol{\Delta}$ is in the direction of $\mathbf{M}_1^{-1}\tilde{\mathbf{x}}_0'$ with the multiple depending on both the prediction error e_2^* and the distance measure Q_2. The increase in error sum of squares is SSE $-$ SSE* $= e_2^{*2}/(\tilde{\mathbf{x}}_0)$, indicating a large increase if the additional observation is far from its predicted value in units of the standard deviation of prediction errors at $\tilde{\mathbf{x}}_0$.

Deleting Observations: We want a formula similar to (4.4.2) which uses $\mathbf{M}$ rather than $\mathbf{M}_1$. It is possible to use formulas for the inverse of partitioned matrices (Section 1.7), but we will avoid this by beginning with (4.4.1). Since $\mathbf{M}\hat{\boldsymbol{\beta}} = (\mathbf{M} - \mathbf{M}_2)\hat{\boldsymbol{\beta}}^* + \mathbf{X}_2'\mathbf{Y}_2$, where

$$\mathbf{M}(\hat{\boldsymbol{\beta}} - \hat{\boldsymbol{\beta}}^*) = \mathbf{X}_2'(\mathbf{Y}_2 - \mathbf{X}_2\hat{\boldsymbol{\beta}}^*) = \mathbf{X}_2'\mathbf{e}_2^*,$$

it follows that

$$\hat{\beta} - \hat{\beta}^* = M^{-1}\mathbf{X}_2'\mathbf{e}_2^*. \tag{4.4.8}$$

This time we should replace $\mathbf{e}_2^*$ by a term depending on $\mathbf{e}_2$. Multiplying by $\mathbf{X}_2$ we get

$$\hat{\mathbf{Y}}_2 - \hat{\mathbf{Y}}_2^* = \mathbf{e}_2^* - \mathbf{e}_2 = (\mathbf{X}_2\mathbf{M}^{-1}\mathbf{X}_2')\mathbf{e}_2^* \tag{4.4.9}$$

The matrix $\mathbf{h} \equiv \mathbf{X}_2\mathbf{M}^{-1}\mathbf{X}_2'$ in parentheses is the lower right $n_2 \times n_2$ submatrix of $\mathbf{X}(\mathbf{X}'\mathbf{X})^{-1}\mathbf{X}' = \mathbf{H}$, projection onto the column space of $\mathbf{X}$, often called the *hat-matrix* because $\mathbf{HY} = \hat{\mathbf{Y}}$. From (4.4.8) we get

$$\mathbf{e}_2^* = [\mathbf{I}_{n_2} - \mathbf{h}]^{-1}\mathbf{e}_2, \tag{4.4.10}$$

so that (4.4.8) becomes

$$\hat{\boldsymbol{\beta}}^* - \hat{\boldsymbol{\beta}} = -\mathbf{M}^{-1}\mathbf{X}_2'[\mathbf{I}_{n_2} - \mathbf{h}]^{-1}\mathbf{e}_2 \tag{4.4.11}$$

Finally then,

$$\hat{\mathbf{Y}}^* - \hat{\mathbf{Y}} = \mathbf{X}(\hat{\boldsymbol{\beta}}^* - \hat{\boldsymbol{\beta}}) = -\mathbf{X}\mathbf{M}^{-1}\mathbf{X}_2'[\mathbf{I}_{n_2} - \mathbf{h}]^{-1}\mathbf{e}_2. \tag{4.4.12}$$

The $n \times n_2$ matrix $\mathbf{X}\mathbf{M}^{-1}\mathbf{X}_2'$ on the right consists of the last n_2 columns of $\mathbf{H}$. Matrix computation then gives

$$\|\hat{\mathbf{Y}}^* - \hat{\mathbf{Y}}\|^2 = \mathbf{e}_2'(\mathbf{I}_{n_2} - \mathbf{h})^{-1}\mathbf{h}(\mathbf{I}_{n_2} - \mathbf{h})^{-1}\mathbf{e}_2 \tag{4.4.13}$$

To express $\text{SSE}^* = \|\mathbf{Y}_1 - \hat{\mathbf{Y}}_1^*\|^2$ in terms of $\text{SSE} = \|\mathbf{Y} - \hat{\mathbf{Y}}\|^2$, we again use (4.4.6):

$$\|\mathbf{Y} - \hat{\mathbf{Y}}^*\|^2 = \text{SSE}^* + \|\mathbf{e}_2^*\|^2$$

From (4.4.12) and (4.4.13) and the Pythagorean Theorem we get

$$\text{SSE}^* = \text{SSE} - \mathbf{e}_2'(\mathbf{I}_{n_2} - \mathbf{h})^{-1}\mathbf{e}_2 \tag{4.4.14}$$

In the special case that $n_2 = 1$, $\mathbf{X}_2 = \tilde{\mathbf{x}}_0$, $h = \tilde{\mathbf{x}}_0\mathbf{M}^{-1}\mathbf{x}_0'$ is the (n, n) term of $\mathbf{H}$. $h = h(\tilde{\mathbf{x}}_0)$ is sometimes called the *leverage* of the observation vector $\tilde{\mathbf{x}}_0$, since by (4.4.8) $\mathbf{e} = (1 - \mathbf{h})\mathbf{e}^*$. Then (4.4.11) becomes

$$\hat{\boldsymbol{\beta}}^* - \hat{\boldsymbol{\beta}} = -\mathbf{M}^{-1}\tilde{\mathbf{x}}_0'\left[\frac{e_2}{1 - h}\right] \tag{4.4.15}$$

and

$$\hat{\mathbf{Y}}^* - \hat{\mathbf{Y}} = -\left[\frac{e_2}{1 - h}\right]\mathbf{X}\mathbf{M}^{-1}\tilde{\mathbf{x}}_0', \tag{4.4.16}$$

where $e_2 = \mathbf{e}_2 = Y_2 - \tilde{\mathbf{x}}_0\hat{\boldsymbol{\beta}}$. From (4.4.14) we get

$$\text{SSE}^* = \text{SSE} - e_2^2/(1 - h) = \text{SSE} - e_2^{*2}(1 - h)$$

It is sometimes useful to study the change vector $\mathbf{\Delta} = \mathbf{\Delta}_i$ for $\tilde{\mathbf{x}}_0 = \tilde{\mathbf{x}}_i$, the ith row of $\mathbf{X}$. In this case, let $\hat{\boldsymbol{\beta}}_{-i} = \hat{\boldsymbol{\beta}}^*$, $\hat{\mathbf{Y}}_{-i} = \hat{\mathbf{Y}}^*$, $\text{SSE}_{-i} = \text{SSE}^*$, $e_{-i} = Y_i - \tilde{\mathbf{x}}_i\hat{\boldsymbol{\beta}}^*$, $h_i = \tilde{\mathbf{x}}_i\mathbf{M}^{-1}\tilde{\mathbf{x}}_i'$. Then

$$\mathbf{\Delta}_i = \hat{\boldsymbol{\beta}}_{-i} - \hat{\boldsymbol{\beta}} = -\mathbf{M}^{-1}\tilde{\mathbf{x}}_i'[e_{-i}/(1 - h_i)], \tag{4.4.17}$$

where h_i is the (ii) diagonal term of the hat-matrix $\mathbf{H}$.

$$\hat{\mathbf{Y}}_{-i} - \hat{\mathbf{Y}} = -\mathbf{X}\mathbf{M}^{-1}\tilde{\mathbf{x}}_i'[e_{-i}/(1 - h_i)], \tag{4.4.18}$$

$$\text{SSE}_{-i} = \text{SSE} - e_i^2/(1 - h_i) \tag{4.4.19}$$

Study of the values given by (4.4.17) to (4.4.19) for each i may suggest omission of one or more observations, or perhaps another model. The residuals e_{-i} are called the *PRESS residuals*, and $\sum_i e_{-i}^2$ is the *PRESS statistic*. One criterion for the choice of a model is to choose the one with the smallest PRESS statistic.

Example 4.4.1: Consider the simple linear regression model $\mathbf{Y} = \beta_0\mathbf{x}_0 + \beta_1\mathbf{x}_1 + \boldsymbol{\varepsilon}$ for $\mathbf{x}_0$ the vector of all ones. It is convenient to suppose $\mathbf{x}_1$ is in mean-deviation form, so that $\mathbf{x}_0 \perp \mathbf{x}_1$. Then

$$\mathbf{M} = \begin{bmatrix} M & 0 \\ 0 & S_{xx} \end{bmatrix}, \qquad S_{xx} = \|\mathbf{x}_1\|^2, \qquad \tilde{\mathbf{x}}_i = (1, x_i - \bar{x}),$$

and from (4.4.17),

$$\hat{\boldsymbol{\beta}}_{-i} - \hat{\boldsymbol{\beta}} = -\begin{bmatrix} 1/n \\ (x_i - \bar{x})/S_{xx} \end{bmatrix}[e_{-i}/(1 - h_i)],$$

for

$$h_i = \frac{1}{n} + (x_i - \bar{x})^2/S_{xx}$$

The jth element of $\hat{\mathbf{Y}}_{-i} - \hat{\mathbf{Y}}$ is

$$-\left[\frac{1}{n} + \frac{(x_i - \bar{x})(x_j - \bar{x})}{S_{xx}}\right][e_{-i}/(1 - h_i)]$$

Example 4.4.2: Consider the eight (x, Y) pairs given in Table 4.4.1 with values of the corresponding components of $\hat{\mathbf{Y}}$ and $\mathbf{e} = \mathbf{Y} - \hat{\mathbf{Y}}$ corresponding to the simple linear regression model. The least squares estimate of (β_0, β_1) is (4.9167, 0.4167). Elimination of each of the observations (x_i, Y_i) in turn produces the eight estimates $\hat{\boldsymbol{\beta}}_{-i}$ of (β_0, β_1) given in Table 4.4.2. See, for example in Table 4.4.3, that elimination of $(x_8, Y_8) = (8, 3)$ causes the estimate of slope

Table 4.4.1

x	$\mathbf{Y}$	$\hat{\mathbf{Y}}$	$\mathbf{e}$
2	5	5.750	−0.750
3	5	6.167	−1.667
3	6	6.167	−0.167
5	8	7.000	1.000
5	8	7.000	1.000
6	10	7.417	2.583
8	11	8.250	2.750
8	3	8.250	−5.250

Table 4.4.2

$\hat{\beta}$	$\hat{\beta}_{-1}$	$\hat{\beta}_{-2}$	$\hat{\beta}_{-3}$	$\hat{\beta}_{-4}$	$\hat{\beta}_{-5}$	$\hat{\beta}_{-6}$	$\hat{\beta}_{-7}$	$\hat{\beta}_{-8}$
4.9167	5.5667	5.5318	5.0045	4.7738	4.7738	4.5950	6.2000	2.4667
0.4167	0.3167	0.3318	0.4046	0.4167	0.4167	0.3198	0.0500	1.1167

Table 4.4.3

x	Y	$\hat{\mathbf{Y}}$	$\hat{\mathbf{Y}}_{-1}$	$\hat{\mathbf{Y}}_{-2}$	$\hat{\mathbf{Y}}_{-3}$	$\hat{\mathbf{Y}}_{-4}$	$\hat{\mathbf{Y}}_{-5}$	$\hat{\mathbf{Y}}_{-6}$	$\hat{\mathbf{Y}}_{-7}$	$\hat{\mathbf{Y}}_{-8}$
2	5	5.750	6.200	6.195	5.814	5.607	5.607	5.623	6.300	4.700
3	5	6.167	6.517	6.527	8.218	6.024	6.024	5.955	6.350	5.817
3	6	6.167	6.517	6.527	6.218	6.024	6.024	5.955	6.350	5.817
5	8	7.000	7.150	7.191	7.027	6.857	6.857	6.619	6.450	8.050
5	8	7.000	7.150	7.191	7.027	6.857	6.857	6.619	6.450	8.050
6	10	7.417	7.467	7.523	7.432	7.274	7.274	6.950	6.500	9.167
8	11	8.250	8.100	8.186	8.241	8.107	8.107	7.615	6.600	11.400
8	3	8.250	8.100	8.186	8.241	8.107	8.107	7.615	6.600	11.400

to be large. Elimination of (x_7, Y_7) causes the estimated regression line to be almost flat, while elimination of observation number 4 or 5 has almost no effect on the slope. This is so because $|x_i - \bar{x}|/S_{xx}$ in (4.4.18) for $x_i = 5$ is small. The residual vectors $\mathbf{e}_{-i} = \mathbf{Y} - \hat{\mathbf{Y}}_{-i}$ are given in Table 4.4.4. Notice, for example, that the eighth component of $\mathbf{e}_{-8}$ is very negative. Error sums of squares and the corresponding estimate S^2_{-i} of variance vary considerably with i (Table 4.4.5).

Problem 4.4.1: Suppose that a simple linear regression analysis has been performed on the following pairs:

Table 4.4.4

x	$\mathbf{e}$	$\mathbf{e}_{-1}$	$\mathbf{e}_{-2}$	$\mathbf{e}_{-3}$	$\mathbf{e}_{-4}$	$\mathbf{e}_{-5}$	$\mathbf{e}_{-6}$	$\mathbf{e}_{-7}$	$\mathbf{e}_{-8}$
2	−0.750	−1.200	−1.200	−0.814	−0.607	−0.607	−0.623	−1.300	0.300
3	−1.170	−1.520	−1.530	−1.220	−1.020	−1.020	−0.955	−1.350	−0.817
3	−0.167	−0.517	−0.527	−0.218	−0.024	−0.024	0.045	−0.350	0.183
5	1.000	0.850	0.809	0.973	1.140	1.140	1.380	1.550	−0.050
5	1.000	0.850	0.809	0.973	1.140	1.140	1.380	1.550	0.050
6	2.580	2.530	2.480	2.570	2.730	2.730	3.050	3.500	0.833
8	2.750	2.900	2.810	2.760	2.890	2.890	3.390	4.400	−0.400
8	−5.250	−5.100	−5.190	−5.240	−5.110	−5.110	−4.610	−3.600	−8.400

Table 4.4.5

S^2	S^2_{-i}								ESS	ESS_{-i}							
7.62	8.97	8.79	9.14	8.92	8.92	7.57	6.73	0.33	45.8	44.9	44.0	45.7	44.6	44.6	37.9	33.7	1.65

x	-3	-2	-1	0	1	2	3
Y	14	7	9	4	5	5	-2

(The x's add to zero, because the original mean was subtracted). Thus

$$\bar{Y} = 6, \quad \sum x_i^2 = 28, \quad \sum x_i Y_i = -56, \quad \sum Y_i^2 = 396$$

(a) Find the last squares estimates $\hat{\boldsymbol{\beta}}$ and $\hat{\mathbf{Y}}$ for the simple linear regression model. Also find ESS.

(b) Use (4.4.11) and (4.4.12) to find the least squares estimates $\hat{\boldsymbol{\beta}}^*$ and $\hat{\mathbf{Y}}^*$ for the case that the observation $(3, -2)$ is omitted. Check your work by computing these from scratch. Also find ESS*.

(c) Suppose an additional observation $(4, 4)$ is obtained (so there are eight (x, Y) pairs). Find $\boldsymbol{\Delta} = \hat{\boldsymbol{\beta}} - \hat{\boldsymbol{\beta}}^*$ and use this to obtain $\hat{\boldsymbol{\beta}} = (\hat{\beta}_0, \hat{\beta}_1)'$, the vector $\hat{\mathbf{Y}}$, and the new error sum of squares, ESS. (The stars correspond now to the smaller data set.)

(d) Suppose two new observations $(-4, 17)$ and $(4, -6)$ are added to the original data. Use (4.4.3) and (4.4.4) to find $\hat{\boldsymbol{\beta}}$ and $\hat{\mathbf{Y}}$, also find ESS. Verify your solution $\hat{\boldsymbol{\beta}}$ by starting from scratch.

Problem 4.4.2: Consider the two sample model with n_1 independent observations from $N(\mu_1, \sigma^2)$, n_2 from $N(\mu_2, \sigma^2)$. Suppose one more observation Y_0 is taken from the distribution. Use (4.4.3) to show that the change in the estimator of (μ_1, μ_2) is $((Y_0 - \bar{Y}_1)/(n_1 + 1), 0)$.

Problem 4.4.3: Suppose the observations $(\mathbf{X}_2, \mathbf{Y}_2)$ are deleted to obtain $\hat{\boldsymbol{\beta}}^*$ from $\hat{\boldsymbol{\beta}}$ using (4.4.11). Then a new estimator $\hat{\boldsymbol{\beta}}^{**}$ is obtained from $\hat{\boldsymbol{\beta}}^*$ by adding $(\mathbf{X}_2, \mathbf{Y}_2)$ to the deleted data set using (4.4.2). Show that $\hat{\boldsymbol{\beta}}^{**} = \hat{\boldsymbol{\beta}}$.

Problem 4.4.4: Fill in the details in the paragraph following (4.4.4) to show that (4.4.7) holds.

4.5 FINDING THE "BEST" SET OF REGRESSORS

Whenever a large set of independent variables $x_1, \ldots, x_k$ is available, particularly when n is not considerably larger, we are faced with the problem of choosing the best subset. For example, we may have $n = 25$ observations $(\tilde{\mathbf{x}}_i, \mathbf{Y}_i)$ where $\tilde{\mathbf{x}}_i$ is a vector with $k = 10$ components. While we could use all 10 components to predict $\mathbf{Y}$, and as a result reduce error sum of squares as much as possible, considerations of precision discussed in Section 4.2, and the wish to find a model with reasonable simplicity suggest a smaller subset of variables.

There are $2^k - 1$ subsets of k-variables with at least one member. Given today's computer power it is often possible to fit models for all such subsets. Algorithms for doing so are available. See, for example, Seber and Wild (1989, ch. 13) on choosing the "best" regression. We would need some criterion for choice of a model from among a large number. Knowledge of the subject matter should almost always be a guide. Such knowledge may suggest, for example, that variables #1, #3, and #7 should be included, and that future observations of variable #9 may be so difficult or expensive that we should look for a model not including #9.

A number of procedures are available for choosing a model. For details see Seber (1977). Usually these are step-up or step-down types. A *step-up* procedure begins with a small set of variables, possibly only the constant term, and adds variable one at a time, depending on the contribution of that variable to the fit. The procedure may allow for elimination of certain variables after addition of another. The *step-down* procedure begins with a more complex model, possibly all x-variables, and eliminates variables one at a time, choices being made in such a way that error sum of squares increases least (or regression sum of squares decreases least). Both procedures usually involve F- or t-tests, but it is usually difficult or impossible to evaluate error probabilities. The *sweep* algorithm, to be described here without proof, facilitates computation. See Kennedy and Gentle (1980) for a complete discussion.

The Sweep Algorithm: (Beaton, 1964, in slightly different form) Let the k columns of $\mathbf{X}$ correspond to the set of explanatory (regressor) variables under consideration. Usually the first column of $\mathbf{X}$ is the column of ones. Let $\mathbf{W} = (\mathbf{X}, \mathbf{Y})$, and let $\mathbf{Q} = \mathbf{W}'\mathbf{W} = \begin{bmatrix} \mathbf{X}'\mathbf{X} & \mathbf{X}'\mathbf{Y} \\ \mathbf{Y}'\mathbf{X} & \mathbf{Y}'\mathbf{Y} \end{bmatrix}$, the inner-product matrix. Then the sweep operator $S(k)$ (sweep on the kth row) is a matrix-valued function of a square matrix $\mathbf{A} = (a_{ij})$ with value $S(k)\,\mathbf{A} = \mathbf{B} = (b_{ij})$ of the same size defined as follows. We assume for simplicity that $\mathbf{A}$ is nonsingular: (1) $b_{kk} = 1/a_{kk}$, (2) $b_{kj} = a_{kj}/a_{kk}$ and $b_{jk} = -a_{jk}/a_{kk}$ for $j \neq k$, (3) $b_{ij} = a_{ij} - a_{jk}a_{kj}/a_{kk}$ for $i \neq k$, $j \neq k$. Let $S(0)$ be the identity sweep, that is, $S(0)\,\mathbf{A} = \mathbf{A}$ for all square $\mathbf{A}$.

For any sequence of integers $(i_1, \ldots, i_r)$, let $S(i_1, \ldots, i_r) = S(i_1)S(i_2)\cdots S(i_r)$. That is, $S(i_1, \ldots, i_r)\,\mathbf{A} = S(i_1)S(i_2)\cdots S(i_r)\,\mathbf{A}$. Then, these sweep operators have the following properties.

(1) The order in which $(i_1, \ldots, i_r)$ is written is irrelevant, i.e. if $(j_1, \ldots, j_r)$ is a permutation of $(i_1, \ldots, i_r)$ then $S(i_1, \ldots, i_r) = S(j_1, \ldots, j_r)$.

(2) $S(i, i) = S(0)$. That is, sweeping $\mathbf{A}$ twice produces $\mathbf{A}$.

(3) Suppose that $\mathbf{A}$ is symmetric. If $\mathbf{A} = \begin{bmatrix} \mathbf{A}_{11} & \mathbf{A}_{12} \\ \mathbf{A}'_{12} & \mathbf{A}_{22} \end{bmatrix}$, and $\mathbf{A}_{11}$ is $n_1 \times n_1$, then $S(1, \ldots, n_1)\,\mathbf{A} = \mathbf{B} = , \begin{bmatrix} \mathbf{B}_{11} & \mathbf{B}_{12} \\ -\mathbf{B}'_{12} & \mathbf{B}_{22} \end{bmatrix}$, where $\mathbf{B}_{11} = \mathbf{A}_{11}^{-1}$, $\mathbf{B}_{12} = \mathbf{A}_{11}^{-1}\mathbf{A}_{12}$, $\mathbf{B}_{22} = \mathbf{A}_{22} - \mathbf{A}'_{12}\mathbf{A}_{11}^{-1}\mathbf{A}_{12}$.

Suppose now that we apply $S(1, \ldots, r)$ to the inner-product matrix $\mathbf{Q}$. Let $\mathbf{X} = (\mathbf{X}_1, \mathbf{X}_2)$, where $\mathbf{X}_1$ consists of the first r columns of $\mathbf{X}$. Let $\mathbf{M}_1 = \mathbf{X}_1'\mathbf{X}_1$, $\mathbf{U}_1 = \mathbf{X}_1'\mathbf{Z}$, where $\mathbf{Z} = (\mathbf{X}_2, \mathbf{Y})$. Then $S(1, \ldots, r)$

$$\mathbf{Q} = \begin{bmatrix} \mathbf{M}_1^{-1} & \mathbf{M}_1^{-1}\mathbf{Z} \\ -\mathbf{Z}'\mathbf{M}_1^{-1} & \mathbf{Z}'\mathbf{Z} - \mathbf{Z}'\mathbf{M}_1^{-1}\mathbf{Z} \end{bmatrix}.$$

The matrix in the upper right is the matrix of regression coefficients when the last $(k - r)$ variables are regressed on the first r. In particular, $\hat{\boldsymbol{\beta}}$, the coefficient vector in the regression of $\mathbf{Y}$ on the first r regressor variables, is the vector consisting of the first r elements in the last column. The matrix on the lower right is the inner-product matrix among the $(k - r)$ residual vectors, after removal of the effects of the first r. ESS for the regression of $\mathbf{Y}$ on the first r regressors is in the lower right corner. The $r \times r$ matrix on the upper left is useful because $D[\hat{\boldsymbol{\beta}}] = \mathbf{M}_1^{-1}\sigma^2$ in the case that $\mathbf{Y} = \mathbf{X}_1\boldsymbol{\beta} + \boldsymbol{\varepsilon}$, $D[\boldsymbol{\varepsilon}] = \sigma^2\mathbf{I}_n$. By sweeping consecutively on rows 1, 2, 1, 3, 2, 1, for example, we get important statistics in the least squares fits of all of the possible linear models involving the first three variables. An algorithm due to Garside (1965) can be used to consecutively fit all possible regressions of $\mathbf{Y}$ against subsets of the regressor variables with a minimum of sweeps.

Example 4.5.1: Let

$$\mathbf{W} = (\mathbf{X}, \mathbf{Y}) = \begin{bmatrix} 1 & 1 & 1 & 18.21 \\ 1 & 1 & 2 & 16.29 \\ 1 & 1 & 3 & 14.37 \\ 1 & 0 & 4 & 12.10 \\ 1 & 0 & 5 & 10.05 \\ 1 & 0 & 6 & 8.13 \end{bmatrix},$$

and

$$\mathbf{Q} = \begin{bmatrix} 6 & 3 & 21 & 79.15 \\ 3 & 3 & 6 & 48.87 \\ 21 & 6 & 91 & 241.32 \\ 79.15 & 48.87 & 241.32 & 1{,}117.04 \end{bmatrix}.$$

$\mathbf{Q}$ is the inner-product matrix corresponding to $\mathbf{W}$. $\mathbf{Y} = \mathbf{X}\boldsymbol{\beta} + \boldsymbol{\varepsilon}$, where $\boldsymbol{\beta} = (20, 0, -2)'$, and $\boldsymbol{\varepsilon}$ was an observation from $N_6(\mathbf{0}, \sigma^2\mathbf{I}_6)$, $\sigma = 0.2$. Refer to the second and third columns of $\mathbf{X}$ by x_2 and x_3. The following sweeps were determined:

$$S(1)\mathbf{Q} = \begin{bmatrix} 0.167 & 0.500 & 3.500 & 13.192 \\ -0.500 & 1.500 & -4.500 & 9.299 \\ -3.500 & -4.500 & 17.500 & -35.707 \\ -13.192 & 9.299 & -35.707 & 72.902 \end{bmatrix}.$$

Sample means are given in the first row. The 3×3 submatrix on the lower right is the sum of squares and cross-products matrix for deviations of the second, third, and fourth columns of $\mathbf{W}$ from their means. For example, the total sum of squares for Y is 72.902.

$$S(2,1)\mathbf{Q} = S(2)S(1)\mathbf{Q} = \begin{bmatrix} 0.333 & -0.333 & 5.000 & 10.092 \\ -0.333 & 0.667 & -3.000 & 6.199 \\ -5.000 & 3.000 & 4.000 & -7.810 \\ -10.092 & -6.199 & -7.810 & 15.257 \end{bmatrix}.$$

This indicates, for example, that simple linear regression of Y on x_2 produces the estimate $\hat{Y} = 10.092 - 6.199x_2$, with ESS = 15.257,

$$S^2(\hat{\boldsymbol{\beta}}) = \begin{bmatrix} 0.333 & -0.333 \\ -0.333 & 0.667 \end{bmatrix} S^2.$$

$$S(1,3)\mathbf{Q} = \begin{bmatrix} 0.867 & 1.400 & -0.200 & 20.333 \\ -1.400 & 0.343 & 0.257 & 0.117 \\ -0.200 & -0.257 & 0.057 & -2.040 \\ -20.333 & 0.117 & 2.040 & 0.046 \end{bmatrix},$$

$$S(1,3,2)\mathbf{Q} = \begin{bmatrix} 6.583 & -4.083 & -1.250 & 19.855 \\ -4.083 & 2.917 & 0.750 & 0.341 \\ 1.250 & -0.750 & 0.250 & -1.953 \\ -19.855 & -0.341 & 1.953 & 0.007 \end{bmatrix}.$$

A 95% confidence interval on β_2, assuming the model corresponding to the design matrix $\mathbf{X}$, is $[0.341 \pm (3.182)\sqrt{(0.007/3)(2.917)}]$. $R_2^2 = 1 - 15.257/72.902$, $R_3^2 = 1 - 0.046/72.902$.

Model fitting is an art, and different good statisticians may arrive at somewhat different models. There is no substitute for a strong interaction between statistical and subject matter knowledge.

In one situation it is possible to determine the probability of the correct choice of a model. Suppose that $V_0 = \Omega$, of dimension n, and that $V_0 \supset V_1 \supset \cdots \supset V_r$ is a decreasing sequence of subspaces of dimensions $k_0 = n > k_1 > k_2 > \cdots > k_r$. Suppose that $\mathbf{Y} = \boldsymbol{\theta} + \boldsymbol{\varepsilon}$ for $\boldsymbol{\theta} \in V_{j_0}$ for some j_0, $1 \le j_0 \le r$, j_0 unknown, and $\boldsymbol{\varepsilon} \sim N(\mathbf{0}, \sigma^2\mathbf{I}_n)$. Suppose also that $\alpha_1, \ldots, \alpha_r$ are chosen error probabilities. Define $\boldsymbol{\theta}_j = p(\boldsymbol{\theta} | V_j)$.

Let $H(i)$ be the null hypothesis: $\boldsymbol{\theta} \in V_i$. Then we choose a subspace by first testing $H(2)$ assuming $H(1)$ holds, using an F-test at level α_1. If $H(2)$ is rejected then we decide that $\boldsymbol{\theta} \in V_1, \boldsymbol{\theta} \notin V_2$. If $H(2)$ is accepted then we test $H(3)$, assuming $H(2)$ holds, using an F-test at level α_2. If $H(3)$ is rejected then we decide that $\boldsymbol{\theta} \in V_2$ but $\boldsymbol{\theta} \notin V_3$. We continue in this way until an F-test rejects $H(j)$ for some j. In this case we decide $\boldsymbol{\theta} \in V_{j-1}$, but $\boldsymbol{\theta} \notin V_j$.

More formally, let $\hat{\mathbf{Y}}_i = p(\mathbf{Y} | V_i)$ and

$$E_i = \|\hat{\mathbf{Y}}_i - \hat{\mathbf{Y}}_{i+1}\|^2 = \|\hat{\mathbf{Y}}_i\|^2 - \|\hat{\mathbf{Y}}_{i+1}\|^2 = \|\mathbf{Y} - \hat{\mathbf{Y}}_{i+1}\|^2 - \|\mathbf{Y} - \hat{\mathbf{Y}}_i\|^2.$$

Define $d_i = n_i - n_{i+1}$ and $F_i = \dfrac{E_i/d_i}{(E_0 + \cdots + E_{i-1})/(n_i - k_i)}$. Let I_i be the event $[F_i \le F_{1-\alpha_i}(d_i, n - k_i))]$. Then we decide that ($\boldsymbol{\theta} \in V_j$, but $\boldsymbol{\theta} \notin V_{j+1}$) if $I_1, \ldots, I_j$ occur but not I_{j+1}.

We therefore want to evaluate the probability of the event $D_j = (I_1 \cap \cdots \cap I_j) \cap I_{j+1}^c$ for $j \le j_0$. We can do so because the events $I_1, \ldots, I_{j+1}$ are independent for $j \le j_0$. To see this, note that

(1) $E_1, \ldots, E_r$ are independent
(2) For $j < j_0$, $E_j \sim \sigma^2\chi^2_{d_j}$
(3) $E_{j_0+1} \sim \sigma^2\chi^2_{d_{j_0}}(\Delta)$ for $\Delta = \|\boldsymbol{\theta} - \boldsymbol{\theta}_{j_0+1}\|^2/\sigma^2$
(4) I_j depends only on $R_j = E_j / \sum_0^{j-1} E_i$. From Problem 2.5.5 $R_1, \ldots, R_{j_0+1}$ are independent.

Thus $P(I_j) = 1 - \alpha_j$ for $j < j_0$ and $P(I_{j+1}^c) = \alpha_{j+1}$ for $j < j_0$, $\gamma(\Delta)$ for $j = j_0$, where $\gamma(\Delta)$ is the power of the F-test for d_{j_0} and $n - k_{j_0}$ d.f. and noncentrality parameter Δ. It follows that

$$\begin{aligned} P(D_j) &= (1 - \alpha_1)(1 - \alpha_2)\cdots(1 - \alpha_{j-1}) && \text{for } j < j_0 \\ &= (1 - \alpha_1)(1 - \alpha_2)\cdots(1 - \alpha_{j_0+1})\gamma(\Delta) && \text{for } j = j_0 \end{aligned}$$

Example 4.5.2: Suppose that $g(x) = E(Y|x) = \sum_0^5 \beta_i x^i$ and we wish to decide whether a polynomial of smaller degree than five is reasonable. We fit each of the models $\mathcal{M}_j$: $g(x) = \sum_0^{6-j} \beta_i x^i$ in turn, and determine the error sum of squares and their differences E_j, for each. We first compute $F_1 = \dfrac{E_1}{E_0/(n-6)}$. If $F_1 > F_{1-\alpha_1}(1, n-6)$ then we decide that the fifth degree polynomial is required. Since the spaces V_0 and V_1 have dimensions 6 and 5, E_1 has one d.f., so that we could use the t-test for H_0: $\beta_5 = 0$. Otherwise we fit the fourth degree model, compute $F_2 = \dfrac{E_1}{(E_0 + E_1)/(n-5)}$, and decide a fourth degree polynomial is required if $F_2 > F_{1-\alpha_2}(1, n-5)$. Continuing this way we decide among six possible models, of degrees 5, 4, 3, 2, 1, 0.

Suppose Y is observed for $x = 0, 1, \ldots, 10$, $\beta_0 = 30$, $\beta_1 = -2$, $\beta_2 = 0.15$, $\beta_3 = \beta_4 = \beta_5 = 0$, $\sigma^2 = 2$, and we choose each $\alpha_i = 0.05$. The probability that we make the correct decision is then $(0.95)^3\gamma(\Delta)$ for $\Delta = \|\boldsymbol{\theta} - \boldsymbol{\theta}_5\|^2/\sigma^2$, where $V_5 = \mathscr{L}(\mathbf{x}_0, \mathbf{x}_1)$, the space spanned by the vector of ones and the $\mathbf{x}$-vector.

In this case $\boldsymbol{\theta}$, $\boldsymbol{\theta}_5 = p(\boldsymbol{\theta} | V_5)$, and $\boldsymbol{\theta} - \boldsymbol{\theta}_{V_5}$ are given in Table 4.5.1. So that $\Delta = 19.31/4 = 4.83$. From the Pearson–Hartley charts, we find for $\phi = \sqrt{4.83/2} = 1.55$, $\nu_1 = 1$, $\nu_2 = 10 - 3 = 7$, $\gamma(\Delta) = 0.47$. Thus, the probability of a correct decision is $(0.95)^3(0.47) = 0.40$. The probability that the simple linear regression model is chosen is $(0.95)^3(0.53) = 0.45$. Table 4.5.1 presents $\theta_i = 30 - 2x_i + 0.15x_i^2$ and the best linear approximation. We obtain $\Delta = \|\boldsymbol{\theta} - \boldsymbol{\theta}_5\|^2/\sigma^2 = 19.31$ and $\theta_{5i} = 27.75 - 0.5x_i$, $\boldsymbol{\theta}_5 = (\theta_{51}, \ldots, \theta_{5,11})' = p(\boldsymbol{\theta} | \mathscr{L}(\mathbf{x}_0, \mathbf{x}_1))$.

A Simulation: $\mathbf{Y} = \boldsymbol{\theta} + \boldsymbol{\varepsilon}$ was generated according to Example 4.5.2. The vectors $\mathbf{Y}$, $\hat{\mathbf{Y}}_j$, $\mathbf{e}_j = \mathbf{Y} - \hat{\mathbf{Y}}_j$ are given in Table 4.5.2 for $j = 1, \ldots, 5$. Then $F_1 = \dfrac{0.366}{6.348/4} = 0.23$, so $H(1)$: $\beta_5 = 0$ is accepted. $F_2 = \dfrac{2.817}{6.714/5} = 2.108 < F_{0.95}(1, 5)$, so $H(2)$: $\beta_4 = \beta_5 = 0$ is accepted. $F_3 = \dfrac{2.339}{9.531/6} = 1.47 < F_{0.95}(1, 6)$,

Table 4.5.1

x	$\boldsymbol{\theta}$	$\boldsymbol{\theta}_5$	$\boldsymbol{\theta} - \boldsymbol{\theta}_5$
0	30.00	27.75	2.25
1	28.15	27.25	0.90
2	26.60	26.75	−0.15
3	25.35	26.25	−0.90
4	24.40	25.75	−1.35
5	23.75	25.25	−1.50
6	23.40	24.75	−1.35
7	23.35	24.25	−0.90
8	23.60	23.75	−0.15
9	24.15	23.25	0.90
10	25.00	22.75	2.25

Table 4.5.2

	$\mathbf{Y}$	$\hat{\mathbf{Y}}_1$	$\hat{\mathbf{Y}}_2$	$\hat{\mathbf{Y}}_3$	$\hat{\mathbf{Y}}_4$	$\hat{\mathbf{Y}}_5$	$\hat{\mathbf{e}}_1$	$\hat{\mathbf{e}}_2$	$\hat{\mathbf{e}}_3$	$\hat{\mathbf{e}}_4$	$\hat{\mathbf{e}}_5$
0	27.95	28.01	27.87	28.46	29.16	26.48	−0.06	0.15	−0.60	−0.70	2.69
1	28.10	27.68	27.97	27.37	27.23	26.18	0.42	−0.30	0.60	0.14	1.07
2	25.69	26.73	26.77	26.18	25.67	25.84	−1.03	−0.04	0.60	0.51	−0.18
3	26.36	25.28	25.09	24.99	24.45	25.53	1.08	0.19	0.10	0.54	−1.07
4	23.35	23.73	23.53	23.93	23.60	25.21	−0.38	0.19	−0.40	0.33	−1.61
5	22.84	22.51	22.51	23.11	23.11	24.90	0.33	0.00	−0.60	0.00	−1.79
6	20.86	22.05	22.25	22.64	22.97	24.58	−1.19	−0.19	−0.40	−0.33	−1.61
7	23.94	22.56	22.75	22.66	23.19	24.27	1.38	−0.19	0.10	−0.54	−1.07
8	23.29	23.90	23.85	23.26	23.77	23.95	−0.62	0.05	0.60	−0.51	−0.18
9	25.51	25.46	25.16	24.57	24.71	23.64	0.06	0.30	0.60	−0.14	1.07
10	25.99	25.96	26.11	26.71	26.01	23.32	0.02	−0.15	−0.60	−0.70	2.69

so $H(3)$: $\beta_3 = \beta_4 = \beta_5 = 0$ is accepted. $F_4 = \dfrac{27.5}{31.085/7} = 6.192 > F_{0.95}(1, 7) =$ 5.59, so we decide that the model should be quadratic. This time, we made the right decision.

Problem 4.5.1: Let

$$\mathbf{X} = \begin{bmatrix} 1 & 1 & 1 & 0 \\ 1 & 1 & -1 & 1 \\ 1 & 1 & 0 & -1 \\ 1 & -1 & -1 & 0 \\ 1 & -1 & 1 & 1 \\ 1 & -1 & 0 & -1 \end{bmatrix} = (\mathbf{x}_0, \mathbf{x}_1, \mathbf{x}_2, \mathbf{x}_3),$$

and consider the model $\mathbf{Y} = \sum_0^3 \beta_j \mathbf{x}_j + \boldsymbol{\varepsilon}$.

(a) Let $\mathbf{Y} = (16, 10, 7, 0, 2, 1)'$. Use the step-down procedure described above, with α_i's all 0.05, to choose an appropriate model. Notice that, in a rare act of kindness, the author has chosen orthogonal regressor vectors. The projection vectors $\hat{\mathbf{Y}}_j$ consist entirely of integers.

(b) Let $\mathcal{M}_i$ be the model $\mathbf{Y} = \sum_0^i \beta_j \mathbf{x}_j + \boldsymbol{\varepsilon}$. Suppose that $\beta_1 = 4$, $\beta_2 = \beta_3 = 0$, and $\sigma^2 = 8.333$. What are the probabilities that the procedure chooses each of the models $\mathcal{M}_i$, $i = 3, 2, 1, 0$? (β_0 is not given because the probability does not depend on β_0). In order to facilitate the computation, we offer Table 4.5.3 (as for the Pearson–Hartley charts, $\phi = \sqrt{\delta/(\nu_1 + 1)}$).

(e) Repeat (b) for the case $\beta_3 = 0$, $\beta_2 = 2$, $\beta_1 = 4$, $\sigma^2 = 8.333$.

Problem 4.5.2: Consider the inner-product matrix $\mathbf{Q}$ of Example 4.5.1.
(a) Verify that $S(2)\mathbf{Q} = S(1)S(1, 2)\mathbf{Q}$.

Table 4.5.3 Power of the $\alpha = 0.05$ F-Test for $\nu_1 = 1$, Small ν_2

	ϕ										
ν_2	0.00	0.40	0.80	1.20	1.60	2.00	2.40	2.80	3.20	3.60	4.00
1	0.05	0.06	0.08	0.11	0.14	0.18	0.21	0.24	0.28	0.31	0.34
2	0.05	0.06	0.11	0.17	0.26	0.36	0.46	0.56	0.65	0.73	0.80
3	0.05	0.07	0.13	0.22	0.35	0.49	0.63	0.75	0.84	0.91	0.95
4	0.05	0.07	0.14	0.26	0.41	0.57	0.72	0.84	0.92	0.96	0.98
5	0.05	0.08	0.15	0.28	0.45	0.62	0.77	0.88	0.95	0.98	0.99

(b) For the simple linear regression model for Y vs. x_3, find $\hat{\boldsymbol{\beta}}$. Use this to find a 95% confidence interval on β_2, and a 95% prediction interval for $x_3 = 7$.

(c) Find the partial correlation coefficient $r_{y3.12}$.

4.6 EXAMINATION OF RESIDUALS

In judging the adequacy of the fit of a model and the distributional assumptions on $\boldsymbol{\varepsilon}$ for these models it is useful to examine the residual vector $\mathbf{e} = \mathbf{Y} - \hat{\mathbf{Y}}$, for $\hat{\mathbf{Y}} = p(\hat{\mathbf{Y}} | V)$, $V = \mathscr{L}(\mathbf{x}_1, \ldots, \mathbf{x}_k)$. Since $\mathbf{e} = \mathbf{P}_{V^\perp}\boldsymbol{\varepsilon} = (\mathbf{I}_n - \mathbf{P}_V)\boldsymbol{\varepsilon}$, $E(\boldsymbol{\varepsilon}) = \mathbf{0}$ implies $E(\mathbf{e}) = \mathbf{0}$ and $D[\boldsymbol{\varepsilon}] = \sigma^2\mathbf{I}_n$ implies $D[\mathbf{e}] = \sigma^2(\mathbf{I}_n - \mathbf{P}_V)$. For $\mathbf{H} = \mathbf{P}_V = (h_{ij})$, the hat-matrix, e_i therefore has variance $\sigma^2(1 - h_i)$, for $h_i = h_{ii}$ the ith diagonal term of $\mathbf{H}$. For simple linear regression $h_{ij} = \dfrac{1}{n} + (x_i - \bar{x})(x_j - \bar{x})/S_{xx}$. Since the trace of a projection matrix is the dimension of the subspace onto which it projects,

$$\sum_1^n \text{Var}(e_i) = \sigma^2 \text{ trace}(\mathbf{I} - \mathbf{P}_V) = \sigma^2[n - \dim V]$$

An observation on Y taken at $\tilde{\mathbf{x}} = \tilde{\mathbf{x}}_i$, the ith row of $\mathbf{X}$, is said to have high *leverage* on the residual e_i if an observation on Y for that $\tilde{\mathbf{x}}$ will tend to cause the prediction error to be small compared to what it otherwise would be. Since, from (4.4.5) $e_i = e_{-i}(1 - h_i)$, $\text{Var}(e_i)/\text{Var}(e_{-i}) = (1 - h_i)^2$, so that large values of h_i imply a significant payoff toward prediction at $\tilde{\mathbf{x}}_i$ by taking an observation at $\tilde{\mathbf{x}}_i$. For example, in simple linear regression h_i will be large for x_i far from $\bar{x}$, and in two variable regression with a constant term h_i will be large for $\tilde{\mathbf{x}}_i = (1, x_{1i}, x_{2i})$ far from $(1, \bar{x}_1, \bar{x}_2)$. Of course, placement of observations too far from corresponding means often leads to nonlinearity of regression. Though the regression of weight on height may be roughly linear for heights near the average, weights for the range (4 feet, 8 feet) are centainly not linear in height.

We obtain "Studentized residuals," often abbreviated as R-student$_i$, by dividing e_i by an estimate of its standard deviation, uncontaminated by Y_i. From (4.4.18) this uncontaminated estimator of σ^2 is $S_{-i}^2 = \text{SSE}_{-i}/(n - k)$, where $\text{SSE}_{-i} = \text{SSE} - e_i^2/(1 - h_i)$. Then

$$R\text{-STUDENT}_i \equiv r_i = e_i/(S_{-i}\sqrt{(1 - h_i)}).$$

Since r_i is approximately distributed as standard normal for reasonably large n, these r_i may be used to decide whether some observations are *outliers*, not consistent with the model. Values of $|r_i|$ greater than 3 or 4 may be labeled as outliers, and some consideration given to discarding them, or to methods not so sensitive to outliers as least squares methods.

Checking for Normality: These r_i, being distributed under our usual model approximately as standard normal, may be used to check for normality of the ε_i. We will not suggest any formal test of hypothesis but will instead describe a graphical technique. As will be discussed in Section 4.7 the normality of the ε_i is not vital to normality of $\hat{\boldsymbol{\beta}}$ or even to the $\hat{\mathbf{Y}}_i$, since these statistics are linear combinations of the ε_j so that, particularly for large n, a form of the Central Limit Theorem holds.

Suppose we have observed a random sample $W_1, \ldots, W_n$ from some continuous distribution F and wonder whether F is a normal distribution; that is, $F(x) = \Phi\left(\frac{x-\mu}{\sigma}\right)$ for some μ and σ. Let $W_{(1)} < W_{(2)} < \cdots < W_{(n)}$ be the corresponding order statistics, the ordered W_i's. Since $F(W_j)$ has a uniform distribution on $[0, 1]$, it can be shown that $E(W_{(j)})$ is approximately $F^{-1}(u_j)$ for $u_j = (j - 1/2)/n$. Since, under normality, $F^{-1}(u_j) = \mu + \sigma\Phi^{-1}(u_j)$, this means that in approximation $E[W_{(j)}] = \mu + \sigma\Phi^{-1}(u_j)$ so that for $Z_j = \Phi^{-1}(u_j)$ the $(Z_j, W_{(j)})$ pairs satisfy the simple linear regression model. Therefore, if we plot the $(Z_j, W_{(j)})$ pairs for $j = 1, \ldots, n$ they should fall approximately on a straight line with slope σ, intercept μ. "Normal" graph paper allows for easy plotting of the pairs. The horizontal Z_j-axis is labeled with u_j values instead of Z_j so that the Z_j need not be computed.

Example 4.6.1: Table 4.6.1 below gives values of $Y_i = 50 - 2x_i + \varepsilon_i$ for 60 x_i's as in the first column, where the ε_i have the double exponential density $f(x) = (1/2\theta)\exp(-|x|/\theta)$ for $\theta = 5\sqrt{2}$. f has mean 0, standard deviation $\sqrt{2}\theta = 10$. Figure 4.9 compares this density with the normal density with the same mean and variance. Figure 4.10 compares corresponding c.d.f.'s. Figure 4.11 presents the scatter diagram and the corresponding least squares line. Figure 4.12 is a plot of the ordered R-Student values r_i vs $Z_i = \Phi^{-1}(u_i)$. For ordered r_i values $r_{(1)} < r_{(2)} < \cdots < r_{(63)}$, $u_i = (i - 1/2)/63$, the points $(u_i, r_{(i)})$

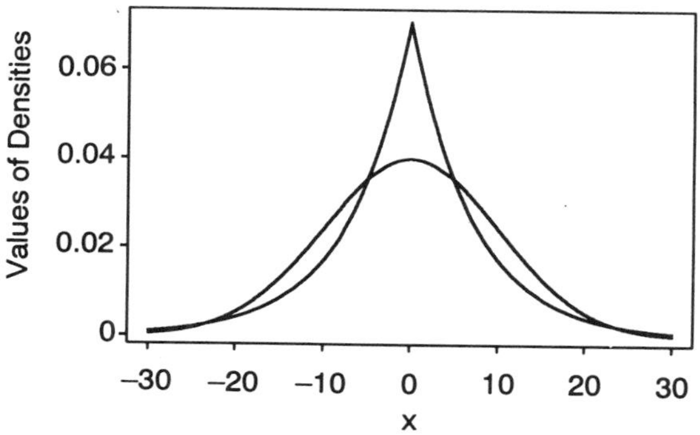

FIGURE 4.9 Normal and double exponential densities.

Table 4.6.1

x	Y	$\hat{Y}$	ε	h	r	x	Y	$\hat{Y}$	ε	h	r
0	49.3	50.0	−0.71	0.059	−0.16	11	25.4	27.6	−2.2	0.016	−0.48
0	47.4	50.0	−2.61	0.059	−0.58	11	24.8	27.6	−2.80	0.016	−0.61
0	51.2	50.0	1.19	0.059	0.26	11	26.4	27.6	−1.20	0.016	−0.26
1	47.7	48.0	−0.27	0.051	−0.06	12	24.9	25.6	−0.66	0.018	−0.14
1	40.5	48.0	−7.47	0.051	−1.69	12	24.8	25.6	−0.76	0.018	−0.16
1	48.5	48.0	0.53	0.051	0.12	12	27.0	25.6	1.44	0.018	0.31
2	46.8	45.9	0.87	0.044	0.19	13	20.9	23.5	−2.63	0.020	−0.57
2	49.8	45.9	3.87	0.044	0.85	13	27.2	23.5	3.67	0.020	0.80
2	45.3	45.9	−0.63	0.044	−0.14	13	20.5	23.5	−3.03	0.020	−0.66
3	50.9	43.9	7.00	0.037	1.56	14	16.4	21.5	−5.09	0.023	−1.11
3	46.0	43.9	2.10	0.037	0.46	14	23.4	21.5	1.91	0.023	0.41
3	44.8	43.9	0.90	0.037	0.20	14	24.6	21.5	3.11	0.023	0.68
4	42.5	41.9	0.64	0.032	0.14	15	20.2	19.4	0.75	0.027	0.16
4	43.2	41.8	1.34	0.032	0.29	15	23.1	19.4	3.65	0.027	0.79
4	44.7	41.8	2.84	0.032	0.62	15	19.0	19.4	−0.45	0.027	−0.10
5	42.8	39.8	2.98	0.027	0.65	16	19.5	17.4	2.08	0.032	0.45

5	38.8	39.8	−0.02	0.027	−0.22	16	25.0	17.4	7.58	0.032	1.69
5	30.4	39.8	−9.42	0.027	−2.13	16	19.4	17.4	1.98	0.032	0.43
6	36.5	37.8	−1.29	0.023	−0.28	17	19.1	15.4	3.72	0.037	0.82
6	37.6	37.8	−0.19	0.023	−0.04	17	21.4	15.4	6.02	0.037	1.33
6	45.9	37.8	8.11	0.023	1.81	17	17.8	15.4	2.42	0.037	0.53
7	39.8	35.8	4.05	0.020	0.88	18	18.4	13.3	5.06	0.044	1.12
7	38.0	35.8	2.25	0.020	0.49	18	13.9	13.3	0.56	0.044	0.12
7	35.9	35.8	0.15	0.020	0.03	18	11.3	13.3	−2.04	0.044	−0.45
8	37.0	33.7	3.29	0.018	0.71	19	14.2	11.3	2.89	0.051	0.64
8	34.7	33.7	0.99	0.018	0.21	19	2.1	11.3	−9.21	0.051	−2.11
8	33.0	33.7	−0.71	0.018	−0.15	19	4.3	10.8	−7.01	0.051	−1.58
9	26.0	31.7	−5.67	0.016	−1.24	20	10.4	9.3	1.13	0.059	0.25
9	29.9	31.7	−1.77	0.016	−0.38	20	4.4	9.3	−4.87	0.059	0.32
9	25.4	31.7	−6.27	0.016	−1.38	20	10.7	9.3	1.43	0.059	0.32
10	15.2	29.6	−14.44	0.016	−3.48						
10	30.6	29.6	0.96	0.016	0.21						
10	30.6	29.6	0.96	0.016	0.21						

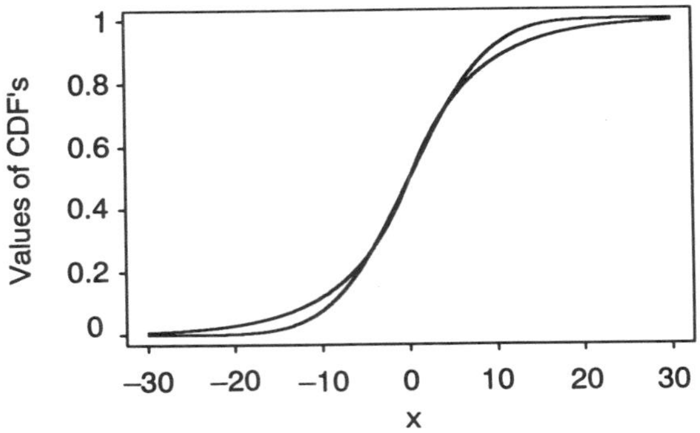

FIGURE 4.10 Normal and double exponential c.d.f.'s.

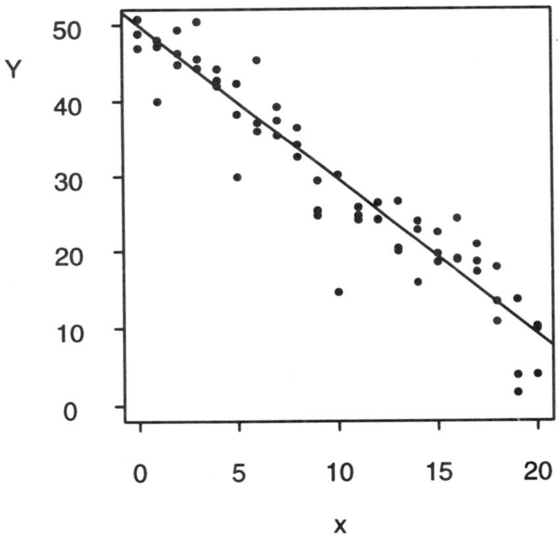

FIGURE 4.11 Scatterplot of Y vs. x and LS fit.

and the best fitting straight line were plotted in Figure 4.12. The plot indicates that the tails of the distribution are spread more widely than would be expected for observations taken from a normal distribution. The intercept and slope were -0.008 and 0.898, near 0 and 1 as would be expected.

In order to investigate the behavior of the estimator $\hat{\beta}_1$ of the slope for the case that the ε_i have a double exponential distribution this same example was repeated 1,000 times. As Figure 4.13 illustrates the distribution of $\hat{\beta}_1$ certainly appears to be normal. The sample standard deviation of these 1,000 values was 0.205 8, very close to the theoretical value 0.208 1. Among 1,000 90% and 95%

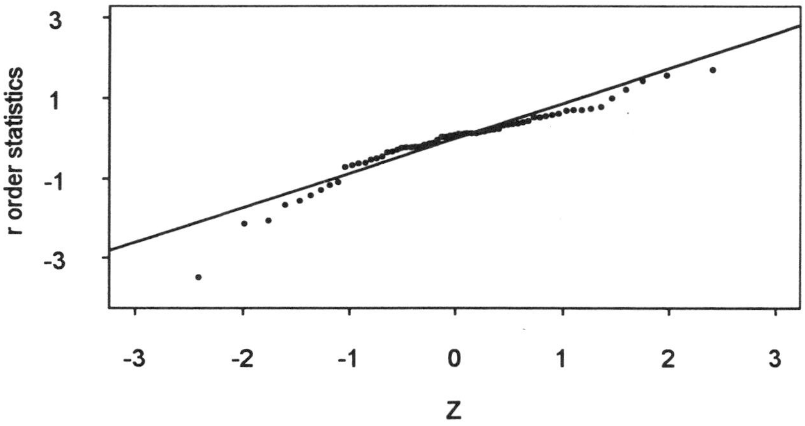

FIGURE 4.12 Plot of the pairs $(j/1000, \hat{\beta}_1^{(j)})$ for 1,000 samples, where $\hat{\beta}_1^{(j)}$ is the jth order statistic.

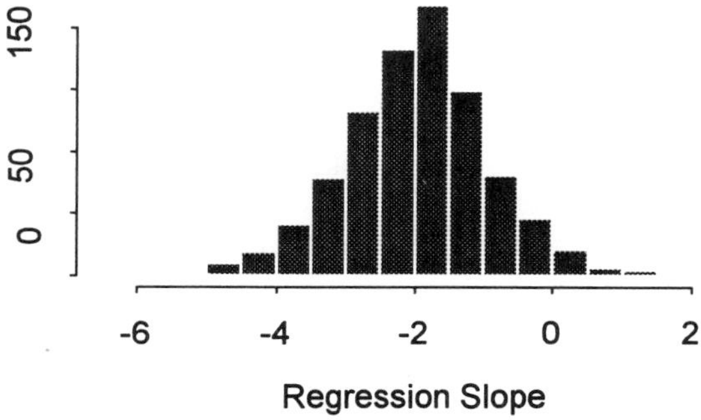

FIGURE 4.13 Histogram of 1,000 sample regression slopes for double exponential errors.

confidence intervals 909 and 955, respectively, covered $\beta_1 = -2$. The experiment was repeated for 10 observations on Y corresponding to $x = 1, \ldots, 10$, again with the ε_i double exponential. Again $\hat{\beta}_1$ seems to have an approximate normal distribution. This time 893 among 1,000 90% confidence intervals covered $\beta_1 = -2$.

4.7 COLLINEARITY

Collinearity in general is a relationship among the vectors $\mathbf{x}_1, \ldots, \mathbf{x}_k$ in which one or more are "almost" a linear combination of the others. Such collinearity can cause (1) inflation in the variance of estimators $\hat{\beta}_j$ of regression coefficients or of linear combinations $\hat{\eta}$ of such coefficients, (2) excessive effects of roundoff errors or measurement errors on the $\mathbf{x}_j$.

Let $\tilde{\mathbf{x}} = (x_1, \ldots, x_k)$ be a point at which we would like to predict $\mathbf{Y}$. If $E(\mathbf{Y}|\tilde{\mathbf{x}}) = g(\tilde{\mathbf{x}}) = \tilde{\mathbf{x}}\boldsymbol{\beta}$, then under our usual model $\hat{g}(\tilde{\mathbf{x}}) = \tilde{\mathbf{x}}\hat{\boldsymbol{\beta}}$ is the BLUE, having variance $h(\tilde{\mathbf{x}})\sigma^2$, for $h(\tilde{\mathbf{x}}) = \tilde{\mathbf{x}}(\mathbf{X}'\mathbf{X})^{-1}\tilde{\mathbf{x}}'$. In general h is smaller for $\tilde{\mathbf{x}}$ near the row vectors of $\mathbf{X}$. For example, in simple linear regression, $h(\tilde{\mathbf{x}}) = h(1, x) = (1/n) + (x - \bar{x})^2/S_{xx}$, so that h is large if x is far from the mean of the x-values used to estimate the regression line.

In order to investigate the variation in $h(\tilde{\mathbf{x}})$ as $\tilde{\mathbf{x}}$ varies, let $(\lambda_i, \tilde{\mathbf{w}}_i)$ for $i = 1, \ldots, k$ be eigenvalue, eigenvector pairs for $\mathbf{M} = \mathbf{X}'\mathbf{X}$. Suppose that the $\tilde{\mathbf{w}}_i$ have been chosen to be orthogonal. Then $\mathbf{M}^{-1}$ has eigenpairs $(1/\lambda_i, \tilde{\mathbf{w}}_i)$. Suppose also that $0 < \lambda_1 \leq \cdots \leq \lambda_k$ and $\|\tilde{\mathbf{w}}_i\|^2 = 1$ for each i. If we let $P_i = \tilde{\mathbf{w}}_i\tilde{\mathbf{w}}_i'$, then P_i is orthogonal projection onto $\mathscr{L}(\tilde{\mathbf{w}}_i)$, $P_iP_j = 0$ for $i \neq j$, $\mathbf{M} = \sum_i \lambda_i P_i = \mathbf{W}\mathbf{d}(\lambda)\mathbf{W}'$ and $\mathbf{M}^{-1} = \sum (1/\lambda_i)P_i = \mathbf{W}\mathbf{d}(1/\lambda)\mathbf{W}'$, where $\mathbf{W} = (\tilde{\mathbf{w}}_1, \ldots, \tilde{\mathbf{w}}_k)$, $\mathbf{d}(\lambda) = \text{diag}(1/\lambda_1, \ldots, \lambda_k)$ and $\mathbf{d}(1/\lambda) = \text{diag}(1/\lambda_1, \ldots, \lambda_k)$. Then $h(\tilde{\mathbf{w}}_i) = 1/\lambda_i$, and h is maximum for $\tilde{\mathbf{x}} = \tilde{\mathbf{w}}_1$, minimum for $\tilde{\mathbf{x}} = \tilde{\mathbf{w}}_k$ (each subject to $\|\tilde{\mathbf{x}}\| = 1$). The ratio

$$[h(\tilde{\mathbf{w}}_1)/h(\tilde{\mathbf{w}}_k)]^{1/2} = [\lambda_k/\lambda_1]^{1/2} = K$$

is usually called the *condition number* for $\mathbf{X}$. $\lambda_1^{1/2}, \ldots, \lambda_k^{1/2}$ are the diagonal elements in the singular value decomposition of $\mathbf{X}$ and of $\mathbf{X}'$.

If K is large then we can do rather poorly in estimating $g(\tilde{\mathbf{x}})$ for some $\tilde{\mathbf{x}}$ relative to that for others at the same distance from the origin. K may not be an interesting number if our interest is in $g(\tilde{\mathbf{x}})$ only for $\tilde{\mathbf{x}}$ near those for which observations have already been taken. For $\tilde{\mathbf{x}} = \tilde{\mathbf{x}}_i$, the ith row of $\mathbf{X}$, $h(\tilde{\mathbf{x}}_i) = h_{ii}$, the ith diagonal element of $\mathbf{H}$, the hat-matrix. Thus, if h_{ii} is large ($\tilde{\mathbf{x}}_i$ has high leverage), then the prediction $\hat{Y}_i$ at $\tilde{\mathbf{x}}_i$ depends heavily on Y_i, and has relatively high variance. Since $\sum h_{ii} = \text{trace}(\mathbf{H}) = k$, the average h_{ii} is k/n.

Scaling of the $\mathbf{x}$-vectors does not affect prediction of $\mathbf{Y}$, since the $\boldsymbol{\beta}$'s are correspondingly inversely scaled, and when a constant term is included in the model, replacement of an $\mathbf{x}$-vector by the corresponding vector of deviations from the mean also does not affect predictions. Such scalar and centering changes do affect the condition number K, however. K therefore has more meaning for comparison purposes if these changes are made. If each $\mathbf{x}$-vector is centered and has length one (except the vector of ones) then $\mathbf{M} = \mathbf{X}'\mathbf{X}$ is of the form

$$\mathbf{M} = \begin{bmatrix} n & 0 \\ 0 & \mathbf{R} \end{bmatrix},$$

where $\mathbf{R}$ is the correlation matrix, and we can consider the ratio of the largest and smallest eigenvalues of $\mathbf{R}$ as a measure of the condition K for $\mathbf{R}$, large values indicating a problem, $K = 1$ being ideal, achieved when all $\mathbf{x}$-vectors are uncorrelated. The eigenvectors $\tilde{\mathbf{w}}_i$ for $\mathbf{R}$ are called the *principal components*.

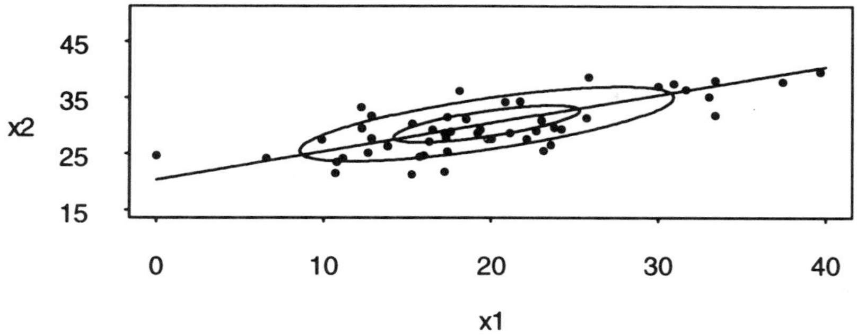

FIGURE 4.14 Contours of constant variance for predicted values.

If the aim is to estimate β_j then we can take $\tilde{\mathbf{x}} = \tilde{\mathbf{u}}_i$, the ith unit vector, so that

$$h(\tilde{\mathbf{x}}) = 1/\|\mathbf{x}_j^{\perp}\|^2$$

If $\mathbf{x}_j$ has been centered, and has length one then this is $1/(1 - R_j^2)$, where R_j is the multiple correlation coefficient of $\mathbf{x}_j$ with the other $\mathbf{x}$-vectors. $1/(1 - R_j^2)$ is called the *variance inflation factor* (Marquardt 1970).

Example 4.7.1: Consider the model $\mathbf{Y} = \beta_0\mathbf{x}_0 + \beta_1\mathbf{x}_1 + \beta_2\mathbf{x}_2 + \boldsymbol{\varepsilon}$, where $\mathbf{x}_0$ is the vector of ones, and 50 (x_{1i}, x_{2i}) pairs are as given in Figure 4.14. For $\tilde{\mathbf{x}} = (1, x_1, x_2)$, $\text{Var}(\tilde{\mathbf{x}}\hat{\boldsymbol{\beta}}) = h(\tilde{\mathbf{x}})\sigma^2$ for $h(\tilde{\mathbf{x}}) = \dfrac{1}{n} + \tilde{\mathbf{d}}\mathbf{M}^{-1}\tilde{\mathbf{d}}'$, $\tilde{\mathbf{d}} = (x_1 - \bar{x}_1, x_2 - \bar{x}_2)$, $\mathbf{M}$ the 2×2 sum of cross products of deviations matrix. The sample covariance matrix among the (x_{1i}, x_{2i}) pairs is $\mathbf{M}/(n-1)$. Contours of constant values of $h(\tilde{\mathbf{x}})$ are as indicated in Figure 4.14. The straight line, the major axis of the ellipse, has the equation $\tilde{\mathbf{d}}\tilde{\mathbf{w}}_2' = 0$, where $\tilde{\mathbf{w}}_2 = (0.867, -0.498)$ is the eigenvector corresponding to the smallest eigenvalue of $\mathbf{M}^{-1}$ (largest for $\mathbf{M}$).

The contours indicate that variances are smallest in the direction of the major axis, and that for $\tilde{\mathbf{x}}$ in a direction away from that axis, variances increase rapidly as distance from $(\bar{x}_1, \bar{x}_2)$ increases. Since $(1, 1)$ is much more in the direction of the major axis than is $(1, -1)$, we could expect $\text{Var}(\hat{\beta}_1 + \hat{\beta}_2)$ to be much smaller than $\text{Var}(\hat{\beta}_1 - \hat{\beta}_2)$. These variances are $(7.95 \times 10^{-4})\sigma^2$ and $(47.70 \times 10^{-4})\sigma^2$.

$$\mathbf{M} = \begin{bmatrix} 2{,}538 & 1{,}169 \\ 1{,}169 & 1{,}181 \end{bmatrix} \qquad \mathbf{M}^{-1} = \begin{bmatrix} 7.238 & -7.165 \\ -7.165 & 15.565 \end{bmatrix} \times 10^{-4}$$

$$\lambda_1 = 3{,}211.31 \qquad \tilde{\mathbf{w}}_1 = \begin{bmatrix} 0.8667 \\ 0.4988 \end{bmatrix}$$

$$\lambda_2 = 507.92 \qquad \tilde{\mathbf{w}}_2 = \begin{bmatrix} 0.4988 \\ -0.8667 \end{bmatrix}$$

One of the problems with use of the condition number K as an indicator of possible difficulty caused by collinearity is that it gives no indication of the cause or of possible solutions. If a major aim of the regression study is the estimation of slopes β_j then a finer analysis can suggest such causes and solutions. Since

$$D[\hat{\boldsymbol{\beta}}] = \mathbf{M}^{-1}\sigma^2 = \sigma^2 \sum (1/\lambda_i)\tilde{\mathbf{w}}_i\tilde{\mathbf{w}}_i', \qquad \mathrm{Var}(\hat{\beta}_j) = \sigma^2 \sum \tilde{\mathbf{w}}_{ij}^2/\lambda_i = \sigma^2 c_{jj}$$

and the *variance proportions*

$$p_{ij} = w_{ij}^2/(\lambda_i c_{jj})$$

are measures of the relative contributions of the ith eigenpair to the variance of $\hat{\beta}_j$, since $\sum_i p_{ij} = 1$. Study of these p_{ij} provides some insight into the causes of large variance inflation.

Example 4.7.2: Consider the following excerpt from *SAS/STAT User/ Guide*: *Volume 2, GLM-VARCOMP 5*, page 141. The collinearity diagnostics (Table 4.7.1) for the fitness data, discussed earlier in Section 3.12, based on the matrix with columns scaled to have length one, but with the means not subtracted, are presented. Tables 4.7.2 and 4.7.3 present corresponding diagnostics based on the correlation matrix $\mathbf{R}$ for two different models. In the author's opinion these last diagnostics are much more useful. The condition number of $\mathbf{R}$ is $\sqrt{42.68}$. The variance proportion $p_{64} = 0.953\,5$ indicates that variability of x_4 (runpulse) in the direction of the eigenvector $\tilde{\mathbf{w}}_6$ corresponding to the smallest eigenvalue (0.060 3) contributes 95.35% of the variability of $\hat{\beta}_4$. $\tilde{\mathbf{w}}_6$ is essentially a multiple of the difference between runpulse and maxpulse. This same eigenvector contributes 95.97% of the variability of $\hat{\beta}_5$. That variations in $\hat{\beta}_4$ and $\hat{\beta}_5$ are affected so strongly by the same eigenvector should not be surprising in light of the correlation 0.929 8 for the corresponding variables. These considerations suggest that runpulse and maxpulse measure substantially the same thing, and that we should try an analysis in which one of these is omitted. Table 4.7.2 presents a regression analysis for the model {1, 2, 3, 4, 6} (with runpulse dropped). The condition number K for this smaller model is $\sqrt{5.320}$, considerably smaller than for the full model. The variance inflation factor for variable 5 (runpulse) has been reduced considerably from that for the full model.

The estimated standard error for $\hat{\beta}_4$ is correspondingly smaller than for the full model. These p_{ij} can be helpful in indicating possible difficulties caused by roundoff or measurement errors in the independent variables. If errors are made observing a row $\tilde{\mathbf{x}}_i$ of $\mathbf{X}$ in the direction of the eigenvector corresponding to the smallest eigenvalue λ_1, $\mathbf{X}'\mathbf{X}$ it can have large effects on the $\hat{\beta}_j$ for which p_{j1} is large.

Table 4.7.1a Parameter Estimates

Variable	DF	Parameter Estimate	Standard Error	T for H_0: Parameter = 0	Prob > |T|	Tolerance	Variance Inflation
Intercept	1	102.934 479	12.403 258 10	8.299	0.000 1		0.000 000 00
Runtime	1	−2.628 653	0.384 562 20	−6.835	0.000 1	0.628 587 71	1.590 867 88
Age	1	−0.226 974	0.099 837 47	−2.273	0.032 2	0.661 010 10	1.512 836 18
Weight	1	−0.074 177	0.054 593 16	−1.359	0.186 9	0.865 554 01	1.155 329 40
Runpulse	1	−0.369 628	0.119 852 94	−3.084	0.005 1	0.118 521 69	8.437 274 18
Maxpulse	1	0.303 217	0.136 495 19	2.221	0.036 0	0.114 366 12	8.743 848 43
Rstpulse	1	−0.021 534	0.066 054 28	−0.326	0.747 3	0.706 419 90	1.415 588 65

Table 4.7.1b Collinearity Diagnostics

			Var Prop						
Number	Eigenvalue	Condition Number	Intercep	Runtime	Age	Weight	Runpulse	Maxpulse	Rstpulse
1	6.949 91	1.000 00	0.000 0	0.000 2	0.000 2	0.000 2	0.000 0	0.000 0	0.000 3
2	0.018 68	19.290 87	0.002 2	0.025 2	0.146 3	0.010 4	0.000 0	0.000 0	0.390 6
3	0.015 03	21.500 72	0.000 6	0.128 6	0.150 1	0.235 7	0.001 2	0.001 2	0.028 1
4	0.009 11	27.621 15	0.006 4	0.609 0	0.031 9	0.183 1	0.001 5	0.001 2	0.190 3
5	0.006 07	33.829 18	0.001 3	0.125 0	0.112 8	0.444 4	0.015 1	0.008 3	0.364 8
6	0.001 02	82.637 57	0.799 7	0.097 5	0.496 6	0.103 3	0.069 5	0.005 6	0.020 3
7	0.000 179 5	196.785 60	0.189 8	0.014 6	0.062 1	0.022 8	0.912 8	0.983 6	0.005 7

Source: Reprinted with permission from *SAS/Stat (R) User's Guide*, Version 6, Fourth Edition, Volume 2. Cary, NC. © 1990 SAS Institute Inc.

Table 4.7.2a Correlation Matrix R for Indepenent Variables $x_1, \ldots, x_6$

	x_1	x_2	x_3	x_4	x_5	x_6
1	1.0000	0.1887	0.1435	0.3136	0.2261	0.4504
2	0.1887	1.0000	−0.2335	−0.3379	−0.4329	−0.1641
3	0.1435	−0.2335	1.0000	0.1815	0.2494	0.0440
4	0.3136	−0.3379	0.1815	1.0000	0.9298	0.3525
5	0.2261	−0.4329	0.2494	0.9298	1.0000	0.3051
6	0.4504	−0.1641	0.0440	0.3525	0.3051	1.0000

x_1 = runtime, x_2 = age, x_3 = weight, x_4 = runpulse, x_5 = maxpulse, x_6 = rstpulse

Table 4.7.2b Eigenvectors $\tilde{\mathbf{w}}_i$ for R

1	2	3	4	5	6
0.2794	0.6670	0.2369	−0.0732	−0.6430	−0.0467
−0.3156	0.5617	0.0759	−0.5341	0.5388	−0.0597
0.2370	−0.1939	0.9210	0.0193	0.2355	0.0475
0.5647	−0.0414	−0.2118	−0.3686	0.1156	0.6967
0.5666	−0.1565	−0.1639	−0.3297	0.1125	−0.7115
0.3555	0.4192	−0.1349	0.6815	0.4635	−0.0206

Table 4.7.2c Variance Inflation Factors $1/\|\mathbf{x}_j^{\perp}\| = c_{jj}$

j	1	2	3	4	5	6
c_{jj}	1.591	1.513	1.55	8.437	8.744	1.416

Table 4.7.2d Collinearity Diagnostics p_{jj} for Model {2, 3, 4, 5, 6, 7}

			j					
i	λ_i	λ_1/λ_i	1	2	3	4	5	6
1	2.575	1.0000	0.0190	0.0256	0.0189	0.0147	0.0143	0.0347
2	1.328	1.9390	0.2106	0.1571	0.0245	0.0002	0.0021	0.0935
3	0.9251	2.7830	0.0382	0.0041	0.7936	0.0057	0.0033	0.0139
4	0.7432	3.4650	0.0045	0.2536	0.0004	0.0217	0.0167	0.4415
5	0.3687	6.9840	0.7408	0.5206	0.1302	0.0043	0.0039	0.4115
6	0.0603	42.6800	0.0228	0.0390	0.0324	0.9535	0.9597	0.0049

Table 4.7.3a Regression Analysis for Model $\{1, 2, 3, 5, 6\}$, $R^2 = 0.9042$

j	$\hat{\beta}_j$	$S_{\hat{\beta}_j}$	t_j	R_j
0	116.5	11.62	10.030	
1	−2.704	0.412	−6.561	0.709
2	−0.285	0.104	−2.753	0.873
3	−0.052	0.058	−0.898	0.901
5	−0.126	0.052	−2.414	0.880
6	−0.027	0.071	−0.382	0.904

Table 4.7.3b Variance Inflation Factors for Model $\{1, 2, 3, 5, 6\}$

j	1	2	3	5	6
c_{jj}	1.579	1.408	1.116	1.389	1.414

Table 4.7.3c Collinearity Diagnostics p_{ij} for Model $\{1, 2, 3, 5, 6\}$

			j				
	λ_i	λ_1/λ_i	1	2	3	5	6
1	1.873	1.000	0.073	0.030	0.030	0.107	0.098
2	1.275	1.469	0.169	0.271	0.070	0.011	0.036
3	0.900	2.081	0.064	0.091	0.704	0.039	0.097
4	0.036	3.153	0.821	0.575	0.099	0.196	0.319
5	0.594	5.230	0.009	0.056	0.032	0.629	0.465

The hyperplane $\hat{\mathbf{Y}} = \hat{\beta}_0 + \hat{\beta}_1 x_1 + \hat{\beta}_2 \hat{x}_2$ above the (x_1, x_2) plane therefore has a *wobble* in the direction of $\tilde{\mathbf{w}}_1 = (0.8671, -0.4981)$. We are trying to balance a thin board sheet on a rough picket fence running in the direction of $\tilde{\mathbf{w}}_2$, the major axis. A few more pickets (observations) taken at points distant from this fence would contribute considerably to the stability of the sheet.

Ridge Regression: Since $\mathbf{M}^{-1} = \sum_i (1/\lambda_i)\tilde{\mathbf{w}}_i\tilde{\mathbf{w}}_i' = \mathbf{W}'\mathbf{d}(1/\lambda)\mathbf{W}$, the error in the estimation of $\boldsymbol{\beta}$ is $\hat{\boldsymbol{\beta}} - \boldsymbol{\beta} = \mathbf{M}^{-1}\mathbf{X}'\boldsymbol{\varepsilon} = \sum_i (1/\lambda_i)D_i\tilde{\mathbf{w}}_i = \sum_i F_i\tilde{\mathbf{w}}_i/\sqrt{\lambda_i}$, where $D_i \equiv \tilde{\mathbf{w}}_i'\mathbf{X}'\boldsymbol{\varepsilon}$, and $F_i \equiv D_i/\sqrt{\lambda_i}$ which have zero mean. Let $\mathbf{F} = (F_1, \ldots, F_k)' = \mathbf{d}(\lambda^{-1/2})\mathbf{W}'\mathbf{X}'\boldsymbol{\varepsilon}$. Then $E(\mathbf{F}) = \mathbf{0}$ and $D[\mathbf{F}] = \sigma^2\mathbf{I}_k$. Of course, it follows that $D[\hat{\boldsymbol{\beta}}] = \sigma^2\mathbf{M}^{-1} = \sigma^2\sum_i (1/\lambda_i)P_i$, so that some of the variances of the $\hat{\beta}_j$ may be large if any of the λ_i are close to zero. The ridge regression estimator of $\boldsymbol{\beta}$ was defined to remove some of this excessive variation by instead accepting a certain amount of bias.

Definition 4.7.1: For $r \geq 0$ the rth order ridge regression estimator of $\boldsymbol{\beta}$ is

$$\hat{\boldsymbol{\beta}}_r = (\mathbf{M} + r\mathbf{I}_k)^{-1}\mathbf{X}'\mathbf{Y}.$$

Note that $\hat{\boldsymbol{\beta}}_r$ is a vector, not the rth component of $\hat{\beta}$. This should not cause confusion. We can write $\boldsymbol{\beta}_r$ in a form which will provide more insight. Since $\mathbf{X}'\mathbf{Y} = \mathbf{M}\hat{\boldsymbol{\beta}}$, $\hat{\boldsymbol{\beta}}_r = \mathbf{Z}_r\hat{\boldsymbol{\beta}}$ for $\mathbf{Z}_r \equiv (\mathbf{M} + r\mathbf{I}_k)^{-1}\mathbf{M}$. But, writing $\mathbf{M}$ in its spectral form, we get

$$\mathbf{Z}_r = \left[\sum_i \left(\frac{1}{\lambda_i + r}\right)P_i\right]\left[\sum_j \lambda_j P_j\right] = \sum_i \left[\frac{\lambda_i}{\lambda_i + r}\right]P_i,$$

so that

$$\hat{\boldsymbol{\beta}}_r = \left[\sum_i \left[\frac{\lambda_i}{\lambda_i + r}\right]P_i\right][\boldsymbol{\beta} + \sum F_i\tilde{\mathbf{w}}_i/\sqrt{\lambda_i}] = \left[\sum_i \frac{\lambda_i}{\lambda_i + r}P_i\right]\boldsymbol{\beta} + \sum_i \frac{\sqrt{\lambda_i}}{\lambda_i + r}F_i\tilde{\mathbf{w}}_i.$$

The bias in $\hat{\boldsymbol{\beta}}_r$ is the first term on the right minus $\boldsymbol{\beta}$, which simplifies to $r\left[\sum_i \frac{1}{\lambda_i + r}P_i\right]\boldsymbol{\beta} = r[\mathbf{M} + r\mathbf{I}_k]^{-1}\boldsymbol{\beta}$. One measure of the bias of the vector $\hat{\boldsymbol{\beta}}_r$ is the squared length of this vector, which is $\gamma_2(r) \equiv r^2\boldsymbol{\beta}'\left[\sum_i \left(\frac{1}{r + \lambda_i}\right)^2 P_i\right]\boldsymbol{\beta} = r^2 \sum_i \alpha_i/(r + \lambda_i)^2$, where $\alpha_i \equiv \boldsymbol{\beta}'P_i\boldsymbol{\beta} = \|p(\boldsymbol{\beta}|\tilde{\mathbf{w}}_i)\|^2$. The function $\gamma_2(r)$ in an increasing function of r, with $\gamma_2(0) = 0$, (corresponding to the least squares estimator) and $\lim_{r\to\infty} \gamma_2(r) = \sum_i \alpha_i = \|\boldsymbol{\beta}\|^2$. Thus, as r becomes larger $\hat{\boldsymbol{\beta}}_r$ is forced to be nearer the origin. In the limit it is the zero vector itself.

The payoff in the use of $\hat{\boldsymbol{\beta}}_r$ is in the decrease in variances. The covariance matrix for $\hat{\boldsymbol{\beta}}_r$ is

$$D[\hat{\boldsymbol{\beta}}_r] = \sigma^2 \sum_i \frac{\lambda_i}{(\lambda_i + r)^2}P_i = \sigma^2(\mathbf{M} + r\mathbf{I}_k)^{-2}\mathbf{M}. \tag{4.7.2}$$

All terms are decreasing as functions of r. An overall measure of the precision of $\hat{\boldsymbol{\beta}}_r$ is the sum of the variances, the trace of $D[\hat{\boldsymbol{\beta}}_r]$. We find $\gamma_1(r) \equiv \text{trace}(D[\hat{\boldsymbol{\beta}}_r]) = \sigma^2 \sum_i \frac{\lambda_i}{(\lambda_i + r)^2}$. $\gamma_1(0) = \sigma^2 \sum_i (1/\lambda_i) = \sigma^2\,\text{trace}(\mathbf{M}^{-1})$ is the variance of the least squares estimator $\hat{\boldsymbol{\beta}}$. It is easy to show that γ_1 has a negative derivative near 0, so that for at least some values of r near 0 this measure of overall variance is smaller than it is for $\hat{\boldsymbol{\beta}}$. Since γ_2 has derivative with limit zero as r approaches zero from the right, it follows that for at least some positive r the sum $\gamma_1(r) + \gamma_2(r)$ of the mean square errors is smaller than it is for $\hat{\boldsymbol{\beta}}$.

Hoerl and Kennard (1970a) show that, subject to the willingness to let the

SSE increase by some fixed amount ϕ_0, the estimator which minimizes the squared length of the estimate $\mathbf{b}$ of $\boldsymbol{\beta}$ is a ridge estimator.

An obvious generalization of the ridge estimator is produced by substituting a vector $\mathbf{r} \equiv (r_1, \ldots, r_k)$, with nonnegative components, for the number r. Define $\hat{\boldsymbol{\beta}}_r = \left[\sum \frac{1}{\lambda_i + r_i} P_i\right] \mathbf{X}'\mathbf{Y} = \sum_i \frac{\sqrt{\lambda_i}}{\lambda_i + r_i} F_i \tilde{\mathbf{w}}_i$. Formulas for biases and covariance matrices are given merely by replacing each $\lambda_i + r$ by $\lambda_i + r_i$. It seems reasonable to use larger r_i whenever λ_i is small. In the extreme, when λ_i is particularly small we could take the limit as r_i approaches infinity, equivalently omitting $\tilde{\mathbf{w}}_i$ from the analysis. We get $\hat{\boldsymbol{\beta}}_A \equiv \left[\sum_{i \in A} (1/\lambda_i) P_i\right] \mathbf{X}'\mathbf{Y}$, where A is the set of indices not omitted. Properties of $\hat{\boldsymbol{\beta}}_A$ are those of the estimator obtained by taking the $\mathbf{r}$-vector to be the vector of zeros and infinities (really taking limits as $r_i \to \infty$), with zeros corresponding to indices in A.

A ridge trace (for the case that $r_i \equiv \mathbf{r}$) is a graph of $\hat{\boldsymbol{\beta}}_j(\mathbf{r})$, the jth component of $\hat{\boldsymbol{\beta}}_\mathbf{r}$ as a function of $\mathbf{r}$. Some insight into the effects of the use of ridge regression on individual regression components is gained. An appropriate choice of $\mathbf{r}$ may be made by studying all ridge traces, though in general it is not an easy choice. Nor is it easy to decide whether to use ridge regression at all. It is tempting to try to estimate the functions γ_1 and γ_2 and to minimize this estimate. This requires an estimate of the bias term, and our reasoning becomes rather circular. The reader might try to replace $\boldsymbol{\beta}$ in γ_2 by $\hat{\boldsymbol{\beta}}_\mathbf{r}$ or by $\hat{\boldsymbol{\beta}}$, to get an estimate $\hat{\gamma}_2$, then compute $E(\hat{\gamma}_2)$.

Example 4.7.3: Let $\mathbf{u}_1 = (1, -1, 0, 0, 0)'/\sqrt{2}$, $\mathbf{u}_2 = (1, 1, -2, 0, 0)/\sqrt{6}$, $\mathbf{u}_3 = (1, 1, 1, -3, 0)/\sqrt{12}$. Let $\mathbf{x}_1 = \mathbf{u}_1$, $\mathbf{x}_2 = \mathbf{u}_2$, and $\mathbf{x}_3 = (a_1\mathbf{u}_1 + a_2\mathbf{u}_2 + \mathbf{u}_3)/\sqrt{a_1^2 + a_2^2 = 1}$. Thus, the $\mathbf{x}$-vectors are orthogonal to (1, 1, 1, 1, 1), and each has length one. As a_1 and a_2 increase in absolute value the matrix $\mathbf{M} = \mathbf{X}'\mathbf{X}$ becomes increasingly ill-conditioned, so that ridge regression becomes more appropriate. To illustrate this let $a_1 = 5$, and $a_2 = 10$. Then

$$\mathbf{M} = \begin{bmatrix} 1 & 0 & 0.4454 \\ 0 & 1 & 0.8909 \\ 0.4454 & 0.8909 & 1 \end{bmatrix},$$

which has eigenvalues 1.996, 0.996, 0.00396, so that $\mathbf{M}$ is somewhat ill-conditioned. The corresponding eigenvectors are $\tilde{\mathbf{w}}_1 = (0.3162, 0.6325, 0.7071)'$, $\tilde{\mathbf{w}}_2 = (0.8944, -0.4472, 0)'$, and $\tilde{\mathbf{w}}_3 = (0.3162, 0.6325, -0.7071)$. For $\boldsymbol{\beta} = (10, 20, 30)'$, and $\boldsymbol{\theta} = (40.65, 40.61, -45.95, -2.31, 0)'$ and $\sigma = 5$ an observation $\mathbf{Y} = (38.00, 12.97, -37.69, -3.54, 0.15)'$ was generated. The ridge estimates of $\boldsymbol{\beta}$ and the resulting error mean squares for each of 10 values of r are given in Table 4.7.4. The least squares estimates correspond to $r = 0$, for which ESS is

Table 4.7.4

r	$\hat{\beta}_{r1}$	$\hat{\beta}_{r2}$	$\hat{\beta}_{r3}$	ESS
0.000	−16.98	−17.41	77.44	23.74
0.010	−0.71	14.60	41.34	34.10
0.020	1.96	19.79	35.25	38.03
0.030	3.06	21.83	32.66	40.10
0.040	3.66	22.88	31.19	41.58
0.050	4.04	23.49	30.22	42.86
0.060	4.30	23.86	29.51	44.09
0.070	4.48	24.10	28.96	45.34
0.080	4.63	24.24	28.51	46.64
0.090	4.74	24.33	28.13	48.01
0.100	4.83	24.37	27.80	49.47

smallest. Notice that the estimates seem to be closer to the corresponding parameters (10, 20, 30) as r increases. As r increases the ridge estimates will begin to be pushed down toward zero, as we accept more bias in the effort to decrease variation (Figure 4.15). The problem the statistician faces, of course, is that $\boldsymbol{\beta}$ is unknown and it is difficult to choose the appropriate r. Knowledge of the subject matter may provide some guidance, so that if too small values of r seem to provide unreasonable estimates, these values can be rejected. Usually the ridge traces will stabilize and gradually shrink toward zero, and a good choice for r may be the a minimum value for which all traces have stabilized. For better understanding see the papers mentioned above.

Problem 4.7.1: Show that the variance inflation factor for the jth component of the ridge estimator $\hat{\boldsymbol{\beta}}_r$ is $\sum_i \frac{\lambda_i \tilde{\mathbf{w}}_{ij}^2}{(\lambda_i + r)^2}$, where $\tilde{\mathbf{w}}_i = (\tilde{\mathbf{w}}_{i1}, \ldots, \tilde{\mathbf{w}}_{ik})'$ is the eigenvector of $\mathbf{M}$ corresponding to λ_i.

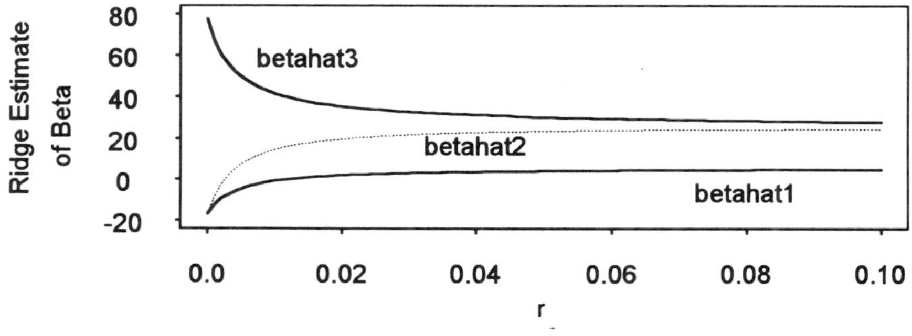

FIGURE 4.15 Plots of ridge estimates of the betas vs. r.

Problem 4.7.2: Use the formula $\hat{\boldsymbol{\beta}}_r = (\mathbf{M} + r\mathbf{I}_k)^{-1}\mathbf{M}\hat{\boldsymbol{\beta}}$ and matrix algebra to show that $D[\hat{\boldsymbol{\beta}}_r] = (\mathbf{M} + r\mathbf{I}_k)^{-2}\mathbf{M}\sigma^2 = \mathbf{M}(\mathbf{M} + r\mathbf{I}_k)^{-2}\sigma^2$.

Problem 4.7.3: Suppose that $\mathbf{M} = \begin{bmatrix} 1 & a \\ a & 1 \end{bmatrix}$ for $0 \le a \le 1$. Find λ_i and P_i for $i = 1, 2$, and use these to write a nonmatrix formula for $D[\hat{\boldsymbol{\beta}}_r]$. Compare the performances of $\hat{\boldsymbol{\beta}}_r$ and $\hat{\boldsymbol{\beta}}$ for the case that $1 - a$ or $1 + a$ is small, say ε.

Problem 4.7.4: What are the variance inflation factors for $\hat{\boldsymbol{\beta}}$ in Example 4.7.1?

4.8 ASYMPTOTIC NORMALITY

In this section we will briefly discuss, without proof, conditions under which the estimator $\hat{\boldsymbol{\beta}}$ is approximately normally distributed, even though the usual assumptions on $\boldsymbol{\varepsilon}$ are not satisfied. Though in any application n is finite, any mathematical treatment of the distributional properties of $\hat{\boldsymbol{\beta}}$ under these relaxed conditions must be asymptotic. In order to investigate the closeness of these distributions to normality some computer simulations will be discussed.

Eicher's Theorem 3.1 (Eicher, 1965) considers a sequence of regression models

$$\mathbf{Y}_n = \mathbf{X}_n\boldsymbol{\beta} + \boldsymbol{\varepsilon}_n,$$

where $\mathbf{Y}_n$ has n components, $E(\boldsymbol{\varepsilon}_n) = \mathbf{0}$, $\boldsymbol{\Sigma}_n = D[\boldsymbol{\varepsilon}_n] = \text{diag}(\sigma_1^2, \ldots, \sigma_n^2)$, and $\mathbf{X}_n$ is an $n \times k$ matrix of constants. Suppose the components of $\boldsymbol{\varepsilon}_n$ are independent. Then, as usual,

$$\hat{\boldsymbol{\beta}}_n = \mathbf{M}_n^{-1}\mathbf{X}_n'\mathbf{Y}_n = \boldsymbol{\beta} + \mathbf{M}_n^{-1}\mathbf{X}_n'\boldsymbol{\varepsilon}_n$$

for $\mathbf{M}_n = \mathbf{X}_n'\mathbf{X}_n$. Then $\hat{\boldsymbol{\beta}}_n$ is an unbiased estimator of $\boldsymbol{\beta}$ with covariance matrix

$$D[\hat{\boldsymbol{\beta}}_n] = \mathbf{M}_n^{-1}\mathbf{X}_n'\boldsymbol{\Sigma}_n\mathbf{X}_n\mathbf{M}_n^{-1} \equiv \mathbf{F}_n$$

For a simple example suppose $k = 1$ and $\mathbf{X}_n = (1, 2, \ldots, n)'$ with $\sigma_1 = \cdots = \sigma_n$. Is $\hat{\beta}_n = \sum_1^n iY_i/\left(\sum_1^n i^2\right)$ asymptotically normally distributed? Since $\hat{\beta}_n$ is β plus a linear combination of the components of $\boldsymbol{\varepsilon}_n$ we might hope that a form of the Central Limit Theorem will cause $\hat{\beta}_n$ to be normally distributed even when the components of $\boldsymbol{\varepsilon}_n$ are not.

Consider an even more extreme example. For $k = 1$ take $\mathbf{X}_n = (1, \ldots, 1, n)'$, the components of $\boldsymbol{\varepsilon}_n$ identically distributed with variance σ^2.

$$\hat{\beta}_n = \left[\sum_1^{n-1} Y_i + nY_n\right]/(n - 1 + n^2)$$

has variance $\sigma^2/(n^2 + n - 1)$ and

$$\mathbf{Z}_n = [\hat{\boldsymbol{\beta}}_n - \boldsymbol{\beta}]/\sqrt{\mathrm{Var}(\hat{\boldsymbol{\beta}}_n)} = \left[\sum_1^{n-1} \varepsilon_i + m\varepsilon_n\right]/\sqrt{(n^2 + n - 1)\sigma^2}$$

$$= \left[\frac{1}{\sigma}\sum_1^{n-1} \varepsilon_i/\sqrt{n-1}\right][(n-1)/(n^2 + n - 1)]^{1/2} + (\varepsilon_n/\sigma)/\left[1 + \frac{1}{n} - \frac{1}{n^2}\right]^{-1/2}.$$

The first factor in the first term on the right is asymptotically $N(0, 1)$, and the second factor converges to 0. Thus the first term converges in probability to zero. The second term converges in distribution to the distribution of ε_n/σ. Thus $\mathbf{Z}_n$ will be asymptotically normally distributed only if ε_n is normally distributed. Too much weight has been put on Y_n, relative to the weight on the other Y_i.

Similarly, if $\mathbf{X}_n$ is the vector of all ones, but the components of $\boldsymbol{\varepsilon}_n$ have variances which differ greatly, then most of the variation in $\hat{\boldsymbol{\beta}}_n$ will be caused by those components with relatively large variance. Eicher's Theorem makes the requirements that not too much weight be given to some components, and that there be not too much relative variation in the components of $\boldsymbol{\varepsilon}_n$.

Let $\tilde{\mathbf{x}}_{i,n}$ be the ith row of $\mathbf{X}_n$ and let $\mathbf{B}_n$ be the symmetric $k \times k$ matrix satisfying $\mathbf{B}_n^2 = \mathbf{M}_n$. Let $\mathscr{F}$ be the collection of all distributions with mean zero, finite positive variance. Suppose that all components of $\boldsymbol{\varepsilon}_n$ have a distribution in $\mathscr{F}$.

Theorem 4.8.1 (Eicher): Let $\mathbf{Z}_n = \mathbf{B}_n^{-1}(\hat{\boldsymbol{\beta}}_n - \boldsymbol{\beta})$. Then $\mathbf{Z}_n$ is asymptotically distributed as $N_k(\mathbf{0}, \mathbf{I}_k)$ for all $G \in \mathscr{F}$ if and only if

(1) For $d_{in} = \tilde{\mathbf{x}}_{in}\mathbf{M}_n^{-1}\tilde{\mathbf{x}}_{in}'$, $h_n \equiv \max_{i=1,\ldots,k} d_{in} \to 0$ as $n \to \infty$.

(2) $\sup_{G \in \mathscr{F}} \int_{|x|>c} x^2 dG(x) \to 0$ as $c \to \infty$.

(3) $\inf_{G \in \mathscr{F}} \int x^2 dG(x) > 0$.

Condition (1) assures that not too much weight be put on any single observation Y_i. d_{in} is the leverage of the observation corresponding to $\tilde{\mathbf{x}}_i$. Condition (2) does not allow components of $\boldsymbol{\varepsilon}_n$ to have probability mass at locations which diverge too far from that of other components. Condition (3) forces all components to a minimum standard of variation.

Consider the first example above with $\mathbf{X}_n = (1, 2, \ldots, n)'$. Then $d_{in} = x_{in}^2/\sum_1^n x_{in}^2 = i^2/\sum_1^n i^2 = 6i^2/[n(n+1)(2n+1)]$, so $h_n = 6n/(n+1)(2n+1) < 6(2n+1)^{-1} \to 0$ as $n \to \infty$. Therefore, if the components of $\boldsymbol{\varepsilon}_n$ have identical distributions with finite positive variance then $\mathbf{Z}_n \equiv (\hat{\boldsymbol{\beta}}_n - \boldsymbol{\beta})/\sqrt{\mathrm{Var}(\hat{\boldsymbol{\beta}}_n)}$, where $\mathrm{Var}(\hat{\boldsymbol{\beta}}_n) = 6/[n(n+1)(2n+1)]$, is asymptotically $N(0, 1)$.

In the second example, with $\mathbf{X}_n' = (1, \ldots, 1, n)$,

$$\tilde{\mathbf{x}}_{jn}\mathbf{M}_n^{-1}\tilde{\mathbf{x}}_{jn}' = 1/[(n-1) + n^2] \qquad \text{for} \quad 1 \le j \le n-1$$
$$= n^2/[(n-1) + n^2)] \qquad \text{for} \quad j = n$$

Since this is maximum for $j = n$ and this maximum converges to one, (1) is not satisfied, so that $\mathbf{Z}_n$ is not asymptotically normal for all $G \in \mathscr{F}$.

Eicher showed, for example, that if the ij element of $\mathbf{X}_n$ is i^{c_j} for $c_1 > \cdots > c_k > -\frac{1}{2}$ then $\mathbf{X}_n$ satisfies (1). Thus polynomial regression ($c_j = j$) on integers $1, \ldots, n$ satisfies I.

One of the problems with Theorem 4.8.1 is that $\mathbf{Z}_n$ depends on the unknown $\mathbf{\Sigma}_n$. Fortunately it is possible to replace $\mathbf{M}_n$ by $\mathbf{C}_n = \mathbf{M}_n^{-1}\mathbf{X}_n'\mathbf{S}_n\mathbf{X}_n\mathbf{M}_n^{-1}$ where $\mathbf{S}_n = \text{diag}(e_1, \ldots, e_n)$ and $\mathbf{e} = (e_1, \ldots, e_n)$ is the usual residual vector.

Theorem 4.6.2 (Eicher): Let $\mathbf{G}_n$ be the symmetric $n \times n$ matrix satisfying $\mathbf{G}_n^2 = \mathbf{C}_n$ and define $\mathbf{W}_n = \mathbf{G}_n^{-1}(\hat{\boldsymbol{\beta}}_n - \boldsymbol{\beta})$. Then $\mathbf{W}_n$ converges in distribution to $N_k(\mathbf{0}, \mathbf{I}_k)$ if all three parts of Theorem 4.8.1 hold.

In the case that the components of $\boldsymbol{\varepsilon}$ have identical distributions F we can state a simpler version of the Central Limit Theorem for least squares estimators for linear models, due to Huber (1981). We follow the presentation of Mammen (1992).

Theorem 4.8.3: Let the components of $\boldsymbol{\varepsilon}$ be independent, with common c.d.f. F. Let h_n be the largest diagonal term (leverage) of the projection matrix $\mathbf{P}_n = \mathbf{X}_n\mathbf{M}_n^{-1}\mathbf{X}_n'$. Let $\mathbf{c} = (c_1, \ldots, c_k)'$ be a vector of constants, let $\eta = \mathbf{c}'\boldsymbol{\beta}$, and $\hat{\eta}_n = \mathbf{c}'\hat{\boldsymbol{\beta}}_n$. Then $\mathbf{Z}_n = [\mathbf{c}'\mathbf{M}_n^{-1}\mathbf{c}]^{1/2}(\hat{\eta}_n - \eta)$ converges in distribution to standard normal if and only if either (1) $h_n \to 0$, or (2) F is a normal distribution.

Example 4.8.1: Suppose $\boldsymbol{\varepsilon}_n$ has independent identically distributed components with mean 0, variance $\sigma^2 > 0$. Let $\mathbf{x}_0$ be the vector of n ones, and let $\mathbf{x}_1$ be the first n components of $(1, \frac{1}{2}, \frac{1}{3}, \ldots)$. Let $\mathbf{x}_1^* = \mathbf{x}_1 - \bar{x}_1\mathbf{x}_0$ and $\mathbf{X} = (\mathbf{x}_0, \mathbf{x}_1^*)$. Then $\mathbf{M}_n = \mathbf{X}_n'\mathbf{X}_n = \text{diag}(1/n, \|\mathbf{x}_1^*\|^2)$, for $\|\mathbf{x}_1^*\|^2 = \sum_1^n (1/i^2) - [\sum (1/i)]^2/n$. Since $\mathbf{U}_n \equiv \sum_1^n (1/i^2) - \Pi^2/6 \to 0$ and $V_n = \sum_1^n (1/i) - \log n \to \eta \doteq 0.577\,23$ (Euler's constant), for $\tilde{\mathbf{x}}_1 = (1, 1 - \bar{x}_n))$,

$$d_{1n} = \tilde{\mathbf{x}}_1 \begin{bmatrix} \frac{1}{n} & 0 \\ 0 & 1/\|\mathbf{x}_1^*\|^2 \end{bmatrix} \tilde{\mathbf{x}}_1' = \frac{1}{n} + \frac{(1 - \bar{x}_n)^2}{\|\mathbf{x}_1^*\|^2} \to 1/(\Pi^2/6) = 6/\Pi^2 \qquad \text{as} \quad n \to \infty.$$

Thus condition (1) is not satisfied. The regression slope $\hat{\beta}_1$ depends too heavily on the observations corresponding to the first few x_i. If F were a normal distribution, then the standardized version $\mathbf{Z}$ of $\hat{\boldsymbol{\beta}}$ has a standard normal distribution for every n. However, if F is not a normal distribution, then $\hat{\boldsymbol{\beta}}_n$ cannot be asymptotically normal.

This result suggests that in the case that the errors ε_i corresponding to a few of the x_i's which are relatively far from the mean have distributions which differ greatly from the normal, that $\hat{\beta}_1$ will also have a distribution differing greatly from normal (though not as much as for the ε_i).

Though limit theorems are useful in that they indicate the conditions under which a distribution may be approximated by the limiting distribution, they do not in general say how large n must be before the approximation is good. The next few examples may provide some understanding of these approximations.

Example 4.8.2: Let Z have the standard normal distribution, Let V be 0 or 1 with probabilities $1 - p$, p independent of Z. Then

$$\varepsilon = [\sigma_1(1 - V) + \sigma_2 V]Z = \begin{cases} \sigma_1 Z, & \text{if} \quad V = 0 \\ \sigma_2 Z, & \text{if} \quad V = 1 \end{cases}$$

has the "contaminated normal distribution," with c.d.f.

$$F(x) = (1 - p)\Phi(x/\sigma_1) + p\Phi(x/\sigma_2).$$

F has mean $E(\varepsilon) = 0$, variance $\mathrm{Var}(\varepsilon) = (1 - p)\sigma_1^2 + p\sigma_2^2$. The density of E is plotted in Figure 4.16(a).

Now consider simple linear regression $Y_i = \beta_0 + \beta_i x_i + \varepsilon_i$ for $i = 1, \ldots, n$, where the ε_i's are a random sample from F above. Take $n = 10$, $x_i = i$ for $i = 1, \ldots, 9$ and $x_{10} = 30$. Take $\sigma_1 = 2$, $\sigma_2 = 20$, $p = 0.2$.

Thus, $d_{10,10}$ as defined in Eicher's Theorem is $\left(\frac{1}{10}\right) + \frac{(30 - 7.5)^2}{622.5} = 0.913$, not very close to 0 as suggested by condition (1). The other $d_{i,10}$ values are considerably smaller. Certainly the ε_i are not normally distributed, and it seems that too much weight on $x_i = 30$ in the determination of $\hat{\beta}_1$ may cause $\hat{\beta}_1$ and the corresponding t-statistic to have distributions differing greatly from the normal and t distributions. Figure 4.16(b), the histogram for $\hat{\beta}_1$ for 1000 simulations, indicates that $\hat{\beta}_1$ takes more extreme values than would be expected under the normal distribution. These values are caused by large $|\varepsilon_i|$ corresponding to $x_i = 30$. Figure 4.16(c) is the corresponding histogram for $T = \hat{\beta}_1/s_{\hat{\beta}_1}$.

The comparison of the distribution of $\hat{\beta}_1$ with the normal distribution is more evident in Figure 4.16(d). Notice that the c.d.f.'s are approximately equal at the 5% and 95% percentile points. Figure 4.16(e) indicates the same things for the t-statistic, showing that T has a 95th percentile somewhat larger than that given by the t_8 distribution. As a result claimed 95% confidence intervals on β_1 would in reality be roughly 90% intervals. Among 1,000 values of T frequencies and nominal probabilities using the t_8 distribution were as follows:

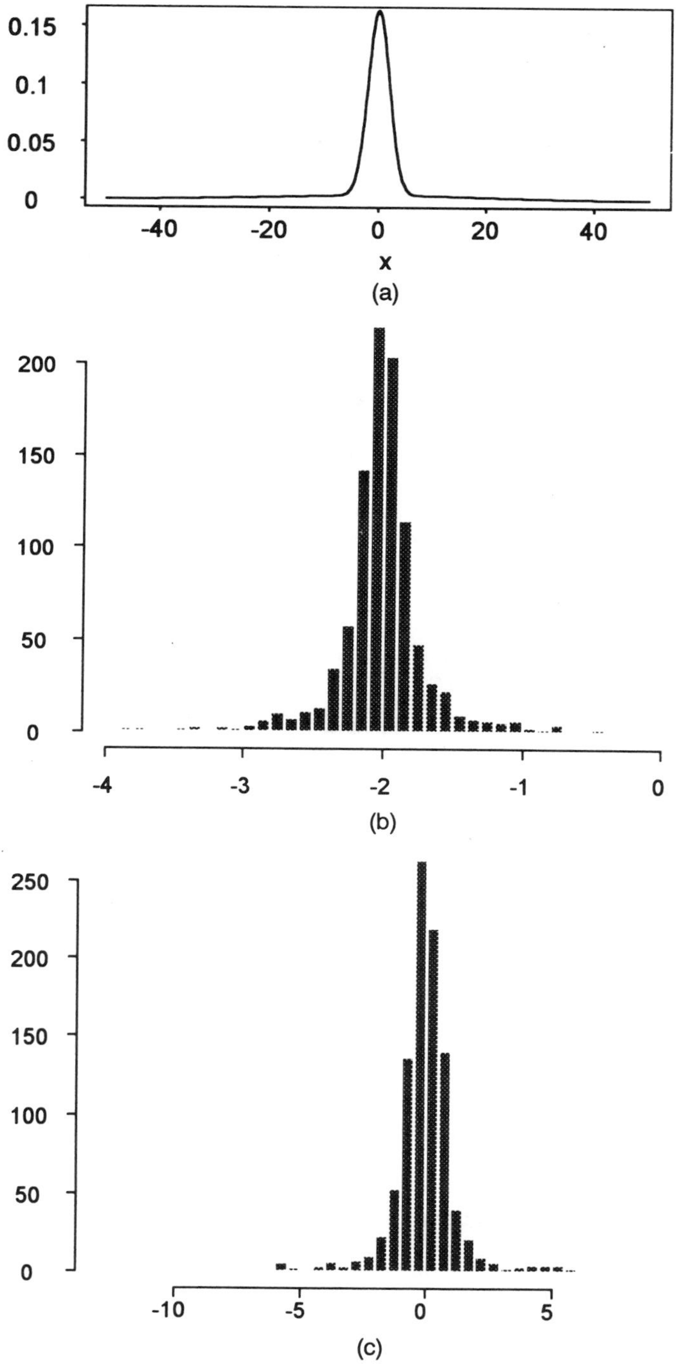

FIGURE 4.16 (a) Contaminated normal density. (b) Histogram of $\hat{\beta}_i$'s. (c) Histogram of t-statistics.

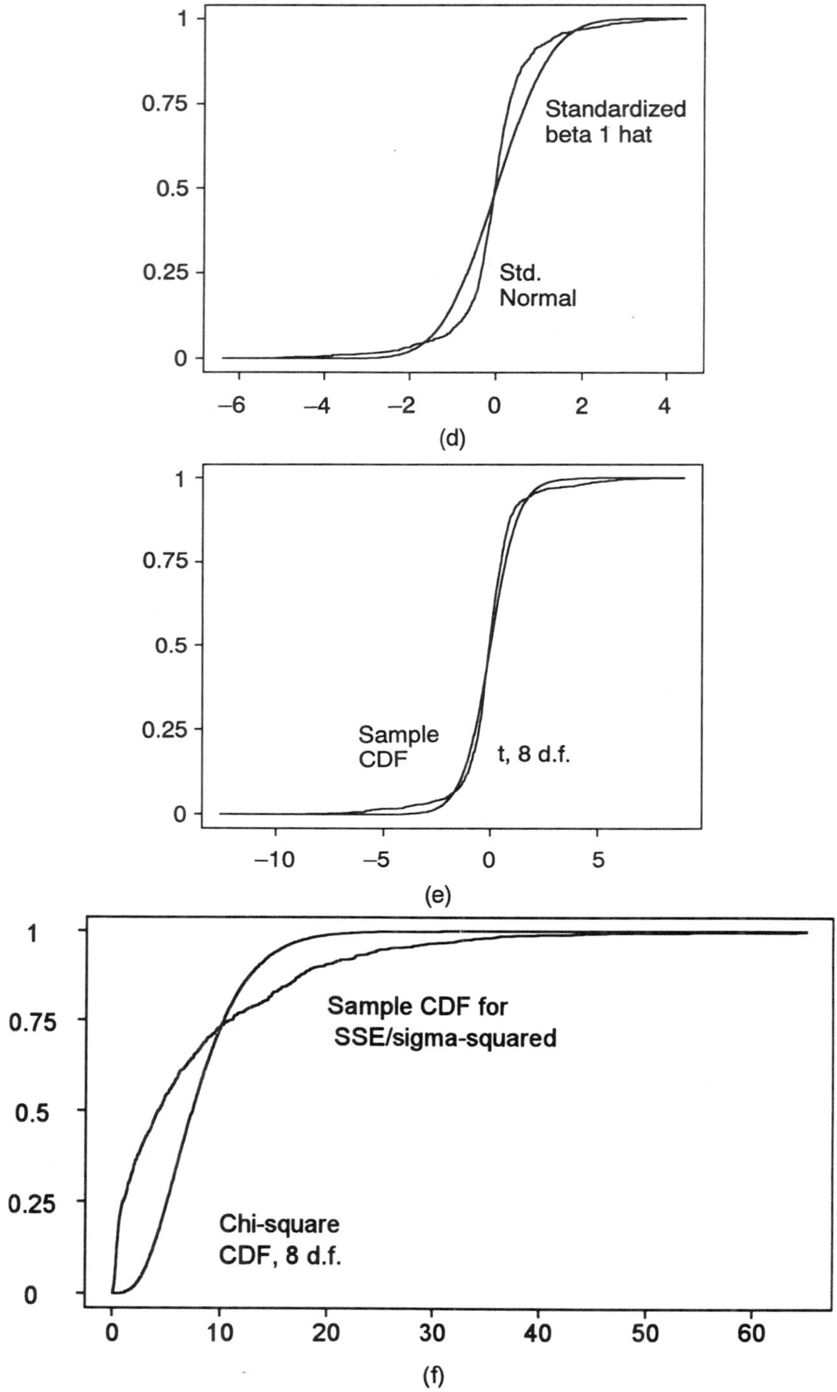

FIGURE 4.16 (d) Sample and normal c.d.f.'s for $\hat{\beta}_1$'s. (e) Sample and t_8 c.d.f.'s for t-statistics. (f) Sample c.d.f. and chi-square c.d.f.

t	-2.897	-2.306	-1.860	-1.397	1.397	1.860	2.306	2.897
$P(T \leq t)$	0.01	0.025	0.05	0.10	0.90	0.95	0.975	0.99
Frequency	30	42	57	72	915	931	947	959

We estimate $P(|T| \leq 2.306)$, for example, to be $(947 - 42)/1{,}000 = 0.905$, while the nominal value is 0.95.

Thus, even for this case, for a distribution F with rather "heavy tails" (kurtosis 13.67), and relatively heavy weight on one observation (corresponding to $x = 30$) the distribution of the t-statistic is not terribly far from the nominal t distribution. Certainly for a larger number of observations or less relative weight on a relatively few observations, or "more normal" distributions for the ε the approximation is better.

Figure 4.16(f), a comparison between the sample c.d.f. for $W = \text{ESS}/\sigma^2 = S^2(8)/\sigma^2$ and the c.d.f. for χ^2_8 indicates that the heavy tail for F has a stronger effect on the distribution of W. In general, nonnormality has a much stronger effect on the distribution of S^2 than on $\hat{\boldsymbol{\beta}}$, so that conclusions concerning σ^2 must remain somewhat tentative in the presence of suspected heavy non-normality. In general, positive kurtosis $(\mu_4/\sigma^4 - 3)$ tends to cause W to have heavier tails than does χ^2.

We will go no further in discussing these approximations. Readers interested in the effects of departure from the assumptions of the usual linear model, especially in the case of the analysis of variance are referred to Chapter 10 of Scheffé's text, *The Analysis of Variance*. At the conclusion of Section 10.2 Scheffé summarizes

> Our conclusions from the examples of this section may be briefly summarized as follows: (i) Nonnormality has little effect on inferences about means but serious effects on inferences about variances of random variables whose kurtosis γ_2 differs from zero. (ii) Inequality of variances in the cells of a layout has little effect on inferences about means if the cell numbers are equal, serious effects with unequal cell numbers. (iii) The effect of correlation in the observations can be serious on inference about means.

The kurtosis of a random variable ε is $\gamma_2 \equiv E(\varepsilon - \mu_\varepsilon)^4/\sigma^4 - 3$. For the normal distribution $\gamma_2 = 0$.

4.9 SPLINE FUNCTIONS

On occasion we may choose to approximate a regression function $g(x) \equiv E(Y|x)$ by functions $h_j(x)$ over nonoverlapping intervals I_j. We could simply treat the data corresponding to x_i's in I_j as separate curve fitting problems. However, it often desirable that the approximating function $h(x) \equiv h_j(x)$ for $x \in I_j$ have certain smoothness properties. For example, the viscosity of a chemical may change continuously with temperature x (degrees Celsius), and linearly for the three intervals $(-100, 0)$, $(0, 100)$, and $(100, 300)$, but the slope may change at $x = 0$ or at $x = 100$.

Let $v_0 < v_1 < \cdots < v_{r+1}$ be fixed known points and let I_j be the closed interval $[v_{j-1}, v_j]$ for $j = 1, \ldots, r + 1$. Suppose that the regression function is $g(x)$ which we hope to approximate by a function

$$h(x) = p_j(x) \qquad \text{for} \quad x \in I_j, \qquad \text{for} \quad j = 1, \ldots, r + 1,$$

where $p_j(x)$ is a polynomial of degree at most m. It is common to choose $m = 3$. In the viscosity example above $m = 1$. Suppose that $h(x)$ has continuous derivatives of order $m - 1$ on the interval (v_0, v_{r+1}). The points v_j for $j = 1, \ldots, r$ are called *knots* and the function $h(x)$ is called a *spline* function.

The word "spline" is taken from the draftsman's spline, a flexible thin rod tied down at certain fixed points (the knots), which physically must then follow a cubic path ($m = 3$) between points. The word was chosen by Schoenberg (1946).

It is possible to represent such spline functions in a simple way, so that least squares computations are facilitated. Define $d_j(x) = p_j(x) - p_{j-1}(x)$ for $j = 1, \ldots, r$ for $x \in (x_0, x_{r+1})$. Then $p_{j+1} = p_1 + d_1 + \cdots + d_j$. Because derivatives of h up to order $m - 1$ are continuous, the first $m - 1$ derivatives of d_j at v_j must all be zero. Since d_j is a polynomial of order at most m, this implies that $d_j(x) = \beta_j(x - v_j)^m$. Thus,

$$h(x) = p_1(x) + \sum_{i=1}^{j} \beta_i(x - v_i)^m \qquad \text{for} \quad x \in I_{j+1}, j = 1, \ldots, r.$$

By defining

$$u_+ = \begin{cases} u & \text{for} \quad u \geq 0 \\ 0 & \text{for} \quad u < 0 \end{cases}$$

we can write g in the form

$$g(x) = p_1(x) + \sum_{i=1}^{r} \beta_i(x - v_i)_+^m, \tag{4.9.1}$$

since for $x \in I_{j+1}$, $i > j$, $(x - v_i)_+ = 0$. Let $a_i(x) = (x - v_i)_+^m$, and $p_1(x) = \alpha_0 + \alpha_1 x + \cdots + \alpha_m x^m$. Then

$$h(x) = \sum_0^m \alpha_j x^j + \sum_1^r \beta_i a_i(x),$$

so that h has been expressed as a linear combination of unknown parameters, with known coefficients (for known knots). Thus, if we observe n pairs (x_i, Y_i), we can represent the model $Y_i = h(x_i) + \varepsilon_i$ in the usual regression form for $\boldsymbol{\beta} = (\alpha_0, \ldots, \alpha_m, \beta_1, \ldots, \beta_r)'$, $\mathbf{X} = (\mathbf{x}_0, \ldots, \mathbf{x}_m, \mathbf{a}_1, \ldots, \mathbf{a}_r)$, with ith row $(1, x_i, x_i^2, \ldots, x_i^m, a_1(x_i), \ldots, a_r(x_i))$.

Example 4.9.1: Let $m = 2$, $r = 2$, $v_0 = 0$, $v_1 = 5$, $v_2 = 10$, $v_3 = 20$, and suppose $p_1(x) = 80 - 10x + 0.5x^2$, $\beta_1 = 0.6$, and $\beta_2 = -2$, so that

$$p_2(x) = p_1(x) + 0.6(x - 5)^2 = 95 - 16x + 1.1x^2$$

and

$$p_3(x) = p_1(x) - 2(x - 10)^2 = -105 + 24x - 0.9x^2.$$

Then $h(x) = p_j(x)$ over I_j, $j = 1, 2, 3$, where $I_1 = [0, 5]$, $I_2 = [5, 10]$, $I_3 = [10, 20]$. For $x_i = 0.5(i - 1)$ and $i = 1, \ldots, 41$ observations $Y_i = h(x_i) + \varepsilon_i$ for $\varepsilon_i \sim H(0, 100)$ where taken independently.

Figure 4.17 presents graphs of the points (x_i, Y_i), $h(x)$, and $\hat{h}(x)$. Table 4.9.1 presents $\mathbf{X}, \boldsymbol{\theta}, \mathbf{Y}, \hat{\mathbf{Y}}, = \mathbf{Y} - \hat{\mathbf{Y}}$, for $\theta_i = h(x_i)$. The parameters and their estimates were

$$\begin{matrix} \alpha_0 \\ \alpha_1 \\ \alpha_2 \\ \beta_0 \\ \beta_1 \end{matrix} \begin{bmatrix} 80 & 86.39 \\ -10 & -13.54 \\ 0.5 & 0.853 \\ 0.6 & 0.280 \\ -2 & -2.026 \end{bmatrix} \qquad \begin{aligned} \sigma^2 &= 100.0 \\ S^2 &= \|\mathbf{Y} - \hat{\mathbf{Y}}\|^2/(41 - 5) = 142.6 \end{aligned}$$

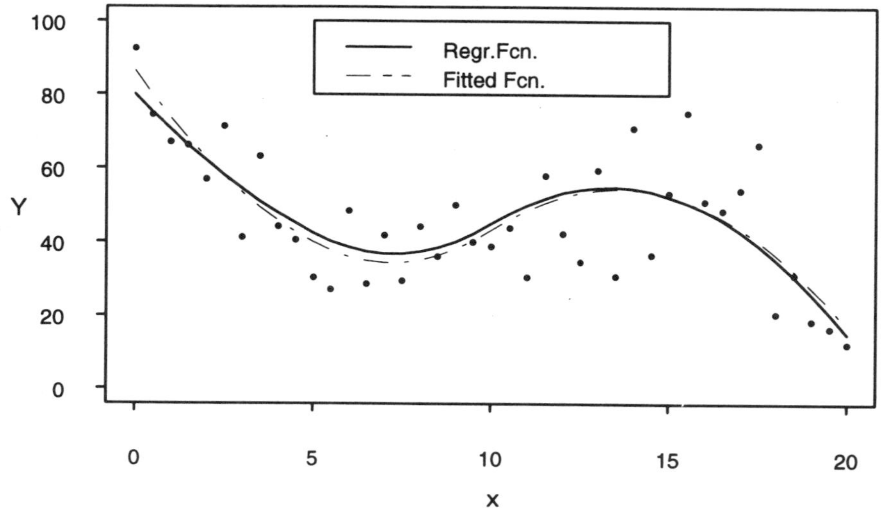

FIGURE 4.17

Table 4.9.1

X					θ	Y	Ŷ	e
1	0.0	0.00	0.00	0.00	80.00	93.24	86.39	6.950
1	0.5	0.25	0.00	0.00	75.12	75.14	79.83	−4.692
1	1.0	1.00	0.00	0.00	70.50	67.79	73.70	−5.913
1	1.5	2.25	0.00	0.00	66.13	66.86	68.00	−1.143
1	2.0	4.00	0.00	0.00	62.00	57.71	72.72	−5.017
1	2.5	6.25	0.00	0.00	58.13	72.10	57.87	14.230
1	3.0	9.00	0.00	0.00	54.50	42.12	53.45	−11.330
1	3.5	12.25	0.00	0.00	51.12	63.96	49.46	14.500
1	4.0	16.00	0.00	0.00	48.00	45.01	45.89	−0.880
1	4.5	20.25	0.00	0.00	45.12	41.45	42.74	−1.290
1	5.0	25.00	0.00	0.00	42.50	31.28	40.03	−8.751
1	5.5	30.25	0.25	0.00	40.27	27.97	37.01	−9.040
1	6.0	36.00	1.00	0.00	38.60	49.26	36.16	13.100
1	6.5	42.25	2.25	0.00	37.48	29.55	35.07	−5.514
1	7.0	49.00	4.00	0.00	36.90	42.69	34.55	8.146
1	7.5	56.25	6.25	0.00	36.88	30.31	34.60	−4.282
1	8.0	64.00	9.00	0.00	37.40	44.99	35.21	9.779
1	8.5	72.25	12.25	0.00	38.48	36.98	36.39	0.596
1	9.0	81.00	16.00	0.00	40.10	50.96	38.14	12.830
1	9.5	90.25	20.25	0.00	42.27	40.87	40.45	0.421
1	10.0	100.00	25.00	0.00	45.00	39.67	43.33	−3.664
1	10.5	110.30	30.25	0.25	47.78	44.69	46.27	−1.579
1	11.0	121.00	36.00	1.00	50.10	31.39	48.76	−17.370
1	11.5	132.30	42.25	2.25	51.97	58.89	50.81	8.078
1	12.0	144.00	49.00	4.00	53.40	43.12	52.41	−9.294
1	12.5	156.20	56.25	6.25	54.37	35.61	53.57	−17.950
1	13.0	169.00	64.00	9.00	54.90	60.35	54.28	6.071
1	13.5	182.20	72.25	12.25	54.98	31.72	54.54	−22.820
1	14.0	196.00	81.00	16.00	54.60	71.76	54.36	17.400
1	14.5	210.20	90.25	20.25	53.78	37.47	53.73	−16.250
1	15.0	225.00	100.00	25.00	52.50	54.21	52.65	1.562
1	15.5	240.20	110.30	30.25	50.77	75.81	51.13	24.680
1	16.0	256.00	121.00	36.00	48.60	51.88	49.16	2.725
1	16.5	272.20	132.30	42.25	45.97	49.50	46.74	2.754
1	17.0	289.00	144.00	49.00	42.50	55.00	43.88	11.120
1	17.5	306.30	156.20	56.25	39.37	67.29	40.57	26.720
1	18.0	324.00	169.00	64.00	35.40	21.48	36.82	−15.340
1	18.5	342.20	182.20	72.25	30.97	32.02	32.62	−0.603
1	19.0	361.00	196.00	81.00	26.10	19.47	27.97	−8.500
1	19.5	380.20	210.20	90.25	20.78	17.42	22.88	−5.465
1	20.0	400.00	225.00	100.00	15.00	13.18	17.34	−4.163

The estimate of $D[\hat{\boldsymbol{\beta}}]$ was

$$S_{\hat{\beta}}^2 = \begin{bmatrix} 69.160 & -39.57 & 4.742 & -5.748 & 1.212 \\ -39.57 & 31.69 & -4.248 & 5.661 & -1.726 \\ 4.742 & -4.248 & 0.596 & -0.831 & 0.294 \\ -5.748 & 5.661 & -0.831 & 1.218 & -0.505 \\ 1.212 & -1.726 & 0.294 & -0.505 & 0.311 \end{bmatrix}.$$

We can test the null hypothesis of no change of regression at knot v_j, equivalently $\beta_j = 0$, using $t_j \equiv \hat{\beta}_j/\hat{S}_{\hat{\beta}_j}$. We obtain $t_1 = 0.280/\sqrt{1.218} = 0.254$, $t_2 = -2.026/\sqrt{0.311} = 3.63$. We conclude that a model with a single knot at $v_2 = 10$ would seem to suffice (an incorrect decision).

The prediction function $\hat{h}$ is an approximation of h, which itself is an approximation of the regression function g. The approximation of g by h can be improved if the number and positions of the knots are chosen carefully. For a discussion of splines in a nonstatistical setting see deBoor (1978).

One problem with the choice of the matrix $\mathbf{X}$ is that the vectors $\mathbf{a}_i$ may be almost dependent so that $\mathbf{X}$ is somewhat ill-conditioned. The functions $a_i(x)$ may be replaced by other functions $B_i(x)$, called B-splines, which are zero outside relatively narrow intervals, so that the column space of $\mathbf{X}$ remains the same. The solution $\hat{\boldsymbol{\beta}}$ remains the same, but computations are likely to be more precise. The condition that the polynomials and their first $m - 1$ derivatives agree at the knots may be weakened by requiring that they agree for fewer derivatives, or the value of m can be made to vary with the knots, with resulting complications in computations. For a thorough discussion of the fitting of surfaces (regression functions of two or more variables), see the book by Lancaster and Salkauskas (1986), or the paper by Friedman (1991).

Problem 4.9.1: Let $0 < v_1$ and let $0 < x_{11} \le x_{12} \le \cdots \le x_{1n_1} \le v_1 \le x_{21} \le \cdots \le x_{2n_2}$. For observations (x_{ij}, Y_{ij}) for $j = 1, \ldots, n_i$, $i = 1, 2$, find the least squares approximation by functions of the form

$$h(x) = \begin{cases} \beta_1 x & \text{for} \quad 0 < x < v_1. \\ \beta_1 v_1 + \beta_2(x - v_1) & \text{for} \quad x \ge v_1. \end{cases}$$

Evaluate $\hat{\beta}_1$, $\hat{\beta}_2$ for $v_1 = 3$, and pairs of observations (1, 3), (2. 5), (4, 3), (5, 5). Sketch the scatter diagram and the function $\hat{h}$.

Problem 4.9.2: (a) Suppose $m = 2, r = 2, v_0 = 0, v_1 = 4, v_2 = 8, v_3 = 10$ and observations on Y are taken for $x = 1, 2, \ldots, 10$. What is the matrix $\mathbf{X}$ needed in order to fit a spline function?

(b) For the following (x, Y) pairs estimate the coefficients $(\alpha_1, \alpha_2, \alpha_3, \beta_1, \beta_2)$.

x	1	2	3	4	5	6	7	8	9	10
Y	12.81	11.73	15.74	28.29	40.44	50.60	59.52	68.11	78.10	97.23

(c) Suppose these Y_i satisfy a spline model for $m = 2$ and v_i as in (a). Find a 95% confidence interval on $g(x) \equiv E(Y|x)$ for $x = 7$.

(d) Test the null hypothesis that g is the same quadratic function on [0, 10] for $\alpha = 0.05$.

4.10 NONLINEAR LEAST SQUARES

Almost all of the models so far considered have been linear in the parameters. Even when nonlinear models were considered in Section 4.1 we made a *linearizing transformation*, in order to take advantage of the mathematical apparatus available for linear models. There are occasions, however, when a nonlinear model cannot be linearized, or when we would greatly prefer to get a better fit to the data than that provided by linearizing.

Suppose, for example, that we have observed n pairs (x_i, y_i), and for theoretical reasons, or simply based on a graphical look at the data, we hope to fit a function of the form $g(\mathbf{x}; \boldsymbol{\beta}) = g(\mathbf{x}; \beta_0, \beta_1, \beta_2) = \beta_0 + \beta_1 \mathbf{x}^{\beta_2}$ to the data. There are no transformations on $\mathbf{x}$ or $\mathbf{y}$ which will result in a function which is linear in the parameters. Instead, we can attempt to use least squares directly. Let $Q(\boldsymbol{\beta}) = \sum_i [y_i - g(x_i; \boldsymbol{\beta})]^2$. The principle of least squares chooses $\boldsymbol{\beta} = \hat{\boldsymbol{\beta}}$, the value which minimizes Q. What makes this problem different from those already considered is the nonlinearity of $g(x; \boldsymbol{\beta})$ in $\boldsymbol{\beta}$.

To emphasize the dependence of $g(x_i; \boldsymbol{\beta})$ on $\boldsymbol{\beta}$, define $g_i(\boldsymbol{\beta}) = g(x_i; \boldsymbol{\beta})$, and let $\mathbf{g}(\boldsymbol{\beta})$ be the corresponding n-component column vector. $\boldsymbol{\beta}$ is also written as a column vector. Define $r(\boldsymbol{\beta}) = \mathbf{y} - \mathbf{g}(\boldsymbol{\beta})$, the vector of residuals. Our task is to choose $\boldsymbol{\beta} = \hat{\boldsymbol{\beta}}$ so that $Q(\boldsymbol{\beta}) = \|r(\boldsymbol{\beta})\|^2$ is minimum. The trick we will use is to suppose that we have a rough estimate, or guess, say $\hat{\boldsymbol{\beta}}^0$ of $\boldsymbol{\beta}$. For $\boldsymbol{\beta}$ reasonably close to $\hat{\boldsymbol{\beta}}^0$ $\mathbf{g}(\boldsymbol{\beta})$ will be approximately linear in $\boldsymbol{\beta}$. If $\mathbf{g}(\boldsymbol{\beta})$ is a reasonably smooth function (that is, all the component functions of $\mathbf{g}(\boldsymbol{\beta})$ are smooth), then we can use a Taylor approximation of $\mathbf{g}(\boldsymbol{\beta})$. Let $g_i^j(\hat{\boldsymbol{\beta}}^0) = \left.\dfrac{\partial g_i(\boldsymbol{\beta})}{\partial \beta_j}\right|_{\boldsymbol{\beta}=\hat{\boldsymbol{\beta}}^0}$, and let $\mathbf{g}^j(\hat{\boldsymbol{\beta}}^0)$ be the corresponding n-component row vector. Let $\mathbf{W} = \mathbf{W}(\hat{\boldsymbol{\beta}}^0)$ be the $n \times 3$ matrix with jth column $\mathbf{g}^j(\hat{\boldsymbol{\beta}}^0)$, with ith row $\mathbf{W}_i$. The linear Taylor approximation (the differential) of $\mathbf{g}(\boldsymbol{\beta})$ at $\hat{\boldsymbol{\beta}}^0$ is $\mathbf{h}(\hat{\boldsymbol{\beta}}^0) = \mathbf{g}(\hat{\boldsymbol{\beta}}^0) + \sum_j \mathbf{g}^j(\hat{\boldsymbol{\beta}}^0)(\beta_j - \hat{\beta}_j^0) = \mathbf{g}(\hat{\boldsymbol{\beta}}^0) + \mathbf{W}(\hat{\boldsymbol{\beta}}^0)(\boldsymbol{\beta} - \hat{\boldsymbol{\beta}}^0)$, which, of course is linear in $\boldsymbol{\beta}$. Let $\boldsymbol{\gamma} = \boldsymbol{\beta} - \hat{\boldsymbol{\beta}}^0$. Replace $Q(\boldsymbol{\beta})$

by $Q^*(\boldsymbol{\beta}) = \sum_i [y_i - h_i(\boldsymbol{\beta})]^2 = \sum_i [(y_i - g_i(\hat{\boldsymbol{\beta}}^0)) - \mathbf{W}_i\boldsymbol{\gamma}]^2$. Letting $z_i = y_i - g_i(\hat{\boldsymbol{\beta}}^0)$, we get $Q^*(\boldsymbol{\beta}) = \sum_i [z_i - \mathbf{W}_i\boldsymbol{\gamma}]^2 = \|\mathbf{z} - \mathbf{W}\boldsymbol{\gamma}\|^2$, where $\mathbf{z} = \mathbf{y} - \mathbf{g}(\hat{\boldsymbol{\beta}}^0)$. The function $Q^*(\boldsymbol{\beta})$ is minimized by $\boldsymbol{\gamma} = \hat{\boldsymbol{\gamma}} = (\mathbf{W}'\mathbf{W})^{-1}\mathbf{W}'\mathbf{z}$, $\boldsymbol{\beta} = \hat{\boldsymbol{\beta}} = \hat{\boldsymbol{\gamma}} + \hat{\boldsymbol{\beta}}^0$. More explicitly,

$$\hat{\boldsymbol{\beta}}^1 = \hat{\boldsymbol{\beta}}^0 + (\mathbf{W}'\mathbf{W})^{-1}\mathbf{W}'(\mathbf{y} - \mathbf{g}(\hat{\boldsymbol{\beta}}^0)). \tag{4.10.1}$$

Once we have obtained the improvement $\hat{\boldsymbol{\beta}}^1$ we can replace $\hat{\boldsymbol{\beta}}^0$ by $\hat{\boldsymbol{\beta}}^1$, then improve on $\hat{\boldsymbol{\beta}}^1$, using (4.10.1) again with $\mathbf{W} = \mathbf{W}(\hat{\boldsymbol{\beta}}^1)$. In this way we get a sequence of vectors $\hat{\boldsymbol{\beta}}^r$, which will under suitable *smoothness conditions*, (which are concerned with the existence of derivatives of the $g_i(\boldsymbol{\beta})$), will converge to a point in 3-space. At the $(r + 1)$th iteration take $\mathbf{W}_r = \mathbf{W}(\hat{\boldsymbol{\beta}}^r)$ and

$$\hat{\boldsymbol{\beta}}^{r+1} = \hat{\boldsymbol{\beta}}_0^r + (\mathbf{W}_r'\mathbf{W}_r)^{-1}\mathbf{W}_r'[\mathbf{y} - \mathbf{g}(\hat{\boldsymbol{\beta}}^r)]. \tag{4.10.2}$$

We will refer to the procedure provided by (4.10.2) as the Newton NLLS method. If the starting point is too far from the minimum point the procedure can fail to converge. It is sometimes worthwhile to choose a collection of points $\mathbf{b}$ in the parameter space at which to evaluate $Q(\mathbf{b})$, then start at the point $\hat{\boldsymbol{\beta}}_0$ at which $Q(\mathbf{b})$ is minimum.

Example 4.10.1: Suppose $Y_i = g(x_i; \boldsymbol{\beta}) + \varepsilon_i$ for $g(\mathbf{x}; \boldsymbol{\beta}) = \beta_0 + \beta_1 x_i^{\beta_2}$ for $i = 1, \ldots, 12$. The following x_i and parameter values β_0, β_1, β_2 were chosen in order to generate Y_i values: $\beta_0 = 1$, $\beta_1 = 3$, and $\beta_2 = 0.5$. Using these parameters and x_i values given in Table 4.10.1, values of $g_i(\boldsymbol{\beta}) = g(x_i; \boldsymbol{\beta})$ were determined. Then a vector $\boldsymbol{\varepsilon} = (\varepsilon_1, \ldots, \varepsilon_{12})'$ was generated, with the ε_i's independent $N(0, 0.01)$ and $\mathbf{Y} = \mathbf{g}(\boldsymbol{\beta}) + \boldsymbol{\varepsilon}$ determined.

Table 4.10.1 A Fit Using Nonlinear Least Squares

i	x_i	$g_i(\boldsymbol{\beta})$	ε_i	Y_i	$\hat{Y}_i$	$e_i = Y_i - \hat{Y}_i$
1	0.05	1.671	−0.032	1.639	1.664	−0.025
2	0.05	1.671	0.010	1.681	1.664	0.017
3	0.10	1.949	0.093	2.042	1.941	0.101
4	0.10	1.949	−0.169	1.780	1.941	−0.161
5	0.20	2.342	0.061	2.402	2.336	0.066
6	0.20	2.342	0.022	2.364	2.336	0.028
7	0.40	2.897	0.200	3.097	2.900	−0.197
8	0.40	2.897	−0.128	2.769	2.900	0.131
9	0.80	3.683	−0.267	3.416	3.704	−0.288
10	0.80	3.683	0.120	3.803	3.704	0.099
11	1.20	4.286	0.222	4.508	4.325	0.183
12	1.20	4.286	−0.048	4.238	4.325	−0.087

The functions $g^0(\boldsymbol{\beta}) \equiv 1$, $g^1(\boldsymbol{\beta}) = x^{\beta_2}$, $g^2(\boldsymbol{\beta}) = \beta_1(\ln x)x^{\beta_2}$ were then determined. Following this, a rather arbitrary starting point $\boldsymbol{\beta}_0 = (0.5, 2, 0.8)'$ was chosen. Formula (4.10.1) was used iteratively. On the rth iteration the matrix $\mathbf{W}_r = \mathbf{W}(\boldsymbol{\beta}^r)$ and $\mathbf{g}(\boldsymbol{\beta}_0)$ had to be determined. The sequence converged rapidly: $\hat{\boldsymbol{\beta}}^1 = (1.228, 2.808, 0.440)'$, $\hat{\boldsymbol{\beta}}^2 = (1.044, 2.986, 0.520)'$, $\hat{\boldsymbol{\beta}}^3 = (1.016, 3.014, 0.513)'$, $\hat{\boldsymbol{\beta}}^4 = (1.015, 3.014, 0.513)$. The difference $\hat{\boldsymbol{\beta}}^4 - \hat{\boldsymbol{\beta}}^3$ had maximum absolute value less than 0.001, so the computer program written to perform these computations ordered the iterations to stop. Table 4.10.1 presents interesting statistics. We find ESS $= \|\mathbf{Y} - \hat{\mathbf{Y}}\|^2 = 0.2319$, $S^2 = \text{ESS}/(12 - 3) = 0.0258$.

Asymptotically, as $n \to \infty$, the statistical properties of $\hat{\boldsymbol{\beta}}$, $\hat{\mathbf{Y}}$, and S^2 are the same as they would be if the model were truly linear, with design matrix $\mathbf{W} = \mathbf{W}(\boldsymbol{\beta})$, which must be estimated by $\hat{\mathbf{W}} = \mathbf{W}(\hat{\boldsymbol{\beta}})$. Thus, for example, for large n, in approximation $\hat{\boldsymbol{\beta}} \sim N(\boldsymbol{\beta}, \sigma^2(\mathbf{W}'\mathbf{W})^{-1})$, even without the normality of $\boldsymbol{\varepsilon}$, (if $g(\mathbf{x}; \boldsymbol{\beta})$ is reasonably smooth, and the x_i are not spread out too much.) We are obviously being somewhat vague here. The student eager for more rigor is referred to *Nonlinear Regression* by Seber and Wild (1989).

Continuation of Example 4.10.1: For these observations with $S^2 = 0.0258$, we estimate the covariance matrix for $\hat{\boldsymbol{\beta}}$ by

$$S^2[\mathbf{W}(\hat{\boldsymbol{\beta}})'\mathbf{W}(\hat{\boldsymbol{\beta}})]^{-1} = \begin{bmatrix} 0.159 & -0.156 & 0.048 \\ -0.156 & 0.160 & -0.046 \\ 0.048 & -0.046 & 0.016 \end{bmatrix}.$$

Since we know $\boldsymbol{\beta} = (1, 3, 0.5)$ and σ^2, we can compute the better approximation

$$\sigma^2[\mathbf{W}(\boldsymbol{\beta})'\mathbf{W}(\boldsymbol{\beta})]^{-1} = \begin{bmatrix} 0.152 & -0.149 & 0.045 \\ -0.149 & 0.152 & -0.043 \\ 0.045 & -0.043 & 0.014 \end{bmatrix}.$$

Even this is an approximation, since it pretends that $\mathbf{g}(\boldsymbol{\beta})$ is linear near the true parameter value. We simulated the experiment 500 times, each time computing $\hat{\boldsymbol{\beta}}$. The sample mean of the 500 values of $\hat{\boldsymbol{\beta}}$ was (0.946, 3.049, 0.497), suggesting that $\hat{\boldsymbol{\beta}}$ is almost unbiased. The mean value for S^2 was 0.0089, so S^2 has a small negative bias. The sample covariance matrix was $\begin{bmatrix} 0.085 & -0.083 & 0.022 \\ 0.083 & 0.084 & -0.021 \\ 0.022 & -0.021 & 0.006 \end{bmatrix}$, indicating that the estimates of the variances of $\hat{\beta}_0$, $\hat{\beta}_1$, and $\hat{\beta}_2$ were somewhat larger than their true values. Since the sample size was relatively small, we did reasonably well. Histograms of the coefficients based on the 500 simulations indicate that the distribution of $\hat{\beta}_0$ is somewhat skewed to the left (long tail on the left, with a few negative values), $\hat{\beta}_1$ is skewed to the right, but that $\hat{\beta}_2$ has a roughly normal distribution.

We have emphasized the example with three β_j's, and one independent variable x. Of course, three may be replaced by any number $k < n$ of independent variables, though $\mathbf{g}(\boldsymbol{\beta})$ must be a 1–1 function on the domain on which $\boldsymbol{\beta}$ takes its values, and unless k/n is small the statistical distributional approximations may be poor. In the following example, a multiplicative model, we compare the solutions given by nonlinear least squares, and those given by a linearizing transformation.

Example 4.10.2: Let $Y_i = g(\tilde{\mathbf{x}}_i; \boldsymbol{\beta}) + \varepsilon_i$, where $\tilde{\mathbf{x}}_i = (x_{i1}, x_{i2})$, and $g(\tilde{\mathbf{x}}; \boldsymbol{\beta}) = \beta_0 x_1^{\beta_1} x_2^{\beta_2}$ for $\tilde{\mathbf{x}} = (x_1, x_2)$. Observations Y_i were taken for 18 values $\tilde{\mathbf{x}}_i$ as given in Table 4.10.2. These were taken for $\boldsymbol{\beta} = (2.7, 0.5, 0.9)'$, and $\varepsilon_i \sim N(0, \sigma^2)$ with $\sigma = 0.5$. Define $g_i(\boldsymbol{\beta}) = g(\tilde{\mathbf{x}}_i, \boldsymbol{\beta})$, and, as before, let $\mathbf{g}(\boldsymbol{\beta})$ be the corresponding 18-component column vector. Let $\mathbf{g}^j(\boldsymbol{\beta})$ be the vector of partial derivatives with respect to β_j for $j = 0$, 1, 2. For example, the ith component of $\mathbf{g}^1(\boldsymbol{\beta})$ is $\beta_0(\ln x_{i1})x_{i1}^{\beta_1}x_{i2}^{\beta_2} = (\ln x_{i1})g_i(\beta)$. Then $\mathbf{W}(\boldsymbol{\beta})$ is the 18×3 matrix with jth column $\mathbf{g}^j(\boldsymbol{\beta})$.

The Newton method was used to minimize $Q(\boldsymbol{\beta}) = \sum_i [Y_i - g_i(\boldsymbol{\beta})]^2$, resulting in the estimate $\hat{\boldsymbol{\beta}} = (2.541, 0.551, 0.799)$, and $\hat{\mathbf{Y}} = \mathbf{g}(\hat{\boldsymbol{\beta}})$ as given in the table. The

Table 4.10.2 Comparison of the Fits Provided by Nonlinear Least Squares and by a Linearizing Transformation

i	x_{i1}	x_{i2}	$g_i(\boldsymbol{\beta})$	ε_i	Y_i	$\hat{Y}_i$	$Y_i - \hat{Y}_i$	Y_i^*	$Y_i - Y_i^*$
1	0.5	0.2	0.449	−0.087	0.361	0.480	−0.119	0.400	−0.039
2	0.5	0.6	1.206	0.256	1.461	1.153	0.308	0.871	0.590
3	0.5	1.0	1.909	−0.622	1.287	1.734	−0.447	1.896	−0.609
4	1.0	0.2	0.634	−0.525	0.109	0.703	−0.594	0.515	−0.406
5	1.0	0.6	1.705	0.193	1.898	1.690	0.209	1.121	0.778
6	1.0	1.0	2.700	0.144	2.844	2.541	0.303	2.439	0.405
7	1.5	0.2	0.777	0.279	1.056	0.879	0.178	0.663	0.394
8	1.5	0.6	2.088	0.015	2.103	2.113	−0.009	1.442	0.662
9	1.5	1.0	3.307	0.003	3.309	3.177	0.132	3.138	0.172
10	2.0	0.2	0.897	0.346	1.243	1.030	0.214	0.852	0.391
11	2.0	0.6	2.411	−0.507	1.904	2.476	0.571	1.855	0.050
12	2.0	1.0	3.818	0.029	3.847	3.723	0.124	4.036	−0.189
13	2.5	0.2	1.003	0.086	1.089	1.164	−0.075	1.096	−0.007
14	2.5	0.6	2.696	0.217	2.913	2.800	0.113	2.386	0.527
15	2.5	1.0	4.269	−0.208	4.061	4.210	−0.149	5.192	−1.131
16	3.0	0.2	1.099	0.250	1.349	1.287	0.062	1.410	−0.061
17	3.0	0.6	2.953	0.359	3.312	3.095	0.217	3.069	0.243
18	3.0	1.0	4.677	−0.144	4.533	4.655	−0.122	6.679	−2.146

function g can be linearized. Taking logs, we get the approximation $Z_i \equiv \ln Y_i \doteq \ln \beta_0 + \beta_1 \ln x_{i1} + \beta_2 \ln x_{i2}$. Least squares, minimizing $\sum_i [Z_i - \ln g_i(\boldsymbol{\beta})]^2$, was used to obtain the estimate $\boldsymbol{\beta}^* = (2.47, 0.711, 0.988)$. The error sums of squares for the two estimates were $Q(\hat{\boldsymbol{\beta}}) = 1.339$ and $Q(\boldsymbol{\beta}^*) = 8.695$. The estimate of σ^2 was $S^2 = 1.339/15 = 0.0893$, substantially below $\sigma^2 = 0.25$. The approximation of $\mathbf{Y}$ by $\hat{\mathbf{Y}}$ is certainly better than that provided by $\hat{\mathbf{Y}}^* = \mathbf{g}(\boldsymbol{\beta}^*)$, as must be the case. We estimate the covariance matrix of $\hat{\boldsymbol{\beta}}$ to be

$$10^{-3}\begin{bmatrix} 20.51 & -7.51 & 3.07 \\ -7.51 & 4.17 & 0.00 \\ 3.07 & 0.00 & 5.27 \end{bmatrix}.$$

The entire experiment was simulated 400 times, providing the estimate

$$10^{-3}\begin{bmatrix} 53.00 & -0.02 & 4.90 \\ -19.82 & 11.46 & 1.31 \\ 4.90 & 1.31 & 16.56 \end{bmatrix}$$

of the covariance matrix, with (mean $\hat{\boldsymbol{\beta}}$) = (2.713, 0.491, 0.899 4), and (mean S^2) = 0.253. Histograms indicated that $\hat{\boldsymbol{\beta}}$ has a distribution which is close to normal (see Table 4.10.2).

Problem 4.10.1: Use least squares to determine $\boldsymbol{\beta}$ so that $g(x; \beta) = e^{\beta x}$ approximates $\mathbf{y}$ for the three (x, y) pairs (1, 2.071), (2, 4.309), (3, 8.955).

Problem 4.10.2: Repeat 4.10.1 for $g(x; \boldsymbol{\beta}) = \beta_1 e^{\beta_2 x}$, for the three (x, y) pairs (1, 2.713), (2, 3.025), (3, 11.731).

Problem 4.10.3: Let $g(\mathbf{x}, \beta) = \mathbf{x}/\beta$, and suppose (x, y) takes the values (1, 2), (2, 4), (3, 6), (4, 8).

(a) Show that if $\hat{\beta}^r = b$, then $\hat{\beta}^{r+1} = b - b^2(2 - 1/b)$.

(b) Use nonlinear least squares beginning with $\hat{\beta}^0 = 0.8$, and iterate until you get tired. What happens if the starting point is $\hat{\beta}^0 = 1$?

(c) What is the estimate of β if we use the linearizing methods of Section 4.1?

Problem 4.10.4: Make one iteration of the Newton NLLS procedure for the starting point $\hat{\boldsymbol{\beta}}^0 = (0.6, 3, 2)$, to try to improve on the least squares fit for the regression function $g(\tilde{\mathbf{x}}; \boldsymbol{\beta}) = g(x_1, x_2; \beta_1, \beta_2, \beta_3) = \beta_0 x_1^{\beta_1} x_2^{\beta_2}$, for (x_1, x_2, y) triples (1, 1, 1.382), (1, 2, 0.192), (2, 1, 27.567), (2, 2, 3.633). Let $\hat{\boldsymbol{\beta}}^1$ be the "improved value." Evaluate $Q(\hat{\boldsymbol{\beta}}_0)$ and $Q(\hat{\boldsymbol{\beta}}^1)$ to see whether you have an improvement. If you have a computer or enough patience also find the least squares estimate $\hat{\boldsymbol{\beta}}$.

Problem 4.10.5: If $Y_i \sim N(g(\tilde{\mathbf{x}}_i; \boldsymbol{\beta}), \sigma^2)$, for $i = 1, \ldots, n$ are independent r.v.'s, what is the maximum likelihood estimator of $\boldsymbol{\beta}$?

Problem 4.10.6: Suppose that $\mathbf{Y} = \mathbf{X}\boldsymbol{\beta} + \boldsymbol{\varepsilon}$, and that $\boldsymbol{\varepsilon} \sim N(0, \sigma^2)$. If the Newton method is used with a starting value $\hat{\boldsymbol{\beta}}^0$, what are $\hat{\boldsymbol{\beta}}^1, \hat{\boldsymbol{\beta}}^2, \ldots$?

4.11 ROBUST REGRESSION

Consider the scatter diagram of Figure 4.18, with straight lines fit to the data using three different methods: least squares, the M-method of Huber, and least median squares. The aim under each method is to minimize the *distance* between the vector $\mathbf{y}$ and the vector of predicted distances $\mathbf{y} = \hat{\beta}_0\mathbf{J} + \hat{\beta}_1\mathbf{x}$. Under the least squares method, the squared distance is $\|\mathbf{y} - \hat{\mathbf{y}}\|^2$. A problem with this measure of distance is that it puts particularly heavy weight on larger deviations. If a relatively few of the error terms (the ε_i's) are exceptionally large, these may have a heavy influence on the estimates of (β_0, β_1), particularly when the leverage $[1/n + (x_i - \bar{x})^2/\sum (x_i - \bar{x})^2]$ of an observation at $x = x_i$ is large, that is, when x_i is far from $\bar{x}$. This sensitivity of least squares to one or a few observations led Box and Andersen (1955) to use the word "robust" in connection with a study of the effects of departures from the usual assumptions of a model. A robust statistical procedure has come to mean that the procedure continues to have desirable properties when the assumptions of the model are not satisfied. For example, a robust procedure would work well in the case that ε_i has the contaminated normal distribution, being an observation from a $N(\mu, \sigma^2)$ distribution with a large probability p, but being an observation from $N(\mu, \sigma^2 K)$, for large K, with probability $1 - p$. Tukey's paper (1962) on data

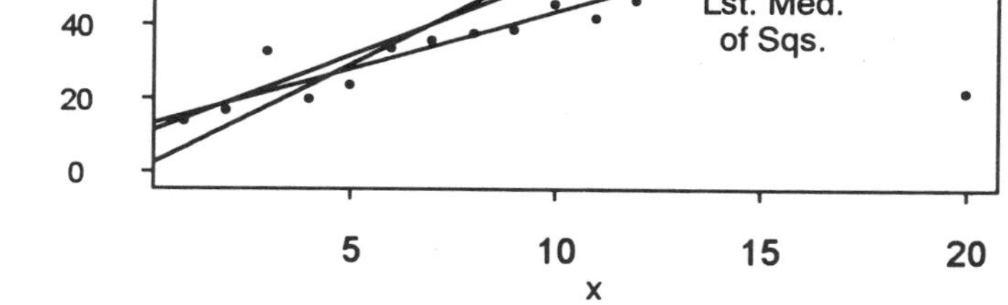

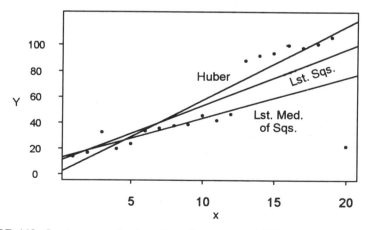

FIGURE 4.18 Least squares, least median of squares, and Huber estimates of a regression function.

analysis, calling for more realistic statistical methodology, less dependent on assumptions, was very important in stimulating the effort of the last 30 years or so on robust methods. Huber's book of 1981 listed 116 papers and books on the subject, and the pace has increased each year, particularly with the rapid increase in computational power. For a thorough discussion of the use of robust methods in regression, both for the M-method discussed here and for the R-method (R for rank) see the monograph by H. Koul (1992).

Huber (1964) suggested the M-estimator for the location problem. In the location problem we observe a random sample $Y_1, \dots, Y_n$ from a distribution $F(y - \theta)$ and wish to estimate θ. The letter M was chosen to remind us of the mean, the median, and the maximum likelihood estimator. Let $\rho(u)$ be a continuous convex function on the real line converging to $+\infty$ as $u \to -\infty$ or $u \to +\infty$. Informally convexity means that a straight line connecting two points of the graph of ρ lies above or on the graph. Simple choices are $\rho_1(u) = u^2/2$, $\rho_2(u) = |u|$, and

$$\rho_3(u) = \begin{cases} u^2/2 & \text{for } |u| \le k \\ |u|k - k^2/2 & \text{for } |u| > k \end{cases}, \qquad \text{for some} \quad k > 0.$$

The M-estimator of θ is the value $\hat{\theta}$ of t which minimizes

$$Q(t) = \sum \rho(Y_i - t).$$

The M-estimators corresponding to ρ_1 and ρ_2 are $\hat{\theta}_1 = \bar{Y}$ and $\hat{\theta}_2 = \text{med}(Y_1, \dots, Y_n)$. The estimator $\hat{\theta}_3$ corresponding to ρ_3, usually called the *Huber estimator*, is more difficult to compute, but may be thought of as a compromise between the mean and the median. Suppose that ρ has the derivative ψ, so that $Q'(t) = \frac{\delta}{\delta t} Q(t) = \sum \psi(Y_i - t)$. For ρ_1, ρ_2, ρ_3 the corresponding ψ are $\psi_1(u) = u$,

$$\psi_2(u) = \begin{cases} -1 & \text{for } u < 0 \\ +1 & \text{for } u > 0 \end{cases},$$

and

$$\psi_3(u) = \begin{cases} -k & u \le -k \\ u & \text{for } -k < u \le k\,. \\ k & u > k \end{cases}$$

Let U have distribution F, and define $\lambda(t) = E[\psi(U - t)]$. Let $\lambda(c) = \min_t |\lambda(t)|$. If $\rho(-u) = \rho(u)$ for all u, as it is for the examples above, and f is symmetric about zero, then $c = 0$. In his 1964 paper Huber proved (in slightly different notation, Lemma 4) that if (1) $\lambda(c) = 0$, (2) λ has a derivative at c and $\lambda'(c) < 0$, (3) $E[\psi^2(U - t)]$ is finite and continuous at c, then $n^{1/2}[\hat{\theta}_n - \theta - c]$ is asymptotically normal with variance $V(\psi, F) = E[\psi^2(U - c)]/[\lambda'(c)]^2$. In the case that ψ is sufficiently smooth, $\lambda'(c) = -E(\psi'(U - c))$. Thus, $V(\psi_1, F) = \text{Var}(U)/1$, $c = 0$, and $E[\psi_3^2(U - c)] = 1$, $\lambda(c) = 1 - 2F(c)$, and $\lambda'(c) = 2f(c)$, so that $V(\psi_3, F) = 1/[4f^2(c)]$. If the distribution of the Y_i's has median θ, so that F has median 0, then this is the usual formula for the asymptotic variance of the median.

Consider the contamination model $F = (1 - \varepsilon)G + \varepsilon H$, where $\varepsilon > 0$ is small, and G and H are c.d.f.'s. G is considered to be fixed but H is allowed to vary over a collection of c.d.f.'s. Huber showed (Theorem 1) that there exist ψ_0, F_0 such that $\sup_H V(\psi_0, F) = V(\psi_0, F_0) = \inf_\psi V(\psi, F_0)$. The first supremum is taken over all c.d.f.'s H for which $E_H[\psi_0(U)] = 0$. The density f_0 corresponding to F_0 is given explicitly in terms of G, and $\psi_0 = -f'_0/f_0$, the choice of ψ corresponding to the maximum likelihood estimator, is a Huber estimator for an explicit choice of k depending on ε and g. This result suggests that if we want an estimator which will perform well when F is G with no more than ε contamination, we should use a Huber estimator. At least asymptotically this will minimize the worst we can do (in a certain sense). As Huber points out, his Lemma 4 is somewhat unsatisfactory in that the supremum over H depends on ψ_0, though this can be avoided by assuming that both G and H define symmetric distributions about 0. See the paper for details.

Figure 4.19 presents $Q'(t)$ for the choices ψ_1, ψ_2, ψ_3 for the sample 1, 2, 3, 12, 17. The parameter k for the Huber estimator is $k = 6$. The median and Huber estimates would not change if 12 and 17 were made arbitrarily larger.

Now consider simple linear regression. Define

$$e_i = e_i(b_0, b_1) = y_i - (b_0 + b_1 x_i) \qquad \text{and} \qquad Q(b_0, b_1) = \sum \rho(e_i(b_0, b_1)).$$

Let

$$Q^0(b_0, b_1) = \frac{\partial}{\partial b_0} Q(b_0, b_1) = \sum \psi(e_i)$$

and

$$Q^1(b_0, b_1) = \frac{\partial}{\partial b_1} Q(b_0, b_1) = \sum \psi(e_i) x_i.$$

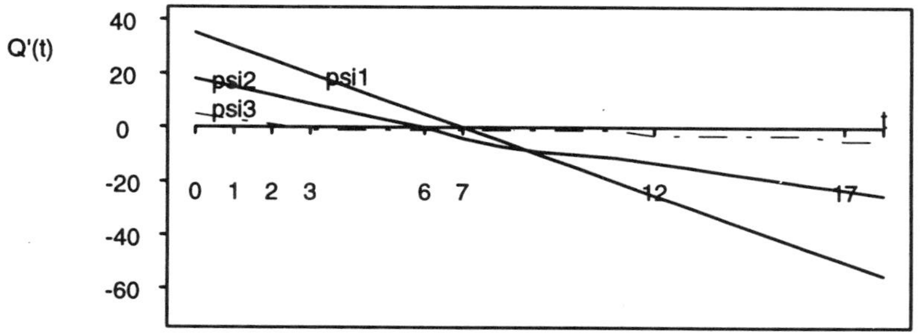

FIGURE 4.19 $Q'(t)$ for ψ_1, ψ_2, ψ_3 for the sample 1, 2, 3, 12, 17.

We seek the pair (b_0, b_1) for which Q^0 and Q^1 are simultaneously 0. The estimates corresponding to ψ_1, ψ_2, and ψ_3 are called, respectively, the least squares, least median of squares, and Huber estimates of (β_0, β_1). The Huber parameter k is usually chosen to be a multiple c of a consistent robust estimator of a scale parameter. For example, for some choice (b_0, b_1), which is considered to be close to (β_0, β_1) with high probability, k might be chosen to be c $\text{med}(|e_i(b_0, b_1)|)$. By defining $w_i = \psi(e_i)/e_i$, the equations $Q^0 = 0$, $Q^1 = 0$, become $\sum w_i e_i = 0$, and $\sum w_i x_i e_i = 0$, the weighted least squares equations. Since the e_i depend on (b_0, b_1) the weights e_i and weights w_i must be recomputed on each of the iterations used to find the solution. For the function rreg in the software package S-Plus c is 1.345. In the S-Plus language the weight function is $\psi(u)/u$. For computational details see Heiberger and Becker (1992).

In Figure 4.18 the straight line estimates provided by these three methods differ considerably, because the observations corresponding to $x = 13, \ldots, 19$ are above the straight line extrapolation suggested by the observations for smaller x, and because the observation (20, 58), which appears to be an *outlier*. The least median of squares (LMS) estimate ignores these last eight observations almost completely. The Huber estimate puts no more weight on this outlier than it would if it were considerably closer to the Huber line. Had all points except the outlier been close to the LMS line, the Huber line would almost coincide with it, and the least squares line would have a smaller slope, since it must account for the outlier.

Suppose now that $Y_i = 10 + 2x_i + \varepsilon_i$, and that two observations Y_i are taken for each integer x_i, $1 \leq x_i \leq 20$. Explicitly, $x_1 = 1$, $x_2 = 1$, $x_3 = 2$, $x_{39} = 20, \ldots, x_{40} = 20$. Suppose also that $\varepsilon_i \sim F$, where $F = (1 - \varepsilon)N(0, 3^2) + \varepsilon N(0, ((5)(3))^2)$, $\varepsilon = 0.1$. That is, the ε_i have the distribution of $(3Z)[(1 - \xi) + 5\xi]$, where Z is has a standard normal distribution, and ξ is 0 or 1 with probabilities 0.9, 0.1. The ε_i's have a contaminated normal distribution with contamination probability 0.1, contamination distribution $N(0, 225)$.

This experiment was performed 200 times, with the coefficient estimates of (β_0, β_1) produced for each of the Huber and LS methods. Figure 4.20 presents the pairs of estimates for both the Huber and LS methods. Notice that there is considerably less variation in the Huber estimates. The corresponding sample covariance matrix, with order $(\hat{\beta}_0, \hat{\beta}_1)$ for least squares, then $(\hat{\beta}_0, \hat{\beta}_1)$ for the Huber estimates) among the four estimates (based this time on 400 experiments) was

$$\begin{bmatrix} 3.431 & -0.249 & 1.865 & -0.138 \\ -0.249 & 0.023 & -0.135 & 0.013 \\ 1.865 & -0.135 & 1.534 & -0.112 \\ -0.138 & 0.013 & -0.112 & 0.010 \end{bmatrix}.$$

The sample means were 10.002, 2.001, 10.049, 1.995, so that all estimators appear

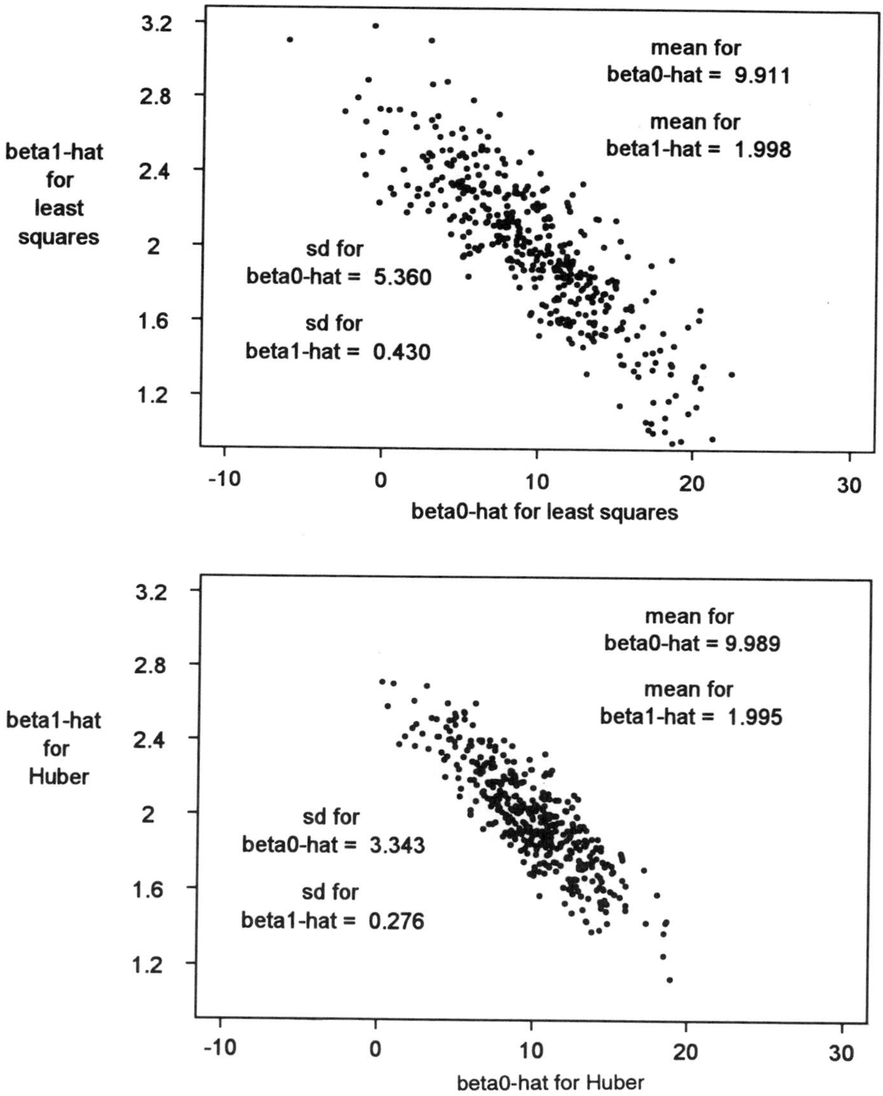

FIGURE 4.20 Least squares estimates and Huber estimates.

to be unbiased. Histograms (Figure 4.21) of $\hat{\beta}_1$ for 400 simulations indicate that the least squares and Huber estimators are normally distributed.

The Gauss–Markov Theorem states that the LS estimators are best among *linear* unbiased estimators (in having smallest variance). The Huber estimator (and other M-estimators, other than the least squares estimator) are not linear,

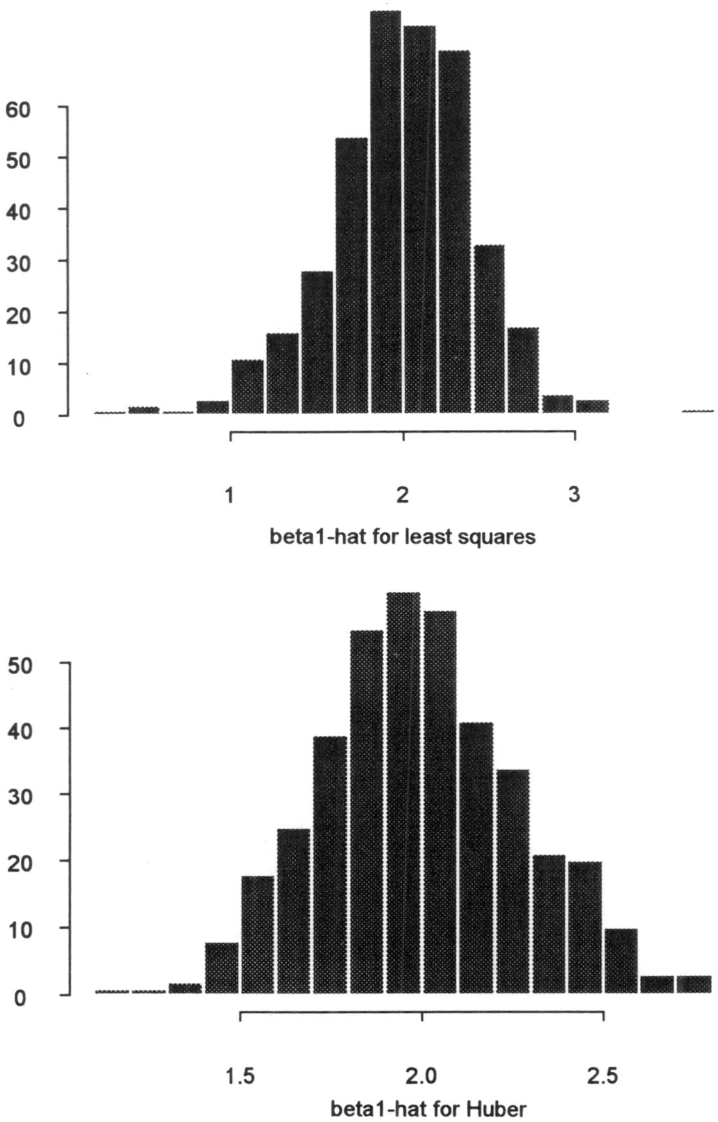

FIGURE 4.21 Histograms of least squares estimates of slope and Huber estimates of slope.

so that they are not eligible for the Gauss–Markov competition. The contaminated normal distribution is not itself a normal distribution, so we cannot call on normal theory to argue that the LS estimator is best. Perhaps we should therefore not be surprised to learn from this simulation, or from the theory described in Huber (1981) or Hampel *et al.* (1986) that we can do better in the presence of contamination than the LS estimator. The problem, of course, for the applied statistician is recognizing whether contamination is present. It is tempting to view the data before deciding whether to use a robust, rather than a LS estimator, and good statisticians will do that, though it is difficult to judge the properties of procedures which use informal judgments.

Let us return now to the more general linear model $\mathbf{Y} = \mathbf{X}\boldsymbol{\beta} + \boldsymbol{\varepsilon}$, where the components of $\boldsymbol{\varepsilon}$ are independent with c.d.f. F. To avoid confusion with the k of Huber's estimator, suppose that $\mathbf{X}$ is $n \times p$, and that $\boldsymbol{\beta}$ has p components. Define $Q(\mathbf{b}) = \sum \rho(Y_i - \tilde{\mathbf{x}}_i\mathbf{b})$, where $\tilde{\mathbf{x}}_i$ is the ith row of $\mathbf{X}$. The M-estimator of $\boldsymbol{\beta}$ in the minimizer $\hat{\boldsymbol{\beta}}$ of $Q(\mathbf{b})$, equivalently, the $\mathbf{b}$ satisfying $\sum \psi(Y_i - \tilde{\mathbf{x}}_i\mathbf{b})x_{ij} = 0$, for $j = 1, \ldots, p$. Let $\mathbf{e} = \mathbf{e}(\mathbf{b}) = \mathbf{Y} - \mathbf{X}\mathbf{b}$. Then this normal equation can be written in the form $\psi(\mathbf{e}) \perp \mathbf{x}_j$, for each j, or $\psi(\mathbf{e}) \perp \mathscr{L}(\mathbf{x}_1, \ldots, \mathbf{x}_p) = V$. By $\psi(\mathbf{e})$ we mean the componentwise application of ψ to the components of $\mathbf{e}$. In order to consider the asymptotic properties of an M-estimator in this regression setting, we must consider, as in Section 4.8, a sequence of matrices $\mathbf{X}_n$. Let $\mathbf{M}_n = \mathbf{X}_n'\mathbf{X}_n$, and let $\mathbf{H}_n = \mathbf{X}_n\mathbf{M}_n^{-1}\mathbf{X}_n'$, projection onto the column space of $\mathbf{X}_n$. Let h_n be the maximum of the diagonal elements of $\mathbf{H}_n$. That is, h_n is the maximum of the *leverages* of all the rows of $\mathbf{X}_n$. For simple linear regression this is $\max_i[1/n + (x_i - \bar{x})^2/S_{xx}] = 1/n + \max(x_i - \bar{x})^2/S_{xx}$. Consider any parameter $\eta = (\mathbf{c}, \boldsymbol{\beta}) = c_1\beta_1 + \cdots + c_p\beta_p$, and the estimator $\hat{\eta}_n = (\mathbf{c}, \hat{\boldsymbol{\beta}}_n)$, where $\hat{\boldsymbol{\beta}}_n$ is the M-estimator of $\boldsymbol{\beta}$. Huber proved (1981, Section 7.4) that $Z_n = (\hat{\eta}_n - \eta)/\sigma_n$ is asymptotically standard normal, under suitable conditions on ψ and F, when $h_n \to 0$ as $n \to \infty$. Here $\sigma_n^2 = [\mathbf{c}'\mathbf{M}_n^{-1}\mathbf{c}]V(\psi, F)$, where $V(\psi, F) = E[\psi(U)^2]/[E(\psi')]^2$ as defined above (not the subspace). $V(\psi, F)$ is minimum if $\psi = f'/f$, producing the maximum likelihood estimator. In practice we will not know F, but can estimate V consistently by $\hat{V}_n = [(1/(n-p))\sum \psi(e_i)^2]/[(1/n)\sum \psi'(e_i)]^2$, as suggested by Huber. If fact, Huber suggests that $\hat{V}_n$ be multiplied by $K^2 = 1 + (p/n)\dfrac{\text{Var}(\psi'(U))}{E(\psi'(U))^2}$. We will ignore this factor, which is close to one if $p \ll n$. In the case $\psi(u) = u$, corresponding to least squares, $\hat{V}_n$ becomes the usual error mean square S^2.

For the contamination example above numerical integration was used to find $E[\psi(U)^2] = 10.30$, and $E[\psi'(U)] = P(-6 < U < 6) = 0.890\,1$. Simulations gave approximately the same values. Thus, $V(\psi, F) \doteq 12.998$, so that the asymptotic variance of the Huber estimator (for $k = 6$) is $12.998/S_{xx} = 12.998/133\,0 = 0.009\,77$, which compares well with 0.010 given in the sample covariance matrix. We were a bit lucky because the value of k used in the S-Plus procedure is itself estimated from the data. The variance of the least squares estimator is $\sigma^2/S_{xx} = [0.9(9) + 0.1(225)]/S_{xx} = 30.6/1{,}330 = 0.023\,0$, which compares well with 0.023 3 obtained in the simulation.

Other simulations indicate that the variances of the Huber estimators remains about the same across a wide ranges of choices of the parameter k. As k becomes larger the Huber estimator becomes more like the least squares estimator. In the case that F is a normal distribution, least squares, corresponding to $k = \infty$, is best. However, for more commonly chosen values of k, variances for Huber estimators increase relatively little over those for LS.

We estimate σ_n^2 by $\hat{\sigma}_n^2 = [\mathbf{c}'\mathbf{M}_n^{-1}\mathbf{c}]\hat{V}_n$. Corresponding $100(1-\alpha)\%$ confidence intervals on η are given by $\hat{\eta}_n \pm z_{1-\alpha}\hat{\sigma}_n$. For large n this is the same formula as given in Section 3.2, with the extra multiplier $\hat{V}_n^{1/2}$, so the asymptotic relative length of confidence intervals produced by the M-estimator to the length of those produced by the method of LS is $V_n^{1/2}$.

Huber showed that asymptotic normality holds even in the case that $p = p_n$ is allowed to increase with n so that $h_n p_n^2 \to 0$. Yohai and Maronna (1979) showed that the power 2 on p_n may be replaced by 3/2.

Huber (1981, Section 7.10) suggests the following procedure for testing $H_0: \boldsymbol{\theta} = \sum \beta_j \mathbf{x}_j \in V_0$, where V_0 is a p_0-dimensional subspace of $V = \mathscr{L}(\mathbf{x}_1, \ldots, \mathbf{x}_p)$. Find the M-estimate $\hat{\mathbf{Y}} = \mathbf{X}\hat{\boldsymbol{\beta}}$ of $\boldsymbol{\theta}$. Find the M-estimate of $\boldsymbol{\theta}$, based on the model $\boldsymbol{\theta} \in V_0$, using $\hat{\mathbf{Y}}$, rather than $\mathbf{Y}$ (the answers will be different). Call this $\hat{\mathbf{Y}}_0$. Define

$$W = \|\hat{\mathbf{Y}} - \hat{\mathbf{Y}}_0\| / \{[K^2\sigma^2(n-p)^{-1} \sum \psi(e_i/\sigma)^2] / [(1/n) \sum \psi'(e_i/\sigma)]^2\}.$$

σ must be replaced by a consistent estimate of a scale parameter based on $\mathbf{e} = \mathbf{Y} - \hat{\mathbf{Y}}$. Under H_0, W is asymptotically distributed as chi-square with $p - p_0$ degrees of freedom.

Problem 4.11.1: (a) For the location parameter problem and the sample 3, 5, 8, 16, 30, plot the function $Q'(t)$ corresponding to the Huber estimator with $k = 5$. To do this first find $Q''(t)$, which is constant on intervals. Use your plot to determine the Huber estimate. Repeat for $k = 10$ and $k = 15$.

(b) For the same sample determine the M-estimate corresponding to

$$\rho(u) = \begin{cases} |u| & \text{for } |u| \le 6 \\ 6 & \text{for } |u| > 6 \end{cases}$$

What is the estimate if 6 is replaced by 1 million?

Problem 4.11.2: Is the Huber estimator with $k = 4$ as applied in simple linear regression the same as least squares estimate of β_1 for the pairs of observations (1, 3), (2, 6), (3, 21)?

Problem 4.11.3: Let $\mathbf{X}$ be an $n \times p$ matrix of rank $p < n$. Let $\hat{\boldsymbol{\beta}}$ and $\hat{\boldsymbol{\beta}}_H$ be the least squares and Huber estimates of $\boldsymbol{\beta}$ in the linear model corresponding to $\mathbf{y}$. Let $\mathbf{e} = \mathbf{y} - \mathbf{X}\hat{\boldsymbol{\beta}} = (e_1, \ldots, e_n)'$. Prove that $\max_i(|e_i|) \le k$ implies that $\hat{\boldsymbol{\beta}}_H = \hat{\boldsymbol{\beta}}$.

Problem 4.11.4: (a) For $\rho = \rho_3$ and the uniform distribution on $[-1, 1]$ for errors, with density $f(x) = (1/2)$ for $-1 < x < 1$, show that $V(\psi, F) =$

$(1 - 2k/3)$ for $k \leq 1$ and $V(\psi, F) = 1/3$ for $k > 1$. What is the optimum choice for k? What is the resulting Huber estimator? Find its variance and compare it to $V(\psi, F)$.

Problem 4.11.5: Let $Y_i = \beta x_i + e_i$ for $i = 1, 2, 3$. Find the least squares and Huber ($k = 5$) estimates of $\boldsymbol{\beta}$ for the (x_i, Y_i) pairs (1, 1), (2, 2), (3, 30). For which values of y_3 does the sample of pairs (1, 1), (2, 2), $(3, y_3)$ produce the Huber estimate ($k = 5$) which is equal to the least squares estimate?

4.12 BOOTSTRAPPING IN REGRESSION

Bradley Efron (1979, 1982) introduced the "bootstrap estimate" as a means of estimating the distribution of a function $R(\mathbf{Y}, F)$ of data and an unknown distribution F, where the components of $\mathbf{Y}$ are a random sample from F. Suppose, for example, that $R = R(\mathbf{Y}, F) = T = (\bar{Y} - \mu)/[S_Y/\sqrt{n}]$, where μ is the mean for F. R is Student's "t-statistic" (not strictly a statistic, since it depends on the unknown μ). T is used as a pivotal quantity in order to determine confidence intervals on μ. This requires that its distribution be known. If F is a normal distribution then R has the t distribution with $n - 1$ d.f. However, if F is not normal then R does not, in general, have a t distribution. If n is large the t distribution may serve as an approximation, but certainly in some applications n may not be large.

For another example let $R(\mathbf{Y}, F) = T_k(\mathbf{Y}) - \theta$, where θ is the median of the symmetric distribution F, and T_k is the trimmed mean, the mean when the smallest k and largest k Y_i's are omitted. In this case R is the error made in using T_k to estimate θ. More explicitly, if $\varepsilon_i = Y_i - \theta$, then, since $T_k(\mathbf{Y}) = \theta + T_k(\boldsymbol{\varepsilon})$, $R(\mathbf{Y}, F) = T_k(\boldsymbol{\varepsilon})$. In still one more example discussed by Efron, $\mathbf{Y}$ was a vector of pairs of observations (U_i, V_i) from a bivariate distribution F, and $R(\mathbf{Y}, F) = r - \rho$, where r and ρ are the sample and population correlation coefficients.

Efron suggested that the conditional distribution of $R^* = R(\mathbf{Y}^*, F_n)$ given $\mathbf{Y}$ serve as an estimator of the distribution of R. Here $\mathbf{Y}^* = (Y_1^*, \ldots, Y_n^*)$ is a random sample (the bootstrap sample, taken with replacement) from the empirical (or sample) distribution function F_n determined by the components of $\mathbf{Y}$. That is, $F_n(y) = (1/n) \sum I[Y_i \leq y]$, the proportion of the Y_i's less than or equal to y. His intuitive argument was that at least for large n, R^*, conditionally on the sample $\mathbf{Y}$, should have a distribution close to that of R.

Consider $R(\mathbf{Y}, F) = \bar{Y} - \mu = \bar{\varepsilon}$, the error in using the sample mean $\bar{Y}$ to estimate the mean μ of the distribution F. Then $R^* = \bar{Y}^* - \bar{Y}$, where $\bar{Y}^*$ is the mean of a bootstrap sample of n from the "population" $\{Y_1, \ldots, Y_n\}$ with mean $\bar{Y}$. If we define $\varepsilon_i^* = Y_i^* - \bar{Y}$, then $R^* = \bar{\varepsilon}^*$. We know that the conditional distribution of R^*, given $\mathbf{Y}$, has mean 0 and variance $\hat{\sigma}_Y^2/n$, where $\hat{\sigma}_Y = (1/n) \sum (Y_i - \bar{Y})^2$. We also know that for large n, and reasonably behaved $\mathbf{Y}$, conditionally on $\mathbf{Y}$, R^* will be approximately normally distributed. We usually use $\hat{\sigma}_Y^2/n$ or $(\hat{\sigma}_Y^2/n)[n/(n-1)]$ as an estimator of the variance of R. We also know that R has an approximate normal distribution if n is large and F is reasonably well behaved. Can we find the exact conditional distribution of R^*? In theory we can, since $\mathbf{Y}$ is known, though the problem may be mathematically intractable. Efron's idea was to take bootstrap samples of n from F_n some large number B times to obtain $R_1^*, \ldots, R_B^*$, then to use these to estimate the distribution of R. The use of F_n, rather than some other estimator $F_{\hat{\lambda}}$, obtained by assuming that $F = F_\lambda$ belongs to some collection of distributions parameterized by λ, causes the bootstrap methods we will discuss to be called nonparametric.

Theory developed over the last 15 years has shown that in a number of circumstances, the "bootstrap" is a better approximation to the distribution of R than is the normal theory. In fact, there are cases in which the normal approximation may be quite poor, but the bootstrap approximation is good. The bootstrap has become practically possible because computing power has increased tremendously over these last 15 years.

In this section we will avoid presenting proofs or even precise statements of the limit theory which justifies the approximations provided by the bootstrap. We are particularly interested, of course, in applications to linear models. Refer to the book by Rousseeuw and Leroy (1987) and the monograph by Mammen (1992). Mammen's references include 79 papers with the name bootstrap in the title. For a less theoretical expository review of the bootstrap see Efron and Tibshirani (1986).

Let $\theta = \theta(F)$ be an unknown parameter, a function of F. Let $\hat{\theta} = \theta(F_n)$ the corresponding value of θ for the distribution F_n. Let $R(\mathbf{Y}, F) = \hat{\theta} - \theta$, the error made when θ is estimated by $\hat{\theta}$. The bootstrap estimate of the variance of R is $\hat{\sigma}_R^2 = \text{Var}(R^* | \mathbf{Y})$, the conditional variance of $R^* = R(\mathbf{Y}^*, F_n)$, given $\mathbf{Y}$. There are n^n possible samples of the components of $\mathbf{Y}$ of size n, all equally likely under "simple" bootstrap sampling. In theory, all we have to do is compute R^* for each such sample, then compute the variance of these n^n values. Though this is in the realm of the possible for $n = 10$ (10 billion samples), for larger n this soon becomes impossible. Instead the bootstrap method requires that we choose some large number B samples at random, determine $R_1^*, \ldots, R_B^*$, then estimate $\text{Var}(R)$ by $\hat{\sigma}_R^2 = (1/B) \sum (R_i^* - \bar{R}^*)^2$. In practice it is often enough to let $B = 200$, though it is usually relatively inexpensive to let $B = 1{,}000$, $10{,}000$, or even $100{,}000$.

Example 4.12.1: Let F be the contaminated normal distribution:

$$F(x) = 0.9\Phi(x) + 0.1\Phi(x/5).$$

Let $\theta(F)$ be the 0.2-trimmed mean. That is, for $F(x_{0.2}) = 0.2$ and $F(x_{0.8}) = 0.8$, $\theta(F) = \int_{x_{0.2}}^{x_{0.8}} xf(x)/0.6$, where f is the density corresponding to F. By symmetry

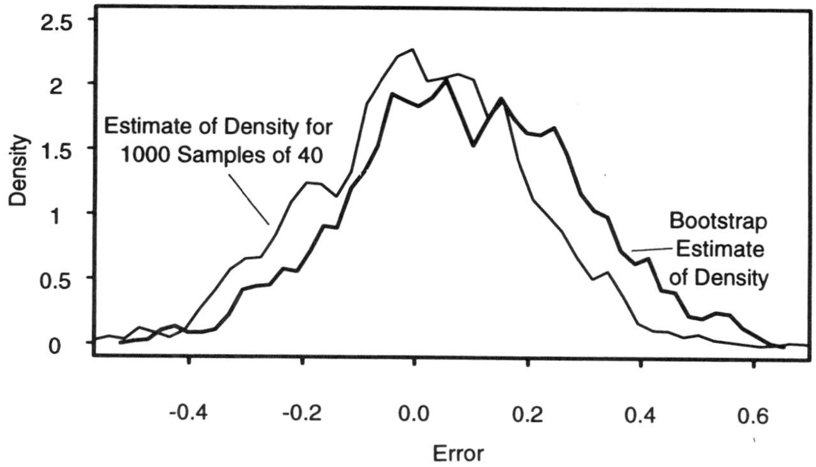

FIGURE 4.22 Density of error for trimmed mean and bootstrap estimate of the density, based on one sample of 40.

$\theta(F) = 0$ and $\hat{\theta} = \theta(F_n)$ is the sample 0.2-trimmed mean. We consider samples of size $n = 40$. To find $\text{Var}(\hat{\theta})$ 500 samples of 40 were chosen. The variance was found to be 0.0349. Then a single sample of 40 was taken and $B = 1{,}000$ bootstrap samples of 40 were taken. The sample variance of these 1,000R^*'s was $\hat{\sigma}_R^2 = 0.030$.

Figure 4.22 presents estimates of the density of R based on this simulation, and also by bootstrapping a single sample of 40 1,000 times.

We will briefly discuss two basic methods for the determination of bootstrap confidence intervals, the naive or percentile method and the t-method. For a full discussion of these and others from a theoretical point of view, see Hall (1988). For a more general review of the bootstrap method with examples, see Efron and Tibshirani (1986, 1993). Let $\theta = \theta(F)$ be an unknown parameter and let $\hat{\theta} = \theta(F_n)$. If, for example, θ is the mean of the distribution F, then $\hat{\theta}$ is the sample mean. If $\theta(F)$ is the median for F then $\hat{\theta}$ is the sample median. Let $R = R(\mathbf{Y}, F) = \theta(F_n) - \theta(F) = \hat{\theta} - \theta$, the error in estimating θ by $\hat{\theta}$. We would like to estimate the distribution $\mathscr{D}(R)$ of R. Let R_γ be the γth percentile of $\mathscr{D}(R)$. Then $1 - \alpha = P(R \geq R_\alpha) = P(\hat{\theta} - \theta \geq R_\alpha) = P(\hat{\theta} - R_\alpha \geq \theta)$, so that $\hat{\theta} - R_\alpha$ is an upper $100(1 - \alpha)\%$ confidence limit on θ. Similarly,

$$1 - \alpha = P(R \leq R_{1-\alpha}) = P(\hat{\theta} - \theta \leq R_{1-\alpha}) = P(\theta \geq \hat{\theta} - R_{1-\alpha}),$$

so that $\hat{\theta} - R_{1-\alpha}$ is a $100(1 - \alpha)\%$ lower confidence limit on θ. Thus, $[\hat{\theta} - R_{1-\alpha_1}, \hat{\theta} - R_{\alpha_2}]$ is a $100(1 - \alpha_1 - \alpha_2)\%$ confidence interval on θ.

The problem with this is that the distribution of R is unknown. For an observation **Y** let $R^* = R(\mathbf{Y}^*, F_n)$, where $\mathbf{Y}^*$ is a bootstrap sample from F_n, the empirical distribution of **Y**. We can estimate $\mathscr{D}(R)$ by $\mathscr{D}(R^*|\mathbf{Y})$, the conditional distribution of R^* given **Y**. We can do this, if we have enough computing power, by determining R^* for all n^n possible bootstrap samples. In general this is impossible so we instead choose some large number B (say 200 or 1,000 or 10,000) of bootstrap samples. For the ith bootstrap sample let $R_i^* = \hat{\theta}_i^* - \hat{\theta}$, the bootstrap estimate of the error. We can now estimate R_γ for any γ by the corresponding sample percentile of $R_1^*, \ldots, R_B^*$. That is, if k_γ is the nearest integer to γn, then we estimate R_γ by $\hat{R}_\gamma = R^*_{(k_\gamma)}$, the k_γth order statistic among the R_i^*. The interval $[\hat{\theta} - \hat{R}_{1-\alpha_1}, \hat{\theta} - \hat{R}_{\alpha_2}]$ is then an approximate $100(1 - \alpha_1 - \alpha_2)\%$ confidence interval on θ. This is a confidence interval obtained by a percentile method, though it does not quite correspond to the $100(1 - 2\alpha)\%$ percentile interval $[\hat{\theta} - \hat{R}_{1-\alpha_2}, \hat{\theta} - \hat{R}_{\alpha_1}]$ of Efron (1982).

In the case that F is symmetric about θ, $R_\alpha = 1 - R_{1-\alpha}$, so that a better confidence interval is given by $\hat{\theta} \pm \hat{A}_{1-\alpha}$, where $\hat{A}_{1-\alpha}$ is the $(1 - \alpha)$th sample quantile of the absolute values $|R_i^*|$. It is reasonable to believe that near symmetry will also cause the absolute percentile bootstrap method to be slightly better.

The bootstrap t-method for confidence intervals requires that some estimator $\hat{\sigma}$ of the standard deviation of $\hat{\theta}$ be available. This could be the bootstrap estimator. Define $T(\mathbf{Y}, F) = (\hat{\theta} - \theta)/\hat{\sigma}$. If we knew the distribution of T then we could determine quantiles $t_{1-\alpha_1}$ and t_{α_2}, so that $1 - \alpha_1 = P(T \le t_{1-\alpha_1}) = P(\hat{\theta} - \theta \le t_{1-\alpha_1}\hat{\sigma}) = P(\hat{\theta} - t_{1-\alpha_1}\hat{\sigma} \le \theta)$, and $1 - \alpha_2 = P(T \ge t_{\alpha_2}) = P(\theta \le t_{\alpha_2}\hat{\sigma})$. Then $[\hat{\theta} - t_{1-\alpha_1}\hat{\sigma}, \hat{\theta} - t_{\alpha_2}\hat{\sigma}]$ would be a $100[1 - \alpha_1 - \alpha_2]\%$ confidence interval on θ.

We can estimate the distribution of T by that of $T^* = (\hat{\theta}^* - \hat{\theta})/\hat{\sigma}^*$, conditionally on the sample **Y**. Here $\hat{\theta}^*$ and $\hat{\sigma}^*$ are the estimates of θ and the standard deviation of $\hat{\theta}$ based on a bootstrap sample $\mathbf{Y}^*$ from F_n. We can generate B values of T^*, then estimate $t_{1-\alpha_1}$ and t_{α_2} by the corresponding percentiles $\hat{t}_{1-\alpha_1}$ and $\hat{t}_{\alpha_2}$ of these B values of T^*. The bootstrap t-confidence interval is then $[\hat{\theta} - \hat{t}_{1-\alpha_1}\hat{\sigma}, \hat{\theta} - \hat{t}_{\alpha_2}\hat{\sigma}]$. If $\hat{\sigma}$ itself is a bootstrap estimator, this would require bootstrap of a bootstrap sample. If $B = 1{,}000$ for both stages of bootstrapping, this would require 1 million bootstrap samples, each of size n. Again in the case of symmetry or near symmetry of F about θ, it might be better to use the interval $\hat{\theta} \pm \hat{A}_{1-\alpha}\hat{\sigma}$, where $\hat{A}_{1-\alpha}$ is the $(1 - \alpha)$th sample quantile of the T_i^*'s.

Example 4.12.2: Consider the contaminated normal example of Example 4.12.1, for samples of size 20. To keep things simple consider the mean $\theta = \theta(F) = 0$. Then $\hat{\theta} = \bar{Y}$, and $\hat{\sigma}^2 = S^2/n$, where S^2 is the usual sample variance. In this case F certainly is not a normal distribution so that T does not have a t distribution. We computed six 90% confidence intervals for each sample of 20, by (1) the usual t-method with interval $\bar{Y} \pm 1.729S/\sqrt{20}$, (2) the bootstrap percentile method described above, (3) the bootstrap percentile method of Efron, (4) the bootstrap absolute percentile method, (5) the bootstrap t-method

Table 4.12.1 Coverage Percentages and Lengths for 90% Confidence Intervals Determined by t-Method and Bootstrap Methods

Method	1	2	3	4	5	6
Percentage	90.09	89.96	84.2	86.8	81.9	93.8
Mean Length	1.75	1.62	1.62	1.62	1.99	2.06

described above, and, finally, (6) the bootstrap absolute t-method. Table 4.12.1 above presents the coverage percentages, and the mean lengths of each of these intervals. The conclusion would seem to be that we are quite well off if we stick to the t-method, and forego bootstrapping. Another simulation, with $F = 0.3N(-1, 1) + 0.7N(3/7, 1)$, $n = 15$, a nonsymmetric distribution, produced essentially the same results. In these examples the variance for F exists, so that the t-statistic is asymptotically normal. Simulations for F with a mean but not a variance, showed that the bootstrap percentile method can defeat the t-method, and that the bootstrap t-method can also be bad. Comparisons of lengths of intervals based on their mean lengths can give false impressions for such distributions, however, since heavy tails for F tend occasionally to produce very long intervals.

Application of the Bootstrap Method to Regression

Suppose that the linear model $\mathbf{Y} = \sum_{1}^{k} \beta_j \mathbf{x}_j + \boldsymbol{\varepsilon}$ holds, where the components of $\boldsymbol{\varepsilon}$ are independent, identically distributed with c.d.f. F. Let the column space of $\mathbf{X}$ be V, of dimension k. By the Gauss–Markov Theorem the least squares estimator $\hat{\boldsymbol{\beta}} = \mathbf{X}^+\mathbf{Y} = \boldsymbol{\beta} + \mathbf{X}^+\boldsymbol{\varepsilon}$, for $\mathbf{M} = \mathbf{X}'\mathbf{X}$, $\mathbf{X}^+ = \mathbf{M}^{-1}\mathbf{X}'$ has minimum variance among all linear unbiased estimators of the components of $\boldsymbol{\beta}$. However, $\hat{\boldsymbol{\beta}}$ is not normally distributed unless F is a normal distribution. We know from Section 4.10 that $\hat{\boldsymbol{\beta}}$ is asymptotically normally distributed if k remains fixed and the maximum h_n of the diagonal elements of the projection matrix $\mathbf{P}_V = \mathbf{X}\mathbf{X}^+ = \mathbf{X}\mathbf{M}^{-1}\mathbf{X}'$ converges to zero as $n \to \infty$. However, we usually do not know the distribution of $\hat{\boldsymbol{\beta}}$ or of a linear combination $\mathbf{c}'\hat{\boldsymbol{\beta}}$ for small n. The normal approximation may not be good. From Section 4.11 on robust estimation we know that in some circumstances it is better to give up the relative simplicity of least squares in order to gain precision. The bootstrap method offers a way of avoiding the assumption of normality and the relative complexity of the analysis required by these robust methods.

Let $\mathbf{R}(\mathbf{Y}, \boldsymbol{\beta}) = \hat{\boldsymbol{\beta}} - \boldsymbol{\beta} = \mathbf{X}^+\boldsymbol{\varepsilon}$ and for fixed $\mathbf{c} = (c_1, \ldots, c_k)'$ let $R_c(\mathbf{Y}, \boldsymbol{\beta}) = \mathbf{c}'(\hat{\boldsymbol{\beta}} - \boldsymbol{\beta}) = \mathbf{c}'\mathbf{R}(\mathbf{Y}, \boldsymbol{\beta})$. We would like to know the distributions of $\mathbf{R}$ and R_c. Let $\mathbf{e} = \mathbf{Y} - \hat{\mathbf{Y}} = (\mathbf{I}_k - \mathbf{X}\mathbf{X}^+)\mathbf{Y} = (\mathbf{I}_k - \mathbf{X}\mathbf{X}^+)\boldsymbol{\varepsilon}$, the residual vector. Let $\mathbf{e}^*$ be an n-component vector obtained by randomly choosing, with replacement, from the components of $\mathbf{e}$ (not $\boldsymbol{\varepsilon}$). That is, conditionally on $\mathbf{Y}$, the components are independent and identically distributed with c.d.f. F_n, the empirical c.d.f. of $\mathbf{e}$. One bootstrap idea is to use the conditional distributions of $\mathbf{R}^* = \mathbf{X}^+\mathbf{e}^*$ and of $R_c^* = \mathbf{c}'\mathbf{X}^+\mathbf{e}^*$, given $\mathbf{Y}$ (and therefore $\mathbf{e}$), to approximate that of $\mathbf{R} = \mathbf{X}^+\boldsymbol{\varepsilon}$ and $R_c = \mathbf{c}'\mathbf{X}^+\boldsymbol{\varepsilon}$. It turns out (Bickel and Freedman 1983) that $k^2/n \to 0$ implies that the conditional distributions of $\mathbf{R}^*$ and of R_c^* will converge (in a certain sense, not to be discussed here) to that of $\mathbf{R}$ and R_c. We will describe the percentile bootstrap and t-bootstrap methods for finding confidence intervals and the F-bootstrap method for testing linear hypotheses. See Mammen (1992) for details of the theory. We are assuming here that the vector of all ones lies in the column space of $\mathbf{X}$; otherwise the components of $\mathbf{e}$ should be adjusted by subtracting their mean, before determining $\mathbf{e}^*$.

Let $\mathbf{e}$ be determined. Let $\mathbf{e}_i^*$, for $i = 1, \ldots, B$, be a random bootstrap sample from the empirical distribution determined by $\mathbf{e}$. That is, each $\mathbf{e}_i^*$ is an n-component vector whose components are independently chosen from the components of $\mathbf{e}$. Let $\mathbf{R}_i^* = \mathbf{X}^+\mathbf{e}_i^*$ and $R_{ci}^* = \mathbf{c}'\mathbf{X}^+\mathbf{e}_i^* = \mathbf{c}'\mathbf{R}_i^*$. Then the empirical distributions of the k-component vectors $\mathbf{R}_1^*, \ldots, \mathbf{R}_B^*$ and of $R_{c1}^*, \ldots, R_{cB}^*$ serve as approximations of the distribution of $\mathbf{R}$ and of R_c^*.

Define $h(\mathbf{c}) = \mathbf{c}'\mathbf{M}^{-1}\mathbf{c}$, $\quad S^2 = \|\mathbf{Y} - \mathbf{X}\hat{\boldsymbol{\beta}}\|^2/(n-k) = \|\mathbf{e}\|^2/(n-k)$,

and

$$T = \mathbf{c}'(\hat{\boldsymbol{\beta}} - \boldsymbol{\beta})/[h(\mathbf{c})S^2]^{1/2} = [R_c/S]h(\mathbf{c})^{-1/2}.$$

In order to determine a confidence interval on $\eta_c = \mathbf{c}'\boldsymbol{\beta}$ we should know the distribution of T. We can approximate the distribution of T from the bootstrap samples $\mathbf{e}_1^*, \ldots, \mathbf{e}_B^*$, to obtain $T_1^*, \ldots, T_B^*$, where $T_i^* = [R_{ci}^*/S_i^*]h(\mathbf{c})^{-1/2}$. Here $S_i^{*2} = \|(\mathbf{I}_n - \mathbf{P}_V)\mathbf{e}_i^*\|^2$. Thus, in order to determine T_i^*, we must perform a regression analysis on $\mathbf{e}^*$. Though $\mathbf{e} \in V^\perp$, because of the random choice of the ε_i^* from the components of $\mathbf{e}$, $\mathbf{e}^*$ is not contained in $V^\perp$. From the ordered values of these T_i^* we can determine estimates $\hat{t}_{1-\alpha_1}$ and $\hat{t}_{\alpha_2}$ of the percentiles $t_{1-\alpha_1}$ and t_{α_2}. One percentile bootstrap $100(1 - \alpha_1 - \alpha_2)\%$ confidence interval on $\mathbf{c}'\boldsymbol{\beta}$ is $[\mathbf{c}'\hat{\boldsymbol{\beta}} - \hat{t}_{1-\alpha_1}Sh(\mathbf{c})^{1/2}, \mathbf{c}'\hat{\boldsymbol{\beta}} - \hat{t}_{\alpha_2}Sh(\mathbf{c})^{1/2}]$.

We can determine bootstrap cutoff points for the F-test of $H_0: \boldsymbol{\theta} = \sum \beta_j \mathbf{x}_j \in V_0$, a $k_0 < k$ dimensional subspace of V. Let $\mathbf{e}_1^*, \ldots, \mathbf{e}_B^*$ be defined as before. Let $F_i^* = [\|(\mathbf{P}_V - \mathbf{P}_{V_0})\mathbf{e}_i^*\|^2/[(k - k_0)]/S_i^*]$, the usual F-statistic for the observation vector $\mathbf{e}_i^*$. Then the proper cutoff point for the F-test can be estimated by the $(1 - \alpha)$th quantile of the empirical distribution of F_i^*'s.

Scheffé simultaneous confidence intervals can be obtained by simply substituting the estimated F-cutoff point for $F_{1-\alpha}$ in the usual normal theory formula. Of course, the Bonferroni method can also be used, finding, for example, five 99% confidence intervals by the t-bootstrap method in order to have 95% overall confidence. In theory the bootstrap Tukey method could also be used by bootstrapping to approximate the distribution of $q = \text{Range}(\bar{Y}_1, \ldots, \bar{Y}_k)/S$.

Fundamental to our discussion of the bootstrap has been the assumption that the ε_i have the same distribution F. There are certainly many applications when that is not a reasonable assumption. Theory has been developed in recent years which allows the bootstrap method, (the "wild bootstrap" for example) to be applied to the case that the ε_i do not have the same distribution. In other applications it may be more reasonable to believe that the pairs $(\tilde{\mathbf{x}}_i, Y_i)$ constitute a random sample form a $(k + 1)$-dimensional distribution. For discussions of applications of the bootstrap in these situations, see Wu (1986) and Tibshirani and Efron (1993).

Example 4.12.3: We looked for a design matrix $\mathbf{X}$ and distribution F for which the usual normal theory method would perform poorly, while the bootstrap method did well. This suggests that we estimate a parameter $\eta = \mathbf{c}'\boldsymbol{\beta}$ for which the estimator $\hat{\eta} = \mathbf{c}'\hat{\boldsymbol{\beta}}$ put heavy weight on one or just a few ε_i's, and that F differ considerably from any normal distribution. As an extreme we chose $\mathbf{X} = (\mathbf{x}_1, \mathbf{x}_2)$, where $\mathbf{x}_1 = (1, 0, \ldots, 0)'$, and $\mathbf{x}_2 = (0, 1, \ldots, 1)'$, each of 50 components, and $\mathbf{c} = (1, 0)'$, so that $\eta = \beta_1$, the coefficient of $\mathbf{x}_1$. Then $\hat{\eta} = \hat{\beta}_1 = \beta_1 + \varepsilon_1$. The distribution of $\hat{\eta} - \eta = \varepsilon_1$ is F. F was chosen to be the 1/2, 1/2 mixture of $N(0, 1)$ and $N(0, 100)$. The t-95%-interval on η: $\hat{\beta}_1 \pm 2.01S$, would be expected to have approximate probability of coverage $F(1.96\sigma) - F(-1.96\sigma) = 2\Phi(1.96(50.5)^{1/2}) - 1 = 0.918$. Using $B = 1{,}000$, six nominal 95% intervals were found for 2,000 repetitions of this experiment (see Table 4.12.2). The methods were as described in Example 4.12.1. Even in this extreme case, chosen to make the bootstrap method look good, the usual t-method (#1) does reasonably well. The two t-bootstrap procedures (#5 and #6) have coverage probabilities closer to the nominal 95%, but they pay the price of larger mean length. An $\alpha = 0.05$ level F-test of H_0: $\beta_1 = \beta_2$ was performed for the case that H_0 was true for each repetition of the experiment, using both the usual method, which rejects for $F > F_{0.95,1,48}$, and the bootstrap method described above. These two methods rejected 169 and 127 times, respectively, so that the true α-levels are estimated to be 0.0845 and 0.0635.

Example 4.12.4: The t-method can be expected to fail when the variance for F does not exist. For that reason we also chose F to be the c.d.f. for the distribution of $\varepsilon = \xi U^{-1/\delta}$, for $\xi = 1$ or -1 with probabilities 1/2, 1/2, and U uniform [0, 1], ξ and U independent. The mean exists for $\delta > 1$, the variance for $\delta > 2$. For $\delta = 1.5$ we simulated the two-sample problem with $n_1 = n_2 = 15$,

Table 4.12.2 Coverage Percentages and Lengths for 95% Confidence Intervals on $\boldsymbol{\beta}_1$ as Determined by t-Method and Bootstrap Methods

Method	1	2	3	4	5	6
Percentage	90.4	92.2	86.7	91.9	93.6	93.8
Mean Length	14.2	15.7	18.8	15.7	17.0	17.0

Table 4.12.3 Coverage Percentages and Lengths for 95% Confidence Intervals on $\mu_2 - \mu_1$ Determined by t-Method and Bootstrap Method

Method	1	2	3	4	5	6
Percentage	95.1	94.6	87.4	81.6	93.4	90.9
Mean Length	7.32	6.31	6.31	8.32	9.76	12.1

again finding six intervals as described in Example 4.12.2. 2,000 samples of $n_1 + n_2 = 30$ were taken, and $B = 1{,}000$ bootstrap samples were taken in each case. Application of the bootstrap regression method requires in this case that the deviations $e_{1i} = Y_{1i} - \bar{Y}_1$ and $e_{21} = Y_{2i} - \bar{Y}_2$, a bootstrap sample $(e_{1i}^*, \ldots, e_{1n_1}^*, e_{21}^*, \ldots, e_{2n_2}^*)$ taken, then the bootstrap distributions of $(\bar{e}_2^* - \bar{e}_1^*) - (\bar{e}_2 - \bar{e}_1) = d^*$ and of $t^* = d^*/S_{d^*}^*$ obtained by B repetitions (Table 4.12.3). The conclusion in this case is that the usual t-method performs surprisingly well, that the Efron percentile method does rather badly, that the absolute t and absolute percentile methods do not not do well as compared to methods #2 and #6, and, as would be expected in a case in which the variance for F does not exist, it is better to use a percentile method rather than a t-method.

In general, the author found in a number of other simulations, that the usual t- and F-methods perform surprisingly well, with respect to confidence intervals and to tests, both for the level of significance, and for the power. In a wide range of problems, it seems to be doubtful that with replacement bootstrapping will do much better than the classical methods.

For discussion of "permutation bootstrapping in regression" see LePage and Podgorski (1992, 1994). For a discussion of bootstrapping for non-normal errors see LePage, Podgorski, and Ryznar (1994).

Problem 4.12.1: (a) For the sample 3, 9, 6 find the exact bootstrap distribution of the sample mean.

(b) Find the bootstrap estimate of the variance of the sample mean. Can you do this without answering (a) first?

(c) Repeat (a) for the sample median.

(d) Repeat (b) for the sample median.

Problem 4.12.2: Define an algorithm which could be used on your favorite computer and software package to find a 95% confidence interval on the mean μ of a distribution F, based on a random sample of n from F. The method should use the sample mean as the estimator, and employ both the percentile-bootstrap and t-bootstrap methods. If possible use the algorithm to carry out a simulation for the case that F is as in example 4.1.2. $n = 10$, $B = 200$. Repeat 500 times in order to estimate the coverage probability and mean lengths of the intervals.

Problem 4.12.3: Let $Y_i = \beta x_i + \varepsilon_i$, $i = 1, 2, 3$, suppose the (x_i, Y_i) pairs (1, 1), (2, 3), (3, 7) are observed.

(a) Find the bootstrap estimate of the distribution of $\hat{\boldsymbol{\beta}} - \boldsymbol{\beta}$. (Since the residual vector will not have components summing to zero, the components of $\mathbf{e}^*$ should be a random sample from the "corrected $\mathbf{e}$").

(b) Use the result of (a) to find an approximate 80% confidence interval on $\boldsymbol{\beta}$.

(c) The *permutation bootstrap* method for regression analysis chooses the elements of $\mathbf{e}^*$ without rather than with replacement, after correcting so the components sum to zero. Use this method to estimate the distribution of $\hat{\boldsymbol{\beta}} - \boldsymbol{\beta}$.

CHAPTER 5

Simultaneous Confidence Intervals

We have discussed methods for the setting of confidence intervals on parameters $\eta = c_1\beta_1 + \cdots + c_k\beta_k$. These are of the form $I_c = [\hat{\eta} \pm tS_{\hat{\eta}}]$ and have the property $P(\eta \in I_c) = 1 - \alpha$ for each choice of $\mathbf{c} = (c_1, \ldots, c_k)$. It is often desirable to be able to make the claim

$$P(\eta \in I_c, \mathbf{c} \in C) = 1 - \alpha, \tag{5.1.1}$$

where C is some finite or infinite collection of such vectors. The collection $\{I_c, \mathbf{c} \in C\}$ is then a family of confidence intervals with confidence coefficient $1 - \alpha$.

For example, consider one-way analysis of variance with four means μ_1, μ_2, μ_3, μ_4 of interest. We may be interested in the 6 differences $\mu_i - \mu_j$ for $i > j$ and might like to have 95% confidence that all intervals of the form of I_c simultaneously hold. Then C is the collection of six coefficient vectors of the form $(1, -1, 0, 0)$ or $(1, 0, -1, 0)$, etc.

For simple linear regression we may be interested in intervals on $\beta_0 + \beta_1 x$ for all x over the entire range of x for which the model holds. Thus each $\mathbf{c}$ is of the form $(1, x)$. Since a confidence interval I_x on $g(x)$ determines an interval aI_x on $ag(x) = a\beta_0 + \beta_1(ax)$ it is therefore equivalent to find intervals $I_{a,b}$ on all linear combination $a\beta_0 + b\beta_1$. Thus C may be taken to be all of R_2.

In this chapter we discuss three simultaneous confidence interval (SCI) methods: (1) Bonferroni, (2) Scheffé, (3) Tukey. The Bonferroni method is fundamentally the simplest method and for small finite C is usually the best in that the resulting intervals are shorter. The Scheffé method is the most mathematically elegant (an opinion of the author), and has applicability to all the linear models we have and will consider. The Tukey method is applicable only to cases with repeated observations for each of several means but for those situations usually provides shorter intervals than the Scheffé method. The Neuman–Kuels (Kuels, 1952) and Duncan procedures are multiple tests of hypotheses, rather than methods for the determination of SCI's. The Duncan procedure is often used because the method finds more significant differences than the simultaneous methods to be discussed, but does not offer the error protection that SCI's do. For example, for $\alpha = 0.05$, with five equal means and large degrees of freedom for error, the Duncan procedure has probability approximately $1 - 0.95^4 = 0.185$ that at least one significant difference will be found. We will not discuss the Duncan and Newman–Kuels procedures.

5.1 BONFERRONI CONFIDENCE INTERVALS

Suppose we want confidence intervals on a fixed number of linear combinations

$$\eta_1 = \mathbf{c}_1'\boldsymbol{\beta}, \qquad \eta_2 = \mathbf{c}_2'\boldsymbol{\beta}, \ldots, \eta_r = \mathbf{c}_r'\boldsymbol{\beta}$$

of the parameters of the linear model. We already have a technique for finding a $100(1-\alpha)\%$ confidence interval on each such linear combination, namely

$$I_j = [\hat{\eta}_j \pm t_{1-\alpha_j/2} S(\hat{\eta}_j)] \qquad \text{for} \quad j = 1, \ldots, r,$$

where $\hat{\eta}_j = \mathbf{c}_j'\hat{\boldsymbol{\beta}}$, $S^2(\hat{\eta}_j) = S^2\mathbf{c}_j'(\mathbf{X}'\mathbf{X})^{-1}\mathbf{c}_j$ and $t_{1-\alpha_j/2}$ has $(n-k)$ d.f..

Let E_j be the event that the confidence interval I_j covers $\eta_j = \mathbf{c}_j'\boldsymbol{\beta}$. We can make use of the *Bonferroni Inequality* to put a lower bound on the probability that all confidence intervals hold simultaneously. Thus

$$P\left(\bigcap_{j=1}^{r} E_j\right) = 1 - P\left(\bigcup_{j=1}^{r} \bar{E}_j\right) \geq 1 - \sum_{j=1}^{r} P(\bar{E}_j) = 1 - \sum_{1}^{r} \alpha_j$$

If we want $P\left(\bigcap_{j=1}^{r} E_j\right) \geq 1 - \alpha$ we therefore need only choose $\alpha_1, \ldots, \alpha_r$ such that $\sum_{1}^{r} \alpha_j \leq \alpha$. The usual choice is $\alpha_j = \alpha/r$.

In one-way analysis of variance treatment #1 might be a control, perhaps the standard seed. Treatments #2, #3, #4 might be new varieties of seed, and we might want to compare each against the control. We may therefore be interested primarily in the linear combinations $\mu_2 - \mu_1$, $\mu_3 - \mu_1$, $\mu_4 - \mu_1$.

Tables 2.1–2.3 in the Appendix, presenting t-distribution quantiles $t_{1-\alpha/2m}$ for various choices of α and m, facilitates use of the Bonferroni inequality. For example, for the three parameters above and $\alpha = 0.05$ we find for $\nu = 20$, $t_{20,1-0.05/6} = 2.61$ so we can use $\bar{Y}_j - \bar{Y}_1 \pm 2.61\sqrt{1/n_j + 1/n_1}S$ for $j = 2, 3, 4$. If we wish simultaneous 95% intervals on $\mu_1 - \mu_2$, $\mu_1 - \mu_3$, $\mu_1 - \mu_4$, $\mu_2 - \mu_3$, $\mu_2 - \mu_4$, $\mu_3 - \mu_4$ we use $t_{20,1-0.05/12} = 2.93$.

5.2 SCHEFFÉ SIMULTANEOUS CONFIDENCE INTERVALS

Let $\mathbf{Y} = \sum_1^k \boldsymbol{\beta}_j \mathbf{x}_j + \boldsymbol{\varepsilon}$ with $\boldsymbol{\varepsilon} \sim N(\mathbf{0}, \sigma^2 \mathbf{I}_n)$. Let C be a subspace of R_k of dimension q, $1 \le q \le k$. Consider the collection of parameters η_c as $\mathbf{c}$ ranges over C. The Scheffé simultaneous confidence interval method provides a collection of intervals I_c on η_c which simultaneously hold for all $\mathbf{c} \in C$ with prescribed probability (confidence).

For example, for the one-way layout we are often interested in "contrasts" only, linear combinations of $\mu_1, \ldots, \mu_k$ with coefficients adding to zero. Thus $C = \mathscr{L}(\mathbf{J}_k)^\perp$. For simple linear regression we are interested in linear combinations $\beta_0 + \beta_1 x$. In this case we take $C = R_2$ and get confidence intervals on all linear combinations $c_0\beta_0 + c_1\beta_1$.

To develop these SCI's, define $\mathbf{a}_c = \mathbf{X}\mathbf{M}^{-1}\mathbf{c}$, where $\mathbf{M} = \mathbf{X}'\mathbf{X}$. Then $\eta_c \equiv (\mathbf{c}, \boldsymbol{\beta}) = (\mathbf{a}_c, \boldsymbol{\theta})$ for all $\boldsymbol{\theta} \in V$ and, since $\mathbf{a}_c\mathbf{X} = \mathbf{c}$, $(\hat{\boldsymbol{\eta}}_c, \mathbf{x}_i) = c_i$ for $i = 1, \ldots, k$. In addition, $\hat{\boldsymbol{\beta}} - \boldsymbol{\beta} = \mathbf{m}^{-1}\mathbf{X}'(\mathbf{Y} - \boldsymbol{\theta}) = \mathbf{M}^{-1}\mathbf{X}'\boldsymbol{\varepsilon}$, so $\hat{\eta}_c = (\mathbf{a}_c, \boldsymbol{\varepsilon})$. We have $E(\hat{\eta}_c) = \eta_c$ and $\text{Var}(\eta_c) = \text{Var}((\mathbf{a}_c, \mathbf{Y})) = \sigma^2\|\mathbf{a}_c\|^2 = \sigma^2\mathbf{c}'\mathbf{M}^{-1}\mathbf{c}$. Let $S_c^2 = S^2\|\mathbf{a}_c\|^2$, an unbiased estimator of $\text{Var}(\hat{\eta}_c)$.

Recall the method used to find a confidence interval on an individual parameter $\eta_c = \mathbf{c}'\boldsymbol{\beta}$. We used the fact that the *pivotal quantity* $T_c = \mathbf{c}'(\hat{\boldsymbol{\beta}} - \boldsymbol{\beta})/[S^2\mathbf{c}'\mathbf{M}^{-1}\mathbf{c}]^{1/2} = (\hat{\eta}_c - \eta_c)/S_c = (\mathbf{a}_c, \boldsymbol{\varepsilon})/[S\|\mathbf{a}_c\|]$ has Student's t distribution with $n - k$ d.f. We can determine simultaneous confidence intervals by considering the random variable $W^* = \sup_{c \in C} T_c^2$.

Define $V_1 = \{\mathbf{a}_c | \mathbf{c} \in C\}$. Then $\dim(V_1) = \dim(C) = q$. Let $\hat{\boldsymbol{\varepsilon}}_1 = p(\boldsymbol{\varepsilon}| V_1)$ and $\hat{\boldsymbol{\varepsilon}}_c = p(\boldsymbol{\varepsilon}|\mathbf{a}_c)$. Then, since $p(\hat{\boldsymbol{\varepsilon}}_1|\mathbf{a}_c) = \boldsymbol{\varepsilon}_c$ and $(\hat{\boldsymbol{\varepsilon}}_1 - \hat{\boldsymbol{\varepsilon}}_c) \perp \hat{\boldsymbol{\varepsilon}}_c$, it follows that $S^2T_c^2 = \|\hat{\boldsymbol{\varepsilon}}_c\|^2 = \|\hat{\boldsymbol{\varepsilon}}_1\|^2 - \|\hat{\boldsymbol{\varepsilon}}_1 - \hat{\boldsymbol{\varepsilon}}_c\|^2 \le \|\hat{\boldsymbol{\varepsilon}}_1\|^2$, with equality only if $\mathbf{a}_c$ is a multiple of $\hat{\boldsymbol{\varepsilon}}_1$. Thus, from Theorem 2.5.6, $W^*/q = [\|\hat{\boldsymbol{\varepsilon}}_1\|^2/q]/S^2 \sim F_{q,n-k}$. Taking $K^2/q = F_{q,n-k,\gamma}$, the 100γth percentile of the $F_{q,n-k}$ distribution, we get $K = (qF_{q,n-k,\gamma})^{1/2}$, and

$$P(T_c^2 \le K \text{ for all } \mathbf{c} \in C) = P(W^*/q \le F_{q,n-k,\gamma}) = \gamma.$$

Finally, we conclude that

$$P(\eta_c \in [\hat{\eta}_c \pm (qF_{q,n-k,\gamma})^{1/2} S(\hat{\eta}_c)] \text{ for all } \mathbf{c} \in C) = \gamma.$$

The intervals I_c within the brackets will therefore contain the corresponding parameters η_c for all $\mathbf{c} \in C$ with probability γ. The intervals I_c are called Scheffé simultaneous confidence intervals after Henry Scheffé (1953).

Application to Simple Linear Regression: Let $g(x) = \beta_0 + \beta_1 x$. Suppose we observe (x_i, Y_i) for $Y_i = \beta_0 + \beta_1 x_i + \varepsilon_i$ for $i = 1, \ldots, n$ and want confidence intervals I_x on $g(x)$ which hold simultaneously for all x for which this simple linear regression model holds.

Let $C = R_2$. Then the Scheffé method provides simultaneous intervals on

$c_0\beta_0 + c_1\beta_1$, hence on $g(x) = \beta_0 + \beta_1 x$. Since $C = R_2$, $V_1 = V = \mathscr{L}(\mathbf{J}, \mathbf{x})$. Thus, $100\gamma\%$ simultaneous confidence intervals on $g(x) = \beta_0 + \beta_1 x$ are given by

$$\hat{g}(x) \pm KS(\hat{\eta}_c) \qquad \text{for} \quad K = \sqrt{2F_{2,n-2,\gamma}},$$

where $\hat{g}(x) = \hat{\beta}_0 + \hat{\beta}_1 x$. Since $\text{Var}(\hat{g}(x)) = h(x)\sigma^2$, where $h(x) = 1/n + (x - \bar{x})^2/S_{xx}$, the simultaneous intervals are

$$\hat{g}(x) \pm [h(x)S^2]^{1/2}K$$

We earlier found that a $100\gamma\%$ confidence interval on $g(x)$, holding for that x only, is

$$\hat{g}(x) \pm th(x)^{1/2}S \qquad \text{for} \quad t = t_{n-2,(1+\gamma)/2}$$

Thus the ratio of the length of the simultaneous interval at x to the individual interval is $(K/t) = (2F_{2,n-2,\gamma}/t^2_{n-2,(1+\gamma)/2})^{1/2}$, which always exceeds one.

Connection Between Scheffé Intervals and Tests of Hypotheses: Let

$$t_c = \frac{\hat{\eta}_c}{\|\mathbf{a}_c\|S} = \frac{(\mathbf{Y}, \mathbf{a}_c)}{\|\mathbf{a}_c\|S},$$

as defined in Section 3.7. t_c is the statistic used to test the null hypothesis that $\eta_c = 0$. Note that T_c was defined similarly, with $\boldsymbol{\varepsilon}$ rather than $\mathbf{Y}$. Then $\hat{\eta}_c - KS\|\mathbf{a}_c\| \le 0 \le \hat{\eta}_c + KS\|\mathbf{a}_c\|$ if and only if $[t_c^2 \le K^2]$. It follows that $0 \in I_c$ for all $\mathbf{c} \in C$ if and only if

$$W \equiv \sup_{\mathbf{c} \in C} t_c^2 \le K^2 = qF_{q,n-k,\gamma}. \tag{5.2.1}$$

It was shown in Section 3.7 that $W/q = F$, where F is the statistic used to test H_0: $\boldsymbol{\theta} \in V_1^{\perp} \Leftrightarrow \boldsymbol{\theta} \perp V_1$, where $V_1 = \{\mathbf{v} = \mathbf{XM}^{-1}\mathbf{c}, \mathbf{c} \in C\}$. Therefore, (5.2.1) holds if and only if H_0 is accepted at level α. We conclude that $0 \in I_c$ for all $\mathbf{c} \in C$ if and only if the α-level F-test for H_0 is accepted. Stated conversely, this means that rejection of H_0 at level α implies the existence of at least one $\mathbf{c} \in C$ for which the interval I_c does not include zero.

One-Way Analysis of Variance: Let $C = \{\mathbf{c} \in R_k | \mathbf{c} \perp \mathbf{J}\}$. Then the collection of linear combinations $(\mathbf{c}, \boldsymbol{\mu}) = \sum_1^k c_i\mu_i$ for $\mathbf{c} \in C$ is the collection of contrasts. For $\mathbf{c} \in C$, $\mathbf{a}_c = \sum_1^k (c_i/n_i)\mathbf{J}_i$ so $V_C = \left\{\sum_1^k \frac{c_i}{n_i}\mathbf{J}_i \,\middle|\, \sum_1^k c_i = 0\right\}$, a $(k-1)$-dimensional subspace of R_n. But $[\mu_1 = \mu_2 = \cdots = \mu_k] \Leftrightarrow [\boldsymbol{\theta} = \mu_1\mathbf{J}$ for some $\mu_1] \Leftrightarrow [(\mathbf{a}, \boldsymbol{\theta}) = 0$ for $\mathbf{a} \in V_C] \Leftrightarrow \left[\sum_i c_i\mu_i = 0 \text{ for all } \mathbf{c} \in C\right]$.

The simultaneous Scheffé confidence intervals on contrast $\sum_1^k c_i\mu_i$ are given by

$$\sum_1^k c_i \bar{Y}_i \pm KS\left[\sum_1^k c_i^2/n_i\right]^{1/2},$$

where $K = [qF_{q,n-k,\alpha}]^{1/2} = [(k-1)F_{k-1,n-k,1-\alpha}]^{1/2}$.

If it is desirable to make an overall statement of confidence on all linear combinations $\sum_1^k c_i\mu_i$, then we can take $C = R_k$. We can include, for example, confidence intervals on $\mu_1, \mu_2, \ldots, \mu_k$ as well as on contrasts. In this case we need only change K to

$$K = (kF_{k,n-k,1-\alpha})^{1/2}.$$

Confidence Ellipsoids: Let $C = R_k$, so that $V_C = V$. Then $\boldsymbol{\varepsilon}_1 = p(\boldsymbol{\varepsilon}|V_1) = p(\boldsymbol{\varepsilon}|V) = \hat{\boldsymbol{\varepsilon}} = \hat{\mathbf{Y}} - \boldsymbol{\theta} = \mathbf{X}(\hat{\boldsymbol{\beta}} - \boldsymbol{\beta})$. It follows that

$$W^* = \|\boldsymbol{\varepsilon}_1\|^2/S^2 = (\hat{\boldsymbol{\beta}} - \boldsymbol{\beta})'\mathbf{M}(\hat{\boldsymbol{\beta}} - \boldsymbol{\beta})/S^2.$$

Let this last term, considered as a function of $\boldsymbol{\beta}$, be $Q(\boldsymbol{\beta})$ and define $A = \{\mathbf{b} \in R_k | Q(\mathbf{b}) \le kF_{k,n-k,\gamma}\}$. A is the convex hull of an ellipsoid in R_k (union of interior and boundary). Since $\boldsymbol{\beta} \in A \Leftrightarrow W^*/k \le F_{k,n-k,\gamma}$, it follows that $P(\boldsymbol{\beta} \in A) = \gamma$, so that A is a $100\gamma\%$ confidence ellipsoid on $\boldsymbol{\beta}$. Since $W^* = \sup_{\mathbf{c} \in R_k} T_c^2$, $\boldsymbol{\beta} \in A \Leftrightarrow \eta_c \in I_c$ for all $\mathbf{c} \in R_k$. If $\boldsymbol{\beta}_0$ is a specified value for the parameter vector, we can test H_0: $\boldsymbol{\beta} = \boldsymbol{\beta}_0$ at level α by rejecting H_0 whenever the $100(1-\alpha)\%$ confidence ellipsoid A does not contain $\boldsymbol{\beta}_0$.

Problem 5.2.1: Consider the weighing Problem 3.1.1 with the two unknown weights β_1 and β_2.

(a) Suppose we want Scheffé simultaneous 95% confidence intervals on all linear combinations of β_1 and β_2. For the four weighings made and for $\mathbf{Y} = (7, 3, 1, 7)'$ find these intervals for the three linear combinations β_1, β_2, and $\beta_1 - \beta_2$.

(b) Use the Bonferroni method to find 95% simultaneous confidence intervals on these same three linear combinations.

(c) Suppose that we wish to test H_0: $\beta_1 = \beta_2 = 0$. Then $\sup_{\mathbf{c} \in C} t_c^2/q = F$ is the corresponding F-statistic. For which value of $\mathbf{c}$ does $t_c^2/q = F$? What is the corresponding $\mathbf{a}_c$?

(d) Find a 95% confidence ellipsoid for $\boldsymbol{\beta} = (\beta_1, \beta_2)'$ for these data.

Problem 5.2.2: For the pairs (x_i, Y_i): (0, 7), (1, 7), (2, 5), (3, 1) and the simple linear regression model sketch the 95% confidence ellipsoid on $\boldsymbol{\beta} = (\beta_0, \beta_1)'$. Suppose that you wished to test H_0: $\boldsymbol{\beta} = (7, -1)'$. Would you reject H_0 at level $\alpha = 0.05$?

Problem 5.2.3: Consider the model for one-way analysis of variance, with $k = 4$, means $\mu_1, \mu_2, \mu_3, \mu_4$, and observations: treatment #1: 3, 5, 7; treatment #2: 6, 8; treatment #3: 8, 10; treatment #4: 8, 10, 12.

(a) Perform the $\alpha = 0.05$ level F-test of H_0: $\mu_1 = \mu_2 = \mu_3 = \mu_4$.

(b) Use the Scheffé method to find 95% simultaneous confidence intervals on the six parameters of the type $\mu_i - \mu_j$.

(c) For which $\mathbf{c} = (c_1, c_2, c_3, c_4)'$ does $t_c^2/(4-1) = F$, the F-statistic you found in (a)?

(d) Suppose that you wish SCI's on all linear combinations of the μ_j. What are the resulting 95% confidence intervals on μ_1 and on $(\mu_1 - \mu_2)$?

Problem 5.2.4: For the fitness data of Section 3.12 consider the simple linear regression of Y(oxygen) vs. x_1 (runtime). Find the least squares estimate $\hat{g}(x_1)$ of $g(x_1) = E(Y|x_1)$ and two functions $k_1(x_1)$ and $k_2(x_1)$ such that for each x_1 $P(g(x_1) \in [\hat{g}(x_1) \pm k_1(x_1)S]) = 0.95$ and $P(g(x_1) \in [\hat{g}(x_1) \pm k_2(x_1)S]$ for all $x_1) = 0.95$.

5.3 TUKEY SIMULTANEOUS CONFIDENCE INTERVALS

The Tukey procedure for finding simultaneous confidence intervals depends on the following definition.

Definition 5.3.1: Let $W_1, \ldots, W_k$ be independent r.v.'s, each $N(\mu, \sigma^2)$. Let $R = \left(\max_i W_i\right) - \left(\min_i W_i\right) = \text{Range}(W_1, \ldots, W_k)$. Let $\nu S^2/\sigma^2$ have a χ^2_ν distribution and be independent of $(W_1, \ldots, W_k)$. Then $q = R/S$ is said to have the *studentized range distribution*. Let $q_{\gamma k \nu}$ be the γ-quantile of this distribution.

Let $M = \max_i (|W_i - \mu|)$. Then $q' = \dfrac{\max(R, M)}{S}$ has the *studentized augmented range distribution*. Let $q'_{\gamma k \nu}$ be the γ-quantile of this distribution.

Comment: For k large it is unlikely that all W_i are on the same side of μ, so that usually $M < R$ and $q' = q$. For any k, $P(q = q') = 1 - 2^{-(k-1)}$. Even when q and q' differ, they will usually be close. For γ large $q'_{\gamma k \nu} \doteq q_{\gamma k \nu}$.

The densities and c.d.f.'s of q and q' cannot be expressed in closed form. See Table 6 in the Appendix for 95 and 99 percentiles for varying values of k and ν. Harter (1960) presents much more complete tables of the c.d.f.

The following theorem, proved by John Tukey (1953), justifies the Tukey simultaneous confidence interval method.

Theorem 5.3.1: Let $\hat{\theta}_1, \ldots, \hat{\theta}_k$ be independent with $\hat{\theta}_i \sim N(\theta_i, a^2\sigma^2)$, where a is a known positive constant. Let $\nu S^2/\sigma^2 \sim \chi^2_\nu$, independent of $(\hat{\theta}_1, \ldots, \hat{\theta}_k)$.

Then

(1) $P((\theta_j - \theta'_j) \in [(\hat{\theta}_j - \hat{\theta}'_j) \pm TS]$ for all j and $j') = \gamma$ for $T = aq_{\gamma k\nu}$, and

(2) $P((\theta_j - \theta'_j) \in [(\hat{\theta}_j - \hat{\theta}'_j) \pm T'S]$ and $\theta_j \in [\hat{\theta}_j \pm T'S]) = \gamma$ for all j and j', for $T' = aq'_{\gamma k\nu}$.

In application to one-way analysis of variance, $\hat{\theta}_j$ will be the treatment mean $\bar{Y}_j$, the constant a will be $1/\sqrt{n_j}$, which we will assume for now is the same for all j. S^2 is error mean square. The theorem makes it possible to: (1) using q, give simultaneous confidence intervals which will cover all differences $\mu_i - \mu_j$ with a prescribed probability γ and (2) using q', give simultaneous confidence intervals which will cover all differences $\mu_i - \mu_j$ and all means μ_j with prescribed probability γ. The proof is easy.

***Proof*:** Let $R = \text{Range}(\hat{\theta}_1 - \theta_1, \ldots, \hat{\theta}_k - \theta_k)$. Since $(\hat{\theta}_j - \theta_j)/a \sim N(0, \sigma^2)$, $R/(Sa)$ has the studentized range distribution. Hence $\gamma = P(R/S \leq aq_{\gamma k\nu})$

$$= P(|(\theta_j - \hat{\theta}_j) - (\theta_{j'} - \hat{\theta}_{j'})| \leq Saq_{\gamma k\nu} \text{ for all } j, j')$$
$$= P([(\theta_j - \theta_{j'}) \in [(\hat{\theta}_j - \hat{\theta}_{j'}) \pm TS] \text{ for all } j, j').$$

The proof of (2) is similar. □

In order to expand the number of linear combinations on $\theta_1, \ldots, \theta_k$ for which confidence intervals are given from those of type $\theta_j - \theta'_j$ to all contrasts $\sum_1^k c_j\theta_j$ for $\sum_1^k c_j = 0$ we need.

Lemma 5.3.1: Let $d_1, \ldots, d_k$ be any real numbers. Then

$$\sup\left\{\left(\frac{1}{2}\sum_1^k |c_j|\right)^{-1}\left(\sum_1^k c_j d_j\right) \middle| \text{ all } c_1, \ldots, c_k \text{ such that } \sum_1^k c_j = 0\right\}$$
$$= \max_{jj'}(d_j - d_{j'}) = \text{Range}(d_1, \ldots, d_k)$$

***Proof*:** Without loss of generality we may suppose that $0 \leq d_1 \leq \cdots \leq d_k$, by a change of notation and the fact that adding the same amount to all d_i does not change $\sum_1^k c_j d_j$, since $\sum_1^k c_j = 0$. Then

$$\sum c_j d_j = \sum_{c_j > 0} c_j d_j + \sum_{c_j \leq 0} c_j d_j \leq d_k \sum_{c_j > 0} c_j + d_1 \sum_{c_j \leq 0} c_j$$
$$= d_k\left(\frac{1}{2}\sum_j |c_j|\right) - d_1\left(\frac{1}{2}\sum_j |c_j|\right) = (d_k - d_1)\left(\frac{1}{2}\sum_j |c_j|\right).$$

Similarly,

$$\sum_j c_j d_j \geq d_1 \sum_{c_j > 0} c_j + d_k \sum_{c_j \leq 0} c_j = (d_1 - d_k)\left(\frac{1}{2}\sum_j \left| c_j \right|\right).$$

Theorem 5.3.2: Under the assumptions of Theorem 5.3.1:

$$P\left(\sum_1^k c_j\theta_j \in \left[\sum_1^k c_j\hat{\theta}_j \pm TS\left(\frac{1}{2}\sum_1^k \left| c_j \right|\right)\right] \text{ for all contrasts}(c_1, \ldots, c_k)\right) = \gamma, \qquad \text{where} \quad T = aq_{\gamma k v}.$$

***Proof*:** Let $d_j = \hat{\theta}_j - \theta_j$ in the inequality of Theorem 5.3.1. Then Theorem 5.3.2 follows immediately from Theorem 5.3.1.

Theorem 5.3.3: Under the assumptions of Theorem 5.3.1:

$$P\left(\sum c_j\theta_j \in \left[\sum_j c_j\hat{\theta}_j \pm T'S \max\left(\sum_{c_j > 0} \left| c_j \right|, \sum_{c_j < 0} \left| c_j \right|\right)\right] \text{ for all}(c_1, \ldots, c_k)\right) = \gamma, \qquad \text{where} \quad T' = aq'_{\gamma k v}$$

Extension to Unequal Sample Sizes: Suppose we want simultaneous confidence intervals on $\sum c_j\theta_j$, but $\text{Var}(\hat{\theta}_j) = \sigma^2/n_j$, with differing n_j. The original Tukey procedure required all $\hat{\theta}_j$ to have the same variance. Extensions have been proposed by many authors, including Dunn (1974, 101–103), Sidak (1967, 626–633), Hochberg (1975, 426–433), Tukey (1953), Kramer (1956, 307–310), and Spjøtvoll and Stoline (1973, 975–978). See the comparison of these and several others by Stoline (1981). Stoline recommends the use of the Tukey–Kramer (T–K) method (Tukey (1953) and Kramer (1956)). The T–K procedure, applied to the one-way model with n_i observations for sample i, $i = 1, \ldots, k$, yields the $100(1 - \alpha)\%$ simultaneous confidence intervals

$$\bar{Y}_{j\cdot} - \bar{Y}_{j'\cdot} \pm q_{1-\alpha, kv} S[(1/n_j + 1/n_{j'})/2]^{1/2}$$

on $\mu_j - \mu_{j'}$. L. D. Brown (1984) showed that for the cases $k = 3, 4, 5$ that the simultaneous coverage probability is at least $1 - \alpha$. Simulation work of Dunnett (1980) indicates that this may be true for all k, or at least that the coverage probability is not much less than the nominal value. In addition, the lengths of these intervals are in general a bit less or equal to those provided by other methods. The T–K method and most of the others reduce to the Tukey method when the sample sizes are equal.

Example 5.3.1: Suppose that 40 pigs were chosen to take part in a study designed to determine the effects of four different feeds on weight gains over a one-month period. The pigs were randomly assigned to the four feeds, 10 to each feed. The weight gains were:

Feed 1	24	20	29	26	29	30	33	27	20	28
Feed 2	37	30	35	41	31	34	33	32	32	32
Feed 3	32	34	31	23	33	31	30	32	33	33
Feed 4	26	22	20	28	28	32	25	27	32	32

The sample means were: feed 1, 26.6; feed 2, 33.7; feed 3, 31.2; feed 4, 27.2; with grand mean $\bar{Y}_{..} = 29.675$, corrected total SSqs. = 840.77, feed SSqs. = 341.07, error SSqs = 499.7, feed MSq. = 113.69, $S^2 = 13.881$. Therefore, the F-statistic for H_0: $\mu_1 = \mu_2 = \mu_3 = \mu_4$ is $F = (\text{feed MSq.})/S^2 = 8.191$. Since $F_{3,36,0.9995} \doteq 7.51$, we reject H_0 at any reasonable α-level. The estimate of the standard error of the feed means is $\sqrt{S^2/10} = 1.178$. Since $q_{0.95,4,36} = 3.81$, Tukey simultaneous 95% confidence intervals on the contrasts $\mu_i - \mu_j$ for $i \neq j$ are given by $(\bar{Y}_{i\cdot} - \bar{Y}_{j\cdot}) \pm 3.81(1.178) = (\bar{Y}_{i\cdot} - \bar{Y}_{j\cdot}) \pm 4.488$. Two sample means which differ by more than 4.488 are said to be significantly different. We have 95% confidence that all these statements are correct simultaneously. Consider Table 5.3.1. Since the interval on $\mu_1 - \mu_3$ is to the left of 0, we can conclude that $\mu_1 < \mu_3$. Similarly, we conclude that $\mu_2 > \mu_4$, and that $\mu_3 > \mu_4$.

Suppose that the eighth, ninth, and tenth observations for feed 2, the ninth and tenth for feed 3, and the tenth for feed 4 were not obtained because the pigs died, or were sick for reasons not connected to the feed they ate. We can still perform an analysis of variance. We find $S^2 = 14.434$, feed MSq. = 121.91, $F = 8.45$, with means: feed 1, 26.60; feed 2, 33.89; feed 3, 30.88; feed 4, 25.86. We can use the T–K method to obtain simultaneous confidence intervals on differences $\mu_j - \mu_{j'}$. For $\mu_2 - \mu_4$, for example, the 95% interval (one of a family) becomes $[(33.89 - 25.86) \pm (3.86)\sqrt{14.434/[(1/7 + 1/9)/2]^{1/2}}] = [8.03 \pm 5.23]$.

Problem 5.3.1: For the contrasts in Table 5.3.1 find the ratio of the lengths of 95% Bonferroni SCI's to those of 95% Scheffé and Tukey intervals for the 40 observations in Example 5.3.1.

Table 5.3.1

Contrast	Estimate	Interval	Contrast	Estimate	Interval
$\mu_1 - \mu_2$	−7.1	−11.59, 2.61	$\mu_2 - \mu_3$	2.5	−1.99, 6.99
$\mu_1 - \mu_3$	−4.6	−9.09, −0.11	$\mu_2 - \mu_4$	6.5	2.01, 10.99
$\mu_1 - \mu_4$	−0.6	−5.09, 3.89	$\mu_3 - \mu_4$	6.0	1.51, 10.49

Problem 5.3.2: In Example 5.3.1 with the missing observations, use the 95% T–K method to find the interval on $\mu_3 - \mu_4$.

Problem 5.3.3: Find C, not depending on γ or v, such that $q_{\gamma,2,v} = Ct_{v,(1+\gamma)/2}$ for all γ and v. Demonstrate this relationship for $v = 10$, $\gamma = 0.95$.

Problem 5.3.4: For the case $k = 1$, define $R = 0$. How is the distribution of q' related to the t-distribution for this case?

Problem 5.3.5: Let $Y_1, \ldots, Y_{20}$ be a random sample from the $N(\mu, 25)$ distribution. Find the probability that at least two of these r.v.'s differ by 28.25 or more.

Problem 5.3.6: For Example 5.3.1, each $n_i = 10$, find a contrast $\sum c_i\mu_i$ for which 95% Scheffé intervals on all contrasts are shorter than corresponding 95% Tukey intervals.

5.4 COMPARISON OF LENGTHS

The lengths of Bonferroni, Scheffé and Tukey intervals are each multiples of S, so that relative lengths are constants which depend on $c_1, \ldots, c_k$ and (in one-way analysis of variance) on $n_1, \ldots, n_k$. Therefore the choice of a method can be made independent of the data, and can be made after α, the n_i, and the set C are chosen. The choice as to method should not depend on the sample means because probability statements on the performances of the methods would no longer be valid.

For equal sample sizes n_1, relative lengths of confidence intervals are

$$r_{\mathrm{T,S}} = \frac{\text{Tukey length}}{\text{Scheffé length}} = \frac{q_{1-\alpha,k,v}(\frac{1}{2}\sum |c_i|)}{\sqrt{(k-1)F_{k-1,v,1-\alpha}(\sum c_i^2)}}$$

where $v = (n_1 - 1)k$ is error d.f.

$$r_{\mathrm{B,S}} = \frac{\text{Bonferroni length}}{\text{Scheffé length}} = \frac{t_{v,1-\alpha/k(k-1)}}{\sqrt{(k-1)F_{k-1,v,1-\alpha}}} \qquad \text{and} \qquad r_{\mathrm{T,B}} = \frac{r_{\mathrm{T,S}}}{r_{\mathrm{B,S}}}$$

For all parameters of the type $\eta = \mu_i - \mu_j$, $\alpha = 0.05$, $k = 3, 5, 7$, $v = 10, \infty$ these relative lengths are as in Table 5.4.1.

Thus, for simultaneous intervals on all $\mu_i - \mu_j$ the Tukey method provides shorter intervals than both the Bonferroni and Scheffé methods. For parameters of the type $\mu_{j_1} - \frac{1}{2}(\mu_{j_2} + \mu_{j_3}) r_{\mathrm{T,S}}$ is $2/\sqrt{3} = 1.155$ times as large, so that Scheffé intervals are sometimes shorter. For even more complex parameters the Scheffé method begins to win the battle. Since inclusion of more confidence intervals

Table 5.4.1 Ratios of Lengths Among Tukey, Bonferroni, and Scheffé SCI's

	$v = 10$				$v = \infty$			
	$k = 3$	$k = 5$	$k = 7$	$k = 10$	$k = 3$	$k = 5$	$k = 7$	$k = 10$
$r_{T,S}$	0.958	0.884	0.824	0.765	0.956	0.886	0.831	0.768
$r_{B,S}$	1.002	0.960	0.919	0.875	0.976	0.913	0.854	0.793
$r_{T,B}$	0.956	0.918	0.897	0.874	0.980	0.970	0.973	0.968

in the family causes all Bonferroni intervals to be longer, while Tukey and Scheffé methods apply to all contrasts, the Bonferroni method is relatively undesirable if many more complex parameters are of interest.

There is some tendency for users of these methods to apply them only when the F-test for equal means rejects. However, as shown by Olshen (1973), the conditional probability that all resulting confidence intervals are correct, given rejection, is always less than the nominal value. This should make it clear that probability interpretations for confidence intervals are relative to the entire sample space, and that it is good practice to present confidence intervals whether or not the F-test rejects.

Suppose we are interested in five hybrids of corn with four observations on the yield for each hybrid, so that we have 15 d.f. for error. Then $t_{15,1-0.05/20} = 3.29$ and $q_{0.95,.5,15} = 4.37$, so $e_{T,B} = 4.37/(3.29)\sqrt{2} = 0.939$. Tukey intervals are shorter.

If hybrid #1 is standard, and comparisons of μ_j for $j > 1$ with μ_1 are desirable then we may be interested only in confidence intervals on $\mu_j - \mu_1$ for $j > 1$. The multiplier of $S/\sqrt{4}$ for the Bonferroni interval is $\sqrt{2}t_{15,0.05/8} = 4.016$, rather than 4.23, so that $r_{T,B} = 4.23/4.016 = 1.05$. Bonferroni intervals are shorter.

If the family of confidence intervals should also include those on the μ_j, we can either replace q by q' (which is very slightly larger) for the Tukey method, or possibly use the Scheffé method with $K = \sqrt{(5)F_{5,15,0.95}} = 3.81$, resulting in the multiplier $\sqrt{2}K = 5.38$. For the Bonferroni method the corresponding multiplier is $\sqrt{2}t_{15,1-0.05/30} = 5.40$, since there are now 15 confidence intervals (10 pairs, 5 individuals). Thus, the Tukey method does considerably better than either the Scheffé or Bonferroni method.

Example 5.4.1: Five hybrids of corn were each planted on 4 half-acre plots, each chosen randomly from 20 available plots (completely randomized design). Yields in bushels for all 20 plots were recorded. Sample means were $\bar{Y}_1 = 49.5$, $\bar{Y}_2 = 58.1$, $\bar{Y}_3 = 53.2$, $\bar{Y}_4 = 51.3$, $\bar{Y}_5 = 56.8$. Error mean square was $S^2 = 8.73$. Hybrid mean square was 52.59, so $F = 6.02 > F_{4,15,0.95} = 3.06$. 95% Bonferroni, Scheffé and Tukey confidence intervals on all differences $\mu_i - \mu_j$ all have the form $\bar{Y}_i - \bar{Y}_j \pm KS/\sqrt{4}$, where $K = \sqrt{2}t_{1-0.05/20} = 5.12$ for Bonferroni, $K =$

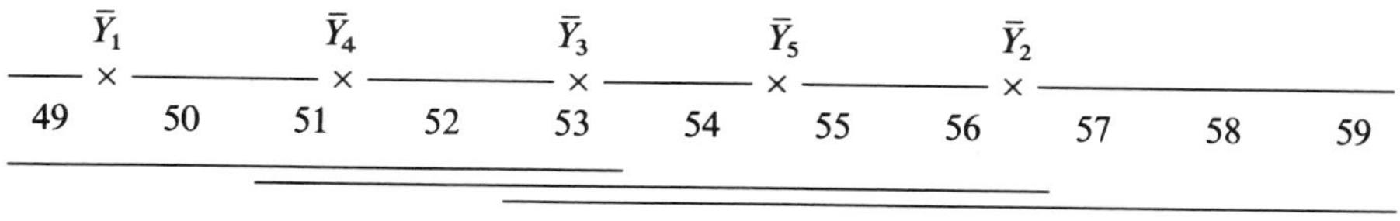

Figure 5.1

$\sqrt{(5-1)F_{4,15,0.95}(2)} = 4.948$ for Scheffé, and $K = q_{5,15,0.95} = 4.37$ for Tukey, Thus, Tukey intervals are best, of the form $\bar{Y}_i - \bar{Y}_j \pm 6.46$.

Figure 5.1 illustrates a convenient graphical procedure for comparison of means. Lines are drawn under any two sample means which differ by less than 6.46, for which the confidence interval on the corresponding difference in μ_j's includes zero. Thus, the overall 95% confidence allows us to say that $\mu_2 > \mu_4$, $\mu_2 > \mu_1$, $\mu_5 > \mu_1$.

Problem 5.4.1: Evaluate $R_{\text{T,S}}$, $R_{\text{B,S}}$ and $R_{\text{T,B}}$ for $\alpha = 0.05$, one-way analysis of variance, for $k = 4$, for common sample sizes $n_1 = 5$ and $n_1 = 10$.

Problem 5.4.2: Consider the model for one-way analysis of variance with observations $Y_{ij} \sim N(\mu_i, \sigma^2)$ for $j = 1, \ldots, i = 1, \ldots, k$. Let $W_i = \bar{Y}_i - \mu_i$ for each i and $S^2 = \left[\sum_{ij} (Y_{ij} - \bar{Y}_{..})^2\right]/(n-k)$. Let $WMS = \left[\sum_i (W_i - \bar{W})n_1\right]/(k-1)$ and $F = WMS/S^2$. Let $q = \text{Range}(W_1, \ldots, W_k)/[S/\sqrt{n_1}]$. Let $\mathbf{c} = (c_1, \ldots, c_k)$ be a contrast. Let $H_c = \left[\sum_i |c_i|\right]^2 \Big/ \left[\sum_i c_i^2\right]$.

(a) Show that $H_c \leq k$.

(b) Prove that $F(k-1) = \sup_{c \in C} T_c^2$, where $T_c = \left(\sum_i c_i W_i\right) \Big/ \left[S^2 \sum_i c_i^2/n_1\right]^{1/2}$ and C is the collection of contrasts.

(c) Prove that $\dfrac{1}{2} \leq \dfrac{F(k-1)}{q^2} \leq \dfrac{k}{4}$.

(d) Use these inequalities to prove that $H_c/(4k) \leq R_{\text{T,S}} \leq H_c/8$.

(e) Compute $R_{\text{T,S}}$ for $\mathbf{c}$ of the forms $(1, -1, 0, \ldots, 0)$ and $(k-1, -1, \ldots, -1)$ for $k = 3, 5$, $n_1 = 10$, and $\gamma = 0.95$ and compare the values with the bounds given in (d).

5.5 BECHHOFER'S METHOD

It is sometimes desirable not only to compare several means but also to choose the largest and offer some measure of assurance that it is the best. A method developed by Bechhofer (1954), by Bechhofer, Dunnett and Sobel (1954) and improved by Fabian (1962) does this.

Definition 5.5.1: Let $(Z_1, \ldots, Z_k) = \tilde{\mathbf{Z}}$ be a r.s. from $N(0, 1)$ and let $W \sim \chi_v^2$ be independent of $\tilde{\mathbf{Z}}$. Let $D = \max_{i=2,\ldots,k} (Z_i - Z_1)/\sqrt{W/v}$. Then D is said to have Bechhofer's distribution $D(k, v)$ with parameters k, v.

Dunnett (1955) prepared the tables, which were reproduced in Fabian and Hannan (1985). As defined in the latter $\kappa_{1-\alpha}(k, v)$ is the $(1-\alpha)100$ percentile of the $D(k, v)$ distribution. See Table 7 in the Appendix.

The following theorem, a direct consequence of the definition, essentially identifies the population with the largest, or nearly the largest mean. The theorem is given as originally stated by Fabian (1962).

Theorem 5.5.1: Let $\bar{X}_i - \mu_i$ for $i = 1, \ldots, k$ be independent $N(0, \sigma^2/n)$. Let $\mu_0 = \max(\mu_1, \ldots, \mu_k)$. Let $\frac{S^2}{\sigma^2} \sim \chi_v^2$ be independent of $(\bar{X}_1, \ldots, \bar{X}_k)$. Let $\bar{X}_I = \max(\bar{X}_1, \ldots, \bar{X}_k)$ (I is the index of the largest $\bar{X}_i$). Define $\delta = \kappa_{1-\alpha}(k, v)(S^2/n)^{1/2} - \left(\bar{X}_I - \max_{i \neq I} \bar{X}_i\right)$ and $\delta_+ = \max(0, \delta)$. Then

$$P(\mu_I \geq \mu_0 - \delta_+) \geq 1 - \alpha.$$

Example 5.5.1: Consider the data of Example 4.3.1. Then $v = 15$, $k = 5$, $\kappa_{0.95}(5, 15) = 3.30$, $I = 2$, $\bar{X}_I = 58.1$, $\sigma = 3.30$, $\delta = \delta_+ = (8.73/4)^{1/2} - (58.1 - 56.8) = 3.58$. Thus, we have 95% confidence that $\mu_2 \geq \mu_0 - 3.58$, that μ_2 is at least as large as the maximum (of the μ_i's, *not* of the $\bar{X}_i$'s) minus 3.58.

***Proof of Theorem 5.5.1*:** For simplicity of notation suppose $\mu_1 = \mu_0$. Define $Z_i = (\bar{X}_i - \mu_i)/(\sigma/\sqrt{n})$ for each i and let $W = S^2 v/\sigma^2$. Then for $\kappa = \kappa_{1-\alpha}(k, v)$ the event

$$A = \left\{ \max_{2 \leq i \leq n} (Z_i - Z_1)/\sqrt{S^2/n} \leq \kappa \right\}$$

has probability $1 - \alpha$, and implies the event

$$\begin{aligned} H &= \{(\bar{X}_I - \mu_I) - (\bar{X}_1 - \mu_I) \leq \kappa\sqrt{S^2/n}\} \cup \{I = 1\} \\ &= \{\mu_1 \leq \mu_I + (\bar{X}_1 - \bar{X}_I) + \kappa\sqrt{S^2/n}\} \cup \{I = 1\} \end{aligned}$$

If $I = 1$ then, since $\delta_+ \geq 0$, certainly $\mu_I \geq \mu_0 - \delta_+$. If $I \neq 1$ then $\bar{X}_1 \leq \max_{i \neq I} \bar{X}_i$ so that when $\bar{X}_1$ is replaced by this maximum the inequality still holds. Thus H implies $\mu_I \geq \mu_0 - \delta_+$. □

Let $\Delta = K\sqrt{S^2/n}$. If $\mu_1 = \mu_0 \geq \mu_j + \Delta$ for $j = 2, \ldots, k$ then, since $\Delta > \delta_+$ with probability one, $\mu_I \geq \mu_1 - \delta_+$ implies $\mu_I \geq \mu_1 - \Delta$, which then implies $I = 1$. Thus, if the maximum μ_j is at least Δ larger than the second largest μ_j then with probability at least $1 - \alpha \mu_I$ is the largest sample mean. This was the original formulation by Bechhofer (1954). In this formulation S^2 was obtained from a first sample, then n was chosen to make Δ equal to a prescribed constant Δ_0.

Problem 5.5.1: Table 5.5.1, taken from Hald (1952, p. 434), presents the measured strength minus 340 of nine cables, with 12 independent measurements on each cable. The last two rows present (sums of squared deviation) $= s_i$, so that the sample variance for the ith sample is $S_i^2 = s_i/11$, and the sample means $x_i \equiv \bar{X}_i$ for each cable. Error sum of square was $\sum s_i = 2{,}626.9$, and $\sum x_i = -6.58$, $\sum x_i^2 = 165.1$.

The usual one-way analysis of variance model with means μ_j seems appropriate.

(a) Fill out an ANOVA table and test $H_0\colon \mu_1 = \cdots = \mu_9$ for $\alpha = 0.05$.

(b) Suppose you wish confidence interval on all pairs $\mu_j - \mu_{j'}$ with simultaneous confidence coefficient 0.95. For the Bonferroni, Scheffé, and Tukey methods find constants $K = K_B$, K_S, and K_T so that $(\bar{X}_j - \bar{X}_{j'}) \pm KS/\sqrt{12}$ are the appropriate confidence intervals.

(c) For the smallest of K_B, K_S, K_T find $KS/\sqrt{12}$ and present a line diagram similar to that of Example 4.3.1.

(d) Suppose cable #1 is the standard cable and you therefore only want simultaneous intervals on $\mu_j - \mu_1$ for $j > 1$. Which method is appropriate?

(e) Use Bechhofer's method to estimate the largest population mean μ_0 and make an appropriate 95% confidence statement.

(f) Suppose these cables are of three types, with cables 1, 2, 3 of type A; 4, 5, 6 of type B; and 7, 8, 9 of type C. Suppose also that you only wish to compare cables of the same type. Thus you want simultaneous confidence intervals on $\mu_j - \mu_{j'}$ where j and j' are of the same type. For coefficient 0.95 use Scheffé's method to do this for all nine such differences. Compare the length with that given by the Bonferroni method. How could the Tukey and Bonferroni method be combined to do this? Additional quantiles for q would be needed, so the tables for q given here are inadequate.

Table 5.5.1 Strength Measurements of Nine Cables

	Strength of Cable i − 340 (kilograms)								
j	1	2	3	4	5	6	7	8	9
1	5	−11	0	−12	7	1	−1	−1	2
2	−13	−13	−10	4	1	0	0	0	6
3	−5	−8	−15	2	5	−5	2	7	7
4	−2	8	−12	10	0	−4	1	5	8
5	−10	−3	−2	−5	10	−1	−4	10	15
6	−6	−12	−8	−8	6	0	2	8	11
7	−5	−12	−5	−12	5	2	7	1	−7
8	0	−10	0	0	2	5	5	2	7
9	−3	5	−4	−5	0	1	1	−3	10
10	2	−6	−1	−3	−1	−2	0	6	7
11	−7	−12	−5	−3	−10	6	−4	0	8
12	−5	−10	−11	0	−2	7	2	5	1
s_i	270.9	532.0	280.9	454.7	300.9	153.7	110.9	180.7	342.2
x_i	−4.08	−7.0	−6.08	−2.67	1.92	0.83	0.92	3.33	6.25

Source: From Hald (1952) with permission of John Wiley & Sons.

CHAPTER 6

Two-Way and Three-Way Analyses of Variance

6.1 TWO-WAY ANALYSIS OF VARIANCE

Hicks' (1982, p. 105) has the following problem:

To determine the effect of two glass types and three phosphor types on the light output of a television tube, light output is measured by the current required to produce 30 foot-lamberts of light output. Thus the higher the current is in microamperes, the poorer the tube is in light output. Three observations were taken under each of the six treatment conditions and the experiment was completely randomized. The following data were recorded.

	Phosphor Type		
Glass Type	*A*	*B*	*C*
1	280	300	270
	290	310	285
	285	295	290
2	230	260	220
	235	240	225
	240	235	230

Do an analysis of variance on these data and test the effect of glass type, phosphor types, and interaction on the current flow.

By "completely randomized" the author means that the 18 tubes were randomly partitioned into six groups of 3 with the glass on the tubes in group i receiving treatment i for $i = 1, \dots, 6$. By the "analysis of variance" in this case we mean that the measure of overall variation (corrected total sum of squares) is expressed as the sum of four measures of variation due to glass type, phosphor type, interaction between glass and phosphor types, and error.

It will be particularly useful in the consideration of such tables to consider them as vectors without reshaping them into column vectors. By keeping the shape of the table, the corresponding linear models, with **x**-vectors which are indicators of rows, columns or cells of the table, will be much more obvious. Mathematically, models remain the same whether or not we reshape into columns. Reshaping into columns does have an intuitive cost, however.

In general suppose we observe Y_{ijk} for $k = 1, \ldots, K$, $j = 1, \ldots, J$, and $i = 1, \ldots, I$. The values of j correspond to the levels of a factor B. The values of i correspond to the levels of a factor A. The values of k correspond to repeated measurements taken for each i,j combination

$$\text{Model:} \quad Y_{ijk} \sim N(\mu_{ij}, \sigma^2) \qquad \text{and these} \quad Y_{ijk} \quad \text{are independent.}$$

Let **Y** be the array of Y_{ijk}. Let $\boldsymbol{\mu}$ (rather than $\boldsymbol{\theta}$) denote $E(\mathbf{Y})$. All the elements in the same cell of $\boldsymbol{\mu}$ are identical (Figure 6.1).

Define

$$\mu = \left(\sum_{ij} \mu_{ij}\right) \Big/ IJ$$

$$\alpha_i = \bar{\mu}_{i\cdot} - \mu = \frac{1}{J} \sum_j \mu_{ij} - \mu$$

		B 1	2	*J*	Means
		Y_{111}	Y_{121}	Y_{1J1}	
	1	$\vdots$	$\vdots$	$\vdots$	$\bar{Y}_{1\cdot\cdot}$
		Y_{11K}	Y_{12K}	Y_{1JK}	
A		$\vdots$	$\vdots$	$\vdots$	
		Y_{I11}	Y_{I21}	Y_{IJ1}	
	I	$\vdots$	$\vdots$	$\vdots$	$\bar{Y}_{I\cdot\cdot}$
		Y_{I1K}	Y_{I2K}	Y_{IJK}	
Means		$\bar{Y}_{\cdot 1\cdot}$	$\bar{Y}_{\cdot 2\cdot}$	$\bar{Y}_{\cdot J\cdot}$	$\bar{Y}_{\cdot\cdot\cdot}$

FIGURE 6.1

$$\beta_j = \bar{\mu}_{\cdot j} - \mu = \frac{1}{I}\sum_i \mu_{ij} - \mu$$

$$(\alpha\beta)_{ij} = \mu_{ij} - [\mu + \alpha_i + \beta_j].$$

Then $\mu_{ij} = \mu + \alpha_i + \beta_j + (\alpha\beta)_{ij}$. The full model then can be written as follows.

$$\text{Full model:} \qquad Y_{ijk} = \mu + \alpha_i + \beta_j + (\alpha\beta)_{ij} + \varepsilon_{ijk},$$

where

$$\sum_1^I \alpha_i = \sum_1^J \beta_j = \sum_i (\alpha\beta)_{ij} = \sum_j (\alpha\beta)_{ij} = 0, \qquad \text{and} \qquad \varepsilon_{ijk} \sim N(0, \sigma^2).$$

The ε_{ijk} are independent. There is a 1–1 correspondence between the parameter vectors $\boldsymbol{\mu} \equiv E(\mathbf{Y})$ and the $(1 + I + J + IJ)$-tuple of parameter vectors $(\mu, \alpha_1, \ldots, \alpha_I, \beta_1, \ldots, \beta_J, (\alpha\beta)_{11}, \ldots, (\alpha\beta)_{IJ})$ satisfying the above equalities. This expression of the full model is simply another way of presenting the model on the previous page. Thus it is not necessary and is employed only because it makes the study of the variability of the μ_{ij} more convenient. The parameters $(\alpha\beta)_{ij}$ are called *interactions.*

Example 6.1.1: Consider a 2×3 table of μ_{ij}'s as follows:

				Mean
	69	65	58	64
	57	59	52	56
Mean	63	62	55	60

Then $\mu = 60$, $\alpha_1 = 4$, $\alpha_2 = -4$, $\beta_1 = 3$, $\beta_2 = 2$, $\beta_3 = -5$, and the $(\alpha\beta)_{ij}$ are

	B		
A	2	−1	−1
	−2	1	1

A graphical display is useful. See Figure 6.2. The fact that the graphs of the means for rows 1 and 2 are almost parallel is a reflection of the fact that the interactions are small. If, in fact, interactions were zero, so that $\mu_{ij} = \mu + \alpha_i + \beta_j$, we would say that means are *additive*, or that there is *additivity of means.*

In vector form, we can write

$$\boldsymbol{\mu} = \sum_{ij} \mu_{ij}\mathbf{C}_{ij} = \sum_{ij} (\mu + \alpha_i + \beta_j + (\alpha\beta)_{ij})\mathbf{C}_{ij}$$

$$= \mu\mathbf{x}_0 + \sum_i \alpha_i\mathbf{A}_i + \sum_j \beta_j\mathbf{B}_j + \sum_{ij} (\alpha\beta)_{ij}\mathbf{C}_{ij},$$

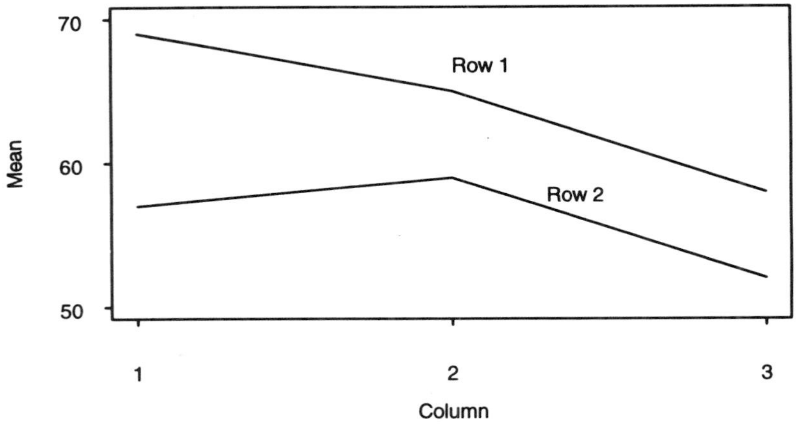

FIGURE 6.2 Cell population means.

where $\mathbf{x}_0$ is the array of all 1's

$$\mathbf{C}_{ij} = \text{indicator of cell } ij$$

$$\mathbf{A}_i = \sum_{j=1}^{J} \mathbf{C}_{ij} = \text{indicator or row } i \text{ (level } i \text{ of } \mathbf{A})$$

$$\mathbf{B}_j = \sum_{i=1}^{I} \mathbf{C}_{ij} = \text{indicator of column } j \text{ (level } j \text{ of } \mathbf{B})$$

Define Ω to be space of $I \times J \times K$ arrays, so that each realization of $\mathbf{Y}$ is in Ω.

Let $V = \mathscr{L}(\mathbf{C}_{11}, \mathbf{C}_{12}, \ldots, \mathbf{C}_{IJ})$ = set of arrays with elements in the same cell equal. Let $V_0 = \mathscr{L}(\mathbf{x}_0)$, $V_R = \mathscr{L}(\mathbf{A}_1, \ldots, \mathbf{A}_I)$, $V_A = V_R \cap V_0^\perp$, $V_C = \mathscr{L}(\mathbf{B}_1, \ldots, \mathbf{B}_J)$, $V_B = V_C \cap V_0^\perp$, $V_{AB} = V \cap (V_0 \oplus V_A \oplus V_B)^\perp$. Then it is easy to show that

$$V_A = \{\sum a_i \mathbf{A}_i | \sum a_i = 0\}, \qquad V_B = \{\sum b_j \mathbf{B}_j | \sum b_j = 0\}$$

$$V_{AB} = \left\{\sum_{ij} c_{ij} \mathbf{C}_{ij} | \sum_i c_{ij} = 0 \text{ for each } j \text{ and } \sum_j c_{ij} = 0 \text{ for each } i\right\}.$$

The subspaces V_0, V_A, V_B, V_{AB} are mutually orthogonal and $V = V_0 \oplus V_A \oplus V_B \oplus V_{AB}$. That $V_A \perp V_B$ follows from a simple computation of inner products. The other orthogonalities follow from the definitions of these subspaces. Thus every vector $\mathbf{y} \in \Omega$ is the sum of its projections onto the five mutually orthogonal subspaces V_0, V_A, V_B, V_{BA}, $V^\perp$. That is, $\mathbf{Y} = \hat{\mathbf{Y}}_0 + \hat{\mathbf{Y}}_A + \hat{\mathbf{Y}}_B + \hat{\mathbf{Y}}_{AB} + \mathbf{e}$, where

$$\hat{\mathbf{Y}}_0 = p(\mathbf{Y} | V_0) = \bar{Y}_{\ldots} \mathbf{x}_0 = \hat{\mu} \mathbf{x}_0,$$

$$\hat{\mathbf{Y}}_A = p(\mathbf{Y} | V_A) = p(\mathbf{Y} | V_R) - p(\mathbf{Y} | V_0) = \sum_i (\bar{Y}_{i\cdot\cdot} - \bar{Y}_{\ldots}) \mathbf{A}_i = \sum_i \hat{\alpha}_i \mathbf{A}_i$$

$$\hat{\mathbf{Y}}_B = p(\mathbf{Y}|V_B) = p(\mathbf{Y}|V_C) - p(\mathbf{Y}|V_0) = \sum_j (\bar{Y}_{.j.} - \bar{Y}_{...})\mathbf{B}_j = \sum \hat{\beta}_j \mathbf{B}_j,$$

$$\hat{\mathbf{Y}}_{AB} = p(\mathbf{Y}|V_{AB}) = p(\mathbf{Y}|V) - (\hat{\mathbf{Y}}_0 + \hat{\mathbf{Y}}_A + \hat{\mathbf{Y}}_B) = \sum_{ij} [\bar{Y}_{ij.} - (\bar{Y}_{...} + \hat{\alpha}_i + \hat{\beta}_j)]\mathbf{C}_{ij},$$

$$= \sum_{ij} \widehat{(\alpha\beta)}_{ij}\mathbf{C}_{ij},$$

$\mathbf{e} = \mathbf{Y} - \hat{\mathbf{Y}}$, where $\hat{\mathbf{Y}} = p(\mathbf{Y}|V) = \sum_{ij} \bar{Y}_{ij.}\mathbf{C}_{ij} = \hat{\mathbf{Y}}_0 + \hat{\mathbf{Y}}_A + \hat{\mathbf{Y}}_B + \hat{\mathbf{Y}}_{AB}$. Substituting $\boldsymbol{\mu}$ for $\mathbf{Y}$ in these formulas we get

$$\boldsymbol{\mu} = \boldsymbol{\mu}_0 + \boldsymbol{\mu}_A + \boldsymbol{\mu}_B + \boldsymbol{\mu}_{AB},$$

where

$$\boldsymbol{\mu}_0 = p(\boldsymbol{\mu}|V_0) = \mu\mathbf{x}_0, \qquad \boldsymbol{\mu}_A = p(\boldsymbol{\mu}|V_A) = \sum_i \alpha_i \mathbf{A}_i$$

$$\boldsymbol{\mu}_B = p(\boldsymbol{\mu}|V_B) = \sum_j \beta_j \mathbf{B}_j, \qquad \boldsymbol{\mu}_{AB} = p(\boldsymbol{\mu}|V_{AB}) = \sum_i (\alpha\beta)_{ij}\mathbf{C}_{ij}.$$

These expressions explain why $\hat{\mu}$, $\hat{\alpha}_i$, $\hat{\beta}_j$, $\widehat{(\alpha\beta)}_{ij}$ were defined as above.

Properties of the Estimators $\hat{\boldsymbol{\mu}}$, $\hat{\boldsymbol{\alpha}}_i$, $\hat{\boldsymbol{\beta}}_j$, and $\widehat{(\boldsymbol{\alpha\beta})}_{ij}$: Each of these estimators is linear in the observations Y_{ijk}, so that their properties are relatively easy to determine. It may be surprising that we should even attempt to estimate any of α_i, β_j, or $(\alpha\beta)_{ij}$, since, as defined in Section 3.3, these parameters are not estimable with respect to the parameter space $R_{1+I+J+IJ}$ of all possible parameter vectors. However, we have restricted our parameter space by forcing row and column sums to be zero, so that relative to the restricted parameter space these parameters are estimable. Since $E(\bar{Y}_{ij.}) = \mu_{ij}$, it follows that any linear function of these μ_{ij} are estimated unbiasedly by the corresponding functions of the $\bar{Y}_{ij.}$. Thus, $\hat{\mu} = \bar{Y}_{...}$, $\hat{\alpha}_i = \bar{Y}_{i..} - \bar{Y}_{...}$, $\hat{\beta}_{.j.} = \bar{Y}_{.j.} - \bar{Y}_{...}$ and $\widehat{(\alpha\beta)}_{ij} = \bar{Y}_{ij.} - [\hat{\mu} + \hat{\alpha}_i + \hat{\beta}_j]$ are all unbiased estimators of the corresponding parameters (the "polite" versions of the same symbols—without the hats). The four arrays $\hat{\mu}$, $\{\hat{\alpha}_1, \ldots, \hat{\alpha}_I\}$, $\{\hat{\beta}_1, \ldots, \hat{\beta}_J\}$, $\{\widehat{(\alpha\beta)}_{ij}\}$ are mutually uncorrelated because they are linear functions of the corresponding four orthogonal projections $\hat{\mathbf{Y}}_0$, $\hat{\mathbf{Y}}_A$, $\hat{\mathbf{Y}}_B$, $\hat{\mathbf{Y}}_{AB}$. By symmetry it is clear that the members of the same array have the same variance. Let us find formulas for these variances: $\sigma_0^2 \equiv \text{Var}(\hat{\mu})$, $\sigma_a^2 \equiv \text{Var}(\hat{\alpha}_i)$, $\sigma_b^2 \equiv \text{Var}(\hat{\beta}_j)$, and $\sigma_{ab}^2 \equiv \text{Var}(\widehat{(\alpha\beta)}_{ij})$.

Since $\hat{\mu} = \bar{Y}_{...}$ is the mean of $n = KIJ$ uncorrelated observations with variance σ^2, we find $\text{Var}(\hat{\mu}) = \sigma^2/KIJ$. Similarly, $\text{Var}(\bar{Y}_{i..}) = \sigma^2/KJ$ and $\text{Var}(\bar{Y}_{.j.}) = \sigma^2/KI$. Since $\bar{Y}_{i..} = \hat{\mu} + \hat{\alpha}_i$, and the two terms on the right are uncorrelated, $\sigma_a^2 = (\sigma^2/JK)\left[1 - \dfrac{1}{I}\right] = \dfrac{\sigma^2(I-1)}{IJK}$. Similarly, $\sigma_b^2 = \dfrac{\sigma^2(J-1)}{IJK}$. This expression for σ_a^2 may also be found from the computation $\text{trace}(D[\hat{\mathbf{Y}}_A]) = IJK\sigma_a^2 = \text{trace}(\mathbf{P}_{V_A}\sigma^2) = \sigma^2 \dim(V_A) = \sigma^2(I-1)$.

Since $\bar{Y}_{ij\cdot} = \hat{\mu} + \hat{\alpha}_i + \hat{\beta}_j + \widehat{(\alpha\beta)}_{ij}$, and the four terms on the right are uncorrelated, we find $\sigma_{ab}^2 = \text{Var}(\bar{Y}_{ij\cdot}) - \sigma_0^2 - \sigma_a^2 - \sigma_b^2 = \dfrac{\sigma^2(I-1)(J-1)}{IJK}$.

Covariances among terms in the same array may be found by exploiting the linear restrictions: $\sum_i \hat{\alpha}_i = \sum_j \hat{\beta}_j = \sum_i \widehat{(\alpha\beta)}_{ij} = \sum_j \widehat{(\alpha\beta)}_{ij} = 0$. For example, suppose that $c_a = \text{cov}(\hat{\alpha}_i, \hat{\alpha}_{i'})$ for $i \neq i'$. Then $\text{Var}\left(\sum_i \hat{\alpha}_i\right) = 0 = I\sigma_a^2 + (I-1)Ic_a$, so that $c_a = -\sigma_a^2/(I-1) = -\sigma^2/IJK$ and $\rho(\hat{\alpha}_i, \hat{\alpha}_{i'}) = -1/(I-1)$. Similarly, $\text{cov}(\hat{\beta}_j, \hat{\beta}_{j'}) = -\sigma_b^2/(J-1)$, $\rho(\hat{\beta}_j, \hat{\beta}_{j'}) = -1/(J-1)$. Covariances among interaction terms can be shown to be

$$\text{cov}(\widehat{(\alpha\beta)}_{ij}, \widehat{(\alpha\beta)}_{i'j'}) = \begin{cases} -(I-1)\sigma^2/IJK & \text{for} \quad i = i', j \neq j' \\ -(J-1)\sigma^2/IJK & \text{for} \quad i \neq i', j = j' \\ \sigma^2/IJK & \text{for} \quad i \neq i', j \neq j'. \end{cases}$$

If $\boldsymbol{\varepsilon}$ has a multivariate normal distribution then these estimators are jointly normally distributed.

The Analysis of Variance: By the Pythagorean Theorem, we can write $\|\mathbf{Y}\|^2$ as the sum of the squared lengths of the five vectors in the decomposition $\mathbf{Y} = \hat{\mathbf{Y}}_0 + \hat{\mathbf{Y}}_A + \hat{\mathbf{Y}}_B + \hat{\mathbf{Y}}_{AB} + \mathbf{e}$ and organize the data into the analysis of variance table, Table 6.1.1. We present these sums of squares in their more intuitive forms and also in their computational form. In the old days, when the author was a student, not long after R. A. Fisher developed these methods, before easy computations were possible, these formulas were necessary to avoid the necessity of first computing means, then sums of squares. These computational formulas are no longer so important. In fact, sums of squares are computed more precisely in their "deviation" form, using the first expression given below. If the computational form is used, precision can be enhanced by first subtracting a convenient constant from all Y_{ijk}. Only $\|\hat{\mathbf{Y}}_0\|^2$ is affected.

$$\|\hat{\mathbf{Y}}_0\|^2 \equiv \text{"Correction Term"} \equiv \text{CT} = \bar{Y}_{\cdots}^2(IJK)$$

$$\|\hat{\mathbf{Y}}_A\|^2 = KJ\sum \hat{\alpha}_i^2 = KJ\sum_i (\bar{Y}_{i\cdot\cdot} - \bar{Y}_{\cdots})^2 = KJ\sum_i \bar{Y}_{i\cdot\cdot}^2 - \text{CT} = \sum_i T_{i\cdot\cdot}^2/(KJ) - \text{CT}$$

where

$$T_{i\cdot\cdot} = \sum_{jk} Y_{ijk}, \qquad T_{\cdot j\cdot} = \sum_{ik} Y_{ijk}, \qquad T_{ij\cdot} = \sum_k Y_{ijk}$$

$$\|\hat{\mathbf{Y}}_B\|^2 = KI\sum_j \hat{\beta}_j^2 = KI\sum_j (\bar{Y}_{\cdot j\cdot} - \bar{Y}_{\cdots})^2 = KI\sum_j \bar{Y}_{\cdot j\cdot}^2 - \text{CT} = \sum_j T_{\cdot j\cdot}^2/(IK) - \text{CT}$$

$$\text{Subtotal SSqs.} = \|\hat{\mathbf{Y}} - \hat{\mathbf{Y}}_0\|^2 = K\sum_{ij} (\bar{Y}_{ij\cdot} - \bar{Y}_{\cdots})^2 = K\sum_{ij} \bar{Y}_{ij\cdot}^2 - \text{CT}$$

$$= \sum_{ij} T_{ij\cdot}^2/K - \text{CT}$$

Table 6.1.1 Analysis of Variance Table

Source	Space	DF	SSqs.	MSq.	Expected MSq.
A	V_A	$I - 1$	$\|\hat{\mathbf{Y}}_A\|^2$		$\sigma^2 + \frac{JK}{I-1}\sum \alpha_i^2$
B	V_B	$J - 1$	$\|\hat{\mathbf{Y}}_B\|^2$		$\sigma^2 + \frac{IK}{J-1}\sum \beta_j^2$
A × **B** Interaction	V_{AB}	$(I-1)(J-1)$	$\|\hat{\mathbf{Y}}_{AB}\|^2$		$\sigma^2 + \frac{K}{(I-1)(J-1)}\sum_{ij} (\alpha\beta)_{ij}^2$
Subtotal	$V \cap V_0^\perp$	$IJ - 1$	$\|\hat{\mathbf{Y}} - \hat{\mathbf{Y}}_0\|^2$		$\sigma^2 + \frac{K}{IJ-1}\sum_{ij} (\mu_{ij} - \mu)^2$
Error	$V^\perp$	$(K-1)IJ$	$\|\mathbf{Y} - \hat{\mathbf{Y}}\|^2$	S^2	σ^2
Corr. Total	$V_0^\perp$	$IJK - 1$	$\|\mathbf{Y} - \hat{\mathbf{Y}}_0\|^2$		
Mean	V_0	1	$\|\hat{\mathbf{Y}}_0\|^2$		
Unadj. Total	Ω	IJK	$\|\mathbf{Y}\|^2$		

$$\text{Adjusted Total SSqs.} = \sum_{ijk} (Y_{ijk} - \bar{Y}_{...})^2 = \sum_{ijk} Y_{ijk}^2 - \text{CT}$$

$$A \times B \text{ Interaction SSqs.} = \text{Subtotal SSqs.} - \text{ASSqs} - \text{BSSqs.}$$

$$\text{Error SSqs} = \sum_{ijk} (Y_{ijk} - \bar{Y}_{ij\cdot})^2 = \text{Adjusted Total SSqs.} - \text{Subtotal SSqs.}$$

$$S^2 = \text{Error Mean Square} = \text{Error SSqs.}/[(K-1)IJ]$$

Suppose we wish to test H_{AB}: $(\alpha\beta)_{ij} = 0$ for all i, j (no interaction). Under H_{AB} $\boldsymbol{\mu}$ lies in $V_0 \oplus V_A \oplus V_B \equiv V_*$. Since $V \cap V_*^\perp = V_{AB}$, the F-statistic needed to test H_{AB} is therefore

$$F_{AB} = \frac{\|\hat{\mathbf{Y}}_{AB}\|^2/(I-1)(J-1)}{S^2} \quad \text{for} \quad (I-1)(J-1) \quad \text{and} \quad (K-1)IJ \text{ d.f.,}$$

Similarly, suppose we wish to test H_A: $\alpha_i = 0 \forall i$ (no **A** effect). Under H_A $\boldsymbol{\mu}$ lies in $V_0 \oplus V_B \oplus V_{AB}$ and, since $V \cap (V_0 \oplus V_B \oplus V_{AB})^\perp = V_A$, the statistic needed to test H_A is

$$F_A = \frac{\|\hat{\mathbf{Y}}_A\|^2/(I-1)}{S^2} \quad \text{for} \quad (I-1) \quad \text{and} \quad (K-1)IJ \text{ d.f..}$$

Similarly, the F-statistic for testing H_B: $\beta_j = 0$ for all j (no **B** effect) is

$$F_B = \frac{\|\hat{\mathbf{Y}}_B\|^2/(J-1)}{S^2} \quad \text{for} \quad (J-1) \quad \text{and} \quad (K-1)IJ \text{ d.f..}$$

Distributional Properties of the Sums of Squares: Being projections on mutually orthogonal subspaces, the random vectors $\hat{\mathbf{Y}}_0$, $\hat{\mathbf{Y}}_A$, $\hat{\mathbf{Y}}_B$, $\hat{\mathbf{Y}}_{AB}$, $\mathbf{e}$ are independent. We also know that for any subspace V^*

$$\|p(\mathbf{Y}|V^*)\|^2/\sigma^2 \sim \chi^2_{\dim V^*}(\|p(\boldsymbol{\mu}|V^*)\|^2/\sigma^2)$$

$\|p(\boldsymbol{\mu}|V^*)\|^2$ may be determined by substituting $\boldsymbol{\mu}$ for $\mathbf{Y}$ in the formula for $\|p(\mathbf{Y}|V^*)\|^2$. Thus, for example,

$$\|p(\boldsymbol{\mu}|V_A)\|^2 = \sum_i (\mu_{i\cdot} - \mu)^2 JK = JK \sum_1^I \alpha_i^2$$

To summarize:

(1) $\|\hat{\mathbf{Y}}_A\|^2/\sigma^2 \sim \chi^2_{I-1}((\sum \alpha_i^2)JK/\sigma^2)$(**A** SSqs.)

(2) $\|\hat{\mathbf{Y}}_B\|^2/\sigma^2 \sim \chi^2_{J-1}((\sum \beta_j^2)IK/\sigma^2)$(**B** SSqs.)

(3) $\|\hat{\mathbf{Y}}_{AB}\|^2/\sigma^2 \sim \chi^2_{(I-1)(J-1)}\left(\sum_{ij} (\alpha\beta)_{ij}^2 K/\sigma^2\right)$(**AB** Inter. SSqs.)

(4) $\|\hat{\mathbf{Y}} - \hat{\mathbf{Y}}_0\|^2/\sigma^2 \sim \chi^2_{IJ-1}\left(\sum_{ij} (\mu_{ij} - \mu)^2 K/\sigma^2\right)$(Subtotal SSqs.)

(5) $\|\mathbf{Y} - \hat{\mathbf{Y}}\|^2/\sigma^2 \sim \chi^2_{(K-1)IJ}(0)$ (Error SSqs.)

(6) $\mathbf{Y} - \hat{\mathbf{Y}}_0\|^2/\sigma^2 \sim \chi^2_{IJK-1}\left(\sum_{ij} (\mu_{ij} - \mu)^2 K/\sigma^2\right)$ (Adj. Total SSqs.)

(7) $\|\hat{\mathbf{Y}}_0\|^2/\sigma^2 \sim \chi^2_1(IJK\mu^2/\sigma^2)$

(8) $\|\mathbf{Y}\|^2/\sigma^2 \sim \chi^2_{IJK}\left(K \sum_{ij} \mu_{ij} /\sigma^2\right)$

The r.v.'s $\|\hat{\mathbf{Y}}_A\|^2$, $\|\hat{\mathbf{Y}}_B\|^2$, $\|\hat{\mathbf{Y}}_{AB}\|^2$, $\|\mathbf{e}\|^2$, $\|\hat{\mathbf{Y}}_0\|^2$ are independent.

Example 6.1.2: Consider the television tube data at the start of the chapter. These vectors are then

$$\mathbf{Y} = \begin{bmatrix} 280 & 300 & 270 \\ 290 & 310 & 285 \\ 285 & 295 & 290 \\ 230 & 260 & 220 \\ 235 & 240 & 225 \\ 240 & 235 & 230 \end{bmatrix} \qquad \mathbf{x}_0 = \begin{bmatrix} 1 & 1 & 1 \\ 1 & 1 & 1 \\ 1 & 1 & 1 \\ 1 & 1 & 1 \\ 1 & 1 & 1 \\ 1 & 1 & 1 \end{bmatrix}$$

$$\mathbf{A}_1 = \begin{bmatrix} 1 & 1 & 1 \\ 1 & 1 & 1 \\ 1 & 1 & 1 \\ 0 & 0 & 0 \\ 0 & 0 & 0 \\ 0 & 0 & 0 \end{bmatrix} \qquad \mathbf{A}_2 = \begin{bmatrix} 0 & 0 & 0 \\ 0 & 0 & 0 \\ 0 & 0 & 0 \\ 1 & 1 & 1 \\ 1 & 1 & 1 \\ 1 & 1 & 1 \end{bmatrix}$$

$$\mathbf{B}_1 = \begin{bmatrix} 1 & 0 & 0 \\ 1 & 0 & 0 \\ 1 & 0 & 0 \\ 1 & 0 & 0 \\ 1 & 0 & 0 \\ 1 & 0 & 0 \end{bmatrix} \qquad \mathbf{B}_2 = \begin{bmatrix} 0 & 1 & 0 \\ 0 & 1 & 0 \\ 0 & 1 & 0 \\ 0 & 1 & 0 \\ 0 & 1 & 0 \\ 0 & 1 & 0 \end{bmatrix} \qquad \mathbf{B}_3 = \begin{bmatrix} 0 & 0 & 1 \\ 0 & 0 & 1 \\ 0 & 0 & 1 \\ 0 & 0 & 1 \\ 0 & 0 & 1 \\ 0 & 0 & 1 \end{bmatrix}$$

$$(\mathbf{AB})_{23} = \begin{bmatrix} 0 & 0 & 0 \\ 0 & 0 & 0 \\ 0 & 0 & 0 \\ 0 & 0 & 1 \\ 0 & 0 & 1 \\ 0 & 0 & 1 \end{bmatrix}$$

$$\hat{\mathbf{Y}}_0 = \begin{bmatrix} 262.2 & 262.2 & 262.2 \\ 262.2 & 262.2 & 262.2 \\ 262.2 & 262.2 & 262.2 \\ 262.2 & 262.2 & 262.2 \\ 262.2 & 262.2 & 262.2 \\ 262.2 & 262.2 & 262.2 \end{bmatrix} \qquad \hat{\mathbf{Y}}_A = \begin{bmatrix} 27.22 & 27.22 & 27.22 \\ 27.22 & 27.22 & 27.22 \\ 27.22 & 27.22 & 27.22 \\ -27.22 & -27.22 & -27.22 \\ -27.22 & -27.22 & -27.22 \\ -27.22 & -27.22 & -27.22 \end{bmatrix}$$

$$\hat{\mathbf{Y}}_B = \begin{bmatrix} -2.22 & 11.11 & -8.889 \\ -2.22 & 11.11 & -8.889 \\ -2.22 & 11.11 & -8.889 \\ -2.22 & 11.11 & -8.889 \\ -2.22 & 11.11 & -8.889 \\ -2.22 & 11.11 & -8.889 \end{bmatrix} \qquad \hat{\mathbf{Y}}_{AB} = \begin{bmatrix} -2.22 & 1.11 & 1.11 \\ -2.22 & 1.11 & 1.11 \\ -2.22 & 1.11 & 1.11 \\ 2.22 & -1.11 & -1.11 \\ 2.22 & -1.11 & -1.11 \\ 2.22 & -1.11 & -1.11 \end{bmatrix}$$

$$\hat{\mathbf{Y}} = \begin{bmatrix} 285 & 301.7 & 281.7 \\ 285 & 301.7 & 281.7 \\ 285 & 301.7 & 281.7 \\ 235 & 245.0 & 225.0 \\ 235 & 245.0 & 225.0 \\ 235 & 245.0 & 225.0 \end{bmatrix} \qquad \mathbf{e} = \begin{bmatrix} -5 & -1.67 & -1.67 \\ 5 & 8.33 & 3.33 \\ 0 & -6.67 & 8.33 \\ -5 & 15.00 & -5.00 \\ 0 & -5.00 & 0.00 \\ 5 & -10.00 & 5.00 \end{bmatrix}$$

Means

$$\bar{Y}_{\cdots} = 262.2 \qquad \bar{Y}_{1\cdot\cdot} = 289.4 \qquad \bar{Y}_{2\cdots} = 235.0$$

$$\bar{Y}_{\cdot 2\cdot} = 260.0 \qquad \bar{Y}_{\cdot 2\cdot} = 273.3 \qquad \bar{Y}_{\cdot 3\cdot} = 253.3$$

$$\bar{Y}_{ij\cdot}\text{'s} \quad \begin{bmatrix} 285 & 301.7 & 281.7 \\ 235 & 245.0 & 225.0 \end{bmatrix}$$

Squared Lengths

$$\|\hat{\mathbf{Y}}_0\|^2 = 1{,}237{,}689 \qquad \|\hat{\mathbf{Y}}\|^2 = 1{,}252{,}317 \qquad \|\hat{\mathbf{Y}}\|^2 = 1{,}253{,}150$$

$$\|\hat{\mathbf{Y}}_A\|^2 = 13{,}339 \qquad \|\hat{\mathbf{Y}} - \hat{\mathbf{Y}}_0\|^2 = 14{,}628 \qquad \|\mathbf{Y} - \hat{\mathbf{Y}}_0\|^2 = 15{,}461$$

$$\|\hat{\mathbf{Y}}_B\|^2 = 1{,}244 \qquad \|\mathbf{e}\|^2 = 833 \qquad \|\hat{\mathbf{Y}}_{AB}\|^2 = 44$$

And the analysis of variance table is Table 6.1.2.

Table 6.1.2 Analysis of Variance Table

Source	Subspace	DF	SSqs.	MSq.	Expected MSq.
Glass	V_A	1	13,339	13,339	$\sigma^2 + 9\sum \alpha_i^2$
Phosphor	V_B	2	1,244	622	$\sigma^2 + 3\sum \beta_j^2$
G × P	V_{AB}	2	44	22	$\sigma^2 + (3/2)\sum (\alpha\beta)_{ij}^2$
Subtotal	$V \cap V_0^\perp$	5	14,628	2925.6	$\sigma^2 + (3/5)\sum (\mu_{ij} - \mu)^2$
Error	$V^\perp$	12	833	69.4	σ^2
Corr. Total	$V_0^\perp$	17	15,461		
Mean	V_0	1	1,237,689		
Total	Ω	18	1,253,150		

F-ratios

$$F_{ST} = \frac{\text{Subtotal MSq.}}{\text{Error MSq.}} = 42.16$$

$$F_{AB} = \frac{\text{G} \times \text{P MSq.}}{\text{Error MSq.}} = 0.32$$

$$F_A = \frac{\text{G MSq.}}{\text{Error MSq.}} = 192.2$$

$$F_B = \frac{\text{P MSq.}}{\text{Error MSq.}} = 8.96$$

F_{ST} may be used to test the null hypotheses that all cell means are equal. Since F_{ST} is so large we certainly reject. F_{AB} is certainly consistent with interactions all zero or small. F_A and F_B indicate strong A (glass) and B (phosphor) effects.

A graphical display of cell means makes the conclusion clear (see Figure 6.3). Cell means have standard errors $\sigma/\sqrt{3}$, which we estimate to be $\sqrt{S^2/3} = \sqrt{69.4/3} = 4.81$.

Tukey 95% confidence intervals on differences $\mu_{ij} - \mu_{i'j'}$ are of the form $\bar{Y}_{ij\cdot} - \bar{Y}_{i'j'\cdot} \pm qS/\sqrt{3}$ for $q = q_{0.95,6,12} = 4.75$. Thus, cell means differing by more than $qS/\sqrt{3} = 22.8$ may be labeled as "significantly different." Any two cells means corresponding to different glass levels are significantly different. Otherwise they are not.

If we believe that no interaction is present, a reasonable belief in this case, then the α_i and β_j are interesting parameters, and we may wish to find confidence intervals on differences $\alpha_i - \alpha_{i'}$ or $\beta_j - \beta_{j'}$. For example, Tukey intervals on these $\beta_j - \beta_{j'}$ are $[\bar{Y}_{\cdot j\cdot} - \bar{Y}_{\cdot j'\cdot} \pm 3.77\sqrt{69.4/6}] = [\bar{Y}_{\cdot j\cdot} - \bar{Y}_{\cdot j'\cdot} \pm 12.82]$, since

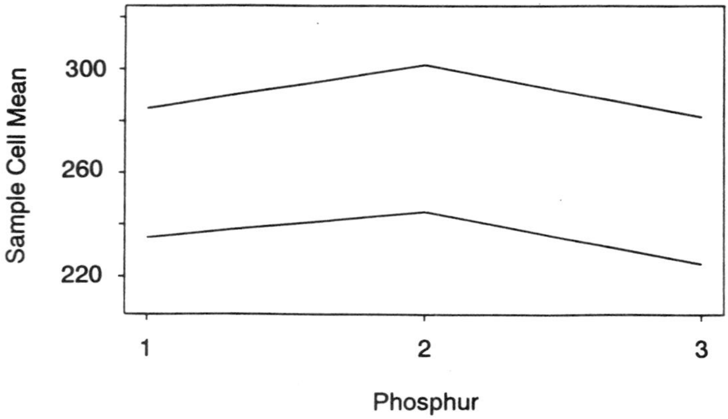

FIGURE 6.3 Sample cell means for combinations of phosphor and glass.

$q_{0.95,3,12} = 3.77$. Thus, levels 2 and 3 of phosphor can be viewed as significantly different at overall level 0.05.

Estimation of Cell Means: In the possible presence of interaction the model $Y_{ijk} \sim N(\mu_{ij}, \sigma^2)$ implies that μ_{ij} should be estimated by the corresponding cell mean $\bar{Y}_{ij}$. Simultaneous confidence intervals on all or some cell means may be obtained by treating the $I \times J$ cells as $k = I \times J$ treatments in one-way analysis of variance. The Bechhofer method may be used in the same way if the object is to estimate the largest μ_{ij}.

If interactions are known to be zero or nearly zero, these methods may be improved. It is tempting to perform an F-test for interaction, then decide upon failure to reject or for small F that interaction is lacking. Such procedures have been advocated. See Yates (1935) and a discussion of this problem by Traxler (1976). It is quite possible that a fair degree of interaction is present, however, but that by chance the F-value is small. This procedure can therefore lead to bias in the estimation of the μ_{ij}.

Under the assumption of no interaction $Y_{ij}^* = \hat{\mu} + \hat{\alpha}_i + \hat{\beta}_j = \bar{Y}_{\cdots} + (\bar{Y}_{i\cdot\cdot} - \bar{Y}_{\cdots}) + (\bar{Y}_{\cdot j\cdot} - \bar{Y}_{\cdots})$ is sometimes used as an estimator of μ_{ij}. Since $E(Y_{ij}^*) = \mu_{ij} - (\alpha\beta)_{ij}$, Y_{ij}^* has bias $-(\alpha\beta)_{ij}$.

$$\text{Var}(\hat{\mathbf{Y}}_{ij}^*) = \sigma^2\left[\frac{1}{n} + \left(\frac{1}{JK} - \frac{1}{n}\right) + \left(\frac{1}{IK} - \frac{1}{n}\right)\right] = \frac{\sigma^2}{n}[I + J - 1]$$

for $n = IJK$, so that the mean square error for Y_{ij}^* is $E(Y_{ij}^* - \mu_{ij})^2 = [-(\alpha\beta)_{ij}]^2 + \text{Var}(Y_{ij}^*)$.

$\text{MSE}(Y_{ij}^*)/\text{MSE}(\bar{Y}_{ij}) = [1 + (\alpha\beta)_{ij}^2/\sigma^2]\dfrac{1}{IJ}(I + J - 1)$, which is always less than 1 if $(\alpha\beta)_{ij} = 0$, considerably so for large I or J. $\hat{\mathbf{Y}}_{ij}^*$ has smaller mean square error than $\bar{Y}_{ij\cdot}$ if and only if $(\alpha\beta)_{ij}^2 < \mathrm{Var}(\bar{Y}_{ij}) - \mathrm{Var}(\hat{\mathbf{Y}}_{ij}^*) = \mathrm{Var}(\widehat{(\alpha\beta)_{ij}}) = [(I-1)(J-1)/IJ](\sigma^2/K)$. If we choose to use $Y^* = (Y_{ij}^*)$ to estimate $\boldsymbol{\mu}$ whenever the sum over i and j of these mean square errors is less than that for $\hat{\mathbf{Y}} = (\bar{Y}_{ij\cdot})$, we are lead to the inequality

$$\sum_{ij} (\alpha\beta)_{ij}^2 < (I-1)(J-1)\sigma^2/K$$

If we then replace these parameters by unbiased estimators in terms of MSAB and MSE, we get the inequality $F < 2$ (see Section 4.2). Thus, if $F < 2$, we might expect to do better (smaller total mean square error) using the $\hat{\mathbf{Y}}_{ij}^*$ rather than the $\bar{Y}_{ij}$. Some computer simulations indicate that such a procedure works better than the procedure which uses $(\bar{Y}_{ij\cdot})$ only in the case that interactions are very close to zero, which we almost never can know.

Fabian (1991) shows that when the goal is to find a confidence intervals on one μ_{ij} or on all μ_{ij} or to find the largest μ_{ij}, a two-stage procedure which uses $\hat{\mathbf{Y}}_{ij}^*$ and bounds on the sizes of interaction terms determined from the $\bar{Y}_{ij\cdot}$ will not in general improve on the direct use of the $\bar{Y}_{ij\cdot}$, treating the problem from a one-way ANOVA point of view.

In the case that interactions are absent $\boldsymbol{\mu}$ ranges over V_*, which has dimension $\nu = I + J - 1$. Scheffé simultaneous confidence intervals on $\eta_c = \sum_{ij} c_{ij}\mu_{ij}$ have the form $\hat{\eta}_c \pm K\sqrt{S^2(\sum c_{ij}^2)}$, where $K = \sqrt{\nu F_{\nu,(K-1)I}}$ and $\hat{\eta}_c = \sum_{ij} c_{ij}\hat{\mathbf{Y}}_{ij}^*$. For contrasts only, ν becomes $I + J - 2$. Since the cell mean estimators $\hat{\mathbf{Y}}_{0ij}$ have unequal covariances, the Tukey method must stick to the estimators $\bar{Y}_{ij\cdot}$, so that Scheffé intervals will be shorter. Recall, however, that if in truth interactions are present, the biases in Y_{ij}^* could cause errors in some of the Scheffé intervals.

Problem 6.1.1: Let the number of observations in cell ij be K_{ij}. Define $V_A = \mathscr{L}(\mathbf{A}_1, \ldots, \mathbf{A}_I) \cap V_0^{\perp}$, $V_B = \mathscr{L}(\mathbf{B}_1, \ldots, \mathbf{B}_J) \cap V_0^{\perp}$. What conditions must the a_i satisfy in order that $\sum a_i\mathbf{A}_i \in V_A$? Prove that $V_A \perp V_B$ if and only if $K_{ij} = K_{i\cdot}K_{\cdot j}/K_{\cdot\cdot}$ (where a dot means that the corresponding subscript has been summed over). Thus the subspaces V_0, V_A, V_B, V_{AB}, $V^{\perp}$ are still orthogonal when the cell frequencies are proportional. *Hint*: First show that the vectors $\mathbf{A}_i^* \equiv \mathbf{A}_i - p(\mathbf{A}_i | V_0)$ span V_A, and define vectors $\mathbf{B}_j^*$ similarly.)

Problem 6.1.2: In order to determine the effects of training on rats the following 2×4 factorial experiment was performed. Rats were trained by forcing them to swim in a tub of water for a given length of time with small weights attached. Four different lengths of training session (10, 20, 30, 40 minutes) were used. Training occurred every day, every second day, or every third day. For each of the 12 treatment combinations, 5 rats were trained. The experiment was completely randomized. That is, 60 rats were used and randomly assigned to the 12 treatment combinations. The measured variable was the log of the length of time the rats were able to swim with added weights after 6 weeks of training. (An experiment similar to this was actually performed. The experimenter called on the author's help because some observations were missing—some rats drowned.)

Length

	10	20	30	40
Every Day	3.26	4.01	2.88	3.76
	3.18	3.33	2.80	5.87
	2.84	4.94	3.87	4.31
	1.94	2.21	3.57	5.33
	2.69	2.73	3.18	4.53
Every Second Day	3.66	5.79	2.98	2.95
	3.07	4.23	4.21	4.34
	2.35	3.25	3.31	3.84
	4.23	3.28	2.44	4.05
	2.50	3.61	3.41	3.58
Every Third Day	3.36	3.35	3.47	6.53
	5.10	5.36	5.86	5.96
	4.53	4.94	3.61	5.32
	4.55	4.46	6.40	5.54
	4.60	5.47	4.43	6.59

Cell Totals

13.91	17.22	16.30	23.80
15.81	20.16	16.35	18.76
22.14	23.58	23.77	29.94

Cell Means

2.782	3.444	3.260	4.760
3.162	4.032	3.270	3.752
4.428	4.716	4.754	5.988

Sum of Squared Deviations From Cell Means

1.108	4.604	0.831	2.812
2.495	4.485	1.678	1.115
1.647	2.965	6.995	1.404

$$Y_{ijk} = 241.709 \qquad \sum Y_{ijk}^2 = 1052.88$$

Training means: 3.561, 3.554, 4.972

Length means: 3.457, 4.064, 3.761, 4.832

(a) State an appropriate model and fill out the analysis of variance table.

(b) Perform appropriate F-tests and state conclusions.

(c) Use Tukey's method to make comparisons among training and also among length effects.

(d) These data were actually generated using computer simulation for $\sigma = 0.8$, $\mu = 4$, $\alpha_1 = -0.7$, $\alpha_2 = 0$, $\alpha_3 = 0.7$, $\beta_1 = -0.6$, $\beta_2 = 0.2$, $\beta_3 = 0.2$, $\beta_4 = 0.6$, $(\alpha\beta)_{ij} \equiv 0$, $\sigma^2 = 0.64$. Determine the powers of the $\alpha = 0.05$ level tests of the null hypotheses of no training and of no length effects.

(e) Suppose that $\beta_j = \beta_0 + 10j\gamma$. That is, log of swimming time is linearly affected by the length of the training session. Give a 95% confidence interval on γ.

(f) Suppose that the observations 3.26 in cell (1, 1) and 5.54 in cell (3, 4) were missing. Determine an unbiased estimator of σ^2, assuming that the loss of these observations was independent of their values, and evaluate it.

Problem 6.1.3: Consider a 2×3 table with two observations per cell. Make up data so that the following conditions are satisfied:

(a) SSA = SSB = SSAB = 0, SSE = 2
(b) SSA > 0, SSB > 0, SSAB = SSE = 0
(c) SSA = SSB = 0, SSAB > 0, SSE = 100
(d) SSA > 0, SSB = 0, SSAB > 0, SSE = 0

Problem 6.1.4: Derive the formulas for the variances and covariances of the $\widehat{(\alpha\beta)}_{ij}$.

6.2 UNEQUAL NUMBERS OF OBSERVATIONS PER CELL

Let the number of observations in cell ij be K_{ij}. Define $V_0 = \mathscr{L}(\mathbf{J})$,

$$V_A = V_R \cap V_0^\perp, \qquad V_B = V_C \cap V_0^\perp, \qquad V_{AB} = V \cap (V_0 \oplus V_A \oplus V_B)^\perp,$$

as before. Then

$$V_A = \{\sum a_i \mathbf{A}_i | \sum a_i K_{i\cdot} = 0\}, \qquad V_B = \left\{\sum_j b_j \mathbf{B}_j | \sum b_j K_{\cdot j} = 0\right\},$$

$$V_{AB} = \left\{\sum_{ij} c_{ij}\mathbf{C}_{ij} \middle| \sum_j K_{ij}c_{ij} = 0 \text{ for each } i \text{ and } \sum_i K_{ij}c_{ij} = 0 \text{ for each } j\right\}.$$

As shown in Problem 6.1.1, $V_A \perp V_B$ if and only if $K_{ij} = K_{i\cdot}K_{\cdot j}/K_{\cdot\cdot}$ for all i and j.

Under the full model $Y_{ijk} \sim N(\mu_{ij}, \sigma^2)$, so that the BLUE for μ_{ij} is $\bar{Y}_{ij\cdot}$, which exists if $K_{ij} > 0$. It follows in the case that $K_{ij} > 0$ for all i and j that the BLUE for μ, α_i, β_j, $(\alpha\beta)_{ij}$ are

$$\hat{\mu} = \frac{1}{IJ}\sum_{ij} \bar{Y}_{ij\cdot}, \qquad \hat{\alpha}_i = \frac{1}{J}\sum_j \bar{Y}_{ij\cdot} - \hat{\mu},$$

$$\hat{\beta}_j = \frac{1}{I}\sum_i \bar{Y}_{ij\cdot} - \hat{\mu}, \qquad \widehat{(\alpha\beta)}_{ij} = \bar{Y}_{ij\cdot} - (\hat{\mu} + \hat{\alpha}_i + \hat{\beta}_j).$$

Variances and covariances are easy to determine. For example, $\text{Var}(\hat{\alpha}_i) = \sigma_i^2(1 - 2/I) + \frac{1}{I^2}\sum_j \sigma_j^2$ for $\sigma_i^2 = \text{Var}(W_{i\cdot}) = \frac{\sigma^2}{J^2}\sum_j \frac{1}{K_{ij}}$, where $W_{i\cdot} = \frac{1}{J}\sum_j \bar{Y}_{ij\cdot}$ and $\text{Var}(\hat{\alpha}_i - \hat{\alpha}_{i'}) = \text{Var}(W_{i\cdot} - W_{i'\cdot}) = \sigma_i^2 + \sigma_{i'}^2$.

Under the full model $\hat{\mathbf{Y}} = p(\mathbf{Y}|\mathbf{V}) = \sum_{ij} \bar{Y}_{ij\cdot}\mathbf{C}_{ij}$. The error space $V^\perp$ is the null space unless at least one $K_{ij} > 1$. As before $\text{SSE} = \|\mathbf{Y} - \hat{\mathbf{Y}}\|^2 = \sum_{ijk} (Y_{ijk} - \bar{Y}_{ij\cdot})^2$.

The statement that all interactions $(\alpha\beta)_{ij}$ are zero is equivalent to the statement that $\boldsymbol{\mu} \in V_* = V_0 \oplus V_A \oplus V_B$. To test the null hypothesis of no interaction, equivalently $\boldsymbol{\mu} \in V_*$, we must find $\hat{\mathbf{Y}}^* = p(\mathbf{Y}|V^*) = p(\hat{\mathbf{Y}}|V^*)$.

Define $\mathbf{A}_i^* = \mathbf{A}_i - \mathbf{A}_I$ for $i = 1, \ldots, I - 1$, and $\mathbf{B}_i^* = \mathbf{B}_i - \mathbf{B}_J$ for $j = 1, \ldots, J$. Then, since $\sum \alpha_i = 0$, and $\sum \beta_j = 0$, $V^* = \mathscr{L}(\mathbf{x}_0, \mathbf{A}_1^*, \ldots, \mathbf{A}_{I-1}^*, \mathbf{B}_1^*, \ldots, \mathbf{B}_{J-1}^*)$. Regression methods may be employed to find the least squares estimates $\hat{\mu}$, $\hat{\alpha}_1, \ldots, \hat{\alpha}_{I-1}, \hat{\beta}_1, \ldots, \hat{\beta}_{J-1}$ and $\hat{\alpha}_I = -\sum_i^{I-1} \hat{\alpha}_i$, $\hat{\beta}_J = -\sum_j^{J-1} \hat{\beta}_j$. Then $\hat{\mathbf{Y}} = \hat{\mu}\mathbf{x}_0 + \sum_i^I \hat{\alpha}_i\mathbf{A}_i + \sum_j^J \hat{\beta}_j\mathbf{B}_j$ and $\|\hat{\mathbf{Y}}^*\|^2 = \hat{\mu}T_{\cdots} + \sum_i^I \hat{\alpha}_i T_{i\cdot\cdot} + \sum_j^J \hat{\beta}_j T_{\cdot j\cdot}$. Only in exceptional cases will there be simple formulas for these estimators and squared lengths. The F-statistic for H_0: $\boldsymbol{\mu} \in V^*$ is $F = \{[\|\hat{\mathbf{Y}}\|^2 - \|\hat{\mathbf{Y}}^*\|^2]/(I - 1)(J - 1)\}/S^2$.

We may be interested in testing the null hypotheses of no A effects (all α_i's 0) in the absence or in the presence of interaction. Absence is equivalent to $\boldsymbol{\mu} \in V_*$, and H_0: (no A effect) then implies $\boldsymbol{\mu} \in V_0 \oplus V_B = \mathscr{L}(\mathbf{B}_1, \ldots, \mathbf{B}_J)$. The numerator sum of squares in the F-statistic is therefore $\|\hat{\mathbf{Y}}^* - \hat{\mathbf{Y}}_0\|^2$, where $\hat{\mathbf{Y}}_0 = \sum_j \bar{Y}_{\cdot j\cdot}\mathbf{B}_j$ for $(I - 1)$ d.f.. The denominator sum of squares is $\|\mathbf{Y}^* - \hat{\mathbf{Y}}_0\|^2$ for $K_{\cdot\cdot} - (I + J - 1)$ d.f.. It may be preferable to use $\|\mathbf{Y} - \hat{\mathbf{Y}}\|^2$, however, since interaction just may be present.

In the presence of interaction it probably does not make much practical sense to test the null hypotheses of no A effect. The null hypothesis then states that the average A effect α_i across all levels of B is 0 for each i. It is rare that such an average is of real interest. Nevertheless, there is nothing which prohibits such a test from a mathematical point of view. Under H_0: $\boldsymbol{\mu} = \mu\mathbf{x}_0 + \sum \beta_j\mathbf{B}_j + \sum_{ij} (\alpha\beta)_{ij}\mathbf{C}_{ij}$, so that H_0 is equivalent to $\boldsymbol{\mu} \in \mathscr{L}(\mathbf{B}_1, \ldots, \mathbf{B}_J) \cap V_{\alpha\beta} = V_{**}$ (say), where $V_{\alpha\beta} = \left\{\sum_{ij} (\alpha\beta)_{ij}\mathbf{C}_{ij} \,\middle|\, \sum_i (\alpha\beta)_{ij} = \sum_j (\alpha\beta)_{ij} = 0 \text{ for all } i \text{ and } j\right\}$. Unless K_{ij} is of the form $g_i f_j$ for all i and j, $V_{\alpha\beta}$ is not V_{AB}. It can be shown that $V_{\alpha\beta}$ is spanned by the vectors $(\mathbf{A}_i - \mathbf{A}_I) \times (\mathbf{B}_j - \mathbf{B}_J)$, where multiplication is componentwise. $V_{\alpha\beta}$ is not, in general, orthogonal to V_B. Using these basis vectors $p(\mathbf{Y}|V_{**})$ can be computed using regression methods. If $K_{ij} > 0$ for all ij then V_{**} has dimension $J + (I - 1)(J - 1)$.

The F-statistic F_B for H_0: (no B effect) is given analogously. The numerators of statistics F_A and F_B are no longer independent.

In the case that the K_{ij} are approximately equal a shortcut approximation is available. Computer experimentation has shown that even in the case of a 2×3 table with two observations in cells (1, 1) and (2, 3), four in the others, the null distributions of F-statistics remain approximately the same.

Define $\bar{K} = \left(\frac{1}{IJ}\sum_{ij}\frac{1}{K_{ij}}\right)^{-1}$, the harmonic mean. Let Error SSqs. $= \sum_{ijk}(Y_{ijk} - \bar{Y}_{ij.})^2$. Since sums of squares higher in the ANOVA table depend only on the means $\bar{Y}_{ij.}$, compute these using the formulas for the equal K_{ij} case using $\bar{K}$ instead of K, using these $\bar{Y}_{ij.}$. Thus, take

$$\bar{Y}_{i..} = \frac{1}{J}\sum_j \bar{Y}_{ij.}, \qquad \bar{Y}_{.j.} = \frac{1}{I}\sum_i \bar{Y}_{ij.}, \qquad \bar{Y}_{...} = \frac{1}{IJ}\sum_{ij} \bar{Y}_{ij.}, \qquad \text{CT} = \bar{Y}^2_{...}(IJ\bar{K}).$$

$$\text{Subtotal SSqs.} = \bar{K}\sum_{ij}(\bar{Y}_{i..} - \bar{Y}_{...})^2 = \bar{K}\sum_{ij}\bar{Y}^2_{i..} - \text{CT},$$

$$SSA = J\bar{K}\sum_i(\bar{Y}_{i..} - \bar{Y}_{...})^2 = J\bar{K}\sum_i \bar{Y}^2_{i..} - \text{CT},$$

$$\text{SSB} = I\bar{K}\sum_j(\bar{Y}_{ij.} - \bar{Y}_{...})^2 = I\bar{K}\sum_j \bar{Y}^2_{.j.} - \text{CT},$$

and

$$\text{SSAB} = \text{Subtotal SSqs.} - \text{SSA} - \text{SSB}.$$

Approximate confidence intervals on linear combinations $\sum_{ij} c_{ij}\mu_{ij}$ can be obtained by using the usual formulas with $\bar{K}$ replacing K.

Example 6.2.1: Suppose that we observe $\mathbf{Y}$ as follows for $I = 2$, $J = 3$.

$$\begin{array}{c} \\ \mathbf{A}_1 \\ \\ \\ \mathbf{A}_2 \end{array} \begin{array}{c} \begin{array}{ccc} \mathbf{B}_1 & \mathbf{B}_2 & \mathbf{B}_3 \end{array} \\ \left[\begin{array}{ccc} 24 & 12 & 9 \\ 22 & & 10 \\ & & 11 \\ \hline 10 & 5 & 4 \\ 12 & 3 & \end{array}\right] \end{array}.$$

Then, for example,

$$\mathbf{B}_2^* = \left[\begin{array}{ccc} 0 & 1 & -1 \\ 0 & & -1 \\ & & -1 \\ \hline 0 & 1 & -1 \\ 0 & 1 & \end{array}\right].$$

The vectors $\mathbf{Y}$, $\mathbf{x}$, $\mathbf{A}_1^*$, $\mathbf{B}_1^*$, $\mathbf{B}_2^*$, were put in column form and multiple regression was used to find $\hat{\mu} = 10.47$, $\hat{\alpha}_1 = 4.52$, $\hat{\beta}_1 = 6.53$, $\hat{\beta}_2 = -2.30$. Therefore $\hat{\alpha}_2 = -4.52$, and $\hat{\beta}_3 = -4.23$. Then $\hat{Y}_{111} = \hat{Y}_{112} = \hat{\mu} + \hat{\alpha}_1 + \hat{\beta}_1 = 21.52$. Similarly, $\hat{Y}_{121} = 12.69$, $\hat{Y}_{131} = \hat{Y}_{132} = 10.76$, $\hat{Y}_{211} = \hat{Y}_{212} = 12.48$, $\hat{Y}_{221} = \hat{Y}_{222} = 3.66$, $\hat{Y}_{231} = 1.72$. We find $\|\mathbf{Y}\|^2 = 1{,}800.0$, $\|\hat{\mathbf{Y}}\|^2 = 1{,}775.6$, ESS $= \|\mathbf{Y} - \hat{\mathbf{Y}}\|^2 = \|\mathbf{Y}\|^2 - \|\hat{\mathbf{Y}}\|^2 = 24.4$, $S^2 = 24.4/(11 - 4) = 3.49$. (SSE and S^2 correspond to the model $\boldsymbol{\mu} \in V_*$; for the model with interaction term SSE $= 8$.)

To test H_0: $\alpha_1 = 0$, we fit the model $\boldsymbol{\mu} \in V_0^* = \mathscr{L}(\mathbf{B}_1, \mathbf{B}_2, \mathbf{B}_3)$. We find $\hat{\mathbf{Y}}_0^* = p(\mathbf{Y} \mid V_0^*) = \sum \bar{Y}_{\cdot j \cdot} \mathbf{B}_j$, where $\bar{Y}_{\cdot 1 \cdot} = 17.00$, $\bar{Y}_{\cdot 2 \cdot} = 6.67$, $\bar{Y}_{\cdot 3 \cdot} = 8.50$, and $\|\hat{\mathbf{Y}}_0^*\|^2 = 1578.3$, $\|\hat{\mathbf{Y}} - \hat{\mathbf{Y}}_0^*\|^2 = 197.5$. Then $F = [197.5/1]/S^2 = 56.6$, for 1 d.f.. Since $\dim(V^* \cap V_0^*) = 1$, $F = t^2$, where $t = \hat{\alpha}_1/[S/\|\mathbf{d}\|]$, $\hat{\alpha}_1 = (\mathbf{d}, \mathbf{Y})/\|\mathbf{d}\|^2$, where $\mathbf{d} = \mathbf{A}_1^* - p(\mathbf{A}_1^* \mid V_0^*)$, and $\|\mathbf{d}\|^2 = \|\mathbf{A}_1^*\|^2 - \|p(\mathbf{A}_1^* \mid V_0^*)\|^2 = 11 - 4/3 = 29/3$. Thus $t = 4.52/[3.49(3/29)]^{1/2} = 7.52 = \sqrt{56.6}$.

Problem 6.2.1: Consider the following 2×3 table:

		Factor B		
		B_1	B_2	B_3
Factor A	A_1	25, 23	15, 17	8, 7, 9
	A_2	10	1, 3	4, 8

(a) Find the least squares estimates of the parameters μ_{ij}, μ, α_i, β_j, $(\alpha\beta)_{ij}$ in the full model.

(b) For the model with interactions zero determine the least squares estimates of the μ_{ij}, and the resulting error sum of squares. Also show that for this model $\hat{\mu} = 74/7$, $\hat{\alpha}_1 = -\hat{\alpha}_2 = 231/42$, $\hat{\beta}_1 = 99/14$, $\hat{\beta}_2 = -22/14$, $\hat{\beta}_3 = -77/14$. Test the null hypothesis H_{AB}: (no interaction effects) at level $\alpha = 0.05$.

(c) Assuming no interaction effects, test H_A: $\alpha_1 = \alpha_2 = 0$ at level $\alpha = 0.05$, using an F-statistic, F_A. The assumption of no interaction is not realistic, based on the test in (b).

(d) Repeat the F-test of (b) using the approximate procedure.

Problem 6.2.2: Suppose that the factors A and B have each have two levels, A_1, A_2 and B_1, B_2. Suppose also that one observation Y_{ij1} is taken in cell ij for all $(i, j) \neq (2, 2)$, and that for cell $(2, 2)$ two observations Y_{221} and Y_{222} are taken. Consider the additive model $Y_{ijk} = \mu + \alpha_i + \beta_j + \varepsilon_{ij}$, with $\alpha_1 + \alpha_2 = 0$, $\beta_1 + \beta_2 = 0$, ε_{ij}'s independent $N(0, \sigma^2)$.

(a) Give explicit nonmatrix formulas for the least squares estimators of μ, α_1, β_1.

(b) Test H_0: $\alpha_1 = 0$, for $\alpha = 0.05$, for $Y_{111} = 19$, $Y_{121} = 13$, $Y_{211} = 11$, $Y_{221} = 3$, $Y_{222} = 7$. Also find a 95% confidence interval on α_1. What is the relationship between this confidence interval and the test?

(c) Consider the approximate method. Let the estimators of μ, α_1, β_1 be $\mu^* = (\mathbf{m}, \mathbf{Y})$, $\alpha_1^* = (\mathbf{a}, \mathbf{Y})$, $\beta_1^* = (\mathbf{b}, \mathbf{Y})$. Find $\mathbf{m}$, $\mathbf{a}$, $\mathbf{b}$. Show that these estimators are unbiased, but that they have larger variances than the least squares estimators.

(d) Show that the results of (c) hold in general for any $I \times J$ table with K_{ij} observations in cell ij.

6.3 TWO-WAY ANALYSIS OF VARIANCE, ONE OBSERVATION PER CELL

If the number K of observations per cell in a two-way layout is just one, the degree of freedom for error is zero. In fact V, the space spanned by the cell indicators, *is* the sample space. If there is another estimator S^2 of σ^2 available, possibly from some previous experiment, then this estimator can serve as the error mean square. Otherwise there is no way to separate σ^2 from the interaction effect. If interaction mean square is used instead of error mean square in the F-tests then power is lost if the interaction effect, as measured by $\frac{K}{(I-1)(J-1)}\sum_{ij} (\alpha\beta)_{ij}^2$, is large relative to σ^2. Similarly, confidence intervals on linear combinations of the μ_{ij} (or on the parameters μ, α_i, β_j, $(\alpha\beta)_{ij}$) will be longer.

It is sometimes reasonable to believe that the model

$$\mu_{ij} = \mu + \alpha_i + \beta_j + \varepsilon_{ij} \tag{6.3.1}$$

for $\sum_i \alpha_i = \sum_j \beta_j = 0$ holds at least in good approximation. Then interaction mean square can serve as a stand in for error mean square in the F-tests and in confidence intervals. The ANOVA table is Table 6.3.1.

The model (6.3.1) is called the *additive* model.

Problem 6.3.1: In an effort to compare the mileages produced by three types G_1, G_2 and G_3 of gasoline, four automobiles A_1, A_2, A_3, A_4 were chosen. Each automobile was driven over a 200 mile course three times, beginning once with a full tank of gasoline of type G_i, $i = 1, 2, 3$. The numbers of gallons of gasoline consumed were:

Gasoline Type	Automobiles A_1	A_2	A_3	A_4
G_1	8.34	9.16	7.82	8.25
G_2	8.07	8.78	7.61	8.95
G_3	8.51	9.41	7.95	8.65

(a) State an appropriate model, determine the corresponding analysis of

Table 6.3.1

Source	DF	SSqs.	MSq.	Error MSq.
A	$I-1$	$J\sum_i (\bar{Y}_{i\cdot} - \bar{Y}_{\cdot\cdot})^2$		$\sigma^2 + \frac{1}{I-1}\sum \alpha_i^2$
B	$J-1$	$I\sum_j (\bar{Y}_{\cdot j} - \bar{Y}_{\cdot\cdot})^2$		$\sigma^2 + \frac{1}{J-1}\sum \beta_j^2$
Residual	$(I-1)(J-1)$	Difference		σ^2
Corr. Total	$IJ-1$	$\sum_{ij} (Y_{ij} - \bar{Y}_{\cdot\cdot})^2$		

variance table, and perform appropriate F-tests. Use the symbols μ, γ_i (for gasoline type) and α_j (for automobiles).

(b) Find 95% simultaneous confidence intervals on $\alpha_j - \alpha_{j'}$ for $j \neq j'$. Use the method which provides the shortest intervals.

(c) Suppose that in previous tests with an automobile similar to these on the same 200 mile test track, the sample variances S^2 were 0.023 7 for G_1 for three trials, 0.034 5 for G_2 for four trials, the 0.019 9 for G_3 for two trials. How could this additional information be used to change the F-tests and the confidence intervals?

Problem 6.3.2: Suppose that automobiles A_4 had engine trouble just before it was to be used with gasoline G_3, so that that observation was missing, though all other observations were obtained.

(a) Show that the least squares estimators of the parameters μ, α_i, β_j are for this case of one missing observation, the same as they would be if the missing observation Y_{34} were replaced by $y_{34} = \bar{Y}_{3\cdot} + \bar{Y}_{\cdot 4} - \bar{Y}_{\cdot\cdot}$, where these means are determined from the observations which were obtained. *Hint*: Pretend that the observation y_{34} *was* available, and add to the model the extra parameter μ_{34}, the mean of cell 34. For each possible selection of y_{34} the least squares estimates of the cell means would be, as functions of y_{34}, the same as discussed for the full data case. But, the estimate for cell 34, with this extra parameter, would equal y_{34}. Thus, $y_{34} = \bar{Y}^*_{3\cdot} + \bar{Y}^*_{\cdot 4} - \bar{Y}^*_{\cdot\cdot}$, where the starred means are expressed in terms of y_{34} and the means of the observations actually obtained.

(b) Generalize the result to the case of any $I \times J$ table with one missing observation.

(c) Carry out the arithmetic for the data in Problem 6.3.1, and determine the estimate S^2. Use it to test H_G: no Gasoline Effect.

(d) Show that $\hat{\gamma}_1 - \hat{\gamma}_3 = (7T'_{1\cdot} + T'_{2\cdot} - 8T'_{3\cdot} + 3Y_{14} - 3Y_{24})/24$, where $T'_{i\cdot} = \sum_{j=1}^{3} Y_{ij}$. Find $\text{Var}(\hat{\gamma}_1 - \hat{\gamma}_3)$. Also determine $\text{Var}(\hat{\gamma}_1 - \hat{\gamma}_2)$.

(e) Find individual 95% confidence intervals on $\gamma_1 - \gamma_2$ and $\gamma_1 - \gamma_3$.

6.4 DESIGN OF EXPERIMENTS

The purpose of this section is to introduce the student to some of the language of the design of experiments. It is the design of the experiment which justifies, or at least makes credible, the models we have and will be considering. The randomness which the experimenter deliberately introduces not only makes the conclusions reached more believable to justifiably suspicious readers, but often makes the distributional assumptions of the models used more realistic.

Definition 6.4.1: An experimental design is a plan for the assignment of treatment levels or combinations of levels of treatments to experimental units and for the taking of measurements on the units under those treatment levels or combinations of levels.

Comment: An experimental unit is an element, thing, material, person, etc., to which treatment levels are applied as a whole. Experimental units are not split; the entire unit must receive the same treatment level or combination of levels.

Definition 6.4.2: A completely randomized design is a design for which the levels of treatments (or combinations of levels) are assigned randomly to the units, i.e., so that if a treatment level t is to be assigned n_t times to the N experimental units available, for $t = 1, \ldots, k$, then all $(n_0, n_1, \ldots, n_k) = N!\Big/\left(\prod_0^k n_t!\right)$ possible assignments are equally likely. Here $n_0 = N - \sum_1^k n_t$ is the number of units receiving no treatment level.

Example 6.4.1: For $k = 3$ treatment levels. Level 1, a control, is assigned to four units, level 2 to three units, level 3 to two units. Then $N = 9$, and there are $\binom{9}{4, 3, 2} = 9!/(4!\,3!\,2!) = 1{,}260$ possible assignments. The three treatments might be methods of heart surgery, the nine experimental units 3-month-old rats.

Definition 6.4.3: A randomized block design is a design for which the experimental units are separated (partitioned) into blocks of units, and treatment levels are then randomly assigned within the separate blocks.

Example 6.4.2: In an agricultural experiment we might be interested in four levels of seed, with the measured variable being yield on half-acre plots. The field might have 40 half-acre plots as follows.

Blocks

	1	2	3	4	5	6	7	8	9	10	

West   East

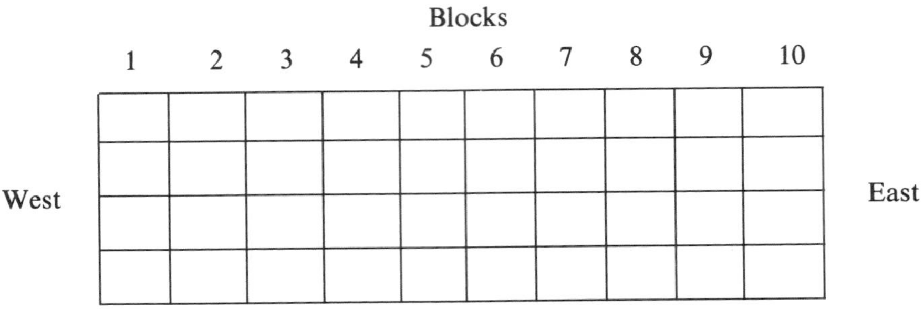

We could allocate each of the four levels to 10 randomly chosen plots from among the 40, a completely randomized design. However, if the land has higher fertility as we move to the east, we might restrict the randomization so that each seed level occurs in each of the 10 blocks.

In general, blocks should be chosen so that units within blocks are relatively homogeneous, while block-to-block variation is as large as possible. In the language of sample surveys blocks are called *strata*.

6.5 THREE-WAY ANALYSIS OF VARIANCE

Consider a three-way complete factorial. Three factors A, B, C have, respectively, a, b, c levels and m observation are taken for each combination of the A, B, C levels, for $abcm$ observations in all.

Factor B

Factor C →	1 1 2 $\cdots$ c	2 1 2 $\cdots$ c	$\cdots$ $\cdots$	b 1 2 $\cdots$ c
1	$\cdots$	$\cdots$	$\cdots$	$\cdots$
2	$\cdots$	$\cdots$	$\cdots$	$\cdots$
Factor A $\vdots$				
a	$\cdots$	$\cdots$	$\cdots$	$\cdots$

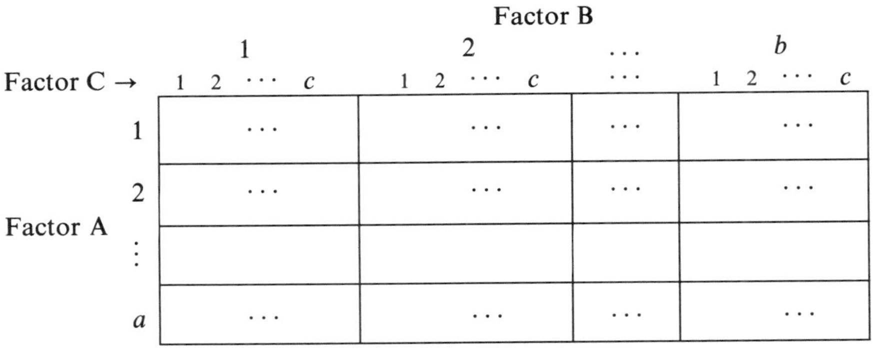

Let the observations corresponding to level i of A, level j of B, level k of C be $Y_{ijk1}, \ldots, Y_{ijkm}$. We suppose $Y_{ijk1} \sim N(\mu_{ijk}, \sigma^2)$ and that the Y_{ijk1} are independent.

Define

$$\mu = \frac{1}{abc}\sum_{ijk} \mu_{ijk},$$

$$\alpha_i = \bar{\mu}_{i\cdot\cdot} - \mu,\ \beta_j = \mu_{\cdot j \cdot} - \mu,\ \gamma_k = \mu_{\cdot\cdot k} - \mu$$

$$\left.\begin{aligned}(\alpha\beta)_{ij} &= \bar{\mu}_{ij\cdot} - [\mu + \alpha_i + \beta_j]\\ (\alpha\gamma)_{ik} &= \bar{\mu}_{i\cdot k} - [\mu + \alpha_i + \gamma_k]\\ (\beta\gamma)_{jk} &= \bar{\mu}_{\cdot jk} - [\mu + \beta_j + \gamma_k]\end{aligned}\right\}\quad \text{First-order interaction terms}$$

$$(\alpha\beta\gamma)_{ijk} = \mu_{ijk} - [\mu + \alpha_i + \beta_j + \gamma_k + (\alpha\beta)_{ij} + (\alpha\gamma)_{ik} + (\beta\gamma)_{ij}]$$

The $(\alpha\beta\gamma)_{ijk}$ are second-order interaction terms. Then the model can be stated as

$$\begin{aligned}Y_{ijk1} &= \mu_{ijk} + \varepsilon_{ijk1}\\ &= \mu + \alpha_i + \beta_j + \gamma_k + (\alpha\beta)_{ij} + (\alpha\gamma)_{ik} + (\beta\gamma)_{jk} + (\alpha\beta\gamma)_{ijk} + \varepsilon_{ijk1}.\end{aligned}$$

where

$$\begin{aligned}\sum \alpha_i = \sum \beta_j = \sum \gamma_k = \sum_i (\alpha\beta)_{ij} = \sum_j (\alpha\beta)_{ij} = \sum_i (\alpha\gamma)_{ik} = \sum_k (\alpha\gamma)_{ik}\\ = \sum_k (\alpha\gamma)_{ik} = \sum_j (\beta\gamma)_{jk} = \sum_k (\beta\gamma)_{jk} = \sum_i (\alpha\beta\gamma)_{ijk} = \cdots = 0\end{aligned}$$

Define the vectors $\mathbf{A}_i$, $\mathbf{B}_j$, $\mathbf{C}_k$, $(\mathbf{AB})_{ij}$, $(\mathbf{AC})_{ik}$, $(\mathbf{BC})_{jk}$, $(\mathbf{ABC})_{ijk}$ to be the indicator arrays suggested by the letters. Thus, for example, $(\mathbf{AB})_{ij}$ is one in cells at level i of A, level j on B.

The sample space Ω of possible values of $\mathbf{Y}$ may be broken into mutually orthogonal subspaces as follows:

$$V_0 = \mathscr{L}(\mathbf{x}_0) \quad V_A = \mathscr{L}(\mathbf{A}_1, \ldots, \mathbf{A}_a) \cap V_0^\perp = \left\{\sum_i a_i \mathbf{A}_i \,\middle|\, \sum_1^a a_i = 0\right\}$$

$$V_B = \mathscr{L}(\mathbf{B}_1, \ldots, \mathbf{B}_b) \cap V_0^\perp = \left\{\sum_j b_j \mathbf{B}_j \,\middle|\, \sum_1^b b_j = 0\right\}$$

$$V_C = \mathscr{L}(\mathbf{C}_1, \ldots, \mathbf{C}_c) \cap V_0^\perp = \left\{\sum_k c_k \mathbf{C}_k \,\middle|\, \sum_1^c C_k = 0\right\}$$

$$V_{AB} = \mathscr{L}((\mathbf{AB})_{11}, \ldots, (\mathbf{AB})_{ab}) \cap V_0^\perp \cap V_A^\perp \cap V_B^\perp$$

V_{AC}, V_{BC} (defined similarly)

$$V_{ABC} = (V_0 \oplus V_A \oplus V_B \oplus V_C \oplus V_{AB} \oplus V_{AC} \oplus V_{BC})^\perp \cap V,$$

and

$$V^\perp \quad \text{for} \quad V = \mathscr{L}((\mathbf{ABC})_{111}, \ldots, (\mathbf{ABC})_{abc}).$$

Thus,

$$\Omega = V_0 \oplus V_A \oplus V_B \oplus V_C \oplus V_{AB} \oplus V_{AC} \oplus V_{BC} \oplus V_{ABC} \oplus V^{\perp}.$$

It is easy to verify that these nine subspaces are mutually orthogonal. For example, a vector $\mathbf{v} \in V_{AB}$ is of the form

$$\mathbf{v} = \sum_{ij} d_{ij}(\mathbf{AB})_{ij} \qquad \text{for} \quad \sum_j d_{ij} = \sum_i d_{ij} = 0.$$

A vector $\mathbf{w} \in V_{BC}$ is of the form

$$\mathbf{w} = \sum_{jk} f_{jk}(\mathbf{BC})_{jk} \qquad \text{for} \quad \sum_j f_{jk} = \sum_k f_{jk} = 0.$$

Then $(\mathbf{v}, \mathbf{w}) = \sum_{ij}\sum_{j'k} d_{ij} f_{j'k}((\mathbf{AB})_{ij}, (\mathbf{BC})_{jk})$. This inner product is 0 if $j \neq j'$, m if $j = j'$. Thus $(\mathbf{v}, \mathbf{w}) = \sum_{ijk} m d_{ij} f_{jk} = m \sum_{ij} d_{ij} \sum_k f_{jk} = 0$.

The model can be written in the form $\mathbf{Y} = \boldsymbol{\mu} + \boldsymbol{\varepsilon}$, where

$$\begin{aligned}\boldsymbol{\mu} = \sum_{ijk} \mu_{ijk}(\mathbf{ABC})_{ijk} &= \mu \mathbf{x}_0 + \sum_i \alpha_i \mathbf{A}_i + \sum_{ij} (\alpha\beta)_{ij}(\mathbf{AB}) \\ &\quad + \sum_{ik} (\alpha\gamma)_{jk}(\mathbf{AC})_{ik} + \sum_{jk} (\beta\gamma)_{jk}(\mathbf{BC})_{jk} \\ &\quad + \sum_{ijk} (\alpha\beta\gamma)_{ijk}(\mathbf{ABC})_{ijk} \qquad \text{and} \qquad \boldsymbol{\varepsilon} \sim N_n(\mathbf{0}, \sigma^2 \mathbf{I}_n).\end{aligned}$$

The projections of $\mathbf{Y}$ onto these subspaces are

$$\begin{aligned}
\hat{\mathbf{Y}}_0 &= p(\mathbf{Y}|V_0) = \bar{Y}_{\cdots\cdot}\mathbf{x}_0 = \hat{\mu}\mathbf{x}_0 \\
\hat{\mathbf{Y}}_A &= p(\mathbf{Y}|V_A) = \sum_i (\bar{Y}_{i\cdots} - \bar{Y}_{\cdots})\mathbf{A}_i = \sum_i \hat{\alpha}_i \mathbf{A}_i \\
\hat{\mathbf{Y}}_B &= p(\mathbf{Y}|V_B) = \sum_j (\bar{Y}_{\cdot j\cdot\cdot} - \bar{Y}_{\cdots\cdot})\mathbf{B}_j = \sum \hat{\beta}_j \mathbf{B}_j \\
\hat{\mathbf{Y}}_C &= p(\mathbf{Y}|V_C) = \sum_k (\bar{Y}_{\cdot\cdot k\cdot} - \bar{Y}_{\cdots\cdot})\mathbf{C}_k = \sum \hat{\gamma}_k \mathbf{C}_k \\
\hat{\mathbf{Y}}_{AB} &= p(\mathbf{Y}|V_{AB}) = \sum_{ij} [\bar{Y}_{ij\cdot\cdot} - (\hat{\mu} + \hat{\alpha}_i + \hat{\beta}_j)](\mathbf{AB})_{ij} = \sum_{ij} \widehat{(\alpha\beta)}_{ij}(\mathbf{AB})_{ij} \\
\hat{\mathbf{Y}}_{AC} &= p(\mathbf{Y}|V_{AC}) = \sum_{ik} [\bar{Y}_{i\cdot k\cdot} - (\hat{\mu} + \hat{\alpha}_i + \hat{\gamma}_k)](\mathbf{AC})_{ik} \\
\hat{\mathbf{Y}}_{BC} &= p(\mathbf{Y}|V_{BC}) = \sum_{ij} [\bar{Y}_{\cdot jk} - (\hat{\mu} + \hat{\beta}_j + \hat{\gamma}_k)](\mathbf{BC})_{jk}
\end{aligned}$$

$$\hat{\mathbf{Y}}_{ABC} = p(\mathbf{Y} \mid V_{ABC}) = \sum_{ijk} [\bar{Y}_{ijk\cdot} - (\hat{\mu} + \hat{\alpha}_i + \hat{\beta}_j + \hat{\gamma}_k + \widehat{(\alpha\beta)}_{ij} + \widehat{(\alpha\gamma)}_{ik} + \widehat{(\beta\gamma)}_{jk}](\mathbf{ABC})_{ijk} = \sum \widehat{(\alpha\beta\gamma)}_{ijk}(\mathbf{ABC})_{ijk}$$

$$\hat{\mathbf{Y}} = p(\mathbf{Y} \mid V) = \sum_{ijk} \bar{Y}_{ijk\cdot}(\mathbf{ABC})_{ijk}, \qquad \text{and} \qquad \mathbf{e} = p(\mathbf{Y} \mid V^{\perp}) = \mathbf{Y} - \hat{\mathbf{Y}}$$

Sums of squares are computed easily:

$$\text{SSA} = \|\hat{\mathbf{Y}}_A\|^2 = \sum_i \hat{\alpha}_i^2 \|\mathbf{A}_i\|^2 = \sum_i (\bar{Y}_{i\cdots} - \bar{Y}_{\cdots\cdot})^2(bcm)$$
$$= \sum_i (\bar{Y}_{i\cdots})^2 bcm - \bar{Y}^2_{\cdots} n \qquad \text{for} \quad n = abcm$$

$$\text{SSAB} = \|\hat{\mathbf{Y}}_{AB}\|^2 = \sum_{ij} \widehat{(\alpha\beta)}_{ij}^2 \|(\mathbf{AB})_{ij}\|^2 = cm \sum_{ij} \widehat{(\alpha\beta)}_{ij}^2$$

For computational purposes, let

$$\mathbf{Y}^* = p(\mathbf{Y} \mid \mathscr{L}((\mathbf{AB})_{11}, \ldots, (\mathbf{AB})_{ab})) = \sum_{ij} \bar{Y}_{ij\cdot}(\mathbf{AB})_{ij}$$

Then $\hat{\mathbf{Y}}_{AB} = \mathbf{Y}^* - (\hat{\mathbf{Y}}_0 + \hat{\mathbf{Y}}_A + \hat{\mathbf{Y}}_B)$ and by the Pythagorean Theorem,

$$\|\hat{\mathbf{Y}}_{AB}\|^2 = \|\hat{\mathbf{Y}}^* - \hat{\mathbf{Y}}_0\|^2 - [\|\hat{\mathbf{Y}}_A\|^2 + \|\hat{\mathbf{Y}}_B\|^2]$$

But $\mathbf{Y}^* - \hat{\mathbf{Y}}_0 = \sum_{ij} (\bar{Y}_{ij\cdot\cdot} - \bar{Y}_{\cdots\cdot})(\mathbf{AB})_{ij}$ so

$$\|\hat{\mathbf{Y}}^* - \hat{\mathbf{Y}}_0\|^2 = \sum_{ij} (\bar{Y}_{ij\cdot\cdot} - \bar{Y}_{\cdots\cdot})^2(cm) = \|\hat{\mathbf{Y}}^*\|^2 - \|\hat{\mathbf{Y}}_0\|^2 = \sum_{ij} \bar{Y}^2_{ij\cdot\cdot}(cm) - \bar{Y}^2_{\cdots\cdot} n$$

This is called the 'AB subtotal'. It is the adjusted total sum of squares when the data is treated as a two-way ANOVA on A and B, with C ignored. Thus

$$\text{SSAB} = (AB \text{ Subtotal}) - (\text{SSA} + \text{SSB}).$$

Similarly,

$$\|\hat{\mathbf{Y}}_{AC}\|^2 = \sum_{ik} (\bar{Y}_{i\cdot k\cdot} - \bar{Y}_{\cdots\cdot})^2(bm) - (\text{SSA} + \text{SSC})$$
$$= AB \text{ Subtotal} - (\text{SSA} + \text{SSC})$$
$$\|\hat{\mathbf{Y}}_{BC}\|^2 = \sum_{jk} (\bar{Y}_{\cdot jk\cdot} - \bar{Y}_{\cdots\cdot})^2(am) - (\text{SSB} + \text{SSC})$$
$$= BC \text{ Subtotal} - (\text{SSB} + \text{SSC})$$

Finally, by defining

$$ABC\text{ Subtotal} = \sum_{ijk} (\bar{Y}_{ijk\cdot} - \bar{Y}_{\cdots})^2$$
$$= \sum_{ijk} \bar{Y}^2_{ijk\cdot} m - \bar{Y}^2_{\cdots} n = \|\hat{\mathbf{Y}} - \hat{\mathbf{Y}}_0\|^2 = \|\hat{\mathbf{Y}}\|^2 - \|\hat{\mathbf{Y}}_0\|^2$$

we get

$$\|\hat{\mathbf{Y}}_{ABC}\|^2 = (ABC\text{ Subtotal}) - (\text{SSA} + \text{SSB} + \text{SSC} + \text{SSAB} + \text{SSAC} + \text{SSBC})$$

Of course,

$$\|\mathbf{Y} - \hat{\mathbf{Y}}\|^2 = \|\hat{\mathbf{e}}\|^2 = \sum_{ijk1} (Y_{ijk1} - Y_{ijk\cdot})^2 = \sum_{ijk1} Y^2_{ijk1} - \left(\sum_{ijk} \bar{Y}_{ijk\cdot}\right)^2 m$$

Summarizing, we get Table 6.5.1. We can then test the null hypothesis that the projection of $\boldsymbol{\mu}$ on any of the subspaces V_A, $V_B, \ldots, V_{ABC}$ is zero, using the F-statistic with numerator the corresponding MSq. and the denominator S^2 = Error MSq. Usually we would want to proceed upward beginning with the more complex model terms. Whenever we reject the hypothesis that the corresponding lower-order terms in the interaction are zero. For example, if we decide AC interaction is present it makes little practical sense to test for A effects or C effects.

Example 6.5.1: Consider a three-way factorial discussed by Cochran and Cox (1957, p. 177):

5.32 Numerical Example: a 4 × 4 × 3 Factorial in Randomized Blocks
A number of experiments have indicated that electrical stimulation may be helpful in preventing the wasting away of muscles that are denervated. A factorial experiment on rats was conducted by Solandt, DeLury, and Hunter (5.8) in order to learn something about the most effective method of treatment. The factors and their levels are shown below.

A: Number of Treatment Periods Daily (Minutes)	B: Length of Treatment	C: Type of Current
1	1	Galvanic
3	2	Faradic
6	3	60 cycle alternating
	5	60 cycle alternating

Treatments were started on the third day after denervating and continued for 11 consecutive days. There are 48 different combinations of methods of treatment, each of which was applied to a different rat. Two replicates were conducted, using 96 rats in all.

Table 6.5.1 Analysis of Variance

Source	Subtotal Space	DF	SSqs.	MSq.	Expected MSq.
A	V_A	$a-1$			$\sigma^2 + \frac{bcm}{a-1}\sum \alpha_i^2$
B	V_B	$b-1$			$\sigma^2 + \frac{acm}{b-1}\sum \beta_j^2$
C	V_C	$c-1$			$\sigma^2 + \frac{abm}{c-1}\sum \gamma_k^2$
AB	V_{AB}	$(a-1)(b-1)$			$\sigma^2 + \frac{cm}{(a-1)(b-1)}\sum_{ij} (\alpha\beta)_{ij}^2$
AC	V_{AC}	$(a-1)(c-1)$	$\vdots$		
BC	V_{BC}	$(b-1)(c-1)$			
ABC	V_{ABC}	$(a-1)(b-1)(c-1)$			$\sigma^2 + \frac{m}{(a-1)(b-1)(c-1)}\sum_{ijk} (\alpha\beta\gamma)_{ijk}^2$
Error	$V^{\perp}$	$abc(m-1)$			σ^2
Corr. Total	$V \cap V_0^{\perp}$	$abcm-1$			$\sigma^2 + \frac{m}{abcm-1}\sum_{ijk} (\mu_{ijk}-\mu)^2$
Mean	V_0	1			$\sigma^2 + \mu^2 n$
Total	Ω	$abcm$			$\sigma^2 + \frac{1}{abcm}\sum_{ijk} \mu_{ijk}^2$

The muscles denervated were the gastronemius-soleus group on one side of the animal, denervation being accomplished by the removal of a small part of the sciatic nerve. The measure used for judging the effects of the treatments was the weight of the denervated muscle at the end of the experiment. Since this depends on the size of the animal, the weight of the corresponding muscle on the other side of the body was included as a covariate.

The data are shown in Table 6.5.2.

Though Cochran and Cox did not describe how Reps. I and II differ, let us assume that they were repetitions of the experiment with 48 rats at different points in time. For an initial analysis we will ignore this Reps. variable. A discussion of the use of the covariate x will be postponed until Section 6.6. Cell means were

		a_1	a_2	a_3
b_1	c_1	59.0	74.0	63.5
	c_2	60.5	62.5	58.5
	c_3	66.5	64.5	70.5
	c_4	69.0	70.5	63.5
b_2	c_1	55.5	55.0	58.0
	c_2	58.5	55.0	55.0
	c_3	63.0	63.0	71.5
	c_4	63.5	75.0	71.0
b_3	c_1	55.0	58.0	66.5
	c_2	64.0	50.0	49.5
	c_3	59.5	61.0	71.5
	c_4	56.0	66.5	80.5
b_4	c_1	51.5	55.5	71.0
	c_2	58.0	59.0	57.5
	c_3	62.5	72.5	65.0
	c_4	66.0	72.0	79.5

Two-way means were

	a_1	a_2	a_3
b_1	63.750	67.875	64.000
b_2	60.125	62.000	63.875
b_3	58.625	58.875	67.000
b_4	59.500	64.750	68.250

	a_1	a_2	a_3
c_1	55.250	60.625	64.750
c_2	60.250	56.625	55.125
c_3	62.875	65.250	69.625
c_4	63.625	71.000	73.625

Table 6.5.2 Weights of Denervated (y) and Corresponding Normal (x) Muscle (unit $= 0.1$ gram)

Length of Treatment (Minutes)		Type of Current	One (a_1)		Three (a_3)		Six (a_6)	
			y	x	y	x	y	x
Rep. I	1 (b1)	G	72	152	74	131	69	131
		F	61	130	61	129	65	126
		60	62	141	65	112	70	111
		25	85	147	76	125	61	130
	2 (b2)	G	67	136	52	110	62	122
		F	60	111	55	180	59	122
		60	64	126	65	190	64	98
		25	67	123	72	117	60	92
	3 (b3)	G	57	120	66	132	72	129
		F	72	165	43	95	43	97
		60	63	112	66	130	72	180
		25	56	125	75	130	92	162
	4 (b4)	G	57	121	56	160	78	135
		F	60	87	63	115	58	118
		60	61	93	79	126	68	160
		25	73	108	86	140	71	120
Rep. II	1 (b1)	G	46	97	74	131	58	81
		F	60	126	64	124	52	102
		60	71	129	64	117	71	108
		25	53	108	65	108	66	108
	2 (b2)	G	44	83	58	117	54	97
		F	57	104	55	112	51	100
		60	62	114	61	100	79	115
		25	60	105	78	112	82	102
	3 (b3)	G	53	101	50	103	61	115
		F	56	120	57	110	56	105
		60	56	101	56	109	71	105
		25	56	97	58	87	69	107
	4 (b4)	G	46	107	55	108	64	115
		F	56	109	55	104	57	103
		60	64	114	66	101	62	99
		25	59	102	58	98	88	135

Number of Treatment Periods Daily

	b_1	b_2	b_3	b_4
c_1	65.500	60.500	67.167	67.667
c_2	56.167	56.167	65.833	69.833
c_3	59.833	54.500	64.000	67.667
c_4	59.333	58.167	66.667	72.500

A means			*B* means				*C* means			
a_1	a_2	a_3	b_1	b_2	b_3	b_4	c_1	c_2	c_3	c_4
60.5	63.375	65.791	65.208	62.0	61.5	64.167	60.208	57.333	65.917	69.417

$$\text{Grand mean} = \bar{Y} = \bar{Y}_{\cdots\cdot} = 63.218$$

Then $\text{CT} = \bar{Y}^2(96) = 383{,}674.6$

$$\text{SSA} = 32[60.5^2 + 63.375^2 + 65.781^2] - \text{CT} = 447.44$$

$$AB \text{ Subtotal} = 8[63.75^2 + \cdots + 68.25^2] - \text{CT} = 1{,}038.531$$

$$\text{SSAB} = (AB \text{ Subtotal}) - (\text{SSA} + \text{SSB}) = 367.98$$

$$ABC \text{ Subtotal} = 2[59^2 + \cdots + 79.5^2] - \text{CT} = 5{,}177.906$$

$$\text{SSABC} = (ABC \text{ Subtotal}) - [\text{SSA} + \text{SSB} + \text{SSC} + \text{SSAD} + \text{SSAC} + \text{SSBC}]$$

$$= 1{,}050.9$$

Other sums of squares were found similarly. The ANOVA Table is Table 6.5.3. Only the *A* and *C* main effects are significantly different at the 0.05 level ($F_{2,48,0.95} = 3.20$). Thus the *A* and *C* means are of particular interest.

Table 6.5.3

Source	DF	SSqs.	MSq.	*F*
A	2	447.44	223.72	3.36
B	3	223.11	74.37	1.12
C	3	2,145.45	715.15	10.73
AB	6	367.98	61.33	0.92
AC	6	644.40	107.40	1.61
BC	9	298.68	33.19	0.50
ABC	18	1,050.85	58.38	0.87
Error	48	3,804.50	66.66	
(Corr. Total)	95	8,992.41		

If a term r_t for $t = 1, 2$ is added to the model for the two levels of replication, then

$$\text{SSRep.} = 48 \sum_{t=1}^{2} (\bar{Y}_{\cdots t} - \bar{Y}_{\cdots\cdot})^2$$

$$= 48(\bar{Y}_{\cdots 1} - \bar{Y}_{\cdots 2})^2/2 = 48\,(65.729 - 60.708)^2/2 = 605.01$$

for one d.f. The error sum of squares becomes 3,804.50 − 605.01 = 3,199.49 for 47 d.f. We reject the null hypothesis of no Rep. effect at reasonable α levels. Conclusions relative to other effects do not change.

Problem 6.5.1: Consider the following $2 \times 2 \times 2$ table of means μ_{ijk}.

$$b_1: \quad \begin{array}{c} \\ a_1 \\ a_2 \end{array} \begin{array}{c} \begin{array}{cc} c_1 & c_2 \end{array} \\ \begin{bmatrix} 35 & 25 \\ 21 & 11 \end{bmatrix} \end{array} \qquad b_2: \quad \begin{array}{c} \\ a_1 \\ a_2 \end{array} \begin{array}{c} \begin{array}{cc} c_1 & c_2 \end{array} \\ \begin{bmatrix} 13 & 7 \\ 7 & 1 \end{bmatrix} \end{array}$$

(a) Find the parameters μ, α_i, $\beta_j, \ldots, (\alpha\beta\gamma)_{ijk}$.

(b) If three observations are taken independently from $N(\mu_{ijk}, \sigma^2)$ for each cell ijk, and $\sigma = 4$, what is the power of the $\alpha = 0.05$ level F-test of H_0: No $A \times B$ interaction?

Problem 6.5.2: Sample of sizes two were taken from each of the eight normal distributions with the means μ_{ijk} as presented in Problem 6.5.1, and variances each $\sigma^2 = 16$, then rounded to the nearest integer.

$$a_1: \quad \begin{array}{c} \\ b_1 \\ \\ b_2 \\ \end{array} \begin{array}{c} \begin{array}{cc} c_1 & c_2 \end{array} \\ \begin{bmatrix} 40 & 33 \\ 31 & 21 \\ 18 & 12 \\ 21 & 14 \end{bmatrix} \end{array} \qquad a_2: \quad \begin{array}{c} \\ b_1 \\ \\ b_2 \\ \end{array} \begin{array}{c} \begin{array}{cc} c_1 & c_2 \end{array} \\ \begin{bmatrix} 17 & 11 \\ 5 & 16 \\ 2 & 3 \\ 10 & 1 \end{bmatrix} \end{array}$$

(a) Estimate the parameters, and fill out the analysis of variance table.

(b) Since the eight subspaces V_0, V_A, $V_B, \ldots, V_{ABC}$, $V^{\perp}$ all have dimension one, the corresponding sums of squares may all be expressed in the form $\|p(\mathbf{Y}|\mathbf{x})\|^2 = (\mathbf{Y}, \mathbf{x})^2/\|\mathbf{x}\|^2$. Give a vector $\mathbf{x}$ for each of these subspaces.

(c) What are the lengths of simultaneous 95% Scheffé and Tukey confidence intervals on differences among all cell means? Draw a diagram indicating which of the means are significantly different.

(d) For the model with all first- and second-order interaction terms zero what is the least squares estimator of μ_{112}? What is its variance under this model? Find a 95% individual confidence interval on μ_{112} under this model.

(e) What is the bias of the estimator of (d) for μ_{ijk} as given in Problem 6.5.1?

Problem 6.5.3: Prove that the subspaces V_A and V_{ABC} are orthogonal.

6.6 THE ANALYSIS OF COVARIANCE

Suppose that Y is observed under varying experimental conditions in a factorial design, but that in addition measurements x_1, or x_1 and x_2, taking values on a continuous scale, are made on each unit. In cases in which it is reasonable to suppose that Y is affected linearly by these *covariates* the data may be analyzed by the method of *analysis of covariance.*

For example, we might measure the yield of corn on 24 plots of land, for 3 varieties of seed, 4 levels of fertilizer, 2 plots for each combination of seed and fertilizer. It is reasonable to suppose that the yield is related also to the fertility x of the soil, and therefore measure x for each plot. A reasonable model might then be:

$$Y_{ijk} = \mu_{ij} + \gamma x_{ijk} + \varepsilon_{ijk},$$

for $k = 1, 2$; $j = 1, 2, 3, 4$; and $i = 1, 2, 3$. We might also observe some other variable w_{ijk} on each plot and then add another term βw_{ijk} to the model.

Define the parameters μ, α_i, β_j, $(\alpha\beta)_{ij}$ as functions of the μ_{ij}, and the vectors $\mathbf{x}_0$, $\mathbf{A}_i$, $\mathbf{B}_j$, $\mathbf{C}_{ij}$ as before. In vector form the model becomes:

$$\mathbf{Y} = \mu\mathbf{x}_0 + \sum_i \alpha_i\mathbf{A}_i + \sum_j \beta_j\mathbf{B}_j + \sum_{ij} (\alpha\beta)_{ij}\mathbf{C}_{ij} + \gamma\mathbf{x} + \boldsymbol{\varepsilon},$$

where $\mathbf{x} = (x_{ijk})$. In the case of several covariates $\gamma\mathbf{x}$ could be replaced by $\sum \gamma_m\mathbf{x}_m$.

There is in theory no difficulty in testing null hypotheses of the form H_0: $(\alpha\beta)_{ij} = 0$ for all ij or H_0: $\alpha_i = 0$ for all i. We need only fit the model with and without these terms present, and express the F-statistic in terms of the appropriate sums of squares.

We will discuss the analysis for the case of several covariates, but will attempt to give explicit formulas for estimators only for the case of a single covariate $\mathbf{x}$.

Consider the model $\mathbf{Y} = \boldsymbol{\theta} + \boldsymbol{\varepsilon}$, for $\boldsymbol{\theta} = \boldsymbol{\mu} + \sum \gamma_m\mathbf{x}_m \in V = V_1 \oplus V_x$, where $V_1 = \mathscr{L}(\{\mathbf{C}_{ij}\})$ and $V_x = \mathscr{L}(\mathbf{x}_1, \ldots, \mathbf{x}_r)$. Suppose V_1 and V_x are linearly independent. This means that for each $\mathbf{x}_m$ at least some cell has values which are not all the same. We will refer to V_x as the *covariate space.* Define $V_{1x} \equiv V_1^\perp \cap V_x$. Then $V = V_1 \oplus V_{1x}$ and $V_1 \perp V_{1x}$. The subspace V_{1x} is spanned by the vectors $\mathbf{x}_m^\perp \equiv \mathbf{x}_m - p(\mathbf{x}_m | V_1)$. When V is spanned by the cell indicators, each $\mathbf{x}_m^\perp$ is the vector obtained by subtracting the corresponding cell mean from each component.

Let $\hat{\mathbf{Y}}_1 \equiv p(\mathbf{Y} | V_1)$, $\hat{\mathbf{Y}} \equiv p(\mathbf{Y} | V)$, and $\hat{\mathbf{Y}}_{1x} \equiv p(\hat{\mathbf{Y}} | V_{1x})$. Then orthogonality implies that $\hat{\mathbf{Y}} = \hat{\mathbf{Y}}_1 + \hat{\mathbf{Y}}_{1x}$, and $(\mathbf{Y} - \hat{\mathbf{Y}}_1) = (\mathbf{Y} - \hat{\mathbf{Y}}) + \hat{\mathbf{Y}}_{1x}$. Orthogonality of $(\mathbf{Y} - \hat{\mathbf{Y}})$ to V and therefore to V_{1x} therefore implies that

$$\|\mathbf{Y} - \hat{\mathbf{Y}}\|^2 = \|\mathbf{Y} - \hat{\mathbf{Y}}_1\| - \|\hat{\mathbf{Y}}_{1x}\|^2.$$

That is,

(SSE under the analysis of covariance model)

= (SSE under the analysis of variance model) $- \|\hat{\mathbf{Y}}_{1x}\|^2$.

In the case that the number of covariates is one we can give explicit formulas for the parameter estimates. From Section 3.5 we have $\hat{\gamma} = (\mathbf{Y}, \mathbf{x}^\perp)/\|\mathbf{x}^\perp\|^2$, and $\text{Var}(\hat{\gamma}) = \sigma^2/\|\mathbf{x}^\perp\|^2$. When V_1 is the space spanned by the cell means we get $\mathbf{x}^\perp = \sum (x_{ijk} - \bar{x}_{ij\cdot})\mathbf{C}_{ij}$, $(\mathbf{Y}, \mathbf{x}^\perp) = \sum_{ijk} Y_{ijk}(x_{ijk} - \bar{x}_{ij\cdot})^2$, and $\|\mathbf{x}^\perp\|^2 = \sum_{ijk} (x_{ijk} - \bar{x}_{ij\cdot})^2$. This last sum of squares is the SSE in an analysis of variance on the x-values. Of course, $\|\hat{\mathbf{Y}}_{1x}\|^2 = (\mathbf{Y}, \mathbf{x}^\perp)^2/\|\mathbf{x}\|^2$.

To get explicit formulas for estimators of the μ_{ij}, μ, α_i, β_j, $(\alpha\beta)_{ij}$, we first notice that $\hat{\mathbf{Y}} = \sum_{ij} \bar{Y}_{ij\cdot}\mathbf{C}_{ij} + \hat{\gamma}\mathbf{x}^\perp = \sum_{ij} (\bar{Y}_{ij\cdot} - \hat{\gamma}\bar{x}_{ij\cdot})\mathbf{C}_{ij} + \hat{\gamma}\mathbf{x}$. Thus, $\hat{\mu}_{ij} = \bar{Y}_{ij\cdot} - \hat{\gamma}\bar{x}_{ij}$, the *corrected cell means*. Since the parameters μ, α_i, β_j, $(\alpha\beta)_{ij}$ are linear functions of the μ_{ij}, their estimators are the corresponding linear functions of the $\hat{\mu}_{ij}$. For example, $\hat{\mu} = \frac{1}{KIJ}\sum \hat{\mu}_{ijk}$, $\hat{\alpha}_i = \frac{1}{KJ}\sum_{jk} \hat{\mu}_{ij}$. The $\bar{Y}_{ij\cdot}$ are uncorrelated with $\hat{\gamma}$, and with each other. Variances and covariances among the $\hat{\mu}_{ij}$ can be determined from the relationship $\hat{\mu}_{ij} = \bar{Y}_{ij\cdot} - \hat{\gamma}\bar{x}_{ij\cdot}$. These can be used to find variances and covariances for linear combinations, such as for the $\hat{\alpha}_i$ or for differences $\hat{\alpha}_i - \hat{\alpha}_{i'}$.

Suppose that we wish to test a null hypothesis H_0: $\boldsymbol{\theta} \in V_2 \oplus V_x$, where $V_2 \subset V_1$. In two-way analysis of variance, if we wish to test for lack of interaction, V_2 would be the subspace spanned by the row and column indicators. We need to fit the model which holds under the null hypothesis. We need only repeat the argument above with V_1 replaced by V_2. For $V_{2x} \equiv V_2 \cap V_x^\perp$, $\hat{\mathbf{Y}}_{2x} \equiv p(\mathbf{Y} | V_{2x})$,

(Error sum of squares under the H_0 analysis of covariance model)

= (Error sum of squares under the H_0 analysis of variance model)
$- \|\hat{\mathbf{Y}}_{2x}\|^2$.

The numerator sum of squares for the F-statistic is therefore

(Numerator sum of squares for test of H_0 for the analysis of variance model)
$- \|\hat{\mathbf{Y}}_{1x}\|^2 + \|\hat{\mathbf{Y}}_{2x}\|^2$.

The degree of freedom for this numerator sum of squares remains the same as it was for the analysis of variance model.

Again, if there is only one covariate, explicit formulas can be given. Let $\mathbf{x}_2^\perp \equiv p(\mathbf{x}|V_2^\perp \cap V_x) = \mathbf{x} - p(\mathbf{x}|V_2)$, the error vector under H_0 when $\mathbf{x}$ is the observed vector. Then $p(\mathbf{Y}|V_2 \oplus V_x) = p(\mathbf{Y}|V_2) + \hat{\mathbf{Y}}_{2x}$, where $\hat{\mathbf{Y}}_{2x} = \hat{\gamma}_2\mathbf{x}_2^\perp$, and $\hat{\gamma}_2 = (\mathbf{Y}, \mathbf{x}_2^\perp)/\|\mathbf{x}_2^\perp\|^2$, and $\|\hat{\mathbf{Y}}_{2x}\|^2 = (\mathbf{Y}, \mathbf{x}_2^\perp)^2/\|\mathbf{x}_2^\perp\|^2$.

An analysis of covariance can be performed in much the same way as analysis of variance is performed. Add two columns to the usual analysis of variance sums of squares column. Each term in the analysis of variance sum of squares column is of the form $H_{yy} \equiv \mathbf{Y}'\mathbf{P}\mathbf{Y}$, where $\mathbf{P}$ is projection onto the corresponding subspace. Make the second and third column entries in that row $H_{xy} \equiv \mathbf{x}'\mathbf{P}\mathbf{Y}$, and $H_{xx} \equiv \mathbf{x}'\mathbf{P}\mathbf{x}$. Let E_{yy}, E_{xy}, and E_{xx} be the corresponding terms for the error row of the table. Then sums of squares for the analysis of covariance are all of the form

$$\begin{aligned} C &\equiv (\text{Error SSqs. under } H_0) - (\text{Error SSqs. under Full Model}) \\ &= [H_{yy} + E_{yy} - (H_{xy} + E_{xy})^2/(H_{xx} + E_{xx})] - [E_{yy} - E_{xy}^2/E_{xx}]. \end{aligned}$$

The estimate of the regression coefficient γ under the null hypothesis corresponding to a given row of the table is $(H_{xy} + E_{xy})/(H_{xx} + E_{xx})$. These sums of squares no longer have the additive properties the terms H_{yy} did, since the corresponding subspaces $V_2^\perp \cap V_x$ are not orthogonal.

Example 6.6.1: Consider the results (Table 6.6.1) of an experiment conducted to study the effects of three feeding treatments on the weight gains of pigs, as reported by Wishart (1950) and analyzed by Ostle (1963, 455). In this

Table 6.6.1 Initial Weights and Gains in Weight of Young Pigs in a Comparative Feeding Trial

		Food					
		A		B		C	
		Male	Female	Male	Female	Male	Female
Pen I	x	38	48	39	48	48	48
	y	9.52	9.94	8.51	10.00	9.11	9.75
Pen II	x	35	32	38	32	37	28
	y	8.21	9.48	9.95	9.24	8.50	8.66
Pen III	x	41	35	46	41	42	33
	y	9.32	9.32	8.43	9.34	8.90	7.63
Pen IV	x	48	46	40	46	42	50
	y	10.56	10.90	8.86	9.68	9.51	10.37
Pen V	x	43	32	40	37	40	30
	y	10.42	8.82	9.20	9.67	8.76	8.57

experiment 15 male pigs and 15 female pigs were randomly assigned to 15 pens, combinations of 5 pens, and three feeding treatments, so that in each pen one male and one female pig received each of the three treatments. The initial weight x and the weight gain were recorded for each pig. Pens were considered to be a blocking variable, with interactions between pens and the sex and treatment variables expected to be relatively small.

For food level i, sex level j (1 for male, 2 for female), pen level k, let the weight gain be Y_{ijk} and the initial weight x_{ijk}. Suppose that $Y_{ijk} = \mu_{ijk} + \gamma x_{ijk} + \varepsilon_{ijk}$, with the usual assumptions on the ε_{ijk}. Suppose that $\mu_{ijk} = \mu + f_i + s_j + (fs)_{ij} + p_k$, where f_i, s_j, $(fs)_{ij}$, and p_k are the food, sex, food $\times$ sex, and pen effects, and these parameters add to zero over each subscript. We have chosen to use the symbols f_i, s_j, and p_k rather than the more generic notation α_i, β_j, and δ_k (say) because it will remind us more readily of the meaning of the effect. Such notation is usually preferable.

The terms $\sum y^2$, $\sum xy$, and $\sum x^2$ for food, for example, are computed as follows:

$$\sum y^2 = \sum_i (5 \times 2)(\bar{Y}_{i\cdot\cdot} - \bar{Y}_{\cdot\cdot\cdot})^2, \qquad \sum x^2 = \sum_i (5 \times 2)(\bar{x}_{i\cdot\cdot} - \bar{X}_{\cdot\cdot\cdot})^2,$$

$$\sum xy = \sum_i (5 \times 2)(\bar{Y}_{i\cdot\cdot} - \bar{Y}_{\cdot\cdot\cdot})(\bar{x}_{i\cdot\cdot} - \bar{x}_{\cdot\cdot\cdot}).$$

Other sums of squares and cross-products are computed similarly, using the formulas for two-way analysis of variance. Then the error SSqs. for the analysis of covariance is $E_{yy} - E_{xy}^2/E_{xx} = 8.414 - (39.367^2/442.93) = 4.815$. The degree of freedom for error is $20 - 1 = 19$, since we have one covariate. Notice that error MSq. has been reduced for $8.314/20 = 0.416$ to $4.815/19 = 0.253$.

The sum of squares for food, for example, in the analysis of covariance was computed using the formulas above with $H_{yy} = 2.269$, $H_{xy} = -0.147$, $H_{xx} = 5.40$. The estimate of γ for the full model is $\hat{\gamma} = E_{xy}/E_{xx} = 0.889$. Table 6.6.2 presents

Table 6.6.2

	Sums of Squares and Cross Products				Analysis of Covariance		
Source	DF	$\sum y^2$	$\sum xy$	$\sum x^2$	SSqs.	Mean Sq.	$\hat{\gamma}$
Pens	4	4.852	39.905	605.87	2.359	0.590	0.076
Food	2	2.269	−0.147	5.40	2.337	1.168	0.087
Sex	1	0.434	−3.730	32.03	1.259	1.259	0.075
Food × sex	2	0.476	3.112	22.47	0.098	0.049	0.091
Error	20	8.314	39.367	442.93	4.815	0.253	0.089
Corr. Total	29	16.345	78.507	1,108.70			

Table 6.6.3

	F	Approx. p-value
Pens	2.33	0.10
Food	4.62	0.025
Sex	4.98	0.040
Food × sex	0.19	0.980

the estimates of γ for the models corresponding to each of the other rows. For example, $\hat{\gamma}_F = 0.087$ is the estimate when the food terms f_i are omitted from the model.

The F-statistics corresponding to the first four rows of Table 6.6.2 are given in Table 6.6.3. The model with the interaction terms omitted seems to be the most appropriate. The parameter μ_{ijk} is estimated by $\hat{\mu}_{ijk} = \hat{Y}_{ijk} - \hat{\gamma}x^{\perp}_{ijk}$, which has variance $\text{Var}(\hat{\mu}_{ijk}) = \text{Var}(\hat{Y}_{ijk}) + \text{Var}(\hat{\gamma})(x^{\perp}_{ijk})^2 = \sigma^2[20/25 + (x^{\perp}_{ijk})^2/E_{xx}]$, with covariances $\text{cov}(\hat{\mu}_{ijk}, \hat{\mu}_{i'j'k'}) = \sigma^2(x^{\perp}_{ijk}x^{\perp}_{i'j'k'})/E_{xx}$.

Table of $\hat{\mu}_{ijk}$

		A		B		C	
		M	F	M	F	M	F
	I	5.71	6.01	5.13	5.71	4.99	5.38
	II	6.24	6.54	5.66	6.23	5.52	5.91
Pen	III	5.52	5.82	4.94	5.52	4.80	5.19
	IV	6.17	6.47	5.59	6.17	5.45	5.85
	V	6.17	6.47	5.59	6.17	5.45	5.85

The estimates of the standard errors of these $\hat{\mu}_{ijk}$ were all between 0.41 and 0.45. There are too many covariances for us to attempt to give them here, though they are small relative to these standard deviations. Estimates of the μ, f_i, s_j, and $(fs)_{ij}$ are given in Table 6.6.4.

To make comparisons of the effects of the levels of the food effects we need to know

$$\begin{aligned}\text{Var}(\hat{f}_i - \hat{f}_{i'}) &= \text{Var}(\bar{Y}_{i\cdot\cdot} - \bar{Y}_{i'\cdot\cdot} - \hat{\gamma}(\bar{x}_{i\cdot\cdot} - \bar{x}_{i'\cdot\cdot}) \\ &= \sigma^2[2/15 + (\bar{x}_{i\cdot\cdot} - \bar{x}_{i'\cdot\cdot})^2/E_{xx}],\end{aligned}$$

which we estimate by replacing σ^2 by $S^2 = \text{Error MSq.} = 0.253$ to get 0.0849 for $i = 1$, $i' = 2$, 0.0845 for $i = 1$, $i' = 3$, and 0.0849 for $i = 2$, $i' = 3$. Similarly, we estimate $\text{Var}(\hat{s}_1 - \hat{s}_2) = \sigma^2[2/15 + (\bar{x}_{\cdot 1\cdot} - \bar{x}_{\cdot 2\cdot})^2/E_{xx}]$ to be 0.0362. These estimates can be used to give confidence intervals on the differences.

Table 6.6.4

μ	5.740	
f_1, f_2, f_3	0.371, −0.070, −0.301	
s_1, s_2	−0.212, 0.212	
$(fs)_{ij}$	0.063	−0.063
	−0.077	0.077
	0.014	−0.014
p_k	−0.254, 0.274, −0.442, 0.211, 0.211	

As is evident from this example, the payoff in using the covariate in the analysis, despite its additional complication, is that it reduces the size of error MSq., providing shorter confidence intervals, and more power. Analysis of covariance should *not* be used if the covariate itself is affected by the factors being studied. In this last example that is definitely not the case because x was the weight of the pig before the experiment began. Thus, in a study of the effects of three different methods of teaching algebra to high school freshman it would be appropriate to use the score on a standardized math exam if the exam were given before the experiment, but not if the exam were given during or after the experiment. (The models discussed in this chapter would not be appropriate for most such experiments because the performances of students in the same classroom could not be considered to be independent, usually being affected by the same teacher and by interaction among students.)

Problem 6.6.1: Suppose that following pairs (x_{ijk}, Y_{ijk}) are observed for $i = 1, 2;\ j = 1, 2;\ k = 1, 2$.

$$\begin{array}{cc} & \begin{array}{cc} B_1 & B_2 \end{array} \\ \begin{array}{c} A_1 \\ \\ A_2 \end{array} & \begin{bmatrix} (10, 25) & (7, 8) \\ (14, 23) & (5, 12) \\ (1, 20) & (0, 7) \\ (7, 12) & (4, 5) \end{bmatrix} \end{array}$$

(a) For the analysis of covariance model $Y_{ijk} = \mu + \alpha_i + \beta_j + (\alpha\beta)_{ij} + \gamma x_{ijk} + \varepsilon_{ijk}$ determine the analysis of covariance table, estimate the parameters, and perform appropriate F-tests.

(b) Plot the scatter diagram and the estimated regression line for each ij cell of the two-way table. (You can do this on just one pair of xy-axes if you use different labels for the points corresponding to different cells.)

(c) Find $\text{Var}((\hat{\alpha}_1 - \hat{\alpha}_2))$, both for the model in (a) and for the same model with the γx_{ijk} term omitted. Use this to find 95% confidence intervals on $\alpha_1 - \alpha_2$ for both models.

(d) Find a 95% confidence interval on γ, for the model of (a).

(e) Let SSA be the sum of squares used to test $H_0\colon \alpha_1 = \alpha_2 = 0$ for this model. Show that $\text{SSA} = (\hat{\alpha}_1 - \hat{\alpha}_2)^2/[\text{Var}(\hat{\alpha}_1 - \hat{\alpha}_2)/\sigma^2]$.

(f) How many covariates $x_1, \ldots, x_r$ could be used for these eight observations on Y?

(g) Suppose that $x_{ijk} = x_{ij}$ for all ijk, so that the x-values within the ij cell were the same. Could an analysis of covariance be performed? Could it be performed if the model did not include interaction terms $(\alpha\beta)_{ij}$?

Problem 6.6.2: The analysis of covariance models we have considered has assumed that the slope γ is the same for every cell of the table. In the case of a 2×3 table with three observations per cell suppose that $Y_{ijk} = \mu_{ij} + \gamma_{ij}x_{ij} + \varepsilon_{ijk}$.

(a) Describe how you could test the null hypothesis that the γ_{ij} are the same for all i and j.

(b) Could you carry out this test if there were only two observations per cell?

(c) Consider the model with γ_{ij} replaced by γ_i. How could you test the null hypothesis that $\gamma_1 = \gamma_2$? Is there a corresponding t-test?

CHAPTER 7

Miscellaneous Other Models

7.1 THE RANDOM EFFECTS MODEL

In this chapter we consider models which do not quite satisfy the general linear model in the sense that they contain two independent random terms, say η and ε rather than only one. In another sense the error term is the sum $\xi = \eta + \varepsilon$, for which different observations will not in general be independent. The special structure of ξ as a sum allows us to develop estimators and tests in computationally simple form.

We will only treat a few of these *random component* models here and will, for example, not even discuss *mixed* models for two-way layouts when one factor has randomly chosen levels. Multivariate analysis of variance is best used in such situations, and we shall not attempt to discuss its techniques. Those interested in multivariate statistical methods are referred to books by Morrison (1976) or Johnson and Wichern (1988).

Suppose we are interested in studying the output in numbers of parts turned out by the workers in a factory. A large number of workers are available, and we choose I of them at random, asking each to work J different two-hour time periods. The measured variable is then Y_{ij}, the number of parts turned out by worker i in time period j. We suppose that the worker has had enough experience so that there is no learning effect. The following model may be appropriate:

$$Y_{ij} = \mu + a_i + \varepsilon_{ij} \qquad \text{for} \quad j = 1, \ldots, J$$
$$i = 1, \ldots, I$$

where $a_i \sim N(0, \sigma_a^2)$, $\varepsilon_{ij} \sim N(0, \sigma^2)$, and the ε_{ij} and a_i are $JI + I$ independent r.v.'s.

This model differs from the usual one-way ANOVA model, called the *fixed effects model*, in that the a_i are considered to be random variables. In fact, conditional on the a_i (on the specific workers chosen in the example) this is the fixed effects model with $\mu_i = \mu + a_i$. This model, unconditional on the a_i, is called the *random components model*, and is appropriate when the levels of the treatment variable (the workers in the example) under study can be considered as a random sample from a large collection of possible levels, and inferences are to be made about this large collection, rather than just those levels on which observations are taken. The example above and a complete discussion of the mathematics involved is found in Scheffé (1959, Ch. 7).

As for the fixed model define $\mathbf{Y}$ to be the $I \times J$ array of Y_{ij}, $\mathbf{A}_i$ to be the $I \times J$ array with 1's in column i, 0's elsewhere,

$$V_0 = \mathscr{L}(\mathbf{x}_0), \qquad \text{for} \quad \mathbf{x}_0 = \sum_1^I \mathbf{A}_i$$

$$V_A = \mathscr{L}(\mathbf{A}_1, \ldots, \mathbf{A}_I) \cap V_0^\perp = \left\{ \sum_1^I b_i \mathbf{A}_i \,\middle|\, \sum_1^I b_i = 0 \right\}$$

$$V = \mathscr{L}(\mathbf{A}_1, \ldots, \mathbf{A}_I) = V_0 \oplus V_A$$

Then

$$\mathbf{Y} = \mu \mathbf{x}_0 + \sum_1^I a_i \mathbf{A}_i + \boldsymbol{\varepsilon}, \qquad \text{and} \qquad \mathbf{Y} \sim N(\mu \mathbf{x}_0, \boldsymbol{\Sigma}_{\mathbf{Y}}),$$

where the elements of $\boldsymbol{\Sigma}_{\mathbf{Y}}$ are given by

$$\begin{aligned}
\operatorname{cov}(Y_{ij}, Y_{i'j'}) &= \operatorname{cov}(a_i + \varepsilon_{ij}, a_{i'} + \varepsilon_{i'j'}) \\
&= \sigma_a^2 + \sigma^2 \qquad \text{for} \quad i = i' \quad \text{and} \quad j = j' \\
&= \sigma_a^2 \qquad \text{for} \quad i = i' \quad \text{and} \quad j \neq j' \\
&= 0 \qquad \text{for} \quad i \neq i'
\end{aligned}$$

If we order the elements of $\mathbf{Y}$ by first going down columns then across rows, then

$$\underset{IJ \times IJ}{\boldsymbol{\Sigma}_{\mathbf{Y}}} = \sigma_a^2 \begin{pmatrix} \mathbf{B}_J^2 & & 0 \\ & \ddots & \\ 0 & & \mathbf{B}_J^2 \end{pmatrix} + \sigma^2 \mathbf{I}_{IJ}$$

where $\mathbf{B}_J$ is a $J \times J$ block of all ones. Thus for $I = 3$, $J = 2$, we get

$$\boldsymbol{\Sigma}_{\mathbf{Y}} = \sigma_a^2 \begin{bmatrix} 1 & 1 & 0 & 0 & 0 & 0 \\ 1 & 1 & 0 & 0 & 0 & 0 \\ 0 & 0 & 1 & 1 & 0 & 0 \\ 0 & 0 & 1 & 1 & 0 & 0 \\ 0 & 0 & 0 & 0 & 1 & 1 \\ 0 & 0 & 0 & 0 & 1 & 1 \end{bmatrix} + \sigma^2 \begin{bmatrix} 1 & 0 & 0 & 0 & 0 & 0 \\ 0 & 1 & 0 & 0 & 0 & 0 \\ 0 & 0 & 1 & 0 & 0 & 0 \\ 0 & 0 & 0 & 1 & 0 & 0 \\ 0 & 0 & 0 & 0 & 1 & 0 \\ 0 & 0 & 0 & 0 & 0 & 1 \end{bmatrix}$$

As for the fixed effects model, let

$$\hat{\mathbf{Y}}_0 = p(\mathbf{Y} \mid V_0) = \bar{Y}_{..}\mathbf{x}_0 = (\mu + \bar{a} + \bar{\varepsilon}_{..})\mathbf{x}_0$$

$$\hat{\mathbf{Y}} = p(\mathbf{Y} \mid V) = \sum_1^I \bar{Y}_{i\cdot}\mathbf{A}_i = \sum_1^I [\mu + a_i + \bar{\varepsilon}_{i\cdot}]\mathbf{A}_i$$

$$\begin{aligned}\mathbf{Y}_A = p(\mathbf{Y} \mid V_A) &= p(\mathbf{Y} \mid V) - p(\mathbf{Y} \mid \mathbf{x}_0)\\ &= \sum_i (\bar{Y}_{i\cdot} - \bar{Y}_{..})\mathbf{A}_i = \sum_1^I [a_i + \bar{\varepsilon}_{..} - (\bar{a} + \bar{\varepsilon}_{..})]\mathbf{A}_i\\ &= \sum_i [(a_i - \bar{a}) + (\bar{\varepsilon}_{i\cdot} - \bar{\varepsilon}_{..})]\mathbf{A}_i\end{aligned}$$

$$\begin{aligned}\mathbf{e} = \mathbf{Y} - \hat{\mathbf{Y}} = p(\mathbf{Y} \mid V^{\perp}) &= p\left(\mu\mathbf{x}_0 + \sum_1^I a_i\mathbf{A}_i + \boldsymbol{\varepsilon} \mid V^{\perp}\right)\\ = p(\boldsymbol{\varepsilon} \mid V^{\perp}) &= \begin{bmatrix} \varepsilon_{11} - \bar{\varepsilon}_{1\cdot} & \cdots & \varepsilon_{I1} - \bar{\varepsilon}_{I\cdot} \\ \vdots & & \\ \varepsilon_{1J} - \varepsilon_{1\cdot} & \cdots & \varepsilon_{IJ} - \bar{\varepsilon}_{I\cdot} \end{bmatrix}\end{aligned}$$

Then

$$\text{SSA} = \|\hat{\mathbf{Y}}_A\|^2 = \sum_1^I [(a_i - \bar{a}) + (\bar{\varepsilon}_{i\cdot} - \bar{\varepsilon}_{..})]^2 J,$$

and

$$\text{SSE} = \|\mathbf{e}\|^2 = \sum_{ij} (\varepsilon_{ij} - \bar{\varepsilon}_{i\cdot})^2$$

Let $W_i = a_i + \bar{\varepsilon}_{i..}$ Then $W_i \sim N\left(0, \sigma_a^2 + \dfrac{\sigma^2}{J}\right)$ and the W_i are independent. It follows that

$$\frac{\sum_i (W_i - \bar{W})^2}{\sigma_a^2 + \dfrac{\sigma^2}{J}} = \frac{\text{SSA}}{\sigma^2 + J\sigma_a^2} \sim \chi_{I-1}^2$$

In addition, the W_i are independent of the vector **e**, and therefore of SSE, which is the same as it is for the fixed effects model. Thus, $\text{SSE}/\sigma^2 \sim \chi^2_{(J-1)I}$. It follows that

$$\frac{\{\text{SSA}/[\sigma^2 + J\sigma_a^2]\}/(I-1)}{[\text{SSE}/\sigma^2]/(J-1)} \sim F(I-1, (J-1)I)$$

The analysis can be summarized by the usual one-way ANOVA table, with the same d.f.'s, sums of squares, and mean squares as for the fixed effects model.

The only change appears in the expected mean squares column, with $E(\text{MSA}) = \sigma^2 + J\sigma_a^2$.

Though we will not show it here, the maximum likelihood estimator of the pair (σ^2, σ_a^2) is $(\hat{\sigma}^2, \text{SSA}/n - J\hat{\sigma}^2)$, where $\hat{\sigma}^2 = \text{SSE}/n$. It is more common to use the estimator (MSE, (MSA − MSE)/J), which is unbiased. However, it makes little sense to estimate σ_a^2 by a negative number, so that it makes more sense to replace (MSA − MSE)/J by the minimum of this and zero. Of course the estimator is then biased. That seems to be a small price to pay to avoid an embarrassing point estimate.

Let

$$F = \frac{\text{SSA}/(I-1)}{\text{SSE}/(J-1)I} = \frac{\text{A MSq.}}{\text{Error MSq.}} \qquad \text{and} \qquad \theta = \frac{\sigma^2 + J\sigma_a^2}{\sigma^2}.$$

We have shown that F/θ has a *central* F distribution. Thus, for $F_{\alpha/2}$ and $F_{1-\alpha/2}$ the $100(\alpha/2)$ and $100(1 - \alpha/2)$ percentiles of the $F_{I-1,(J-)I}$ distribution,

$$1 - \alpha = P(F_{\alpha/2} \le F/\theta \le F_{1-\alpha/2}) = P\left(\frac{F}{F_{1-\alpha/2}} \le \theta \le \frac{F}{F_{\alpha/2}}\right).$$

We have a $100(1-\alpha)\%$ confidence interval on $\theta = 1 + J(\sigma_a^2/\sigma^2)$. Manipulating the inequalities still further, we get

$$P\left(\frac{1}{J}\left(\frac{F}{F_{1-\alpha/2}} - 1\right) \le \frac{\sigma_a^2}{\sigma^2} \le \frac{1}{J}\left(\frac{F}{F_{\alpha/2}} - 1\right)\right) = 1 - \alpha.$$

An approximate confidence interval on σ_a^2 is obtained by substituting S^2 for σ^2, to get the interval

$$\frac{S^2}{J}\left(\frac{F}{F_{1-\alpha/2}} - 1\right) \le \sigma_a^2 \le \frac{S^2}{J}\left(\frac{F}{F_{\alpha/2}} - 1\right)$$

The approximation is good if $I(J-1) =$ Error d.f. is large. A $100(1 - \alpha/2)\%$ one-sided confidence interval may be obtained by leaving off either end of the interval.

We can test H_0: $\sigma_a^2/\sigma^2 \le r_0$ vs. H_1: $\sigma_a^2/\sigma^2 > r_0$ for r_0 a known constant as follows. Since

$$F/\theta = F\left(\frac{1}{1 + J\sigma_a^2/\sigma^2}\right) \sim F_{I-1,\, I(J-1)},$$

$$\alpha = P\left(F \ge \left(1 + J\frac{\sigma_a^2}{\sigma^2}\right)F_{1-\alpha}\right) \ge P(F \ge (1 + Jr_0)F_{1-\alpha})$$

for $\sigma_a^2/\sigma^2 \le r_0$, so that the test which rejects H_0 for $F \ge (1 + Jr_0)F_{1-\alpha}$ is an

α-level test. Its power function for $r = \sigma_a^2/\sigma^2$ is

$$P(F \geq (1 + Jr_0)F_{1-\alpha}) = P\left(F\left(\frac{1}{1 + Jr}\right) \geq \left(\frac{1 + Jr_0}{1 + Jr}\right)F_{1-\alpha}\right)$$

$$= \text{area under } F_{I-1, I(J-1)} \text{ density to the right of } \frac{(1 + Jr_0)}{(1 + Jr)} F_{1-\alpha}.$$

For the case $r_0 = 0$ the test rejects for $F \geq F_{1-\alpha}$, the usual ANOVA test. In practice it is not reasonable to expect $\sigma_a^2 = 0$, however.

Problem 7.1.1: For the random effects model, what is the c.c. $\rho(Y_{ij}, Y_{ij'})$ for $j \neq j'$?

Problem 7.1.2: Let $U_1, V_1, U_2, V_2, \ldots, U_k, V_k$ be independent r.v.'s with $U_i \sim N(0, \sigma_U^2)$ and $V_i \sim N(0, \sigma_V^2)$. Let $\bar{U} = \sum U_i/k$ and $\bar{V} = \sum V_i/k$, and let $Q = \sum [(U_i + V_i) - (\bar{U} + \bar{V})]^2$.

(a) Describe the conditional distribution of Q, given $V_1 = v_1, \ldots, V_k = v_k$.
(b) Describe the unconditional distribution of Q.
(c) Apply the results of (a) and (b) to $\sum (W_i - \bar{W})^2$, where $W_i = a_i + \bar{\varepsilon}_i$.

Problem 7.1.3: In order to determine the contamination by dioxin of land formerly used as a dumpsite, the land was divided into 20,000 one foot by one foot squares. Fifty of these squares were then chosen randomly for analysis. Five samples of soil, each of one cubic inch, were then taken from each sample square. Measurements, in parts per billion of dioxin, were then obtained; for example,

Sample Square

1	2	3	4	5		50
195	323	257	332	328	...	262
180	295	248	263	284	...	259
187	306	261	281	264	...	267
196	320	282	292	320	...	268
149	344	263	326	262	...	265

Computations gave:

$$\sum_{ij} Y_{ij} = 96{,}936 \qquad \sum_{ij} Y_{ij}^2 = 40{,}835{,}618 \qquad \sum_j \left(\sum_i Y_{ij}\right)^2 = 203{,}375{,}212.$$

(a) State an appropriate random effects model, with variances σ^2 and σ_s^2 (for squares).
(b) Determine the analysis of variance table.

(c) Find 95% confidence intervals on σ^2, $R = \sigma_s^2/\sigma^2$, and σ_s^2.

(d) For $R = \sigma_s^2/\sigma^2$ test $H_0\colon \theta \le 5$ vs. $H_1\colon \theta > 5$, for $\alpha = 0.05$.

(e) These data were actually generated on a computer using $\mu = 400$, $\sigma = 30$, $\sigma_s = 100$. Did the confidence intervals include these parameter values? Find the power of the test of (d).

(f) What were $\text{Var}(\bar{Y}_{\cdot,17})$, $\text{Var}(\bar{Y}_{\cdot,5} - \bar{Y}_{\cdot,50})$, $\text{Var}(\bar{Y}_{\cdot\cdot})$? What were their estimates?

(g) Suppose that the cost of observations is \$90 for each square and \$3 for each measurement of dioxin, so that the total cost was $C = (\$90)(50) + (\$3)(250) = \$5{,}250$. Suppose that the purpose of the study was to estimate the overall mean per square foot as precisely as possible. Find a choice of (I, J) which would cause $\text{Var}(\bar{Y}_{\cdot\cdot})$ to be as small as possible subject to the cost being no larger, and compute $\text{Var}(\bar{Y}_{\cdot\cdot})$ for the experiment performed and for the better experiment.

7.2 NESTING

Suppose we expand the experiment of Section 7.1 as follows. We are interested in I different machines (or machine types). J_i workers are chosen randomly to work on machine i, for $i = 1, \ldots, I$. Then each worker is assigned to K different two-hour periods, all on the same machine. The production in time period k is Y_{ijk} for $k = 1, \ldots, K$. Then, since an individual worker works only on one machine, workers are said to be "nested within machines."

Example 7.2.1: $K = 3$, $I = 4$, $J_1 = 3$, $J_2 = 3$, $J_3 = 2$, $J_4 = 4$

Workers											
Machine 1			Machine 2			Machine 3		Machine 4			
1	2	3	4	5	6	7	8	9	10	11	12
31	41	36	48	39	50	29	37	45	57	50	53
35	38	39	48	42	50	32	39	46	54	50	55
30	37	38	45	41	53	30	39	48	55	54	49

A reasonable model is

$$Y_{ijk} = \mu + m_i + w_{j(i)} + \varepsilon_{ijk}$$

The nesting is indicated by the notation $w_{j(i)}$, since the values which j takes depend on i, with j "nested" within machine i. Let J_i be the number of workers nested within machine i, for $i = 1, \ldots, I$. Let $M_0 \equiv 0$, and $M_i = J_1 + \cdots + J_i$

for each i. Then for each i, j takes only the values $M_{i-1}+1,\ldots,M_i$. For each j, k takes the values $1,\ldots,K$. Suppose $m_1,\ldots,m_I$ are fixed effects with $\sum_i m_i = 0$, and the $w_{j(i)}$ are random variables, independent, with $N(0,\sigma_w^2)$ distributions. Then

$$Y_{ijk} \sim N(\mu + m_i, \sigma_w^2 + \sigma^2)$$

and

$$\operatorname{cov}(Y_{ijk}, Y_{i'j'k'}) = \begin{cases} \sigma_w^2 + \sigma^2 & \text{for} \quad i=i', j=j', k=k' \\ \sigma_w^2 & \text{for} \quad i=i', j=j', k\neq k' \\ 0 & \text{for} \quad j \neq j' \qquad \text{for all} \quad i, k, k'. \end{cases}$$

Let $\mathbf{Y}$ be the array of $K(\sum J_i)$ observations Y_{ijk}. Let $\mathbf{M}_i$ be the indicator of machine i, let $\mathbf{w}_j(i)$ be the indicator of worker j, who uses machine i. Define

$$V_0 = \mathscr{L}(\mathbf{x}_0) \qquad \text{for} \quad \mathbf{x}_0 = \sum_i \mathbf{M}_i, \text{ the array of all 1's.}$$

$$V_M = \mathscr{L}(\mathbf{M}_1,\ldots,\mathbf{M}_I) \cap V_0^\perp = \left\{\sum_1^I b_i\mathbf{M}_i \,\middle|\, \sum_1^I b_i J_i = 0\right\}$$

$$V = \mathscr{L}(\mathbf{w}_{1(1)},\ldots,\mathbf{w}_{J_I(I)}) \qquad V_W = V \cap (V_0 \oplus V_m)^\perp = V \cap V_0^\perp \cap V_M^\perp$$

$$\hat{\mathbf{Y}}_0 = p(\mathbf{Y}\,|\,\mathbf{x}_0) = \bar{Y}_{\ldots}\mathbf{x}_0 \qquad \hat{\mathbf{Y}}_M = \sum_i \bar{Y}_{i..}\mathbf{M}_i - \bar{Y}_{\ldots}\mathbf{x}_0 = \sum (\bar{Y}_{i\ldots} - \bar{Y}_{\ldots})\mathbf{M}_i$$

$$\hat{\mathbf{Y}}_W = \sum_{ij} \bar{Y}_{ij\cdot}\mathbf{w}_{j(i)} - \sum_i \bar{Y}_{i..}\mathbf{M}_i = \sum_{ij} (\bar{Y}_{ij\cdot} - \bar{Y}_{i..})\mathbf{w}_{j(i)}$$

Then

$$\text{SSM} = \|\hat{\mathbf{Y}}_M\|^2 = \sum_i (\bar{Y}_{i..} - \bar{Y}_{\ldots})^2 J_i K = K\sum_i \bar{Y}_{i..}^2 J_i - \bar{Y}_{\ldots}^2 n \qquad \text{for} \quad n = K\sum_1^I J_i$$

$$\text{SSW} = \|\mathbf{Y}_w\|^2 = \sum_{ij} (\bar{Y}_{ij\cdot} - \bar{Y}_{i..})^2 K = \sum_{ij} \bar{Y}_{ij\cdot}^2 K - K\sum_i \bar{Y}_{i..}^2 J_i$$

$$\text{SSE} = \|\mathbf{Y} - \hat{\mathbf{Y}}\|^2 = \sum_{ij} (Y_{ijk} - \bar{Y}_{ij\cdot})^2 = \sum_{ijk} Y_{ijk}^2 - \sum_{ij} \bar{Y}_{ij\cdot}^2 K$$

$$\text{SST} = \|\mathbf{Y} - \hat{\mathbf{Y}}_0\|^2 = \sum_{ijk} (Y_{ijk} - \bar{Y})^2 = \sum_{ijk} Y_{ijk}^2 - \bar{Y}_{\ldots}^2 n$$

The analysis of variance is given in Table 7.2.1. For each (i, j)

$$\bar{Y}_{ij\cdot} = \mu + m_i + w_{j(i)} + \bar{\varepsilon}_{ij\cdot} \sim N\left(\mu + m_i, \sigma_w^2 + \frac{\sigma^2}{K}\right)$$

Table 7.2.1 Analysis of Variance

Source	Subspace	DF	SSqs.	MSq.	Expected Msq.
Machine	V_M	$I-1$	SSM		$\sigma^2 + K\sigma_w^2 + \frac{K}{I-1}\sum m_i^2 J_i$
Worker	V_W	$\sum_1^I (J_i - 1)$	SSW		$\sigma^2 + K\sigma_w^2$
Error	$V^\perp$	$(K-1)\sum J_i$	SSE		σ^2
Adj. Total	$V_0^\perp$	$K(\sum J_i) - 1$	SST		

and the $\bar{Y}_{ij\cdot}$ are independent. Then

$$\frac{\text{SSM}}{K} = \sum_i (Y_i - Y_{\ldots})^2 J_i \qquad \text{and} \qquad \frac{\text{SSW}}{K} = \sum_{ij} (\bar{Y}_{ij\cdot} - \bar{Y}_{i\cdot\cdot})^2$$

are among means SSqs. and error SSqs. for one-way analysis of variance on the $\bar{Y}_{ij\cdot}$. Thus, they are independent,

$$\frac{\text{SSM}}{K\left(\sigma_w^2 + \dfrac{\sigma^2}{K}\right)} = \frac{\text{SSM}}{\sigma^2 + K\sigma_w^2} \sim \chi_{I-1}^2 \qquad \text{for} \quad \delta = \frac{K\sum_i m_i^2 J_i}{\sigma^2 + K\sigma_w^2}$$

and

$$\frac{\text{SSW}}{K\left(\sigma_w^2 + \dfrac{\sigma^2}{K}\right)} = \frac{\text{SSW}}{\sigma^2 + K\sigma_w^2} \sim \chi_{\sum J_i - I}^2 \text{ (central } \chi^2\text{)}.$$

Thus

$$F_M = \frac{\text{Machine MSq.}}{\text{Worker Msq.}} \sim F_{I-1,\sum J_i - I}(\delta).$$

Confidence intervals on σ_w^2/σ^2 can be obtained by the same method used in the one-way ANOVA random effects model. Thus (L, U) is a $100(1-\alpha)\%$ confidence interval on

$$\sigma_w^2/\sigma^2 \text{ for } L = \left(\frac{F_w}{F_{1-\alpha/2}} - 1\right)\frac{1}{K} \qquad \text{and} \qquad U = \left(\frac{F_w}{F_{\alpha/2}} - 1\right)\frac{1}{K},$$

where $F_{\alpha/2}$ and $F_{1-\alpha/2}$ are percentiles of the $F(\nu_1, \nu_2)$ distribution for $\nu_1 = \sum J_i - I$ and $\nu_2 = (K-1)(\sum J_i)$ and $F_w = (W\,\text{MSq.})/S^2$. Since $[E(\text{MSW}) - \sigma^2]/K = \sigma_w^2$, $\hat{\sigma}_w^2 = [\text{MSW} - S^2]/K$ is an unbiased point estimator of σ_w^2. It makes sense, however, to replace $\hat{\sigma}_w^2$ by 0 whenever it is less than 0, so the unbiasedness is lost.

Problem 7.2.1: Hicks (1982, Example 11.1) described the following experiment:

> In a recent in-plant training course the members of the class were assigned a final problem. Each class member was to go to the plant and set up an experiment using the techniques that had been discussed in the class. One engineer wanted to study the strain readings of glass cathode supports from five different machines. Each machine had four "heads" on which the glass was formed, and she decided to take four samples from each head. She treated this experiment as a 5 × 4 factorial with four replications per cell. Complete randomization of the testing for strain readings presented no problem. Her model was
>
> $$Y_{ij} = \mu + M_i + \eta_j + MH_{ij} + \varepsilon_{k(ij)}$$
>
> with
>
> $$i = 1, 2, \ldots, 5 \qquad j = 1, \ldots, 4 \qquad k = 1, \ldots, 4$$
>
> Her data and analysis appear in Table 7.2.2. In this model she assumed that both machines and heads were fixed, and used the 10 percent significance level. The results indicated no significant interaction at the 10 percent level of significance.
>
> The question was raised as to whether the four heads were actually removed from machine A and mounted on machine B, then on C, and so on. Of course, the answer was no, as each machine had its own four heads. Thus machines and heads did not form a factorial experiment, as the heads on each machine were unique for that particular machine. In each case the experiment is called a nested experiment: levels of one factor are nested within, or are subsamples of, levels of another factor. Such experiments are called hierarchical experiments.

Table 7.2.2 Data for Strain Problem in a Nested Experiment

Machine Head																			
A				B				C				D				E			
1	2	3	4	5	6	7	8	9	10	11	12	13	14	15	16	17	18	19	20
6	13	1	7	10	2	4	0	0	10	8	7	11	5	1	0	1	6	3	3
2	3	10	4	9	1	1	3	0	11	5	2	0	10	8	8	4	7	0	7
0	9	0	7	7	1	7	4	5	6	0	5	6	8	9	6	7	0	2	4
8	8	6	9	12	10	9	1	5	7	7	4	4	3	4	5	9	3	2	0
Head Totals																			
16	33	17	27	38	14	21	8	10	34	20	18	21	26	22	19	21	16	7	14
Machine Totals																			
93				81				82				88				58			

(a) Define a more appropriate model than that chosen by the student.

(b) Determine the appropriate analysis of variance table and test the hypotheses $\sigma_H^2 = 0$ and (all $m_i = 0$) at level $\alpha = 0.05$.

(c) Find a 95% confidence intervals on σ_H^2/σ^2 and (approximately) on σ_H^2.

Problem 7.2.2: Express $P(\hat{\sigma}_w^2 > 0)$ in terms of the central F c.d.f. and the parameter $\theta = [\sigma^2 + K\sigma_w^2]/\sigma^2$. Evaluate it for the case $I = 3$, $J_1 = J_2 = J_3 = 5$, $K = 4$, $R = \sigma_w^2/\sigma^2 = 0.355$.

Problem 7.2.3: For the worker–machine model find:

(a) The c.c. $\rho(\bar{Y}_{ij\cdot}, \bar{Y}_{ij'\cdot})$ and $\text{Var}(\bar{Y}_{ij\cdot} - \bar{Y}_{ij'\cdot})$ for $j \neq j'$.

(b) The c.c. $\rho(Y_{ijk}, Y_{ijk'})$ and $\text{Var}(Y_{ijk} - Y_{ijk'})$ for $k \neq k'$.

(c) The c.c. $\rho(\bar{Y}_{i\cdot\cdot}, \bar{Y}_{i'\cdot\cdot})$ and $\text{Var}(\bar{Y}_{i\cdot\cdot} - \bar{Y}_{i'\cdot\cdot})$ for $i \neq i'$.

7.3 SPLIT PLOT DESIGNS

Ostle (1963) describes an experiment designed to determine the effects of temperature and electrolyte on the lifetime of thermal batteries. The electrolytes were A, B, C, D, and the temperatures were low, medium and high. The temperature chamber had positions for four batteries. On six consecutive days (replicates) the chamber was used three times (*whole-plots*, so that there were 18 whole-plots). The three temperatures were randomly assigned to these whole-plots. Within each whole-plot one battery with each of the electrolytes was randomly chosen for a position, split-plot, within the chamber. The measured variable was the activated life of the battery (Table 7.3.1). For an agricultural example suppose four hybrids of corn H_1, H_2, H_3, H_4 and three levels of fertilizer F_1, F_2, F_3 are of interest. Three farms (replicates) each have four acres available for use. On each farm the land is divided into one-acre whole plots, and one of each of the hybrids assigned randomly to these whole plots. Then each whole plot is divided into three split-plots and the three fertilizers randomly assigned to these split-plots.

Consider the battery example again. Let i index replicate, j index temperature and k index electrolyte. A reasonable model is then

$$Y_{ijk} = \mu + \rho_i + \tau_j + p_{ij} + \gamma_k + (\tau\gamma)_{jk} + \varepsilon_{ijk}$$

where ρ_i is the fixed replication effect, τ_j is the fixed temperature effect, p_{ij} is the random whole-plot effect, γ_k is the fixed electrolyte effect, $(\tau\gamma)_{jk}$ is the temperature–electrolyte interaction effect, and ε_{ijk} is the split-plot effect. ε_{ijk} also contains other random errors. In addition, $\sum \rho_i = \sum \tau_j = \sum \gamma_k = \sum_j (\tau\gamma)_{jk} = 0$, $p_{ij} \sim N(0, \sigma_p^2)$, $\varepsilon_{ijk} \sim N(0, \sigma^2)$ and these random variables p_{ij}, ε_{ijk} are all independent.

The sums of squares for each of the terms of the model are determined as before for a three-way factorial with one observation per cell assuming no

Table 7.3.1 Activated Lives in Hours of 72 Thermal Batteries Tested in a Split Plot Design Which Used Temperatures as Whole Plots and Electrolytes as Split Plots

Electrolyte	Replicate 1	2	3	4	5	6
			Low Temperature			
A	2.17	1.88	1.62	2.34	1.58	1.66
B	1.58	1.26	1.22	1.59	1.25	0.94
C	2.29	1.60	1.67	1.91	1.39	1.12
D	2.23	2.01	1.82	2.10	1.66	1.10
			Medium Temperature			
A	2.33	2.01	1.70	1.78	1.42	1.35
B	1.38	1.30	1.85	1.09	1.13	1.06
C	1.86	1.70	1.81	1.54	1.67	0.88
D	2.27	1.81	2.01	1.40	1.31	1.06
			High Temperature			
A	1.75	1.95	2.13	1.78	1.31	1.30
B	1.52	1.47	1.80	1.37	1.01	1.31
C	1.55	1.61	1.82	1.56	1.23	1.13
D	1.56	1.72	1.99	1.55	1.51	1.33

Source: Reprinted with permission from *Statistics in Research* by Bernard Ostle. © 1963 Iowa State Press.

interaction between replicates and electrolyte or between temperature and electrolyte. In general, suppose there are R replicates, T temperatures, L electrolytes. Then

$$\text{SSR} = (TL)\sum_i (\bar{Y}_{i..} - \bar{Y}_{...})^2 \qquad \text{(Replicates)}$$

$$\text{SST} = (RL)\sum_j (\bar{Y}_{.j.} - \bar{Y}_{...})^2 \qquad \text{(Temperature)}$$

$$RT\,\text{Subtotal} = L\sum_{ij} (\bar{Y}_{ij.} - \bar{Y}_{...})^2$$

$$\text{SSRT} = (RT\,\text{Subtotal}) - \text{SSR} - \text{SST} \qquad (RT\ \text{Interaction})$$

$$\text{SSP} = L\sum_{ij} (\bar{Y}_{ij.} - \bar{Y}_{...})^2 \qquad \text{(Whole-Plots)}$$

$$\text{SSL} = (RT)\sum_k (\bar{Y}_{..k} - \bar{Y}_{...})^2 \qquad \text{(Electrolytes)}$$

$$TL\,\text{Subtotal} = R\sum_{jk} (\bar{Y}_{.jk} - \bar{Y}_{...})^2$$

$$\text{SSTL} = (TL\,\text{Subtotal}) - \text{SST} \qquad (TL\ \text{interaction})$$

$$\text{(Corr.) Total SS} = \sum_{ijk} (\bar{Y}_{ijk} - \bar{Y}_{...})^2$$

$$\text{SSE} = \text{Total SS} - (RT\ \text{Subtotal}) - \text{SSL (Error)} - \text{SSTL}$$

Each sum of squares is the squared length of the projection of **Y** on a subspace, the subspaces for R, T, P, L, error being orthogonal. SSRT is RT interaction sum of squares, which we have chosen to call whole-plot error and labeled as p_{ij}.

Since $\bar{Y}_{ij\cdot} \sim N(\mu + \rho_i + \tau_j, \sigma_p^2 + \sigma^2/L)$, by arguments similar to those for nested designs we get

$$\frac{\text{SSR}}{L\sigma_p^2 + \sigma^2} \sim \chi^2_{R-1}(\delta) \qquad \text{for} \quad \delta = TL(\sum \rho_i^2)/(L\sigma_p^2 + \sigma^2)$$

$$\frac{\text{SST}}{L\sigma_p^2 + \sigma^2} \sim \chi^2_{T-1}(\delta) \qquad \text{for} \quad \delta = RL(\sum \tau_j^2)/(L\sigma_p^2 + \sigma^2)$$

$$\frac{\text{SSP}}{L\sigma_p^2 + \sigma^2} \sim \chi^2_{(R-1)(T-1)}$$

$$\frac{\text{SSL}}{\sigma^2} \sim \chi^2_{L-1}(\delta) \qquad \text{for} \quad \delta = RT(\sum \gamma_k^2)/\sigma^2$$

$$\frac{\text{SSTL}}{\sigma^2} \sim \chi^2_{(T-1)(L-1)}(\delta) \qquad \text{for} \quad \delta = R\sum (\tau\gamma)_{jk}^2/\sigma^2$$

$$\frac{\text{SSE}}{\sigma^2} \sim \chi^2_{R(T-1)(L-1)}$$

These sums of squares are independent.

F-tests for replicate and temperature effects use the mean square for whole plots = SSP/$(R-1)(T-1)$ in the denominator. F-tests for the split-plot factor, electrolyte, use error mean square. The ANOVA table presented by Ostle for these data is Table 7.3.2.

Table 7.3.2 Abbreviated ANOVA

Source	DF	SSqs.	MSq.	F	Expected MSq.
		Whole Plots			
Replicates	5	4.1499	0.8300	6.09*	$\sigma^2 + 4\sigma_p^2 + 12\sum \rho_i^2/5$
Temperatures	2	0.1781	0.0890	0.65	$\sigma^2 + 4\sigma_p^2 + 24\sum \tau_j^2/2$
Whole plot error	10	1.3622	0.1362		$\sigma^2 + 4\sigma_p^2$
		Split Plots			
Electrolytes	3	1.9625	0.6542*	23.4	$\sigma^2 + 18\sum \gamma_k^2/3$
Temperature × electrolyte	6	0.2105	0.0351	1.25	$\sigma^2 + 6\sum (\tau\gamma)_{jk}^2/6$
Split plot error	45	1.2586	0.0280		σ^2

* Significant at the $\alpha = 0.01$ level.

Since $F_{5,10,0.99} = 5.64$, $F_{3,45,0.99} = 4.24$ both replicate and electrolyte effects are significantly different from zero at the 0.01 level. There seems to be little effect due to temperature or temperature × electrolyte interaction. It would be appropriate to compare electrolyte means, which have variances $\sigma^2/18$, estimated to be 0.028 0/18.

Problem 7.3.1: Analysis of the battery data produced: $\sum Y_{ijk} = 114.97$, $\sum Y_{ijk}^2 = 192.7$, and means as follows:

$$\text{Temperature}\begin{array}{c}\text{Replicates}\\ \begin{bmatrix} 2.068 & 1.688 & 1.583 & 1.985 & 1.470 & 1.205 \\ 1.960 & 1.705 & 1.842 & 1.453 & 1.383 & 1.088 \\ 1.595 & 1.688 & 1.935 & 1.565 & 1.265 & 1.268 \end{bmatrix}\end{array}$$

$$\text{Temperature}\begin{array}{c}\text{Electrolytes}\\ \begin{bmatrix} 1.875 & 1.307 & 1.663 & 1.820 \\ 1.765 & 1.302 & 1.577 & 1.643 \\ 1.703 & 1.413 & 1.483 & 1.610 \end{bmatrix}\end{array}$$

Rep. Means					
1.874	1.693	1.787	1.667	1.372	1.187

Temp. Means		
1.662	1.572	1.552

Electrolyte Means			
1.781	1.341	1.574	1.691

$$\sum \bar{Y}_{ij\cdot}^2 = 47.319 \qquad \sum \bar{Y}_{i\cdot\cdot}^2 = 15.645$$

$$\sum \bar{Y}_{\cdot jk}^2 = 30.989 \qquad \sum \bar{Y}_{\cdot j\cdot}^2 = 7.647$$

(a) Verify the sums of squares in Ostle's table.
(b) Find an individual 95% confidence interval $\tau_1 - \tau_2$.
(c) Find an individual 95% confidence interval on $\gamma_1 - \gamma_2$.

Problem 7.3.2: Prove that SST, SSR, and SSL are independent r.v.'s.

7.4 BALANCED INCOMPLETE BLOCK DESIGNS

Consider the following experiment described by Mendenhall (1968, p. 325):

> An experiment was conducted to compare the effect of $p = 7$ chemical substances on the skin of male rats. The area of experimentation on the animal's skin was confined

to a region which was known to be relatively homogeneous, but this restricted the experimenter to three experimental units (patches of skin) per animal. Hence to eliminate the rat-to-rat variability for the comparison of treatments, the experiment was blocked on rats using the balanced incomplete block design shown below ($k = 3$, $r = 3$, $b = 7$, $\lambda = 1$). The seven blocks correspond to 7 rats.

Blocks

1	2	3	4	5	6	7
A	D	C	E	B	E	A
10.2	12.9	11.7	9.1	8.8	9.2	11.3
B	F	B	G	G	F	C
6.9	14.1	12.1	7.7	8.6	15.2	9.7
D	C	E	D	F	A	G
14.2	9.9	8.6	14.3	16.3	13.1	6.2

This experimental design is called an *incomplete block design* because not all treatment levels are represented in each block. There are $b = 7$ blocks (rats), $k = 3$ experimental units (patches) within each block, $t = 7$ levels of the chemical factor (the treatment), each level is replicated $r = 3$ times, and each pair of levels is together in the same block $\lambda = 1$ time. The experiment is called balanced because r does not depend on the treatment level, block size k is constant, and the number λ does not depend on a combination ii' of treatment levels.

A second example, taken from Scheffé (1959, p. 189) has $b = 10$ blocks (which corresponds to time), each block has $k = 3$ treatment levels (detergents) from among $t = 5$ treatment levels, each treatment level is replicated $r = 6$ times, and each pair of treatment levels is contained within the same block $\lambda = 3$ times.

In a test to compare detergents with respect to a certain characteristic a large stack of dinner plates soiled in a specified way is prepared and the detergents are tested in blocks of three, there being in each block three basins with different detergents and three dishwashers who rotate after washing each plate. The measurements in the table are the numbers of plates washed before the foam disappears from the basin. Use the T-method with 0.90 confidence coefficient on the intrablock estimates to decide which pairs of detergents differ significantly.

Block		1	2	3	4	5	6	7	8	9	10
	A	27	28	30	31	29	30				
	B	26	26	29				30	21	26	
Detergent	C	30			34	32		34	31		33
	D		29		33		34	31		33	31
	E			26		24	25		23	24	26

The T-method to which Scheffé referred was Tukey's method for simultaneous confidence intervals. He referred to his own method as the S-method.

Other examples may be found in Cochran and Cox (1957, 475–6, 480, 482). The largest has $t = 28, k = 7, r = 9, b = 36, \lambda = 2$. For any $k < t$, we can always take $b = \binom{t}{k}$, $r = bk/t$, $\lambda = bk(k-1)/t(t-1)$, which may be a larger experiment than desired.

Such experiments can be useful in cases in which blocks are not large enough to accommodate all treatment levels. In an agricultural experiment with many locations with at most four plots each of which must receive only one hybrid of corn, an experiment to compare six hybrids must use incomplete blocks. In comparing five cake mixes using ovens with only three positions for cakes, the blocks (baking periods) must be incomplete. In an industrial experiment to compare three procedures, workers may only be able to use two procedures on any day. Thus a block (worker-day) must be incomplete.

Before we discuss these balanced incomplete block designs, let us first consider a model for incomplete block designs which may not be balanced. If we let the observations at treatment level i, block j be $Y_{ij1}, \ldots, Y_{ijK_{ij}}$, the model

$$Y_{ijk} \sim N(\mu_{ij}, \sigma^2)$$

with independence, may be fit using standard regression methods as described in Section 6.2.

Unless K_{ij} is at least one for each ij, however, we have no estimator for μ_{ij}, particularly for an interaction parameter $(\alpha\beta)_{ij}$. Therefore we are led to consideration of the additive model $\boldsymbol{\mu} \in V = \mathscr{L}(\mathbf{A}_1, \ldots, \mathbf{A}_t, \mathbf{B}_1, \ldots, \mathbf{B}_k)$. To test H_0: $\boldsymbol{\mu} \in \mathscr{L}(\mathbf{B}_1, \ldots, \mathbf{B}_b) = V_B$, let $V_A = V \cap V_B^{\perp}$, $\hat{\mathbf{Y}} = p(\mathbf{Y} \mid V)$, $\hat{\mathbf{Y}}_B = p(\mathbf{Y} \mid V_B)$, $\hat{\mathbf{Y}}_A = p(\mathbf{Y} \mid V_A) = \hat{\mathbf{Y}} - \hat{\mathbf{Y}}_B$, $\mathbf{e} = \mathbf{Y} - \hat{\mathbf{Y}}$. Since $\mathbf{B}_1, \ldots, \mathbf{B}_k$ is an orthogonal basis for V_B, $\hat{\mathbf{Y}}_B = \sum_1^b \bar{Y}_{\cdot j}\mathbf{B}_j$. Since $V_A \perp V_B$, $\hat{\mathbf{Y}} = \hat{\mathbf{Y}}_A + \hat{\mathbf{Y}}_B$. V_A is spanned by the vectors

$$\mathbf{A}_i^* = \mathbf{A}_i - p(\mathbf{A}_i \mid V_B) = \mathbf{A}_i - \sum_j (K_{ij}/K_{\cdot j})\mathbf{B}_j \tag{7.4.1}$$

for $i = 1, \ldots, t$.

These vectors sum to the zero vector, so that V_A has dimension at most $t - 1$. In general, without some restrictions on the K_{ij} it may have smaller dimension.

In the case that the design is a balanced incomplete block design (BIBD), special relationships among the parameters t, r, b, k, λ make it possible to solve explicitly for $\hat{\mathbf{Y}}$. First, the total number of observations is

$$n = \sum_i K_{i\cdot} = \sum_i r = rt \qquad \text{and} \qquad n = \sum_j K_{\cdot j} = \sum_j k = bk$$

so that (1) $rt = bk$. Secondly, the number of pairs of different treatment levels in the same block, summed across all blocks, is $\lambda\binom{t}{2}$ and also $\binom{k}{2}b$, so that (2) $\lambda t(t-1) = k(k-1)b$.

For BIBD's all K_{ij} are zero or one. Let $\mathscr{I}$ be the collection of pairs (i, j) for which $K_{ij} = 1$. Let $\mathscr{I}(i) = \{j \mid (i, j) \in \mathscr{I}\}$.

From (7.4.1) we get $\mathbf{A}_i^* = \mathbf{A}_i - (1/k)\sum_j K_{ij}\mathbf{B}_j = \mathbf{A}_i - \sum_{j \in \mathscr{I}(i)} \mathbf{B}_j$. For $(i', j) \in \mathscr{I}$, the (i', j) element of $\mathbf{A}_i^*$ is $1 - 1/k$ if $j \in \mathscr{I}(i)$ and $i = i'$, $-1/k$ if $j \in \mathscr{I}(i)$ and $i \neq i'$, 0 for $j \notin \mathscr{I}(i)$. Notice that $\sum_i \mathbf{A}_i^* = \mathbf{0}$. These $\mathbf{A}_i^*$ have inner product matrix $\mathbf{M}^* \equiv (m_{ij})$, where

$$\begin{aligned} m_{ij} &= r - (1/k^2)kr = r(k-1)/k = (t-1)\lambda/k \qquad \text{for} \quad i = j \qquad \text{and} \\ &= -(1/k^2)\lambda k = -\lambda/k \qquad \text{for} \quad i \neq j. \end{aligned} \tag{7.4.2}$$

Thus $\mathbf{M}^* = (\lambda t/k)[\mathbf{I}_t - (1/t)\mathbf{J}_t]$, where $\mathbf{J}_t$ is the $t \times t$ matrix of all ones.

Let $c = (\lambda t/k)$ and suppose $\hat{\mathbf{Y}}_A = \sum a_i \mathbf{A}_i$. Since $\hat{\mathbf{Y}}_A \perp \mathbf{x}_0$, $\sum a_i = 0$. To find these a_i, compute

$$\begin{aligned} (\mathbf{Y}, \mathbf{A}_i^*) = (\hat{\mathbf{Y}}_\mathbf{A}, \mathbf{A}_i^*) &= \sum_{i' \neq i} a_{i'}(\mathbf{A}_{i'}^*, \mathbf{A}_i) + a_i(\mathbf{A}_i^*, \mathbf{A}_i^*) \\ &= (-\lambda/k)(-a_i) + a_i(t-1)(\lambda/k) = ca_i \end{aligned}$$

It follows that $a_i = (\mathbf{Y}, \mathbf{A}_i^*)c^{-1}$. More explicitly,

$$a_i = c^{-1}\left[Y_{i\cdot} - \sum_j K_{ij}\bar{Y}_{\cdot j} \right], \qquad \text{where} \quad Y_{i\cdot} = \sum_{j \in \mathscr{I}(i)} Y_{ij\cdot}.$$

The term $B(i) = \sum_j K_{ij}\bar{Y}_{\cdot j} = \sum_{j \in \mathscr{I}(i)} \bar{Y}_{\cdot j}$ is the correction for blocks for treatment level i.

If $\mu_{ij} = \mu + \alpha_i + \beta_j$ with $\sum \alpha_i = \sum \beta_j = 0$, then, substituting μ_{ij} for Y_{ij}, we get $a_i = \alpha_i$, the effect of treatment level i. Thus a_i is an unbiased estimator of α_i.

The coefficient vector $\mathbf{a} = (a_1, \ldots, a_t)$ has covariance matrix $\sigma^2 c^{-2}\mathbf{M}^* = (\sigma^2/c)[\mathbf{I}_t - (1/t)\mathbf{J}_t]$, so that $\mathrm{Var}(a_i - a_{i'}) = 2\sigma^2/c = 2k\sigma^2/\lambda t$. More generally, for a contrast $\eta = \sum c_i\alpha_i$, $\mathrm{Var}(\hat{\eta}) = (\sum c_i^2)(k\sigma^2)/t$.

For $(i, j) \in \mathscr{I}$ the (i, j) term of $\hat{\mathbf{Y}}_A$ is $a_i - \bar{b}_j$, where $\bar{b}_j = (1/k)\sum_{(i,j) \in \mathscr{I}} a_i$. (see Problem 7.4.1). Of course, the (i, j) term of $\hat{\mathbf{Y}}_B$, for $(i, j) \in \mathscr{I}$, is $\bar{Y}_{\cdot j}$.

Explicit formulas for sums of squares may be determined as follows:

$$\begin{aligned} &\mathrm{SSA} = \|\hat{\mathbf{Y}}_A\|^2 = c\sum a_i^2, \text{ (see Problem 7.1.2)}, \\ &\qquad \|\hat{\mathbf{Y}}_B\|^2 = k\sum \bar{Y}_{\cdot j}^2,\ \|\hat{\mathbf{Y}}\|^2 = \|\hat{\mathbf{Y}}_A\|^2 + \|\hat{\mathbf{Y}}_B\|^2,\ \mathrm{SSE} = \|\mathbf{e}\|^2 = \|\hat{\mathbf{Y}}\|^2 - \|\hat{\mathbf{Y}}\|^2. \end{aligned}$$

$S^2 = \|\mathbf{e}\|^2/v$, for $v = n - \dim(V) = n - t - b + 1$ is the usual estimator of σ^2. The F-statistic for the test of H_0: ($\alpha_i = 0$ for all i) is $F = [\|\hat{\mathbf{Y}}_A\|^2/(t-1)]/S^2$. F has the noncentral F distribution with noncentrality parameter $(c \sum \alpha_i^2)/\sigma^2$, and $(t-1, v)$ d.f.

Notice that the definitions of $\hat{\mathbf{Y}}_A$ and $\hat{\mathbf{Y}}_B$ are not symmetric. A test of H_0: (no block effect) $\Leftrightarrow$ (all $\beta_j = 0$) may be constructed analogously to the test of H_0: (no treatment effect) $\Leftrightarrow$ (all $\alpha_i = 0$). The numerator sum of squares is not $\|\hat{\mathbf{Y}}_B\|^2$.

The joint distribution of all differences $a_i - a_{i'}$ is the same as it would be if the a_i were independent, with variances σ^2/c. It follows that Tukey simultaneous confidence intervals on the $\alpha_i - \alpha_{i'}$ are given by $[a_i - a_{i'} \pm q_{1-\alpha,t,v}(S^2/c)^{1/2}]$.

For a randomized block design with all treatment levels within each block, r blocks, the estimator for η is $\hat{\eta}_{\text{BL}} = \sum c_i \bar{Y}_{i\cdot}$, which has $\text{Var}(\hat{\eta}_{\text{BL}}) = \sigma^2(\sum c_i^2)/r$. The *efficiency* of the balanced incomplete block design relative to the randomized block design is therefore

$$e = \frac{\text{Var}(\hat{\eta}_{\text{BL}})}{\text{Var}(\hat{\eta})} = \frac{\lambda t}{kr} = \frac{t(k-1)}{(t-1)k} = c/r,$$

so that $e < 1$ for $t > k$. For the dishwasher example $e = 5/6$, so the design is relatively efficient, even though there are three basins rather than five.

Example 7.4.1: Consider the dishwashing example with $k = 3$, $t = 5$, $b = 10$, $r = 6$, $\lambda = 3$. In order to represent the sample space Ω conveniently take Ω to be the collection of 5×10 matrices with rows corresponding to levels of treatment, columns to blocks, with zeros where no observations was taken. Thus

$$\mathbf{Y} = \begin{bmatrix} 27 & 28 & 30 & 31 & 29 & 30 & 0 & 0 & 0 & 0 \\ 26 & 26 & 29 & 0 & 0 & 0 & 30 & 21 & 26 & 0 \\ 30 & 0 & 0 & 34 & 32 & 0 & 34 & 31 & 0 & 33 \\ 0 & 29 & 0 & 33 & 0 & 34 & 31 & 0 & 33 & 31 \\ 0 & 0 & 26 & 0 & 24 & 25 & 0 & 23 & 24 & 26 \end{bmatrix}$$

$$\mathbf{A}_1 = \begin{bmatrix} 1 & 1 & 1 & 1 & 1 & 1 & 0 & 0 & 0 & 0 \\ 0 & 0 & 0 & 0 & 0 & 0 & 0 & 0 & 0 & 0 \\ 0 & 0 & 0 & 0 & 0 & 0 & 0 & 0 & 0 & 0 \\ 0 & 0 & 0 & 0 & 0 & 0 & 0 & 0 & 0 & 0 \\ 0 & 0 & 0 & 0 & 0 & 0 & 0 & 0 & 0 & 0 \end{bmatrix}$$

$$\mathbf{A}_2 = \begin{bmatrix} 0 & 0 & 0 & 0 & 0 & 0 & 0 & 0 & 0 & 0 \\ 1 & 1 & 1 & 0 & 0 & 0 & 1 & 1 & 1 & 0 \\ 0 & 0 & 0 & 0 & 0 & 0 & 0 & 0 & 0 & 0 \\ 0 & 0 & 0 & 0 & 0 & 0 & 0 & 0 & 0 & 0 \\ 0 & 0 & 0 & 0 & 0 & 0 & 0 & 0 & 0 & 0 \end{bmatrix}$$

$$\mathbf{B}_2 = \begin{bmatrix} 0 & 1 & 0 & 0 & 0 & 0 & 0 & 0 & 0 & 0 \\ 0 & 1 & 0 & 0 & 0 & 0 & 0 & 0 & 0 & 0 \\ 0 & 0 & 0 & 0 & 0 & 0 & 0 & 0 & 0 & 0 \\ 0 & 1 & 0 & 0 & 0 & 0 & 0 & 0 & 0 & 0 \\ 0 & 0 & 0 & 0 & 0 & 0 & 0 & 0 & 0 & 0 \end{bmatrix}$$

$$p(\mathbf{A}_2 \mid V_{\mathbf{B}}) = \begin{bmatrix} 1 & 1 & 1 & 0 & 0 & 0 & 0 & 0 & 0 & 0 \\ 1 & 1 & 1 & 0 & 0 & 0 & 1 & 1 & 1 & 0 \\ 1 & 0 & 0 & 0 & 0 & 0 & 1 & 1 & 0 & 0 \\ 0 & 1 & 0 & 0 & 0 & 0 & 1 & 0 & 1 & 0 \\ 0 & 0 & 1 & 0 & 0 & 0 & 0 & 1 & 1 & 0 \end{bmatrix} \frac{1}{3}$$

$$\mathbf{A}_2^* = \begin{bmatrix} -1 & -1 & -1 & 0 & 0 & 0 & 0 & 0 & 0 & 0 \\ 2 & 2 & 2 & 0 & 0 & 0 & 2 & 2 & 2 & 0 \\ -1 & 0 & 0 & 0 & 0 & 0 & -1 & -1 & 0 & 0 \\ 0 & -1 & 0 & 0 & 0 & 0 & -1 & 0 & -1 & 0 \\ 0 & 0 & -1 & 0 & 0 & 0 & 0 & -1 & -1 & 0 \end{bmatrix} \frac{1}{3}$$

Notice that $p(\mathbf{A}_2 \mid V_B)$ has ones in blocks in which treatment level i appears, and, of course, each $\mathbf{A}_i^* \perp V_B$.

Example 7.4.2: For the dishwasher data block sums and means are given Table 7.4.1.

Table 7.4.1

Block	1	2	3	4	5	6	7	8	9	10
Sum mean	83	83	85	98	85	89	95	75	83	90
	27.67	27.67	29.33	32.67	28.33	29.67	31.67	25.00	27.67	30.00

Table 7.4.2

Treatment	1	2	3	4	5	
Sum mean	175 29.17	158 26.33	194 32.33	191 31.83	143 24.67	$c = 5$
$B_{(i)}$	174.3	168.0	175.3	179.3	169.0	
a_i	0.13	−2.00	3.73	2.33	−4.20	

The treatment totals, means, correlations, and a_i are given in Table 7.4.2.

The projections are

$$\hat{\mathbf{Y}} = \begin{bmatrix} 27.18 & 27.64 & 30.49 & 30.73 & 28.58 & 30.38 & 0 & 0 & 0 & 0 \\ 25.04 & 25.51 & 28.36 & 0 & 0 & 0 & 28.31 & 23.82 & 26.96 & 0 \\ 30.78 & 0 & 0 & 34.33 & 32.18 & 0 & 34.04 & 29.56 & 0 & 33.11 \\ 0 & 29.84 & 0 & 32.93 & 0 & 32.58 & 32.64 & 0 & 31.29 & 31.71 \\ 0 & 0 & 26.16 & 0 & 24.24 & 26.04 & 0 & 21.62 & 24.76 & 25.18 \end{bmatrix}$$

$$\hat{\mathbf{Y}}_B = \begin{bmatrix} 27.67 & 27.67 & 28.33 & 32.67 & 28.33 & 29.67 & 0 & 0 & 0 & 0 \\ 27.67 & 27.67 & 28.33 & 0 & 0 & 0 & 31.67 & 25 & 27.67 & 0 \\ 27.67 & 0 & 0 & 32.67 & 28.33 & 0 & 31.67 & 25 & 0 & 30 \\ 0 & 27.67 & 0 & 32.67 & 0 & 29.67 & 31.67 & 0 & 27.67 & 30 \\ 0 & 0 & 28.33 & 0 & 28.33 & 29.67 & 0 & 25 & 27.67 & 30 \end{bmatrix}$$

$$\hat{\mathbf{Y}}_A = \begin{bmatrix} -0.489 & -0.022 & 2.156 & -1.933 & 0.244 & 0.711 & 0 & 0 & 0 & 0 \\ -2.622 & -2.156 & 0.022 & 0 & 0 & 0 & -3.356 & -1.178 & -0.711 & 0 \\ 3.111 & 0 & 0 & 1.667 & 3.844 & 0 & 2.378 & 4.556 & 0 & 3.111 \\ 0 & 2.178 & 0 & 0.267 & 0 & 2.911 & 0.978 & 0 & 3.622 & 1.711 \\ 0 & 0 & -2.178 & 0 & -4.089 & -3.622 & 0 & -3.378 & -2.911 & -4.822 \end{bmatrix}$$

Directly from these vectors or from the formulas, we get

$$\|\hat{\mathbf{Y}}\|^2 = \sum_{ij} Y_{ij}^2 = 25{,}366, \qquad \|\hat{\mathbf{Y}}\|^2 = 25{,}335.9, \qquad \|\hat{\mathbf{Y}}_B\|^2 = k \sum \bar{Y}_{\cdot j}^2 = 25{,}130.7,$$

$$\|\hat{\mathbf{Y}}_A\|^2 = c \sum a_i^2 = 205.2, \qquad \|\mathbf{e}\|^2 = \|\mathbf{Y} - \hat{\mathbf{Y}}\|^2 = \|\mathbf{Y}\|^2 - \|\hat{\mathbf{Y}}\|^2 = 30.1$$

Thus $S^2 = 30.1/(30 - 14) = 1.88$, and the F-statistic is $F = \dfrac{205.2/4}{1.88} = 27.3$ for $(4 - 1)$ and 16 d.f. We reject the null hypothesis of no detergent effect at any reasonable α-level.

Since $\text{Var}(a_i - a_{i'}) = 2\sigma^2 c^{-1}$, we estimate these variances to be $S^2(a_i - a_{i'}) = 2(1.88)/5 = 0.75$, so that individual 95% confidence intervals on $\alpha_i - \alpha_{i'}$ are given by $a_i - a_{i'} \pm (2.131)\sqrt{0.75}$.

Problem 7.4.1: Prove that for $(i, j) \in \mathscr{I}$, the (i, j) term of $\mathbf{Y}_A$ is $a_i - \bar{b}_j$, where $\bar{b}_j = (1/k) \sum_{(i, j) \in \mathscr{I}} a_i$.

Problem 7.4.2: Prove that $\|\hat{\mathbf{Y}}_A\|^2 = c \sum a_i^2$.

Problem 7.4.3: For the rat experiment with seven levels of chemical as described in this section:

(a) Estimate the chemical effects α_i and find a random variable Q so that $(a_i - a_{i'}) \pm Q$ for all $i \neq i'$ are Tukey simultaneous 90% confidence intervals on all $\alpha_i - \alpha_{i'}$.

(b) Test H_0: all $\alpha_i = 0$ at level $\alpha = 0.05$.

(c) Determine $\mathbf{A}_2^*$, $\hat{\mathbf{Y}}_A$, $\hat{\mathbf{Y}}_B$, and $\hat{\mathbf{Y}}$, or for those with less time, at least the terms corresponding to $j = 3$.

(d) Find the efficiency e of this experiment relative to a randomized block design with 21 observations.

Problem 7.4.4: Consider the case $k = 2$, $b = 3$, $t = 3$, $r = 2$, $\lambda = 1$, with observations as indicated.

$$\mathbf{Y} = \begin{bmatrix} Y_{11} & Y_{12} & - \\ Y_{21} & - & Y_{23} \\ - & Y_{32} & Y_{33} \end{bmatrix}.$$

Let Ω be the collection of all possible $\mathbf{Y}$.

(a) Find $\mathbf{A}_1^*$, $\mathbf{A}_2^*$, $\mathbf{A}_3^*$ and $\mathbf{M}^*$.

(b) Give formulas for a_1, a_2, a_3 without using summation or matrix notation.

(c) Consider the estimator $\hat{\eta} = Y_{11} - Y_{21}$ of $\eta = \alpha_1 - \alpha_2$. Compare its variance with that of $a_1 - a_2$.

(d) For any BIBD compare the variance of $a_1 - a_2$ with that of the estimator obtained only using differences between observations in the same block (the intrablock estimator).

Problem 7.4.5: The following experiment to measure the effects of cold storage on the tenderness of beef was conducted by Dr. Pauline Paul at Iowa State University (Paul, 1943; see Table 7.4.1). An analysis was described by Cochran and Cox (1957). The six periods of storage were 0, 1, 2, 4, 9, and 18 days. Thirty muscles were used in 15 pairs. Members of the same pairs were the left and right versions of the same muscle. The five replicates were types of muscle. For this analysis we ask the student to ignore the replicate effect, and to consider the block (pair) and treatment (storage time) effects only.

All pieces of meat were roasted. The measured variable was the total tenderness score given by four judges on a 0–10 scale. Treatment numbers are indicated in parentheses, followed by the observation.
Treat this as a BIBD, with nine blocks of $k = 2$.

(a) Find the parameters b, r, t, λ, c, and efficiency e.

(b) Find the statistics $a_1, \ldots, a_6$ and use these to determine an analysis of variance table.

(c) Test H_0: (no storage effect) for $\alpha = 0.05$.

(d) Use the Tukey method to produce a line diagram which describes significant ($\alpha = 0.05$) differences among storage effects.

(e) *For braver students*: Test H_0: (no replicate effect) for $\alpha = 0.05$, assuming a model in which block effects are random, replicate effects fixed.

Table 7.4.1 Scores for Tenderness of Beef

Rep. I		Rep. II		Rep. III		Rep. IV		Rep. V	
(1) 7	(2) 17	(1) 17	(3) 27	(1) 10	(4) 25	(1) 25	(5) 40	(1) 11	(6) 27
(3) 26	(4) 25	(2) 23	(5) 27	(2) 26	(6) 37	(2) 25	(4) 34	(2) 24	(3) 21
(5) 33	(6) 29	(4) 29	(6) 30	(3) 24	(5) 26	(3) 34	(6) 32	(4) 26	(5) 32

CHAPTER 8

Analysis of Frequency Data

In this chapter we will discuss methodology for the analysis of *count* or *frequency* data, for which the observation $\mathbf{Y}$ is a table for which the ith component is the number of occurrences of some event A_i. The names categorical data analysis, analysis of contingency tables, frequency table analysis, log-linear models, and discrete multivariate analysis have also been used to describe the subject. Though the probability models we will discuss are quite different than those of the first seven chapters, many of the linearity properties developed can still be exploited to give insight into this somewhat more difficult theory.

The theory to be discussed is more difficult for two major reasons. First, the mean vector $\mathbf{m} \equiv E(\mathbf{Y})$ can no longer be assumed to lie in a known linear subspace V. Instead we will discuss models in which $\boldsymbol{\mu} = \log(\mathbf{m})$ lies in V. The function $\log(\cdot)$, linking $\mathbf{m}$ to $\boldsymbol{\mu} = \log(\boldsymbol{\mu}) \in V$, is often called the *link* function. Fortunately, the need for this link function is not too difficult to overcome, and we will be able to draw vector space pictures which offer intuitive understanding which the author (and, he thinks, at least some of his students) finds invaluable. Secondly, the theory concerning the sampling distributions of the estimators of β, $\boldsymbol{\mu}$, and $\mathbf{m}$ is asymptotic, depending for good approximation on large total frequencies. We will discuss some of this theory, but will omit many proofs. For thorough discussions of this theory a student should see the books by Haberman (1974), Bishop, Fienberg, and Holland (1975), Aicken (1983), Agresti (1990), Santner and Duffy (1989), and Christensen (1990). Some books emphasizing application are Everitt (1977), Haberman (1978, 1979), Fienberg (1977), and Hosmer and Lemeshow (1989). Also of interest, though we will discuss relatively little of this very general theory, is the book by McCullagh and Nelder (1990) on generalized linear models.

We begin by giving some examples for which the methodology to be discussed will be useful, postponing the analysis until the theory has been discussed. Section 8.2 is devoted to a study of the Poisson, binomial, multinomial, and generalized hypergeometric distributions and their interrelationships. We will also discuss the Multivariate Central Limit Theorem (MCLT) as it applies to these distributions, and the δ-method which will be needed in order to develop the asymptotic distributions of our estimators. Section 8.3 is concerned with inference on binomial and Poisson parameters p, $p_1 - p_2$, λ, and λ_1/λ_2. Section 8.4 introduces some log-linear models, develops the notation needed for their analysis, and defines log-odds. Section 8.5 concerns the maximum likelihood estimation of the parameters. Section 8.6 discusses goodness-of-fit statistics. Section 8.7 is devoted to the asymptotic theory for the parameter estimators. And Section 8.8 discusses logistic regression, the case of one dichotomous dependent variable.

8.1 EXAMPLES

Example 8.1.1 (Haberman, 1974, p. 5): The drug digitalis was injected into the lymph nodes of 45 frogs, with each of the drug dosages d_1, d_2, d_3, where $\log(d_1 . d_2 . d_3) = (0.75, 0.85, 0.95)$. Each dosage was assigned to 15 frogs, chosen randomly. The numbers dying for these three dosages were respectively 2, 5, and 8. While it is fairly obvious that increasing dosage tends to kill more frogs, what can be said about the kill-rates for these or other dosages?

Example 8.1.2: A report of the police department of East Lansing for 1990 gave the numbers of fights and assaults for downtown and nondowntown for each of the months of the year.

	Jan.	Feb.	Mar.	Apr.	May	June
Downtown	62	44	46	46	64	43
Nondowntown	24	25	19	30	34	40

	July	Aug.	Sept.	Oct.	Nov.	Dec.
Downtown	42	32	40	29	39	11
Nondowntown	40	48	47	62	30	20

East Lansing is the home of Michigan State University. It had about 25,000 nonstudent residents, and 44,000 students, of which about 35,000 live in East Lansing in university dormitories and apartments and in rooms and apartments in the city. The university was in session Sept. 18 to Dec. 10, Jan. 3 to Mar. 17, and Mar. 24 to June 10. A summer session June 21 to Aug. 30 had about 15,000 students. Obviously the numbers downtown seemed to vary with the number of students in East Lansing, while being somewhat steady for non-downtown areas (though the number in October seems strangely high). Is there a relatively simple few-parameter model which fits these data?

Example 8.1.3: A doctor did a study of osteopathic hospitals in the Detroit area to see whether there was any relationship between cancer and multiple sclerosis. He found

	Cancer	Not Cancer
MS	5	225
Not MS	14,286	119,696

The usual chi-square statistic was 17.4, so the observed significance level for the null hypothesis of independence was extremely small. Was he correct in suspecting that there might be something in the biochemistry of the two diseases which prevents the other? He argued that the age distributions for the two diseases seemed to be about the same.

Example 8.1.4: A report of the National Center for Health Statistics for 1970 classified 13,832 homicides in the U.S. by the race and sex of the victim and by the murder weapon used (Table 8.1.1). Is instrument used independent of the sex and race of the victim, or of either? If the instrument does depend on race or sex, how strong is the relationship?

Example 8.1.5 (Bickel, Hammel, and O'Connel, 1975): The authors studied the rates of admission to graduate school by sex and department at the University of California at Berkeley. To make their point they invented the following data for the departments of "Machismatics" and "Social Warfare." For the combined departments their data were

	Admit	Deny	Percentage
Men	250	300	45.5
Women	250	400	38.5

Table 8.1.1 Type of Assault

Race	Sex	Firearms and Explosives	Cutting and Piercing Instruments	Total
White	Male	3,910	808	4,718
	Female	1,050	234	1,284
Black	Male	5,218	1,385	6,603
	Female	929	298	1,227
Total		11,107	2,725	13,832

Assuming relatively equal ability for men and women, there seems to be discrimination against women. Frequencies for individual departments were

	Machismatics			Social Warfare		
	Admit	Deny	% Admitted	Admit	Deny	% Admitted
Men	200	200	50.0	50	100	33.3
Women	100	100	50.0	150	300	33.3

These data seem to indicate that the two departments are each acting fairly, yet the university seems to be acting unfairly. Which is true? Or are both true?

8.2 DISTRIBUTION THEORY

In this section we study three discrete distributions, Poisson, multinomial, and generalized hypergeometric, which serve as models for frequency data. We also discuss interrelationships among these distributions given by conditioning, the Multivariate Central Limit Theorem, the approximations of these discrete distributions it provides, and the multivariate delta method, which provides approximations for the distributions of functions of the vector $\mathbf{Y}$ of observed frequencies.

The observation vector $\mathbf{Y}$ will always have T components, with each having a discrete distribution taking only nonnegative integer values. T will be fixed throughout any discussion on the properties of $\mathbf{Y}$ and functions of $\mathbf{Y}$. In most of the asymptotic theory we will discuss, certain other parameters will change but not T. For that reason we have chosen T rather than n to represent the number of components. The index set will be called $\mathscr{I}$. We can always take $\mathscr{I}$ to be $\{1, 2, \ldots, T\}$, but in the case of two or multiway tables we will let $\mathscr{I}$ be a Cartesian product. For a $2 \times 3 \times 4$ table we could, for example, take $\mathscr{I} = \{1, 2\} \times \{1, 2, 3\} \times \{1, 2, 3, 4\}$.

Definition 8.2.1: Let $\mathbf{Y}$ have T components indexed by $\mathscr{I}$, and let $\mathbf{p}$ be a probability vector with components indexed by $\mathscr{I}$ with component i denoted by p_i. Let $\mathbf{u}_i$ be the indicator of component i for each $i \in \mathscr{I}$, and suppose that $\mathbf{Y}$ takes the value $\mathbf{u}_i$ with probability p_i for each $i \in \mathscr{I}$. Then $\mathbf{Y}$ is said to have the *generalized Bernoulli distribution* with parameter $\mathbf{p}$.

Thus, for example, if $\mathscr{I} = \{1, 2, 3\}$, and $\mathbf{p} = (0.3, 0.2, 0.5)$, then $\mathbf{Y}$ takes the values $(1, 0, 0)$, $(0, 1, 0)$ and $(0, 0, 1)$ with probabilities 0.3, 0.2, and 0.5. If $T = 2$, we can let $\mathbf{Y}$ take the values $(1, 0)$ and $(0, 1)$ with probabilities p_1 and $p_2 = 1 - p_1$, or, for simplicity, only record the first component Y_1 of $\mathbf{Y}$. In that case we say that Y_1 has the *Bernoulli distribution* with parameter p_1.

It is, of course, easy to determine the moments of the generalized Bernoulli distribution. Since the kth power of any unit vector is still the same unit vector $E(\mathbf{Y}^k) = E(\mathbf{Y}) = \mathbf{p}$. (By $\mathbf{x}^k$ for any vector $\mathbf{x}$ we mean the vector obtained by replacing each component by its kth power.) If Y_i and Y_j are components of $\mathbf{Y}$ for $i \neq i'$, then $Y_i Y_{i'} \equiv 0$, so $\text{cov}(Y_i, Y_{i'}) = 0 - p_i p_{i'} = -p_i p_{i'}$. Of course, $\text{Var}(Y_i) = p_i - p_i^2 = p_i(1 - p_i)$. If $\mathbf{p}$ is a column vector we can write the covariance matrix for $\mathbf{Y}$ in a convenient way: $\mathbf{\Sigma_Y} \equiv D[\mathbf{Y}] \equiv d(\mathbf{p}) - \mathbf{pp}'$, where $d(\mathbf{u})$ for any vector $\mathbf{u}$ is the square matrix with diagonal $\mathbf{u}$. Since the components of $\mathbf{p}$ sum to one we have $\mathbf{\Sigma_Y J} = \mathbf{p} - \mathbf{p}(\mathbf{p}'\mathbf{J}) = \mathbf{p} - \mathbf{p} = \mathbf{0}$. Thus, $\mathbf{\Sigma_Y}$ has rank at most $T - 1$. We will show later that the rank of $\mathbf{\Sigma_Y}$ is always one less than the number of positive components of $\mathbf{p}$. Since the writing of $\mathbf{Y}$ and $\mathbf{p}$ as column vectors was merely a notational convenience, the same statements remain true when these vectors are written in other shapes. If $\mathbf{p}$ is not written as a column, then simply interpret $\mathbf{pp}'$ as the $T \times T$ matrix with ij element $p_i p_j$, the outer product of $\mathbf{p}$ and $\mathbf{p}$ under multiplication.

Just as a binomial r.v. is the sum of independent Bernoulli r.v.'s with the same parameter p, the multinomial distribution is defined similarly as the sum of independent generalized Bernoulli random vectors.

Definition 8.2.2: Let $\mathbf{Y}_1, \ldots, \mathbf{Y}_n$ be independent generalized Bernoulli random vectors, all with the same parameter vector $\mathbf{p}$. Then $\mathbf{Y} = \sum_{j=1}^{n} \mathbf{Y}_i$ is said to have the *multinomial distribution* with parameters n and $\mathbf{p}$. We will denote this distribution by $\mathcal{M}_T(n, \mathbf{p})$.

If a fair die is tossed 10 times, the vector $(Y_1, \ldots, Y_6)$ denoting the frequencies of occurrence of the six numbers has the multinomial distribution with parameters $n = 10$, and $\mathbf{p} = (1/6, \ldots, 1/6)$. If a pair of fair dice are thrown 20 times and the total of the two dice recorded for each then $\mathcal{I} = \{2, \ldots, 12\}$, and $\mathbf{Y} = (Y_2, \ldots, Y_{12})$, the vector of frequencies of occurrence of these possible totals, has the multinomial distribution with parameters $n = 20$, and $\mathbf{p} = (p_2, \ldots, p_{12})$, where $p_i = [6 - |i - 7|]/36$.

If $\mathcal{I} = \{4, 7, 9\}$, $n = 4$, and $\mathbf{p} = (0.2, 0.3, 0.5)$, then such a random vector $\mathbf{Y}$ takes the value $(1, 2, 1)$ with probability $\binom{4}{1, 2, 1}(0.2^1, 0.3^2, 0.5^1) = 6(0.009) = 0.054$. The coefficient $\binom{4}{1, 2, 1}$ is the number of ways in which the four trials can be assigned to receive one 4, two 7's, and one 9. The other factor in parentheses is the probability four trials will produce one 4, two 7's, and one 9 in a particular order. Of course, the components of $\mathbf{Y}$ must sum to n. In general, if $\mathbf{y}$ is any vector of nonnegative integers adding to n,

$$m_T(\mathbf{y}; n, \mathbf{p}) \equiv P(\mathbf{Y} = \mathbf{y}) = \binom{y}{y_1, \ldots, y_T} p_1^{y_1} \cdots p_T^{y_T}.$$

Recall that the multinomial coefficient may be evaluated by

$$\binom{n}{y_1, \ldots, y_T} = n!/(y_1! \cdots y_T!)$$

In the case that $T = 2$, we need not keep a record of the value of Y_2, since $T_2 = n - Y_1$. We therefore say that Y_1 has a binomial distribution with parameters n and p_1. Thus, if $\mathbf{p} = (p, 1 - p)$, then $\mathscr{B}(n, p)$ is the distribution of the first component of the $\mathscr{M}_2(n, \mathbf{p})$ distribution. The probability function for the binomial distribution specializes from the multinomial to

$$b(k; n, p) = m_2((k, n - k); n, (p, 1 - p)) = \binom{n}{k} p^k (1 - p)^{n-k} \quad \text{for } k = 0, \ldots, n.$$

The mean vector and covariance matrix for the multinomial can be computed easily from the representation as a sum. Thus,

$$E(\mathbf{Y}) = n\mathbf{p} \qquad \text{and} \qquad D[\mathbf{Y}] = n[d(\mathbf{p}) - \mathbf{p}\mathbf{p}'].$$

(Recall that $d(\mathbf{u})$ is the diagonal matrix with $\mathbf{u}$ on the diagonal.) For the binomial distribution (the marginal distribution of the first component of the $\mathscr{M}_2(n, (p, 1 - p))$ distribution),

$$E(Y) = np \qquad \text{and} \qquad \text{Var}(Y) = np(1 - p).$$

We will often use models in which $\mathbf{Y}$ is a k-tuple of independent multinomial random vectors. Consider, for example, a study in which random samples of 100 each are taken from the six combinations of the two sexes and three age-groups in order to determine the opinions of these six groups on abortion, with the opinion having three possible values. If $\mathbf{Y}_{ij}$ is the 3-tuple of frequencies of opinion for sex i and age-group j, then a reasonable model would suppose that $\mathbf{Y}$ is the 6-tuple of independent multinomial vectors $\mathbf{Y}_{11}, \ldots, \mathbf{Y}_{23}$.

Definition 8.2.3: Let $\mathbf{Y}_i \sim \mathscr{M}_{r_i}(n_i, \mathbf{p}_i)$ for $i \in \mathscr{I}$, be independent random vectors. Then $\mathbf{Y} = (\mathbf{Y}_i, i \in \mathscr{I})$ is said to satisfy the product (or independent) multinomial model.

Such a random vector $\mathbf{Y}$ has $T = \sum r_i$ components, with mean vector the T-tuple with ith component vector $n_i\mathbf{p}_i$. The covariance matrix consists of blocks of size $r_i \times r_i$ on the diagonal. Usually the r_i will be the same, though that is not necessary. We will refer to the case that $\mathscr{I}$ has only one element as the single multinomial model, to distinguish it from the product multinomial model, for which $\mathscr{I}$ will have at least two elements.

A random variable X is said to have a Poisson distribution if it takes only nonnegative integer values with probabilities $p(k; \lambda) \equiv e^{-\lambda}\lambda^k/k!$ for $k = 0, 1, \ldots$.

We will refer to this as the $\mathscr{P}(\lambda)$ distribution. Recall that $E(X) = \lambda$, and $\text{Var}(X) = \lambda$. The moment-generating function is $m(t) = e^{-\lambda(e^t - 1)}$. The m.g.f. or an inductive argument can be used to prove that the sum of independent Poisson r.v.'s is itself Poisson, with parameter equal to the sum of the parameters of the r.v.'s summed. Our first limit theorem suggests the approximation of binomial by Poisson probabilities in some situations.

Theorem 8.2.1: Let $\{p_n\}$ be a sequence of probabilities satisfying $np_n^2 \to 0$ as $n \to \infty$. Then $\lim_{n\to\infty} [b(k; n, p_n)/p(k; np_n)] = 1$ for $k = 0, 1, \ldots$.

***Proof*:** $r(k; n) = b(k; n, p_n)/p(k; np_n)$ is the product of the three factors:

$$F_{1n} = [n(n-1)\cdots(n-k+1)]/n^k, \qquad F_{2n} = (1 - p_n)^{-k}$$

and

$$F_{3n} = (1 - p_n)^n/e^{-np_n} = [(1 - p_n)e^{p_n}]^n.$$

These three factors each have the limit one. To see that $F_{3n} \to 1$ as $n \to \infty$, note that $\log F_{3n} = n[\log(1 - p_n) + p_n)] = n[-p_n + o(p_n^2) + p_n] \to 0$ as $n \to \infty$, since $np_n^2 \to \infty$ as $n \to \infty$. (The notation $o(p_n^2)$ denotes a function of p_n^2 having the property $o(p_n^2)/p_n^2 \to 0$ as $n \to \infty$.) □

We can therefore expect the approximation of a binomial distribution by the Poisson distribution with the same mean to be good if np^2 is small. A more general result, and of great practical value is the following theorem, due to LeCam (1960). For a very interesting and relatively simple discussion of this see the paper by T. W. Brown (1984). We do not prove the theorem here.

Theorem 8.2.2 (LeCam): Let $Y_1, \ldots, Y_n$ be independent, with $Y_i \sim \mathscr{B}(1, p_i)$. Let $T = \sum X_i$. Let W have the Poisson distribution with parameter $\lambda = \sum p_i$. Then, for any subset A of the real line, $G(A) \equiv |P(T \in A) - P(W \in A)| \leq \sum p_i^2$.

Comments: Since $\text{Var}(T) = \sum p_i(1 - p_i)$ and $\text{Var}(W) = \lambda = \sum p_i$, $\sum p_i^2 = \text{Var}(W) - \text{Var}(T)$. If all p_i are equal to $p = \lambda/n$, we get the upper bound λ^2/n on $G(A)$, so that as $n \to \infty$, $G(A) \to 0$ uniformly in A. If, for example, we observe the number of deaths due to cancer over a large population, it may be reasonable to adopt a model in which different people die from cancer independently, with small probabilities which differ across people. Still, the Poisson distribution can serve as a good approximation of the distribution of T, the total number of people who die.

Example 8.2.1: Let Y_1, Y_2, Y_3 be independent Bernoulli r.v.'s with $p_1 = 0.01$, $p_2 = 0.02$, $p_3 = 0.03$. Then $T = \sum Y_i$ and a Poisson r.v. W with mean $\lambda = 0.06$

have probability distributions, accurate to five decimal places, as follows

k	0	1	2	3
$P(T = k)$	0.941 09	0.057 82	0.001 08	0.000 01
$P(W = k)$	0.941 76	0.056 51	0.001 70	0.000 03

$P(W > 3)$ is positive but less than 10^{-6}. $|P(T \in A) - G(A)|$ is maximum for $A = \{1\}$, with value 0.001 31. The upper bound given by LeCam's theorem is 0.001 40. In general, the approximating Poisson distribution puts greater mass on the left and right tails.

The simplest model we will consider, and therefore the starting point for the discussion of estimation for log-linear models will be the independent Poisson model.

Definition 8.2.4: Let the components of $\mathbf{Y}$ be independent, $Y_i \sim \mathscr{P}(\lambda_i)$, $i = 1, \ldots, T$. Then $\mathbf{Y}$ is said to satisfy the independent Poisson model with parameter $\boldsymbol{\lambda} \equiv E(\mathbf{Y})$, where $\boldsymbol{\lambda} = (\lambda_1, \ldots, \lambda_T)$.

The models we have considered so far are tied together through conditioning. Beginning with the simplest model, the independent Poisson model, we condition on the total to get the multinomial. By conditioning the multinomial random vector on the totals of subsets of components, we get the product multinomial model.

Theorem 8.2.3: Let $\mathbf{Y}$ satisfy the independent Poisson model with parameter vector $\boldsymbol{\lambda}$. Let S be the sum of the components of $\mathbf{Y}$, and λ be the sum of the components of $\boldsymbol{\lambda}$. Then, conditional on $S = s$, $\mathbf{Y}$ has the multinomial distribution with parameters $n = s$, and $\mathbf{p} = \boldsymbol{\lambda}/\lambda$.

***Proof*:** Details of the proof are left to the student. Consider any fixed vector $\mathbf{y}$ with components adding to s, and express the conditional probability function as the ratio of two probabilities. The denominator is determined using the fact that the sum of independent Poisson r.v.'s also has a Poisson distribution. □

In particular, if Y_1 and Y_2 are independent Poisson r.v.'s with means λ_1 and λ_2, then, conditionally on $Y_1 + Y_2 = s$, Y_1 has a binomial distribution with parameters $n = s$ and $p = \lambda_1/(\lambda_1 + \lambda_2)$. We will exploit this to develop a formula for a confidence interval on the ratio $R = \lambda_1/\lambda_2$ on Poisson parameters.

The following theorem implies that we can construct the product multinomial model by conditioning the single multinomial model.

Theorem 8.2.4: Let $\mathbf{Y}$ have index set $\mathscr{I} = \mathscr{I}_1 \cup \mathscr{I}_2 \cup \cdots \cup \mathscr{I}_k$, where index set $\mathscr{I}_i$ has T_i elements and the $\mathscr{I}_i$ are disjoint. Let $T = \sum T_i$. Let $\mathbf{Y} \sim \mathscr{M}_T(n, \mathbf{p})$, and partition $\mathbf{p}$ in the same way that $\mathscr{I}$ is partitioned, so that $\mathbf{p}$ is the k-tuple $(\mathbf{p}_1, \ldots, \mathbf{p}_k)$, the components in $\mathscr{I}_i$ having corresponding probability vector $\mathbf{p}_i$. Let S_i be the sum of the components of $\mathbf{Y}$ corresponding to index set $\mathscr{I}_i$. Let $\mathbf{S}$ be the k-vector of sums, and let $\mathbf{s}$ be a k-vector of nonnegative integers with sum n. Then, the conditional distribution of $\mathbf{Y}$, given $\mathbf{S} = \mathbf{s}$, is product multinomial with ith component vector $\mathbf{Y}_i \sim \mathscr{M}_{T_i}(s_i, \mathbf{p}_i/P_i)$, where P_i is the sum of the components of $\mathbf{p}_i$).

Comments: To see that this theorem says, consider the index sets $\mathscr{M} = R \times C$ for $R = \{1, 2, 3\}$ and $C = \{1, 2, 3, 4\}$. Then $\mathbf{Y}$ is a 3×4 table of frequencies. Let $\mathscr{M}_i$ be the set of indices corresponding to the ith row. Suppose that $\mathbf{Y}$ has the multinomial distribution with $n = 20$, and

$$\mathbf{p} = \begin{bmatrix} 0.1 & 0.2 & 0.15 & 0.05 \\ 0.1 & 0.2 & 0 & 0 \\ 0 & 0.1 & 0.1 & 0 \end{bmatrix}.$$

Then, conditionally on the three row sums S_1, S_2, S_3 being 12, 5, and 3, $\mathbf{Y}$ satisfies the product multinomial model with the ith row $\mathbf{Y}_i$ having the multinomial distribution with parameter n_i, where $n_1 = 12$, $n_2 = 5$, $n_3 = 3$, and probability vector $\mathbf{p}_i$ for $\mathbf{p}_1 = (0.2, 0.4, 0.3, 0.1)$, $\mathbf{p}_2 = (1/3, 2/3, 0, 0)$, $\mathbf{p}_3 = (0, 0.5, 0.5, 0)$. Here $P_1 = 0.5$, $P_2 = 0.3$, and $P_3 = 0.2$.

***Proof*:** Again we will avoid the messy details by leaving them to the student. The crucial point is that the vector $\mathbf{S}$ of sums has a multinomial distribution with parameters n and $\mathbf{p}$-vector, having ith component P_i. This follows directly from the definition of the multinomial distribution, since S_i is the frequency of occurrence of observations in index set $\mathscr{M}_i$. □

One more discrete multivariate distribution arises frequently in the analysis of categorical data. To make the definition to follow more intuitive, consider a box of 20 marbles of three colors, with 8 red, 7 white, and 5 blue. Suppose that four people A, B, C, and D randomly partition the marbles by carefully mixing and drawing with eyes shut tightly, with A drawing 7, B drawing 6, C drawing 4, D drawing 3. Let $\mathbf{Y}_i = (Y_{i1}, Y_{i2}, Y_{i3})$ be the numbers of marbles of the three colors drawn by person i, and let $\mathbf{Y} = (Y_{ij})$ be the 4×3 matrix of counts. What is the distribution of $\mathbf{Y}$? The generalized hypergeometric distribution.

Anyone who has played a card game in which all the 52 cards are dealt can think of other examples. If Y_{ij} is the the number of rank j cards received by player i for $i = 1, 2, 3, 4$ then the table $\mathbf{Y}$ also has a generalized hypergeometric distribution, with index set $\{1, 2, 3, 4\} \times \{\text{Ace}, 2, \ldots, 10, \text{J}, \text{Q}, \text{K}]$.

Definition 8.2.5: Let a set B of N elements be partitioned into k subsets $B_1, \ldots, B_k$, with $N_j = N(B_j)$. Suppose that B is randomly partitioned into subsets $A_1, \ldots, A_r$, with $M_i = N(A_i)$. Let Y_{ij} be the number of elements in $A_i \cap B_j$. Then the random vector $\mathbf{Y} = (Y_{ij})$ is said to have the generalized hypergeometric distribution with parameter vectors $\mathbf{N} = (N_1, \ldots, N_k)$ and $\mathbf{M} = (M_1, \ldots, M_r)$.

In the marble example above $k = 3$, $r = 4$, $\mathbf{N} = (8, 7, 5)$ and $\mathbf{M} = (7, 6, 4, 3)$.

A possible observation on $\mathbf{Y}$ is $\begin{bmatrix} 3 & 2 & 2 \\ 2 & 3 & 1 \\ 1 & 2 & 1 \\ 2 & 0 & 1 \end{bmatrix}$, so, for example, person B drew 2 red, 3 white, and 1 blue marble.

Since the number of ways in which a set of $N = \sum_i M_i = \sum_j N_j$ elements can be partitioned into the subsets A_i of sizes $M_1, \ldots, M_r$ is $\binom{N}{M_1, \ldots, M_r}$, the number of ways in which this can be done so that y_{ij} elements are contained in $A_i \cap B_j$ for all i and j is

$$\prod_i \binom{M_i}{y_{i1}, \ldots, y_{ik}} = \left[\prod_i M_i!\right] \Big/ \left[\prod_{ij} y_{ij}!\right].$$

Thus

$$P(\mathbf{Y} = \mathbf{y}) = \prod_i \binom{M_i}{y_{i1}, \ldots, y_{ik}} \Big/ \binom{N}{M_1, \ldots, M_r}$$
$$= \left[\prod_i M_i!\right]\left[\prod_j N_j!\right] \Big/ [n!]\left[\prod_{ij} y_{ij}!\right],$$

for all tables $\mathbf{y}$ for which the ith row total is M_i and the jth column is N_j for all i and j.

For the 4×3 table with $\mathbf{y}$ as indicated

$$P(\mathbf{Y} = \mathbf{y}) = [(8!)(7!)(5!)][(7!)(6!)(4!)(3!)]/[20!]\left[\prod_{ij} y_{ij}!\right]$$
$$= 1.2746 \times 10^{19}/[2.4329 \times 10^{18}][1{,}152] = 0.0045.$$

In order to perform certain tests for two-way tables, we may wish to compute these probabilities, rather than relying on the asymptotic results.

The marginal distribution of Y_{ij} is the *hypergeometric distribution*, with parameters N_j, M_i, and N, with

$$P(Y_{ij} = y) = \left[\binom{M_i}{y}\binom{N - M_i}{M_i - y}\right] \Big/ \binom{N}{M_i} = \left[\binom{N_j}{y}\binom{N - N_j}{N_j - y}\right] \Big/ \binom{N}{N_j}.$$

Indicators may be used to show that

$$E(Y_{ij}) = (M_i N_j)/N, \quad \mathrm{Var}(Y_{ij}) = M_i N_j (N - M_i)(N - N_j)/[N^2(N - 1)]$$
$$= [M_i][p_j(1 - p_j)][(N - M_i)/(N - 1)], \qquad \text{for} \quad p_j = N_j/n.$$

The factors within the first two brackets of the last term give the variance for the case that a sample of M_i is taken with replacement. The third factor is the finite correction factor, since sampling is without replacement. More generally, the ith row $\mathbf{Y}_i$ of $\mathbf{Y}$ has covariance matrix $[d(\mathbf{p}) - \mathbf{pp}']$, where $\mathbf{p} = (p_1, \ldots, p_k)'$, and $d(\mathbf{p})$ is the $k \times k$ diagonal matrix with $\mathbf{p}$ on the diagonal (see Problem 2.2.4).

If $\mathbf{Y}$ satisfies the product multinomial model, with rows independent, with the ith row $\mathbf{Y}_i \sim \mathscr{M}_k(M_i, \mathbf{p})$, $\mathbf{p}$ the same for each i, then conditioned on column totals $(N_1, \ldots, N_k)$, $\mathbf{Y}$ has the generalized hypergeometric distribution with parameter vectors given by row and column sums. Again, details are left to the hard-working student.

Looking back at the discussion of the Poisson, multinomial, product multinomial, and generalized hypergeometric models, we see that by conditioning in various ways and determining the parameters from the conditions, we are lead from the relatively simple independent Poisson model to these more "dependent" models.

We now turn to the limit theory which we will need for statistical inference for log-linear models. We remind students of the meaning of convergence in distribution.

Definition 8.2.6: Let $\{\mathbf{Z}_n = (Z_{n1}, \ldots, Z_{nT})\}$ be a sequence of random vectors. Let F be a c.d.f. defined on R_T. $\{\mathbf{Z}_n\}$ is said to converge in distribution to F if

$$\lim_{n \to \infty} P(Z_{ni} \le z_i \text{ for } i = 1, \ldots, T) = F(\mathbf{z})$$

for every $\mathbf{z} = (z_1, \ldots, z_T)$ at which F is continuous. If $\mathbf{Z}$ has c.d.f. F, then we also say that $\{\mathbf{Z}_n\}$ converges in distribution to $\mathbf{Z}$. We will write $\mathbf{Z}_n \xrightarrow{D} F$, or $\mathbf{Z}_n \xrightarrow{D} \mathbf{Z}$.

The definition does not demand convergence for all $\mathbf{z}$, but only for those $\mathbf{z}$ at which F, the limiting distribution, is continuous. To see the need for this definition, consider the uniform distribution on the interval $(0, 1/n)$. A reasonable definition would allow this sequence to converge to the distribution with mass one at 0. Under this definition it does, though $F_n(0) \equiv 0$ does not converge to $F(0) = 1$. In the case that the limiting distribution is continuous, as it is for the multivariate normal, for example, the convergence must hold for all $\mathbf{z}$.

Theorem 8.2.5: $\mathbf{Z}_n \xrightarrow{D} \mathbf{Z}$ if and only if $E[g(\mathbf{Z}_n)]$ to $E[g(\mathbf{Z})]$ for all continuous real-valued functions g which are zero outside a bounded set in T-space.

Convergence in probability of $\{\mathbf{Z}_n\}$ to $\mathbf{Z}$, which demands that

$$\lim_{n \to \infty} P(\|\mathbf{Z}_n - \mathbf{Z}\| > \varepsilon) = 0 \qquad \text{for all} \quad \varepsilon > 0,$$

implies convergence in distribution, but the converse is not true.

This definition immediately produces the following useful limit therems, which we give without proof.

Theorem 8.2.6: Let $\{\mathbf{Z}_n\}$ be a sequence of random vectors of T components, converging in distribution to $\mathbf{Z}$, which has c.d.f. F. Let g be a function on R_T into R_k, which is continuous on a set A, such that $P(\mathbf{Z} \in A) = 1$. Let G be the c.d.f. of $g(\mathbf{Z})$. Then $\{g(\mathbf{Z}_n)\} \xrightarrow{D} G$.

Theorem 8.2.7: Let $\{\mathbf{Z}_n\} \xrightarrow{D} \mathbf{Z}$, where each $\mathbf{Z}_n$ has T components. Then, for any Borel subset A of R_T for which

$$P(\mathbf{Z} \in \text{Bdy}(A)) = 0, \qquad \lim_{n \to \infty} P(\mathbf{Z}_n \in A) = P(\mathbf{Z} \in A).$$

Students who have not studied real analysis should interpret a Borel subset of R_T as a reasonable subset, certainly including all those in which we would normally be interested. Bdy(A) is the boundary of A, the closure of A minus the interior of A.

Moment-generating functions (or characteristic functions for those who understand complex variables) may be used to establish convergence in distribution. Recall that the m.g.f. of an r.v. X is $m_X(t) = E(e^{Xt})$, defined at least for t in some neighborhood of 0. Not all r.v.'s possess m.g.f.'s. If t on the right is replaced by $\sqrt{-1}\,t$, we get the characteristic function, which is always defined for all t. If two r.v.'s possess the same m.g.f., defined for all t in some neighborhood of the origin, then the two r.v.'s must have the same distribution. (Actually it is enough that two m.g.f.'s agree on any interval.) The m.g.f. of the standard normal distribution is $e^{t^2/2}$, defined for all t. The log-normal distribution (the distribution of e^Z for Z normal), does not possess a m.g.f.

Theorem 8.2.8, The Continuity Theorem (Billingsley 1986, 408): Let F_n have m.g.f. m_n. Let F be a c.d.f. with m.g.f. m. Then convergence of $m_n(t)$ to $m(t)$ for t in a neighborhood of 0 implies $F_n \xrightarrow{D} F$.

In the case that a Poisson parameter is large we will want a convenient approximation. Moment-generating functions can be used to prove the following limit theorem for Poisson r.v.'s for which the parameter λ converges to infinity.

Theorem 8.2.9: Let $\{Y_n\}$ be a sequence of Poisson r.v.'s with $E(Y_n) = \lambda_n$. Define $Z_n = (Y_n - \lambda_n)/\sqrt{\lambda_n}$. Suppose that $\{\lambda_n\} \to \infty$. Then $Z_n \xrightarrow{D} N(0, 1)$.

***Proof*:** The m.g.f. of Z_n is $m_n(t) = e^{-t\sqrt{\lambda_n}} m_{Y_n}(t/\sqrt{\lambda_n})$, so that $\log(m_n(t)) = -t\sqrt{\lambda_n} + \lambda_n(e^{t/\sqrt{\lambda_n}} - 1)$. Expanding the second term in a power series about 0, we find the limit $t^2/2$ as $n \to \infty$. Continuity of the exponential function (inverse of the log) then implies that $m_n(t) \to e^{t^2/2}$ for every t. Since this is the m.g.f. of the standard normal distribution, Theorem 8.2.8 then implies the conclusion of Theorem 8.2.9. □

The fact that Z_n as defined above converges to standard normal is useful because the probabilities provided by the normal are close to those provided by the Poisson for even moderate λ. We can improve the approximation by using the 1/2 correction. Thus, if $Y \sim$ Poisson, mean λ, then we approximate

$$P(Y \le k) = P\left(\frac{Y - \lambda}{\sqrt{\lambda}} \le \frac{k + 1/2 - \lambda}{\sqrt{\lambda}}\right) \quad \text{by} \quad \Phi\left(\frac{k + 1/2 - \lambda}{\sqrt{\lambda}}\right),$$

where Φ is the c.d.f. for the standard normal distribution. Consider the approximations of Table 8.2.1 given for some selected values of λ and k. We

Table 8.2.1 Poisson Probabilities and Their Normal Approximations*

k	$P(Y \le k)$	Normal Approx.	$P(Y = k)$	Normal Approx.
		$\lambda = 16$		
8	0.021 99	0.030 40	0.011 99	0.021 68
12	0.193 12	0.190 79	0.066 13	0.075 19
16	0.565 96	0.549 74	0.099 22	0.096 43
20	0.868 17	0.869 71	0.055 92	0.045 73
24	0.977 68	0.983 21	0.014 37	0.008 02
		$\lambda = 64$		
48	0.022 59	0.026 34	0.006 43	0.008 61
56	0.174 78	0.174 25	0.031 58	0.034 00
64	0.533 18	0.524 92	0.049 80	0.049 45
72	0.855 57	0.856 00	0.029 05	0.026 49
80	0.977 37	0.980 42	0.007 00	0.005 23

* The normal approximations for $P(Y = k)$ were found by taking differences of approximations of $P(Y \le k)$ and $P(Y \le k - 1)$.

will need to study the behavior of a sequence of random vectors. Fortunately, we have the Multivariate Central Limit Theorem, which follows directly from the univariate theorem by considering linear combinations of components. Recall (Section 2.4) that a random vector $\mathbf{Z}$ has the $N_T(\mathbf{0}, \mathbf{\Sigma})$ distribution if and only if its m.g.f. is $m(\mathbf{t}) = e^{\mathbf{t}'\mathbf{\Sigma}\mathbf{t}/2}$.

Theorem 8.2.10 (The Multivariate Central Limit Theorem): Let $\{\mathbf{Y}_k\}$ be a sequence of T-component independent identically distributed random vectors, with means $\boldsymbol{\mu}$ and common covariance matrices $\mathbf{\Sigma}$. Let

$$\mathbf{S}_n = \sum_{k=1}^{n} \mathbf{Y}_k \qquad \text{and} \qquad \mathbf{Z}_n = (\mathbf{S}_n - n\boldsymbol{\mu})/\sqrt{n}$$

Then $\mathbf{Z}_n \xrightarrow{D} N_T(\mathbf{0}, \mathbf{\Sigma})$.

Comment: We will refer to this theorem as the MCLT. The sample mean vector is $\bar{\mathbf{Y}}_n \equiv \frac{1}{n}\sum_{k=1}^{n} \mathbf{Y}_k = \mathbf{S}_n/n$. Then $\mathbf{Z}_n = (\bar{\mathbf{Y}}_n - \boldsymbol{\mu})\sqrt{n}$. If $\mathbf{\Sigma}$ has rank r, and $\mathbf{B}$ is an $r \times T$ matrix such that $\mathbf{B\Sigma B}' = \mathbf{I}_r$, then $\mathbf{W}_n = \mathbf{BZ}_n \xrightarrow{D} N_r(\mathbf{0}, \mathbf{I}_r)$.

Our most important application of the MCLT is to the multinomial distribution.

Theorem 8.2.11: Let $\mathbf{Y}_n \sim \mathcal{M}_T(n, \mathbf{p})$. Then $\mathbf{Z}_n = (\mathbf{Y}_n - n\mathbf{p})/\sqrt{n} \xrightarrow{D} N_T(\mathbf{0}, \mathbf{Q}_p)$, where $\mathbf{Q}_p = d(\mathbf{p}) - \mathbf{pp}'$.

Comment: Let $\hat{\mathbf{p}}_n = \mathbf{Y}_n/n$, the vector of proportions. Then $\mathbf{Z}_n = (\hat{\mathbf{p}}_n - \mathbf{p})\sqrt{n}$. In particular, the ith component of $\mathbf{Z}_n$ converges in distribution to $N(0, p_i(1 - p_i))$.

Proof: $\mathbf{Y}_n$ has the distribution of the sum of n independent generalized Bernoulli random vectors, each with parameter $\mathbf{p}$. These Bernoulli r.v.'s have mean $\mathbf{p}$ and covariance matrix $\mathbf{Q}_p$. The result follows by the MCLT. □

We are now in position to consider Karl Pearson's chi-square goodness-of-fit statistic $C_n = \sum_{k=1}^{n} (Y_{ni} - np_i)^2/(np_i)$. Karl Pearson, the leading statistician of the period 1890–1910 and the founder of the journal *Biometrika* in 1901, was the father of Egon Pearson, who along with Jerzy Neyman developed the theories of testing hypotheses and confidence intervals in about 1933. Karl Pearson invented the statistic C_n as a measure of the deviation of an observed vector of frequencies from their expectations.

Theorem 8.2.12: Let $\mathbf{Y}_n = (Y_{n1}, \ldots, Y_{nT}) \sim \mathcal{M}_T(n, \mathbf{p})$, where all the components of $\mathbf{p}$ have positive components. Let C_n be defined as above. Then $C_n \xrightarrow{D} \chi^2_{T-1}$.

***Proof*:** Let $\mathbf{Z}_n$ be defined as in Theorem 8.2.11. Let $\mathbf{p}^{1/2}$ be the vector of square roots of the elements of $\mathbf{p}$. Then $\mathbf{W}_n = d(\mathbf{p}^{-1/2})\mathbf{Z}_n$ has covariance matrix $\mathbf{M}_p \equiv d(\mathbf{p}^{-1/2})\mathbf{Q}_p\mathbf{d}(\mathbf{p}^{-1/2}) = \mathbf{I}_T - \mathbf{p}^{1/2}\mathbf{p}^{1/2\prime}$. Thus, $\mathbf{W}_n \xrightarrow{D} N(\mathbf{0}, \mathbf{M}_p)$. $\mathbf{M}_p$ is the projection matrix onto $V^\perp$, for $V = \mathscr{L}(\mathbf{p}^{1/2})$. If $\mathbf{Z} \sim N(\mathbf{0}, \mathbf{I}_T)$, and $\hat{\mathbf{Z}} = p(\mathbf{Z}|V^\perp)$, then $\hat{\mathbf{Z}} \sim N_T(\mathbf{0}, \mathbf{M}_p)$, and, from Theorem 2.5.3, $\|\hat{\mathbf{Z}}\|^2 \sim \chi^2_{\dim(V^\perp)} = \chi^2_{T-1}$. Since $\mathbf{W}_n \xrightarrow{D} \hat{\mathbf{Z}}$, and squared length is a continuous function of its argument, it follows from Theorem 8.2.6 that $\|\mathbf{W}_n\|^2 = C_n$ converges in distribution to χ^2_{T-1}. □

There is a rough rule often suggested that the chi-square approximation of the distribution of C_n is adequate if all expectations are at least 5. Actually, the approximation seems to be quite good even in cases in which some of these expectations are considerably smaller. Consider the case that $T = 3$, $\mathbf{p} = (0.2, 0.3, 0.5)$. Figure 8.1 presents the cumulative chi-square distribution for 2 d.f. (which is the exponential with mean 2), and the c.d.f. of the Pearson chi-square statistics for $n = 6$ and $n = 10$. Most of the expectations are less than 5. Notice the closeness of the approximations. The 95th percentile of the χ^2_2 distribution is $-2 \log 0.05 = 5.99$, while the the true probabilities of exceeding 5.99 are 0.0527 for $n = 6$, and 0.0502 for $n = 10$. The approximation is not always quite this good.

We will be interested in the estimation of $\mathbf{p}$ for the case that $\mathbf{p} = \mathbf{p}(\boldsymbol{\beta})$, where $\boldsymbol{\beta}$ is a vector with fewer than T parameters. If $\hat{\mathbf{p}} = \mathbf{p}(\hat{\boldsymbol{\beta}})$ replaces $\mathbf{p}$ in the definition of C_n, the distribution of C_n changes. For some functions $\mathbf{p}(\boldsymbol{\beta})$ and estimators $\hat{\boldsymbol{\beta}}$, C_n is asymptotically distributed as chi-square with $T - 1 - (\#$ components of $\boldsymbol{\beta})$.

We will also be interested in the distributions of logs and exponentials of random variables and random vectors whose asymptotic distribtions we know. For example, we know that $\mathbf{Z}_n = \sqrt{n}(\mathbf{Y}_n - n\mathbf{p}) \xrightarrow{D} N_T(\mathbf{0}, d(\mathbf{p}) - \mathbf{pp}')$, if $\mathbf{Y}_n \sim \mathcal{M}_T(n, \mathbf{p})$. What happens to the distribution of $\mathbf{W}_n \equiv \log(\mathbf{Y}_n/n)$ as $n \to \infty$? (If any component of $\mathbf{Y}_n$ is 0, replace it by 1/2, so that $\mathbf{W}_n$ is defined.) Fortunately, every smooth function is approximately linear over small intervals, and with high probability the random vector of interest ($\mathbf{Y}_n/n$ in this case) will for large n be confined to a small interval. This, together with the fact that linear functions of normally distributed random vectors are still normally distributed, provides us with the very useful multivariate *δ-method.*

Theorem 8.2.13: Let $\mathbf{U}_n$ be a sequence of T-component random vectors, with the same mean $\boldsymbol{\mu}_0 = E(\mathbf{U}_n)$. Suppose that $\mathbf{Z}_n \equiv \sqrt{n}(\mathbf{U}_n - \boldsymbol{\mu}_0) \xrightarrow{D} N_T(\mathbf{0}, \boldsymbol{\Sigma})$. Let $g = (g_1, \ldots, g_k)$ be a function from R_T into R_k. Suppose that the partial

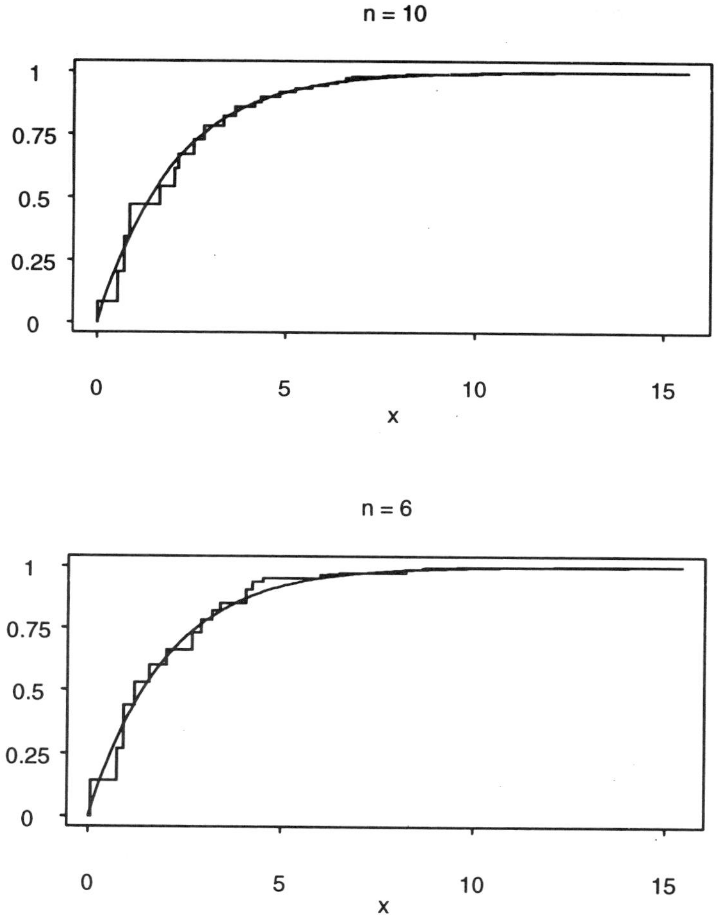

FIGURE 8.1 Sample c.d.f. for 1,000 chi-square statistics and the chi-square c.d.f.

derivatives $g_j^i(\boldsymbol{\mu}) \equiv \frac{\partial}{\partial \mu_i} g_j(\boldsymbol{\mu})$ exist at $\boldsymbol{\mu}_0$ for each i and j. Let $\mathbf{A}$ be the $T \times k$ matrix with ij element $g_j^i(\boldsymbol{\mu}_0)$. Then

$$\mathbf{W}_n = \sqrt{n}[g(\mathbf{U}_n) - g(\boldsymbol{\mu}_0)] \xrightarrow{D} N_k(\mathbf{0}, \mathbf{A}\boldsymbol{\Sigma}\mathbf{A}').$$

We will only outline a proof here. The essential idea is that by Taylor's Theorem $g(\mathbf{u}) = h(\mathbf{u}) + e(\mathbf{u} - \boldsymbol{\mu}_0)$, where $h(\mathbf{u}) = g(\boldsymbol{\mu}_0) + \mathbf{A}(\mathbf{u} - \boldsymbol{\mu}_0)$, $e(\mathbf{0}) = 0$, and $\|e(\mathbf{x})\|/\|\mathbf{x}\| \to 0$ as $\mathbf{x} \to \mathbf{0}$. Then

$$\mathbf{W}_n = \sqrt{n}[h(\mathbf{U}_n) - h(\boldsymbol{\mu}_0)] + \sqrt{n}e(\mathbf{U}_n - \boldsymbol{\mu}_0) = \mathbf{A}\mathbf{Z}_n + \sqrt{n}e(\mathbf{Z}_n/\sqrt{n}).$$

The second term on the right converges in probability to 0, so that the first term $\mathbf{AZ}_n$ has the same limiting distribution as does $\mathbf{W}_n$. By Theorem 8.2.6 it follows that $\mathbf{W}_n \xrightarrow{D} N_k(\mathbf{0}, \mathbf{A\Sigma A'})$.

The existence of first partial derivatives guarantees the breakup of g into the sum of the linear function h and the error function e, with satisfactory properties.

Example 8.2.2: Let $\mathbf{Y}_n \sim \mathscr{M}_T(n, \mathbf{p}_0)$, with all components of $\mathbf{p}$ positive, and let $\hat{\mathbf{p}}_n = \mathbf{Y}_n/n$. Let $g(\mathbf{y}) = \log \mathbf{y}$ for $\mathbf{y} \in R_T^+$. Then $\mathbf{U}_n = \hat{\mathbf{p}}_n$ satisfies the conditions of Theorem 8.2.11 with $\boldsymbol{\mu}_0 = \mathbf{p}_0$, $\boldsymbol{\Sigma} = d(\mathbf{p}_0) - \mathbf{p}_0\mathbf{p}_0'$, and $k = T$. The matrix of partial derivatives defined in Theorem 8.2.12 is $\mathbf{A} = d(\mathbf{p}^{-1})$, the diagonal matrix of reciprocals. Thus, $\mathbf{W}_n = \sqrt{n}[\log \hat{\mathbf{p}}_n - \log \mathbf{p}_0] \xrightarrow{D} N_T(\mathbf{0}, d(\mathbf{p}^{-1}) - \mathbf{J}_T)$, where $\mathbf{J}_T$ is the $T \times T$ matrix of all 1's. A less rigorous but more intuitive way to say this is that $\log \hat{\mathbf{p}}_n$ is approximately $N_T\left(\log \mathbf{p}, \frac{1}{n}[d(\mathbf{p}^{-1}) - \mathbf{J}_T]\right)$.

We will be particularly interested in contrasts among these logs, inner products of the form $\eta = (\mathbf{c}, \log \mathbf{p})$, where the components of c add to one. Our estimator of η will be $\hat{\eta}_n \equiv (\mathbf{c}, \hat{\mathbf{p}}_n)$. Then the estimator $\hat{\eta}_n$ is asymptotically normally distributed with mean η, and variance

$$\frac{1}{n}\mathbf{c}[d(\mathbf{p}^{-1}) - \mathbf{J}_T]\mathbf{c} = \frac{1}{n}\mathbf{c}'d(\mathbf{p}^{-1})\mathbf{c} = \frac{1}{n}\sum c_i^2/p_i = \sum c_i^2/m_i,$$

where $m_i = E(Y_{ni}) = np_i$.

Often a random variable whose distribution we are able to determine for finite n or asymptotically depends on one or more unknown parameters and we would like to replace one or more of the unknown parameters by an estimator, which we expect to be close to the unknown parameter if the sample size is large. Consider the r.v.

$$Z_n = (\bar{X}_n - \mu)/(\sigma/n^{1/2}),$$

where $\bar{X}_n$ is the mean for a random sample from a distribution with mean μ, variance σ^2. If n is large then by the CLT Z_n is approximately distributed as standard normal, no matter what the distribution sampled. If the distribution sampled is normal then Z_n has a standard normal distribution for every positive n. These facts allow us to make probability statements about the error $(\bar{X} - \mu)$, and to give confidence intervals on μ. However, σ is usually unknown, and it is tempting to simply substitute the sample standard deviation S_n for σ, and to assume that the distribution of Z_n is not changed. The following theorem implies that under certain circumstances the substitution is valid in approximation.

Slutsky's Theorem (Fabian and Hannan, 1985, p. 144): Let c be a constant and let $h(x, y)$ be a function on a subset of $R_1 \times R_1$, continuous on the straight line $\{(x, c)|x \in A \subset R_1\}$. Let $\{T_n\}$ and $\{W_n\}$ be sequences of r.v.'s. Suppose that T_n converges in distribution to a r.v. T, that $h(T_n, c)$ converges in distribution, and that $\{W_n\}$ converges in probability to c. Suppose $P(T \in A) = 1$. Then $h(T_n, W_n)$ converges in distribution to $h(T, c)$.

In the example above take $c = \sigma$, $h(x, y) = x/y$, $A = R_1$, $T_n = n^{1/2}(\bar{X}_n - \mu)$, and $W_n = S_n$. $\{S_n\}$ is consistent for σ. We conclude that $h(T_n, W_n) = (\bar{X}_n - \mu)/(S_n/n^{1/2}) \xrightarrow{D} N(0, 1)$. Since a sample proportion is a special case of a sample mean, with $\mu = p$, $\bar{X}_n = \hat{p}_n$, and $\sigma^2 = p(1 - p)$, with a change in notation we get $(\hat{p}_n - p)/[\hat{p}_n(1 - \hat{p}_n)/n]^{1/2} \xrightarrow{D} N(0, 1)$. We can use n or $n - 1$ in the denominator of the denominator with impunity.

In Example 8.2.2, we showed that for $\hat{\eta}_n = (\mathbf{c}, \log \hat{\mathbf{p}}_n)$

$$\sqrt{n}(\hat{\eta} - \eta)/\sqrt{\sum c_i^2/p_i} \xrightarrow{D} N(0, 1).$$

We can replace the p_i in the denominator by the consistent estimators $\hat{p}_{in}$, and, by Slutsky's Theorem, get the same limiting distribution. Hence $\hat{\eta}_n \pm z_{(1+\gamma)/2}/\sqrt{\sum c_i^2/(n\hat{p}_{in})}$ is an approximate $100\gamma\%$ confidence interval on η.

We have not given limit theorems for the generalized or univariate hypergeometric distribution, and will only present results without proof. Consider a finite population B of N elements, with disjoint subsets $B_1, \ldots, B_k$ of sizes $N_1, \ldots, N_k$. Suppose a simple random sample (without replacement) of size n is taken. Let Y_j be the number of elements chosen from subset B_j and let $\mathbf{Y} = (Y_1, \ldots, Y_k)$. In order to apply limit theory we must let the population size grow as well as the sample size. To indicate this growth add the superscript N to n and the N_j, Y_j, and to $\mathbf{Y}$. Thus, for example, N_j^N is the size of B_j when N is the population size. Define $\mathbf{N}^N = (N_1^N, \ldots, N_k^N)$. The superscript N is *not* an exponent. Then as $N \to \infty$:

(1) $\mathbf{Y}^N \xrightarrow{D} \mathscr{M}_K(n, \mathbf{p})$ as $\mathbf{N}^N/N \to \mathbf{p} = (p_1, \ldots, p_k)$, and $n^N \equiv n$ remains fixed.

(2) $Y_j^N \xrightarrow{D} \mathscr{P}(\lambda)$ as $n^N \to \infty$, $n^N N_j^N/N \to \lambda > 0$ for $N \to \infty$.

(3) $\mathbf{Z}^N = (\mathbf{Y}^N - n^N\mathbf{N}^N/N)/\sqrt{n^N} \xrightarrow{D} N_k(\mathbf{0}, [d(\mathbf{p}) - \mathbf{p}'\mathbf{p}](1 - r))$ as $\mathbf{N}^N/N \to \mathbf{p}$, $n^N/N \to r$, $0 \le r < 1$.

r is the *asymptotic sampling fraction.* The limit would be the same if $\mathbf{N}^N/N$ in the numerator of $\mathbf{Z}^N$ were replaced by $\mathbf{p}$.

These limit theorems become useful when we replace the distribution of the r.v. on the left by the more tractable distribution on the right for finite N. The approximations provided are surprisingly good, particularly if the 1/2 correction is used. Consider, for example, the normal approximation of the hypergeometric distribution with $N = 10$, $N_1 = 6$, and sample size $n = 5$. Then $E(Y_1) = nN_1/N = 3$, $\text{Var}(Y_1) = [nN_1/N]\dfrac{N_1 - n}{N - 1} = 2/3$. We find $P(Y_1 = 2) = 0.238\,10$, and by the normal approximation of $P(1.5 \le Y_1 \le 2.5)$, 0.237. Similarly, $P(Y_1 \le 2) = 0.261\,9$ and the normal approximation gives 0.270. It would be silly to use the normal approximation in such a case, of course, but such calculations should give us great faith that these limit theorems are indeed useful.

Problem 8.2.1: Let Y_1, Y_2, Y_3 be independent, with $Y_i \sim \mathscr{P}(\lambda_i)$.

(a) Prove that $S_3 = Y_1 + Y_2 + Y_3 \sim \mathscr{P}(\lambda_1 + \lambda_2 + \lambda_3)$.

(b) Show that, conditionally on $S_3 = s$, $\mathbf{Y} = (Y_1, Y_2, Y_3)$ has a multinomial distribution.

(c) For $\lambda_1 = 2$, $\lambda_2 = 5$, $\lambda_3 = 3$, find $P(Y_2 \ge 5 | S_3 = 6)$, $P(Y_1 = 2,\ Y_2 = 3 | S_3 = 6)$, and $P(Y_1 + Y_2 \ge 5 | S_3 = 6)$.

(d) Find $D[\mathbf{Y}]$ and $D[\mathbf{Y} | S_3 = s]$, the conditional covariance matrix.

Problem 8.2.2: Perform calculations similar to those of Example 8.2.1, illustrating LeCam's upper bound (Theorem 8.2.2) for the case $n = 2$, $p_1 = 0.02$, $p_2 = 0.03$. Determine $G(A)$ and the upper bound given by the LeCam theorem for the case $A = \{1, 2, \ldots\}$.

Problem 8.2.3: Let $\mathscr{I} = \{1, 2\} \times \{1, 2, 3\}$, and let $\mathbf{Y}$ be a random vector indexed by $\mathscr{I}$. Suppose that $\mathbf{Y}_i$ is the ith row of $\mathbf{Y}$, $\mathbf{Y}_i \sim \mathscr{M}_3(n_i, \mathbf{p}_i)$, and $\mathbf{Y}_1$, $\mathbf{Y}_2$ are independent.

(a) For $n_1 = 10$, $n_2 = 20$, $\mathbf{p}_1 = (0.2, 0.3, 0.5)$, and $\mathbf{p}_2 = (0.4, 0.5, 0.1)$, give the mean vector and covariance matrix for $\mathbf{Y}$.

(b) Let $\mathbf{W} \sim \mathscr{M}_6(30, \mathbf{p})$, where $\mathbf{W}$ has the same index set as does $\mathbf{Y}$, and $\mathbf{p} = \begin{bmatrix} 0.08 & 0.12 & 0.20 \\ 0.24 & 0.30 & 0.06 \end{bmatrix}$. Show that the conditional distribution of $\mathbf{W}$, given the row sums for $\mathbf{W}$ are 10 and 20, is the distribution of $\mathbf{Y}$ in (a).

Problem 8.2.4: The members of a large population of voters were asked to select among candidates A, B, C. For the population 40% favored A, 30% favored B, and 30% favored C. A pollster took a random sample of 200. Assume for simplicity that sampling was with replacement.

(a) What is the distribution of $\mathbf{Y} = (Y_A, Y_B, Y_C)$, the numbers in the sample voting for the three candidates?

(b) Find $E(\mathbf{Y})$ and $D[\mathbf{Y}]$.

(c) Find an approximation for the probability that A loses in the sample (that $Y_A < Y_B$ or $Y_A < Y_C$). *Hints*: The event of interest can be written in the form $E_1 \cup E_2$, where $E_1 = [W_1 \equiv a_1 Y_A + b_1 Y_B + c_1 Y_C < 0]$ and $E_2 = [W_2 \equiv a_2 Y_A + b_2 Y_B + b_2 Y_C < 0]$. Find an approximation for the distribution of (W_1, W_2). You will have to use tables of or a computer program for the bivariate normal distribution.

Problem 8.2.5: A class of 14 students sit in four rows of 2, 3, 4 and 5. The instructor decides to grade the class randomly, giving three A's, five C's, and six F's. Find the probability that the observed table of frequencies is $\begin{bmatrix} 1 & 2 & 0 & 0 \\ 1 & 1 & 2 & 1 \\ 0 & 0 & 2 & 4 \end{bmatrix}$. What is the probability that both of the A's are given to students in the same row?

Problem 8.2.6: A, B, and C each throw two coins three times, resulting in 9 throws of two coins. Among these 9 throws, 3 resulted in two heads, 5 in one head, and 1 in no heads. What is the conditional probability that A had two heads each time, and B had one head each time?

Problem 8.2.7: Let $Y \sim \mathscr{B}(n, p)$. Let $\hat{p} = Y/n$. Use the δ-method to find an approximation for the distribution of arcsin $\sqrt{\hat{p}}$. *Hints*: The variance of the limiting distribution does not depend on p. And $\frac{d}{du}$ arcsin $u = (1 - u^2)^{-1/2}$. Use this limiting distribution to find a 95% confidence interval on p for $n = 1{,}000$ and $Y = 84$. Compare its length to that of the interval $\hat{p} \pm 1.96\sqrt{\hat{p}(1 - \hat{p})/n}$.

Problem 8.2.8: In 250 days the number of accidents at a large automobile manufacturing plant was 579. It seems reasonable to suppose that the number of accidents on each day has a Poisson distribution with mean λ, and that the numbers on different days are independent. Use the asymptotic normality of Poisson r.v.'s to find a 98% confidence interval on the daily rate.

Problem 8.2.9: Let $Y_n \sim$ Poisson$(n\lambda)$, for $\lambda > 0$ fixed. Define $R_n = Y_n/n$, and for a smooth function g define $U_n = g(R_n)$. Find a function g such that the asymptotic distribution of $W_n = \sqrt{n}[g(R_n) - g(\lambda)]$ does not depend on λ. The function g is often called a *variance-stabilizing transformation.* What is the variance-stabilizing transformation for the binomial distribution? See Problem 8.2.7).

Problem 8.2.10: (a) Let U have the uniform distribution on [0, 1]. Show that $X = -\log U$ has the exponential distribution with mean 1.

(b) Let $X_1, X_2, \ldots$ be independent, each with the exponential distribution with mean one. For $\lambda > 0$ let Y be the smallest value n satisfying $S_n = \sum_{i=1}^{n} X_i > \lambda$. Show that Y has the Poisson distribution with mean λ. *Hint*: Use the properties of the Poisson process, or prove directly using the fact that S_n has a gamma distribution and integration by parts that $P(Y \geq k) = \sum_{k}^{\infty} e^{-\lambda}\lambda^i/i!$. (a) and

(b) together may be used to generate an observation Y from the Poisson distribution with mean λ.

Problem 8.2.11: Let $U_1, \ldots, U_n$ be a random sample from the uniform distribution on [0, 1]. Let $M_n = \max(X_1, \ldots, X_n)$. Does M_n, or $z_n = (M_n - a_n)/b_n$ for some a_n and b_n, converge in distribution? To what distribution?

Problem 8.2.12: Suppose that events occur in time in a Poisson process with mean λ. n nonoverlapping intervals of time, each of length T, are chosen and the number of intervals Y, the *Hansen frequency*, for which there is no occurrence is recorded. Then Y has the binomial distribution with parameters n and $p = e^{-\lambda T}$. Since the maximum likelihood estimator (MLE) of p, based on Y, is $\hat{p} = Y/n$, the MLE for λ is the solution to $e^{-\hat{\lambda} T} = \hat{p}$, or $\hat{\lambda} = -[\log \hat{p}]/T$.

(a) Find an approximation for $P(|\hat{\lambda} - \lambda| < 0.3)$ if $n = 100$, $T = 0.4$, and $\lambda = 2$.

(b) Suppose the actual numbers $X_1, \ldots, X_n$ of occurrences in these intervals were observed. Let $\hat{\lambda}^*$ be the MLE of λ, based on these X_i. Find an approximation for $P(|\hat{\lambda} - \lambda| < 0.3)$.

(c) The asymptotic relative efficiency of $\hat{\lambda} = \hat{\lambda}_n$ to that of $\hat{\lambda}^* = \hat{\lambda}_n^*$ is $e_T = \lim_{n \to \infty} \mathrm{Var}(\hat{\lambda}_n)/\mathrm{Var}(\hat{\lambda}_n^*)$. Show that $e_T \to 1$ as $T \to 0$, and $e_T \to 0$ as $T \to \infty$.

Problem 8.2.13: A tree has unknown height h. In order to estimate h, a surveyor writes $h = \delta \tan \alpha$, where δ is the distance on the ground from the base of the tree to the surveying instrument, and α is the angle between ground level and the top of the tree. The surveyor measures the distance and the angle independently, with estimators $d \sim N(\delta, \sigma_d^2)$ and $a \sim N(\alpha, \sigma_a^2)$. Use the δ-method to find an approximation to the distribution of $\hat{h} = d \tan a$, and to $P(|\hat{h} - h| \leq 0.2)$ if $\alpha = \pi/6$, $\delta = 50$, $\sigma_a = 0.002$ radians, and $\sigma_d = 0.02$ meters? *Hint*: In 1,000 computer simulations the mean was 28.881 7, the sample s.d. was 0.132 3, the largest was 29.29 and the smallest 28.46. The event of interest occurred 867 times.

8.3 CONFIDENCE INTERVALS ON POISSON AND BINOMIAL PARAMETERS

In this section we will be concerned exclusively with the estimation of binomial and Poisson parameters. Situations arise frequently in which the observed random variables may reasonably be assumed to have one of these relatively simple distributions, yet, sadly, many introductory texts and courses provide only cursory discussion. The methodology discussed will depend on both small and large sample theory. It is this small sample theory (really *any* size sample theory), which provides confidence intervals on the binomial parameter p, on the Poisson parameter λ, and on a ratio λ_1/λ_2 of two Poisson parameters, which will probably be new to many students.

We will use $\mathscr{B}(n, p)$ and $\mathscr{P}(\lambda)$ respectively to denote the binomial and Poisson distributions. We begin by developing confidence intervals on the parameters p and λ. We need two very simple inequalities.

Lemma 8.3.1: Let X be a random variable with c.d.f. F. Define $\bar{F}(x) = P(X \geq x)$ for each x. Then, for $0 < \alpha < 1$, (1) $P(F(X) \leq \alpha) \leq \alpha$ and (2) $P(\bar{F}(X) \leq \alpha) \leq \alpha$.

***Proof*:** Let $M_\alpha = \{x | F(x) \leq \alpha)$. M_α is an interval (see Figure 8.2). Let x_α be the least upper bound of M_α, the right endpoint of M_α. If $x_\alpha \in M_\alpha$, then $P(F(X) \leq \alpha) = P(X \leq x_\alpha) = F(x_\alpha) = \alpha$. If $x_\alpha \notin M_\alpha$, then there exists a monotone increasing sequence of points x_n, with $F(x_n) \leq \alpha$, converging to x_α. Then $M_\alpha = \bigcup_{n=1}^{\infty} (-\infty, x_n]$, so that

$$P(F(X) \leq \alpha) = P(X \in M_\alpha) = \lim_{n \to \infty} P(X \in (-\infty, x_n]) = \lim_{n \to \infty} F(x_n) \leq \alpha.$$

To prove the other inequality consider the random variable $Y = -X$, which has c.d.f. $G(y) = P(Y \leq y) = P(X \geq -y) = \bar{F}(-y)$. Then $P(\bar{F}(X) \leq \alpha) = P(G(-X) \leq \alpha) = P(G(Y) \leq \alpha) \leq \alpha$. The last inequality follows from (1) by replacing F by G. □

The lemma allows us to use $F(X; p)$ and $\bar{F}(X; p)$ as pivotal quantities in order to find confidence limits on p, since the probability inequalities hold for all p.

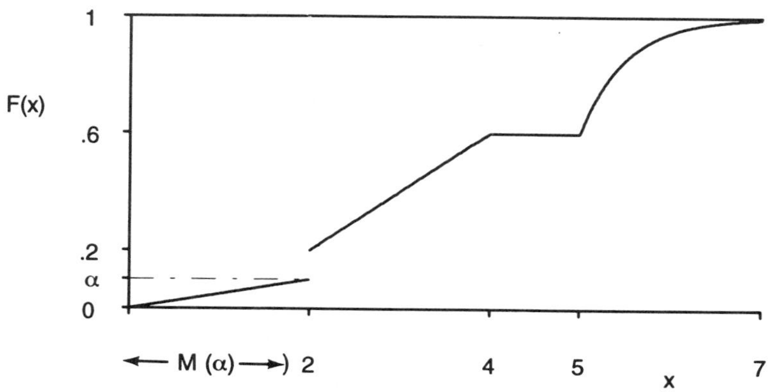

FIGURE 8.2 Cumulative distribution function.

Let $F(k; p)$ be the $\mathscr{B}(n, p)$ cumulative distribution. For fixed k, $F(k; p)$ is a monotone decreasing continuous function of p. To prove monotinicity, let $U_1, \ldots, U_n$ be independent $U(0, 1)$ random variables. Then, for any p, $X_p = \sum_i I[U_i \leq p] \sim \mathscr{B}(n, p)$. For $p' > p$, $X'_p \geq X_p$, so that $F(k; p) = P(X_p \leq k) \geq P(X'_p \leq k) = F(k; p')$.

For $k < n$, let $p_2(k)$ be the solution p of $F(k; p) = \alpha$. The subscript 2 is used because we will later define p_1, which will be less than p_2. The solution exists because $F(k; p)$ is continuous in p, $F(k; 0) = 1$, and $F(k; 1) = 0$. Define $p_2(n) = 1$. Then $p_2(k) \leq p$ if and only if $\alpha = F(k; p_2(k)) \geq F(k; p)$, so that $P(p_2(X) \leq p) = P(\alpha \geq F(X; p)) \leq \alpha$. The last inequality follows from Lemma 8.3.1. The r.v. $p_2(X)$ is therefore an upper $100(1 - \alpha)\%$ confidence limit for p, in the sense that the probability is *at least* $1 - \alpha$ that p_2 exceeds p.

Since the distribution of X is discrete, the probability of coverage will be exactly $1 - \alpha_2$ only for those p for which there is a k such that $F(k; p) = \alpha$.

Example 8.3.1: Suppose $n = 20$ and we observe $X = 0$. Since $F(0; p) = (1 - p)^{20} = \alpha$, $p = p_2(0) = 1 - \alpha^{1/20}$ is an upper $100(1 - \alpha)\%$ confidence limit on p. For $\alpha = 0.05$, we find $p_2 = 0.139\,1$. If we instead observed $X = 1$, then p_2 is the solution to $F(1; p) = q^{20} + 20pq^{19}$ for $q = 1 - p$. We find $p_2 = 0.216\,1$, so that we have 95% confidence that $p \leq 0.216\,1$. For $X = 2$, we solve $F(2; p) = 0.05$ to find $p_2 = 0.282\,6$. Graphs of the functions $F(0; p)$, $F(1; p)$ and $F(2; p)$ are given in Figure 8.3.

The function $\bar{F}(k; p)$, giving right tail probabilities, is a monotone increasing continuous function of p for each k. For $k > 0$ let $p_1 = p_1(k)$ be the solution to $\bar{F}(k; p) = \alpha$. Let $p_1(0) = 0$. Then in a argument similar to that for p_2 we can show that $P(p_1(X) \geq p) \leq \alpha$. Therefore, p_1 is a lower $100(1 - \alpha)\%$ confidence

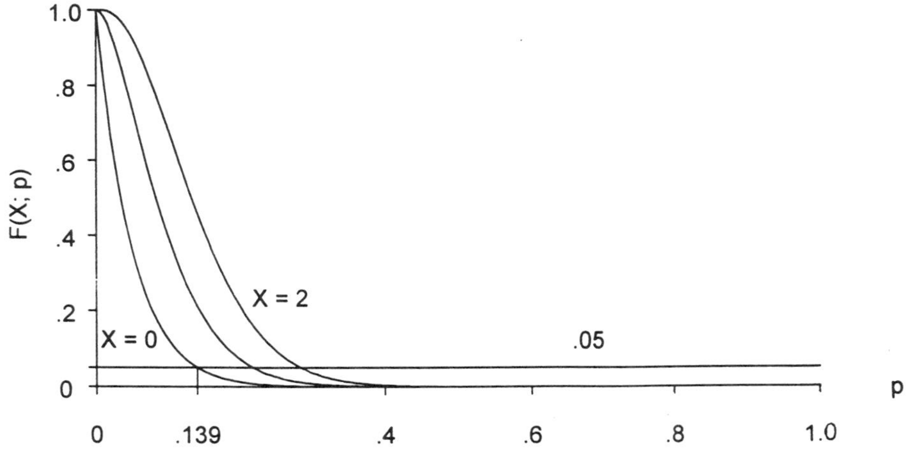

FIGURE 8.3 The functions $F(0; p)$, $F(1; p)$, $F(2; p)$.

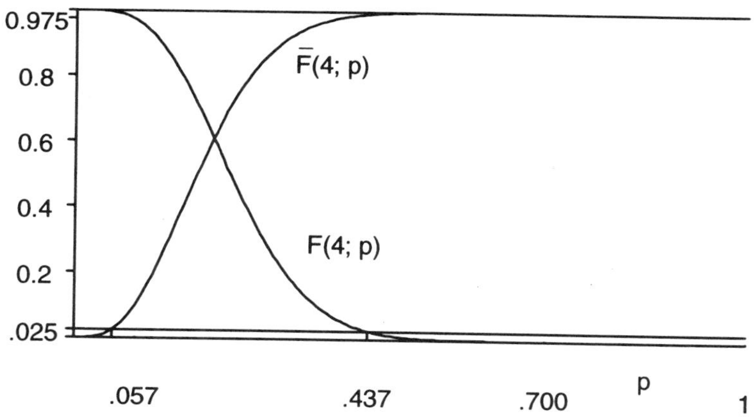

FIGURE 8.4 $F(4; p)$, $\bar{F}(4; p)$ and corresponding 95% confidence interval.

limit for p. For example, if $n = 20$, and we observe $X = 20$, then p_1 is the solution to $\bar{F}(20; p) = p^{20} = \alpha$, so $p_1 = \alpha^{1/20}$. For $\alpha = 0.05$ we get $p_1 = 0.860\,89$, so that we have 95% confidence that $p > 0.860\,89$.

If we want a two-sided confidence interval on p we can use both p_1 and p_2. Let p_2 be the solution to $F(X; p) = \alpha_2$, and let p_1 be the solution to $\bar{F}(X; p) = \alpha_1$. Then $P(p_1 < p < p_2) \geq 1 - (\alpha_1 + \alpha_2)$. If α_1 and α_2 are chosen to add to α, then the interval (p_1, p_2) is a $(1 - \alpha)100\%$ confidence interval on p. For example, if $n = 20$, we observe $X = 4$, and we want a 95% confidence interval on p, then we can choose $\alpha_1 = \alpha_2 = 0.025$, and we find $p_1 = 0.057\,3$, $p_2 = 0.436\,6$, so that we have 95% confidence that $0.057\,3 < p < 0.436\,6$ (Figure 8.4). These values p_1 and p_2 can be found with a computer program generating binomial probabilities, or by using a connection to the F distribution which we give now.

Binomial tail probabilities are related to the beta distribution through the equality

$$\sum_{j=k}^{n} \binom{n}{j} p^j (1 - p)^{n-j} = \frac{\Gamma(n + 1)}{\Gamma(k + 1)\Gamma(n + 1 - k)} \int_0^p x^{k-1}(1 - x)^{n+1-k}\,dx.$$

This can be proved by integrating by parts on the right $n - k$ times. The right side is the c.d.f. of a Beta$(k, n + 1 - k)$ r.v. U. There is a connection between the beta and F distributions: If $U \sim$ Beta(v_1, v_2), then $F \equiv \frac{v_2}{v_1}\frac{U}{1 - U}$ has an $F(2v_1, 2v_2)$ distribution. This ıelationship can then be exploited to give, for observed $X = k$:

$$p_1 = 1\Big/\left[1 + \frac{v_1}{v_2} F_{1-\alpha_1}(v_1, v_2)\right] \qquad \text{for} \quad v_1 = 2(n + 1 - k), \quad v_2 = 2k.$$

Take

$$p_2 = W/(1 + W) \quad \text{for} \quad W = (\nu_1/\nu_2)F_{1-\alpha_2}(\nu_1, \nu_2),$$
$$\text{for} \quad \nu_1 = 2(k+1), \quad \nu_2 = 2(n-k).$$

F-table values may not be easily available for large ν_1 or ν_2, though most statistical computer packages now provide them. If ν_2 is large and ν_1 relatively small then $F_\gamma(\nu_1, \nu_2) \doteq \chi^2_{\nu_1,\gamma}/\nu_1$. If ν_1 is large and ν_2 small then $F_\gamma(\nu_1, \nu_2) \doteq \nu_2/\chi^2_{\nu_2,1-\gamma}$. If both ν_1 and ν_2 are large we can instead use the fact that $Z = (\hat{p} - p)/\sigma(\hat{p})$ for $\sigma(\hat{p}) = \sqrt{p(1-p)/n}$, is approximately distributed as standard normal. Take p_1 and p_2 to be the solutions to $Z = z_{\alpha_1}$ and to $Z = z_{1-\alpha_2}$. The solution offered in most introductory texts on statistics is obtained by replacing p under the square root by its estimator $\hat{p} = X/n$, to obtain $\hat{p} - z_{\alpha_1}\hat{\sigma}_{\hat{p}}$ and $\hat{p} + z_{1-\alpha_2}\hat{\sigma}_{\hat{p}}$. Usually, too often in the author's opinion, people take, $\alpha_1 = \alpha_2 = \alpha/2$, so that the interval is symmetric about $\hat{p}$. In many applications it makes more sense to take $\alpha_1 = 0$ to get an interval $(p_1, 1]$ or $\alpha_2 = 0$ to get an interval $[0, p_2)$.

We sometimes are interested on the *odds* for success $\theta = p/(1-p)$, or *log-odds* $\mu \equiv \log\theta = \log p - \log(1-p) \equiv g(p)$. As will be shown as we develop the theory and applications over the next few sections, this scale turns out to be very convenient for the analysis of frequency data. We begin with one p only, though the principal application will be to the comparison of two or many p's. Since the log-odds function has derivatives of all orders, except at zero and one, with $g'(p) = 1/[p(1-p)]$, we can apply the δ-method to conclude that $W_n \equiv \sqrt{n}[g(\hat{p}_n) - g(p)] \xrightarrow{D} N(0, 1/p(1-p))$. That is, for large n, the *sample log-odds* $\hat{\mu} \equiv g(\hat{p}_n) = \log[\hat{p}_n/(1-\hat{p}_n)]$ is approximately normally distributed with mean $\mu = \log[p/(1-p)]$ and variance $1/[np(1-p)] \equiv \sigma^2(\hat{\mu})$ for large n, p not too close to 0 or 1. Since p is unknown, with consistent estimator $\hat{p}_n$, we can replace p by $\hat{p}_n$ in $\sigma(\hat{\mu})$, let $\hat{\sigma}(\hat{\mu}) = 1/[n\hat{p}_n(1-\hat{p}_n)]$, and use $[\hat{\mu} - \mu]/\hat{\sigma}(\hat{\mu})$ as a pivotal quantity to obtain the approximate $100\gamma\%$ confidence interval $(\mathrm{L}, \mathrm{U}) = [\mu \pm z_{(1+\gamma)/2}\hat{\sigma}(\hat{\mu})]$ on μ. Since $\theta = e^\mu$, and $\theta = p/(1-p)$, $p = e^\mu/(1+e^\mu)$, we therefore have the 100% confidence interval $(e^L/(1+e^L), e^U I(1+e^U))$ on p. This interval will not be symmetric about $\hat{p}_n$. These intervals are shown in Figure 8.5.

Suppose $n = 20$ and we observe $X = 4$. Then $\hat{p}_n = 0.20$, $\hat{\mu} = -1.3863$, $\hat{\sigma}(\hat{\mu}) = 0.5590$ and, for $\gamma = 0.95$, $L = -2.4819$, $U = -0.2907$. The 95% confidence interval on p is (0.077, 0.428). The corresponding intervals found by the exact and more direct large-sample method are (0.0866, 0.4366) and (0.025), 0.375), all roughly of the same length, but of different *shape*.

These large sample methods work quite well, even for small n, and p surprisingly close to 0 or 1. Table 8.3.1 presents probabilities of coverage of p for various p for nominal 95% confidence intervals on p found by the direct (method #1, P_1) and log-odds (method #2, P_2) large sample methods for samples of size 20. Also given are mean lengths L_1 and L_2 of these confidence intervals. Intervals with endpoints less than 0 or greater than 1 were truncated

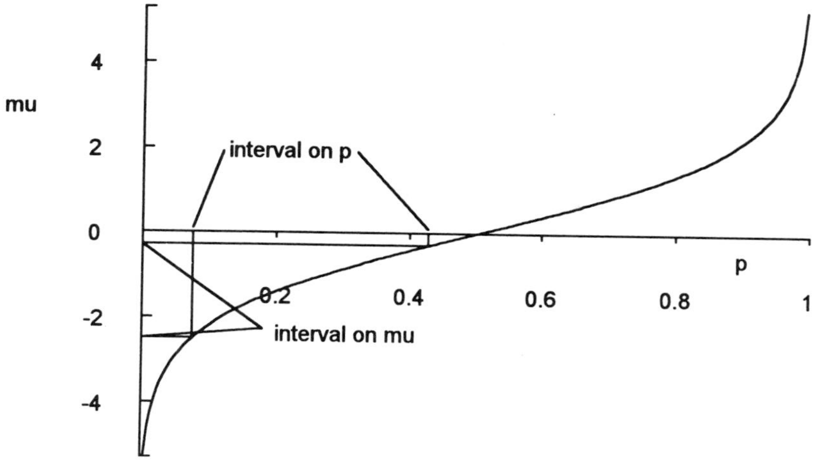

FIGURE 8.5 95% confidence intervals on $\mu = \log(p/(1-p))$ and on p.

back to the interval [0, 1]. Values for $p > 0.50$ are the same, except that $\mu(1-p) = -\mu(p)$.

For the case of two independent binomial random variables $X_1 \sim \mathscr{B}(n_1, p_1)$ and $X_2 \sim \mathscr{B}(n_2, p_2)$, there is no good small sample confidence interval on $\Delta = p_1 - p_2$. For large n_1 and n_2, with p_1 and p_2 not too close to 0 or 1, we can use the fact that $Z \equiv (\hat{\Delta} - \Delta)/\hat{\sigma}(\hat{\Delta}) \xrightarrow{D} N(0, 1)$, for $\hat{\Delta} = \hat{p}_1 - \hat{p}_2$, and

$$\hat{\sigma}^2(\hat{\Delta}) = \frac{\hat{p}_1(1-\hat{p}_1)}{n_1} + \frac{\hat{p}_2(1-\hat{p}_2)}{n_2}.$$

A $100(1-\alpha)\%$ confidence interval on Δ is given by $\hat{\Delta} \pm z_{1-\alpha/2}\hat{\sigma}(\hat{\delta})$. A slightly better approximation can be obtained by adjusting $\hat{p}_i(1-\hat{p}_i)$ in the estimate of the variance to $(X_i + 0.5)(n_i - X_i + 0.5)/n_i^2$ for $i = 1, 2$. The approximation works surprisingly well, even when n_1 and n_2 are very small. For example, for

Table 8.3.1

p	μ	P_1	P_2	L_1	L_2
0.01	−4.60	1.000	0.983	0.104	0.293
0.05	−2.94	0.997	0.984	0.156	0.292
0.10	2.20	0.876	0.957	0.223	0.305
0.20	−1.39	0.921	0.968	0.327	0.345
0.30	−0.85	0.947	0.975	0.387	0.378
0.40	0.41	0.928	0.963	0.418	0.398
0.50	0.00	0.959	0.959	0.427	0.404

$n_1 = 8$, $n_2 = 7$, $p_1 = 0.5$, $p_2 = 0.3$ the adjusted and unadjusted coverage probabilities for nominal 95% confidence intervals on $\Delta = 0.2$ are 0.967 and 0.895. The mean lengths are 0.70 and 0.89, very long, and therefore of not much value. For the same p-values, but $n_1 = n_2 = 20$, these probabilities are 0.947 and 0.942, with average lengths 0.48 and 0.46. The lesson is that we should not fear the use of the large sample approximation for relatively small sample sizes, though, their usefulness is limited because of their excessive length.

We can extend the usefulness of log-odds to the comparison of two proportions p_1 and p_2. Let $q_i = 1 - p_i$, $\theta_i = p_i/q_i$, $\mu_i = \log \theta_i$ for $i = 1, 2$. Then $R \equiv \theta_1/\theta_2 = \dfrac{p_1 q_2}{p_2 q_1}$ is the *odds-ratio* and $\delta \equiv \mu_1 - \mu_2$ is the *log odds-ratio*. The *sample odds-ratio* is $\hat{R}$ is obtained by replacing each p_i by $\hat{p}_i = X_i/n_i$. $\hat{\mu} \equiv \log \hat{R}$ is the *sample log odds-ratio*. Using the δ-method again, we find that $[\hat{\mu} - \mu]/\hat{\sigma}(\hat{\mu}) \xrightarrow{D} N(0, 1)$, where $\hat{\sigma}^2(\hat{\mu}) = \dfrac{1}{n_1\hat{p}_1\hat{q}_1} + \dfrac{1}{n_2\hat{p}_2\hat{q}_2}$. It follows that $\hat{\mu} \pm z_{(1+\gamma)/2}\hat{\sigma}(\hat{\mu})$ is a $100\gamma\%$ confidence interval on μ. Of course, this interval can be transformed into an interval on R.

Estimation of Poisson Parameters: If X has a Poisson distribution the functions $F(X; \lambda)$ and $\bar{F}(X; \lambda)$ can again be used as pivotal quantities. For observed X, the solution $\lambda_2 = \lambda_2(X)$ to $F(X; \lambda) = 0.05$ is an upper 95% confidence limit on λ. Similarly, the solution $\lambda_1 = \lambda_1(X)$ to $\bar{F}(X; \lambda) = 0.05$ is a lower 95% confidence limit for λ. (If $X = 0$, the lower limit is taken to be 0.) The relationship between the Poisson distribution and the chi-square distribution can be exploited to find explicit formulas for λ_1 and λ_2. If $Y \sim \chi^2_{2k}$ then $P(Y \geq 2\lambda) = F(k; \lambda)$, the Poisson c.d.f. (see Problem 3.8.5).

Take $\lambda_1(X) = \chi^2_{2X, \alpha_1}/2$ and $\lambda_2(X) = \chi^2_{2(X+1), 1-\alpha_2}/2$. Then λ_1 is a $100(1 - \alpha_1)\%$ lower confidence limit on λ and λ_2 is a $100(1 - \alpha_2)\%$ upper confidence limit on λ. For example, if we observe $X = 2$, and we want a 95% confidence interval on λ, take $\alpha_1 = \alpha_2 = 0.025$. Then $\lambda_1 = \chi^2_{4, 0.025}/2 = 0.242$ and $\lambda_2 = \chi^2_{6, 0.975}/2 = 7.22$, so that (0.242, 7.22) is a 95% confidence interval on λ.

For large X, we can use the cube-root transformation to find quantiles of the chi-square distribution (Section 2.5) for which table values are not available, or we can use the fact that $Z \equiv (X - \lambda)/\sqrt{\lambda}$ is approximately standard normal. We find that λ_1 and λ_2 are the solutions to $Z = z_{\alpha_1}$ and $Z = z_{1-\alpha_2}$. A still rougher approximation is given by replacing λ under the square root by X to give $\lambda_1 = X + z_{\alpha_1}\sqrt{X} = X - z_{1-\alpha_2}\sqrt{X}$ and $\lambda_2 = X + z_{1-\alpha_2}\sqrt{X}$. For large X the three methods will provide approximately the same answers.

Suppose now that X_1 and X_2 have Poisson distributions with parameters λ_1 and λ_2, and X_1 and X_2 are independent. We would like to construct a confidence interval on $R = \lambda_1/\lambda_2$. We will use the fact that, conditionally on their sum, $X_1 + X_2 = n$, X_1 has a $\mathscr{B}(n, p)$ distribution with parameters n and $p = \lambda_1/(\lambda_1 + \lambda_2) = R/(1 + R)$ (see Theorem 8.2.3).

Let $(p_1 = p_1(X_1), p_2 = p_2(X_1))$ be a $100\gamma\%$ confidence interval on p. That is,

$$P(p_1 < p < p_2 | X_1 + X_2 = n) \geq \gamma \qquad \text{for all} \quad p \quad \text{and} \quad n.$$

Replacing p by $R/(1 + R)$, and manipulating the inequalities, we get

$$P(p_1/(1 - p_1) < R < p_2/(1 - p_2) | X_1 + X_2 = n) \geq \gamma.$$

Since this is true conditionally for every n, it must therefore hold unconditionally, so that $(p_1/(1 - p_1), p_2/(1 - p_2))$ is a $100\gamma\%$ confidence interval on R.

Example 8.3.2: Suppose that the number of highway deaths in July of 1998 in Michigan was 145. After a concerted safety campaign, the number of deaths in 1999 in July was 121. Assuming that the numbers of miles driven in the two years were the same, find a 95% confidence interval on the ratio $R = \lambda_1/\lambda_2$ of the rates for the two years.

We will suppose that the numbers of deaths X_1 and X_2 by highway accident have Poisson distributions with parameters λ_1 and λ_2. This model may not be realistic, since accidents often kill more than one person. It would probably be better to deal with accidents in which deaths occur rather than with numbers of deaths. Conditionally on the total number $n = X_1 + X_2 = 266$ deaths, X_1 has a binomial distribution with parameters $n = 266$, and $p = \lambda_1/(\lambda_1 + \lambda_2)$. We first find a 95% confidence interval on p, given by

$$(p_1, p_2) = (\hat{p} \pm 1.96\sqrt{\hat{p}(1 - \hat{p})/n}) = (0.545\,11 \pm 0.059\,84) = (0.485\,27, 0.604\,95).$$

Then a 95% confidence interval on R is $(p_1/(1 - p_1), p_2/(1 - p_2)) = (0.942\,77, 1.531\,33)$. We are presenting more decimal places than are warranted by the methods. In a report to the possibly statistically naive, it would be better to give (0.94, 1.53). The fact that the interval includes 1 should lead us to be cautious about claiming that the safety campaign was a success.

Suppose now that X_1 and X_2 are independent Poisson r.v.'s with parameters $\lambda_1 = \theta_1 t_1$, and $\lambda_2 = \theta_2 t_2$, where t_1 and t_2 and known constants. To find a confidence interval on the ratio $\rho = \theta_1/\theta_2$ we can simply first find a confidence interval on $R = \lambda_1/\lambda_2$ and, since $\rho = \theta_1/\theta_2 = (\lambda_1/t_1)/(\lambda_2/t_2) = (t_2/t_1)R$, multiply the confidence interval for λ_1/λ_2 through by t_2/t_1 to get an interval for ρ.

We can use the log method to find confidence intervals on $R = \lambda_1/\lambda_2$. Define $\mu = \log R$, $\hat{R} = X_1/X_2$, $\hat{\mu} = \log \hat{R}$. Then, for R fixed, with $\mu_2 = R\mu_1$, and $\mu_1 \to \infty$, $[\hat{\mu} - \mu]/\hat{\sigma}(\hat{\mu}) \xrightarrow{D} N(0, 1)$, where $\hat{\sigma}^2(\hat{\mu}) = \dfrac{1}{X_1} + \dfrac{1}{X_2}$. The approximation improves as λ_1 and λ_2 become large. The resulting $100\gamma\%$ confidence interval on μ is $\hat{\mu} \pm z_{(1+\gamma)/2}\hat{\sigma}(\hat{\mu})$.

For the highway accident example above with $X_1 = 145$, and $X_2 = 121$, we find $\hat{\mu} = 0.078\,58$, $\hat{\sigma}(\hat{\mu}) = 0.123\,13$, and the 95% confidence interval on μ $(-0.163, 0.320)$. The resulting 95% confidence interval on $R = \lambda_1/\lambda_2$ is (0.850, 1.377).

Odds and Log-Odds: Suppose that you, as a statistician, are asked to design a study to determine whether residence near a high-voltage power line raises the probability that a child will have cancer. Let populations #1 and #2 be the collections of children living (#1) and not living (#2) within 400 yards of a power line. Let p_i be the conditional probability that a child in population i is diagnosed with cancer during a three-year period. We would like to compare p_1 to p_2. Let us ignore for the moment the possibility that a lurking variable, say poverty, may cause children both to live near power lines and also to have cancer. Suppose also for simplicity that children do not change residence during the three-year period of interest.

Consider the two-way table:

	Cancer	No Cancer
Population 1	p_{11}	p_{12}
Population 2	p_{21}	p_{22}

Here p_{ij} is the proportion of children in the entire population who would fall in row i, column j of the table. You would like to do a prospective study. That is, you would like to choose random samples of children from each of the two population, or one from the entire population of children in the region of interest, then estimate both $p_i = p_{i1}/(p_{i1} + p_{i2})$ for $i = 1, 2$, and compare these estimates. However, the usual cancer rate is 0.2 per 1,000 children per year, and in order to estimate probabilities and to make comparisons which have any reasonable chance of separating real from chance differences, samples of the order of 100,000 and more are required. Since no records are kept of residence near power lines, identification of a large number of children in population #1 seems practically very difficult.

If a random sample were taken from the population of all children in the region of interest then p_{ij} is the probability that a child would fall in cell ij. The conditional probabilities of interest are p_1 and p_2. Since $p_{i1} \ll p_{i2}$, $p_i \doteq p_{i1}/p_{i2} \equiv R_i = p_i/(1 - p_i)$. In fact, $R_i/p_i = 1/(1 - p_i) = (p_{i1}/p_{12}) + 1$. R_i is the *odds* for cancer in population i. The ratio $R \equiv R_1/R_2 = p_{11}p_{22}/p_{12}p_{21} = p_1(1 - p_2)/[p_2(1 - p_1)]$ is called the odds-ratio. This odds-ratio, for the case that each p_i is small, can serve as a stand-in for $\theta \equiv p_1/p_2$, since $R/\theta = [(p_{11}/p_{12}) + 1]/[(p_{21}/p_{22}) + 1] = (1 - p_2)/(1 - p_1)$.

The benefit of the use of R, rather than θ, to compare rates for cancer in the two populations is most evident when we consider that we can estimate R by doing a *retrospective study*. That is, we can randomly sample the cancer and noncancer populations and still estimate R, since R is symmetric in row and column probabilities. Since such records are kept, we may have access to files of addresses of such children. Suppose that after three years we choose a random sample of n_C (say 400) such children, and another random sample of n_{NC} (say 500) children from the population NC who were not diagnosed to have cancer. Identifying the children in this noncancer population may not be easy, and we may need to confine the study to school-age children, since school records could then be used. Suppose we then use maps to identify whether each of the $n_C + n_{NC}$ children live near a power line.

Let Y_j be the number among the sample of n_j who live near a power line. Then $Y_1 \sim \mathscr{B}(n_1, p_C)$ and $Y_2 \sim \mathscr{B}(n_2, p_{NC})$ in good approximation, where $p_C = p_{11}/(p_{11} + p_{21})$ is the conditional probability that a child with cancer is in population #1, and $p_{NC} = p_{12}/(p_{12} + p_{22})$ is the conditional probability that a noncancer child is in population #1. Let $q_C = 1 - p_C$, $q_{NC} = 1 - p_{NC}$, $\hat{p}_C = Y_1/n_1$, $\hat{q}_C = 1 - \hat{p}_C$, $\hat{p}_{NC} = Y_2/n_2$, and $\hat{q}_{NC} = 1 - \hat{p}_{NC}$. Then $\hat{p}_C$ and $\hat{p}_{NC}$ are independent unbiased estimators of p_C and p_{NC}. Notice that $R = p_C q_{NC}/p_{NC} q_C$. It is this feature of R that allows us to estimate R, despite the fact that sampling is retrospective, rather than prospective.

Define $\hat{R} = \hat{p}_C \hat{q}_{NC}/\hat{p}_{NC}\hat{q}_C = Y_1(n_2 - Y_2)/[Y_2(n_1 - Y_1)]$, and $\hat{\eta} = \log \hat{R} = \hat{\eta}_C - \hat{\eta}_{NC}$, where $\hat{\eta}_C = [\log \hat{p}_C - \log \hat{q}_C]$ and $\hat{\eta}_{NC} = [\log \hat{p}_{NC} - \log \hat{q}_{NC}]$. From Section 8.2 $\hat{\eta}_C$ is approximately distributed as $N(\eta_C \equiv \log(p_C/q_C),\ V_C \equiv (1/n_1 p_C q_C))$, and $\hat{\eta}_{NC}$ is approximately distributed as $N(\eta_{NC} \equiv \log(p_{NC}/q_{NC}),\ V_{NC} \equiv (1/n_2 p_{NC} q_{NC}))$, if n_1p_1, n_1q_1, n_2p_2, n_2q_2 are not too close to zero. An approximate $100\gamma\%$ confidence interval on R is therefore given by $(\hat{\eta}_L, \hat{\eta}_U) = (\hat{\eta} \pm z_{(1+\gamma)/2}\sqrt{\hat{V}_C + \hat{V}_{NC}})$, where $\hat{V}_C$ and $\hat{V}_{NC}$ are obtained by replacing the proportions by their estimates. Note that $\hat{V}_C = 1/Y_1 + 1/(n_1 - Y_1)$ and $\hat{V}_{NC} = 1/Y_2 + 1/(n_2 - Y_2)$. The corresponding confidence interval on R is $(e^{\hat{\eta}_L}, e^{\hat{\eta}_U})$.

For example, suppose we observe the table $\mathbf{Y} = (Y_{ij}) = \mathbf{y} = \begin{bmatrix} 47 & 34 \\ 353 & 466 \end{bmatrix}$. That is, we sampled 400 children with cancer, 47 lived near power lines, and we sampled 500 children who did not have cancer, and determined that 34 lived near power lines. Then, $\hat{p}_C = 0.1175$, $\hat{p}_{NC} = 0.068$, $\hat{R} = 1.8249$, $\hat{\eta} = 0.6015$, $\hat{\eta}_L = 0.1391$, $\hat{\eta}_U = 1.0639$ for $\gamma = 0.95$, so that $(1.1492, 2.8978)$ is a 95% confidence interval on R. There seems to be some relationship between residence near a power line and the incidence of cancer. We are *not* justified in saying that power lines *cause* cancer.

If we know the overall rate of cancer in these children is 0.2 per 1,000 per year (60 per 100,000 over three years), then we can estimate the two-way probability table (p_{ij}). The estimate of p_{11} is $\hat{p}_{11} = \hat{p}_C(0.0006) = 0.0000705$ (7 per 100,000). Similarly, we obtain,

$$\hat{p} = (\hat{p}_{ij}) = \frac{1}{100{,}000}\begin{bmatrix} 7.05 & 6{,}740 \\ 52.95 & 92{,}660 \end{bmatrix}.$$

If the overall rate were unknown, then we could not estimate the conditional probabilities p_1, p_2, or the unconditional probabilities p_{ij}.

Problem 8.3.1: (a) Let X have a binomial distribution with $n = 50$, and p. Suppose $X = 0$ is observed. Find an upper 99% confidence limit U on p.

(b) If $X = 0$, how large must n be in order to have upper 99% confidence limit less than 0.001? This sort of question is vital to the developers of vaccines, who fear recipients will acquire a disease from the vaccine, or automobile manufacturers, who must guarantee the safety of airbags.

(c) To be extra careful, perhaps the automobile maker should prepare for the event that $X = 1$ of the airbags fails. What should n be in order to have upper 99% confidence less than 0.001?

Problem 8.3.2: Suppose X has a binomial distribution with parameters $n = 30$ and p. If $X = 4$. Find a 90% confidence interval on p.

Problem 8.3.3: (a) Let X have a Poisson distribution with parameter λ. Suppose we observe $X = 10$. Find a 90% confidence interval on λ, using the exact method.

(b) Suppose $X = 383$. Find a 90% confidence interval on λ using three methods. (i) Use $Z = (X - \lambda)/\sqrt{\lambda}$ as a pivotal quantity. (ii) Use $\hat{Z} = (X - \lambda)/\sqrt{X}$ as a pivotal quantity. (iii) Use the fact that the chi-square distribution for v d.f. is close to the normal with mean v and variance $2v$.

Problem 8.3.4: The number of cases of lung cancer reported among 8,791 men of ages 50–59 living in a county in which a nuclear reactor was located over a three-year period was 81. During that same time period in other counties in that same state there were 62,547 men of ages 50–59, of which 483 were reported to have lung cancer.

(a) State a reasonable model.

(b) Give point estimates of the rates θ_1 and θ_2 per 1,000 such men per year for the county and for the other counties. Estimate the standard error of your estimators and use these to find a 90% confidence intervals on $\theta_1 - \theta_2$.

(c) Find 90% confidence interval on θ_1/θ_2 using the binomial method.

(d) Find a 90% confidence interval on $R = \theta_1/\theta_2$.

Problem 8.3.5: The following properties of the Poisson process are often established in introductory courses in probability. Suppose that events occur at random points in time $0 < X_1 < X_2 < \cdots$. Let $Y(0) \equiv 0$, and let $Y(t)$ be the number of occurrences in the time interval $(0, t\}$. $Y(t)$ is said to be a Poisson process with parameter $\lambda > 0$ if

(1) For each $t > 0$, $Y(t) \sim \mathscr{P}(\lambda t)$.

(2) The numbers of occurrences in nonoverlapping time intervals are independent r.v.'s.

Let $X_0 \equiv 0$, and $D_j = X_j - X_{j-1}$ for $j = 1, 2, \ldots$ Then $X_j = D_1 + \cdots + D_j$ and $Y(t) = \max\{n | X_n \le t\}$ for $t \ge 0$. The waiting times $D_1, D_2, \ldots$ are independent, each with an exponential distribution with mean $1/\lambda$. The process $\{Y(t), t \ge 0\}$ can therefore be simulated by first generating the waiting times D_j between events.

(a) Prove that $D_j \sim$ exponential, with mean $1/\lambda$. *Hint*: $P(D_j > d) = P$(no occurrences in an interval of length d).

(b) Use induction or moment-generating functions to prove that $X_j/\lambda \sim$ gamma, with scale parameter 1, power parameter j.

(c) Show that $U \sim$ gamma, with scale parameter 1, power parameter ν, for ν a positive integer, implies that $2U \sim \chi^2_{2\nu}$.

(d) For U as in (c), prove that $P(U > \lambda) = \sum_{j=0}^{\nu-1} p(j; \lambda)$, where $p(j; \lambda)$ is the Poisson probability function. (Either differentiate by parts on the left $\nu - 1$ times or use the relationships among U, X_ν, and $Y(\lambda)$ to rewrite $P(U > \lambda)$. The second method is more elegant.)

(e) Derive the formulas for the lower and upper confidence limits λ_1 and λ_2 on λ.

Problem 8.3.6: In order to investigate the effects of smoking on lung cancer, the files of the hospitals in a large metropolitan area were searched. It was found that 867 patients (all adults) had been diagnosed for lung cancer during the year 1990. From these 867, a random sample of 393 was chosen, of which 261 patients were found to have been smokers for at least 10 years in their lifetimes. Another random sample of 612 adults (the controls) was taken from among the residents of the area, using telephone directories. Among these, 197 were found to have been smokers according to the same definition.

(a) Find a 95% confidence interval on the odds-ratio R for cancer–smoking.

(b) The lung cancer rate in this area was known to be 1.2 per 1,000 adults per year. Estimate the probability table $p = (p_{ij})$ and the conditional probabilities for cancer among the smoking and nonsmoking populations. Give a 95% confidence interval on p_{11}.

(c) Suppose that the control population used was the collection of people who were admitted to one of these hospitals in 1990. Does that cause any problems in the interpretation?

Problem 8.3.7: The Doll and Hill (1950) study of 709 lung cancer patients and 709 patients without lung cancer in 20 London hospitals in 1948–49 was one the most important in determining government policy with respect to smoking. In that study only 69 were women smokers (at least once a day for a year), of whom 41 had cancer. Among 51 nonsmoking women, 19 had cancer. Give a 95% confidence interval on the odds-ratio for cancer among women, and state your conclusions. Sir Ronald Fisher warned strongly against the conclusion that smoking caused cancer, though $\hat{R}$ for men was even more extreme. Among other things he pointed out that inhaling seemed to result in lower rates of cancer. See "Smoking and Lung Cancer" in Fienberg and Hinckley (1980).

8.4 LOG-LINEAR MODELS

We will begin by considering some relatively simple log-linear models, delaying their analysis until Section 8.5. These models will be written in the vector space form. We will use such notation as log $\mathbf{Y}$ or $e^{\hat{\mu}}$ to mean that these functions operate componentwise, so that, for example, $\log(Y_1, Y_2) = (\log Y_1, \log Y_2)$. Differences between the theories for linear and log-linear models occur largely because (1) the log of the mean vector $\mathbf{m} = E(\mathbf{Y})$, rather than $\mathbf{m}$ itself, will be assumed to lie in a linear subspace, and (2) the distributional properties of $\mathbf{Y}$ are more complex for frequency data. Most of the difficulties imposed by (1) and (2) will be postponed to later sections.

There are interesting correspondences between the explanatory (also design, independent, or regressor) vectors $\mathbf{x}_j$ which we will choose and independence or conditional independence. We will wish to test for independence or conditional independence, but we shall also wish to measure the strength of the dependencies which do occur. These measures of dependence will usually be odds, odds-ratios, or log odds-ratios.

Example 8.4.1: In order to determine the effect of the length of traffic light cycle on the accident rate at an intersection, four cycle lengths were used, each for one year. The numbers of accidents were

Year	1	2	4	4
Cycle length (s)	40	50	60	70
Number of accidents	149	129	112	112

Assume for simplicity that the traffic each year is approximately the same. It seems reasonable to suppose that the number Y_i of accidents in year i has a Poisson distribution with mean m_i, and that Y_1, Y_2, Y_3, Y_4 are independent. Can we find a simpler model? Longer cycle times might be expected to decrease the numbers of accidents, with m decreasing with increasing cycle time t. Suppose that $m = m(t) = \exp(\beta_0 + \beta_1 t)$, or equivalently, $\mu = \mu(t) \equiv \log m(t) = \beta_0 + \beta_1 t$. Writing $\mathbf{t} = (40, 50, 60, 70)$, $\mathbf{J} = (1, 1, 1, 1)$, $\mathbf{Y} = (Y_1, Y_2, Y_3, Y_4)$, $\boldsymbol{\mu} = \beta_0 \mathbf{J} + \beta_1 \mathbf{t}$, we can state the model as follows: $\mathbf{Y}$ satisfies the independent Poisson model with $\mathbf{m} \equiv E(\mathbf{Y}) = \exp(\boldsymbol{\mu})$, $\boldsymbol{\mu} \in V = \mathscr{L}(\mathbf{J}, \mathbf{t})$. Increasing the cycle time by d will multiply the accident rate by the factor $e^{\beta_1 d}$. If we decide that β_1 is positive or only slightly negative, we might wish to keep t near 40 or even less. More negative values of β_1 suggest that we should be willing to put up with some of the inconveniences of longer cycles in the interest of safety.

Actually these frequencies were generated by a computer with $\beta_0 = 8$, $\beta_1 = -0.8$, so that $\mathbf{m} = (155.9, 130.4, 112.7, 99.6)$.

Example 8.4.2: In order to investigate the effect of a poison on rats, the poison was fed to the rats in four different dosages: $0 < d_1 < d_2 < d_3 < d_4$. The numbers of rats and the numbers dying at these dosages were

Dosage	d_1	d_2	d_3	d_4
Log-dosage	$x_1 = 0.5$	$x_2 = 1$	$x_3 = 1.5$	$x_4 = 4$
Number of rats	15	17	19	16
Number dying	2	6	11	13

Let

$$Y_{ij} = \begin{cases} \#\text{ living when the dosage is } d_j \text{ and } i = 1 \\ \#\text{ dying when the dosage is } d_j \text{ and } i = 2. \end{cases}$$

$\mathbf{Y} = (Y_{ij})$, the 2×4 table of observed frequencies. Suppose that $(Y_{1j}, Y_{2j}) \sim \mathscr{M}_2(\mathbf{p}_j, n_j)$ and that these columns of $\mathbf{Y}$ are independent, where $\mathbf{p}_j = (1 - p_j, p_j)$ for $j = 1, 2, 3, 4$ and $n_1 = 15$, $n_2 = 17$, $n_3 = 19$, $n_4 = 16$. Thus, $\mathbf{Y}$ satisfies the independent multinomial model. Of course, it is equivalent to say that $(Y_{21}, Y_{22}, Y_{23}, Y_{24})$ are independent with $Y_{2j} \sim \mathscr{B}(n_j, p_j)$, and $Y_{1j} = n_j - Y_{2j}$. Define $\mathbf{m} = E(\mathbf{Y}) = (m_{ij})$, where $m_{ij} = n_j(1 - p_j)$ for $i = 1$ and $m_{ij} = n_j p_j$ for $i = 2$. Let $\boldsymbol{\mu} = \log \mathbf{m} = (\mu_{ij})$. Suppose that the log-odds for death under dosage j is γx_j, for $x_j = \log(d_j)$ for each j. The odds for death under dosage d_j are $\exp(\gamma x_j) = d_j^\gamma$. Positive values of γ correspond to increasing probability of dying with increasing dosage. Solving for p_j, we get $p_j = \dfrac{e^{\gamma x_j}}{1 + e^{\gamma x_j}} = \dfrac{d_j^\gamma}{1 + d_j^\gamma}$. Notice that when $x_j = 0$, equivalently when $d_j = 1$, it follows that $p_j = 1/2$. This model forces the probability of death at dosage 1.0 to be 1/2.

Let $\mathbf{J}_j$ be the 2×4 indicator of column j, and let $\mathbf{w}$ be the array with zeros in the first row, and x_j as the jth term in the second row. Then $\log \mathbf{m} = \boldsymbol{\mu} = \sum_{j=1}^{4} \mu_{1j}\mathbf{J}_j + \gamma \mathbf{w}$. Thus, $\mathbf{m}$ satisfies the log-linear model. However, not all vectors $\boldsymbol{\mu} \in \mathscr{L}(\mathbf{J}_1, \ldots, \mathbf{J}_4, \mathbf{w}) \equiv V$ are possible. In fact, the requirement that the column sums of $\mathrm{M} = E(\mathbf{Y})$ be the constants n_j determines the restrictions $e^{\mu_{1j}}(1 + e^{\gamma x_j}) = n_j$, so that $\mu_{1j} = \log[n_j/(1 + e^{\gamma x_j})]$. Since we are interested in γ, rather than the μ_j, we can reduce the dimensionality of the model by concentrating on the *logits* $L_j = \log[p_j/(1 - p_j)] = \gamma x_j$. The two statements

$\boldsymbol{\mu} \in V$, and $L_j = \gamma x_j$ for each j are equivalent, but the second statement, called a *logit model*, seems to be simpler.

This model requires that the death rate for $d = 0$ be zero. We could relax this by taking $L_j = \gamma_0 + \gamma_1 x_j$, or, equivalently, adding the indicator $\mathbf{R}_2$ of the second row to the subspace V.

Example 8.4.3: Consider the index set $\mathscr{I} = \{1, 2, 3\} \times \{1, 2, 3, 4\}$ for 3×4 tables of frequencies. Suppose that Y_{ij} is the observed frequency in cell ij, and let $\mathbf{Y} = (Y_{ij})$. Suppose that $\mathbf{Y}$ has a Poisson distribution with parameter $\mathbf{m} = (m_{ij})$. With no further restrictions on $\mathbf{m}$, this is the *saturated model* because the model allows the estimates $\hat{m}_{ij} = Y_{ij}$, so that the model fits with no residuals. We can state the model in its vector space form by defining $\mathbf{C}_{ij}$ to be the indicator of cell ij. Then $\boldsymbol{\mu} = \log \mathbf{m} = \sum_{ij} \mu_{ij}\mathbf{C}_{ij} \in V = \mathscr{L}(\mathbf{C}_{11}, \ldots, \mathbf{C}_{34})$, a 12-dimensional subspace of 12-space.

We should always seek simpler models, for which the subspace V has smaller dimension. Let $\mathbf{R}_i$ and $\mathbf{C}_j$ be the indicators of the ith row and jth column. One such model supposes instead that $\boldsymbol{\mu} \in \mathscr{L}(\mathbf{R}_1, \mathbf{R}_2, \mathbf{R}_3, \mathbf{C}_1, \mathbf{C}_2, \mathbf{C}_3, \mathbf{C}_4) = V_6$, a 6-dimensional subspace of 12-space. This model implies that there exist parameters ρ_i and γ_j such that $\mu_{ij} = \rho_i + \gamma_j$ and $m_{ij} = e^{\rho_i}e^{\gamma_j}$, so that the Poisson parameters m_{ij} satisfy a multiplicative model.

If we replace the Poisson model by $\mathbf{Y} \sim \mathscr{M}_{12}(n, \mathbf{p})$, then the mean vector $\mathbf{M} = n\mathbf{p}$ and the observation vector $\mathbf{Y}$ must have inner product n with the vector $\mathbf{J}$ of all ones. For both the saturated and multiplicative models $\boldsymbol{\mu}$ may take only those values in V for which $(e^{\boldsymbol{\mu}}, \mathbf{J}) = n$. If we begin with the independent Poisson model, but condition on $\sum_{ij} Y_{ij} = (\mathbf{Y}, \mathbf{J}) = n$, then, conditionally, $\mathbf{Y} \sim \mathscr{M}_{12}(n, \mathbf{p})$, with $p_{ij} = \lambda_{ij}/\sum_{ij} \lambda_{ij}$ (see Theorem 8.2.2). The multiplicative model $\boldsymbol{\mu} \in V_6$ is equivalent to $p_{ij} = p_{i\cdot}\, p_{\cdot j}$, the independence of the row and column factors.

By expanding the μ_{ij} as we did in two-way analysis of variance, we can more systematically study two-way tables. Define $\mu = \frac{1}{12}\sum_{ij} \mu_{ij}$, $\mu_{i\cdot} = \frac{1}{4}\sum_j \mu_{ij}$, $\bar{\mu}_{\cdot j} = \frac{1}{3}\sum_i \mu_{ij}$, $\alpha_i = \bar{\mu}_{i\cdot} - \mu$, $\beta_j = \bar{\mu}_{\cdot j} - \mu$, $(\alpha\beta)_{ij} = \mu_{ij} - [\mu + \alpha_i + \beta_j]$. Then $\mu_{ij} = \mu + \alpha_i + \beta_j + (\alpha\beta)_{ij}$, and the parameters α_i, β_j, $(\alpha\beta)_{ij}$ satisfy the familiar zero-sum restrictions of the analysis of variance. The statement that $(\alpha\beta)_{ij} = 0$ for all i and j is equivalent to the multiplicative model for Poisson $\mathbf{Y}$ or independence of row and column effects for the multinomial model.

In the case that one of the factors has ordered levels, it may be possible to find a model which has interaction effects, but is still smaller than the saturated model. Suppose, for example, that a random sample of 400 adults was chosen from the telephone subscribers in Frequency City. Those sampled were asked their view on a law before Congress which would increase social security (SS) benefits. Their choices were (1) favor, (2) neutral, (3) against. They were classified by age: (1) 18–35, (2) 36–50, (3) 51–65, (4) 66–99. The results were

		Age 1	2	3	4	
View	Agree	27	39	47	58	171
on SS	Neutral	44	25	20	10	99
Bill	Against	56	48	18	8	130
		127	112	85	76	400

It is reasonable to expect that as people age their view towards increases in SS benefits should become increasingly favorable. We can quantify this by replacing the interaction term $(\alpha\beta)_{ij}$ (which might better be called $(as)_{ij}$ or $(\alpha\sigma)_{ij}$ for this example) by a multiplicative term $\gamma(i-2)(j-2.5) = \gamma w_{ij}$, chosen so that the vector $\mathbf{w} = (w_{ij})$ is orthogonal to the row and column indicators. Since we expect frequencies to be higher for small i and large j, and for large i, small j, we should expect to obtain an estimate $\gamma < 0$. The model can now be written as $\boldsymbol{\mu} \in V_7 = V_6 \oplus \mathscr{L}(\mathbf{w})$, a subspace of dimension 7. We will later develop means of fitting this and the other models, and discuss measures of their goodness-of-fit.

We will be interested in odds-ratios and log odds-ratios:

$$R(i_1, i_2, j_1, j_2) = [m_{i_1j_1}/m_{i_1j_2}]/[m_{i_2j_1}/m_{i_2j_2}] = (m_{i_1j_1}m_{i_2j_2})/(m_{i_1j_2}m_{i_2j_1})$$

and

$$L(i_1, i_2, j_1, j_2) = \log R(i_1, i_2, j_1, j_2) = \mu_{i_1j_1} - \mu_{i_1j_2} - \mu_{i_2j_1} - \mu_{i_2j_1} + \mu_{i_2j_2}.$$

L is the inner product of $\boldsymbol{\mu}$ with the vector $\mathbf{v}$ having ones at indices (i_1, j_1) and (i_2, j_2), minus ones at indices (i_1, j_2) and (i_2, j_1). The vector $\mathbf{v} \in V_{\alpha\beta}$, the interaction subspace, and the collection of all such vectors corresponding to all possible choices of $i_1 \neq i_2$ and $j_1 \neq j_2$ span $V_{\alpha\beta}$. For the independence model L is zero for all choices of the indices. That is, independence is equivalent to $\boldsymbol{\mu} \perp V_{\alpha\beta}$. If the interaction term is γw_{ij}, L reduces to $\gamma(i_1 - i_2)(j_1 - j_2)$. For example, for the four extreme corners of the table $L(1, 3, 1, 4) = 6\gamma$, which can be expected to be quite negative, corresponding to a small odds-ratio

$$[P(\text{Favor}|\text{Young})/P(\text{Oppose}|\text{Young})]/[P(\text{Favor}|\text{Elderly})/P(\text{Oppose}|\text{Elderly})].$$

Example 8.4.4: Consider the Table 8.4.1 of frequencies and percentages, taken originally from National Opinion Research Center, 1975 General Social Survey, University of Chicago, excerpted from Haberman (1978, p. 183).

Table 8.4.1 Subjects in the 1975 General Social Survey, Cross-Classified by Attitude Toward Women Staying Home, Sex of Respondent, and Education of Respondent

Respondent		Agree		Disagree		
Sex	Education (Years)	No.	Percent.	No.	Percent.	Total
Male	≤ 8	72	60.5	47	39.5	119
	9–12	110	35.9	196	64.1	306
	≥ 13	44	19.7	179	80.3	223
	Total	226	34.9	422	65.1	648
Female	≤ 8	86	69.4	38	30.6	124
	9–12	173	37.9	283	62.1	456
	≥ 13	28	13.0	187	87.0	215
	Total	287	36.1	508	63.9	795
Total	≤ 8	158	65.0	85	35.0	243
	9–12	283	37.1	479	62.9	762
	≥ 13	72	16.4	366	83.6	438
	Total	513	35.6	930	64.4	1,443

Subjects were asked the question, "Do you agree with this statement—Women should take care of running their homes and leave running the country up to men?"

This is a three-way table, with three categorical variables: sex at two levels, education at three levels, and response at two levels. Sampling was done by choosing independent random samples of 648 men and 795 women, then determining their ages and responses. Let Y_{ijk} be the frequency observed for sex level i, education level j, response level k. Then the index set is $\mathscr{I} = \{1, 2\} \times \{1, 2, 3\} \times \{1, 2\}$. The observation vector $\mathbf{Y} = (Y_{ijk})$ is made up of the two random vectors $\mathbf{Y}_1 = (Y_{1jk})$ for men and $\mathbf{Y}_2 = (Y_{2jk})$ for women. A reasonable model is: $\mathbf{Y}_1$, $\mathbf{Y}_2$ are independent with $\mathbf{Y}_i \sim \mathscr{M}_6(n_i, \mathbf{p}_i)$, for $i = 1, 2$, $n_1 = 648$, $n_2 = 795$. Then $\mathbf{m} = E(\mathbf{Y}) = E(\mathbf{Y}_1, \mathbf{Y}_2) = (n_1\mathbf{p}_1, n_2\mathbf{p}_2)$.

We would like to find a simple model for $\mathbf{m}$. Let $\boldsymbol{\mu} = \log \mathbf{m} = (\mu_{ijk})$. As for the three-way analysis of variance we can write μ_{ijk} as the sum of its effects:

$$\mu_{ijk} = \mu + s_i + e_j + r_k + (se)_{ij} + (sr)_{ik} + (er)_{jk} + (ser)_{ijk}. \qquad (8.4.1)$$

As for the ANOVA, the 12-dimensional sample space V can be broken into the mutually orthogonal subspaces V_0, V_s, V_e, V_r, V_{se}, V_{sr}, V_{er}, V_{ser}. Because of the restriction that $\sum_{jk} m_{ijk} = n_i$, $\boldsymbol{\mu}$ cannot take all possible values in V. In fact, μ and s_1, $s_2 = 1 - s_1$ are uniquely determined by the other parameters and these two linear restrictions on the m_{ijk}.

We have chosen to use symbols s_i, e_j, r_k, etc., which remind us of the meaning of these variables. It is common in the literature of log-linear models, to use the symbol λ_{ijk}, rather than μ_{ijk}, and write $\lambda_{ijk} = \lambda + \lambda_i^1 + \lambda_j^2 + \lambda_k^3 + \lambda_{ij}^{12} + \lambda_{ik}^{13} + \lambda_{jk}^{23} + \lambda_{ijk}^{123}$, where the meanings are the same as for the corresponding symbols in (8.4.1).

With all terms of the representation of μ_{ijk} present the model is saturated. We would like to find a simpler model. The model with the three-way interaction term $(ser)_{ijk}$ missing is at least a little bit simpler. This model is often indicated in shorthand form as (1 2 3 12 13 23), corresponding to the three main effects and the three two-way interactions. It is easy to show that the log odds-ratios for men and for women in the saturated model are

$$\begin{aligned} L_i(j_1, j_2, k_1, k_2) &\equiv \log[(m_{ij_1k_1}/m_{ij_1k_2})/(m_{ij_2k_1}/m_{ij_2k_2})] \\ &= [(er)_{j_1k_1} - (er)_{j_1k_2} - (er)_{j_2k_1} + (er)_{j_2k_2}] \\ &\quad + [(ser)_{ij_1k_1} - (ser)_{ij_1k_2} - (ser)_{ij_2k_1} + (ser)_{ij_2k_2}]. \end{aligned}$$

For this example (k_1, k_2) may be taken to be (1, 2), since response has only two levels. The more general notation is used so that the ideas may be generalized to factors with more than two levels. These log-odds ratios are the same for men and women if and only if the three-way interaction terms are all zero. If this were the case then the interrelationship between education and response, as measured by odds-ratios, are the same for men and women. By the symmetry of the roles of the indices, we could also conclude that the interrelationship between sex and response is the same for each level of education.

The difference $D = D(j_1, j_2, k_1, k_2) = L_1(j_1, j_2, k_1, k_2) - L_2(j_1, j_2, k_1, k_2)$ is zero if and only if the two corresponding odds ratios are equal. D is the inner product of $\boldsymbol{\mu}$ with the vector $\mathbf{v}$ of ones and minus ones corresponding to the indices. The vector $\mathbf{v} \in V_{ser}$, the three-way interaction subspace, and the collection of all such vectors, for all choices of subscripts, span V_{ser}. Thus, equality of the odds-ratios for men and women is equivalent to $\boldsymbol{\mu} \perp V_{ser}$. For these data $\hat{L}_1(1, 2, 1, 2) = \log(72 \times 196)/(110 \times 47) = 1.00415$ and $\hat{L}_2(1, 2, 1, 2) = \log(86 \times 283)/(173 \times 38) = 1.30892$, so that $D(1, 2, 1, 2) = -0.30477$. Similarly, we find $\hat{D}(1, 3, 1, 2) = 1.82971 - 2.71567 = -0.88596$. We will have to decide later whether these is too far from zero for us too discard the three-way interaction term in the model.

Both the $(ser)_{ijk}$ and the $(er)_{jk}$ terms are missing (corresponding to the (1 2 3 12 13) model if and only if the log odds-ratios $L_i(j_1, j_2, k_1, k_2)$ are both zero, equivalently the odds-ratios $R_i(j_1, j_2, k_1, k_2)$ are all one. This, in turn, is equivalent to conditional independence of education and response, separately for men and for women. The estimates from the data given above indicate that this model surely would be a poor fit. The model with the terms $(ser)_{ijk}$ and $(sr)_{ik}$ missing, called the (1 2 3 12 23) model, would therefore correspond to conditional independence of sex and response for each level of education. Perusal of the data indicates that this model may fit well.

Let $R_{ij}(k_1, k_2) = m_{ijk_1}/m_{ijk_2} = p_{ijk_1}/p_{ijk_2}$ be the odds for level k_1 of the response factor for a given combination ij of the levels of sex and education. Let

$$\log R_{ij}(k_1, k_2) \equiv L_{ij}(k_1, k_2)$$
$$= r_{k_1} - r_{k_2} + (sr)_{ik_1} - (sr)_{ik_2} + (er)_{jk_1} - (er)_{jk_2} + (ser)_{ijk_1} - (ser)_{ijk_2}.$$

The functions L_{ij} are the same for all ij if and only if the interaction terms $(ser)_{ijk}$, $(er)_{jk}$, and $(sr)_{ik}$ are zero for all i, j, k. But $L_{i'j'}(k_1, k_2) = L_{ij}(k_1, k_2)$ for all i, i', j, j', k_1, k_2 corresponds to independence of factor 3 from the combination of factors 1, 2. For our example that would mean sex and education do not affect the probability of agreement, obviously not the case. Similarly, absence of the terms $(ser)_{ijk}$, $(sr)_{ik}$, and $(se)_{ij}$, the model (1 2 3 23) corresponds to independence of the sex factor from the combination of education and response. Since sampling was done independently for men and women, it would be better to say that the vectors $\mathbf{p}_i$ (3 × 2 arrays) are identical. Had sampling been done instead by taking one random sample of 1,443 people, with 648 turning out to be the number of men, then $\mathbf{p}_i$ would represent the conditional probability vector for the categories of education and response, given level i of the sex factor.

Absence of all interaction terms is equivalent to $\boldsymbol{\mu} \in V_0 \oplus V_s \oplus V_e \oplus V_r$, to $\boldsymbol{\mu} \perp (V_{se} \oplus V_{sr} \oplus V_{er} \oplus V_{ser})$, and to the representation of m_{ijk} as a product $f_i g_j h_k$. In the case of independent sampling for men and women this means $\mathbf{p}_1 = \mathbf{p}_2$ and independence of the factors education and response. With respect to the one multinomial model, the absence of any interaction terms implies independence of all three factors.

Complete absence of a subscript, say j, in the model, implies that the conclusions of the preceding paragraph hold, plus the equality of expectations and probabilities with respect to the levels of j, education. The same proportion of the population would have to belong to each of the three levels of education.

Table 8.4.1 summarizes the relationships among the terms in the log-linear model and independence or conditional independence in a three-way table. Suppose that the log of the expected frequency in cell ijk is

$$\mu_{ijk} = \log m_{ijk} = \lambda + \lambda_i^1 + \lambda_j^2 + \lambda_k^3 + \lambda_{ij}^{12} + \lambda_{ik}^{13} + \lambda_{jk}^{23} + \lambda_{ijk}^{123} \quad \text{for} \begin{cases} i = 1, \dots, I \\ j = 1, \dots, J \\ k = 1, \dots, K \end{cases}$$

This is the unique representation of μ_{ijk} as the sum of terms, each of which sums to zero over any one of its subscripts. We will use the notation [123] = 0 to mean that all the terms λ_{ijk}^{123} are zero. Similarly, {[123] = 0, [23] = 0} means that all the terms $\lambda_{ijk}^{123} = 0$, $\lambda_{jk}^{23} = 0$. In Table 8.4.1 each row corresponds to a set of model terms which are zero, as indicated in the first column. Equivalently, each row corresponds to the statement that $\boldsymbol{\mu}$ lies in a certain subspace, with subspaces becoming smaller as additional terms become zero. These are the same subspaces defined in Chapter 6 for the three-way analysis of variance. The second column gives the equivalent statement in terms of odds ratios, which hold for any selection of subscripts. We define

$$R_i(j, j', k, k') = [m_{ijk}/m_{ijk'}]/[m_{ij'k}/m_{ij'k'}]$$

and

$$L_i(j, j', k, k') = \log R_i(j, j', k, k').$$

For all but the first row these smaller models produce representations of m_{ijk} in terms of sums of m_{ijk} across one or more subscripts. Replacement of a subscript by a "+" means that the subscript has been summed over. Thus, $m_{i+k} = \sum_j m_{ijk}$ and $m_{+j+} = \sum_{ik} m_{ijk}$. Column 4 gives the interpretation of the model in terms of independence or conditional independence. Equivalent statements for models not considered in Table 8.4.2 may be found by interchanging subscripts. Let us prove the statements of the second row of the table. Others are left to students. $[123] = 0$, $[23] = 0$ is equivalent to $\mu_{ijk} = \lambda + \lambda_i^1 + \lambda_j^2 + \lambda_k^3 + \lambda_{ij}^{12} + \lambda_{ik}^{13}$. Computation gives $L_i(j, j', k, k') = \lambda_{ijk} - \lambda_{ij'k} - \lambda_{ijk'} + \lambda_{ij'k'} = (\boldsymbol{\mu}, \mathbf{x})$, where $\mathbf{x} = \mathbf{C}_{ijk} - \mathbf{C}_{ij'k} - \mathbf{C}_{ijk'}\mathbf{C}_{ij'k'}$, and $\mathbf{C}_{ijk}$ is the indicator of cell ijk. If $V = V_0 \oplus V_1 \oplus V_2 \oplus V_3 \oplus V_{12} \oplus V_{13}$ is the subspace in which $\boldsymbol{\mu}$ lies under this model then such vectors $\mathbf{x}$, for all choices of i, j, j', k, k' span $V^\perp$. Applying the function $\exp(\cdot)$ on each side of $L_i(j, j', k, k') \equiv 0$, we get $R_i(j, j', k, k') \equiv 1$. This establishes the equivalence between the first and second columns. In fact, each of the statements in column 2 is simply a translation of the statement $\boldsymbol{\mu} \perp V^\perp$, where V is the subspace in which $\boldsymbol{\mu}$ lies under the model.

To get the representation given in the columns, write $m_{ijk}m_{ij'k'} = m_{ij'k}m_{ijk'}$. Summing across both j' and k', we get $m_{ijk}m_{i++} = m_{ij+}m_{i+j}$. This representation implies the identity of column 2, so the statements of the first three columns are equivalent. To demonstrate the interpretation of the column 4, let $p_{ijk} = m_{ijk}/m_{+++}$. Then conditional independence of the second and third factors, given the level of the first means that

$$p_{ijk}/p_{i++} = [p_{ij+}/p_{i++}][p_{i+k}/p_{i++}],$$

equivalent to $m_{ijk}m_{i++} = m_{ij+}m_{i+k}$, which is the identity of column 3.

Example 8.4.5: Consider Example 8.1.5 again, which presents frequencies of admission to graduate school for men and women for two fictitious departments. Supposing an equal distribution of credentials for men and women, is there discrimination against women? When the admission rates are the same for men and women in each department, why is the admission rate lower for women in the university? The answer of course is that women applied in larger numbers to the department which admits a smaller percentage of students. This tended to be the case at Berkeley, with men tending to apply to departments which are more technical. The higher admission rates in more

Table 8.4.2 Relationships Among λ-Terms and Means m_{ijk} in Three-Way Contingency Tables

Model Terms Set to Zero	Corresponding Equalities for for Odds Ratios	Equivalent Expression for m_{ijk}	Interpretation
[123]	$\frac{R_i(j,j',k,k')}{R_{i'}(j,j',k,k')} = 1$		None
[123], [23]	$R_i(j,j',k,k') = 1$	$[m_{ij+}m_{i+k}]/m_{i++}$	Independence of factors 2, 3, conditionally on levels of factor 1
[123], [23], [13]	$\frac{m_{ijk}/m_{ijk'}}{m_{i'j'k}/m_{i'j'k'}} = 1$	$\frac{m_{++k}m_{ij+}}{m_{+++}}$	Independence of factor 3 and combination of factors 1 and 2
[123], [23], [13], [12]	$\frac{m_{ijk}/m_{ijk'}}{m_{i'j'k}/m_{i'j'k'}} = 1$ $\frac{m_{ijk}/m_{ijk'}}{m_{i'jk'}/m_{i'j'k'}} = 1$	$\frac{m_{i++}m_{+j+}m_{++k}}{m_{+++}^2}$	Independence of factors 1, 2, 3
All above plus [3]	$m_{ijk}/m_{ijk'} = 1$ $\frac{m_{ijk}/m_{ij'k}}{m_{ijk}/m_{i'j'k}} = 1$	$K\frac{m_{+j+}m_{i++}}{m_{+++}^2}$	Factor 3 has no effect and factors 2 and 3 are independent
[123], [23], [13], [12], [3], [2]	$m_{ijk}/m_{ijk'} = 1$ and $m_{ijk}/m_{ij'k} = 1$	$\frac{m_{i++}}{JK}$	Factors 2 and 3 have no effect
All terms except λ	$m_{ijk} = \lambda$ or $m_{ijk}/m_{i'j'k'} = 1$	m_{+++}/IJK	None of the factors have an effect

technical departments seems either to indicate that such departments take the view that students should have the "right to fail," or that only students with high ability in those subjects apply to such departments. What does this example say about two- and three-way contingency tables?

This is an example of *Simpson's Paradox* (Simpson 1951). Let M, W, A, D_1 and D_2 be the events that a person is a man, a woman, admitted, applies to Dept. #1, and to Dept. #2, respectively. In our example $m_1 \equiv P(A|MD_1) = P(A|WD_1) \equiv w_1$ and $m_2 \equiv P(A|MD_2) = P(A|WD_2) \equiv w_2$, but $m \equiv P(A|M) > P(A|W) \equiv w$. Since $m = m_1P(D_1|M) + m_2P(D_2|M) = w_1P(D_1|M) + w_2P(D_2|M)$ and $w = w_1P(D_1|W) + w_2P(D_2|W)$,

$$m - w = w_1[P(D_1|M) - P(D_1|W)] + w_2[P(D_1|W) - P(D_1|M)]$$
$$= (w_1 - w_2)[P(D_1|M) - P(D_1|W)].$$

The second term before the last equality follows because $P(D_2|M) = 1 - P(D_1|M)$, and $P(D_2|W) = 1 - P(D_1|W)$. Since both factors of the last term are positive, $m > w$. Men simply applied to the department with the higher admission rate.

This is also an example of the danger in *collapsing tables.* A table of frequencies is collapsed across a factor F if frequencies for each category of the other variables are added across all levels of F. The relationships among the other variables in the collapsed table, as measured by odds-ratios may change completely, as they did for the Berkeley admission data. A factor, say 1, in a three-way table with factors 1, 2, and 3 is said to be *collapsible* with respect to the 23 interaction term if the 23 interaction term in the collapsed two-way table (determined by summing across factor 1) is the same as it was for the three-way table. In general, this will hold for either of the models (1 2 3 13 23), which is conditional independence of factors 1 and 2, given factor 3, or (1 2 3 12 23), which is conditional independence of factor 1 and 3, given factor 2. This can be verified by computing log odds-ratios for the collapsed table for these models (see Problem 8.4.3). For the Berkeley data, we have conditional independence of the factors sex and admission given department, so we can collapse across sex, while still preserving the interaction term for department with admission, or we can collapse across admission, while preserving the interaction term for department with sex. We cannot collapse with respect to department without changing the interaction term for sex with admission, and it is this term in which we are interested.

In general, the lesson is that tables are to be collapsed with great care. Students may recall the height—reading score example used to demonstrate the need for a partial correlation coefficient in Section 3.7. Age was said to be a lurking variable. In this example, department is the lurking variable, and we should study the odds-ratios for the separate departments, rather than the odds-ratios for the collapsed table, the frequency table for the entire university.

Problem 8.4.1: Prove the implications of the third row of Table 8.4.1.

Problem 8.4.2: Make up some data for the three-way table of Example 8.4.5 so that there seems to be bias against females within each department, but, when the tables are collapsed across departments, there seems to be bias against men.

Problem 8.4.3: (a) Prove that a three-way table can be collapsed across factor 3, with the interaction terms λ_{ij}^{12} preserved, if the model of the second line of Table 8.4.1 holds, conditional independence of factors 2 and 3, given the levels of factor 1. Suppose that, instead, the model (1 2 3 12 23) holds (indicated by [123] = [13] = 0). Does this also imply that the terms λ_{ij}^{12} are preserved by collapsing across factor 3?

(b) Give an example of a $2 \times 2 \times 2$ table which is collapsible across factor 1, but does not satisfy the model (1 2 3 13 23) or the model (1 2 3 12 23).

Hint: To make things easier let $\lambda = \lambda_i^1 = \lambda_j^2 = \lambda_k^3 = 0$. This reduces the model (1 2 3 13 23) so that it can be expressed in terms of just two parameters.

Problem 8.4.4: Consider four tennis players—Abe, Bob, Carl, and Dan—numbered 1, 2, 3, 4 for simplicity. Suppose that there exist numbers λ_1, λ_2, λ_3, and λ_4, the strengths of these four players, so that the probability that player i beats player j is p_{ij}, where $\mu_{ij} = \log[p_{ij}/(1 - p_{ij})] = \lambda_i - \lambda_j$ for each i and j. Each pair of players plays one set on five different occasions, so that a total of 30 sets are played, with the outcomes of different sets being independent. Let Y_{ij} be the number of times that player i beats j, and let

$$\mathbf{Y} = \begin{bmatrix} - & Y_{12} & Y_{13} & Y_{14} \\ Y_{21} & - & Y_{23} & Y_{24} \\ Y_{31} & Y_{32} & - & Y_{34} \\ Y_{41} & Y_{42} & Y_{43} & - \end{bmatrix}$$

(a) Write this as a log-linear model. What is the dimension of the subspace V? Note that $Y_{ij} = 5 - Y_{ji}$.

(b) Suppose that an expert has determined that $p_{ij} = w_i^\beta/(w_i^\beta + w_j^\beta)$, for $w_i = 5 - i$, though the expert does not know what β should be. Write this as a log-linear model. (The author has applied this model with reasonable success to analyze the records of college basketball teams playing in the National Collegiate Athletic Association Tournament each year. In that case w_i was $17 - s_i$, where s_i was the *seed* of a team. In each of four regions, one team receives each possible seed number $j = 1, \ldots, 16$. Over nine seasons and 567 games the best estimate of β seems to be about 1.34.)

Problem 8.4.5: Consider a three-way model with three factors, 1 at two levels, 2 at three levels, and 3 at four levels.

(a) Which model corresponds to conditional independence of factors 1 and 3, for each level of factor 2?

(b) For which model are factors 1 and 3 jointly independent of factor 2?

(c) Give two $2 \times 3 \times 4$ tables $\mathbf{x}_1$ and $\mathbf{x}_2$, consisting only of -1's, 0's and 1's, which span the interaction space V_{12}, corresponding to the terms λ_{ij}^{12}. Express λ_{11}^{12}, λ_{12}^{12}, and λ_{13}^{12} in terms of $(\boldsymbol{\mu}, \mathbf{x})$ and $(\boldsymbol{\mu}, \mathbf{x}_2)$.

Problem 8.4.6: Let $\mathbf{Y}$ be a $k \times k$ table. The following discussion is particularly useful in the situation in which row and column classifications are the same, though that need not be the case. We might, for example, classify 1,000 father–son pairs, drawn at random from the population in which the sons graduate from the high schools of a large city in 1980. The education of father and son might be classified into E_1, E_2, E_3, E_4, E_5, where a person in E_j has more education than a person in E_i if $i < j$. Obviously the education of fathers and their sons are not independent. The saturated model has k^2 parameters. We would like to find a model in which the number of parameters is smaller. Suppose $\mu_{ij} = \log m_{ij} = \lambda + \lambda_i^1 + \lambda_j^2 + \lambda_{ij}$.

As will be evident the definitions and relationships to be demonstrated in this problem are applicable any time the classifications of the rows and columns and rows of a two-way table are the same. Examples: (1) The members of a panel of people are asked their opinions on some issue at two points in time; (2) matched pairs, say husbands and wives, each classified by religion; (3) people, animals, or things are paired so that they might be expected to produce similar results when treatments are applied. One member of each pair is chosen randomly to receive treatment #1, the other to receive treatment #2. All pairs are then classified according to the reactions of their members, rows corresponding to the member receiving treatment #1, columns to the other.

(a) The table **m** of expected frequencies is said to be *symmetric* if $m_{ij} = m_{ji}$ (or $\mu_{ij} = \mu_{ji}$) for all i and j. Show that **m** is symmetric if and only if $\lambda_i^1 = \lambda_i^2$ and $\lambda_{ij} = \lambda_{ji}$ for all i and j. Let V_s be the collection of vectors $\boldsymbol{\mu}$ corresponding to symmetric tables **m**.

(b) Let $\mathbf{B}_{ij}$ for $i < j$ be the indicator of the pair of cells (i, j) and (j, i) and let $\mathbf{D}_j$ be the indicator of cell (j, j). Express a symmetric table $\boldsymbol{\mu}$ as a linear combination of the $\mathbf{D}_j$, and $\mathbf{B}_{ij}$.

(c) The table **m** is *quasi-symmetric* if $\lambda_{ij} = \lambda_{ji}$ for all $i < j$. Let V_{qs} be the collection of all vectors $\boldsymbol{\mu}$ corresponding to quasi-symmetric tables **m**. Give an example of a quasi-symmetric table which is not symmetric.

(d) Let $\mathbf{R}_i$ and $\mathbf{C}_j$ be row and column indicators. Show that

$$V_{qs} = \mathscr{L}(\mathbf{D}_1, \ldots, \mathbf{D}_k, \mathbf{R}_1, \ldots, \mathbf{R}_k, \mathbf{B}_{12}, \ldots, \mathbf{B}_{k-1,k}).$$

What is $\dim(V_{qs})$? *Hint*: Let the subspace on the right be V^*. First show that a vector $\mathbf{v} \in V_{qs}$ is in V^*. To show this, show that each $\mathbf{C}_j \in V^*$. Next, show that $\mathbf{v} \in V^*$ implies $\mathbf{v} \in V_{qs}$. To do this let $\boldsymbol{\mu} = \sum_j \gamma_j \mathbf{D}_j + \sum_i \alpha_i \mathbf{R}_i + \sum_{i<j} \beta_{ij} \mathbf{B}_{ij}$, and express λ, λ_i^1, λ_j^2, and λ_{ij} in terms of the γ_j, α_i, and β_{ij}.

(e) A table **m** satisfies *marginal homogeneity* if $m_{i+} = m_{+i}$ holds for each i. Marginal homogeneity does not correspond to a log-linear model. However, if a table is quasi-symmetric *and* has marginal homogeneity then it must be symmetric. Prove this.

(f) Give an example of a 4×4 table which has (1) marginal homogeneity but does not have symmetry, and (2) $\lambda_i^1 = \lambda_i^2$ for all i. Give another table which satisfies (2) but not (1).

(g) A table is *quasi-independent* with respect to a subset S of index pairs (i, j) if there exist constants a_i, b_j such that $m_{ij} = a_i b_j$ for all $(i, j) \in S$. Show that quasi-independence with respect to the off-diagonal terms implies quasi-symmetry.

8.5 ESTIMATION FOR THE LOG-LINEAR MODEL

We will begin our discussion of estimation with the simplest of our models, those for which $\mathbf{Y}$ satisfies the independent Poisson model, with mean $\mathbf{m}$ satisfying $\boldsymbol{\mu} = \log \mathbf{m} = \sum_{j=1}^{k} \beta_j \mathbf{x}_j$, where $\mathbf{x}_1, \ldots, \mathbf{x}_k$ are fixed known vectors of constants, chosen by the analyst (or statistician, or student, or political scientist, or ...). All vectors indicated in the discussion have T components, indexed by a set $\mathscr{I}$, which is fixed throughout the discussion. Let $V = \mathscr{L}(\mathbf{x}_1, \ldots, \mathbf{x}_k)$. We will always assume that the vector $\mathbf{J}$ of all ones is in V. We will not always assume that these $\mathbf{x}_j$ are linearly independent. They might, for example, be the row and column indicators for a two-way table. Later we will wish to consider various possible multinomial models, but they cause some complications. We must learn to walk before we can run.

For convenience we will sometimes want to think of $\mathbf{Y}$ and the $\mathbf{x}_j$ as T-component column vectors. In this case we can write $\boldsymbol{\mu} = \mathbf{X}\boldsymbol{\beta}$, where $\boldsymbol{\beta} = (\beta_1, \ldots, \beta_k)'$. We will see that the maximum likelihood estimators (MLEs) of $\boldsymbol{\beta}$, $\boldsymbol{\mu}$, and $\mathbf{m}$ satisfy certain geometric properties, so much of the intuitive appeal of linear models remains.

We will confine ourselves to MLEs for which we have nice asymptotic properties. For the Poisson model the likelihood function is $L(\boldsymbol{\beta}; \mathbf{y}) = \prod_i [e^{-m_i} m_i^{y_i}/y_i!]$. The log likelihood function is

$$l(\boldsymbol{\beta}; \mathbf{y}) = \log(L(\boldsymbol{\beta}; \mathbf{y})) = \sum_i [-m_i + y_i \mu_i] + C = -(\mathbf{J}, \mathbf{m}) + (\mathbf{y}, \boldsymbol{\mu}) + C,$$

where C does not depend on $\boldsymbol{\beta}$. The partial derivative of $\mathbf{m}$ with respect to β_j is $\mathbf{x}_j\mathbf{m}$, where multiplication of two vectors is componentwise. That is, $\mathbf{x}_j\mathbf{m}$ has ith term $x_{ji}m_i$. Since $(\mathbf{J}, \mathbf{x}_j\mathbf{m}) = (\mathbf{m}, \mathbf{x}_j)$, and the partial derivative of $(\mathbf{y}, \boldsymbol{\mu})$ with respect to β_j is $(\mathbf{y}, \mathbf{x}_j)$, we find that $\frac{\partial}{\partial \beta_j} l(\boldsymbol{\beta}; \mathbf{y}) = (\mathbf{y} - \mathbf{m}, \mathbf{x}_j)$, for $j = 1, \ldots, k$. We seek a solution $\boldsymbol{\beta} = \hat{\boldsymbol{\beta}}$ to the likelihood equations:

$$\frac{\partial}{\partial \boldsymbol{\beta}} l(\boldsymbol{\beta}; \mathbf{y}) = (\mathbf{y} - \mathbf{m}, \mathbf{x}_j) = 0 \qquad \text{for} \quad j = 1, \ldots, k, \tag{8.5.1}$$

where $\mathbf{m} = \exp(\sum \beta_j \mathbf{x}_j)$. Let $\boldsymbol{\beta} = \hat{\boldsymbol{\beta}}$ be a solution to (8.5.1), and define $\hat{\mathbf{m}} = \exp(\sum \hat{\beta}_j \mathbf{x}_j)$. Equation (8.5.1) requires that the residual vector $\mathbf{e} = \mathbf{y} - \hat{\mathbf{m}}$ be orthogonal to the subspace V. This, of course, was the condition required of the least squares solution for linear models. The difference is that in this case $\mathbf{X}\hat{\boldsymbol{\beta}} = \hat{\boldsymbol{\mu}} = \log \hat{\mathbf{m}}$, rather than $\hat{\mathbf{m}}$, must lie in V (see Figure 8.6).

We will need to demonstrate that a solution $\hat{\boldsymbol{\beta}}$ to (8.5.1) exists (it usually does) and that the likelihood function is maximized for this choice. We find

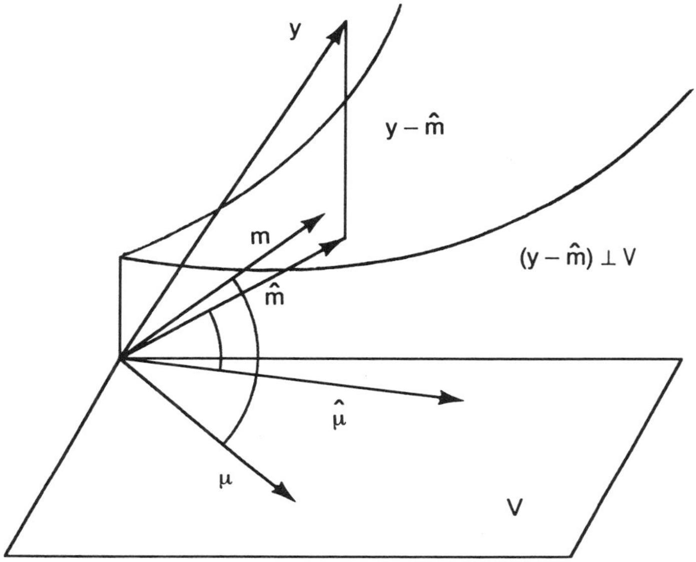

FIGURE 8.6 The log–linear model and estimates.

$\frac{\partial^2}{\partial \beta_j \, \partial \beta_{j'}} l(\boldsymbol{\beta}; \mathbf{y}) = -(\mathbf{x}_j, \mathbf{x}_{j'}\mathbf{m})$. If we momentarily write $\mathbf{m}$ and the $\mathbf{x}_j$ as column vectors, and let $\mathbf{X} = (\mathbf{x}_1, \ldots, \mathbf{x}_k)$, then the matrix of second partial derivatives is $-\mathbf{X}'d(\mathbf{m})\mathbf{X}$, where $d(\mathbf{m})$ is the $T \times T$ diagonal matrix with $\mathbf{m}$ on the diagonal. $\mathbf{X}'d(\mathbf{m})\mathbf{X}$ is nonnegative definite in general, and is positive definite if the $\mathbf{x}_j$ are linearly independent and each component of $\mathbf{m}$ is positive. Hence, if a solution to the likelihood equation exists, it is unique.

Example 8.5.1: Consider the accident data of Example 8.4.1. For cycle lengths of 40, 50, 60, and 70 the numbers of accidents were 149, 129, 112, 112. Represent $\mathbf{Y}$ as a 4-component column vector, and suppose that $\mathbf{Y}$ satisfies the independent Poisson model with $\boldsymbol{\mu} \in V(\mathbf{J}, \mathbf{x})$, where $\mathbf{x} = (40, 50, 60, 70)'$. We seek $\hat{\boldsymbol{\beta}} = (\hat{\beta}_0, \hat{\beta}_1)$ such that for $\mathbf{y} = (149, 129, 112, 112)'$, $(\mathbf{y} - \hat{\mathbf{m}}, \mathbf{J}) = \sum y_i - e^{\hat{\beta}_0} \sum_i e^{\hat{\beta}_1 x_i} = 0$, and $(\mathbf{y} - \hat{\mathbf{m}}, \mathbf{x}) = \sum_i x_i y_i - e^{\hat{\beta}_0} \sum_i x_i e^{\hat{\beta}_1 x_i} = 0$. Letting $b = \hat{\beta}_1$, solving the first equation for $\hat{\beta}_0$, and substituting in the second, we get $26{,}970 - 502\left[\sum_i x_i e^{b x_i}\right] \Big/ \left[\sum_i e^{b x_i}\right] = 0$. Since $\mathbf{x} = (40, 50, 60, 70)'$, and $\eta = e^b$, we get

$$26{,}970 - 502[40\eta^{40} + 50\eta^{50} + 60e^{60} + 70\eta^{70}]/[\eta^{40} + \eta^{50} + \eta^{60} + \eta^{70}] = 0.$$

This equation cannot be solved explicitly for η, but can be solved with patience and a \$20 calculator. We find, with the aid of a personal computer (the author's calculator costs \$65, so that it could not be used), $\eta = 0.989\,82$, $b = \hat{\beta}_1 = -0.010\,23$ and $\hat{\beta}_0 = 5.388\,4$. Then $\hat{\boldsymbol{\mu}} = \hat{\beta}_0\mathbf{J} + \hat{\beta}_1\mathbf{x} = (4.979\,2, 4.876\,9, 4.774\,6, 4.672\,3)'$, and $\hat{\mathbf{m}} = (145.36, 131.23, 118.44, 106.95)'$. The residual vector $\mathbf{e} = \mathbf{y} - \hat{\mathbf{m}} = (3.639\,5, -2.226\,2, -6.466\,3, 5.052\,9)'$ may easily be verified to be orthogonal to $\mathbf{J}$ and $\mathbf{x}$. Pearson and log chi-square statistics for $(4 - 2)$ d.f., to be introduced in Section 8.6 as measures of the distance between $\hat{\mathbf{m}}$ and $\mathbf{y}$, are 0.721 and 0.723, so the fit of the model is quite good. We were a little lucky.

The Newton–Raphson Algorithm: This algorithm provides a technique which will almost always converge to the unique solution $\hat{\boldsymbol{\beta}}$. The idea is to find a sequence of approximate solutions $\{\hat{\boldsymbol{\beta}}^{(r)}\}$ which will eventually change so little with r that we can be confident that $\hat{\boldsymbol{\beta}}^{(r)}$ is close to $\hat{\boldsymbol{\beta}}$. For each $\hat{\boldsymbol{\beta}}^{(r)}$ the function $h(\boldsymbol{\beta}) = \frac{\partial}{\partial \boldsymbol{\beta}} l(\boldsymbol{\beta}; \mathbf{y})$ is approximated by its Taylor linear approximation about $\hat{\boldsymbol{\beta}}^{(r)}$ (its differential). We have already shown that $h(\boldsymbol{\beta}) = \mathbf{X}'(\mathbf{y} - \mathbf{m})$. Where each vector is in column vector form, and, of course, $\mathbf{X}$ is the $T \times k$ design matrix. The matrix of first partials of the vector $h(\boldsymbol{\beta})$ (second partials of $l(\mathbf{y}, \boldsymbol{\beta})$, the *Hessian*) is $-I(\boldsymbol{\beta}) = -\mathbf{X}'d(\mathbf{m})\mathbf{X}$. We therefore approximate $h(\boldsymbol{\beta})$ at the $(r + 1)$th iteration by $h_{r+1}(\boldsymbol{\beta}) = h(\hat{\boldsymbol{\beta}}^{(r)}) - I(\hat{\boldsymbol{\beta}}^{(r)})(\boldsymbol{\beta} - \hat{\boldsymbol{\beta}}^{(r)})$. We then define $\hat{\boldsymbol{\beta}}^{(r+1)}$ to be that value of $\boldsymbol{\beta}$ for which $h_{r+1}(\boldsymbol{\beta}) = 0$. We find $\boldsymbol{\beta} - \hat{\boldsymbol{\beta}}^{(r)} = I(\hat{\boldsymbol{\beta}}^{(r)})^{-1}h_r(\hat{\boldsymbol{\beta}}^{(r)})$,

$$\hat{\boldsymbol{\beta}}^{(r+1)} = \hat{\boldsymbol{\beta}}^{(r)} + [\mathbf{X}'d(\hat{\mathbf{m}}^{(r)})\mathbf{X}]^{-1}\mathbf{X}'(\mathbf{y} - \hat{\mathbf{m}}^{(r)}),$$

where $\hat{\mathbf{m}}^{(r)} = \exp(\hat{\boldsymbol{\mu}}^{(r)})$, $\hat{\boldsymbol{\mu}}^{(r)} = \mathbf{X}\hat{\boldsymbol{\beta}}^{(r)}$, and $d(\mathbf{m}^{(r)})$ is the corresponding diagonal matrix. Sometimes the sequence may fail to converge because the jumps are too big. A good algorithm can produce shorter increments by multiplying them by constants $\alpha^{(r)} < 1$. The criterion for stopping can be small changes in $\hat{\boldsymbol{\beta}}^{(r)}$, in $\hat{\boldsymbol{\mu}}^{(r)}$, or in $\hat{\mathbf{m}}^{(r)}$. A good starting point $\hat{\boldsymbol{\beta}}^{(0)}$ is usually obtained by use of least squares on $\log \mathbf{y}$. That is, $\hat{\boldsymbol{\beta}}^{(0)} = (\mathbf{X}'\mathbf{X})^{-1}\mathbf{X}' \log \mathbf{y}$. To avoid zeros in $\mathbf{y}$, replace any zeros by 1/2.

The method produced by the Newton–Raphson procedure is often called iterative weighted least squares. The change $\boldsymbol{\beta}^{(r+1)} - \boldsymbol{\beta}^{(r)}$ is the generalized least squares estimate of the coefficient vector corresponding to observation vector $(\mathbf{y} - \hat{\mathbf{m}}^{(r)})$, with weight matrix $d(\hat{\mathbf{m}}^{(r)})$, design matrix $\mathbf{X}$.

Example 8.5.2: Babe Ruth ("the Sultan of Swat") was probably the most famous baseball player of all time. He began as a pitcher with the Boston Red Sox at 19 in 1914, was traded to the New York Yankees in 1919, and, because of his home run hitting, became a full-time outfielder in 1920, at the same time that the baseball was made more "lively," to increase the number of home runs. His at bats (AB's) and home runs until the end of his career were

Year	AB	HR
1920	458	54
1921	540	59
1922	406	35
1923	522	41
1924	529	46
1925	359	25
1926	495	47
1927	540	60
1928	536	54
1929	499	46
1930	518	49
1931	534	46
1932	457	41
1933	459	34
1934	365	22

It seems reasonable to suppose that the number of home runs Y_i in the ith year should have a Poisson distribution with mean m_i, which is a multiple of the number z_i of AB's. Ruth was 38 years old in 1933, and it is not surprising that his HR production decreased in 1933 and 1934. How can we model this to allow for some deterioration with time?

Let i be the index for year $1920 + i$. If $m_i = z_i \exp(\beta_0 + \beta_1 i)$, for $i = 0, 1, \ldots, 14$, then $\mu_i \equiv \log m_i = \log z_i + \beta_0 + \beta_1 i$, so that, strictly speaking, these m_i do not obey a log-linear model. However, we can, put the model in a form which will allow us to use the methods developed for the log-linear model. Let $\mathbf{x}_0$ be the vector of ones. Let $\mathbf{x} = (1, \ldots, 15)$, $\boldsymbol{\mu}^* = \beta_0 \mathbf{x}_0 + \beta_1 \mathbf{x}$, and $\mathbf{m}^* = \exp(\boldsymbol{\mu}^*)$. Then the log likelihood function is $l(\boldsymbol{\beta}; \mathbf{y}) = (\mathbf{y}, \log \mathbf{z} + \boldsymbol{\mu}^*) - (\mathbf{m}^*, \mathbf{z}) + C$, where C does not depend on $\boldsymbol{\beta}$. The ML equations are therefore $(\mathbf{y}, \mathbf{x}_j) - (\mathbf{m}^*\mathbf{x}_j, \mathbf{z}) = (\mathbf{y} - \mathbf{z}\mathbf{m}^*, \mathbf{x}_j) = 0$. The matrix of second partial derivatives, the Hessian, is $-I(\boldsymbol{\beta}) = (\mathbf{X}'d(\mathbf{mz})\mathbf{X})$. The Newton–Raphson algorithm defines

$$\hat{\boldsymbol{\beta}}^{(r+1)} = \hat{\boldsymbol{\beta}}^{(r)} + [\mathbf{X}'d(\hat{\mathbf{m}}^{*(r)}\mathbf{z})\mathbf{X}]^{-1}\mathbf{X}'(\mathbf{y} - \mathbf{z}\hat{\mathbf{m}}^{*(r)}).$$

Using an APL function of the author, checked using S-Plus, we have: $\hat{\boldsymbol{\beta}} = (-2.269, -0.018\,41)$. The coefficient $\hat{\beta}_1 = -0.018\,41$ can be interpreted to mean that the model predicts that Ruth's HR production per time at bat in year $(i + 1)$ could be expected to be $100e^{-0.018\,41}\% = 98.18\%$ of that predicted for year i.

A commonly used statistic in baseball is the number of times at bat per HR, or $w_i = z_i/Y_i$. The *smoothed* w_i for year i is $\hat{w}_i = z_i/\hat{m}_i = \exp(-\hat{\beta}_0 - \hat{\beta}_1 i)$. This is analysed in Table 8.5.1. Pearson and log chi-square values are: 12.90 and 13.24 for $(15 - 2)$ d.f., so the fit is quite good. The estimate of the standard error

Table 8.5.1 Analysis of the Babe Ruth Home Run Data

Year	AB	HR	$\hat{\mathbf{m}}$	$\hat{\mathbf{e}}$	w_i	$\hat{w}_i$
1920	458	54	47.36	6.64	8.48	9.67
1921	540	59	54.83	4.18	9.15	9.85
1922	406	35	40.47	−5.47	11.60	10.03
1923	522	41	51.08	10.08	12.73	10.22
1924	529	46	50.82	−4.82	11.50	10.41
1925	359	25	33.86	−8.86	14.36	10.60
1926	495	47	45.84	1.16	10.53	10.80
1927	540	60	49.09	10.91	9.00	11.00
1928	536	54	47.84	6.16	9.93	11.20
1929	499	46	43.72	2.28	10.85	11.41
1930	518	49	44.56	4.44	10.57	11.62
1931	534	46	45.10	0.91	11.61	11.84
1932	457	41	37.89	3.11	11.15	12.06
1933	459	34	37.36	−3.36	13.50	12.28
1934	365	22	29.17	−7.17	16.59	12.51

of $\hat{\beta}_1$ is 0.009 24, so a 95% confidence interval on β_1 is $\hat{\beta}_1 \pm 1.96(0.009\,24) = -0.018\,41 \pm 0.018\,11$, just missing zero.

Sufficiency for the Poisson Model: For the independent Poisson model the log likelihood function is

$$l(\boldsymbol{\beta}; \mathbf{y}) = -(\mathbf{J}, \mathbf{m}) + (\mathbf{y}, \boldsymbol{\mu}) + C = -(\mathbf{J}, \mathbf{m}) + (\mathbf{P}_V \mathbf{y}, \boldsymbol{\mu}) + C,$$

where $\mathbf{P}_V$ is orthogonal projection onto V, since $\boldsymbol{\mu} \in V$. Therefore, $\mathbf{P}_V \mathbf{y}$ is sufficient for $\boldsymbol{\beta}$, $\boldsymbol{\mu}$, and $\mathbf{m}$. Since $\mathbf{P}_V \mathbf{y}$ is a function of the inner products $(\mathbf{y}, \mathbf{x}_j)$, the vector of these inner products is sufficient for $\boldsymbol{\beta}$, $\boldsymbol{\mu}$, and $\mathbf{m}$.

Example 8.5.3: Let $\mathscr{I} = \{1, 2, \ldots, r\} \times \{1, 2, \ldots, c\}$, so that $\mathbf{Y}$ is an $r \times c$ table of frequencies. Let $\mathbf{R}_i$ and $\mathbf{C}_j$ be the indicators of the ith row and jth column. Let $\mathbf{x}_0$ be the vector of all ones, let $V_0 = \mathscr{L}(\mathbf{x}_0)$, $V_R = \mathscr{L}(\mathbf{R}_1, \ldots, \mathbf{R}_r) \cap V_0^{\perp}$, $V_C = \mathscr{L}(\mathbf{C}_1, \ldots, \mathbf{C}_c) \cap V_0^{\perp}$, and $V = V_0 \oplus V_R \oplus V_C$. Suppose that $\boldsymbol{\mu} = \log E(\mathbf{Y}) \in V$. As noted in Example 8.4.3, $\boldsymbol{\mu} \in V$ is equivalent to the multiplicative model: $m_{ij} = (m_{i\cdot}\, m_{\cdot j})/m_{\cdot\cdot}$ for all i and j. The vector of inner products of any set of spanning vectors for V is sufficient for the parameters of this model. The inner products of the row and column indicators are the corresponding row and column sums. If we were to suppose instead that $\boldsymbol{\mu} \in \mathscr{L}(\mathbf{R}_1, \ldots, \mathbf{R}_r)$, then the vector of row sums would be sufficient.

For observed $\mathbf{y} = (y_{ij})$ it is easy to verify that the array $\left(\hat{m}_{ij} = \dfrac{y_{i\cdot}\, y_{\cdot j}}{y_{\cdot\cdot}}\right) = \hat{\mathbf{m}}$ is the MLE of $\mathbf{m}$, and that $\hat{\boldsymbol{\mu}} = \sum_i [\log y_{i\cdot}]\mathbf{R}_i + \sum_j [\log y_{\cdot j}]\mathbf{C}_j - y_{\cdot\cdot}\mathbf{J}$.

We have yet to prove that the MLE always exist. In fact, the MLE need not exist if some component of $\mathbf{Y}$ is zero. For all the models we will consider there is always a positive probability that the MLE does not exist. Fortunately, the probability is usually extremely small. For a simple example, let $\mathbf{Y} = (Y_1, Y_2)$ satisfy the independent Poisson model with $\boldsymbol{\mu} = (\beta_1, \beta_2)$. The likelihood function is $l(\boldsymbol{\beta}, \mathbf{y}) = -(\mathbf{J}, \mathbf{m}) + (\mathbf{y}, \boldsymbol{\mu}) + C = -[e^{\beta_1} + e^{\beta_2}] + y_1\beta_1 + y_2\beta_2 + C$. If $y_2 = 0$, then $l(\beta, \mathbf{y})$ is a decreasing function of β_2, taking its maximum at $-\infty$. We insist that the values of estimators be real numbers. Fortunately, this is the only kind of situation for which the MLE does not exist.

Whenever all components of $\mathbf{m}$ are positive, as we always assume, the vector $\boldsymbol{\mu}$ exists. The vector $\boldsymbol{\beta}$ is not uniquely defined unless the vectors $\mathbf{x}_j$ are linearly independent. To avoid this assumption we will show that the MLE for $\boldsymbol{\mu}$ (and therefore for $\mathbf{m}$) exists whenever all components of $\mathbf{y}$ are positive. If the $\mathbf{x}_j$ are linearly independent then $\hat{\boldsymbol{\beta}} = (\mathbf{X}'\mathbf{X})^{-1}\mathbf{X}\hat{\boldsymbol{\mu}}$ is the MLE for $\boldsymbol{\beta}$.

Theorem 8.5.1: If every component of $\mathbf{y}$ is positive, then the MLE of $\boldsymbol{\mu}$ exists. More generally, if there exists a vector $\boldsymbol{\delta} \perp V$, such that $\mathbf{y} + \boldsymbol{\delta}$ has components which are all positive, then the MLE for $\boldsymbol{\mu}$ exists.

***Proof*:** Suppose all components of $\mathbf{y}$ are positive. The likelihood function minus C is $g(\boldsymbol{\mu}) = -(\mathbf{J}, e^{\boldsymbol{\mu}}) + (\mathbf{y}, \boldsymbol{\mu}) = \sum_i [-e^{\mu_i} + y_i\mu_i] = \sum_i h_i(\mu_i)$, for $h_i(a) = y_i a - e^a$. Each h_i is continuous, $\lim_{a \to -\infty} h_i(a) = -\infty$, $\lim_{a \to \infty} h_i(a) = \infty$, so that each h_i has a finite maximum (at $\log y_i$). It follows that there exists a constant c such that, whenever $|\mu_i| < c$ for all i, $g(\boldsymbol{\mu}) < g(\mathbf{0}) = -T$, where T is the number of components of $\mathbf{y}$. Thus, $F \equiv \{\boldsymbol{\mu} | g(\boldsymbol{\mu}) \geq -T, \boldsymbol{\mu} \in R_T\}$ is closed and bounded. It follows that $G \equiv \{\boldsymbol{\mu} | g(\boldsymbol{\mu}) \geq -T, \boldsymbol{\mu} \in V\}$ is closed and bounded. Since g is continuous this implies that there exists a $\hat{\boldsymbol{\mu}}$ at which g takes its maximum on C. Since g is smaller for all $\boldsymbol{\mu} \in (V - G)$, this proves that g is maximized on V by $\hat{\boldsymbol{\mu}}$.

If $\boldsymbol{\delta} \perp V$, then $g(\boldsymbol{\mu}) = -(\mathbf{J}, \mathbf{m}) + (\mathbf{y} + \boldsymbol{\delta}, \boldsymbol{\mu}) = -(\mathbf{J}, \mathbf{m}) + (\mathbf{y}, \boldsymbol{\mu})$. If all components of $\mathbf{y} + \boldsymbol{\delta}$ are positive then g has a maximum at some point $\hat{\boldsymbol{\mu}}$ by the first part of the theorem. Since the likelihood functions for $\mathbf{y} + \boldsymbol{\delta}$ and for $\mathbf{y}$ are the same, $\hat{\boldsymbol{\mu}}$ also is the MLE corresponding to $\mathbf{y}$. □

Continuation of Example 8.5.3: Suppose that $r = 2$, $c = 3$, and that we observe $\mathbf{y} = \begin{bmatrix} 6 & 3 & 0 \\ 0 & 1 & 2 \end{bmatrix}$. Let $\boldsymbol{\delta} = \begin{bmatrix} -1 & 0 & 1 \\ 1 & 0 & -1 \end{bmatrix}$. Then $\boldsymbol{\delta} \perp V$, since it is orthogonal to the row and column indicators. Since all the components of $\mathbf{y} + \boldsymbol{\delta}$ are positive, the MLE for $\boldsymbol{\mu}$ and $\mathbf{m}$ exists. It is easy to verify that $\hat{\mathbf{m}} = \frac{1}{4}\begin{bmatrix} 18 & 9 & 9 \\ 6 & 3 & 3 \end{bmatrix}$, has the same inner products with these indicators as does $\mathbf{y}$. A two-way table has an MLE for the multiplicative model $\boldsymbol{\mu} \in V$ if and only if all row and column sums are positive. (Why?)

The Independent Multinomial Model: Suppose that $\mathscr{I} = \mathscr{I}_1 \cup \mathscr{I}_2 \cup \cdots \cup \mathscr{I}_k$, where these index sets $\mathscr{I}_j$ are disjoint. Let $\mathscr{I}_j$ have T_j elements, and let $T = \sum_j T_j$. For each j let $\mathscr{I}_j$ be the index set for $\mathbf{Y}_j$, and let $\mathbf{Y} = (\mathbf{Y}_1, \ldots, \mathbf{Y}_k)$. Suppose that these $\mathbf{Y}_j$ are independent, and that $\mathbf{Y}_j \sim \mathscr{M}_{T_j}(n_j, \mathbf{p}_j)$. That is, $\mathbf{Y}$ satisfies the product multinomial model. Let $\mathbf{m}_j = E(\mathbf{Y}_j) = n_j\mathbf{p}_j$ and $\mathbf{m} = E(\mathbf{Y}) = (\mathbf{m}_1, \ldots, \mathbf{m}_k)$. Let $\mathbf{w}_j$ be the indicator of index set $\mathscr{I}_j$, and suppose the subspace V of the sample space $R_{\mathscr{I}}$ of possible vectors $\mathbf{y}$ includes each of these $\mathbf{w}_j$.

Example 8.5.4: Suppose a random sample of 400 adults is chosen from among the residents of Lansing, Michigan, and that they are classified into four age-groups, corresponding to the rows, and three political categories: Republican, Democratic, and Independent, corresponding to the columns. In this case $k = 1$. Then V must include the vector $\mathbf{J} = \mathbf{x}_1$ of all ones. On the other hand, if we sample by choosing 100 people randomly from each of the four age-groups, $k = 4$, $\mathscr{I}_i = \{(i, 1), (i, 2), (i, 3)\}$, each n_i is 100, and $\mathbf{w}_i$ is the indicator of the ith row.

If we classify by sex as well, with sex as the first factor, age-group as the second, political party the third, then $\mathscr{I}_1 = \{1, 2\}$, $\mathscr{I}_2 = \{1, 2, 3, 4\}$, $\mathscr{I}_3 = \{1, 2, 3\}$, $\mathscr{I} = \mathscr{I}_1 \times \mathscr{I}_2 \times \mathscr{I}_3$. If we sample again by taking 100 people randomly from the jth age-group, then $\mathbf{w}_j$, the indicator of the jth level of age-group, must be included in V. If we sample by taking 50 of each sex–age-group combination, then we must include the indicator $\mathbf{w}_{ij}$ of the indices with level i for sex and level j for age-group in V.

If $\boldsymbol{\mu} = \log \mathbf{m} = \sum \beta_j \mathbf{x}_j = \mathbf{X}\boldsymbol{\beta}$ for the product multinomial model, the likelihood function is

$$L_{mn}(\boldsymbol{\beta}; \mathbf{y}) = \sum_{j=1}^{k} \left[\binom{n_j}{\mathbf{y}_j} \sum_{i=1}^{n_j} p_{ji}^{y_{ji}} \right],$$

where $\mathbf{y}_j = (y_{j1}, \ldots, y_{jn_j})$ is the vector of values taken by $\mathbf{Y}_j$. The log likelihood function is

$$l_{mn}(\boldsymbol{\beta}, \mathbf{y}) = \sum_j \sum_i y_{ji} \log(p_{ji}) + C^*,$$

where C^* does not depend on $\mathbf{p}$. Since $m_{ji} = n_j p_{ji}$, and $\sum_i y_{ji} = n_j$, this is

$$l_{mn}(\boldsymbol{\beta}, \mathbf{y}) = \sum_j \sum_i y_{ji} \log(m_{ji}) - \sum_j \sum_i y_{ji} \log n_j + C^* = (\mathbf{y}, \boldsymbol{\mu}) - \sum_j n_j \log n_j + C^*.$$

The log likelihood for the Poisson model was:

$$l_p(\boldsymbol{\beta}, \mathbf{y}) = (\mathbf{y}, \boldsymbol{\mu}) - (\mathbf{J}, \mathbf{m}) + C.$$

Hence

$$l_{mn}(\boldsymbol{\beta}, \mathbf{y}) = l_p(\boldsymbol{\beta}, \mathbf{y}) + (\mathbf{J}, \mathbf{m}) - \sum_j n_j \log n_j + C^* - C$$

$$= l_p(\boldsymbol{\beta}, \mathbf{y}) + \left[\sum_j n_j - \sum_j n_j \log n_j + C^* - C\right] = l_p(\boldsymbol{\beta}, \mathbf{y}) + C^{**},$$

where C^{**} does not depend on $\boldsymbol{\beta}$. Since the conditional distribution of $\mathbf{Y}_j$, given $S_j = \sum_i Y_{ji} = n_j$ is multinomial with parameters n_j and $\mathbf{p}_j = \mathbf{m}_j/n_j$, C^{**} is the negative of the log likelihood of $(n_1, \ldots, n_k)$. Since $\hat{\boldsymbol{\beta}}$ maximizes $l_p(\boldsymbol{\beta}, \mathbf{y})$ and C^{**} does not depend on $\boldsymbol{\beta}$,

$$l_{mn}(\boldsymbol{\beta}; \mathbf{y}) \le l_p(\hat{\boldsymbol{\beta}}; \mathbf{y}) + C^{**} \qquad \text{for all} \quad \boldsymbol{\beta}.$$

$\boldsymbol{\beta}$ must be chosen so that $\sum_i m_{ji} = (\mathbf{m}, \mathbf{w}_j) = n_j$ for each j. However, since $\mathbf{w}_j \in V$ for each j, $(\hat{\mathbf{m}}, \hat{\mathbf{w}}_j) = (\mathbf{y}, \hat{\mathbf{w}}_j) = n_j$ for the MLE $\hat{\mathbf{m}}$ corresponding to $\hat{\boldsymbol{\beta}}$ under the Poisson model. Therefore, the Poisson solution $\hat{\boldsymbol{\beta}}$ automatically satisfies the restrictions of the multinomial model, and is a solution which makes the inequality an equality. That is, the Poisson solution is also the multinomial solution.

Though the solution $\hat{\boldsymbol{\beta}}$'s for the Poisson and multinomial models are the same, $\hat{\boldsymbol{\beta}}$, $\hat{\boldsymbol{\mu}}$, and $\hat{\mathbf{m}}$ will have different distributions under different models. As we will indicate, $\hat{\boldsymbol{\beta}}$, $\hat{\boldsymbol{\mu}}$, and $\hat{\mathbf{m}}$ are all (in a certain sense) asymptotically unbiased. Each will be less variable under the multinomial model than under the Poisson model.

Example 8.5.5: Consider the rat data of Example 8.4.2:

Dosage	d_1	d_2	d_3	d_4
Log-dosage	x_1	x_2	x_3	x_4
Number of rats	15	17	19	16
Number dying	2	6	11	13

For Y_{ij} as defined in that example the observed $\mathbf{Y}$ is $\mathbf{y} = \begin{bmatrix} 13 & 11 & 8 & 3 \\ 2 & 6 & 11 & 13 \end{bmatrix}$. The model states that $Y_{2j} \sim \mathscr{B}(n, p_{2j})$, where $\log[p_{2j}/(1 - p_{2j})] = \gamma x_j$. Equi-

the same equality holds if $\hat{\gamma}$ and $\hat{\mu}_j$ are substituted for γ and μ_j. We will use the log-linear approach rather than the logit approach to fit the model. This means that we will fit the model with the five parameters $\mu_1, \ldots, \mu_4, \gamma$ even though the first four are functions of the last. In this sense the model is overparameterized. In the logit approach we consider instead only the second row of $\mathbf{Y}$. This second row, conditionally on the column totals, has a distribution which depends only on the parameter γ. The estimators $\hat{\mu}_j$, being simple functions of $\hat{\gamma}$, are easily determined from $\hat{\gamma}$.

The Newton–Raphson method produced the MLE

$$\hat{\mathbf{m}} = \begin{bmatrix} 11.98 & 8.50 & 5.87 & 3.22 \\ 3.02 & 8.50 & 13.13 & 12.78 \end{bmatrix}$$

for $d_1 = 0.5$, $d_2 = 1.0$, $d_3 = 1.5$, $d_4 = 2.0$, and therefore

$$\mathbf{x} = \begin{bmatrix} 0 & 0 & 0 & 0 \\ -0.6931 & 0 & 0.4055 & 0.6931 \end{bmatrix}.$$

The estimate of $\boldsymbol{\beta}$ was $\hat{\boldsymbol{\beta}} = (2.483, 2.140, 1.770, 1.171, 1.986)'$, and

$$\hat{\boldsymbol{\mu}} = \sum_i \hat{\mu}_j \mathbf{J}_j + \hat{\gamma}\mathbf{x} = \begin{bmatrix} 2.483 & 2.140 & 1.770 & 1.171 \\ 1.106 & 2.140 & 2.575 & 2.548 \end{bmatrix}.$$

The estimate of the probability matrix is obtained by dividing the jth column of $\hat{\mathbf{m}}$ by n_j, to obtain $\hat{P} = \begin{bmatrix} 0.799 & 0.500 & 0.309 & 0.201 \\ 0.201 & 0.500 & 0.691 & 0.799 \end{bmatrix}$. As noted earlier, this model forces the probability of death for dosage 1.0 to be 1/2.

The residual vector $\mathbf{e} = \mathbf{y} - \hat{\mathbf{m}} = \begin{bmatrix} 1.023 & 2.5 & 2.132 & -0.224 \\ -1.023 & -2.5 & -2.132 & 0.224 \end{bmatrix}$ is orthogonal to the column indicators and to $\mathbf{x}$. The Pearson and log chi-square values, measures of the distance of $\mathbf{y}$ from $\hat{\mathbf{m}}$ were $\chi^2 = \sum (y_{ij} - \hat{m}_{ij})^2/\hat{m}_{ij} = 3.044$ and $G^2 = 2\sum y_{ij}\log(y_{ij}/\hat{m}_{ij}) = 3.053$. We will discuss the properties of these statistics later. Under the hypothesis that this model holds, these statistics should each be approximately distributed as χ^2 with $(8 - 5) = 3$ d.f. Thus, the model fits quite well. Each of $\hat{\boldsymbol{\beta}}$, $\hat{\boldsymbol{\mu}}$, $\hat{\mathbf{m}}$ has an approximate multivariate normal distribution with mean given by the corresponding parameter, and covariance matrix given by formulas presented in Section 8.6. Estimates of the standard deviation of the elements of $\hat{\mathbf{m}}$ are $\begin{bmatrix} 0.956 & 0 & 0.940 & 1.020 \\ 0.956 & 0 & 0.940 & 1.020 \end{bmatrix}$. Since $Y_{2j} = n_j - Y_{1j}$, $\text{Var}(\hat{m}_{1j}) = \text{Var}(\hat{m}_{2j})$. Since $x_2 = \log d_2 = 0$, $\hat{m}_{12} = \hat{m}_{22} = (Y_{12} + Y_{22})/2 = n_2/2 = 8.5$, so that $\text{Var}(\hat{m}_{12}) = \text{Var}(\hat{m}_{22}) = 0$.

Estimates of the standard deviations of the terms of $\hat{\boldsymbol{\beta}}$ are given by taking the square roots of the diagonal elements of the estimate of the covariance matrix: (0.270 3, 0.242 5, 0.279 8, 0.403 1, 0.571 3). Only $\hat{\gamma}$ is of real interest. $\hat{\gamma} \pm 1.96\, \hat{\sigma}(\hat{\gamma}) = 1.986 \pm 1.120$ is an approximate 95% confidence interval on γ. Since this interval does not include zero, we can be quite confident that γ is greater than zero.

The model $\log[p_{2j}/(1 - p_{2j})] = \gamma_0 + \gamma_1 x_j$ was also fit, resulting in $\hat{\beta} = (2.595, 2.352, 1.984, 1.341, -0.482, 2.367)'$, $\hat{\mathbf{m}} = \begin{bmatrix} 13.40 & 10.51 & 7.27 & 3.82 \\ 1.60 & 6.49 & 11.73 & 12.18 \end{bmatrix}$, $\hat{\mathbf{p}} = \begin{bmatrix} 0.893 & 0.618 & 0.404 & 0.239 \\ 0.107 & 0.382 & 0.596 & 0.761 \end{bmatrix}$. Pearson chi-square and Log chi-square values were 0.520, 0.525, certainly smaller, indicating a very good fit. The smaller model with γ_0 necessarily zero fits quite well and the improvement in the fit probably does not warrant the increased complexity. The estimate of the standard deviation of $\hat{\gamma}_0$ is 0.316 5, so that -0.482 ± 0.620. If we let $g_i(d)$ be the probability of death for dosage d for model $i (i = 1, 2)$ then $\hat{g}_1(d) = d^{\hat{\gamma}}/(1 + d^{\hat{\gamma}}) = d^{1.986}/(1 + d^{1.986})$ and $\hat{g}_2(d) = e^{\hat{\gamma}_0} d^{\hat{\gamma}_1}/[1 + e^{\hat{\gamma}_0} d^{\hat{\gamma}_1}] = 0.617\,6 d^{2.367}/(1 + 0.617\,6 d^{2.367})$. These functions are graphed in Figure 8.7.

Sufficient Statistics for the Independent Multinomial Model: Suppose that $\mathbf{Y} = (\mathbf{Y}, \ldots, \mathbf{Y}_{k_0})$ satisfies the independent multinomial model with

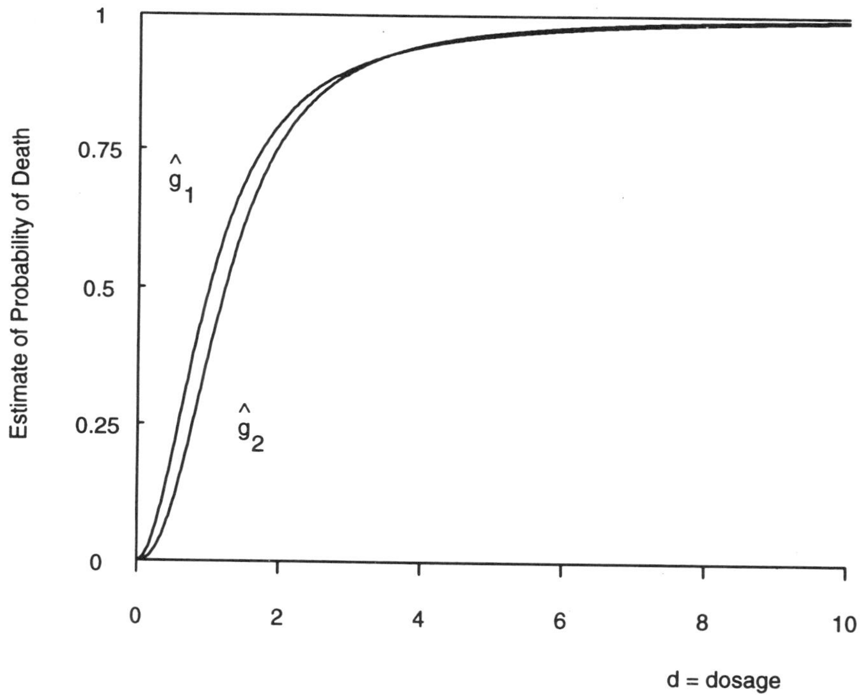

FIGURE 8.7 Estimates of the probability of death.

$\mathbf{Y}_j = (Y_{1j}, \ldots, Y_{T_j j}) \sim \mathcal{M}_{T_j}(n_j, \mathbf{p}_j)$. Let

$$\boldsymbol{\mu} = (\boldsymbol{\mu}_1, \ldots, \boldsymbol{\mu}_{k_0}) = \log(\mathbf{m}) = \log(\mathbf{m}_1, \ldots, \mathbf{m}_{k_0}) = \log(n) + \log(\mathbf{p}).$$

Let $T = \sum_j T_j$. $\mathbf{Y}$ takes values in R_T. Let $\mathbf{J}_j$ be the indicator of the cells $(1, j), \ldots, (k_j, j)$. Then $\sum_j \mathbf{J}_j = \mathbf{J}$, the T-component vector of all ones.

Let $V_0 = \mathcal{L}(\mathbf{J}_1, \ldots, \mathbf{J}_{k_0})$, and suppose that $\boldsymbol{\mu} \in V = V_0 \oplus \mathcal{L}(\mathbf{x}_{k_0+1}, \ldots, \mathbf{x}_k)$, where $\mathbf{x}_j$ is not contained in V_0 for $j > k_0$. Let $V_1 = V \cap V_0^\perp$ and $\boldsymbol{\mu}_\perp = p(\boldsymbol{\mu} | V_\perp) = \boldsymbol{\mu} - p(\boldsymbol{\mu} | V_0) = \boldsymbol{\mu} - \sum_j \mu_j \mathbf{J}_j$, where $\mu_j = \left(\sum_i \mu_{ij}\right)/T_j$. Then $\boldsymbol{\mu} = \sum_j \mu_j \mathbf{J}_j + \boldsymbol{\mu}_\perp$. The kernel of the log likelihood function (the part which depends on $\boldsymbol{\beta}$) for the product multinomial model is

$$l_{mn}(\boldsymbol{\beta}) - C^* = (\mathbf{y}, \boldsymbol{\mu}) = \sum_j \mu_j(\mathbf{y}, \mathbf{J}_j) + (\mathbf{y}, \boldsymbol{\mu}_\perp) = \sum_j \mu_j n_j + (\mathbf{y}, \boldsymbol{\mu}_\perp).$$

Let $\mathbf{y}_1 = p(\mathbf{y} | V_1)$. Then $(\mathbf{y}, \boldsymbol{\mu}_\perp) = (\mathbf{y}_1, \boldsymbol{\mu}_\perp)$, so that the likelihood function may be expressed as a function of $\mathbf{y}_1$. Thus, we have Theorem 8.5.2.

Theorem 8.5.2: $\hat{\mathbf{Y}}_1 = p(\mathbf{Y} | V_1)$ is a sufficient statistic for $\boldsymbol{\beta}$. If $\mathbf{x}_{k_0+1}, \ldots, \mathbf{x}_k$ span V_1, then the $(k - k_0)$-tuple $((\mathbf{x}_j, \mathbf{Y}), j = k_0 + 1, \ldots, k)$ is sufficient for $\boldsymbol{\beta}$.

***Proof*:** The second sentence follows from the fact that $\mathbf{Y}_1$ is a function of the inner products $(\mathbf{Y}, \mathbf{x}_j)$ for $j > k_0$. We are supposing that V has dimension k, so that $\boldsymbol{\beta}$ and $\hat{\boldsymbol{\beta}}$ are uniquely defined. □

The representation $\boldsymbol{\mu} = \boldsymbol{\mu}_0 + \boldsymbol{\mu}_\perp$, with $\boldsymbol{\mu}_0 = \sum_j \mu_j \mathbf{J}_j \in V_0$ is also useful in providing an understanding of the relationship among the parameters for the multinomial model. Since

$$\mathbf{m} = \exp(\boldsymbol{\mu}) = \exp(\boldsymbol{\mu}_0 + \boldsymbol{\mu}_\perp) = \left[\sum_j e^{\mu_j} \mathbf{J}_j\right] e^{\boldsymbol{\mu}_\perp}, \tag{8.5.2}$$

we have $n_j = (\mathbf{m}, \mathbf{J}_j) = e^{\mu_j} \sum_i e^{\mu_{\perp ij}}$, so that $\mu_j = \log n_j - \log\left[\sum_i e^{\mu_{\perp ij}}\right]$. Thus, the coefficients μ_j of the indicators $\mathbf{J}_j$ are determined by $\boldsymbol{\mu}_\perp$, which in turn is a function of the parameters $\beta_{k_0+1}, \ldots, \beta_k$, and $\mathbf{p}_j = \exp(\boldsymbol{\mu}_{\perp j}) / \left[\sum_i \exp(\mu_{\perp ij})\right]$.

Put another way, all the vectors in $\mathcal{M} = \{\mathbf{m} | \boldsymbol{\mu} = \log \mathbf{m} \in V, (\mathbf{m}, \mathbf{J}_j) = n_j, j = 1, \ldots, k\}$, the set of all possible $\mathbf{m}$, have the same projection $\mathbf{m}_0 = \sum (n_j/T_j)\mathbf{J}_j$ onto V_0. The orthogonal part $\mathbf{m} - \mathbf{m}_0$ is determined uniquely by $\boldsymbol{\mu}_\perp$. Figure 8.8 may provide some intuition. In Figure 8.8, of T-space, $\hat{\boldsymbol{\mu}} = \log(\hat{\mathbf{m}}) \in V_0$ is

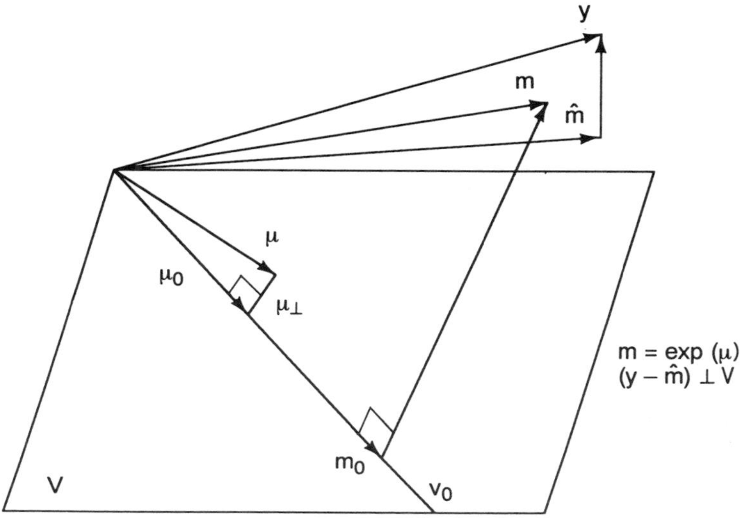

FIGURE 8.8

not pictured so as not to cause even more clutter. The case $\boldsymbol{\mu}_\perp = \mathbf{0}$ corresponds to the case $\boldsymbol{\mu} = \sum \mu_j \mathbf{J}_j$. In this case all the cells of the table corresponding to the same j have the same expected value $\exp(\mu_j) = n_j/T_j$.

In the rat poison example, we considered the model $\boldsymbol{\mu} = \sum_j \beta_j \mathbf{C}_j + \gamma \mathbf{x}$, for $\mathbf{x}$ the 2×4 array with 0's in the first row and $(x_1, \ldots, x_4) = \log(d_1, d_2, d_3, d_4)$ in the second row. For the notation above $\mu_j = \beta_j$ for $1 \le j \le 4$, and

$$\boldsymbol{\mu}_\perp = \gamma \mathbf{x}_\perp = \gamma \begin{bmatrix} -x_1 & -x_2 & -x_3 & -x_4 \\ x_1 & x_2 & x_3 & x_4 \end{bmatrix} \frac{1}{2}.$$

Then $\mu_j = \log n_j - \log[e^{-\gamma x_j} + e^{\gamma x_j}]$.

Example 8.5.6: Reconsider Example 8.4.4. Here we will discuss only two models, using S-Plus, rather than the procedure CATMOD in the statistical computer language SAS, or the command LOGLINEAR in SPSS.

Usually we will wish to fit the saturated model, for which $\hat{\mathbf{m}} = \mathbf{Y}$, then do an ANOVA-type breakup of the $\hat{\mu}_{ijk} = \log \hat{m}_{ijk}$ to obtain estimates of $\boldsymbol{\mu}$, s_i, e_j, etc. Estimates of the standard errors of these parameter estimators and χ^2 statistics can then be used to decide if these estimates are sufficiently close to zero to omit them from the model They will require estimates of the standard errors of these terms, and some chi-square statistics, so we postpone that until these topics have been discussed. We first discuss the model which differs from the saturated model only in that the term $(ser)_{ijk}$ is missing, the (1 2 3 12 13 23) model. The **x**-vectors can be taken to be

$$\mathbf{x}_0, \text{ the vector of all ones,} \quad \mathbf{s} = \begin{cases} 1 & \text{for men} \\ -1 & \text{for women} \end{cases}, \quad \mathbf{e}_1 = \begin{cases} 1 & \text{for age level 1} \\ 0 & \text{for age level 2} \\ -1 & \text{for age level 3} \end{cases}$$

$$\mathbf{e}_2 = \begin{cases} 0 & \text{for age level 1} \\ 1 & \text{for age level 2}, \\ -1 & \text{for age level 3} \end{cases} \quad \mathbf{r} = \begin{cases} 1 & \text{for response agree} \\ -1 & \text{otherwise} \end{cases}$$

and vectors corresponding to interaction terms, which are their componentwise product: $(\mathbf{se})_{11}$, $(\mathbf{se})_{12}$, $(\mathbf{sr})_{11}$, $(\mathbf{er})_{11}$, $(\mathbf{er})_{21}$. For example,

$$s = \begin{bmatrix} 1 & 1 \\ 1 & 1 \\ 1 & 1 \\ -1 & -1 \\ -1 & -1 \\ -1 & -1 \end{bmatrix}, \quad \mathbf{e}_1 = \begin{bmatrix} 1 & 1 \\ 0 & 0 \\ -1 & -1 \\ 1 & 1 \\ 0 & 0 \\ -1 & -1 \end{bmatrix}, \quad (\mathbf{se})_1 = \begin{bmatrix} 1 & 1 \\ 0 & 0 \\ -1 & -1 \\ -1 & -1 \\ 0 & 0 \\ 1 & 1 \end{bmatrix},$$

$$(\mathbf{er})_2 = \begin{bmatrix} 0 & 0 \\ 1 & -1 \\ -1 & 1 \\ 0 & 0 \\ 1 & -1 \\ -1 & 1 \end{bmatrix}, \quad \text{and} \quad \mathbf{y} = \begin{bmatrix} 72 & 47 \\ 110 & 196 \\ 44 & 179 \\ 86 & 38 \\ 173 & 283 \\ 28 & 187 \end{bmatrix}.$$

The choices of the spanning vectors are somewhat arbitrary. These have been chosen to be linearly independent and to span the orthogonal subspaces V_0, V_s, V_e, V_r, V_{se}, V_{sr}, V_{er}. The coefficients $\beta_0, \ldots, \beta_9$ are, in the usual notation of the analysis of variance μ, s_1, e_1, e_2, r_1, $(se)_{11}$, $(se)_{12}$, $(sr)_{11}$, $(er)_{11}$, and $(er)_{21}$. The other terms can be found from the additive property. For example, $(se)_{11} + (se)_{12} + (se)_{13} = 0$, so that $(se)_{13} = -(se)_{11} - (se)_{12}$.

This is the only model for three categorical models with no explicit formulas for the $\hat{m}_{ijk}$. We seek $\hat{\mathbf{m}}$ having the same inner products with these independent vectors as does **y**. That is, the marginal totals for $\hat{\mathbf{m}}$ and **y** across all combinations of any two factors must be the same. For example, the 2×2 table of sex and response combinations determined by adding across education must be $\begin{bmatrix} 226 & 422 \\ 287 & 508 \end{bmatrix}$. The Newton–Raphson method was used to find the solution. The coefficient vector is $\beta = (\mu, s_1, e_1, e_2, r_1, (se)_{11}, (se)_{12}, (sr)_{11}, (er)_{11}, (er)_{21})'$.

To facilitate the use of matrix algebra these $2 \times 3 \times 2$ arrays were rewritten as columns of 12, so that the design matrix $\mathbf{X}$ was 12×10. After the computations for $\hat{\boldsymbol{\beta}}$ were completed, $\hat{\boldsymbol{\mu}} = \mathbf{X}\boldsymbol{\beta}$ was determined as a column vector of 12 components, then rewritten as a $2 \times 3 \times 2$ array for improved understanding. All of this was carried out using the S-Plus function "glm." That function provides options (using a *contrast* option) which determine various spanning vectors. The choice made here deviates in sign only from those given by the contrast = "contr.sum" option. Other contrast choices provides the same $\hat{\mathbf{m}}$, but different $\hat{\boldsymbol{\beta}}$. The solution was

$$\hat{\boldsymbol{\beta}} = (4.550, -0.068, -0.491, -0.645, -0.256, 0.050, -0.132, -0.006, 0.566, -0.009)'.$$

$$\hat{\boldsymbol{\mu}} = \begin{bmatrix} 4.344 & 3.736 \\ 4.724 & 5.265 \\ 3.592 & 5.229 \\ 4.394 & 5.229 \\ 5.138 & 5.655 \\ 3.575 & 5.189 \end{bmatrix}, \qquad \hat{\mathbf{m}} = \begin{bmatrix} 77.05 & 41.95 \\ 112.64 & 193.36 \\ 36.31 & 186.69 \\ 80.95 & 43.05 \\ 170.36 & 285.64 \\ 35.69 & 179.31 \end{bmatrix},$$

and

$$\mathbf{e} = \mathbf{y} - \hat{\mathbf{m}} = \begin{bmatrix} -5.05 & 5.05 \\ -2.64 & 2.64 \\ 7.69 & -7.69 \\ 5.05 & -5.05 \\ 2.64 & -2.64 \\ -7.69 & 7.69 \end{bmatrix}.$$

For example,

$$\begin{aligned} \hat{\mu}_{111} &= \hat{\mu} + \hat{s}_1 + \hat{e}_1 + \hat{r}_1 + (\widehat{se})_{11} + (\widehat{sr})_{11} + (\widehat{er})_{11} \\ &= \hat{\beta}_0 + \hat{\beta}_1 + \hat{\beta}_2 + \hat{\beta}_4 + \hat{\beta}_5 + \hat{\beta}_7 + \hat{\beta}_8. \end{aligned}$$

To verify that $\hat{\mathbf{m}} \in V$, check all the two-way marginal totals. For example, summing $\hat{\mathbf{m}}$ across educational levels, we get $\begin{bmatrix} 226 & 422 \\ 287 & 179 \end{bmatrix}$, as for $\mathbf{y}$. To see that $\hat{\boldsymbol{\mu}} \in V$, check that $\hat{\boldsymbol{\mu}}$ is orthogonal to each of the vectors in the three-way interaction subspace V_{ser}. We can do this by showing that $\hat{\boldsymbol{\mu}}$ is orthogonal to the two spanning vectors

$$\mathbf{I}_1 \equiv \mathbf{se}_1\mathbf{r} = \begin{bmatrix} 1 & -1 \\ 0 & 0 \\ -1 & 1 \\ -1 & 1 \\ 0 & 0 \\ 1 & -1 \end{bmatrix}, \quad \text{and} \quad \mathbf{I}_2 = \mathbf{se}_2\mathbf{r} = \begin{bmatrix} 0 & 0 \\ 1 & -1 \\ -1 & 1 \\ 0 & 0 \\ -1 & 1 \\ 1 & -1 \end{bmatrix}.$$

These vectors define contrasts, $L_i \equiv (\mathbf{I}_i, \boldsymbol{\mu})$, whose estimates under the saturated model are $\hat{L}_1 = (\mathbf{I}_1, \boldsymbol{\mu}) = -0.3048$ and $\hat{L}_2 = (\mathbf{I}_2, \boldsymbol{\mu}) = -0.8860$. Estimates of the standard errors of these $\hat{L}_i$, assuming only the saturated model, are $\hat{\sigma}_1 = (\mathbf{I}_1, 1/\mathbf{y})^{1/2} = 0.3175$, and $\hat{\sigma}_2 = (\mathbf{I}_2, 1/\mathbf{y})^{1/2} = 0.3775$. The z-values are therefore $z_1 = 0.960$ and $z_2 = 2.347$, and the Pearson χ^2 value is 5.948 for 2 d.f. ($\chi^2_{0.95} = 5.991$), indicating that this model may be barely believable. Still, the fit is quite good, and we may be satisfied with it. The contrasts L_1 and L_2 are zero for the model with three-way interactions zero. The same is therefore true for the estimates $\hat{L}_1$ and $\hat{L}_2$ under this model.

Estimates of odds-ratios are obtained by taking the antilogs of these contrast estimates $\hat{L}_i$. Confidence limits may be obtained from the antilogs of the endpoints of confidence intervals on the L_i.

The estimate $\hat{p}_{ij}$ for the conditional probabilities $p_{ij} = m_{ij1}/(m_{ij1} + m_{ij2})$ of agreement for the sex–education classification ij is obtained by substituting $\hat{m}_{ijk}$ for m_{ijk}. We obtain: $\hat{\mathbf{p}} = \begin{bmatrix} 0.647 & 0.368 & 0.197 \\ 0.653 & 0.374 & 0.166 \end{bmatrix}$. Notice that sex seems to play little role, and that the probability p_{ij} is a decreasing function of j. This suggests that we try a model with no sex–response interaction, and that we replace the terms $(er)_{jk}$ by a covariate which is linear in k.

Consider the vector

$$\mathbf{w} = \begin{bmatrix} -1 & 1 \\ 0 & 0 \\ 1 & -1 \\ -1 & 1 \\ 0 & 0 \\ 1 & -1 \end{bmatrix}.$$

This vector should serve as a good stand-in for the two vectors $(\mathbf{er})_{11}$ and $(\mathbf{er})_{21}$. The resulting subspace $V = \mathscr{L}(\mathbf{x}_0, \mathbf{s}_1, \mathbf{e}_1, \mathbf{e}_2, \mathbf{r}_1, (\mathbf{se})_{11}, (\mathbf{se})_{12}, \mathbf{w})$ is 9-dimensional. We obtain, using the Newton–Raphson method, after three iterations:

$$\hat{\boldsymbol{\mu}} = \begin{bmatrix} 4.339 & 3.746 \\ 4.731 & 5.260 \\ 3.581 & 5.232 \\ 4.388 & 3.774 \\ 5.145 & 5.651 \\ 3.562 & 5.192 \end{bmatrix}, \qquad \hat{\mathbf{m}} = \begin{bmatrix} 76.65 & 42.35 \\ 113.46 & 192.54 \\ 35.89 & 187.11 \\ 80.47 & 43.53 \\ 171.31 & 284.69 \\ 35.22 & 179.78 \end{bmatrix}$$

and

$$\mathbf{e} = \mathbf{y} - \hat{\mathbf{m}} = \begin{bmatrix} -4.65 & 4.65 \\ -3.46 & 3.46 \\ 8.11 & -8.11 \\ 4.65 & -4.65 \\ 3.46 & -3.46 \\ -8.11 & 8.11 \end{bmatrix}.$$

	$\hat{\mu}$	$\hat{s}_1$	$\hat{e}_1$	$\hat{e}_2$	$\hat{r}_1$
$\hat{\beta}_j$	4.550	−0.068	−0.488	−0.647	−0.259
$\hat{\sigma}(\hat{\beta}_j)$	0.034	0.030	0.050	0.039	0.029
Z_j	135.7	−2.28	−9.85	−16.45	−8.81

	$\widehat{(se)}_{11}$	$\widehat{(se)}_{12}$	$\widehat{(sr)}_{11}$	$\hat{\beta}_w$
$\hat{\beta}_j$	0.049	−0.132	−0.005	−0.561
$\hat{\sigma}(\hat{\beta}_j)$	0.050	0.036	0.029	0.047
Z_j	1.00	3.65	−0.18	−12.04

The standard error estimates given here were obtained under the assumption that the observed frequencies are independent Poisson r.v.'s. For the independent multinomial model, the standard errors for $\hat{s}_1$ is actually a bit smaller. However, since $\hat{s}_1$ is of no interest, its value being determined by the numbers of men and women sampled, no harm is done by the inclusion of a poor estimate. This will be discussed further in Section 8.7.

Under the null hypothesis that $\beta_j = 0$, $Z_j = \hat{\beta}_j/\hat{\sigma}(\hat{\beta}_j)$ is approximately distributed as standard normal. The only Z_j not significantly far from zero corresponds to $(se)_{12}$. We are not tempted to drop this term from the model because $(se)_{11}$ obviously should be included. Of course, the coefficient -0.561 of $\mathbf{w}$ is of most interest. We estimate that the logit $\log(p_{ij}/(1 - p_{ij}))$ decreases 0.561 for each step upward in educational level. That is, the odds for agreement are multiplied by the factor $e^{-0.561} = 0.571$ for each step upward in educational level. The Pearson and log chi-square values, measures of the distance from $\mathbf{Y}$ to $\hat{\mathbf{m}}$ for this model were 6.03 and 6.06 for (12 − 8) d.f., indicating a reasonably good fit. Estimates of the probability of the response "Yes" are given, under this model, by $\hat{p}_{ij} = \hat{m}_{ij1}/(\hat{m}_{ij1} + \hat{m}_{ij2}) = \hat{m}_{ij1}/Y_{ij+}$. These are

Education Level	Men	Women
1	0.644	0.649
2	0.371	0.376
3	0.161	0.164

Problem 8.5.1: Let $\mathbf{Y} = (Y_1, Y_2, Y_3)$ have a multinomial distribution with $n = 150$, $\mathbf{p} = (p_1, p_2, p_3)$, where $\log p_1 = \mu_1 = \beta_1 + \beta_2$, $\log p_2 = \mu_2 = \beta_1 + \beta_2$, $\log p_3 = \mu_3 = \beta_1$.

(a) Give explicit expressions for the MLE's $\hat{\boldsymbol{\beta}}$ of $\boldsymbol{\beta}$, $\hat{\boldsymbol{\mu}}$ of $\boldsymbol{\mu}$, and $\hat{\mathbf{m}}$ of $\mathbf{m}$, $\hat{\mathbf{p}}$ of $\mathbf{p}$. *Hint*: β_1 and β_2 may be expressed very simply in terms of m_1, m_2, m_3. The relationship between $\hat{\boldsymbol{\beta}}$ and $\hat{\mathbf{m}}$ must be the same. Remember that $(\mathbf{Y}, \mathbf{x}_1) = (\hat{\mathbf{m}}, \mathbf{x}_1)$ and $(\mathbf{Y}, \mathbf{x}_2) = (\hat{\mathbf{m}}, \mathbf{x}_2)$, where $\boldsymbol{\mu} \in V = \mathscr{L}(\mathbf{x}_1, \mathbf{x}_2)$.

(b) For $\mathbf{Y} = (80, 60, 10)$ find $\hat{\boldsymbol{\beta}}$, $\hat{\boldsymbol{\mu}}$, $\hat{\mathbf{m}}$, $\hat{\mathbf{p}}$.

(c) Use the Newton–Raphson algorithm to find $\hat{\boldsymbol{\beta}}$, beginning with $\hat{\boldsymbol{\beta}}^{(0)} = (0, 1)$.

Problem 8.5.2: Find the MLE of (β_1, β_2) in Example 8.4.1.

Problem 8.5.3: Find the MLE of $\boldsymbol{\beta} = (\mu_1, \mu_2, \mu_3, \mu_4, \gamma)'$ in Example 8.4.2. Also find the MLE of $\boldsymbol{\beta} = (\mu_1, \mu_2, \mu_3, \mu_4, \gamma_0, \gamma_1)$. Compare the corresponding estimates of $\mathbf{m}$.

Problem 8.5.4: Suppose the 1,000 father–son pairs of Problem 8.4.6 are classified as follows:

Education of Father	E_1	E_2	E_3	E_4	E_5	
E_1	52	45	21	9	4	131
E_2	64	87	45	13	7	216
E_3	32	59	83	40	28	242
E_4	23	19	51	82	42	217
E_5	10	12	38	50	84	194
	181	222	238	194	165	1,000

(a) Suppose that $\boldsymbol{\mu} \in V_s$. That is, $\boldsymbol{\mu}$ and $\mathbf{m}$ are symmetric. Find the MLE's of $\boldsymbol{\mu}$ and $\mathbf{m}$.

(b) Suppose that $\boldsymbol{\mu} \in V_{qs}$. That is, $\boldsymbol{\mu}$ is quasi-symmetric (see Problem 8.4.6). Show that for this model the MLE is

$$\hat{\mathbf{m}} = \begin{bmatrix} 52.00 & 45.56 & 19.35 & 10.07 & 4.02 \\ 63.44 & 87.00 & 46.25 & 12.49 & 6.83 \\ 33.65 & 57.75 & 83.00 & 40.41 & 27.18 \\ 21.93 & 19.51 & 50.59 & 82.00 & 42.97 \\ 9.98 & 12.17 & 38.82 & 49.03 & 84.00 \end{bmatrix}.$$

(c) Determine the Pearson and log likelihood goodness-of-fit statistics $\chi^2 = \sum_{ij} (Y_{ij} - \hat{m}_{ij})/\hat{m}_{ij}$ and $G^2 = 2 \sum_{ij} Y_{ij} \log(Y_{ij}/\hat{m}_{ij})$, for the models of (a) and (b). If $\boldsymbol{\mu} \in V_s$ these χ^2 statistics are approximately distributed as χ^2 with d.f. $= (n - \dim(V_s)) = (25 - 15) = 10$. Similarly, if $\boldsymbol{\mu} \in V_{qs}$, then d.f. $= (25 - 19) = 6$. Would you reject either of the null hypotheses $\boldsymbol{\mu} \in V_s$ or $\boldsymbol{\mu} \in V_{sq}$ at level $\alpha = 0.05$?

Problem 8.5.5: The numbers of cases of a rare cancer among the residents of a state in one year, broken down by county were as follows:

	1	2	3	4	5	6
County population in 1,000's	213	147	89	190	284	126
No. of cases	14	25	26	22	38	45
No. of people over 50 in 1,000's	58	49	34	56	83	49

Let Y_i be the number of cases in county i. Suppose that $Y_i \sim \mathscr{P}(m_i)$, independent in different counties.

(a) Suppose $m_i = \theta z_i$, where z_i is the number of people (in 1,000's) in the population. Find the MLE's of θ and $\mathbf{m}$, both in symbolic and numerical form.

(b) Suppose that $m_i = z_i \exp(\beta_0 + \beta_1 p_i)$, where p_i is the proportion of people over 50 in the population. Find the MLE's of $\boldsymbol{\beta} = (\beta_0, \beta_1)$ and $\mathbf{m}$.

Problem 8.4.6: Consider a 2×4 table $\mathbf{Y}$ with row vectors $\mathbf{Y}_1$, $\mathbf{Y}_2$ satisfying the independent multinomial model, $\mathbf{Y}_i \sim \mathscr{M}(\mathbf{p}_i, n_i)$ for $i = 1, 2$. Let $R(j) = p_{1j}/p_{2j}$ and $H(j_1, j_2) = R(j_1)/R(j_2)$. Let $\mathscr{M}$ be the model for which $\log[H(j_1, j_2)] = \beta(j_2 - j_1)$ for some parameter β, and each j_1, j_2.

(a) Express the interaction terms λ_{ij} in terms of β for the model $\mathscr{M}$. Use this to write the model in vector form, using just one vector to represent interaction. Why must row effects be included in the model? What is $\dim(V)$?

(b) For $\mathbf{Y} = \begin{bmatrix} 37 & 47 & 83 & 133 \\ 62 & 52 & 68 & 68 \end{bmatrix}$ find the MLE of $\mathbf{m}$ under model $\mathscr{M}$. You will need a computer, or much patience. For those without a computer, we give $\hat{m}_{11} = 36.833$, $\hat{m}_{12} = 46.195$. Thought and a \$10 calculator should be enough now. Verify that $G^2 = 0.176$, indicating a very good fit.

(c) Why must each $\mathbf{m}$ under this model satisfy

$$[m_{11}m_{22}m_{23}m_{14}]/[m_{21}m_{12}m_{13}m_{24}] = 1,$$

and

$$[m_{11}m_{22}^3m_{13}^3m_{24}]/[m_{21}m_{12}^3m_{23}^3m_{14}] = 1?$$

(d) Let the model $\mathscr{M}^*$ correspond to $\mathbf{p}_1 = \mathbf{p}_2$. Find the MLE of $\mathbf{m}$ for this model. Which model, $\mathscr{M}$ or $\mathscr{M}^*$, seems most appropriate? What is the subspace V^* corresponding to $\mathscr{M}^*$ and its dimension? Verify that $G^2 = 24.97$ for $\mathscr{M}^*$, indicating a rather poor fit.

8.6 THE CHI-SQUARE STATISTICS

For linear models F-statistics were used to measure the adequacy of the fit of a model. We always had to begin with a model, called the full model, and used the F-statistic to help decide whether some smaller linear model was adequate. For log-linear models, with Poisson or multinomial sampling, we use chi-square statistics, the general name for statistics which are asymptotically distributed as χ^2 under the null hypothesis. These statistics are measures of distance between two vectors, either between $\mathbf{Y}$ and $\hat{\mathbf{m}}$, or between $\hat{\mathbf{m}}$ and another estimate of $\mathbf{m}$, say $\hat{\mathbf{m}}_0$.

We will be particularly interested in two distance measures:

Pearson chi-square: $\chi^2(\mathbf{x}, \mathbf{y}) = \sum_i (y_i - x_i)^2/x_i$, defined for all $y_i \geq 0$, $x_i > 0$.

Log chi-square or deviance: $G^2(\mathbf{x}, \mathbf{y}) = 2 \sum_i y_i \log(y_i/x_i)$,

defined for all $y_i > 0$, $x_i > 0$.

Let $\mathbf{J}$ be the vector of all ones. In most applications $\sum_i x_i = (\mathbf{J}, \mathbf{x}) = (\mathbf{J}, \mathbf{y}) = \sum_i y_i = n$. In this case these statistics can be written as inner products.

$$\chi^2(\mathbf{x}, \mathbf{y}) = \sum_i y_i^2/x_i - n = (\mathbf{y}/\mathbf{x}, \mathbf{y}) - (\mathbf{J}, \mathbf{y}) = (\mathbf{y}/\mathbf{x} - \mathbf{J}, \mathbf{y}),$$

and

$$G^2(\mathbf{x}, \mathbf{y}) = 2(\mathbf{y}, \log(\mathbf{y}/\mathbf{x})) = 2[(\mathbf{y}, \log \mathbf{y}) - (\mathbf{y}, \log \mathbf{x})].$$

$\chi^2(\mathbf{x}, \mathbf{y})$ obviously takes its minimum value for $\mathbf{x} = \mathbf{y}$. To see that this is also true for G^2, use Lagrangian multipliers. For fixed $\mathbf{y}$ let $H(\mathbf{x}, \lambda) =$

$(\mathbf{y}, \log \mathbf{y} - \log \mathbf{x}) + \lambda[(\mathbf{x}, \mathbf{J}) - n]$. Then $\frac{\partial}{\partial \mathbf{x}} H(\mathbf{x}, \lambda) = -\mathbf{y}/\mathbf{x} + \lambda \mathbf{J}$. Setting this equal to the zero vector, and using the fact that $(\mathbf{x}, \mathbf{J}) = n$, we get $\mathbf{x} = \mathbf{y}$. The matrix of second partial derivatives is diag($\mathbf{y}/\mathbf{x}$), which is positive definite, so that this is a minimum. Since $G^2(\mathbf{x}, \mathbf{x}) = 0$, $G^2(\mathbf{x}, \mathbf{y}) \geq 0$ whenever $\sum x_i = \sum y_i$.

It is easy to show that for constants c_1, c_2: $\chi^2(c_1\mathbf{x}, c_2\mathbf{y}) = c_1\chi^2(\mathbf{x}, \mathbf{y}) + (c_1 - c_2)n$ and $G^2(c_1\mathbf{x}, c_2\mathbf{y}) = c_2 G^2(\mathbf{x}, \mathbf{y}) + c_2 n \log(c_1/c_2)$. Notice that if $c = c_1 = c_2$ then $\chi^2(c\mathbf{x}, c\mathbf{y}) = c\chi^2(\mathbf{x}, \mathbf{y})$ and $G^2(c\mathbf{x}, c\mathbf{y}) = cG^2(\mathbf{x}, \mathbf{y})$.

Another useful identity:

$$\chi^2(\mathbf{x}, \mathbf{y}) - \chi^2(\mathbf{y}, \mathbf{x}) = \sum_i (y_i - x_i)^2[1/x_i - 1/y_i] = \sum (y_i - x_i)^3/x_i y_i. \quad (8.6.1)$$

This last term will be much smaller than either χ^2 if $\mathbf{x}$ and $\mathbf{y}$ are close, so that $\chi^2(\mathbf{x}, \mathbf{y})$ and $\chi^2(\mathbf{y}, \mathbf{x})$ will often be relatively close.

The distance measures $G^2(\mathbf{x}, \mathbf{y})$ and $\chi^2(\mathbf{x}, \mathbf{y})$ will be close whenever $\mathbf{x}$ and $\mathbf{y}$ are close in a sense to be discussed. By Taylor's Theorem for $|\delta| < 1$, $\log(1 - \delta) = -\delta - (1/2)\delta^2 + o(\delta^2)$, where $\lim_{\delta \to 0} o(\delta^2)/\delta^2 = 0$. For a pair of numbers (x, y) let $\Delta = (y - x)/y = 1 - x/y$. Then $\log(y/x) = -\log(x/y) = -\log(1 - \Delta) = \Delta + \Delta^2/2 + o(\Delta^2)$.

Let $\Delta_i = (y_i - x_i)/y_i$. Then $G^2(\mathbf{x}, \mathbf{y}) = 2(\mathbf{y}, \log(\mathbf{y}/\mathbf{x})) = 2[(\mathbf{y}, \mathbf{\Delta}) + (\mathbf{y}, \mathbf{\Delta}^2/2) + \sum_i o(\Delta_i^2)y_i] = \sum_i (y_i - x_i) + \chi^2(\mathbf{y}, \mathbf{x}) + \sum_i y_i o(\Delta_i^2)$. In most application $\sum y_i = \sum x_1 = n$, so that the first term on the right is zero, and

$$G^2(\mathbf{x}, \mathbf{y}) = \chi^2(\mathbf{y}, \mathbf{x}) + \sum_i y_i o(\Delta_i^2). \quad (8.6.2)$$

After all this preparatory work we are ready to replace $\mathbf{x}$ and $\mathbf{y}$ by estimators. Let $\mathbf{p}$ be a T-component probability vector and let $\{\hat{\mathbf{p}}_n = \hat{\mathbf{m}}_n/n\}$ and $\{\hat{\mathbf{p}}_n^* = \hat{\mathbf{m}}_n^*/n\}$ be two sequences of estimators of $\mathbf{p}$, with components summing to one for each n. Let

$$\mathbf{Z}_n = (\hat{\mathbf{m}}_n - n\mathbf{p})/\sqrt{n} = (\hat{\mathbf{p}}_n - \mathbf{p})\sqrt{n}, \quad \text{and} \quad \mathbf{Z}_n^* = (\hat{\mathbf{m}}_n^* - n\mathbf{p})/\sqrt{n} = (\hat{\mathbf{p}}_n^* - \mathbf{p})\sqrt{n}.$$

Then $\hat{\mathbf{m}}_n - \hat{\mathbf{m}}_n^* = (\mathbf{Z}_n - \mathbf{Z}_n^*)\sqrt{n}$. Suppose that the sequences $\{\mathbf{Z}_n\}$ and $\{\mathbf{Z}_n^*\}$ are each tight. A sequence of random variables $\{W_n\}$ is *tight* if for every $\varepsilon > 0$ there exists a constant K_ε such that $P(|W_n| > K_\varepsilon) < \varepsilon$ for all n. A sequence of random vectors is tight if each component is tight. The tightness of the sequences $\{\mathbf{Z}_n\}$ and $\{\mathbf{Z}_n^*\}$ implies that there is a cube in T-space within which all these random vectors will lie with probability close to one. Tightness is implied by the convergence in distribution of the sequence. For our two sequences it implies that $\{\hat{\mathbf{p}}_n\}$ and $\{\hat{\mathbf{p}}_n^*\}$ both converge in probability to $\mathbf{p}$.

Theorem 8.6.1

(1) $D_1(\hat{\mathbf{m}}_n, \hat{\mathbf{m}}_n^*) \equiv [\chi^2(\hat{\mathbf{m}}_n, \hat{\mathbf{m}}_n^*) - \chi^2(\hat{\mathbf{m}}_n^*, \hat{\mathbf{m}}_n)]$
$= n[\chi^2(\hat{p}_n, \hat{p}_n^*) - \chi^2(\hat{p}_n^*, \hat{p}_n)]$ converges in probability to zero.

(2) $D_2(\hat{\mathbf{m}}_n, \hat{\mathbf{m}}_n^*) \equiv [G^2(\hat{\mathbf{m}}_n, \hat{\mathbf{m}}_n^*) - \chi^2(\hat{\mathbf{m}}_n, \hat{\mathbf{m}}_n^*)]$ converges in probability to zero.

(3) $D_3(\hat{\mathbf{m}}_n^*, \hat{\mathbf{m}}_n) \equiv [G^2(\hat{\mathbf{m}}_n, \hat{\mathbf{m}}_n^*) - G^2(\hat{\mathbf{m}}_n^*, \hat{\mathbf{m}}_n)]$ converges in probability to zero.

***Proof*:** From (8.6.1)

$$D_1(\hat{\mathbf{m}}_n, \hat{\mathbf{m}}_n^*) = \sum_i (\hat{m}_{ni} - m_{ni}^*)^3/m_{ni}m_{ni}^* = \sum_i (Z_{ni} - Z_{ni}^*)^3 n^{3/2}/[n^2 \hat{p}_{ni}\hat{p}_{ni}^*],$$

which converges in probability to zero. From (8.6.2)

$$D_2(\hat{\mathbf{m}}_n, \hat{\mathbf{m}}_n^*) = \chi^2(\hat{\mathbf{m}}_n, \hat{\mathbf{m}}_n^*) + D_1(\hat{\mathbf{m}}_n, \hat{\mathbf{m}}_n^*) + \sum_i \hat{m}_{ni}(\hat{m}_{ni} - \hat{m}_{ni}^*)^3/\hat{m}_{ni}^3.$$

The second term converges in probability to zero by (1). The third term is $\sum_i n\hat{p}_{ni}(Z_{ni} - Z_{ni}^*)^3 n^{3/2}/n^3\hat{p}_{ni}^3$, which also converges in probability to zero.

To prove (3) note that

$$D_3(\hat{\mathbf{m}}_n, \hat{\mathbf{m}}_n^*) = [G^2(\hat{\mathbf{m}}_n, \hat{\mathbf{m}}_n^*) - \chi^2(\hat{\mathbf{m}}_n, \hat{\mathbf{m}}_n^*)]$$
$$+ [\chi^2(\hat{\mathbf{m}}_n, \hat{\mathbf{m}}_n^*) - \chi^2(\hat{\mathbf{m}}_n^*, \hat{\mathbf{m}})] + [\chi^2(\hat{\mathbf{m}}_n^*, \hat{\mathbf{m}}_n) - G^2(\hat{\mathbf{m}}_n^*, \hat{\mathbf{m}}_n)],$$

and that each of the terms within brackets converges in probability to zero by (1) and (2). □

For simplicity of the discussion and proof we have defined $\mathbf{Z}_n = (\hat{\mathbf{p}}_n - \mathbf{p})\sqrt{n}$, and $\mathbf{Z}_n^* = (\hat{\mathbf{p}}_n^* - \mathbf{p})\sqrt{n}$. However, $\mathbf{p}$ can be replaced by $\mathbf{p}_n = \mathbf{p} + \mathbf{d}/\sqrt{n}$, for a constant vector $\mathbf{d}$ with $(\mathbf{d}, \mathbf{J}) = 0$, so that $\mathbf{p}_n$ remains a probability vector, and the theorem still holds. The limiting distributions of χ^2 and G^2 will be depend on $\mathbf{d}$, in a way to be discussed later.

Notice that we can take $\hat{\mathbf{m}}_n^* \equiv n\mathbf{p}$, and, from (1), conclude that $p\lim_n[\chi^2(\hat{\mathbf{m}}_n, \mathbf{p}) - \chi^2(\mathbf{p}, \hat{\mathbf{m}}_n)] = 0$, (limit in probability) if $\{\mathbf{Z}_n\}$ is tight. In particular, if $\mathbf{Y}_n \sim \mathcal{M}_T(n, \mathbf{p})$, then, as shown in Section 8.2, $\chi^2(\mathbf{p}, \mathbf{Y}_n)$ is asymptotically distributed as χ^2_{T-1}, so that $\chi^2(\mathbf{Y}_n, \mathbf{p}) = \sum_i (Y_{ni} - np_i)^2/Y_{ni}$ differs from $\chi^2(\mathbf{p}, \mathbf{Y}_n)$ by a random amount which converges in probability to zero as $n \to \infty$.

G^2 and χ^2 are measures of distance, though they are not metrics because they are not symmetric in their two arguments. Theorem 8.6.1 shows that when $\{\mathbf{Z}_n\}$ and $\{\mathbf{Z}_n^*\}$ are tight, as will be the case if the model used to determine $\hat{\mathbf{m}}_n$ and $\hat{\mathbf{m}}_n^*$ is true, G^2 and χ^2 are almost symmetric, and therefore "almost" metrics.

G^2 possesses two other useful properties. Consider the product multinomial model, with parameter vectors $\mathbf{p} = (\mathbf{p}_1, \ldots, \mathbf{p}_k)$ and $\mathbf{n} = (n_1, \ldots, n_k)$ as defined earlier. Suppose that $\log \mathbf{m} = \boldsymbol{\mu} = \log(n_1\mathbf{p}_1, \ldots, n_k\mathbf{p}_k)$. The log likelihood function for the multinomial model is $l_{mn}(\boldsymbol{\beta}, \mathbf{y}) = (\mathbf{y}, \boldsymbol{\mu}) - \sum n_j \log(n_j) + C^*$, where C^* does not depend on $\boldsymbol{\beta}$. But, $G^2(\mathbf{m}, \mathbf{y}) = 2(\mathbf{y}, \log \mathbf{y}/\mathbf{m}) = -2(\mathbf{y}, \boldsymbol{\mu} - \log \mathbf{y}) = -2l_{mn}(\boldsymbol{\beta}, \mathbf{y}) + 2(\mathbf{y}, \log \mathbf{y}) - 2\sum n_j \log n_j + 2C^*$. As a function of $\boldsymbol{\beta}$, $G^2(\mathbf{m}, \mathbf{y})$ takes its minimum value when $\boldsymbol{\beta} = \hat{\boldsymbol{\beta}}$, the MLE for $\boldsymbol{\beta}$. Thus, in this sense, $(\hat{\boldsymbol{\beta}}, \hat{\boldsymbol{\mu}}, \hat{\mathbf{m}})$ is the minimum distance estimator, as well as the MLE of $(\boldsymbol{\beta}, \boldsymbol{\mu}, \mathbf{m})$.

Another useful property of G^2 is its *additivity* as a measure of distance. First consider any subspace V of R_T and let $(\hat{\boldsymbol{\mu}}, \hat{\mathbf{m}})$ be the MLE for $(\boldsymbol{\mu}, \mathbf{m})$ corresponding to an observation $\mathbf{Y} = \mathbf{y}$ and the subspace V. Then $(\mathbf{y} - \hat{\mathbf{m}}) \perp V$, so that $(\hat{\mathbf{m}}, \mathbf{v}) = (\mathbf{y}, \mathbf{v})$ for all $\mathbf{v} \in V$. If $V_1 \supset V_2$, with corresponding MLE's $\hat{\mathbf{m}}_1$, $\hat{\mathbf{m}}_2$ then for any $\mathbf{v} \in V_2$, $(\hat{\mathbf{m}}_1, \mathbf{v}) = (\mathbf{y}, \mathbf{v}) = (\hat{\mathbf{m}}_2, \mathbf{v})$. Now consider three subspaces $V_1 \supset V_2 \supset V_3$ of R_T of dimensions $k_1 > k_2 > k_3$. Let $(\hat{\boldsymbol{\mu}}_i, \hat{\mathbf{m}}_i)$ be the maximum likelihood estimator of $(\boldsymbol{\mu}, \mathbf{m})$ under the model which states that $\boldsymbol{\mu}_i \in V_i$. (We will shorten this to "the model V_i" and will write G_{ij}^2 to denote $G^2(\hat{\mathbf{m}}_i, \hat{\mathbf{m}}_j)$). Then

$$
\begin{aligned}
G_{31}^2 &= 2(\hat{\mathbf{m}}_1, \hat{\boldsymbol{\mu}}_1 - \hat{\boldsymbol{\mu}}_3) = 2(\hat{\mathbf{m}}_1, \hat{\boldsymbol{\mu}}_1 - \hat{\boldsymbol{\mu}}_2) + 2(\hat{\mathbf{m}}_1, \hat{\boldsymbol{\mu}}_2 - \hat{\boldsymbol{\mu}}_3) \\
&= G_{21}^2 + 2(\hat{\mathbf{m}}_2, \hat{\boldsymbol{\mu}}_2 - \hat{\boldsymbol{\mu}}_3) = G_{21}^2 + G_{32}^2.
\end{aligned}
$$

The third equality holds because $(\hat{\boldsymbol{\mu}}_2 - \hat{\boldsymbol{\mu}}_3) \in V_2$ and, from the previous sentence, $\hat{\mathbf{m}}_1$ and $\hat{\mathbf{m}}_2$ have the same inner product with all vectors in V_2.

If we wish to test the hypothesis that $\boldsymbol{\mu} \in V_3$, assuming the model V_2, we can take $V_1 = R_T$, so the model V_1 is saturated and $\hat{\mathbf{m}}_1 = \mathbf{y}$. Then $G^2(\hat{\mathbf{m}}_3, \hat{\mathbf{m}}_2) = G^2(\hat{\mathbf{m}}_3, \mathbf{y}) - G^2(\hat{\mathbf{m}}_2, \mathbf{y})$. For this reason, the "distances" $G^2(\hat{\mathbf{m}}, \mathbf{y})$ can be conveniently combined to test various hypotheses as appropriate models are sought. The same additivity does not hold for χ^2, though Theorem 8.6.1 indicates that when a null hypothesis $\boldsymbol{\mu} \in V_3$ holds, the additivity holds in approximation. See Figure 8.9, where $\mathscr{M}_i = \{\mathbf{m} \mid \mathbf{m} = \exp(\boldsymbol{\mu}), \boldsymbol{\mu} \in V_i\}$.

Example 8.6.1: Consider the sample space of 2×3 tables. Suppose that the random variable $\mathbf{Y} = (Y_{ij}) \sim \mathscr{M}_6(40, \mathbf{p})$ for $\mathbf{p} = \begin{bmatrix} 0.30 & 0.18 & 0.12 \\ 0.20 & 0.12 & 0.08 \end{bmatrix}$. Notice that rows and columns are independent, so that $\boldsymbol{\mu} \in V = \mathscr{L}(\mathbf{R}_1, \mathbf{R}_2, \mathbf{C}_1, \mathbf{C}_2, \mathbf{C}_3)$, a 4-dimensional subspace. Then $\mathbf{m} = 40\mathbf{p}$ and

$$
\boldsymbol{\mu} = \log \mathbf{m} = \log 40 + \begin{bmatrix} -1.204 & -1.715 & -2.120 \\ -1.609 & -2.120 & -2.526 \end{bmatrix}.
$$

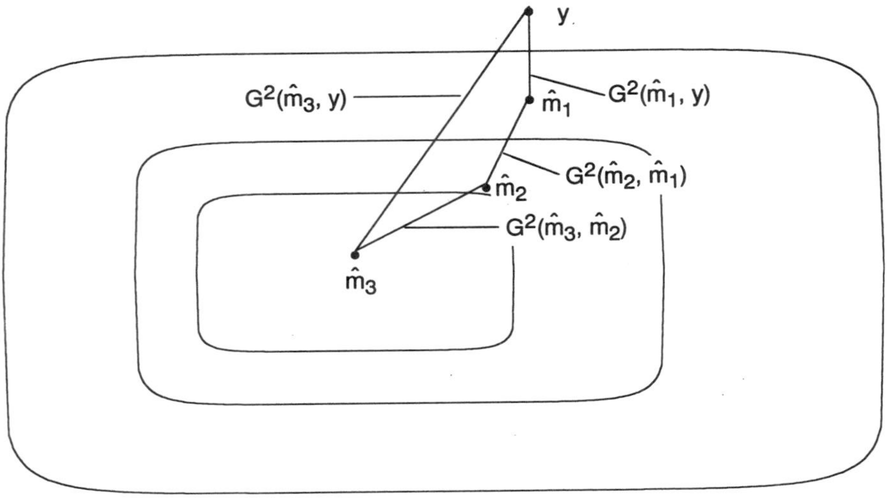

FIGURE 8.9 *T*-space.

Notice that $\boldsymbol{\mu}$ has additivity. That is, $\mu_{ij} - \mu_{ij'} = \mu_{i'j} - \mu_{i'j'}$, which is equivalent to the corresponding odds-ratios being one.

One thousand observations were taken on $\mathbf{Y}$, then $\chi^2 = \chi^2(\hat{\mathbf{m}}, \mathbf{Y})$ and $G^2 = G^2(\hat{\mathbf{m}}, \mathbf{Y})$ were determined for each. Figure 8.10 contains a histogram of the χ^2 values obtained and a scatterplot of the pairs (χ^2, G^2). These pairs had mean (2.009, 2.147), variances 3.480 and 4.503, and correlation 0.987. The limiting distribution has 90th and 95th percentiles 4.605 and 5.991. The χ^2-statistic exceeded these values 93 and 44 times, while the G^2-statistic exceeded them 111 and 65 times.

Continuing with the example, suppose that we let $\mathbf{p} = \begin{bmatrix} 0.35 & 0.13 & 0.12 \\ 0.15 & 0.17 & 0.08 \end{bmatrix}$. Independence of row and columns no longer holds. In fact, $\boldsymbol{\mu} = \log 40 + \begin{bmatrix} -1.050 & -2.040 & -2.120 \\ -1.897 & -1.772 & -2.526 \end{bmatrix}$. The vector of interaction terms is $\boldsymbol{\mu}_\perp = \begin{bmatrix} 0.260 & -0.298 & 0.039 \\ -0.260 & 0.298 & -0.039 \end{bmatrix}$. Again we fit the independence model and determine $\chi^2 = \chi^2(\hat{\mathbf{m}}, \mathbf{Y})$ and $G^2 = G^2(\hat{\mathbf{m}}, \mathbf{Y})$ for 1,000 observations on $\mathbf{Y}$. As shown in Figure 8.11, χ^2 and G^2 are still close. The histogram for χ^2 has moved to the right. We will indicate later than both χ^2 and G^2 are asymptotically distributed as noncentral χ^2 under certain conditions. χ^2 and G^2 had means 4.288 and 4.482, variances 11.85 and 13.97, and correlation 0.995. They exceeded $\chi^2_{0.90} = 4.61$ by 365 and 377 times among the 1,000 $\mathbf{Y}$'s, and exceeded $\chi^2_{0.95} = 5.99$ by 262 and 275 times, indicating that the power for each test is approximately 0.37 for the $\alpha = 0.10$ test and 0.27 for the $\alpha = 0.05$ test. As

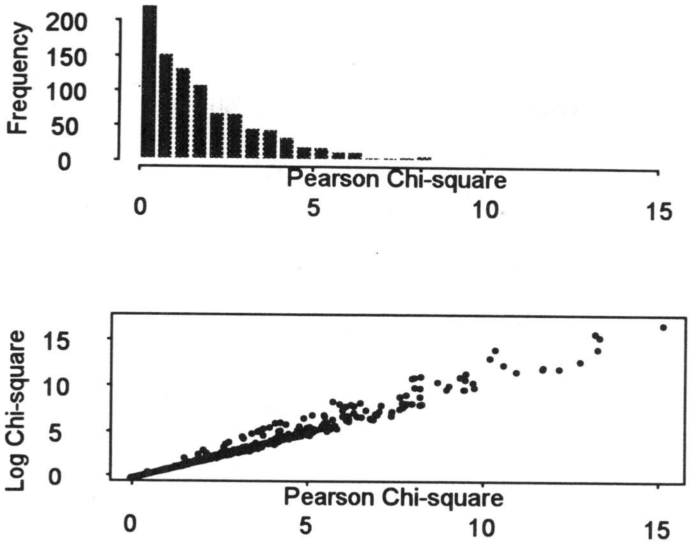

FIGURE 8.10 Histogram of Pearson chi-squares and scatterplot of (Pearson, log) chi-squares for 1,000 experiments for the case of independence.

will be shown later, the power may be determined in approximation from the Pearson–Hartley charts for the noncentrality parameter $\delta = G^2(\mathbf{m}_1, \mathbf{m}) = 2.222$, where $\mathbf{m}_1$ is the MLE of $\mathbf{m}$ corresponding to the observation $\mathbf{m} = 40\mathbf{p}$, and $\nu_1 = 2$, $\nu_2 =$ infinity.

The Wald Statistic: Still another goodness-of-fit statistic is sometimes used. Let $\mathbf{Y}$ satisfy a log-linear model with $\boldsymbol{\mu} = \log \mathbf{m} = \mathbf{X}\boldsymbol{\beta}$, where $\mathbf{X}$ is $T \times k$, of rank k. Let V be the column space of $\mathbf{X}$. Let V_0 and V_1 be subspaces of V of dimensions k_0 and k_1, $V_0 \subset V_1$, where $k_0 < k_1 \leq k$. Let $\hat{\mathbf{m}}_0$ and $\hat{\mathbf{m}}_1$ be the MLE's of $\mathbf{m}$ corresponding to V_0 and V_1. Let $\hat{\mathbf{m}}$ be a consistent estimator of $\mathbf{m}$. $\hat{\mathbf{m}}$ could be the MLE of $\mathbf{m}$ corresponding to V, or $\hat{\mathbf{m}}_0$, or $\hat{\mathbf{m}}_1$ as long as $\hat{\mathbf{m}}$ is consistent. Define $I(\hat{\mathbf{m}}) = \mathbf{X}'d(\hat{\mathbf{m}})\mathbf{X}$, the estimator of information matrix for the model $\mathbf{m} \in V$. Define $W(\mathbf{u}, \mathbf{v}) = \mathbf{u}'I(\mathbf{v})^{-1}\mathbf{u}$. Then

$$W = W(\hat{\mathbf{m}}_0 - \hat{\mathbf{m}}_1, \hat{\mathbf{m}}) = (\hat{\mathbf{m}}_1 - \hat{\mathbf{m}}_0)'\mathbf{X}I(\hat{\mathbf{m}})^{-1}\mathbf{X}'(\hat{\mathbf{m}}_1 - \hat{\mathbf{m}}_0) \qquad (8.6.3)$$

is called *Wald's statistic*. In more generality, Wald's statistic W is of the form $W(\hat{\boldsymbol{\gamma}}) = \hat{\boldsymbol{\gamma}}'\hat{\mathbf{I}}^{-1}\hat{\boldsymbol{\gamma}}$, where $\hat{\boldsymbol{\gamma}}$ is the vector of first partial derivatives of the log likelihood, and $\hat{\mathbf{I}}$ is a consistent estimator of the information matrix, the negative of the matrix of the expectations of second partial derivatives of the likelihood function. If $\hat{\boldsymbol{\gamma}}$ is asymptotically multivariate normal with mean vector $\boldsymbol{\gamma}$, covariance matrix $\mathbf{C}$, then W is asymptotically noncentral χ^2 with k degrees of freedom and noncentrality parameter $\boldsymbol{\gamma}'\mathbf{C}\boldsymbol{\gamma}$. It can be shown that $W(\hat{\mathbf{m}}_0 - \hat{\mathbf{m}}_1, \hat{\mathbf{m}})$

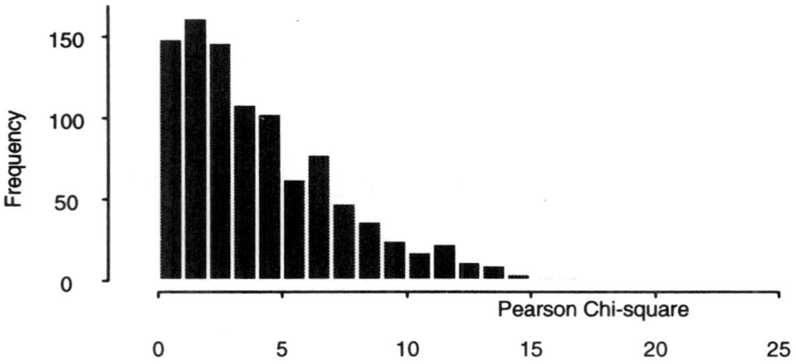

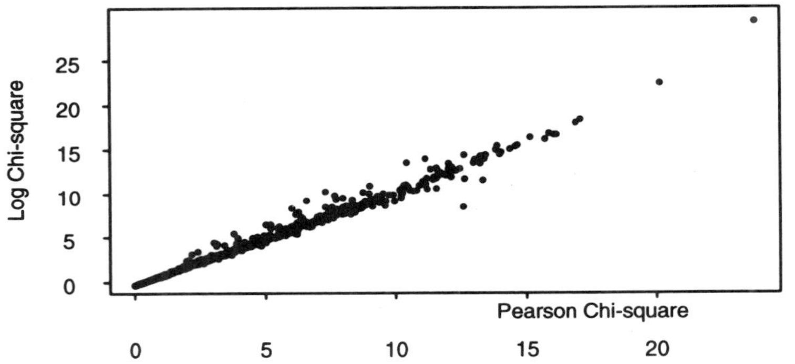

FIGURE 8.11 Histogram of Pearson chi-squares and scatterplot of (Pearson, log) chi-squares for 1,000 experiments for a case of dependence.

is close to $G^2(\hat{\mathbf{m}}_0, \hat{\mathbf{m}}_1)$ and $\chi^2(\hat{\mathbf{m}}_0, \hat{\mathbf{m}}_1)$ if $\hat{\mathbf{m}}_1 - \hat{\mathbf{m}}_0$ is *reasonably close* to $\mathbf{0}$, and that $W(\hat{\mathbf{m}}_0 - \hat{\mathbf{m}}_1, \hat{\mathbf{m}})$ has the same asymptotic distribution as do $G^2(\hat{\mathbf{m}}_0, \hat{\mathbf{m}}_1)$ and $\chi^2(\hat{\mathbf{m}}_0, \hat{\mathbf{m}}_1)$ under the conditions given in Theorem 8.6.1. Notice that when $\hat{\mathbf{m}} = \mathbf{Y}$, $W(\hat{\mathbf{m}}_0 - \hat{\mathbf{m}}_1, \hat{\mathbf{m}}) = \chi^2(\hat{\mathbf{m}}_1, \hat{\mathbf{m}}_0)$. Note also that if $V = V_1$, then $W(\hat{\mathbf{m}}_0 - \hat{\mathbf{m}}_1, \hat{\mathbf{m}}) = W(\hat{\mathbf{m}}_0 - \mathbf{Y}, \hat{\mathbf{m}})$. We omit proofs.

The Power Divergence Statistic: Read and Cressie (1988), in their book on goodness-of-fit statistics, discuss a statistic they introduced in 1984, the power divergence statistic:

$$I^{\lambda}(\mathbf{x}, \mathbf{y}) = [2/\lambda(1 + \lambda)](\mathbf{y}, (\mathbf{y}/\mathbf{x})^{\lambda} - 1).$$

I^{λ} is defined for all real λ by assigning the values $I^0 = \lim_{\lambda \to 0} I^{\lambda}$ and $I^{-1} = \lim_{\lambda \to -1} I^{\lambda}$. Then, $I^0 = G^2$ and $I^{-1} = \chi^2$ (see Problem 8.6.6). $I^{-1/2} = 4 \sum (y_i^{1/2} - x_i^{1/2})^2 \equiv K(\mathbf{x}, \mathbf{y})$ is called the Freeman–Tukey statistic. Read and

Cressie show that I^λ behaves asymptotically as do G^2 and χ^2. They recommend use of $I^{2/3}$ as a test against general alternatives.

Problem 8.6.1: Let $\mathbf{y}_1 = (25, 20, 15, 10)$, $\mathbf{y}_2 = (22.5, 22.5, 15, 10)$, $\mathbf{y}_3 = (20, 20, 20, 10)$.

(a) Verify that $G^2(\mathbf{y}_3, \mathbf{y}_1) = G^2(\mathbf{y}_2, \mathbf{y}_1) + G^2(\mathbf{y}_3, \mathbf{y}_2)$. Is this true for any three vectors $\mathbf{y}_1$, $\mathbf{y}_2$, $\mathbf{y}_3$? If not, why is it true in this case?

(b) Determine whether the equality of (a) holds when G^2 is replaced by χ^2.

Problem 8.6.2: Let $\mathbf{y}_1 = (358, 245, 417)$, $\mathbf{y}_2 = (367, 233, 420)$. Compare the values of $G^2(\mathbf{y}_1, \mathbf{y}_2)$, $G^2(\mathbf{y}_2, \mathbf{y}_1)$, $\chi^2(\mathbf{y}_1, \mathbf{y}_2)$, $\chi^2(\mathbf{y}_2, \mathbf{y}_1)$, $K(\mathbf{y}_2, \mathbf{y}_1)$, $K(\mathbf{y}_1, \mathbf{y}_2)$.

Problem 8.6.3: Consider Problem 8.5.6. Let $\hat{\mathbf{m}}$ be the MLE of $\mathbf{m}$ under model $\mathscr{M}$, and let $\hat{\mathbf{m}}^*$ be the MLE of $\mathbf{m}$ under $\mathscr{M}^*$ (equivalently, $\boldsymbol{\beta} = 0$).

(a) Why should $G^2(\hat{\mathbf{m}}^*, \mathbf{Y}) = G^2(\hat{\mathbf{m}}, \mathbf{Y}) + G^2(\hat{\mathbf{m}}^*, \hat{\mathbf{m}})$?

(b) Verify this equality for $\mathbf{Y} = \mathbf{y}$, $\hat{\mathbf{m}}$, $\hat{\mathbf{m}}^*$ as determined in Problem 8.5.6.

Problem 8.6.4: Prove that $\chi^2(\mathbf{x}, \mathbf{y}) = ((\mathbf{y}/\mathbf{x} - \mathbf{J}), \mathbf{y}) = \sum y_i^2/x_i - n$ whenever $\sum x_i = \sum y_i = n$.

Problem 8.6.5: Prove the statements concerning the Wald statistic W made in the next to last sentence and the previous sentence of the paragraph preceding the discussion of the power divergence statistics.

Problem 8.6.6: Show that $\lim_{\lambda \to 0} I^\lambda(\mathbf{x}, \mathbf{y}) = G^2(\mathbf{x}, \mathbf{y})$, $\lim_{\lambda \to -1} I^\lambda(\mathbf{x}, \mathbf{y}) = \chi^2(\mathbf{x}, \mathbf{y})$, $I^{-2}(\mathbf{x}, \mathbf{y}) = \chi^2(\mathbf{y}, \mathbf{x})$, and $I^{-1/2}(\mathbf{x}, \mathbf{y}) = K(\mathbf{x}, \mathbf{y})$.

8.7 THE ASYMPTOTIC DISTRIBUTIONS OF $\hat{\boldsymbol{\beta}}$, $\hat{\boldsymbol{\mu}}$, AND $\hat{\mathbf{m}}$

Under suitable conditions each of the parameter vectors $\hat{\boldsymbol{\beta}}$, $\hat{\boldsymbol{\mu}}$, and $\hat{\mathbf{m}}$ are asymptotically distributed as multivariate normal with mean vectors $\boldsymbol{\beta}$, $\boldsymbol{\mu}$, and $\mathbf{m}$, and covariance matrices which depend on the space V, and on the subspace V_0 of V corresponding to the probability model chosen: Poisson, multinomial, or product multinomial. We will present the results of this asymptotic theory with only a hint of the proofs. In general, the proofs depend on the asymptotic normality of the multinomial distribution plus the delta method, which exploits the fact that for large n smooth functions are *almost linear*. With the exception of a few exact probability statements (the packages StatXact and LogXact are exceptions), all probability statements found in computer software packages for the analysis of frequency data use the approximations given by this asymptotic theory.

Let $\{\mathbf{Y}^{(n)}, n = 1, 2, \ldots\}$ be a sequence of T-component random vectors with respective mean vectors $\mathbf{m}^{(n)}$ and log means $\boldsymbol{\mu}^{(n)} = \log \mathbf{m}^{(n)}$. We suppose that $\boldsymbol{\mu}^{(n)}$ lies in a subspace V of the space of all T-component vectors and that the vector $\mathbf{J}$ lies in V. To get asymptotic results for multinomial and Poisson models, we suppose that $\mathbf{m}^{(n)}/n \to \mathbf{m}^*$ as $n \to \infty$. Thus, $\log \mathbf{m}^{(n)} - \log n = \boldsymbol{\mu}^{(n)} - \log n \to \log \mathbf{m}^* \equiv \boldsymbol{\mu}^* \in V$.

For multinomial models we have the following. Let $\mathscr{I}_1, \ldots, \mathscr{I}_{k_0}$ be a partitioning of the index set $\mathscr{I}$ of the vector $\mathbf{Y}^{(n)}$. Let $\mathbf{w}_1, \ldots, \mathbf{w}_{k_0}$ be the indicators of $\mathscr{I}_1, \ldots, \mathscr{I}_{k_0}$ and suppose that these k_0 vectors are contained in V. Let $\mathbf{Y}_i^{(n)}$ be the vector of components of $\mathbf{Y}^{(n)}$ corresponding to $\mathscr{I}_i$ and suppose $\mathbf{Y}_i^{(n)} \sim \mathscr{M}_{k_i}(n_i, \mathbf{p}_i)$, where k_i is the number of elements of $\mathscr{I}_i$. Then $E(\mathbf{Y}_i^{(n)}) = n_i^{(n)}\mathbf{p}_i$, so that $\mathbf{m}^{(n)} = (n_1^{(n)}\mathbf{p}_1, \ldots, n_{k_0}^{(n)}\mathbf{p}_{k_0})$. The sequence $\mathbf{m}^{(n)}/n$ converges to a constant vector $\mathbf{m}^*$ if and only if $n_i^{(n)}/n$ converges to a constant for each i. The approximations given by the asymptotic theory to be discussed here will be best when all n_i are large. The index n in the superscript can usually be considered to be the total sample size, though this cannot be the case for the Poisson model. If n is the total sample size, then $\mathbf{m}^* = (\mathbf{p}_1, \ldots, \mathbf{p}_{k_0})$, so that the components of $\mathbf{m}^*$ add to k_0. We suppose also that $\mathbf{Y}_1^{(n)}, \ldots, \mathbf{Y}_{k_0}^{(n)}$ are independent.

In the following let each vector be written as a T-component column vector, so that we can use matrix algebra. Suppose that $\mathbf{X} = (\mathbf{x}_1, \ldots, \mathbf{x}_k)$, and $\mathbf{X}_0 = (\mathbf{w}_1, \ldots, \mathbf{w}_{k_0})$, and suppose that both matrices have full column rank. Often, but not necessarily, $\mathbf{w}_i = \mathbf{x}_i$ for $i = 1, \ldots, k_0$. We do suppose that $V_0 \equiv \mathscr{L}(\mathbf{w}_1, \ldots, \mathbf{w}_{k_0}) \subset \mathscr{L}(\mathbf{x}_1, \ldots, \mathbf{x}_k) \equiv V$.

Define $\mathbf{D}^* = \operatorname{diag}(\mathbf{m}^*) = d(\mathbf{m}^*)$, the $T \times T$ diagonal matrix with diagonal elements $\mathbf{m}^*$. We will not in general know $\mathbf{m}^*$, but will be able to estimate it. Let

$$\mathbf{H} = [\mathbf{X}'\mathbf{D}^*\mathbf{X}]^{-1} \quad \text{and} \quad \mathbf{H}_0 = [\mathbf{X}_0'\mathbf{D}^*\mathbf{X}_0]^{-1}.$$

$$\mathbf{P}_V = \mathbf{X}\mathbf{H}\mathbf{X}'\mathbf{D}^* \quad \text{and} \quad \mathbf{P}_{V_0} = \mathbf{X}_0\mathbf{H}_0\mathbf{X}_0'\mathbf{D}^*.$$

$\mathbf{H}$ is the negative inverse of the Hessian matrix of the Poisson likelihood function $l(\mathbf{y}, \boldsymbol{\beta})$. $\mathbf{P}_V$ and $\mathbf{P}_{V_0}$ are orthogonal projections onto V and onto V_0 with respect to the inner product $((\mathbf{x}, \mathbf{y})) \equiv \mathbf{x}'\mathbf{D}^*\mathbf{y}$. For example, if $\mathbf{x} \in V$, then $\mathbf{x} = \mathbf{X}\mathbf{b}$ for some $\mathbf{b}$, so that $((\mathbf{x}, \mathbf{P}_V\mathbf{y})) = \mathbf{b}'\mathbf{X}'\mathbf{D}^*\mathbf{X}\mathbf{H}\mathbf{X}'\mathbf{D}^*\mathbf{y} = \mathbf{b}'\mathbf{X}'\mathbf{D}^*\mathbf{y} = \mathbf{x}'\mathbf{D}^*\mathbf{y} = ((\mathbf{x}, \mathbf{y}))$. We use the subscripts V and V_0 because these projections depend on the subspaces and not upon the spanning vectors for these subspaces. To see this, replace $\mathbf{X}$ by $\mathbf{XA}$ for $\mathbf{A}$ a $k \times k$ nonsingular matrix.

In the case that $\mathbf{X}_0 = \mathbf{J}$, it follows that $\mathbf{P}_{V_0} = \mathbf{J}\mathbf{p}^{*\prime}$, where $\mathbf{p}^* = \mathbf{m}^*/\mathbf{J}'\mathbf{m}^*$ is a probability vector. In the case of 3×4 tables with row sums fixed, the product multinomial model, the columns of $\mathbf{X}_0$ span the row space, $\mathbf{P}_{V_0}$ is the 3-block diagonal matrix with ith block $\mathbf{J}\mathbf{p}_i^{*\prime}$, where $\mathbf{p}_i = \mathbf{m}_i^*/\mathbf{J}'\mathbf{m}_i^*$, a 4-component probability vector. For the Poisson model $V_0 = \mathscr{L}(\mathbf{0})$ and $\mathbf{X}_0$ is the $T \times 1$ $\mathbf{0}$ matrix.

In the following $o_p^{(n)}(1)$ for $n = 1, 2, \ldots$ is a sequence of random vectors converging to zero in probability. Then $\hat{\mathbf{m}}^{(n)}/n \to \mathbf{m}^*$ in probability implies that

(1)
$$\hat{\boldsymbol\mu}^{(n)} - \boldsymbol\mu^{(n)} \to 0 \text{ in probability},$$

(2)
$$\frac{1}{\sqrt{n}}(\mathbf{Y}^{(n)} - \mathbf{m}^{(n)}) \xrightarrow{D} N_T(\mathbf{0}, \mathbf{D}^*(\mathbf{I}_T - \mathbf{P}_{V_0})),$$

(3)
$$\sqrt{n}(\boldsymbol\beta^{(n)} - \boldsymbol\beta) = \frac{1}{\sqrt{n}}\mathbf{C}(\mathbf{Y}^{(n)} - \mathbf{m}^{(n)}) + o_p^{(n)}(1),$$

where
$$\mathbf{C} = (\mathbf{X}'\mathbf{X})^{-1}\mathbf{X}'(\mathbf{P}_V - \mathbf{P}_{V_0})\mathbf{D}^{*-1},$$
so that
$$\sqrt{n}(\hat{\boldsymbol\beta}^{(n)} - \boldsymbol\beta) \xrightarrow{D} N_k(\mathbf{0}, \mathbf{H} - \mathbf{Q}),$$
for
$$\mathbf{Q} = \mathbf{M}^{-1}\mathbf{X}'\mathbf{P}_{V_0}\mathbf{D}^{*-1}\mathbf{X}\mathbf{M}^{-1}, \qquad \mathbf{M} = \mathbf{X}'\mathbf{X}.$$
$\mathbf{Q}$ reduces to
$$\begin{bmatrix} \mathbf{H}_0 & 0 \\ 0 & 0 \end{bmatrix} \text{ if } \mathbf{X} = (\mathbf{X}_0, \mathbf{X}_1).$$

(4)
$$\sqrt{n}(\hat{\boldsymbol\mu}^{(n)} - \boldsymbol\mu^{(n)}) = \frac{1}{\sqrt{n}}(\mathbf{P}_V - \mathbf{P}_{V_0})\mathbf{D}^{*-1}(\mathbf{Y}^{(n)} - \mathbf{m}^{(n)}) + o_p^{(n)}(1),$$
so that
$$\sqrt{n}(\hat{\boldsymbol\mu}^{(n)} - \boldsymbol\mu^{(n)}) \xrightarrow{D} N_T(\mathbf{0}, (\mathbf{P}_V - \mathbf{P}_{V_0})\mathbf{D}^{*-1}),$$

(5)
$$\frac{1}{\sqrt{n}}(\hat{\mathbf{m}}^{(n)} - \mathbf{m}^{(n)}) = \frac{1}{\sqrt{n}}\mathbf{D}^*(\mathbf{P}_V - \mathbf{P}_{V_0})\mathbf{D}^{*-1}(\mathbf{Y}^{(n)} - \mathbf{m}^{(n)}) + o_p^{(n)}(1),$$
so that
$$\frac{1}{\sqrt{n}}(\hat{\mathbf{m}}^{(n)} - \mathbf{m}^{(n)}) \xrightarrow{D} N_T(\mathbf{0}, \mathbf{D}^*(\mathbf{P}_V - \mathbf{P}_{V_0})),$$

(6)
$$\frac{1}{\sqrt{n}}(\mathbf{Y}^{(n)} - \hat{\mathbf{m}}^{(n)}) = \frac{1}{\sqrt{n}}(\mathbf{I}_T - \mathbf{P}'_V)(\mathbf{Y}^{(n)} - \mathbf{m}^{(n)}) + o_p^{(n)}(1),$$
so that
$$\frac{1}{\sqrt{n}}(\mathbf{Y}^{(n)} - \hat{\mathbf{m}}^{(n)}) \xrightarrow{D} N_T(\mathbf{0}, \mathbf{D}^*[\mathbf{I}_T - \mathbf{P}_V]),$$

(7) $\dfrac{1}{\sqrt{n}}(\mathbf{Y}^{(n)} - \hat{\mathbf{m}}^{(n)})$ and $\dfrac{1}{\sqrt{n}}(\hat{\mathbf{m}}^{(n)} - \mathbf{m}^{(n)})$ are asymptotically independent.

These results are proved in Haberman (1974, Ch. 4) and a less abstract proof is given in Cox (1984); the original result in a less general form is due to Birch (1964).

Part (2) follows from the multivariate central limit theorem. Part (3) can be credited to Birch, though it follows from general results on MLEs (Cox 1984). The other results follow from (3) through the relations $\hat{\boldsymbol{\mu}}^{(n)} = \mathbf{X}\hat{\boldsymbol{\beta}}^{(n)}$, $\hat{\mathbf{m}}^{(n)} = \exp(\hat{\boldsymbol{\mu}}^{(n)})$, and the multivariate δ-method. Notice from (3) that the variances of the $\hat{\beta}_j$ are proportional to the diagonal elements of $\mathbf{H} - \mathbf{Q}$. If $\mathbf{X} = (\mathbf{X}_0, \mathbf{X}_1)$, then $\mathbf{Q} = \begin{bmatrix} \mathbf{H}_0 & \mathbf{0} \\ \mathbf{0} & \mathbf{0} \end{bmatrix}$, so that the coefficients corresponding to the vectors making up the columns of $\mathbf{X}_0$ have variances which become smaller as the sampling is confined to smaller subsets of the populations. If the model is independent Poisson, then $\mathbf{H}_0$ is the zero matrix, so there is no reduction. Under the single multinomial model, $\mathbf{H}_0$ is the 1×1 matrix $1/\text{trace}(\mathbf{D}^*) = 1/\sum p_i^* = 1$.

The coefficients corresponding to the columns of $\mathbf{X}_1$, those for which the inner products with $\mathbf{m}$ are not fixed by the sampling, have distributions which are asymptotically unaffected by the sampling scheme. That is, their asymptotic variances are the same whether or not sampling is Poisson, or single multinomial, or even product multinomial. Since we are usually interested in these coefficients, we can in a sense be a bit careless in specifying $\mathbf{X}_0$, so long as the vectors corresponding to these coefficients are not included in $\mathbf{X}_0$. In fact, some software packages ignore the restrictions implied by multinomial models and present estimates of standard errors which are those given by the Poisson model. These estimates are therefore too large in the case that sampling is multinomial for the coefficients corresponding to $\mathbf{X}_0$, but are the correct estimates for other coefficients.

Example 8.7.1: Let $\mathbf{Y}_1 \sim \mathscr{M}_3(100, \mathbf{p})$ and $\mathbf{Y}_2 \sim \mathscr{M}_3(200, \mathbf{p})$ be independent, with $\mathbf{p} = (0.2, 0.3, 0.5)$. Let $\mathbf{Y} = \begin{bmatrix} \mathbf{Y}_1 \\ \mathbf{Y}_2 \end{bmatrix}$, a 2×3 table. Then $\mathbf{m} = \begin{bmatrix} 20 & 30 & 50 \\ 40 & 60 & 80 \end{bmatrix}$, $\boldsymbol{\mu} = \begin{bmatrix} 2.996 & 3.401 & 3.912 \\ 3.689 & 4.094 & 4.605 \end{bmatrix}$, and in the usual notation for log-linear models, $\lambda = 3.783$, $\lambda_1^1 = -0.346$, $\lambda_1^2 = -0.441$, $\lambda_2^2 = -0.035$. For example, $\mu_{21} = \lambda + \lambda_2^1 + \lambda_1^2 = 3.783 + 0.346 - 0.035 = 4.094$. These are the coefficients of $\mathbf{x}_0$, the table of ones, $\mathbf{r} = \begin{bmatrix} 1 & 1 & 1 \\ -1 & -1 & -1 \end{bmatrix}$, $\mathbf{c}_1 = \begin{bmatrix} 1 & 0 & -1 \\ 1 & 0 & -1 \end{bmatrix}$, and $\mathbf{c}_2 = \begin{bmatrix} 0 & 1 & -1 \\ 0 & 1 & -1 \end{bmatrix}$ in the representation of $\boldsymbol{\mu}$. That is, $\boldsymbol{\mu} = \lambda \mathbf{x}_0 + \lambda_1^1 \mathbf{r} + \lambda_1^2 \mathbf{c}_1 + \lambda_2^2 \mathbf{c}_2$. Let $\mathbf{X}$ be the 6×4 matrix obtained by writing these vectors as columns. The second column of $\mathbf{X}$ becomes $(1, 1, 1, -1, -1, -1)'$. Since $n = 300$, $\mathbf{m}^* = \begin{bmatrix} 20 & 30 & 50 \\ 40 & 60 & 100 \end{bmatrix}/300$, and $\mathbf{D}^*$ is the diagonal matrix with diagonal

(20, 30, 50, 40, 60, 100)/300. Then

$$\mathbf{H} = \begin{bmatrix} 1.273 & 0.375 & 0.519 & -0.037 \\ 0.375 & 1.125 & 0 & 0 \\ 0.519 & 0 & 2.815 & -1.630 \\ -0.037 & 0 & -1.630 & 2.259 \end{bmatrix}.$$

Since row totals are fixed by the multinomial sampling, $\mathbf{X}_0 = (\mathbf{x}_0, \mathbf{r})$, written as a 6×2 matrix. Then $\mathbf{H}_0 = \begin{bmatrix} 1 & 1/3 \\ 1/3 & 1 \end{bmatrix}^{-1} = \begin{bmatrix} 8/9 & 3/8 \\ 3/8 & 8/9 \end{bmatrix}$, and $\mathbf{Q}$ is the 4 by 4 matrix with $\mathbf{H}_0$ in the upper-left corner, zeros elsewhere. The asymptotic covariance matrix for $\hat{\boldsymbol\beta}$ is

$$(1/300)(\mathbf{H} - \mathbf{Q}) = \begin{bmatrix} 0.000\,49 & 0 & 0.001\,73 & -0.000\,12 \\ 0 & 0 & 0 & 0 \\ 0.001\,73 & 0 & 0.009\,38 & -0.005\,43 \\ -0.000\,12 & 0 & -0.005\,43 & 0.007\,53 \end{bmatrix}.$$

$\mathbf{Y}$ was observed independently 500 times. The observed sample variances were 0.000 51, 3.6×10^{-16}, 0.009 04, 0.008 21, in close approximation to that given by the theory. The observe mean vector was (3.780, −0.347, − 0.440, −0.033), indicating that the estimators are almost unbiased. The variances for $\hat{\lambda}$ and $\hat{\lambda}_1^1$ are considerably smaller than those given by (1/300)$\mathbf{H}$, the asymptotic covariance matrix for Poisson sampling. The asymptotic variances for $\hat{\lambda}_1^2$ and $\hat{\lambda}_2^2$ are the same for Poisson and multinomial sampling. The asymptotic variances for the components of $\hat{\mathbf{m}}$, the diagonal elements of 300 $\mathbf{D}^*(\mathbf{P}_V - \mathbf{P}_{V_0})$, are considerably smaller for the multinomial model.

The asymptotic distribution of the G^2 and χ^2 statistics are easy to determine from parts (6) and (7). Let $\mathbf{W}_n = \mathbf{D}^{*-1/2}(\mathbf{X}_n - \mathbf{m}^{(n)})/\sqrt{n}$. Then by (6) $\mathbf{W}_n \xrightarrow{D} N_T(\mathbf{0}, \mathbf{I}_T - \mathbf{P}_B)$, where $\mathbf{B} = \mathbf{D}^{*1/2}\mathbf{X}$, and $\mathbf{P}_B = \mathbf{B}(\mathbf{B}'\mathbf{B})^{-1}\mathbf{B}$ is projection onto the column space V_B of $\mathbf{B}$. Theorem 2.5.2 then implies that $\|\mathbf{W}_n\|^2 = \chi^2(\mathbf{m}^{(n)}, \mathbf{Y}^{(n)})$ converges in distribution to χ^2_ν, where $\nu = \dim(V_B^\perp) = T - k$. Applying Theorem 8.6.1 with $\mathbf{m}_n = \mathbf{Y}^{(n)}$ and $\mathbf{m}_n^* = \mathbf{m}^{(n)}$, we conclude from (2) that $G^2(\mathbf{m}^{(n)}, \mathbf{Y}^{(n)})$ has the same limiting distribution.

Similarly from (4), we conclude that $\chi^2(\mathbf{m}^{(n)}, \hat{\mathbf{m}}^{(n)})$ and $G^2(\mathbf{m}^{(n)}, \hat{\mathbf{m}}^{(n)})$ are each asymptotically distributed as χ^2 with $(k - k_0)$ degrees of freedom.

From (7) we conclude that the statistics in this and the previous paragraph are asymptotically independent. Further, under the notation of Theorem 8.6.1, with $V_1 \supset V_2 \supset V_3$, the statistics $\chi^2(\hat{\mathbf{m}}_1^{(n)}, \mathbf{Y}^{(n)})$, $\chi^2(\hat{\mathbf{m}}_1^{(n)}, \hat{\mathbf{m}}_2^{(n)})$, and $\chi^2(\hat{\mathbf{m}}_2^{(n)}, \hat{\mathbf{m}}_3^{(n)})$ are asymptotically independent, and the same result holds if G^2 is substituted for χ^2. Even more generally, we can consider a sequence of nested subspaces

$V_0 = R_T \supset V_1 \supset \cdots \supset V_r$, and the resulting χ^2-statistics $Q_j^{(n)} = \chi^2(\hat{\mathbf{m}}_j^{(n)}, \hat{\mathbf{m}}_{j-1}^{(n)})$ are asymptotically independent, with Q_j asymptotically χ^2 with $\dim(V_j) - \dim(V_{j-1})$ d.f. if $\mathbf{m}^* \in V_{j-1}$. Of course, the same result holds if G^2 replaces χ^2.

In applications the covariance matrices given by (1) to (7) are unknown. For example, we will suppose that $\hat{\boldsymbol{\beta}} - \boldsymbol{\beta}$ is approximately distributed as $N_k(\mathbf{0}, (1/n)(\mathbf{H} - \mathbf{Q}))$. $\mathbf{H}$ and $\mathbf{Q}$ depend on $\mathbf{m}^* = \lim_{n\to\infty} \mathbf{m}^{(n)}/n$. However, $\mathbf{a}_n \equiv \hat{\mathbf{m}}^{(n)}/n$ is consistent for $\mathbf{m}^*$, so that we can replace $(1/n)(\mathbf{H} - \mathbf{Q})$ by $(1/n)(\hat{\mathbf{H}} - \hat{\mathbf{Q}})$, where $\hat{\mathbf{H}}$ and $\hat{\mathbf{Q}}$ are obtained by replacing $\mathbf{m}^*$ by $\mathbf{a}_n$ in the definition of $\mathbf{D}^*$. $\hat{\mathbf{H}}/n$ is obtained in the last step of the Newton–Raphson algorithm used to compute $\hat{\boldsymbol{\beta}}$. The constant $(1/n)$ may be absorbed into $\hat{\mathbf{H}} - \hat{\mathbf{Q}}$, if we replace $\mathbf{D} = d(\mathbf{m}^*)$ by $d(\hat{\mathbf{m}})$. Similarly, for large n, we can approximate the distribution of $(\hat{\mathbf{m}}^{(n)} - \mathbf{m}^{(n)})$ by the $N_T(\mathbf{0}, n\mathbf{D}^*(\mathbf{P}_V - \mathbf{P}_{V_0}))$ distribution, and replace each occurrence of $\mathbf{D}^* = d(\mathbf{m}^*)$ by $d(\hat{\mathbf{m}})$, to get estimators $\hat{\mathbf{P}}_V$ and $\hat{\mathbf{P}}_{V_0}$ of $\mathbf{P}_V$ and $\mathbf{P}_{V_0}$. Then the distribution of $\hat{\mathbf{m}}^{(n)} - \mathbf{m}^{(n)}$ is approximated by $N_T(\mathbf{0}, d(\hat{\mathbf{m}})(\hat{\mathbf{P}}_V - \hat{\mathbf{P}}_{V_0}))$. In estimating the covariance matrix of any of the estimators $\hat{\boldsymbol{\beta}}^{(n)}$, $\hat{\boldsymbol{\mu}}^{(n)}$, or $\hat{\mathbf{m}}^{(n)}$ the rule is therefore simple: replace $\mathbf{D}^*$ whenever it occurs by $d(\hat{\mathbf{m}})$, and don't worry about the factor n.

Proportional Iterated Fitting (The Stephan–Deming Method: Consider the four-way table indexed by i_1, i_2, i_3, i_4, with $1 \le i_j \le T_i$, $T_1 = 2$, $T_2 = 3$, $T_3 = 4$, $T_4 = 5$. The model (1 2 3 4 12 13 124). The model (1 2 3 4 12 13 124) may be written in the reduced form [13], [124], since, assuming the model is heierarchical, the presence of the terms 1, 2, 3, 4, 12 is implied by the presence of 13 and 124. Similarly, the saturated model may be written as [1234], and the model (1 2 3 4 12 23 34 124 234) may be written as [124], [234]. The reduced form makes it easy to determine a set of sufficient statistics. From Theorem 8.7.1 for the model [13], [124] the collection of sums $\{Y_{i+k+}, Y_{ij+1}\}$ is sufficient. For the model [124], [234] the collection $\{Y_{ij+l}, Y_{+jkl}\}$ is sufficient. For the saturated model $\{Y_{ijkl}\}$ is sufficient, as it is for all smaller models. If a collection of statistics is sufficient for a model V, then it is sufficient for a smaller model $V_1 \subset V$.

In general, consider a d-dimensional table indexed by $i_1, i_2, \ldots, i_d$, with $1 \le i_j \le T_j$ for $j = 1, \ldots, d$. Let $B_1, \ldots, B_r$ be subsets of the integers $\{1, \ldots, d\}$ corresponding to the reduced form of the model under consideration. For the four-way table with the model [13], [124] above $r = 2$, $B_1 = \{1, 3\}$, $B_2 = \{1, 2, 4\}$.

Let $\mathscr{P}_i = \{A_{ij}, j = 1, \ldots, d_i\}$ be the partitioning of the cells defined by the indices in B_i. All cells in any A_{ij} have the same levels for all indices in B_i. B_1 defines a partitioning $\mathscr{P}_1$ of the 120 cells into $d_1 = 2 \times 4 = 8$ subsets, each of $3 \times 5 = 15$ cells. B_2 defines a partitioning $\mathscr{P}_2$ of the 120 cells into $d_2 = 2 \times 3 \times 5 = 30$ subsets, each of 4 cells. A_{17} is the collection of cells for which

$i_1 = 2$, and $i_3 = 3$. Let $\mathbf{x}_{ij}$ be the indicator of A_{ij}. In general, $d_i = \prod_{j \in B_i} T_j$, and $\sum_{j \in B_i} \mathbf{x}_{ij} = \mathbf{J}$.

The likelihood equations (also called the normal equations) are

$$\hat{\mathbf{m}}_{ij+} \equiv (\hat{\mathbf{m}}, \mathbf{x}_{ij}) = (\mathbf{Y}, \mathbf{x}_{ij}) \equiv Y_{ij+} \qquad \text{(condition } C_{ij}\text{)}$$

for all i and j. Proportional iterated fitting adjusts each $\hat{\mathbf{m}}_{ij+}$ in turn by multiplication so that condition C_{ij} is satisfied. These adjustments continue until all conditions are at least approximately satisfied.

Suppose $\hat{\mathbf{m}}$ has been chosen as a starting point. This means that $\hat{\mu} = \log \hat{\mathbf{m}} \in V$, the space spanned by the $\mathbf{x}_{ij}$. If $\mathbf{J} \in V$ then one choice is $\mathbf{J}$ or any multiple. Suppose also that $\hat{\mathbf{m}}$ does not satisfy C_{ij}. Consider a new approximation $\mathbf{v}_b = \hat{\mathbf{m}}e^{b\mathbf{x}_{ij}}$. $\mathbf{v}_b$ differs from $\hat{\mathbf{m}}$ only in the cells in A_{ij}. For the cells in A_{ij}, the components of $\hat{\mathbf{m}}$ have been multiplied by e^b. But $\log \mathbf{v}_b = \log \hat{\mathbf{m}} + b\mathbf{x}_{ij}$, so that $\mathbf{v}_b$ is still contained in V. Thus $\mathbf{v}_b$ remains a possible solution to the likelihood equations.

Proportional iterated fitting chooses $b = b_{ij}$, so that condition C_{ij} is satisfied. To determine the value of b_{ij} let $c = e^b$. In order that $\mathbf{v}_b$ satisfy condition C_{ij} we need $(\mathbf{v}_b, \mathbf{x}_{ij}) = (\mathbf{Y}, \mathbf{x}_{ij}) = Y_{ij+}$. But the first inner product is $c(\hat{\mathbf{m}}, \mathbf{x}_{ij}) = c\hat{m}_{ij+}$. Thus, we should take $c_{ij} = c = (\mathbf{Y}, \mathbf{x}_{ij})/(\hat{\mathbf{m}}, \mathbf{x}_{ij}) = Y_{ij+}/\hat{m}_{ij+}$. That is, multiply all components of $\hat{\mathbf{m}}$ in the cells in A_{ij} by $Y_{ij+}/\hat{m}_{ij+}$. Continue across all combinations of i and j. Later adjustments, as i changes, will in general cause conditions C_{ij} to fail. If so, perform another round of adjustments.

If the MLE has a closed form, then one round of adjustments will suffice. Otherwise, several round will be necessary. In general, however, proportional iterated fitting will converge relatively quickly, with the procedure stopping whenever the likelihood equations hold in good approximation.

In terms of $\hat{\boldsymbol{\mu}}$, the adjustment is one of addition, as indicated by $\mathbf{v}_b$ above. $\mathbf{v}_b$ is adjusted in one of the fixed directions given by the $\mathbf{x}_{ij}$. This is the idea behind the proportional iterated fitting, which, considered as a technique in numerical analysis, is called the Deming–Stephan algorithm.

Example 8.7.2: Consider a $2 \times 2 \times 2$ table, and the model $(1\ 2\ 3\ 12\ 13\ 23) \equiv ([12], [13], [23])$. Then $B_1 = \{1, 2\}$, $B_2 = \{1, 3\}$, $B_3 = \{2, 3\}$, and $d_1 = d_2 = d_3 = 4$. Written in their table form, with the first factor corresponding to layers, the second to rows, and the third to columns, some of the twelve $\mathbf{x}_{ij}$ are

$$\mathbf{x}_{11} = \begin{bmatrix} 1 & 1 \\ 0 & 0 \\ \\ 0 & 0 \\ 0 & 0 \end{bmatrix}, \qquad \mathbf{x}_{12} = \begin{bmatrix} 0 & 0 \\ 1 & 1 \\ \\ 0 & 0 \\ 0 & 0 \end{bmatrix}, \qquad \mathbf{x}_{14} = \begin{bmatrix} 0 & 0 \\ 1 & 1 \\ \\ 0 & 0 \\ 1 & 1 \end{bmatrix},$$

$$\mathbf{x}_{21} = \begin{bmatrix} 1 & 0 \\ 1 & 0 \\ 0 & 0 \\ 0 & 0 \end{bmatrix}, \quad \mathbf{x}_{23} = \begin{bmatrix} 0 & 0 \\ 0 & 0 \\ 1 & 0 \\ 1 & 0 \end{bmatrix}, \quad \mathbf{x}_{32} = \begin{bmatrix} 0 & 1 \\ 0 & 0 \\ 0 & 1 \\ 0 & 0 \end{bmatrix}.$$

Suppose, we observe $\mathbf{Y} = \mathbf{y} = \begin{bmatrix} 10 & 20 \\ 8 & 12 \\ 6 & 10 \\ 16 & 4 \end{bmatrix}$. Begin with $\hat{\mathbf{m}}$ the vector of all ones. After adjustments for $\mathbf{x}_{1j}$, $j = 1, \ldots, 4$, we get $\begin{bmatrix} 15 & 15 \\ 10 & 10 \\ 8 & 8 \\ 10 & 10 \end{bmatrix}$. After adjustments for $\mathbf{x}_{2j}$, $j = 1, \ldots, 4$, we get $\begin{bmatrix} 10.8 & 19.2 \\ 7.2 & 12.8 \\ 88/9 & 56/9 \\ 110/9 & 70/9 \end{bmatrix}$. After adjusting for $\mathbf{x}_{31}, \ldots, \mathbf{x}_{34}$ we get $\begin{bmatrix} 8.397 & 22.657 \\ 8.897 & 9.952 \\ 7.603 & 7.343 \\ 15.103 & 6.048 \end{bmatrix}$. Notice that after the third cycle of adjustments, inner products which had previously been adjusted to be the same as for $\mathbf{y}$, no longer have this property. For example, $8.397 + 22.657 = 31.053 \neq 30$, and $8.397 + 8.897 = 17.294 \neq 18$. After another complete round of adjustments, the vector is $\begin{bmatrix} 8.163 & 21.811 \\ 9.806 & 10.216 \\ 7.837 & 8.189 \\ 14.194 & 5.734 \end{bmatrix}$. After still another round it is $\begin{bmatrix} 8.187 & 21.805 \\ 9.813 & 10.195 \\ 7.813 & 8.195 \\ 14.187 & 5.805 \end{bmatrix}$.

Notice that all 12 inner products are almost the same as for $\mathbf{y}$. After seven rounds the change from one round to another is less than 10^{-6} in any component of $\hat{\mathbf{m}}$. The final solution, to three decimal places, is

$$\begin{bmatrix} 8.191 & 21.809 \\ 9.809 & 10.191 \\ & \\ 8.809 & 8.191 \\ 14.191 & 5.809 \end{bmatrix}.$$

The extra rounds after the first or second were hardly worth the effort. The chi-square values were $G^2 = 2.872$ and $\chi^2 = 2.816$, roughly the 58th percentile for 1 d.f., indicating a reasonably good fit.

Example 8.7.3: Suppose a study is conducted to determine the opinions of men and of women on a proposed abortion law, which would limit the freedom of women to choose abortion. Random samples of 300 women and 200 men were chosen. Each person was then asked to choose one of the options. (1) strongly favor, (2) slightly favor, (3) slightly against, (4) strongly against. Let Y_{ij} be the number choosing response j, $i = 1$ for women, and $i = 2$ for men. Let $\mathbf{Y} = (Y_{ij})$, let $\mathbf{x}_0$ be the identity vector, let

$$\mathbf{r} = \begin{bmatrix} 1 & 1 & 1 & 1 \\ -1 & -1 & -1 & -1 \end{bmatrix}, \quad \mathbf{c}_1 = \begin{bmatrix} 1 & 0 & 0 & -1 \\ 1 & 0 & 0 & -1 \end{bmatrix}, \quad \mathbf{c}_2 = \begin{bmatrix} 0 & 1 & 0 & -1 \\ 0 & 1 & 0 & -1 \end{bmatrix},$$

$$\mathbf{c}_3 = \begin{bmatrix} 0 & 0 & 1 & -1 \\ 0 & 0 & 1 & -1 \end{bmatrix}, \qquad \mathbf{w} = \begin{bmatrix} 3 & 1 & -1 & -3 \\ -3 & -1 & 1 & 3 \end{bmatrix}.$$

The vector $\mathbf{w}$ has been chosen in order to model the different opinions of men and women. Suppose that $\boldsymbol{\mu} = \mu\mathbf{x}_0 + s_1\mathbf{r} + \beta_1\mathbf{c}_1 + \beta_2\mathbf{c}_2 + \beta_3\mathbf{c}_3 + \gamma\mathbf{w}$, where $\beta_1 = -0.2, \beta_2 = 0.2, \beta_3 = 0.1, \gamma = 0.10$. Let $\mu = 4.080\,378$ and $s_1 = 0.206\,254$ be chosen so that the 2×4 matrix $\mathbf{m} = \exp(\boldsymbol{\mu})$ has row sums 300 and 200.

Then $\mathbf{m} = \begin{bmatrix} 80.37 & 98.16 & 72.72 & 48.75 \\ 29.20 & 53.20 & 58.80 & 58.80 \end{bmatrix}$. Let $\mathbf{p}_1$ and $\mathbf{p}_2$ be the probability vectors obtained by dividing the first and second rows of $\mathbf{m}$ by 300 and 200, respectively. Let $\mathbf{Y}_1 \sim \mathcal{M}_4(\mathbf{p}_1, 300)$ and $\mathbf{Y}_2 \sim \mathcal{M}_4(\mathbf{p}_2, 200)$ be independent. Then $\mathbf{Y}$, the 2×4 table formed by $\mathbf{Y}_1$ and $\mathbf{Y}_2$ as the first and second row, has the product multinomial distribution, and $E(\mathbf{Y}) = \mathbf{m}$. The coefficients μ, s_1, β_1, β_2, β_3, $\beta_4 = -(\beta_1 + \beta_2 + \beta_3)$ may be determined from the usual ANOVA expansion of $\boldsymbol{\mu} - \gamma\mathbf{w}$.

Since $\mu_{1j} - \mu_{2j} = 2s_1 - 2\gamma(4 - j)$, log odds–ratios are $L(j_1, j_2) = \log[(m_{2j_2}/m_{1j_2})/(m_{2j_1}/m_{1j_1})] = \mu_{2j_2} - \mu_{1j_2} - \mu_{1j_1} + \mu_{1j_1} = 2\gamma(j_2 - j_1)$. Since $\gamma > 0$, men are more likely to favor such a law.

An observation was simulated on a computer, producing $\mathbf{Y} = \mathbf{y} = \begin{bmatrix} 95 & 91 & 66 & 48 \\ 29 & 57 & 59 & 55 \end{bmatrix}$. Such multinomial random vectors may be generated by summing independent generalized Bernoulli random vectors (GBRVs). Thus, the first row of $\mathbf{Y}$ is the sum of 300 such GBRVs, with, for example (0, 0, 1, 0) being taken with probability 72.72/300. The author used S-Plus to do this.

Using the S-Plus function "glm" the model $\boldsymbol{\mu} \in V = \mathscr{L}(\mathbf{x}_0, \mathbf{r}, \mathbf{c}_1, \mathbf{c}_2, \mathbf{c}_3)$ (equivalent to $\mathbf{p}_1 = \mathbf{p}_2$) was fit, providing the chi-square statistics $G^2 = 25.62$, $\chi^2 = 24.80$, for $(8 - 5) = 3$ d.f., indicating a rather poor fit, as should be expected.

The model $\boldsymbol{\mu} \in \mathscr{L}(\mathbf{x}_0, \mathbf{r}, \mathbf{c}_1, \mathbf{c}_2, \mathbf{c}_3, \mathbf{w})$, the correct model, was then fit, giving $G^2 = 1.657$, $\chi^2 = 1.651$, for $(8 - 6) = 2$ d.f., indicating a good fit, as should be expected. S-Plus provided Table 8.7.1.

The matrix $\mathbf{X}$ corresponding to this model is the 8×6 matrix formed by writing spanning vectors $\mathbf{x}, \mathbf{r}, \mathbf{c}_1, \mathbf{c}_2, \mathbf{c}_3, \mathbf{w}$ as columns. We can estimate $\mathbf{H} = \mathbf{X}'\mathbf{D}^*\mathbf{X}$ by $\hat{\mathbf{H}} = \mathbf{X}'\hat{\mathbf{D}}^*\mathbf{X}$, where $\hat{\mathbf{D}}^* = d(\hat{\mathbf{m}}/500)$. Similarly, let $\mathbf{X}_0$ consist of the first two columns of $\mathbf{X}$, corresponding to $\mathbf{x}_0$ and $\mathbf{r}$. Let $\hat{\mathbf{H}}_0 = \mathbf{X}_0'\hat{\mathbf{D}}^*\mathbf{X}_0$. By replacing $\mathbf{H}$ and $\mathbf{H}_0$ by $\hat{\mathbf{H}}$ and $\hat{\mathbf{H}}_0$ in the definitions of $\mathbf{P}_V$ and $\mathbf{P}_{V_0}$ we get estimates $\hat{\mathbf{P}}_V$ and $\hat{\mathbf{P}}_{V_0}$ of these projection matrices. The estimate of Q is $\hat{Q}$, formed by placing the 2×2 matrix $\hat{\mathbf{H}}_0$ in the upper left corner, zeros in the other 32 places. The estimate of the covariance matrix of $\hat{\boldsymbol{\beta}}$ is $\hat{\mathbf{D}}[\boldsymbol{\beta}] = (1/300)(\hat{\mathbf{H}} - \hat{\mathbf{Q}})$. The square roots of the diagonal of $\hat{\mathbf{D}}[\hat{\boldsymbol{\beta}}]$, estimates of the standard errors, as well as components of $\hat{\boldsymbol{\beta}}$ are given in Table 8.7.2. Notice that the estimates of the standard errors of $\hat{\mu}$ and $\hat{s}_1$ are considerably smaller than those given by S-Plus as shown in Table 8.7.1. This is because the S-Plus routine used assumes Poisson, rather than multinomial sampling, and the

Table 8.7.1

Coef.	Est.	Est. of Std. Error	z = (Est./(Est. of Std. Error)
μ	4.081	0.0472	86.471
s_1	0.194	0.0468	4.155
β_1	−0.080	0.0816	−0.985
β_2	0.178	0.0739	2.411
β_3	0.050	0.0789	0.632
γ	0.106	0.0221	4.795

Table 8.7.2

$\hat{\mu}$	$\hat{s}_1$	$\hat{\beta}_1$	$\hat{\beta}_2$	$\hat{\beta}_3$	$\hat{\gamma}$
4.081	0.194	−0.084	0.178	0.050	0.106
0.012	0.011	0.082	0.074	0.079	0.022

matrix $\mathbf{Q}$ is therefore omitted in the computation of covariance matrices. The estimates of other standard errors are the same.

Estimates of the covariance matrices of $\hat{\mathbf{m}}$, $\hat{\boldsymbol{\mu}}$, and $\mathbf{e}$ are given by $\hat{\mathbf{D}}[\hat{\mathbf{m}}] = 300\ \hat{\mathbf{D}}^*(\hat{\mathbf{P}}_V - \hat{\mathbf{P}}_{V_0})$, $\hat{\mathbf{D}}[\hat{\boldsymbol{\mu}}] = (1/300)(\hat{\mathbf{P}}_V - \hat{\mathbf{P}}_{V_0})\hat{\mathbf{D}}^{*-1}$, and $\hat{\mathbf{D}}[\mathbf{e}] = 300\ \hat{\mathbf{D}}^*(\mathbf{I}_8 - \hat{\mathbf{P}}_V)$. All these estimates were close to those obtained using $\mathbf{D}^* = d(\mathbf{m})/300$ rather than $\hat{\mathbf{D}}^*$.

An approximate 95% confidence interval on γ is given by $[\hat{\gamma} \pm 1.96\ \hat{\sigma}(\hat{\gamma})] = [0.106 \pm 1.96(0.022)] = [0.063, 0.149]$. Since we know $\gamma = 0.10$, we were correct this time. Since the log odds–ratios are multiples of γ, confidence intervals on these or on odds–ratios are easy to determine. Individual 95% confidence intervals on a contrast $\eta = \sum c_j\beta_j$ among $\beta_1, \beta_2, \beta_3$, and $\beta_4 = -(\beta_1 + \beta_2 + \beta_3)$ are determined by $[\hat{\eta} \pm 1.96\hat{\sigma}(\hat{\eta})]$, where $\hat{\sigma}(\hat{\eta}) = \mathbf{c}'\hat{\mathbf{D}}[\hat{\boldsymbol{\beta}}]\mathbf{c}$, where $\mathbf{c}$ is the vector of coefficients. For example, for $\eta = \beta_1 - \beta_3$, $\mathbf{c} = (0, 0, 1, 0, -1, 0, 0)'$, we find $[-0.130 \pm (1.96)(0.132)] = [-0.391, 0.131]$. To get a confidence interval on $\beta_1 - \beta_4$, we can either express β_4 in terms of the other β's or we can determine the covariance matrix of all four $\hat{\beta}$'s by using the fact that row and column sums must be zero.

Table 8.7.3 gives some of the estimates, together with estimates of their standard errors, obtained from the estimates of the covariance matrices. The Pearson chi-square statistic is $\chi^2 = \sum_{ij} e_{ij}^2/\hat{m}_{ij} = 1.651$.

Table 8.7.3

i	j	Y_{ij}	$\hat{m}_{ij}$	$\hat{\sigma}(\hat{m}_{ij})$	e_{ij}	$\hat{\mu}_{ij}$	$\hat{\sigma}(\hat{\mu}_{ij})$	$\hat{\sigma}(e_{ij})$	$e_{ij}/\hat{\sigma}(e_{ij})$
1	1	95	91.27	7.40	3.73	4.51	0.0047	2.95	1.263
1	2	91	95.59	6.60	−4.59	4.56	0.0040	4.64	−0.989
1	3	66	68.00	5.64	−2.00	4.22	0.0048	4.55	−0.438
1	4	48	45.14	5.47	2.86	3.81	0.0070	2.91	0.985
2	1	29	32.73	4.32	−3.73	3.49	0.0076	2.95	−1.263
2	2	57	52.41	4.14	4.59	3.96	0.0046	4.64	0.989
2	3	59	57.00	4.47	2.00	4.04	0.0045	4.55	0.438
2	4	55	57.86	5.72	−2.86	4.06	0.0057	2.91	−0.985

Confidence Intervals: We should be particularly interested in estimating odds and log-odds. Since log-odds are of the form $\eta = c_1\mu_1 + \cdots + c_T\mu_T = (\mathbf{c}, \boldsymbol{\mu})$, with $\mathbf{c} \perp \mathbf{x}_0$, so that η is a contrast, we can first determine a confidence interval (L, U) on η, then determine an interval (e^L, e^U) on the corresponding odds. We are usually interested in coefficient vectors $\mathbf{c}$ for which $\mathbf{c}$ is orthogonal to the columns of $\mathbf{X}_0$. For example, for two-way tables and an independent multinomial model for row vectors $\mathbf{Y}_i$, the columns of $\mathbf{X}_0$ are the row indicators, and we will want the rows of $\mathbf{c}$ to add to zero. Since the row totals of $\mathbf{m}$ are fixed, the row effects λ_i^1 are nuisance parameters, there to adjust the rows of $\mathbf{m}$ so that the row totals are correct. We therefore do not want comparisons among the row effects λ_i^1.

In the case that $\mathbf{c}$ is orthogonal to the columns of $\mathbf{X}_0$, the second term of $D[\hat{\boldsymbol{\mu}}]$ disappears, and we get in approximation: $\hat{\eta} - \eta = (\mathbf{c}, \hat{\boldsymbol{\mu}} - \boldsymbol{\mu}) \sim N(\mathbf{0}, \sigma^2(\hat{\eta}))$ for $\sigma^2(\hat{\eta}) = \mathbf{c}'\mathbf{M}\mathbf{c}$, and $\mathbf{M} = \mathbf{X}(\mathbf{X}'d(\mathbf{m})\mathbf{X})^{-1}\mathbf{X}'$. Therefore,

$$\hat{\eta} \pm z_{1-\alpha/2}\hat{\sigma}(\hat{\eta}) \tag{8.7.1}$$

is an approximate $100(1-\alpha)\%$ confidence interval on $\eta = (\mathbf{c}, \boldsymbol{\mu})$, in the case that $\mathbf{c} \perp V_0$, the subspace spanned by the columns of $\mathbf{X}_0$. The Scheffé method may be used to provide simultaneous confidence intervals on $\eta = \eta_c$ for all $\mathbf{c} \in C$, where C is a subspace orthogonal to V_0:

$$\hat{\eta}_c \pm K\hat{\sigma}(\hat{\eta}_c), \qquad \text{for all} \quad c \in C, \tag{8.7.2}$$

where $K = \sqrt{\chi^2_{\nu, 1-\alpha}}$ and $\nu = \dim(C)$. The proof is only outlined here. See Haberman (1974, 131). Let $Z_c = (\hat{\eta}_c - \eta_c)/\hat{\sigma}(\hat{\eta}_c) = (\mathbf{c}, \hat{\boldsymbol{\mu}} - \boldsymbol{\mu})/[\mathbf{c}'\hat{\mathbf{M}}\mathbf{c}]^{1/2}$. Now apply Theorem 3.8.2, with $\mathbf{b} = \hat{\boldsymbol{\mu}} - \boldsymbol{\mu}$, $\mathbf{M} = \hat{\mathbf{M}}$. We conclude that $\sup_{c \in C} Z_c^2 = (\hat{\boldsymbol{\mu}} - \boldsymbol{\mu})'\hat{\mathbf{M}}^{-1}(\hat{\boldsymbol{\mu}} - \boldsymbol{\mu})$. By Theorem 3.8.1 and Slutsky's Theorem this last r.v. is asymptotically distributed as χ^2 with $q = \dim(C)$ d.f.

In the case that only a few confidence intervals are desired, they will be shorter if the Bonferroni inequality is exploited. If, for example, we wish intervals on $\eta_1 = (\mathbf{c}_1, \boldsymbol{\mu}), \ldots, \eta_5 = (\mathbf{c}_5, \boldsymbol{\mu})$ then we use (8.7.1) on each for $\alpha = 0.01$, and then have 95% confidence that all are correct.

In the case that the model is saturated, so that $V = R^T$, $\mathbf{X} = \mathbf{I}_T$, $\hat{\mathbf{m}} = \mathbf{Y}$, $\hat{\mathbf{M}} = d(\mathbf{Y})^{-1}$, and $\hat{\sigma}^2(\hat{\eta}) = \mathbf{c}'d(\mathbf{Y})^{-1}\mathbf{c} = \sum_i c_i^2/Y_i$. This is an upper bound on $\hat{\sigma}^2(\hat{\eta})$, which will be smaller when nonsaturated models are considered. Of course, when non-saturated models are considered, the risk is always present that $\hat{\eta}$ is biased for η.

Example 8.7.4: Consider 13,832 homocides in 1970, as reported by the National Center for Health Statistics (Table 8.7.4), in which the victim was classified by race (white or black), sex, and by the method (firearms and explosives, or cutting and piercing instruments).

Table 8.7.4 Reported Homocides in 1970, Classified by Race of Victim, Sex, and Type of Assault

		Type of Assault		
Race	Sex	Firearms and Explosives	Cutting and Piercing Instruments	Total
White	Male	3,910	808	4,718
	Female	1,050	234	1,284
Black	Male	5,218	1,385	6,603
	Female	929	298	1,227
Total		11,107	2,725	13,832

Within each of the races let us estimate the log-odds for use of these two methods for males and females. Index race by i, sex by j, method by k. The independent Poisson model seems to be appropriate. Let $\mu_{ijk} = \lambda + \lambda_i^1 + \lambda_j^2 + \lambda_k^3 + \lambda_{ij}^{12} + \lambda_{ik}^{13} + \lambda_{jk}^{23} + \lambda_{ijk}^{123}$, where each of the terms add to zero across any subscript. Define $R_i = (m_{i11}/m_{i12})/(m_{i21}/m_{i22}) = (m_{i11}m_{i22})/(m_{i12}m_{i21})$ and $\eta_i = \log R_i = \mu_{i11} - \mu_{i12} - \mu_{i21} + \mu_{i22} = 4(\lambda_{11}^{23} + \lambda_{i11}^{123})$. The estimate of η_1 under the saturated model is $\hat{\eta}_1 = \log[(3{,}910)(234)]/[(1{,}050)(808)] = 0.075\,5$. Similarly, $\hat{\eta}_2 = 0.189\,4$. Under the saturated model $\hat{\sigma}(\hat{\eta}_1) = (1/Y_{111}) + (1/Y_{112}) + 1/Y_{121}) + (1/Y_{122}) = 0.006\,72$ so that a 95% confidence interval on η_1 is $0.075\,5 \pm (1.96)(0.006\,72)^{1/2} = 0{\cdot}075\,5 \pm 0.160\,7$. A 95% confidence interval of η_2 is $0.189\,4 \pm (1.96)(0.005\,35)^{1/2} = 0.189\,4 \pm 0.143\,31$. Since this confidence interval is to the right of zero, we can conclude with 95% confidence that when blacks were the victims, males were more likely than were females to have been killed by firearms and explosives. For whites, though the estimate $\hat{\eta}_1 = 0.075\,5$ indicates a slight tendency in that direction, the interval covers zero, so that it is possible that η_1 is negative.

We should compare the log-odds ratios by estimating $\eta = \eta_1 - \eta_2 = 8\lambda_{111}^{123}$. We get $\hat{\eta} = \hat{\eta}_1 - \hat{\eta}_2 = -0.113\,9$. Since the vectors (Y_{1jk}) and (Y_{2jk}) are independent, we get $\hat{\sigma}^2(\hat{\eta}) = \sum_{ijk} (1/Y_{ijk}) = 0.006\,72 + 0.005\,35 = 0.012\,07$, so that the 95% confidence interval on η is $-0.113\,9 \pm (1.96)(0.012\,07)^{1/2} = -0.113\,9 \pm 0.215\,3$. The fact that this interval includes zero suggests that we fit the model (1 2 3 12 13 23), equivalently $\lambda_{111}^{123} = 0$, or $R_1 = R_2$. A more formal test of $H_0\colon R_1 = R_2 \Leftrightarrow \lambda_{111}^{123} = 0$ rejects H_0 at level 0.05 for $Z = \hat{\eta}/\hat{\sigma}(\hat{\eta})$, when $|Z| > 1.96$. In this case we get $Z = 1.037$, so we do not reject H_0. The corresponding χ^2 test statistic is $Z^2 = 1.076$, for one d.f.

For the model (1 2 3 12 13 23) we find $\hat{\mathbf{m}} = \begin{bmatrix} 3{,}919.5 & 798.5 \\ 1{,}040.5 & 243.5 \\ \\ 5{,}208.5 & 1{,}394.5 \\ 938.5 & 288.5 \end{bmatrix}$. The corresponding goodness-of-fit statistics are $G^2 = 1.077$, and $\chi^2 = 1.075$, as was promised in the last paragraph above. The log-odds ratio for blacks and for whites is $\eta^* = \eta_i = \log R_i = 4\lambda_{11}^{23}$, which is the same for $i = 1, 2$. The estimate is $\hat{\eta}^* = \hat{\mu}_{i11} - \hat{\mu}_{i12} - \hat{\mu}_{i21} + \hat{\mu}_{i22} = 0.138\,6$, which is the same for each i (why?). $\hat{\mathbf{M}}$ was used to determine $\hat{\sigma}(\hat{\eta}^*) = [\mathbf{c}'\hat{\mathbf{M}}\mathbf{c}]^{1/2} = 0.043\,9$, so that a 95% confidence interval on η^* is $0.138\,6 \pm (1.96)(0.043\,9) = 0.138\,6 \pm 0.086\,1$.

The z-statistic for H_0: $\eta^* = 0 \Leftrightarrow \lambda_{11}^{23} = 0$, is $Z = 0.138\,6/0.043\,9 = 3.16$, so that it does not seem reasonable to drop the 23 interaction term from the model, equivalently to suppose independence of method and sex, conditionally on race. Similar tests on other log-odds provided Z-statistics even further from zero, so no smaller model than (1 2 3 12 13 23) seems appropriate.

Power: For given parameter values, we can determine the approximate power for any hypothesis of the form H_0: $\boldsymbol{\mu} \in V_2$, as follows. Consider the model $\boldsymbol{\mu} \in V_1$, with $\dim(V_1) = d_1$, and let V_2 be a subspace of V_1, with $\dim(V_2) = d_2 < d_1$. Let $\mathbf{m}_1$ and $\mathbf{m}_2$ be the MLEs corresponding to the observation $\mathbf{m}$ for the models V_1 and V_2. Then $G^2 \equiv G^2(\hat{\mathbf{m}}_2, \hat{\mathbf{m}}_1)$ and $\chi^2 \equiv \chi^2(\hat{\mathbf{m}}_2, \hat{\mathbf{m}}_1)$ are approximately distributed as noncentral chi-square with $d_1 - d_2$ degrees of freedom, and noncentrality parameter $\delta = G^2(\mathbf{m}_2, \mathbf{m}_1)$. A more precise statement of the limit theory justifying the approximation is contained in Bishop, Fienberg, and Holland (1975, Ch. 14) and Haberman (1974, 103). The asymptotic theory requires that as sample sizes increase the probability vectors converge to V_2 at the rate proportional to the square roots of sample sizes. If the logs of the probability vectors remain fixed, not contained in V_2, then the powers of the χ^2 and G^2 tests converge to one. That is, the tests are consistent.

Pearson–Hartley charts, or computer packages producing noncentral χ^2 cumulative probability values may be used to evaluate power. Consider the example concerning opinion on an abortion law, Example 8.7.2. Let $V_1 = \mathscr{L}(\mathbf{x}_0, \mathbf{r}, \mathbf{c}_1, \mathbf{c}_2, \mathbf{c}_3)$, the subspace corresponding to equality of the probability vectors for men and women of dimension $d_2 = 5$. Let $V_1 = V^* \otimes \mathscr{L}(\mathbf{w})$, which has dimension $d = 6$. If $\mathbf{m} \in V_1$, the MLE of $\mathbf{m}$ corresponding to observation $\mathbf{m}$ and subspace V_1 is $\mathbf{m}$. The MLE corresponding to $\mathbf{m}$ and V_2 is $\mathbf{m}_2 = m_{i+}m_{+j}/m_{++} = \begin{bmatrix} 65.74 & 90.82 & 78.91 & 64.53 \\ 43.82 & 60.54 & 52.61 & 43.02 \end{bmatrix}$. The "distance" of $\mathbf{m}_2$ from $\mathbf{m} = \mathbf{m}_1$ may be measured by $G^2(\mathbf{m}_2, \mathbf{m}_1) = 20.69$. We conclude that the statistic $G^2(\hat{\mathbf{m}}_2, \hat{\mathbf{m}}_1)$ has an approximate noncentral χ^2 distribution with one d.f. and noncentrality parameter 20.69, indicating that the power of the test H_0: $\gamma = 0$ is approximately 0.995 2. The experiment was repeated using S-Plus 500 times, each time with 300 men and 200 women, resulting in estimates $\hat{\gamma}$ with mean 0.100 1, variance 0.000 86, indicating no or very little bias, and that the estimate of the variance provided by the asymptotic theory (0.000 497) is quite good. The 500 trials resulted in 496 rejections for the $\alpha = 0.05$ level test, indicating approximately the same power as that given by the theory.

The 500 trials were repeated with γ set equal to 0.05. The 500 $\hat{\gamma}$ values averaged 0.051, with variance 0.000 480, again very close to that given by the theory. The noncentral χ^2-statistic for the test of $H_0: \gamma = 0$ was 3.851, indicating power of 0.501 for the $\alpha = 0.05$ level test. The test rejected 323 times among the 500, indicating somewhat larger power than that suggested by the theory. The $\alpha = 0.05$ level tests of the correct model rejected 30 and 35 times for $\gamma = 0.10$ and $\gamma = 0.05$, indicated that the significance levels are reasonably close to those claimed.

Whenever the number of degrees of freedom is one, the chi-square statistic is the square of a $N(\theta, 1)$ r.v., (of the form $(Z + \theta)^2$) with $\theta^2 = \delta$ at least in approximation. In this case the r.v. is $U = \hat{\gamma}/\hat{\sigma}(\hat{\gamma})$, and $\theta = \gamma/\sigma(\hat{\gamma})$. Thus, in the case $\gamma = 0.10$, the test is equivalent to rejection for $|U| \geq 1.96$, and the power is approximately $P(|U| \geq 1.96) = 1 - P(-1.96 - \gamma/\sigma(\hat{\gamma}) < (\hat{\gamma} - \gamma)/\sigma(\hat{\gamma}) < 1.96 - \gamma/\sigma(\hat{\gamma})) \doteq 1 - \Phi(1.96 - \gamma/\sigma(\hat{\gamma})) - \Phi(-1.96 - \gamma/\sigma(\hat{\gamma})) \doteq \Phi(-1.96 + \gamma/\sigma(\hat{\gamma})) = \Phi(2.527) = 0.994$. However, since $E(\hat{\gamma})$ seems to be approximately 0.088 6, computations with that value replacing γ, give power approximately 0.978 1, close to the power achieved. Results for $\gamma = 0.05$ were similar.

If one were to first test for the adequacy of the model corresponding to V_1 at level $\alpha = 0.05$, then test $H_0: \gamma = 0$, the correct decision, nonrejection, then rejection, occurs with probability approximately 0.95(0.995 2) = 0.945, since the statistics $G^2(\mathbf{Y}, \hat{\mathbf{m}}_1)$ and $G^2(\hat{\mathbf{m}}_1, \hat{\mathbf{m}}_2)$ are asymptotically independent. The 500 trials with $\gamma = 0.10$ resulted in 466 correct decisions. In general, if a sequence of nested models $\mathscr{M}_1 \subset \mathscr{M}_2 \subset \cdots \subset \mathscr{M}_k$ is chosen for consideration, with corresponding subspaces $V_1 \subset V_2 \subset \cdots \subset V_k$, of dimensions $d_1 > d_2 > \cdots > d_k$, and differences $\nu_1 = d_{i+1} - d_i$, if H_i: $\mathscr{M}_i$ is rejected for $G_i = G^2(\hat{\mathbf{m}}_i, \hat{\mathbf{m}}_{i+1}) > \chi^2_{\nu_i, 1-\alpha_i}$, the probability of choosing the correct model can be determined in approximation because the statistics G_i are asymptotically independent (see Section 4.5).

Example 8.7.5: Consider a three-way table with three factors of 2, 3, and 4 levels. Suppose the model (1 2 3 12 13), independence of factors 2 and 3, given the level of factor 1, and that the parameter values are as follows:

$$\lambda_1^1 = -0.5, \qquad \lambda_2^1 = 0.5, \qquad \lambda_1^2 = -0.2, \qquad \lambda_2^2 = 0, \qquad \lambda_3^2 = 0.2,$$

$$\lambda_1^3 = -0.2, \qquad \lambda_2^3 = -0.1, \qquad \lambda_3^3 = 0.1, \qquad \lambda_4^3 = 0.2,$$

$$(\lambda_{ij}^{12}) = \begin{bmatrix} -0.1 & 0 & 0.1 \\ 0.1 & 0 & -0.1 \end{bmatrix},$$

and

$$(\lambda_{ik}^{13}) = \begin{bmatrix} -0.12 & -0.04 & 0.04 & 0.12 \\ 0.12 & 0.04 & -0.04 & -0.12 \end{bmatrix}.$$

Then $\mu_{ijk} = \lambda + \lambda_i^1 + \lambda_j^2 + \lambda_k^3 + \lambda_{ij}^{12} + \lambda_{ik}^{13}$, with λ chosen to be 3.598 29 so that

$\sum_{ij} m_{ijk} = 1{,}000$. Then

$$\mathbf{m} = \exp(\boldsymbol{\mu}) = \begin{bmatrix} 11.81 & 14.14 & 18.71 & 22.40 \\ 15.94 & 19.08 & 25.25 & 30.23 \\ 21.52 & 25.76 & 34.09 & 40.81 \\ \\ 49.84 & 50.85 & 57.33 & 58.49 \\ 55.08 & 56.20 & 63.36 & 64.64 \\ 60.88 & 62.11 & 70.03 & 71.44 \end{bmatrix}.$$

Suppose that $\mathbf{Y}$ has the $\mathcal{M}(\mathbf{p}, 1{,}000)$ distribution with $\mathbf{p} = \mathbf{m}/1{,}000$. If we observe $\mathbf{Y}$ we could, not knowing $\mathbf{m}$, or any of the parameters, proceed by performing sequential χ^2 tests on the models $\mathcal{M}_4 = (1\ 2\ 3\ 12\ 13\ 23)$, then $\mathcal{M}_3 = (1\ 2\ 3\ 12\ 13)$, then $\mathcal{M}_2 = (1\ 2\ 3\ 12)$, then $\mathcal{M}_1 = (1\ 2\ 3)$. For simplicity we will suppose that we would not proceed further. Let $\mathcal{M}_5$ be the saturated model. We could, then, decide upon any of the five models $\mathcal{M}_5, \ldots, \mathcal{M}_1$. Let $\mathbf{m}_i$ be the MLE corresponding to the observation $\mathbf{m}$, and model $\mathcal{M}_1$. Since $\mathbf{m}$ satisfies models $\mathcal{M}_3$, $\mathcal{M}_4$, $\mathcal{M}_5$, it follows that $\mathbf{m}_i = \mathbf{m}$ for $i = 3, 4, 5$. The vectors $\mathbf{m}_2$ and $\mathbf{m}_1$ were computed using proportional iterated fitting, though a hand calculator can be used for models $\mathcal{M}_2$ and $\mathcal{M}_1$ since, $\mathbf{m}_2 = (m_{ij+}m_{++k}/1{,}000)$, and $\mathbf{m}_1 = (m_{i++}m_{+j+}m_{++k}/1{,}000^2)$. Let $\gamma_i = G^2(\mathbf{m}_i, \mathbf{m})$, and $\delta_i = \gamma_i - \gamma_{i+1}$. Similarly, let $\hat{\mathbf{m}}_i$ be the MLE corresponding to observation $\mathbf{Y}$ and model $\mathcal{M}_i$, $G_i = G^2(\hat{\mathbf{m}}_i, \mathbf{Y})$, and $D_i = G_i - G_{i+1}$. Then, asymptotically $G_i \sim \chi^2_{\nu_i}(\gamma_i)$, and $D_i \sim \chi^2_{d_i}(\delta_i)$, where $\nu_5 = 0$, $\nu_4 = 6$, $\nu_3 = 12$, $\nu_2 = 15$, $\nu_1 = 17$, and $d_i = \nu_i - \nu_{i+1}$. Of course, $\gamma_5 = \gamma_4 = \gamma_3 = 0$ and computation determined $\gamma_2 = 6.29$, $\gamma_1 = 11.53$. Thus $\delta_4 = \delta_3 = 0$, $\delta_2 = 6.29$, and $\delta_1 = 5.24$, $d_4 = d_3 = 6$, $d_2 = 3$, $d_1 = 2$. These d_i's are the degrees of freedom in the usual three-way analysis of variance, and the D_i correspond to the sums of squares. If V_i is the subspace corresponding to the model $\mathcal{M}_i$ and $W_i = V_i \cap V_{i+1}^{\perp}$, then $d_i = \dim(W_i)$. The vectors $\mathbf{m}_i$ may be considered as (nonlinear) "projections" of $\mathbf{m}$ on the spaces V_i. Similarly for the $\hat{\mathbf{m}}_i$ relative to $\mathbf{Y}$. In this sense $\hat{\mathbf{m}}_i - \hat{\mathbf{m}}_{i+1}$ is the projection of $\mathbf{Y}$ on W_i, and $\hat{\mathbf{m}}_i - \hat{\mathbf{m}}_{i+1}$ is the projection of $\mathbf{m}$ on W_i. Asymptotically the measures D_i of the "lengths" of these projections are independent.

The hypothesis H_i: ($\mathbf{m}$ satisfies $\mathcal{M}_i$) $\Leftrightarrow$ ($\log \mathbf{m}$) $\in V_i$ is rejected when $D_i > c_i = \chi^2_{d_i, 1-\alpha_i}$. We choose $\alpha_i = 0.10$ for each i, so that $c_4 = c_3 = 10.64$, $c_2 = 6.25$, and $c_1 = 4.61$. The sequential procedure chooses model $\mathcal{M}_i$ if H_j is accepted for $j \geq i$, but H_{i-1} is rejected. For example, the correct model $\mathcal{M}_3$ is chosen only if $D_4 \leq 10.64$, $D_3 \leq 10.64$, and $D_2 > 6.25$. Computer computations showed that $P(\chi^2_3(6.29) > 6.25) = 0.662\,3$, and $P(\chi^2_2(5.24) > 4.61) = 0.715\,3$. Thus, the probability that $\mathcal{M}_3$ is chosen is $(0.9)^2(0.662\,3) = 0.536$. Similarly, $P(\mathcal{M}_5$ chosen$) = 0.10$, $P(\mathcal{M}_4$ chosen$) = (0.90)(0.10) = 0.09$, $P(\mathcal{M}_1$ chosen$) = (0.9)^2$

$(0.337\,7)(0.715\,3) = 0.195\,7$, $P(\mathscr{M}_1 \text{ chosen}) = (0.9)^2(0.337\,7)(0.284\,7) = 0.077\,9$.

The experiment was simulated 500 times using Manugistics APL-Plus, of which 257 resulted in the choice of the correct model $\mathscr{M}_2$, the sample proportion 0.514, close to 0.536. Similarly, the proportions of choices other models were 0.092 for $\mathscr{M}_0$, 0.206 for $\mathscr{M}_1$, 0.086 for $\mathscr{M}_3$, 0.102 for $\mathscr{M}_4$. The asymptotic theory seems to provide very good approximations, at least for $n = 1{,}000$.

Problem 8.7.1: Let $\mathbf{Y} = (Y_1, Y_2, Y_3)$, where $\mathbf{Y} \sim \mathscr{M}_3(\mathbf{p}, n)$ and $\log \mathbf{m} = \log(n\mathbf{p}) = (\beta_1, \beta_1 - \beta_2, \beta_1 + \beta_2)$.

(a) Suppose that $\beta_2 = 0.8$ and $n = 2{,}000$. Find β_1, $\mathbf{m}$, and $\mathbf{p}$.

(b) What are the asymptotic covariance matrices for $\hat{\boldsymbol{\beta}}$, and $\hat{\mathbf{m}}$, $\hat{\mathbf{p}}$?

(c) Suppose we observe $\mathbf{Y} = \hat{\mathbf{y}} = (514, 255, 123\,1)$. Give 95% confidence intervals on β_2 and on m_2, supposing β_2 to be unknown. *Hint*: $\hat{m}_1 = 541.742\,4$.

(d) Estimate $P(|\hat{\beta}_2 - \beta_2| < 0{\cdot}05)$ for $\beta_2 = 0.8$.

(e) Suppose that $\mathbf{Y}$ satisfies the independent Poisson model rather than the multinomial model. How is the asymptotic distribution of $\hat{\boldsymbol{\beta}}$ affected?

Problem 8.7.2: Consider the $2 \times 2 \times 2$ table

$$\begin{array}{cc|cc|cc} & j \rightarrow & 1 & & 2 & \\ & k \rightarrow & 1 & 2 & 1 & 2 \\ \mathbf{Y} = \mathbf{y}:\ i & 1 & 200 & 300 & 500 & 800 \\ & 2 & 300 & 400 & 300 & 400 \end{array}$$

(a) Find a 95% confidence interval on λ_{111}^{123}, assuming the saturated independent Poisson model.

(b) Use proportional iterated fitting until $\hat{\mathbf{m}}$ does not change in any component by more than 0.5 to fit the model (1 2 3 12 13 23).

(c) Find a 95% confidence interval on λ_{11}^{12}. For simplicity use $\hat{\mathbf{m}}$ as found in (b) and $d(\hat{\mathbf{m}})$ as for the saturated model.

Problem 8.7.3: (a) For the homocide data of Table 8.7.1 find a 95% confidence interval on λ_{11}^{12}, assuming the saturated model.

(b) Verify that $\hat{\mathbf{m}}$ as given is the MLE of $\mathbf{m}$ under the model (1 2 3 12 13 23).

Problem 8.7.4: Suppose that $\mathbf{Y}$ is 2×2 and has the $\mathscr{M}_4(\mathbf{p}, 500)$ distribution, with $\mathbf{p} = \begin{bmatrix} 0.36 & 0.24 \\ 0.20 & 0.20 \end{bmatrix}$. Find an approximation of the power of the 0.05 level χ^2-test for independence.

Problem 8.7.5: Suppose that you wanted 95% simultaneous confidence intervals on all odds-ratios $R(i_1, i_2, j_1, j_2) = (p_{i_1 j_1} p_{i_2 j_2})/(p_{i_1 j_2} p_{i_2 j_1})$ for a 4×5 table. Describe how you could do this. Would it be better to use the Bonferroni or the Scheffé method?

Problem 8.7.6: Consider a stationary Markov chain $W = \{W(t) | t = 0, 1, \ldots\}$ having just two states 0 and 1. W is said to be a Markov chain of order k if, conditionally on $W(t-1)$, $W(t-2), \ldots, W(0)$, the distribution of $W(t)$ depends only on $W(t-1), \ldots, W(t-k)$. Let $W(t)$ have order 2, and $p_{ijk} \equiv P(W(t) = k \mid W(t-2) = i,\ W(t-1) = j)$, for $i, j, k = 0$ or 1. Suppose we observe $W(t)$ for $t = 0, 1, 2, \ldots, 200$. We would like to decide whether W is of order one.

(a) Show that W is of order one if and only if $p(k \mid i, j)$ is the same for $i = 0$ and 1.

(b) Show that W is of order one if and only if $W(t-1)$ and $W(t+1)$ are conditionally independent given $W(t)$ for each t.

(c) Classify a 3-tuple $(W(t-2), W(t-1), W(t))$ in cell (i, j, k) if $W(t-2) = i$, $W(t-1) = j$, $W(t) = k$ for $t = 2, 3, \ldots, 200$. If Y_{ijk} is the number of 3-tuples classified into cell (i, j, k), then $\mathbf{Y} = (Y_{ijk})$ satisfies the independent binomial model with parameters $p_{ij} = p_{ij1}$ and $n_{ij} \equiv Y_{ij+}$ for every ij pair, conditionally on $Y_{ij+} = n_{ij}$. See Bishop, Fienberg, and Holland (1975), page 267. The result of (b) implies that W is of order 1 if and only if the log-linear model for $\mathbf{m} = 200\ \mathbf{p}$ is (1 2 3 12 23). The following table was produced by simulating the process for the case that

$$P = (p_{ijk}) = \begin{array}{c} \\ \left[\begin{array}{cc} 0.2 & 0.8 \\ 0.3 & 0.7 \\ & \\ 0.3 & 0.7 \\ 0.5 & 0.5 \end{array}\right] \end{array} \begin{array}{l} j = 0 \\ j = 1 \\ \\ j = 0 \\ j = 1 \end{array} \quad \begin{array}{l} i = 0 \\ \\ \\ i = 1 \end{array} \qquad (\text{columns } k = 0,\ k = 1)$$

Thus, for example, $p_{101} = 0.7$, and $p_{001} = 0.8$, so that W is not of order one. This can also be verified through odds ratios. $R_j = \dfrac{p_{0j0}/p_{001}}{p_{1j0}/p_{1j1}}$ is one for each j if W has order one. In this case $R_0 = 7/12$, and $R_1 = 3/7$. The stationary probabilities that $(W(t-1), W(t)) = (i, j)$ are

$$\begin{array}{c} \\ i = 0 \\ i = 1 \end{array} \begin{array}{cc} j = 0 & j = 1 \\ \left[\begin{array}{cc} 0.0993 & 0.2649 \\ 0.2649 & 0.3709 \end{array}\right. & \left.\right] \end{array} \equiv P_{12}$$

Beginning with $W(0) = 0$ and $W(1) = 0$, the first 26 observations on W were 0 01100 11110 11010 11101 1110. The frequencies of consecutive

3-tuples (i, j, k) were

$$\mathbf{Y} = \begin{bmatrix} 4 & 19 \\ 15 & 39 \\ \\ 17 & 37 \\ 39 & 29 \end{bmatrix}.$$

Test at level $\alpha = 0.05$ the null hypothesis H_0 that W is of order one.

(d) Find $\mathbf{m}$, $\boldsymbol{\mu}$, and the λ-terms: λ, λ_1^1, λ_1^2, λ_1^3, λ_{11}^{12}, λ_{11}^{13}, λ_{11}^{23}, λ_{111}^{123}, supposing that the stationary probabilities were used to determine $W(0)$ and $W(1)$.

(e) Find the noncentrality parameter and an approximation of the power of the test used in (c). (In 500 simulations H_0 was rejected 293 times, indicating that the power is approximately 0.586.)

8.8 LOGISTIC REGRESSION

As in multiple regression we often wish to study the effects of one or more explanatory variables $x_1, \ldots, x_k$ on a dependent variable Y, which takes only two values. Suppose that for $i = 1, \ldots, T$, we independently observe $Y_i \sim \mathscr{B}(n_i, p_i)$, where p_i depends on $\tilde{x}_i = (x_{i1}, \ldots, x_{ik})$, in the following way. Recall that the logit of any p, $0 < p < 1$, is $L(p) = \log[p/(1-p)]$. It has the inverse $L^{-1}(u) = e^u/(1 + e^u)$. The logistic regression model supposes that $L(p_i) \equiv \eta_i = \sum_j \beta_j x_{ij}$. Let $\mathbf{Y}_1$ be the column vector $(Y_1, \ldots, Y_T)'$, $\mathbf{n} = (n_1, \ldots, n_T)'$, $\mathbf{x}_j = (x_{1j}, \ldots, x_{Tj})'$. Since $\log E(Y_i) = \log n_i + \log \mathbf{p}_i = \log n_i + \log L^{-1}(\eta_i)$ is not linear in the β's, $\mathbf{Y}_1$ does not satisfy a log-linear model. However, the $T \times 2$ array $\mathbf{Y} = (\mathbf{Y}_1, \mathbf{n} - \mathbf{Y}_1)$ does, and we will show how that fact can be exploited.

As usual let $\mathbf{m} = E(\mathbf{Y})$ and $\boldsymbol{\mu} = \log \mathbf{m}$. Write $p_i = e^{\eta_i}/(1 + e^{\eta_i}) = e^{\omega_i}/(e^{\omega_i} + e^{-\omega_i})$, where $\omega_i = \eta_i/2$. Then $\mu_{i1} = \log n_i + \omega_i - h_i$ and $\mu_{i2} = \log n_i - \omega_i - h_i$, where $h_i = \log[e^{\omega_i} + e^{-\omega_i}]$. Letting $\alpha_i = \log n_i - h_i$, we get $\mu_{i1} = \alpha_i + \omega_i$ and $\mu_{i2} = \alpha_i - \omega_i$. Define $\mathbf{x}_j = (x_{1j}, \ldots, x_{Tj})'$, and let $\mathbf{x}_j^*$ be the $T \times 2$ array with $\mathbf{x}_j$ in the first column, $-\mathbf{x}_j$ in the second column. Let $\mathbf{R}_i$ be the indicator of the ith row. Then $\boldsymbol{\mu} = \sum_i \alpha_i \mathbf{R}_i + \sum_j \beta_j \mathbf{x}_j^*$, and $(\mathbf{Y}, \mathbf{R}_i) = Y_{i1} + Y_{i2} = n_i = m_{i1} + m_{i2}$. Therefore, $\mathbf{Y}$ satisfies the log-linear model. Let $V = \mathscr{L}(\mathbf{R}_1, \ldots, \mathbf{R}_T, \mathbf{x}_1^*, \ldots, \mathbf{x}_k^*)$.

The only difficulty caused by the consideration of $\mathbf{Y}$, rather than $\mathbf{Y}_1$ is that we introduced T additional parameters $\alpha_1, \ldots, \alpha_T$. In many applications T may be quite large. We will show that we can avoid use of numerical algorithms which solve for all $T + k$ parameters. First notice that $\alpha_i = \log n_i + h_i$, so that α_i is determined by ω_i, which depends only on $\beta_1, \ldots, \beta_k$ and n_i. We seek the

$T \times 2$ array $\hat{\mathbf{m}}$ with ith row $(\hat{m}_{i1}, \hat{m}_{i2})$, such that $\hat{\mu} = \log \hat{\mathbf{m}} \in V$, and $(\mathbf{Y} - \hat{\mathbf{m}}, \mathbf{x}_j^*) = 0$ for each j. Let $\hat{\mathbf{m}}_1$ be the first column of $\hat{\mathbf{m}}$. Then $\hat{\mathbf{m}} = (\hat{\mathbf{m}}_1, \mathbf{n} - \hat{\mathbf{m}}_1)$, and $\mathbf{Y} = (\mathbf{Y}_1, \mathbf{n} - \mathbf{Y}_1)$. It follows that $(\mathbf{Y} - \hat{\mathbf{m}}, \mathbf{x}_j^*) = (\mathbf{Y}_1 - \hat{\mathbf{m}}_1, \mathbf{x}_j) + (\mathbf{n} - \mathbf{Y} - \hat{\mathbf{m}}_1 + \mathbf{n}, \mathbf{x}_j) = 2(\mathbf{Y}_1 - \hat{\mathbf{m}}_1, \mathbf{x}_j)$. We seek $\hat{\boldsymbol{\beta}} = (\hat{\beta}_1, \ldots, \hat{\beta}_k)'$ so that $\hat{\mathbf{m}}_1 = \mathbf{n}\exp(\hat{\boldsymbol{\eta}})/[1 + \exp(\hat{\boldsymbol{\eta}})]$, $\hat{\eta} = \sum_j \hat{\beta}_j \mathbf{x}_j$, makes this last inner product zero.

We can apply the Newton–Raphson method to find $\hat{\boldsymbol{\beta}}$. Let $G(\boldsymbol{\beta}) = \mathbf{X}'(\mathbf{Y}_1 - \mathbf{m}_1)$, the vector of inner products. The matrix of partial derivatives $H(\boldsymbol{\beta})$ has j, j' element

$$h_{jj'} = \left(\mathbf{x}_j, \frac{\partial \mathbf{m}_1}{\partial \beta_j'}\right) = (\mathbf{x}_j, \mathbf{x}_j'\mathbf{m}(\mathbf{n} - \mathbf{m})/\mathbf{n}),$$

so that $H(\boldsymbol{\beta}) = \mathbf{X}'\mathbf{D}\mathbf{X}$, where $\mathbf{D} = \mathbf{D}(\boldsymbol{\beta})$ is the $T \times T$ diagonal matrix with ith term $n_i p_i(1 - p_i)$, where $p_i = m_i/n_i$. The Newton–Raphson algorithm therefore starts with some initial estimate $\boldsymbol{\beta}^{(0)}$, and iteratively takes

$$\boldsymbol{\beta}^{(r+1)} = \boldsymbol{\beta}^{(r)} - (\mathbf{X}'\mathbf{D}(\boldsymbol{\beta}^{(r)})\mathbf{X})^{-1}(\mathbf{Y}_1 - m(\boldsymbol{\beta}^{(r)})),$$

where $m(\boldsymbol{\beta}) = \exp(\mathbf{X}\boldsymbol{\beta})/[1 + \exp(\mathbf{X}\boldsymbol{\beta})]$. When $\boldsymbol{\beta}^{(r+1)} - \boldsymbol{\beta}^{(r)}$ is small, iterations stop and $\hat{\boldsymbol{\beta}} = \boldsymbol{\beta}^{(r+1)}$, $\hat{\boldsymbol{\eta}} = \mathbf{X}\hat{\boldsymbol{\beta}}$, and $\hat{\mathbf{m}} = \exp(\hat{\boldsymbol{\eta}})/[1 + \exp(\hat{\boldsymbol{\eta}})]$.

Example 8.8.1: In an experiment designed to determine the affect of poison on rats, male and female rats were fed various levels x of poison, and the numbers dying and surviving observed.

	Males		Females	
x	Die	Live	Die	Live
0.5	3	17	1	18
1.0	4	20	3	22
2.0	11	18	12	18
3.0	20	2	14	9

Let $\mathbf{Y}_1 = (Y_{11}, \ldots, Y_{18})$ be the 8-component column vector numbers of rats dying, the first 4 components corresponding to the males. Suppose that these 8 components are independent, with corresponding parameters which are the components of $\mathbf{n} = (20, 24, 29, 22, 19, 25, 30, 23)'$, and $\mathbf{p} = (p_1, \ldots, p_8)'$. Let $\mathbf{x}_0 = (1, \ldots, 1)'$, let $\mathbf{x}_1$ be the indicator for males, and let $\mathbf{x}_2$ be the vector of poison dosages. Suppose also that $L(\mathbf{p}) = \log(\mathbf{p}/(1 - \mathbf{p})) \equiv \boldsymbol{\eta} = \beta_0\mathbf{x}_0 + \beta_1\mathbf{x}_1 + \beta_2\mathbf{x}_2$. The vector $\mathbf{Y}$ was generated on a computer for $\mathbf{n}$ as given, $\beta_0 = -3$, $\beta_1 = 0.5$, $\beta_2 = 1.2$, so that $\mathbf{p} = (0.130, 0.214, 0.475, 0.750, 0.083, 0.142, 0.354, 0.646)'$. The Newton–Raphson algorithm was used to find $\hat{\boldsymbol{\beta}} = (-3.457, 0.0579, 1.405)'$, with corresponding estimates

$$\hat{\mathbf{m}}_1 = (2.045, 4.487, 14.031, 17.437, 1.139, 2.853, 10.330, 15.678),$$

and

$$\hat{\mathbf{p}} = \hat{\mathbf{m}}_1/\mathbf{n} = (0.102, 0.187, 0.484, 0.793, 0.060, 0.114, 0.344, 0.682)'.$$

As before, let $d(\mathbf{u})$ be the diagonal matrix with diagonal $\mathbf{u}$ and let $\mathbf{X} = (\mathbf{x}_0, \mathbf{x}_1, \mathbf{x}_2)$. Note that $D[\mathbf{W}]$, with square brackets, indicates the covariance matrix of a random vector $\mathbf{W}$. Then the asymptotic covariance matrices were $D[\hat{\boldsymbol{\beta}}] = [\mathbf{X}'d(\mathbf{np}(1-\mathbf{p}))\mathbf{X}]^{-1}$, $D[\hat{\boldsymbol{\mu}}] = \mathbf{X}D[\hat{\boldsymbol{\beta}}]\mathbf{X}'$, $D[\hat{\mathbf{m}}] = d(\mathbf{np})D[\hat{\boldsymbol{\mu}}]d(\mathbf{np})$, and $D[\mathbf{p}] = d(\mathbf{p})D[\hat{\boldsymbol{\mu}}]d(\mathbf{p})$. The corresponding estimates are obtained by substituting $\hat{\mathbf{p}}$ for $\mathbf{p}$. For these data we find

$$\hat{D}[\hat{\boldsymbol{\beta}}] = \begin{bmatrix} 0.2683 & -0.0874 & -0.0988 \\ -0.0874 & 0.1287 & 0.0109 \\ -0.0988 & 0.0109 & 0.0481 \end{bmatrix}.$$

The estimates of the standard errors of the components of $\hat{\mathbf{p}}$ were (0.0367, 0.0386, 0.0422, 0.0621, 0.0308, 0.0339, 0.0399, 0.0604).

An approximate 95% confidence interval on β_2 is given by $(L, U) = (1.4054 \pm (1.96)(0.0481)^{1/2}) = (0.9755, 1.9355)$. If $p_m(x)$ is the probability of death of a male rat for dosage x, then $\hat{p}_m(x) = G(\hat{\eta}(x))$, where $G(u) = \exp(u)/[1 + \exp(u)]$, and $\hat{\eta} = \hat{\beta}_0 + \hat{\beta}_1 + \hat{\beta}x$. The asymptotic variance of $\hat{\eta}(x)$ is $\text{Var}(\hat{\eta}(x)) = \mathbf{d}'D[\hat{\boldsymbol{\beta}}]\mathbf{d}$, where $\mathbf{d} = (1, 1, x)'$. Using the δ-method, we find that the asymptotic variance of $\hat{p}_m(x)$ is $\text{Var}(\hat{p}_m(x)) \equiv h(x) = [p_m(x)(1 - p_m(x)]$ $\text{Var}(\hat{\eta}(x))$. To determine a confidence interval $(L(x), U(x))$ on $p_m(x)$ for any x, first determine one on $\eta(x)$. Since the transformation $\eta(x) \to G(\eta(x)) = p_m(x)$ is monotone, $(G(L(x)), G(U(x))$ is then a confidence interval on $p_m(x)$. Corresponding confidence intervals on the probability $p_f(x)$ of a female dying are given by considering $\eta(x) = \beta_0 + \beta_2 x$, and $\hat{\eta}(x) = \hat{\beta}_0 + \hat{\beta}_2 x$. For these data and six choices of x, 95% individual confidence intervals were found on $p_m(x)$ and $p_f(x)$ as presented in Table 8.8.1. Simultaneous Scheffé 95% confidence intervals on $p_m(x)$ and $p_f(x)$ for all six choices of x were also found by replacing $z_{0.975} = 1.96$ by $\sqrt{\chi^2_{2,0.95}} = \sqrt{5.99} = 2.447$. If the Bonferroni method were used we would instead substitute $z = z_{1-0.05/24} = z_{0.99792} = 2.86$, so Scheffé intervals are shorter. We were lucky; all the intervals covered the corresponding parameters. The goodness-of-fit statistics were $G^2 = 2.004$, the *residual deviance*, and $\chi^2 = 1.996$ for 2 d.f., so that the fit was good, as we should expect.

The limit theory we have described has required that T be held fixed, while n or a vector of sample sizes approach infinity. In many applications we observe $(\mathbf{x}_i, Y_i)$, for $i = 1, \ldots, T$, where $\mathbf{x}_i = (x_{i1}, \ldots, x_{ik})$ is a vector of constants, and

Table 8.8.1 Estimates and 95% Confidence Intervals on $p_m(x)$ and $p_f(x)$

x	$p(x)$	$\hat{p}_m(x)$	Lower Individual	Upper Individual Confidence Limit	Lower Simultaneous Confidence Limit	Upper Simultaneous Confidence Limit
				Males		
0.5	0.130	0.110	0.056	0.205	0.047	0.236
1.0	0.214	0.188	0.113	0.295	0.100	0.326
2.0	0.475	0.448	0.335	0.567	0.309	0.596
3.0	0.750	0.740	0.594	0.847	0.554	0.867
4.0	0.909	0.787	0.787	0.964	0.742	0.972
5.0	0.970	0.972	0.899	0.933	0.864	0.995
				Females		
0.5	0.083	0.075	0.035	0.151	0.029	0.178
1.0	0.142	0.131	0.073	0.223	0.063	0.252
2.0	0.354	0.346	0.245	0.462	0.223	0.493
3.0	0.646	0.649	0.497	0.777	0.458	0.803
4.0	0.858	0.867	0.717	0.943	0.667	0.955
5.0	0.953	0.953	0.860	0.988	0.816	0.991

a reasonable model, as before, states that $Y_i \sim \mathscr{B}(p_i, n_i)$, independently for differing i, and that $\log(p_i/(1 - p_i)) = \mathbf{x}_i\boldsymbol{\beta}$. However, the asymptotic theory would not seem appropriate in cases for which n_i is small for most i, particularly, as is often the case, all n_i are one. This will occur when some of the components of $\mathbf{x}_i$ take values on a continuous scale. Fortunately, theory has been developed over the last 20 years which shows that even in the case, if T becomes large, while the individual $\mathbf{x}_i$ are not too far from the others and the matrix $\mathbf{X}'d(\mathbf{pq})\mathbf{X}$ not too close to singularity. The conditions are similar to those for linear models, as stated in Eicher's Theorem 4.8.1. See Santner and Duffy (1989, Section 5.3) and the paper by Fahrmeir and Kaufmann (1985).

Example 8.8.2: Reconsider Examples 8.4.4, and 8.5.7. As before let p_{ij} be the probability that a subject of sex i (1 for men, 2 for women), and educational level $j (j = 1, 2, 3)$ would respond "Yes." Suppose that $Y_{ij1} \sim \mathscr{B}(n_{ij}, p_{ij})$ for $n_{ij} = Y_{1j1} + Y_{ij2}$, are independent for the six combinations of i and j. Corresponding to the model of (8.4.4) the model (1 2 3 12 13 23), $p_{ij} = \exp(\mu_{ij1})/[\exp(\mu_{ij1}) + \exp(\mu_{ij2})] = \exp(\eta_{ij})/[1 + \exp(\eta_{ij})]$, where $\eta_{ij} = \mu_{ij1} - \mu_{ij2} = 2[r_1 + (sr)_{i1} + (er)_{j1}]$. In terms of the β_j defined in Example 8.5.7, we can write $\eta_{ij} = \beta_4 + \beta_7\delta_{i1} + \beta_8\delta_{j1} + \beta_9\delta_{j2}$, where δ_{uv} is 1 if $u = v$, zero otherwise. Therefore the design matrix for the logistic model is

$$\mathbf{X} = \begin{bmatrix} 1 & 1 & 1 & 0 \\ 1 & 1 & 0 & 1 \\ 1 & 1 & 0 & 0 \\ 1 & 0 & 1 & 0 \\ 1 & 0 & 0 & 1 \\ 1 & 0 & 0 & 0 \end{bmatrix},$$

and the vector $\boldsymbol{\beta}^*$ of the model (we use the symbol * to distinguish it from the $\boldsymbol{\beta}$ of the log-linear form of the model in Example 8.5.7) is $(\beta_5, \beta_7, \beta_8, \beta_9)$.

j	$\hat{\beta}_j$	$\hat{\sigma}(\hat{\beta}_j)$	Z_j
5	−1.614	0.142	11.36
7	−0.0235	0.118	0.20
8	2.245	0.150	15.00
9	1.097	0.150	7.33

The z-statistics indicate that a model without the sex variable might be adequate. We found $G^2(\hat{\mathbf{m}}, \mathbf{Y}) = 6.22$, and $\chi^2(\hat{\mathbf{m}}, \mathbf{Y}) = 5.95$ for $6 - 4 = 2$ d.f., with corresponding p-values 0.045 and 0.051. We fit the model with the last two columns of $\mathbf{X}$ replaced by the single column $\mathbf{x}_5 = (0, 1, 2, 0, 1, 2)'$, allowing for the log-linear effect of education, getting $G^2 = 6.27$, $\chi^2 = 6.02$ for 3 d.f., indicating a reasonably good fit. A z-value of -0.0179 again suggested that the sex variable might be omitted. We then fit the model with $\mathbf{J}_6$ and $\mathbf{x}_5$ only, getting $G^2 = 6.30$ and $\chi^2 = 6.05$ for 4 d.f. This resulted in the estimates 0.647, 0.373, 0.162 for the probabilities of the "Yes" answer for educational levels 1, 2, 3. These are, of course, the same as those obtained under the log-linear model discussed in Example 8.5.7. Responses seemed to be little affected by the sex of the subject.

Example 8.8.3: This example is taken from Lee (1974) and from SAS (1990, 1101–1108). The data (Table 8.8.2) consist of 27 vectors $(Y, x_1, \ldots, x_6)$ of observations on 27 cancer patients, where Y is the indicator of remission, and $x_1, \ldots, x_6$ are patient characteristics. The same example is discussed by Santner and Duffy (1989, 230). For the purposes of this example, we will confine the analysis to the explanatory variables Li, Temp., and Cell, which we call x_1, x_2, and x_3, respectively. Actually, these variables were the first three chosen in applying stepwise regression from among six explanatory variables, including these three. We will discuss the fits produced using these three variables only, ignoring the fact they were obtained by stepwise procedures.

We consider four models $\mathcal{M}_k$ for $k = 0, 1, 2, 3$. For each model $Y_i \sim \mathcal{B}(1, p_i)$, independent for $i = 1, \ldots, 27$. Let $\eta_i = \log(p_i/(1 - p_i))$. For $\mathcal{M}_0$, $\eta_i \equiv \beta_0$. For

Table 8.8.2

Patient i	Remiss Y_i	Li x_{i1}	Temp. x_{i2}	Cell x_{i3}
1	1	1.9	0.996	0.80
2	1	1.4	0.992	0.90
3	0	0.8	0.982	0.80
4	0	0.7	0.986	1.00
5	1	1.3	0.980	0.90
6	0	0.6	0.982	1.00
7	1	1.0	0.982	0.95
8	0	1.9	1.020	0.95
9	0	0.8	0.999	1.00
10	0	0.5	1.038	0.95
11	0	0.7	0.988	0.85
12	0	1.2	0.982	0.70
13	0	0.4	1.006	0.80
14	0	0.8	0.990	0.20
15	0	1.1	0.990	1.00
16	1	1.9	1.020	1.00
17	0	0.5	1.014	0.65
18	0	1.0	1.004	1.00
19	0	0.6	0.990	0.55
20	1	1.1	0.986	1.00
21	0	0.4	1.010	1.00
22	0	0.6	1.020	0.90
23	1	1.0	1.002	1.00
24	0	1.6	0.988	0.95
25	1	1.7	0.990	1.00
26	1	0.9	0.986	1.00
27	0	0.7	0.986	1.00

$\mathcal{M}_k$, $k = 1, \ldots, 6$, $\eta_i = \beta_0 + \sum_{j=1}^{k} \beta_j x_{ij}$. Table 8.8.3 contains estimates of these parameters, and Wald's goodness-of-fit statistic $W_k = W(\hat{\mathbf{m}}_k - \hat{\mathbf{m}}_6, \hat{\mathbf{m}}_k)$ as given in the SAS manual. $\hat{\mathbf{m}}_k$ is the MLE of $\mathbf{m}$ for the model $\mathcal{M}_k$. The $\hat{\beta}$'s given here have the opposite sign than those given in the SAS manual because the Y-values were coded as 1's and 2's there rather than as 1's and 0's, as they are here.

If we accept the model $\mathcal{M}_6$, the Wald statistics provide measures of the adequacy of these smaller models. It is tempting to use $W(\mathbf{m}_k - \mathbf{Y}, \mathbf{Y})$ or $W(\hat{\mathbf{m}}_k - \mathbf{Y}, \hat{\mathbf{m}}_k)$ as measures of the adequacy of the model $\mathcal{M}_k$. However, when $\mathbf{Y}$ is a vector of ones and zeros, or even when $\mathbf{Y}$ is a vector of binomial r.v.'s with very small n-values, the statistics $\chi^2(\hat{\mathbf{m}}_k, \mathbf{Y})$, $G^2(\hat{\mathbf{m}}_k, \mathbf{Y})$, $W(\hat{\mathbf{m}}_k - \mathbf{Y}, \mathbf{Y})$, and $W(\hat{\mathbf{m}}_k - \mathbf{Y}, \hat{\mathbf{m}}_k)$ are not approximately distributed as χ^2 under the model $\mathcal{M}_k$. The asymptotic theory discussed earlier applies only for the case n large, T fixed. The statistic $W_3 = 0.183$ indicates that relative to the model $\mathcal{M}_6$, the model

Table 8.8.3

j	$\hat{\beta}_j$	$\hat{\sigma}(\hat{\beta}_j)$	Z_j	$\hat{\alpha}_j$
		Model $\mathcal{M}_0: \eta_i \equiv \beta_0$		
0	0.692	0.408	−1.6946	0.09
		$W_0 = 9.46$ for 6 d.f.		
		Model $\mathcal{M}_1: \eta_i = \beta_0 + \beta_1 x_{i1}$		
0	−3.77	1.38	2.74	0.0061
1	2.90	1.19	2.44	0.0146
		$W_1 = 3.1174$ for 5 d.f.		
		Model $\mathcal{M}_2: \eta_i = \beta_0 + \beta_1 x_{i1} + \beta_2 x_{i2}$		
0	47.86	46.44	1.03	0.303
1	3.30	1.359	2.43	0.015
2	−52.43	47.49	1.10	0.270
		$W_2 = 2.1431$ for 4 d.f.		
		Model $\mathcal{M}_3: \eta_i = \beta_0 + \beta_1 x_{i1} + \beta_{i2} x_{i2} + \beta_3 x_{i3}$		
0	−67.63	56.89	1.19	0.234
1	−9.65	7.75	1.25	0.213
2	−3.87	1.78	2.17	0.030
3	2.07	61.71	1.33	0.184
		$W_3 = 0.1831$ for 3 d.f.		

$\mathcal{M}_3$ fits almost as well. The statistic $W_2 - W_3$ is (in approximation) independent of W_3 and is approximately distributed as χ_1^2 if $\mathbf{m} \in V_2$. This statistic can therefore be used to test $\beta_3 = 0$. In approximation $(W_3 - W_3) = Z_3^2$, where $Z_3 = \hat{\beta}_3/\hat{\sigma}(\hat{\beta}_3)$, another possible test statistic. In this case the model $\mathcal{M}_1$ seems to be quite adequate, with the resulting estimate $\hat{p}(x_1) = \exp(3.7771 + 2.8973x_1)/[1 + \exp(3.771 + 2.8973x_1)]$ of the probability of remission. The positive coefficient for x_1 indicates that increasing amounts of Li tend to increase the probability of remission.

The Case of a Multinomial Response Variable: We have been studying the case in which the response variable Y_i has a binomial distribution. The response variable may instead take $r > 2$ values. The rats in Example 8.8.1 might be classified as dead, sick, and well. As with the case $r = 2$ we can again reduce the dimensionality from the full log-linear model with $T + \eta_k$ parameters to $(r - 1)k$ parameters by using a logistic approach. Let $\mathbf{Y}_i \sim \mathcal{M}_r(\mathbf{p}_i, n_i)$, independent for $i = 1, \ldots, T$, with $\mathbf{p}_i = (p_i^1, \ldots, p_i^r)$. Suppose that $\log(p_i^h/p_i^r) = \sum_j \beta_j^h x_{ij}$. The rth response category has been chosen as a baseline. As for the case $r = 2$, this model can again be shown to be log-linear. Another response category could be chosen as the baseline. The log likelihood is a function $(r - 1)k$ parameters, and can be fit using the Newton–Raphson method. The corresponding vector space is spanned by the vectors $\mathbf{x}_j^h$, $h = 1, \ldots, r-1$, defined to take the value x_{ij} in cell (i, h), $-x_{ij}$ in cell (i, r), zero elsewhere. For a full discussion of this and other models see Agresti (1990 Ch. 9).

Problem 8.8.1: In his first four years as a major league baseball player, Hank Aaron, the leading home run hitter of all time, had the following record:

Year	No. of At Bats	No. of Home Runs
1954	468	13
1955	602	27
1956	609	26
1957	615	44

(a) State an appropriate model, using only two parameters to model the probability p_j of a home run in a time at bat in year j, $j = 1, 2, 3, 4$. One x-variable should reflect experience.

(b) Fit the model, and determine the goodness-of-fit statistics $G^2(\hat{\mathbf{m}}, \mathbf{y})$ and $\chi^2(\hat{\mathbf{m}}, \mathbf{y})$. Estimate the covariance matrix of $\hat{\boldsymbol{\beta}}$ for the model of (a) and use the estimate to test the null hypothesis that p_j was the same every year. Also test the hypothesis using the G^2-statistic.

Problem 8.8.2: In an experiment to determine the effectiveness of insecticide XXX, 60 cockroaches were divided randomly into three sets of 100. The sets of 100 were exposed to three different doses: $d_1 = 1.0$, $d_2 = 1.5$, and $d_3 = 2.0$. The numbers of deaths were 15, 36, and 79. Let $p(d)$ be the probability of death with dosage d. Suppose that the log-odds for death at dosage d is $\beta_0 + \beta_1 d$.

(a) Find the MLE of $\boldsymbol{\beta} = (\beta_0, \beta_1)'$ and estimate its covariance matrix.

(b) Sketch your estimate $\hat{p}(d)$ of $p(d)$ as a function of d.

(c) Give a 95% confidence interval on $p(3.0)$.

(d) Estimate the dosage $d_{0.5}$ for which $p(d) = 1/2$ and give an approximate 95% confidence interval.

Problem 8.8.3: Let $Y_i \sim \mathscr{B}(n_i, p_i)$ independently for $i = 1, \ldots, T$. Let $\mathscr{M}$ be a log-linear model for the table $\mathbf{Y} = (Y_{ij}, i = 1, \ldots, T, \text{ and } j = 1, 2)$, where $Y_{i1} = Y_i$ and $Y_{i2} = n_i - Y_i$. Let $\hat{\mathbf{m}}$ be the MLE of $\mathbf{m} = E(\mathbf{Y})$ under model $\mathscr{M}$. Let $\hat{p}_i = m_{1j}/n_i$ and $\hat{p}_i^* = Y_i/n_i$. Show that

(a)
$$\chi^2(\hat{\mathbf{m}}, \mathbf{Y}) = \sum n_i(Y_i - \hat{m}_{i1})^2/[\hat{m}_{i1}(n_i - \hat{m}_{i1})]$$
$$= \sum n_i(\hat{p}_i^* - \hat{p}_i)^2/[\hat{p}_i(1 - \hat{p}_i)],$$

and

(b)
$$G^2(\hat{\mathbf{m}}, \mathbf{Y}) = \sum Y_i \log[(Y_i(n_i - \hat{m}_i)/(n_i - Y_i)\hat{m}_i]$$
$$= \sum n_i \hat{p}_i^* \log[p_i^*(1 - \hat{p}_i)/(1 - \hat{p}_i^*)\hat{p}_i].$$

Problem 8.8.4: The experiment described in Problem 8.8.2 was also conducted for the insecticide Super-XXX, using 93, 97, and 95 cockroaches with the results that 18, 43, and 83 cockroaches died. Give an appropriate model, then use it to give a confidence interval on a parameter which measures the difference in effectiveness in the two insecticides.

Problem 8.8.5: Questions concerning Example 8.8.1:

(a) Use the results to give 95% confidence intervals on the dosages x_m and x_f for which 50% of rats will die for males and for females.

(b) Find a 95% confidence interval on β_1, the "male effect." Do males and females seem to respond in the same way?

(c) How could you test the null hypothesis that the regression effect of the dosage is the same for females as for males?

(d) Give 95% confidence intervals on the dosages d_m and d_f necessary to kill 99% of all male and female rats.

Problem 8.8.6: (See Problem 8.4.4.) The following table contains the results of the games played among four teams in a basketball league. Each team played 16 games against each of the opponent, 8 on their home court and 8 at the other team's court. The table below presents the number of games won by the home team. For example, team #2 won 3 games over team #3 while playing on team #2's court, and team #3 won 5 games over team #2 when the games were played on #3's court.

$$
\begin{array}{ccc}
 & & \text{Away Team} \\
 & & \begin{array}{cccc} 1 & 2 & 3 & 4 \end{array} \\
\text{Home Team} & \begin{array}{c} 1 \\ 2 \\ 3 \\ 4 \end{array} & \begin{bmatrix} & 7 & 1 & 4 \\ 3 & & 3 & 1 \\ 7 & 5 & & 4 \\ 6 & 7 & 8 & \end{bmatrix}.
\end{array}
$$

Thus, Team #1 won 20 games, #2 won 12, #3 won 28, and #4 won 36.

(a) Let p_{ij} be the probability that team i wins over team j in games played on team i's court. Let $q_{ij} = 1 - p_{ij}$. One possible model $\mathscr{M}^*$ assumes the existence of *strength parameters* $\lambda_1, \lambda_2, \lambda_3, \lambda_4$ such that $\log(p_{ij}/q_{ij}) = \lambda_i - \lambda_j$. Define the vector $\mathbf{Y}$ and matrix $\mathbf{X}$ corresponding to this model.

(b) Show that the matrix $\mathbf{X}$ has rank 3, so that one of the strength parameters, say λ_1, can be arbitrarily set to zero, so that $\mathbf{X}$ becomes a 12×3 matrix.

(c) Fit the model in (a), comment on how well it fits, and draw conclusions.

(d) Actually the data presented were generated on a computer for the case that $\mu_{ij} = \beta_0 + \lambda_i - \lambda_j$, so that β_0 is the "home field effect." The parameters chosen were $\beta_0 = 0.6$, $\lambda_2 = -0.5$, $\lambda_3 = 0.5$, $\lambda_4 = 1.0$. Determine $\mathbf{p} = (p_{ij})$, $\mathbf{m}$ (a 12×2 matrix), and $D[\hat{\boldsymbol{\beta}}]$ (in 500 simulations the standard deviations of the components of $\hat{\boldsymbol{\beta}}$ were 0.254, 0.420, 0.418, 0.433; the means were 0.623, -0.543, 0.536, 1.034).

(e) Fit the model $\mathscr{M}$ actually used to determine the data. Present $\hat{\boldsymbol{\beta}} = (\hat{\beta}_0, \hat{\lambda}_2, \hat{\lambda}_3, \hat{\lambda}_4)$ and $G^2(\hat{\mathbf{m}}, \mathbf{y})$, the residual deviance.

(f) Assuming the model $\mathscr{M}$, test the null hypothesis that $\mathscr{M}^*$ holds ($\alpha = 0.05$, as usual).

(g) Find an approximation for the power of the test in (f). For 500 simulations the test rejected 371 times.

References

Agresti, A. (1990). *Categorical Data Analysis.* New York: John Wiley.

Aicken, M. (1983). *Linear Statistical Analysis of Discrete Data.* New York: John Wiley.

Akaike, H. (1973). Information theory and an extension of the maximum likelihood principle. In *Proceedings of 2nd International Symposium on Information Theory* (eds. B. N. Petrov and F. Csaki). Budapest: Academia Kiado, pp. 267–281.

Akaike, H. (1978). A Bayesian analysis of the AIC procedure. *Ann. Inst. Stat. Math.*, **30**, 9–14.

Albert, A. (1972). *Regression and the Moore–Penrose Pseudoinverse.* New York: Academic Press.

Bailey, B. J. R. (1977). Tables of the Bonferroni *t* statistic. *J. Am. Stat. Assoc.*, **72**, 469–477.

Bartlett, M. S. (1947). The use of transformations. *Biometrics*, **3**, 39–52.

Beaton, A. (1964). *The Use of Special Matrix Operators in Statistical Calculus.* Princeton NJ: Educational Testing Service, pp. RB64–51.

Bechhofer, R. (1954). A single-sample multiple decision procedure for ranking means of normal populations with known variances. *Ann. Math. Stat.*, **25**, 16–39.

Bechhofer, R., Dunnett, C., and Sobel, M. (1954). A two-sample multiple decision procedure for ranking means of normal populations with a common unknown variance. *Ann. Math. Stat.*, **30**, 102–119.

Belsley, D. A., Kuh, E., and Welsch, R. E., (1980). *Regression Diagnostics.* New York: John Wiley.

Bickel, P. and Freedman, D. (1983). *Bootstrap Regression Models with Many Parameters, A Festschrift for Erich Lehmann.* New York: Wadsworth, pp. 28–48.

Bickel, P., Hammel, E., and O'Connell, J. (1975). Sex bias in graduate admissions. *Science*, **187**, 398–403.

Billingsley, P. (1986). *Probability and Measure*, 2d Ed. New York: John Wiley.

Birch, M. (1964). A new proof of the Pearson–Fisher theorem. *Ann. Math Stat.*, **35**, 818–824.

Bishop, Y., Fienberg, S., and Holland, P. (1975). *Discrete Multivariate Analysis: Theory and Practice.* Cambridge MA: MIT Press.

Bock, M. E., Yancey, J. A., and Judge, G. G. (1973). The statistical consequences of preliminary test estimators in regression. *J. Am. Stat. Assoc.*, **68**, 109–116.

Box, G. E. P. and Andersen, S. L. (1955). Permututation theory in the derivation of robust criteria and the study of departure from assumption. *J. Roy. Stat. Soc. B*, **17**, 1–34.

Box, J. F. (1987). Guinness, Gosset, and Fisher. *Statistical Science*, **2**, 45–42.

Brown, L. D. (1984). A note on the Tukey–Kramer procedure for pairwise comparisons of correlated means. In *Design of Experiments: Ranking and Selection*—Essays in Honor of Robert E. Bechhofer (eds. T. J. Santner and A. C. Tamhane), New York: Marcel Dekker.

Brown, T. (1984). Poisson approximations and the definition of the Poisson process. *Am. Math. Monthly*, **91**, 116–123.

Carroll, R. and Ruppert, D. (1988). *Transformation and Weighting in Regression.* New York: Chapman and Hall.

Chatterjee, S. and Hadi, A. (1986). Influential observations. *Statistical Science*, **1**, 379–392.

Chatterjee, S. and Hadi, A. (1988). *Sensitivity Analysis in Linear Regression.* New York: John Wiley.

Christensen, R. (1990). *Log-Linear Models.* New York: Springer-Verlag.

Cochran, W. G. and Cox, G. M. (1957). *Experimental Designs.* New York: John Wiley.

Cook, R. D. and Weisberg, S. (1982). *Residuals and Influence in Regression.* New York: Chapman and Hall.

Cox, C. (1984). An elementary introduction to maximum likelihood estimation for multinomial models: *Birch's theorem and the delta method. The American Statistician*, **38**, 283–287.

Cramer, H. (1951). *Mathematical Methods of Statistics.* Princeton NJ: Princeton University Press.

deBoor, C. (1978). *A Practical Guide to Splines*, New York: Springer-Verlag.

Dixon, W. J. and Massey, F. J. (1957). *Introduction to Statistical Analysis*, 2d Ed. New York: McGraw-Hill.

Doll, R. and Hill, B. (1950). Smoking and carcinoma of the lung. *Br. Med. J.*, **2**, 739–748.

Draper, N. R. and Smith, H. (1981). *Applied Regression Analysis*, 2d Ed. New York: John Wiley.

Dunn, O. J. (1961). Multiple comparisons among means. *J. Am. Stat. Assoc.*, **56**, 52–64.

Dunn, O. J. (1974). On multiple tests and confidence intervals. *Commun. Stat.*, **3**, 101–103.

Dunnett, C. W. (1955). A multiple comparision procedure for comparing several procedures with a control. *J. Am. Stat. Assoc.*, **50**, 1096–1121.

Dunnett, C. W. (1980). Pairwise multiple comparisons in the homogeneous variance, unequal sample size case. *J. Am. Stat. Assoc.*, **75**, 789–795.

Dunnett, C. W. (1982). Robust multiple comparisons. *Commun. Stat.*, **11**, 2611–2629.

Durbin, J. and Watson, J. S. (1950). Testing for zero correlation in least squares regression, I. *Biometrika*, **37**, 409–428.

Durbin, J. and Watson, J. S. (1951). Testing for zero correlation in least squares regression, II. *Biometrika*, **38**, 159–178.

Eaton, M. (1983). *Multivariate Statistics, A Vector Space Approach.* New York: John Wiley.

Efron, B. (1979). Bootstrap methods: another look at the jackknife. *Ann. Stat.*, **7**, 1–26.

Efron, B. (1979). *The Jackknife, the Bootstrap and Other Resampling Methods.* CBMS-NSF 38, SIAM.

Efron, B. and Tibshirani, R. (1986). Bootstrap methods for standard errors, confidence intervals, and other methods of statistical accuracy. *Stat. Sci.*, **1**, 54–75.

Efron, B. and Tibshirani, R. (1993). *An Introduction to the Bootstrap.* New York: Chapman and Hall.

Eicher, F. (1965). Limit theorems for regression with unequal and dependent errors. In *Proceedings of 5th Berkeley Symposium on Mathematical Statistics and Probability Statistics*, **1**, 59–82.

Everitt, B. (1977). *The Analysis of Contingency Tables.* New York: Chapman and Hall.

Fabian, V. (1962). On multiple decision methods for ranking population means. *Ann. Math. Stat.*, **33**, 248–254.

Fabian, V. (1991). On the problem of interactions in the analysis of variance. *J. Am. Stat. Assoc.*, **86**, 362–367.

Fabian V. and Hannan, J. F. (1985). *Introduction to Probability and Mathematical Statistics.* New York: John Wiley.

Fahrmeir, L. and Kaufmann, H. (1985). Consistency and asymptotic normality of the maximum likelihood estimator in generalized linear models. *The Annals of Statistics*, **13**, 342–368.

Fahrmeir, L. and Tutz, G. (1994). *Multivariate Statistical Modelling Based on Generalized Linear Models*, New York: Springer-Verlag.

Fienberg, S. (1977). *The Analysis of Cross-Classified Categorical Data*, 2d Ed. Cambridge MA: MIT Press.

Fienberg, S. and Hinkley, D., eds. (1980). *R. A. Fisher: An Appreciation.* New York: Springer-Verlag.

Friedman, J. (1991). Multivariate adaptive regression splines. *Ann. Math. Stat.*, **19**, 1–66.

Galton, F. (1889). *Natural Inheritance.* London: Macmillan.

Galton, F. (1890). Kinship and correlation. *North American Review*, **150**, 419–431.

Garside, M. J. (1965). The best subset in multiple regression analysis. *Appl. Stat.*, **14**, 196–200.

Gosset, W. (1907). On the probable error of the mean. *Biometrika*, **5**, 315. Under the name "A Student."

Haberman, S. (1974). *The Analysis of Frequency Data.* Chicago: University of Chicago Press.

Haberman, S. (1978, 1979). *Analysis of Qualitative Data*, Vols. 1 and 2. New York: Academic Press.

Hald, A. (1952). *Statistical Theory, with Engineering Applications.* New York: John Wiley.

Hall, P. (1988). Theoretical comparisons of bootstrap confidence intervals. *Ann. Stat.*, **16**, 927–953.

Hampel, F. R., Ronchetti, E. M., Rousseeuw, P. J., and Stahel, W. A. (1986). *Robust Statistics The Approach Based on Influence Functions.* New York: John Wiley.

Harter, H. L. (1960). Tables of range and studentized range. *Ann. Math. Stat.*, **31**, 1122–47.

Hastie, T. J. and Tibshirani, R. J. (1990). *Generalized Additive Models.* New York: Chapman and Hall.

Hayes, J. G. and Halliday, J. (1974). The least squares fitting of cubic spline surfaces to general data sets. *J. Inst. Math. & Appl.*, **14**, 89–104.

Hedayat, A. and Robson, D. S. (1970). Independent stepwise residuals for testing homoscedasticity. *J. Am. Stat. Assoc.*, 1573.

Heiberger, R. and Becker, R. A. (1992). Design of an *S* function for robust regression using iteratively reweighted least squares. *J. Comp & Graph. Stat.*, **1**, 181–196.

Herr, D. (1980). On the history of the use of geometry in the general linear model. *The American Statistician*, 43–47.

Hicks, C. (1982). *Fundamental Concepts in the Design of Experiments.* New York: Holt, Rinehart & Winston.

Hochberg, Y. (1975). An extension of the *T*-method to general unbalanced models of fixed effects. *J. Roy. Stat. Soc. B*, **37**, 426–433.

Hochberg, Y. and Tamhane, A. C. (1987). *Multiple Comparison Procedures.* New York: John Wiley.

Hocking, R. R. (1985). *The Analysis of Linear Models.* New York: Brooks/Cole.

Hoerl, A. E. and Kennard, R. W. (1970a). Ridge regression: biased estimation for nonorthogonal problems. *Technometrics*, **12**, 55–67.

Hoerl, A. E. and Kennard, R. W. (1970b). Ridge regression: applications to nonorthogonal problems. *Technometrics*, **12**, 69–82.

Hosmer, D. and Lemeshow, S. (1989). *Applied Logistic Regression.* New York: John Wiley.

Huber, P. J. (1964). Robust estimation of a location parameter. *Ann. Math. Stat.*, **35**, 73–101.

Huber, P. J. (1981). *Robust Statistics.* New York: John Wiley.

Hurvich, C. M. and Tsai, Chih-Ling (1990). The impact of model selection on inference in linear regression. *The American Statistician*, **44**, 214–217.

Johnson, R. A. and Wichern, D. W. (1988). *Applied Multivariate Statistical Analysis*, 2d Ed. Englewood Cliffs NJ: Prentice Hall.

Kempthorne, O. (1952). *The Design and Analysis of Experiments.* New York: John Wiley.

Kennedy, W. J. and Gentle, J. E. (1980). *Statistical Computing.* New York: Marcel Dekker.

Koul, H. (1992). *Weighted Empiricals and Linear Models.* I.M.S. Lecture Notes Monograph Series, No. 21.

Kramer, C. (1956). Extension of multiple range tests to groups means with unequal numbers of replications, *Biometrics*, **12**, 307–310.

Kuels, M. (1952). The use of the "studentized range" in connection with an analysis of variance. *Euphytica*, **1**, 112–122.

Lancaster, P. and Salkauskas, K. (1986). *Curve and Surface Fitting: An Introduction.* New York: Academic Press.

LeCam, L. (1960). An approximation of the Poisson binomial distribution. *Pacific J. Math.*, **10**, 1181–1197.

Lee, E. (1974). A computer program for linear logistic regression analysis. *Computer Programs in Biomedicine*, 80–92.

Legendre, A. M. (1805). *Nouvelles méthodes pour la détermination des orbites des comètes.* Paris: Courcier.

LePage, R. and Podgorski, K. (1992). Resampling Permutations in Regression with Exchangeable Errors. *Computing and Statistics*, **24**, 546–553.

LePage, R. and Podgorski, K. (1994). Giving the boot, block, and shuffle to statistics. *Scientific Computing and Automation*, **10**, 29–34.

LePage, R., Podgorski, K. and Ryznar, M. (1994). Strong and conditional principles for samples attracted to stable laws, Technical Report, No. 425, Center for Stochastic Processes, Univ. of North Carolina.

Linhart, H. and Zucchini, W. (1986). *Model Selection.* New York: John Wiley.

Mallows, C. L. (1964). Choosing variables in linear regression: a graphical aid. Paper presented at the Central Regional Meeting of the Institute of Mathematical Statistics, Manhattan KS.

Mammen, E. (1992). *When Does Bootstrap Work? Asymptotic Results and Simulations.* New York: Springer-Verlag.

Marquardt, D. W. (1970). Generalized inverses, ridge regression, biased linear estimation, and nonlinear estimation. *Technometrics*, **12**, 591–612.

McCullagh, P. and Nelder, J. A. (1989). *Generalized Linear Models*, 2d Ed. New York: Chapman and Hall.

Mendenhall, W. (1968). *Design and Analysis of Experiments.* Belmont CA: Duxbury Press.

Miller, R. (1981). *Simultaneous Statistical Inference*, 2d Ed. New York: Springer-Verlag.

Morrison, D. (1976). *Multivariate Statistical Methods*, 2d Ed. New York: McGraw-Hill.

Myers, R. (1986). *Classical and Modern Regression with Applications.* Boston: PWS Publishers.

Neter, J. and Wasserman, W. (1974). *Applied Linear Statistical Models.* Homewood IL: Richard Irwin.

Odeh, R. E. and Fox, M. (1991). *Sample Size Choice*, 2d Ed. New York: Marcel Dekker.

Olshen, R. (1973). The conditional level of the F-test. *J. Am. Stat. Assoc.*, **68**, 692.

Ostle, B. (1963). *Statistics in Research.* Ames IA: Iowa State College Press.

Patnaik, P. B. (1949). The noncentral χ^2 and F distributions and their applications. *Biometrika*, **36**, 202–232.

Paul, P. C. (1943). Ph.D. Thesis, Iowa State College: Ames, Iowa.

Pearson, E. and Hartley, H. O. (1966). *Biometrika Tables for Statisticians*, 3d Ed. Cambridge UK: Cambridge University Press.

Rao, C. R. (1989). *Statistics and Truth, Putting Chance to Work.* Fairland MD: International Co-operative Publishing House.

Read, T. and Cressie, N. (1988). *Goodness-of-Fit Statistics for Discrete Multivariate Data.* New York: Springer-Verlag.

Rousseeuw, P. J. and Leroy, A. M. (1987). *Robust Regression and Outlier Detection.* New York: John Wiley.

Santner, T. and Duffy, D. (1989). *The Statistical Analysis of Discrete Data.* New York: Springer-Verlag.

SAS (1985). *SAS User's Guide: Statistics Version 5.* SAS Institute Inc.

SAS (1990). *SAS/Stat User's Guide*, Ver. 6, Vol. 2, 4th Ed. SAS Institute Inc.

Scheffé, H. (1953). A method for judging all contrasts in the analysis of variance. *Biometrika*, **40**, 87–104.

Scheffé, H. (1959). *The Analysis of Variance.* New York: John Wiley.

Scheffé, H. (1973). A statistical theory of calibration. *Ann. Stat.*, **1**, 1–37.

Schoenberg, I. J. (1946). Contributions to the approximation of equidistant data by analytic functions. *Quart. Appl. Math.*, **4**, 45–99 and 112–141.

Schwarz, G. (1978). Estimating the dimension of a model. *Ann. Stat.*, **6**, 461–464.

Scott, A. and Wild. C. (199). Transformations and R^2. *The American Statistician*, **45**, 127–129.

Searle, S. R. (1971). *Linear Models.* New York: John Wiley.

Searle, S. R. (1987). *Linear Models for Unbalanced Data.* New York: John Wiley.

Seber, G. (1977). *Linear Regression Analysis.* New York: John Wiley.

Seber, G. (1980). *The Linear Hypothesis: A General Theory*, 2d Ed. London: Griffin's Statistical Monograph, No. 19.

Seber, G. and Wild, C. J. (1989). *Nonlinear Regression.* New York: John Wiley.

Sidak, Z. (1967). Rectangular confidence intervals for means of multivariate normal distributions. *J. Am. Stat. Assoc.*, **62**, 626–633.

Simpson, E. H. (1951). The interpretation of interaction in contingency tables, *J. Roy. Stat. Soc. B*, **13**, 238–241.

Snedecor, G. (1967). *Statistical Methods*, 6th Ed. Ames IA: Iowa State College Press.

Snedecor, G. W. and Cochran, W. G. (1980). *Statistical Methods*, 7th Ed. Ames IA: Iowa State University Press.

Spjotvøll, E. (1972). On the optimality of some multiple comparison procedures. *Ann. Math. Stat.*, **43**, 398–411.

Spjotvøll, E. and Stoline, M. R. (1973). An extension of the *T* method of multiple comparison to include the cases with unequal sample sizes. *J. Am. Stat. Assoc.*, **68**, 975–978.

Srivastava, M. S. and Khatri, C. G. (1979). *An Introduction to Multivariate Statistics.* Amsterdam: North Holland.

Stigler, S. (1986). *The History of Statistics: The Measurement of Uncertainty before 1900.* Cambridge MA: Harvard University Press.

Stigler, S. (1989). Galton's account of the invention of correlation. *Stat. Sci.*, 73–86.

Stoline, M. R. (1981). The Status of Multiple Comparisons: Simultaneous Estimation of All Pairwise Comparisons in One-Way ANOVA Designs, *Am. Statistician*, **35**, 134–141.

Theil, H. (1972). *Principles of Econometrics.* New York: John Wiley.

Thorn, J. and Palmer, P. (1989). *Total Baseball.* Warner Books.

Tibshirani, R. and Efron, B. (1993). *An Introduction to the Bootstrap.* New York: Chapman and Hall.

Traxler, R. H. (1976). A snag in the history of factorial experiments. In *On the History of Statistics and Probability* (ed. D. B. Owen). New York: Marcel Dekker.

Tukey, J. (1953). The Problem of Multiple Comparisons. Mimeographed Manuscript, Princeton University, New Jersey.

Tukey, J. (1962). The future of data analysis. *Ann. Math. Stat.*, **33**, 1–67.

Wahba, G. (1990). *Spline Models for Observational Data.* CBMS-NSF 59, SIAM.

Wang, Y. (1971). Probabilities of type I errors of the Welch tests for the Behrens–Fisher problem. *J. Am. Stat. Assoc.*, **66**, 605–608.

Welch, B. L. (1947). The generalization of Students' problem when several different population variances are involved. *Biometrika*, **34**, 28.

Welsch. R. (1977). Stepwise multivariate comparison procedures. *J. Am. Stat. Assoc.*, **72**, 566–575.

Wishart, J. (1950). *Field Trials II: The analysis of Covariance.* Tech. Comm. No. 15, Commonwealth Bureau of Plant Breeding and Genetics, School of Agriculture, Cambridge, England.

Wu, C. F. J. (1986). Jackknife bootstrap, and other resampling methods in regression analysis. *Ann. Math. Stat.*, **14**, 1261–1294.

Yates, F. (1935). Complex experiments. *J. Roy. Stat. Soc. Suppl.*, **2**, 181–247.

Yohai, V. J. and Maronna, R. A. (1979). Asymptotic behavior of M-estimators for the linear model. *Ann. Stat.*, **7**, 258–268.

Appendix

Tables

Table 1.1

Standard Normal C.D.F.

Φ(z)

z	0.00	,01	0.02	0.03	0.04	0.05	0.06	0.07	0.08	0.09
0.0	0.5000	0.5040	0.5080	0.5120	0.5160	0.5199	0.5239	0.5279	0.5319	0.5359
0.1	0.5398	0.5438	0.5478	0.5517	0.5557	0.5596	0.5636	0.5675	0.5714	0.5753
0.2	0.5793	0.5832	0.5871	0.5910	0.5948	0.5987	0.6026	0.6064	0.6103	0.6141
0.3	0.6179	0.6217	0.6255	0.6293	0.6331	0.6368	0.6406	0.6443	0.6480	0.6517
0.4	0.6554	0.6591	0.6628	0.6664	0.6700	0.6736	0.6772	0.6808	0.6844	0.6879
0.5	0.6915	0.6950	0.6985	0.7019	0.7054	0.7088	0.7123	0.7157	0.7190	0.7224
0.6	0.7257	0.7291	0.7324	0.7357	0.7389	0.7422	0.7454	0.7486	0.7517	0.7549
0.7	0.7580	0.7611	0.7642	0.7673	0.7704	0.7734	0.7764	0.7794	0.7823	0.7852
0.8	0.7881	0.7910	0.7939	0.7967	0.7995	0.8023	0.8051	0.8078	0.8106	0.8133
0.9	0.8159	0.8186	0.8212	0.8238	0.8264	0.8289	0.8315	0.8340	0.8365	0.8389
1.0	0.8413	0.8438	0.8461	0.8485	0.8508	0.8531	0.8554	0.8577	0.8599	0.8621
1.1	0.8643	0.8665	0.8686	0.8708	0.8729	0.8749	0.8770	0.8790	0.8810	0.8830
1.2	0.8849	0.8869	0.8888	0.8907	0.8925	0.8944	0.8962	0.8980	0.8997	0.9015
1.3	0.9032	0.9049	0.9066	0.9082	0.9099	0.9115	0.9131	0.9147	0.9162	0.9177
1.4	0.9192	0.9207	0.9222	0.9236	0.9251	0.9265	0.9279	0.9292	0.9306	0.9319
1.5	0.9332	0.9345	0.9357	0.9370	0.9382	0.9394	0.9406	0.9418	0.9429	0.9441
1.6	0.9452	0.9463	0.9474	0.9484	0.9495	0.9505	0.9515	0.9525	0.9535	0.9545
1.7	0.9554	0.9564	0.9573	0.9582	0.9591	0.9599	0.9608	0.9616	0.9625	0.9633
1.8	0.9641	0.9649	0.9656	0.9664	0.9671	0.9678	0.9686	0.9693	0.9699	0.9706
1.9	0.9713	0.9719	0.9726	0.9732	0.9738	0.9744	0.9750	0.9756	0.9761	0.9767
2.0	0.9772	0.9778	0.9783	0.9788	0.9793	0.9798	0.9803	0.9808	0.9812	0.9817
2.1	0.9821	0.9826	0.9830	0.9834	0.9838	0.9842	0.9846	0.9850	0.9854	0.9857
2.2	0.9861	0.9864	0.9868	0.9871	0.9875	0.9878	0.9881	0.9884	0.9887	0.9890
2.3	0.9893	0.9896	0.9898	0.9901	0.9904	0.9906	0.9909	0.9911	0.9913	0.9916
2.4	0.9918	0.9920	0.9922	0.9925	0.9927	0.9929	0.9931	0.9932	0.9934	0.9936
2.5	0.9938	0.9940	0.9941	0.9943	0.9945	0.9946	0.9948	0.9949	0.9951	0.9952
2.6	0.9953	0.9955	0.9956	0.9957	0.9959	0.9960	0.9961	0.9962	0.9963	0.9964
2.7	0.9965	0.9966	0.9967	0.9968	0.9969	0.9970	0.9971	0.9972	0.9973	0.9974
2.8	0.9974	0.9975	0.9976	0.9977	0.9977	0.9978	0.9979	0.9979	0.9980	0.9981
2.9	0.9981	0.9982	0.9982	0.9983	0.9984	0.9984	0.9985	0.9985	0.9986	0.9986

Table 1.2

The Standard Normal

$10^6 * [1 - \Phi(z)]$

z	0.0	0.1	0.2	0.3	0.4	0.5	0.6	0.7	0.8	0.9
2.5	6209.67	6036.56	5867.74	5703.13	5542.62	5386.15	5233.61	5084.93	4940.02	4798.80
2.6	4661.19	4527.11	4396.49	4269.24	4145.30	4024.59	3907.03	3792.56	3681.11	3572.60
2.7	3466.97	3364.16	3264.10	3166.72	3071.96	2979.76	2890.07	2802.82	2717.95	2635.40
2.8	2555.13	2477.08	2401.18	2327.40	2255.68	2185.96	2118.21	2052.36	1988.38	1926.21
2.9	1865.81	1807.14	1750.16	1694.81	1641.06	1588.87	1538.20	1489.00	1441.24	1394.89
3.0	1349.90	1306.24	1263.87	1222.77	1182.89	1144.21	1106.69	1070.29	1035.00	1000.78
3.1	967.60	935.44	904.26	874.03	844.74	816.35	788.85	762.20	736.38	711.36
3.2	687.14	663.68	640.95	618.95	597.65	577.03	557.06	537.74	519.04	500.94
3.3	483.42	466.48	450.09	434.23	418.89	404.06	389.71	375.84	362.43	349.46
3.4	336.93	324.81	313.11	301.79	290.86	280.29	270.09	260.23	250.71	241.51
3.5	232.63	224.05	215.77	207.78	200.06	192.62	185.43	178.49	171.80	165.34
3.6	159.11	153.10	147.30	141.71	136.32	131.12	126.11	121.28	116.62	112.13
3.7	107.80	103.63	99.61	95.74	92.01	88.42	84.96	81.62	78.41	75.32
3.8	72.35	69.48	66.73	64.07	61.52	59.06	56.69	54.42	52.23	50.12
3.9	48.10	46.15	44.27	42.47	40.74	39.08	37.48	35.94	34.46	33.04
4.0	31.67	30.36	29.10	27.89	26.73	25.61	24.54	23.51	22.52	21.57
4.1	20.66	19.78	18.94	18.14	17.37	16.62	15.91	15.23	14.58	13.95
4.2	13.35	12.77	12.22	11.69	11.18	10.69	10.22	9.77	9.35	8.93
4.3	8.54	8.16	7.80	7.46	7.12	6.81	6.50	6.21	5.93	5.67
4.4	5.41	5.17	4.94	4.71	4.50	4.29	4.10	3.91	3.73	3.56

Φ(z)	z
0.800	0.84162
0.900	1.28156
0.950	1.64485
0.975	1.95996
0.990	2.32635

z	1 - Φ(z)
4.5	3.3977 E-6
4.6	2.1125 E-6
4.7	1.3008 E-6
4.8	7.9333 E-7
4.9	4.7918 E-7

z	1 - Φ(z)
5.0	2.8665 E-7
6.0	9.8659 E-10
7.0	1.2798 E-12
8.0	6.2210 E-16
9.0	1.1286 E-19

Table 2.1

Student's-t γ-Quantiles for ν d.f.

ν	γ = 0.550	0.600	0.650	0.700	0.750	0.800	0.850	0.900	0.950	0.975	0.990	0.995	0.999
1	0.158	0.325	0.510	0.727	1.000	1.376	1.963	3.078	6.314	12.71	31.82	63.66	318.3
2	0.142	0.289	0.445	0.617	0.817	1.061	1.386	1.886	2.920	4.303	6.965	9.925	22.33
3	0.137	0.277	0.424	0.584	0.765	0.979	1.250	1.638	2.353	3.182	4.541	5.841	10.21
4	0.134	0.271	0.414	0.569	0.741	0.941	1.190	1.533	2.132	2.776	3.747	4.604	7.173
5	0.132	0.267	0.408	0.559	0.727	0.920	1.156	1.476	2.015	2.571	3.365	4.032	5.893
6	0.131	0.265	0.404	0.553	0.718	0.906	1.134	1.440	1.943	2.447	3.143	3.707	5.208
7	0.130	0.263	0.402	0.549	0.711	0.896	1.119	1.415	1.895	2.365	2.998	3.500	4.785
8	0.130	0.262	0.400	0.546	0.706	0.889	1.108	1.397	1.860	2.306	2.897	3.355	4.501
9	0.129	0.261	0.398	0.544	0.703	0.883	1.100	1.383	1.833	2.262	2.821	3.250	4.297
10	0.129	0.260	0.397	0.542	0.700	0.879	1.093	1.372	1.813	2.228	2.764	3.169	4.144
11	0.129	0.260	0.396	0.540	0.697	0.876	1.088	1.363	1.796	2.201	2.718	3.106	4.025
12	0.128	0.259	0.395	0.539	0.696	0.873	1.083	1.356	1.782	2.179	2.681	3.055	3.930
13	0.128	0.259	0.394	0.538	0.694	0.870	1.080	1.350	1.771	2.160	2.650	3.012	3.852
14	0.128	0.258	0.393	0.537	0.692	0.868	1.076	1.345	1.761	2.145	2.625	2.977	3.787
15	0.128	0.258	0.393	0.536	0.691	0.866	1.074	1.341	1.753	2.131	2.603	2.947	3.733
16	0.128	0.258	0.392	0.535	0.690	0.865	1.071	1.337	1.746	2.120	2.584	2.921	3.686
17	0.128	0.257	0.392	0.534	0.689	0.863	1.069	1.333	1.740	2.110	2.567	2.898	3.646
18	0.127	0.257	0.392	0.534	0.688	0.862	1.067	1.330	1.734	2.101	2.552	2.878	3.611
19	0.127	0.257	0.391	0.533	0.688	0.861	1.066	1.328	1.729	2.093	2.540	2.861	3.579
20	0.127	0.257	0.391	0.533	0.687	0.860	1.064	1.325	1.725	2.086	2.528	2.845	3.552
21	0.127	0.257	0.391	0.533	0.686	0.859	1.063	1.323	1.721	2.080	2.518	2.831	3.527
22	0.127	0.256	0.390	0.532	0.686	0.858	1.061	1.321	1.717	2.074	2.508	2.819	3.505
23	0.127	0.256	0.390	0.532	0.685	0.858	1.060	1.320	1.714	2.069	2.500	2.807	3.485
24	0.127	0.256	0.390	0.531	0.685	0.857	1.059	1.318	1.711	2.064	2.492	2.797	3.467
25	0.127	0.256	0.390	0.531	0.684	0.856	1.058	1.316	1.708	2.060	2.485	2.787	3.450
26	0.127	0.256	0.390	0.531	0.684	0.856	1.058	1.315	1.706	2.056	2.479	2.779	3.435
27	0.127	0.256	0.389	0.531	0.684	0.855	1.057	1.314	1.703	2.052	2.473	2.771	3.421
28	0.127	0.256	0.389	0.530	0.683	0.855	1.056	1.313	1.701	2.048	2.467	2.763	3.408
29	0.127	0.256	0.389	0.530	0.683	0.854	1.055	1.311	1.699	2.045	2.462	2.756	3.396
30	0.127	0.256	0.389	0.530	0.683	0.854	1.055	1.310	1.697	2.042	2.457	2.750	3.385
35	0.127	0.255	0.389	0.529	0.682	0.852	1.052	1.306	1.690	2.030	2.438	2.724	3.340
40	0.127	0.255	0.388	0.529	0.681	0.851	1.050	1.303	1.684	2.021	2.423	2.705	3.307
50	0.126	0.255	0.388	0.528	0.679	0.849	1.047	1.299	1.676	2.009	2.403	2.678	3.261
60	0.126	0.255	0.387	0.527	0.679	0.848	1.046	1.296	1.671	2.000	2.390	2.660	3.232
90	0.126	0.254	0.387	0.526	0.677	0.846	1.042	1.291	1.662	1.987	2.369	2.632	3.183
120	0.126	0.254	0.386	0.526	0.677	0.845	1.041	1.289	1.658	1.980	2.358	2.617	3.160
Infinity	0.1256	0.2533	0.3853	0.5244	0.6745	0.8416	1.0364	1.2816	1.6449	1.9560	2.3263	2.576	3.090

Table 2.2 Student's-t γ-Quantiles for $\gamma = 0.05/(2k)$, ν d.f.

ν \ k	2	3	4	5	6	7	8	9	10
1	25.452	38.189	50.923	63.657	76.390	89.123	101.856	114.589	127.321
2	6.205	7.649	8.860	9.925	10.886	11.769	12.590	13.360	14.089
3	4.177	4.857	5.392	5.841	6.232	6.580	6.895	7.185	7.453
4	3.495	3.961	4.315	4.604	4.851	5.068	5.261	5.437	5.598
5	3.163	3.534	3.810	4.032	4.219	4.382	4.526	4.655	4.773
6	2.969	3.288	3.521	3.707	3.863	3.997	4.115	4.221	4.317
7	2.841	3.128	3.335	3.500	3.636	3.753	3.855	3.947	4.029
8	2.752	3.016	3.206	3.355	3.479	3.584	3.677	3.759	3.833
9	2.685	2.933	3.111	3.250	3.364	3.462	3.547	3.622	3.690
10	2.634	2.870	3.038	3.169	3.277	3.368	3.448	3.518	3.581
11	2.593	2.820	2.981	3.106	3.208	3.295	3.370	3.437	3.497
12	2.560	2.780	2.935	3.055	3.153	3.236	3.308	3.371	3.428
13	2.533	2.746	2.896	3.012	3.107	3.187	3.257	3.318	3.373
14	2.510	2.718	2.864	2.977	3.069	3.146	3.214	3.273	3.326
15	2.490	2.694	2.837	2.947	3.036	3.112	3.177	3.235	3.286
16	2.473	2.673	2.813	2.921	3.008	3.082	3.146	3.202	3.252
17	2.458	2.655	2.793	2.898	2.984	3.056	3.119	3.174	3.222
18	2.445	2.639	2.775	2.878	2.963	3.034	3.095	3.149	3.197
19	2.433	2.625	2.759	2.861	2.944	3.014	3.074	3.127	3.174
20	2.423	2.613	2.744	2.845	2.927	2.996	3.055	3.107	3.153
21	2.414	2.601	2.732	2.831	2.912	2.980	3.038	3.090	3.135
22	2.406	2.591	2.720	2.819	2.899	2.966	3.023	3.074	3.119
23	2.398	2.582	2.710	2.807	2.886	2.953	3.010	3.060	3.104
24	2.391	2.574	2.700	2.797	2.875	2.941	2.997	3.047	3.091
25	2.385	2.566	2.692	2.787	2.865	2.930	2.986	3.035	3.078
26	2.379	2.559	2.684	2.779	2.856	2.920	2.975	3.024	3.067
27	2.373	2.553	2.676	2.771	2.847	2.911	2.966	3.014	3.057
28	2.369	2.547	2.670	2.763	2.839	2.902	2.957	3.005	3.047
29	2.364	2.541	2.663	2.756	2.832	2.895	2.949	2.996	3.038
30	2.360	2.536	2.657	2.750	2.825	2.887	2.941	2.988	3.030
35	2.342	2.515	2.633	2.724	2.797	2.858	2.910	2.955	2.996
40	2.329	2.499	2.616	2.705	2.776	2.836	2.887	2.931	2.971
50	2.311	2.477	2.591	2.678	2.747	2.805	2.855	2.898	2.937
60	2.299	2.463	2.575	2.660	2.729	2.786	2.834	2.877	2.915
90	2.280	2.440	2.549	2.632	2.698	2.753	2.800	2.841	2.878
120	2.270	2.428	2.536	2.617	2.683	2.737	2.784	2.824	2.860

Table 2.3 Student's-t γ-Quantiles for $\gamma = 1 - 0.05/[k(k - 1)]$, ν d.f.

ν \ k	3	4	5	6	7	8
1	38.189	76.390	127.321	190.984	267.379	356.506
2	7.649	10.886	14.089	17.277	20.457	23.633
3	4.857	6.232	7.453	8.575	9.624	10.617
4	3.961	4.851	5.598	6.254	6.847	7.392
5	3.534	4.219	4.773	5.247	5.667	6.045
6	3.288	3.863	4.317	4.698	5.030	5.326
7	3.128	3.636	4.029	4.355	4.636	4.884
8	3.016	3.479	3.833	4.122	4.370	4.587
9	2.933	3.364	3.690	3.954	4.179	4.374
10	2.870	3.277	3.581	3.827	4.035	4.215
11	2.820	3.208	3.497	3.728	3.923	4.091
12	2.780	3.153	3.428	3.649	3.833	3.993
13	2.746	3.107	3.373	3.584	3.760	3.912
14	2.718	3.069	3.326	3.530	3.699	3.845
15	2.694	3.036	3.286	3.484	3.648	3.788
16	2.673	3.008	3.252	3.444	3.604	3.740
17	2.655	2.984	3.222	3.410	3.565	3.698
18	2.639	2.963	3.197	3.380	3.532	3.661
19	2.625	2.944	3.174	3.354	3.503	3.629
20	2.613	2.927	3.153	3.331	3.477	3.601
21	2.601	2.912	3.135	3.310	3.453	3.575
22	2.591	2.899	3.119	3.291	3.432	3.552
23	2.582	2.886	3.104	3.274	3.413	3.531
24	2.574	2.875	3.091	3.258	3.396	3.513
25	2.566	2.865	3.078	3.244	3.380	3.496
26	2.559	2.856	3.067	3.231	3.366	3.480
27	2.553	2.847	3.057	3.219	3.353	3.465
28	2.547	2.839	3.047	3.208	3.340	3.452
29	2.541	2.832	3.038	3.198	3.329	3.440
30	2.536	2.825	3.030	3.189	3.319	3.428
35	2.515	2.797	2.996	3.150	3.276	3.382
40	2.499	2.776	2.971	3.122	3.244	3.347
50	2.477	2.747	2.937	3.083	3.201	3.300
60	2.463	2.729	2.915	3.057	3.173	3.270
90	2.440	2.698	2.878	3.016	3.127	3.220
120	2.428	2.683	2.860	2.995	3.104	3.195
Infinity	2.394	2.638	2.807	2.935	3.038	3.124

Table 3

Chi-Square Quantiles

ν = d.f.	0.10	0.20	0.30	0.40	0.50	0.60	0.70	0.80	0.90	0.95	0.975	0.990	0.995	0.999
1	0.02	0.06	0.15	0.27	0.45	0.71	1.07	1.64	2.71	3.84	5.02	6.63	7.88	10.83
2	0.21	0.45	0.71	1.02	1.39	1.83	2.41	3.22	4.61	5.99	7.38	9.21	10.60	13.82
3	0.58	1.01	1.42	1.87	2.37	2.95	3.66	4.64	6.25	7.81	9.35	11.34	12.84	16.27
4	1.06	1.65	2.19	2.75	3.36	4.04	4.88	5.99	7.78	9.49	11.14	13.28	14.86	18.47
5	1.61	2.34	3.00	3.66	4.35	5.13	6.06	7.29	9.24	11.07	12.83	15.09	16.75	20.52
6	2.20	3.07	3.83	4.57	5.35	6.21	7.23	8.56	10.64	12.59	14.45	16.81	18.55	22.46
7	2.83	3.82	4.67	5.49	6.35	7.28	8.38	9.80	12.02	14.07	16.01	18.48	20.28	24.32
8	3.49	4.59	5.53	6.42	7.34	8.35	9.52	11.03	13.36	15.51	17.53	20.09	21.95	26.12
9	4.17	5.38	6.39	7.36	8.34	9.41	10.66	12.24	14.68	16.92	19.02	21.67	23.59	27.88
10	4.87	6.18	7.27	8.30	9.34	10.47	11.78	13.44	15.99	18.31	20.48	23.21	25.19	29.59
11	5.58	6.99	8.15	9.24	10.34	11.53	12.90	14.63	17.28	19.68	21.92	24.72	26.76	31.26
12	6.30	7.81	9.03	10.18	11.34	12.58	14.01	15.81	18.55	21.03	23.34	26.22	28.30	32.91
13	7.04	8.63	9.93	11.13	12.34	13.64	15.12	16.98	19.81	22.36	24.74	27.69	29.82	34.53
14	7.79	9.47	10.82	12.08	13.34	14.69	16.22	18.15	21.06	23.68	26.12	29.14	31.32	36.12
15	8.55	10.31	11.72	13.03	14.34	15.73	17.32	19.31	22.31	25.00	27.49	30.58	32.80	37.70
16	9.31	11.15	12.62	13.98	15.34	16.78	18.42	20.47	23.54	26.30	28.85	32.00	34.27	39.25
17	10.09	12.00	13.53	14.94	16.34	17.82	19.51	21.61	24.77	27.59	30.19	33.41	35.72	40.79
18	10.86	12.86	14.44	15.89	17.34	18.87	20.60	22.76	25.99	28.87	31.53	34.81	37.16	42.31
19	11.65	13.72	15.35	16.85	18.34	19.91	21.69	23.90	27.20	30.14	32.85	36.19	38.58	43.82
20	12.44	14.58	16.27	17.81	19.34	20.95	22.77	25.04	28.41	31.41	34.17	37.57	40.00	45.31
21	13.24	15.44	17.18	18.77	20.34	21.99	23.86	26.17	29.62	32.67	35.48	38.93	41.40	46.80
22	14.04	16.31	18.10	19.73	21.34	23.03	24.94	27.30	30.81	33.92	36.78	40.29	42.80	48.27
23	14.85	17.19	19.02	20.69	22.34	24.07	26.02	28.43	32.01	35.17	38.08	41.64	44.18	49.73
24	15.66	18.06	19.94	21.65	23.34	25.11	27.10	29.55	33.20	36.42	39.36	42.98	45.56	51.18
25	16.47	18.94	20.87	22.62	24.34	26.14	28.17	30.68	34.38	37.65	40.65	44.31	46.93	52.62

Table 3 (*contd*)

Chi-Square Quantiles

ν = d.f.	0.10	0.20	0.30	0.40	0.50	0.60	0.70	0.80	0.90	0.95	0.975	0.990	0.995	0.999
26	17.29	19.82	21.79	23.58	25.34	27.18	29.25	31.79	35.56	38.89	41.92	45.64	48.29	54.05
27	18.11	20.70	22.72	24.54	26.34	28.21	30.32	32.91	36.74	40.11	43.19	46.96	49.64	55.48
28	18.94	21.59	23.65	25.51	27.34	29.25	31.39	34.03	37.92	41.34	44.46	48.28	50.99	56.89
29	19.77	22.48	24.58	26.48	28.34	30.28	32.46	35.14	39.09	42.56	45.72	49.59	52.34	58.30
30	20.60	23.36	25.51	27.44	29.34	31.32	33.53	36.25	40.26	43.77	46.98	50.89	53.67	59.70
31	21.43	24.26	26.44	28.41	30.34	32.35	34.60	37.36	41.42	44.99	48.23	52.19	55.00	61.10
32	22.27	25.15	27.37	29.38	31.34	33.38	35.66	38.47	42.58	46.19	49.48	53.49	56.33	62.49
33	23.11	26.04	28.31	30.34	32.34	34.41	36.73	39.57	43.75	47.40	50.73	54.78	57.65	63.87
34	23.95	26.94	29.24	31.31	33.34	35.44	37.80	40.68	44.90	48.60	51.97	56.06	58.96	65.25
35	24.80	27.84	30.18	32.28	34.34	36.47	38.86	41.78	46.06	49.80	53.20	57.34	60.27	66.62
40	29.05	32.34	34.87	37.13	39.34	41.62	44.16	47.27	51.81	55.76	59.34	63.69	66.77	73.40
50	37.69	41.45	44.31	46.86	49.33	51.89	54.72	58.16	63.17	67.50	71.42	76.15	79.49	86.66
60	46.46	50.64	53.81	56.62	59.33	62.13	65.23	68.97	74.40	79.08	83.30	88.38	91.95	99.61
80	64.28	69.21	72.92	76.19	79.33	82.57	86.12	90.41	96.58	101.9	106.6	112.3	116.3	124.8
100	82.36	87.95	92.13	95.81	99.33	102.95	106.91	111.67	118.50	124.3	129.6	135.8	140.2	149.5
200	174.84	183.00	189.05	194.32	199.33	204.43	209.99	216.61	226.02	233.99	241.06	249.45	255.26	267.54

The Cube Root Approximation

For large n the γth quantile is given in good approximation by $u = \nu[a + z\,b]^3$, where $a = 1 - 2/(9\nu)$, $b = (2/(9\nu)$, and $\Phi(z) = \gamma$. For example, for $\gamma = 0.99$, $\nu = 100$, $a = 0.99778$, $b = 0.04714$, $z = 2.32635$, and $u = 135.82$.

Table 4.1 0.90 - Quantiles of the F-Distribution

v_2 \ v_1	1	2	3	4	5	6	7	8	9	10	12	15	20	24	30	40	60	120	Inf.
1	39.9	49.5	53.6	55.8	57.2	58.2	58.9	59.4	59.9	60.2	60.7	61.2	61.7	62.0	62.3	62.5	62.8	63.1	63.3
2	8.53	9.00	9.16	9.24	9.29	9.33	9.35	9.37	9.38	9.39	9.41	9.42	9.44	9.45	9.46	9.47	9.47	9.48	9.49
3	5.54	5.46	5.39	5.34	5.31	5.28	5.27	5.25	5.24	5.23	5.22	5.20	5.18	5.18	5.17	5.16	5.15	5.14	5.13
4	4.54	4.32	4.19	4.11	4.05	4.01	3.98	3.95	3.94	3.92	3.90	3.87	3.84	3.83	3.82	3.80	3.79	3.78	3.76
5	4.06	3.78	3.62	3.52	3.45	3.40	3.37	3.34	3.32	3.30	3.27	3.24	3.21	3.19	3.17	3.16	3.14	3.12	3.11
6	3.78	3.46	3.29	3.18	3.11	3.05	3.01	2.98	2.96	2.94	2.90	2.87	2.84	2.82	2.80	2.78	2.76	2.74	2.72
7	3.59	3.26	3.07	2.96	2.88	2.83	2.78	2.75	2.72	2.70	2.67	2.63	2.59	2.58	2.56	2.54	2.51	2.49	2.47
8	3.46	3.11	2.92	2.81	2.73	2.67	2.62	2.59	2.56	2.54	2.50	2.46	2.42	2.40	2.38	2.36	2.34	2.32	2.29
9	3.36	3.01	2.81	2.69	2.61	2.55	2.51	2.47	2.44	2.42	2.38	2.34	2.30	2.28	2.25	2.23	2.21	2.18	2.16
10	3.29	2.92	2.73	2.61	2.52	2.46	2.41	2.38	2.35	2.32	2.28	2.24	2.20	2.18	2.16	2.13	2.11	2.08	2.06
11	3.23	2.86	2.66	2.54	2.45	2.39	2.34	2.30	2.27	2.25	2.21	2.17	2.12	2.10	2.08	2.05	2.03	2.00	1.97
12	3.18	2.81	2.61	2.48	2.39	2.33	2.28	2.24	2.21	2.19	2.15	2.10	2.06	2.04	2.01	1.99	1.96	1.93	1.90
13	3.14	2.76	2.56	2.43	2.35	2.28	2.23	2.20	2.16	2.14	2.10	2.05	2.01	1.98	1.96	1.93	1.90	1.88	1.85
14	3.10	2.73	2.52	2.39	2.31	2.24	2.19	2.15	2.12	2.10	2.05	2.01	1.96	1.94	1.91	1.89	1.86	1.83	1.80
15	3.07	2.70	2.49	2.36	2.27	2.21	2.16	2.12	2.09	2.06	2.02	1.97	1.92	1.90	1.87	1.85	1.82	1.79	1.76
16	3.05	2.67	2.46	2.33	2.24	2.18	2.13	2.09	2.06	2.03	1.99	1.94	1.89	1.87	1.84	1.81	1.78	1.75	1.72
17	3.03	2.64	2.44	2.31	2.22	2.15	2.10	2.06	2.03	2.00	1.96	1.91	1.86	1.84	1.81	1.78	1.75	1.72	1.69
18	3.01	2.62	2.42	2.29	2.20	2.13	2.08	2.04	2.00	1.98	1.93	1.89	1.84	1.81	1.78	1.75	1.72	1.69	1.66
19	2.99	2.61	2.40	2.27	2.18	2.11	2.06	2.02	1.98	1.96	1.91	1.86	1.81	1.79	1.76	1.73	1.70	1.67	1.63
20	2.97	2.59	2.38	2.25	2.16	2.09	2.04	2.00	1.96	1.94	1.89	1.84	1.79	1.77	1.74	1.71	1.68	1.64	1.61
21	2.96	2.57	2.36	2.23	2.14	2.08	2.02	1.98	1.95	1.92	1.87	1.83	1.78	1.75	1.72	1.69	1.66	1.62	1.59
22	2.95	2.56	2.35	2.22	2.13	2.06	2.01	1.97	1.93	1.90	1.86	1.81	1.76	1.73	1.70	1.67	1.64	1.60	1.57
23	2.94	2.55	2.34	2.21	2.11	2.05	1.99	1.95	1.92	1.89	1.84	1.80	1.74	1.72	1.69	1.66	1.62	1.59	1.55
24	2.93	2.54	2.33	2.19	2.10	2.04	1.98	1.94	1.91	1.88	1.83	1.78	1.73	1.70	1.67	1.64	1.61	1.57	1.53
25	2.92	2.53	2.32	2.18	2.09	2.02	1.97	1.93	1.89	1.87	1.82	1.77	1.72	1.69	1.66	1.63	1.59	1.56	1.52
30	2.88	2.49	2.28	2.14	2.05	1.98	1.93	1.88	1.85	1.82	1.77	1.72	1.67	1.64	1.61	1.57	1.54	1.50	1.46
40	2.84	2.44	2.23	2.09	2.00	1.93	1.87	1.83	1.79	1.76	1.71	1.66	1.61	1.57	1.54	1.51	1.47	1.42	1.38
60	2.79	2.39	2.18	2.04	1.95	1.87	1.82	1.77	1.74	1.71	1.66	1.60	1.54	1.51	1.48	1.44	1.40	1.35	1.29
120	2.75	2.35	2.13	1.99	1.90	1.82	1.77	1.72	1.68	1.65	1.60	1.55	1.48	1.45	1.41	1.37	1.32	1.26	1.19
Inf.	2.71	2.30	2.09	1.95	1.85	1.78	1.72	1.67	1.63	1.60	1.55	1.49	1.42	1.39	1.34	1.30	1.24	1.17	1.00

Table 4.2 0.95 - Quantiles of the F-Distribution

ν_2 \ ν_1	1	2	3	4	5	6	7	8	9	10	12	15	20	24	30	40	60	120	Inf.
1	162	200	216	225	230	234	237	239	241	242	244	246	248	249	250	251	252	253	254
2	18.5	19.0	19.2	19.3	19.3	19.3	19.4	19.4	19.4	19.4	19.4	19.4	19.5	19.5	19.5	19.5	19.5	19.5	19.5
3	10.1	9.55	9.28	9.12	9.01	8.94	8.89	8.85	8.81	8.79	8.74	8.70	8.66	8.64	8.62	8.59	8.57	8.55	8.53
4	7.71	6.94	6.59	6.39	6.26	6.16	6.09	6.04	6.00	5.96	5.91	5.86	5.80	5.77	5.75	5.72	5.69	5.66	5.63
5	6.61	5.79	5.41	5.19	5.05	4.95	4.88	4.82	4.77	4.74	4.68	4.62	4.56	4.53	4.50	4.46	4.43	4.40	4.37
6	5.99	5.14	4.76	4.53	4.39	4.28	4.21	4.15	4.10	4.06	4.00	3.94	3.87	3.84	3.81	3.77	3.74	3.70	3.67
7	5.59	4.74	4.35	4.12	3.97	3.87	3.79	3.73	3.68	3.64	3.57	3.51	3.44	3.41	3.38	3.34	3.30	3.27	3.23
8	5.32	4.46	4.07	3.84	3.69	3.58	3.50	3.44	3.39	3.35	3.28	3.22	3.15	3.12	3.08	3.04	3.01	2.97	2.93
9	5.12	4.26	3.86	3.63	3.48	3.37	3.29	3.23	3.18	3.14	3.07	3.01	2.94	2.90	2.86	2.83	2.79	2.75	2.71
10	4.96	4.10	3.71	3.48	3.33	3.22	3.14	3.07	3.02	2.98	2.91	2.85	2.77	2.74	2.70	2.66	2.62	2.58	2.54
11	4.84	3.98	3.59	3.36	3.20	3.09	3.01	2.95	2.90	2.85	2.79	2.72	2.65	2.61	2.57	2.53	2.49	2.45	2.40
12	4.75	3.89	3.49	3.26	3.11	3.00	2.91	2.85	2.80	2.75	2.69	2.62	2.54	2.51	2.47	2.43	2.38	2.34	2.30
13	4.67	3.81	3.41	3.18	3.03	2.92	2.83	2.77	2.71	2.67	2.60	2.53	2.46	2.42	2.38	2.34	2.30	2.25	2.21
14	4.60	3.74	3.34	3.11	2.96	2.85	2.76	2.70	2.65	2.60	2.53	2.46	2.39	2.35	2.31	2.27	2.22	2.18	2.13
15	4.54	3.68	3.29	3.06	2.90	2.79	2.71	2.64	2.59	2.54	2.48	2.40	2.33	2.29	2.25	2.20	2.16	2.11	2.07
16	4.49	3.63	3.24	3.01	2.85	2.74	2.66	2.59	2.54	2.49	2.42	2.35	2.28	2.24	2.19	2.15	2.11	2.06	2.01
17	4.45	3.59	3.20	2.96	2.81	2.70	2.61	2.55	2.49	2.45	2.38	2.31	2.23	2.19	2.15	2.10	2.06	2.01	1.96
18	4.41	3.55	3.16	2.93	2.77	2.66	2.58	2.51	2.46	2.41	2.34	2.27	2.19	2.15	2.11	2.06	2.02	1.97	1.92
19	4.38	3.52	3.13	2.90	2.74	2.63	2.54	2.48	2.42	2.38	2.31	2.23	2.16	2.11	2.07	2.03	1.98	1.93	1.88
20	4.35	3.49	3.10	2.87	2.71	2.60	2.51	2.45	2.39	2.35	2.28	2.20	2.12	2.08	2.04	1.99	1.95	1.90	1.84
21	4.32	3.47	3.07	2.84	2.68	2.57	2.49	2.42	2.37	2.32	2.25	2.18	2.10	2.05	2.01	1.96	1.92	1.87	1.81
22	4.30	3.44	3.05	2.82	2.66	2.55	2.46	2.40	2.34	2.30	2.23	2.15	2.07	2.03	1.98	1.94	1.89	1.84	1.78
23	4.28	3.42	3.03	2.80	2.64	2.53	2.44	2.37	2.32	2.27	2.20	2.13	2.05	2.01	1.96	1.91	1.86	1.81	1.76
24	4.26	3.40	3.01	2.78	2.62	2.51	2.42	2.36	2.30	2.25	2.18	2.11	2.03	1.98	1.94	1.89	1.84	1.79	1.73
25	4.24	3.39	2.99	2.76	2.60	2.49	2.40	2.34	2.28	2.24	2.16	2.09	2.01	1.96	1.92	1.87	1.82	1.77	1.71
30	4.17	3.32	2.92	2.69	2.53	2.42	2.33	2.27	2.21	2.16	2.09	2.01	1.93	1.89	1.84	1.79	1.74	1.68	1.62
40	4.08	3.23	2.84	2.61	2.45	2.34	2.25	2.18	2.12	2.08	2.00	1.92	1.84	1.79	1.74	1.69	1.64	1.58	1.51
60	4.00	3.15	2.76	2.53	2.37	2.25	2.17	2.10	2.04	1.99	1.92	1.84	1.75	1.70	1.65	1.59	1.53	1.47	1.39
120	3.92	3.07	2.68	2.45	2.29	2.18	2.09	2.02	1.96	1.91	1.83	1.75	1.66	1.61	1.55	1.50	1.43	1.35	1.25
Inf.	3.84	3.00	2.60	2.37	2.21	2.10	2.01	1.94	1.88	1.83	1.75	1.67	1.57	1.52	1.46	1.39	1.32	1.22	1.00

Table 4.3 0.975 - Quantiles of the F-Distribution

ν_2 \ ν_1	1	2	3	4	5	6	7	8	9	10	12	15	20	24	30	40	60	120	Inf.
1	648	800	864	900	922	937	948	957	963	969	977	985	993	249	250	251	252	253	254
2	38.5	39.0	39.2	39.3	39.3	39.3	39.4	39.4	39.4	39.4	39.4	39.4	39.5	19.5	19.5	19.5	19.5	19.5	19.5
3	17.4	16.0	15.4	15.1	14.9	14.7	14.6	14.5	14.5	14.4	14.3	14.3	14.2	8.64	8.62	8.59	8.57	8.55	8.53
4	12.2	10.7	9.98	9.60	9.36	9.20	9.07	8.98	8.90	8.84	8.75	8.66	8.56	5.77	5.75	5.72	5.69	5.66	5.63
5	10.0	8.43	7.76	7.39	7.15	6.98	6.85	6.76	6.68	6.62	6.52	6.43	6.33	4.53	4.50	4.46	4.43	4.40	4.37
6	8.81	7.26	6.60	6.23	5.99	5.82	5.70	5.60	5.52	5.46	5.37	5.27	5.17	3.84	3.81	3.77	3.74	3.70	3.67
7	8.07	6.54	5.89	5.52	5.29	5.12	4.99	4.90	4.82	4.76	4.67	4.57	4.47	3.41	3.38	3.34	3.30	3.27	3.23
8	7.57	6.06	5.42	5.05	4.82	4.65	4.53	4.43	4.36	4.30	4.20	4.10	4.00	3.12	3.08	3.04	3.01	2.97	2.93
9	7.21	5.71	5.08	4.72	4.48	4.32	4.20	4.10	4.03	3.96	3.87	3.77	3.67	2.90	2.86	2.83	2.79	2.75	2.71
10	6.94	5.46	4.83	4.47	4.24	4.07	3.95	3.85	3.78	3.72	3.62	3.52	3.42	2.74	2.70	2.66	2.62	2.58	2.54
11	6.72	5.26	4.63	4.28	4.04	3.88	3.76	3.66	3.59	3.53	3.43	3.33	3.23	2.61	2.57	2.53	2.49	2.45	2.40
12	6.55	5.10	4.47	4.12	3.89	3.73	3.61	3.51	3.44	3.37	3.28	3.18	3.07	2.51	2.47	2.43	2.38	2.34	2.30
13	6.41	4.97	4.35	4.00	3.77	3.60	3.48	3.39	3.31	3.25	3.15	3.05	2.95	2.42	2.38	2.34	2.30	2.25	2.21
14	6.30	4.86	4.24	3.89	3.66	3.50	3.38	3.29	3.21	3.15	3.05	2.95	2.84	2.35	2.31	2.27	2.22	2.18	2.13
15	6.20	4.77	4.15	3.80	3.58	3.41	3.29	3.20	3.12	3.06	2.96	2.86	2.76	2.29	2.25	2.20	2.16	2.11	2.07
16	6.12	4.69	4.08	3.73	3.50	3.34	3.22	3.12	3.05	2.99	2.89	2.79	2.68	2.24	2.19	2.15	2.11	2.06	2.01
17	6.04	4.62	4.01	3.66	3.44	3.28	3.16	3.06	2.98	2.92	2.82	2.72	2.62	2.19	2.15	2.10	2.06	2.01	1.96
18	5.98	4.56	3.95	3.61	3.38	3.22	3.10	3.01	2.93	2.87	2.77	2.67	2.56	2.15	2.11	2.06	2.02	1.97	1.92
19	5.92	4.51	3.90	3.56	3.33	3.17	3.05	2.96	2.88	2.82	2.72	2.62	2.51	2.11	2.07	2.03	1.98	1.93	1.88
20	5.87	4.46	3.86	3.51	3.29	3.13	3.01	2.91	2.84	2.77	2.68	2.57	2.46	2.08	2.04	1.99	1.95	1.90	1.84
21	5.83	4.42	3.82	3.48	3.25	3.09	2.97	2.87	2.80	2.73	2.64	2.53	2.42	2.05	2.01	1.96	1.92	1.87	1.81
22	5.79	4.38	3.78	3.44	3.22	3.05	2.93	2.84	2.76	2.70	2.60	2.50	2.39	2.03	1.98	1.94	1.89	1.84	1.78
23	5.75	4.35	3.75	3.41	3.18	3.02	2.90	2.81	2.73	2.67	2.57	2.47	2.36	2.01	1.96	1.91	1.86	1.81	1.76
24	5.72	4.32	3.72	3.38	3.15	2.99	2.87	2.78	2.70	2.64	2.54	2.44	2.33	1.98	1.94	1.89	1.84	1.79	1.73
25	5.69	4.29	3.69	3.35	3.13	2.97	2.85	2.75	2.68	2.61	2.51	2.41	2.30	1.96	1.92	1.87	1.82	1.77	1.71
30	5.57	4.18	3.59	3.25	3.03	2.87	2.75	2.65	2.57	2.51	2.41	2.31	2.20	1.89	1.84	1.79	1.74	1.68	1.62
40	5.42	4.05	3.46	3.13	2.90	2.74	2.62	2.53	2.45	2.39	2.29	2.18	2.07	1.79	1.74	1.69	1.64	1.58	1.51
60	5.29	3.93	3.34	3.01	2.79	2.63	2.51	2.41	2.33	2.27	2.17	2.06	1.94	1.70	1.65	1.59	1.53	1.47	1.39
120	5.15	3.80	3.23	2.89	2.67	2.52	2.39	2.30	2.22	2.16	2.05	1.94	1.82	1.61	1.55	1.50	1.43	1.35	1.25
Inf.	5.02	3.69	3.12	2.79	2.57	2.41	2.29	2.19	2.11	2.05	1.94	1.83	1.71	1.52	1.46	1.39	1.32	1.22	1.00

Table 4.4 0.99 - Quantiles of the F-Distribution

ν_2 \ ν_1	1	2	3	4	5	6	7	8	9	10	12	15	20	24	30	40	60	120	Inf.
1	4052	5000	5403	5625	5764	5859	5928	5981	6022	6056	6106	6157	6209	6235	6261	6287	6313	6339	6366
2	98.5	99.0	99.2	99.3	99.3	99.3	99.4	99.4	99.4	99.4	99.4	99.4	99.5	99.5	99.5	99.5	99.5	99.5	99.5
3	34.1	30.8	29.5	28.7	28.2	27.9	27.7	27.5	27.4	27.2	27.1	26.9	26.7	26.6	26.5	26.4	26.3	26.2	26.1
4	21.2	18.0	16.7	16.0	15.5	15.2	15.0	14.8	14.7	14.6	14.4	14.2	14.0	13.9	13.8	13.8	13.7	13.6	13.5
5	16.3	13.3	12.1	11.4	11.0	10.7	10.5	10.3	10.2	10.1	9.89	9.72	9.55	9.47	9.38	9.29	9.20	9.11	9.46
6	13.8	10.9	9.78	9.15	8.75	8.47	8.26	8.10	7.98	7.87	7.72	7.56	7.40	7.31	7.23	7.14	7.06	6.97	6.88
7	12.3	9.55	8.45	7.85	7.46	7.19	6.99	6.84	6.72	6.62	6.47	6.31	6.16	6.07	5.99	5.91	5.82	5.74	5.65
8	11.3	8.65	7.59	7.01	6.63	6.37	6.18	6.03	5.91	5.81	5.67	5.52	5.36	5.28	5.20	5.12	5.03	4.95	4.86
9	10.6	8.02	6.99	6.42	6.06	5.80	5.61	5.47	5.35	5.26	5.11	4.96	4.81	4.73	4.65	4.57	4.48	4.40	4.31
10	10.0	7.56	6.55	5.99	5.64	5.39	5.20	5.06	4.94	4.85	4.71	4.56	4.41	4.33	4.25	4.17	4.08	4.00	3.91
11	9.65	7.21	6.22	5.67	5.32	5.07	4.89	4.74	4.63	4.54	4.40	4.25	4.10	4.02	3.94	3.86	3.78	3.69	3.60
12	9.33	6.93	5.95	5.41	5.06	4.82	4.64	4.50	4.39	4.30	4.16	4.01	3.86	3.78	3.70	3.62	3.54	3.45	3.36
13	9.07	6.70	5.74	5.21	4.86	4.62	4.44	4.30	4.19	4.10	3.96	3.82	3.66	3.59	3.51	3.43	3.34	3.25	3.17
14	8.86	6.51	5.56	5.04	4.69	4.46	4.28	4.14	4.03	3.94	3.80	3.66	3.51	3.43	3.35	3.27	3.18	3.09	3.00
15	8.68	6.36	5.42	4.89	4.56	4.32	4.14	4.00	3.89	3.80	3.67	3.52	3.37	3.29	3.21	3.13	3.05	2.96	2.87
16	8.53	6.23	5.29	4.77	4.44	4.20	4.03	3.89	3.78	3.69	3.55	3.41	3.26	3.18	3.10	3.02	2.93	2.84	2.75
17	8.40	6.11	5.18	4.67	4.34	4.10	3.93	3.79	3.68	3.59	3.46	3.31	3.16	3.08	3.00	2.92	2.83	2.75	2.65
18	8.29	6.01	5.09	4.58	4.25	4.01	3.84	3.71	3.60	3.51	3.37	3.23	3.08	3.00	2.92	2.84	2.75	2.66	2.57
19	8.18	5.93	5.01	4.50	4.17	3.94	3.77	3.63	3.52	3.43	3.30	3.15	3.00	2.92	2.84	2.76	2.67	2.58	2.49
20	8.10	5.85	4.94	4.43	4.10	3.87	3.70	3.56	3.46	3.37	3.23	3.09	2.94	2.86	2.78	2.69	2.61	2.52	2.42
21	8.02	5.78	4.87	4.37	4.04	3.81	3.64	3.51	3.40	3.31	3.17	3.03	2.88	2.80	2.72	2.64	2.55	2.46	2.36
22	7.95	5.72	4.82	4.31	3.99	3.76	3.59	3.45	3.35	3.26	3.12	2.98	2.83	2.75	2.67	2.58	2.50	2.40	2.31
23	7.88	5.66	4.76	4.26	3.94	3.71	3.54	3.41	3.30	3.21	3.07	2.93	2.78	2.70	2.62	2.54	2.45	2.35	2.26
24	7.82	5.61	4.72	4.22	3.90	3.67	3.50	3.36	3.26	3.17	3.03	2.89	2.74	2.66	2.58	2.49	2.40	2.31	2.21
25	7.77	5.57	4.68	4.18	3.85	3.63	3.46	3.32	3.22	3.13	2.99	2.85	2.70	2.62	2.54	2.45	2.36	2.27	2.17
30	7.56	5.39	4.51	4.02	3.70	3.47	3.30	3.17	3.07	2.98	2.84	2.70	2.55	2.47	2.39	2.30	2.21	2.11	2.01
40	7.31	5.18	4.31	3.83	3.51	3.29	3.12	2.99	2.89	2.80	2.66	2.52	2.37	2.29	2.20	2.11	2.02	1.92	1.80
60	7.08	4.98	4.13	3.65	3.34	3.12	2.95	2.82	2.72	2.63	2.50	2.35	2.20	2.12	2.03	1.94	1.84	1.73	1.60
120	6.85	4.79	3.95	3.48	3.17	2.96	2.79	2.66	2.56	2.47	2.34	2.19	2.03	1.95	1.86	1.76	1.66	1.53	1.38
Inf.	6.64	4.61	3.78	3.32	3.02	2.80	2.64	2.51	2.41	2.32	2.19	2.04	1.88	1.79	1.70	1.59	1.48	1.33	1.00

Table 5.1 PEARSON AND HARTLEY CHARTS* FOR THE POWER OF THE *F*-TEST

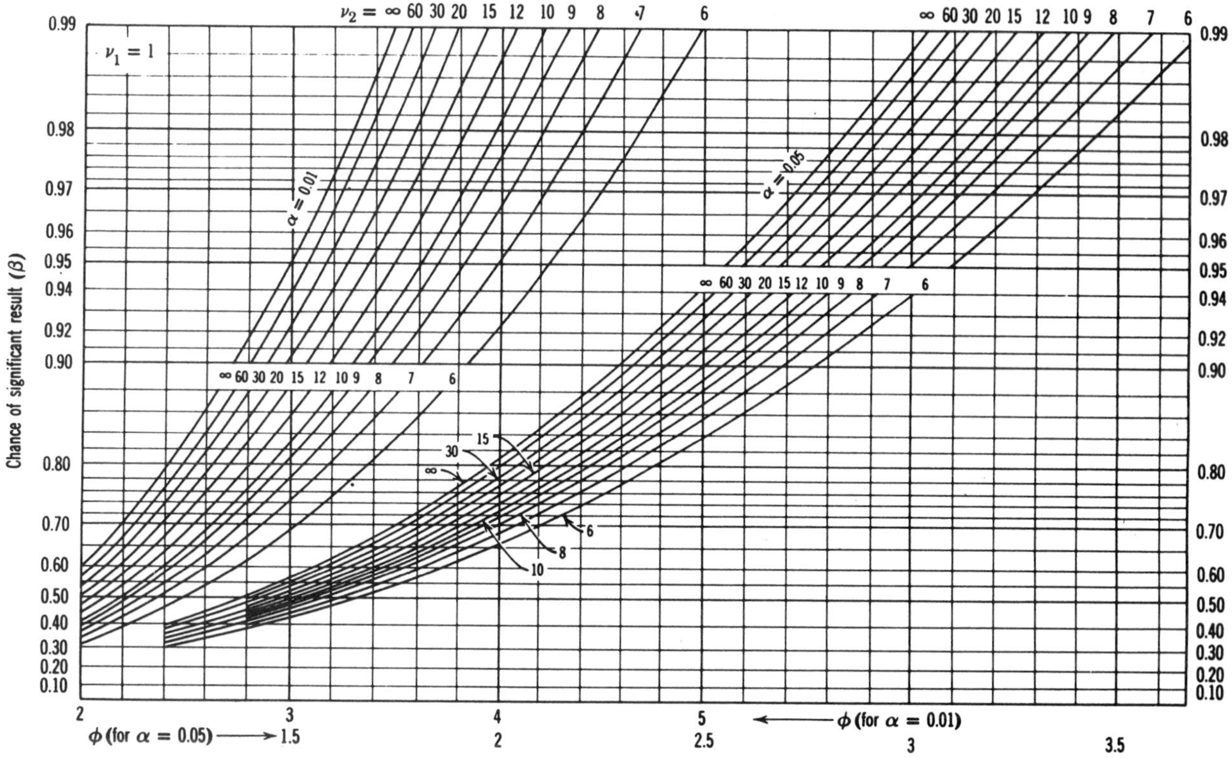

* By E. S. Pearson and H. O. Hartley in *Biometrika*, Vol. 38, pp. 115–122 (1951). Reproduced with the kind permission of the authors and the editor.

Table 5.2 PEARSON AND HARTLEY CHARTS* FOR THE POWER OF THE *F*-TEST

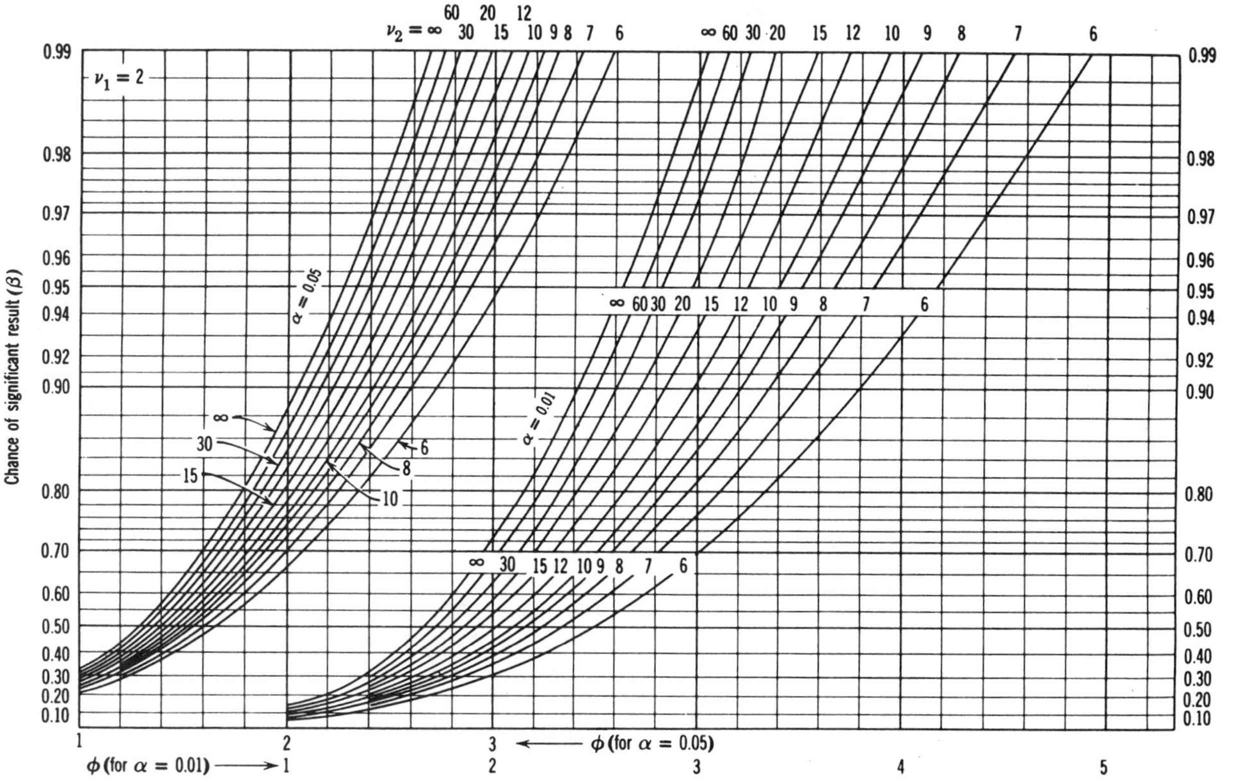

* By E. S. Pearson and H. O. Hartley in *Biometrika*, Vol. 38, pp. 115–122 (1951). Reproduced with the kind permission of the authors and the editor.

Table 5.3 PEARSON AND HARTLEY CHARTS* FOR THE POWER OF THE *F*-TEST

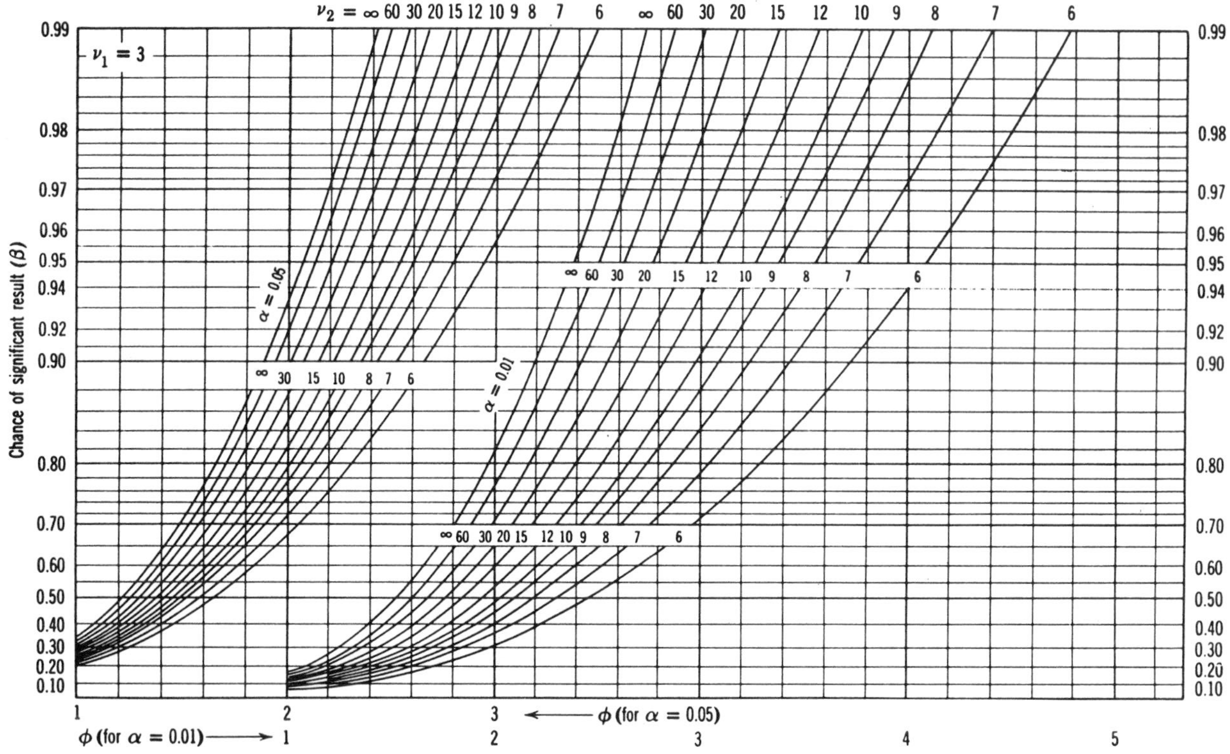

* By E. S. Pearson and H. O. Hartley in *Biometrika*, Vol. 38, pp. 115–122 (1951). Reproduced with the kind permission of the authors and the editor.

Table 5.4 PEARSON AND HARTLEY CHARTS* FOR THE POWER OF THE *F*-TEST

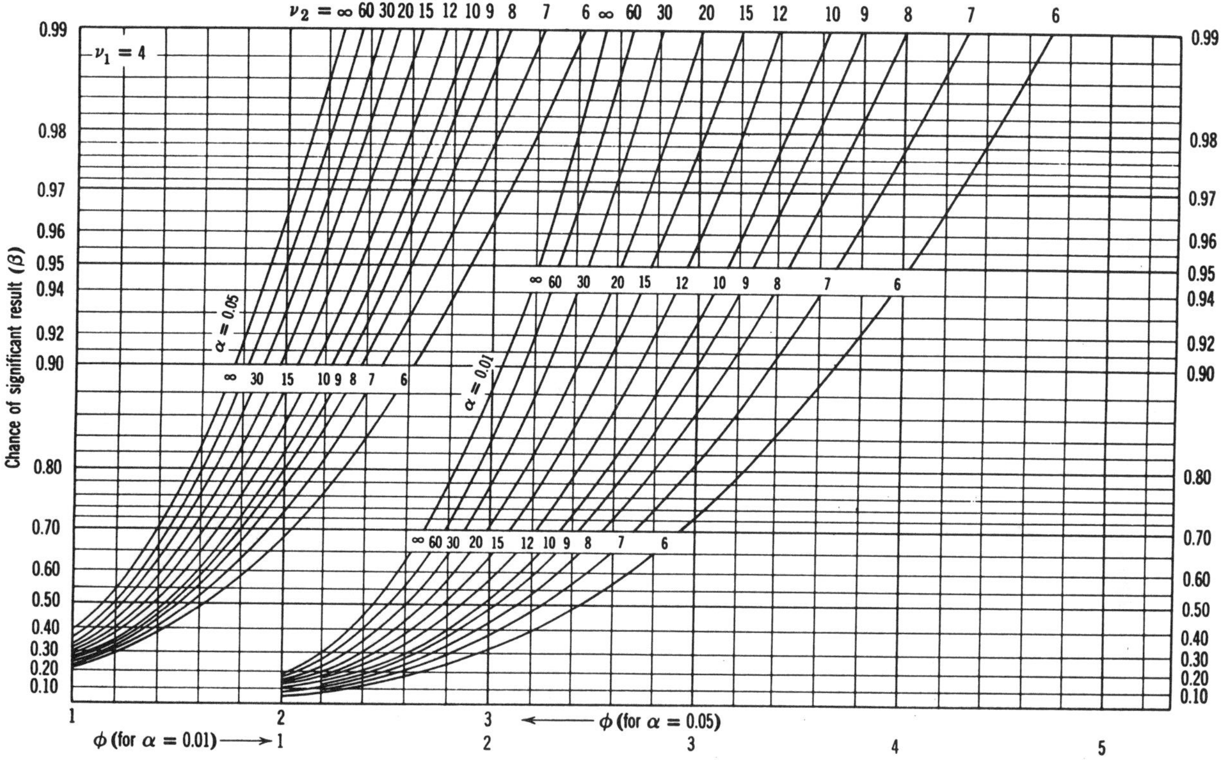

* By E. S. Pearson and H. O. Hartley in *Biometrika*, Vol. 38, pp. 115–122 (1951). Reproduced with the kind permission of the authors and the editor.

Table 5.5 PEARSON AND HARTLEY CHARTS* FOR THE POWER OF THE *F*-TEST

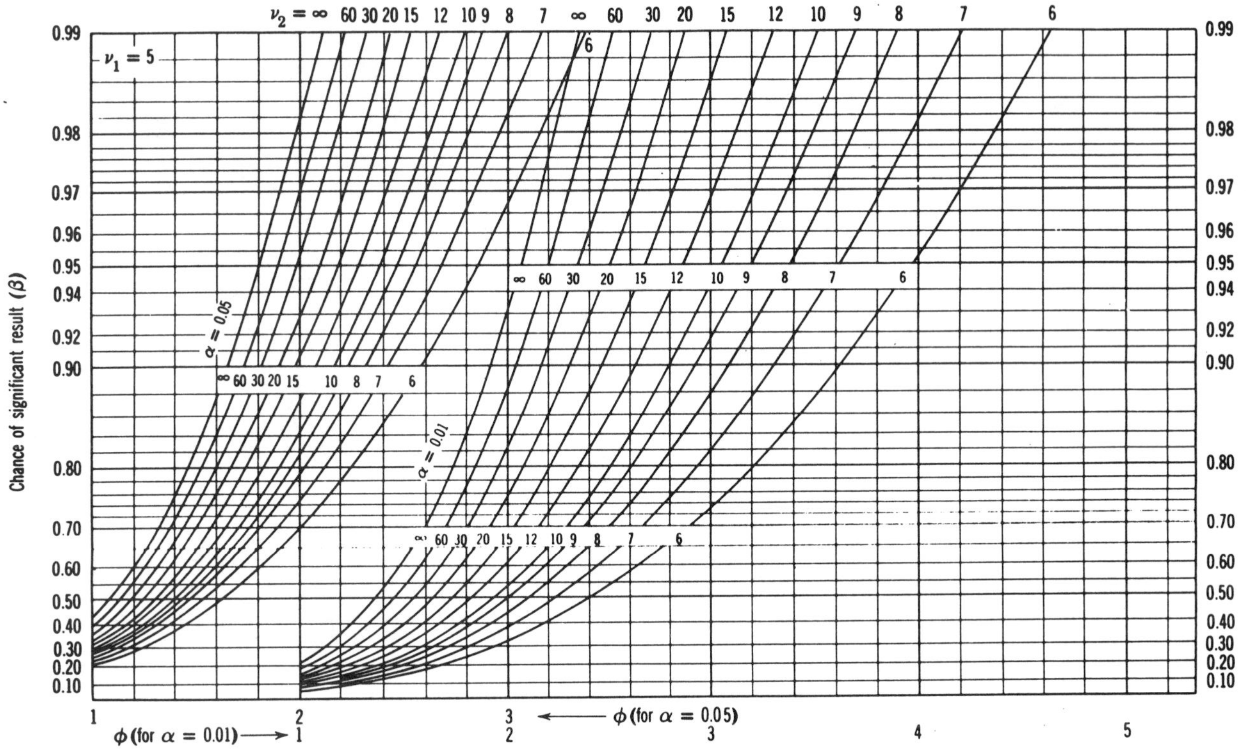

* By E. S. Pearson and H. O. Hartley in *Biometrika*, Vol. 38, pp. 115–122 (1951). Reproduced with the kind permission of the authors and the editor.

Table 5.6 PEARSON AND HARTLEY CHARTS* FOR THE POWER OF THE F-TEST

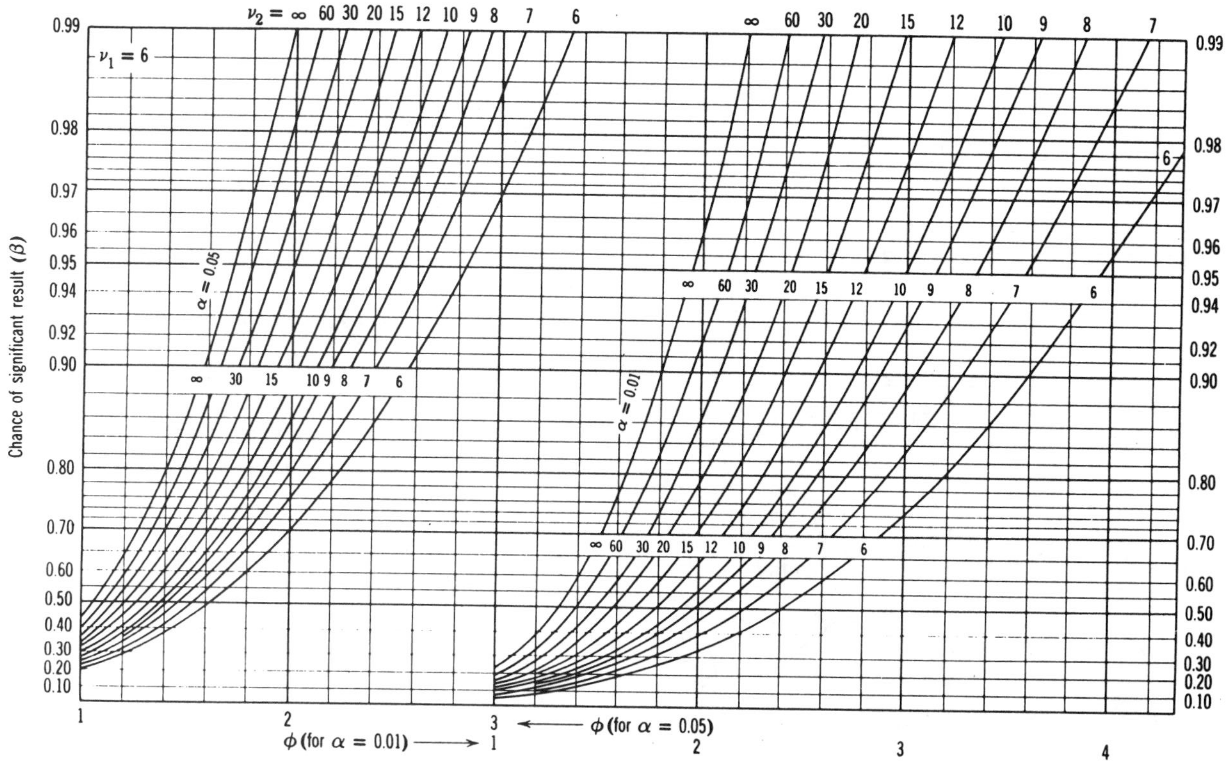

* By E. S. Pearson and H. O. Hartley in *Biometrika*, Vol. 38, pp. 115–122 (1951). Reproduced with the kind permission of the authors and the editor.

Table 5.7 PEARSON AND HARTLEY CHARTS* FOR THE POWER OF THE *F*-TEST

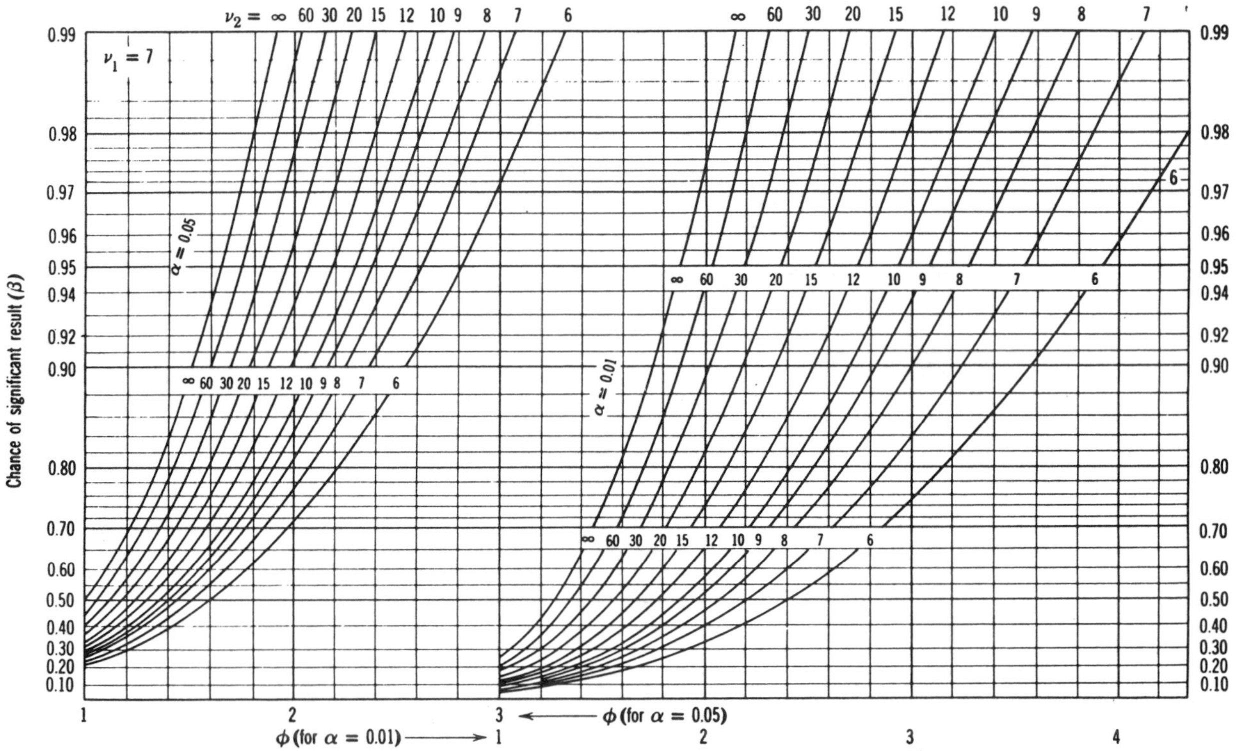

* By E. S. Pearson and H. O. Hartley in *Biometrika*, Vol. 38, pp. 115–122 (1951). Reproduced with the kind permission of the authors and the editor.

Table 5.8 PEARSON AND HARTLEY CHARTS* FOR THE POWER OF THE *F*-TEST

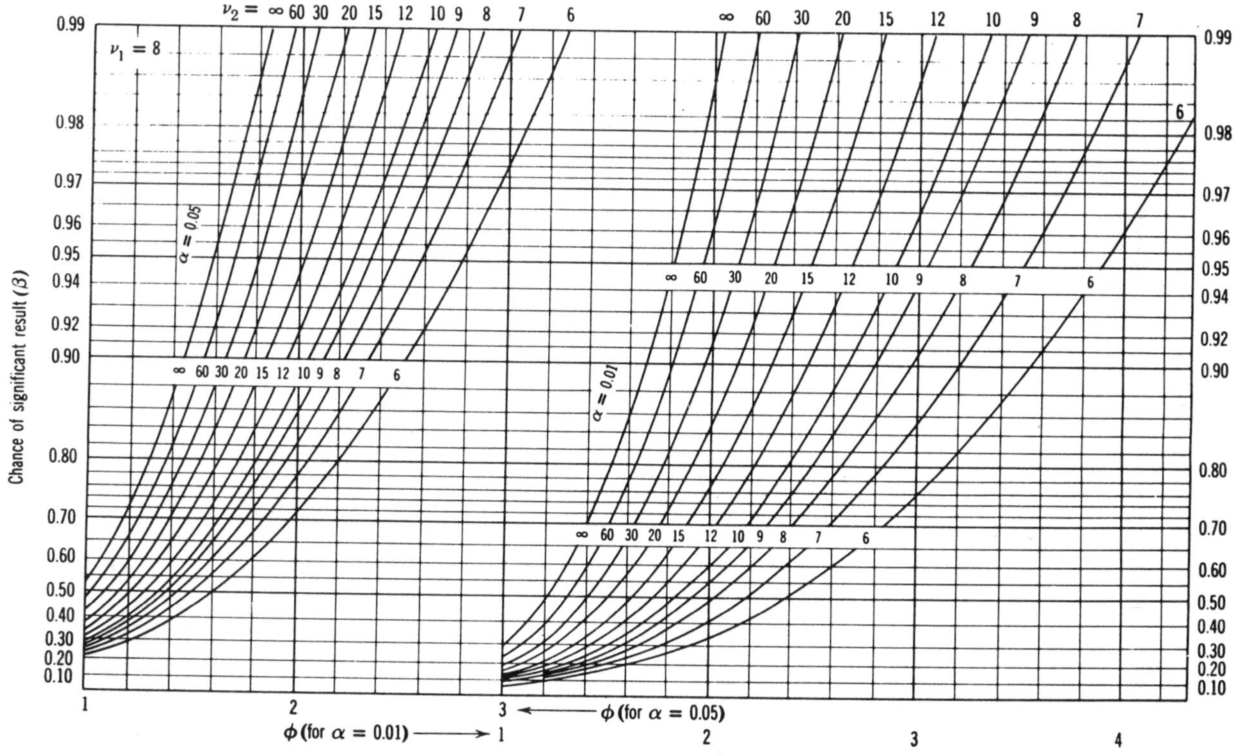

* By E. S. Pearson and H. O. Hartley in *Biometrika*, Vol. 38, pp. 115–122 (1951). Reproduced with the kind permission of the authors and the editor.

Table 6

0.95 and 0.99 Quantiles of the Studentized Range Distribution for Parameters k, ν

ν		k = 2	3	4	5	6	7	8	9	10	11	12	13	14	15
1	0.95	18.00	27.00	32.80	37.10	40.40	43.10	45.40	47.40	49.10	50.60	52.00	53.20	54.30	55.40
	0.99	90.0	135.0	164.0	186.0	202.0	216.0	227.0	237.0	246.0	253.0	260.0	266.0	272.0	277.0
2	0.95	6.09	8.30	9.80	10.90	11.70	12.40	13.00	13.50	14.00	14.40	14.70	15.10	15.40	15.70
	0.99	14.00	19.00	22.30	24.70	26.60	28.20	29.50	30.70	31.70	32.60	33.40	34.10	34.80	35.40
3	0.95	4.50	5.91	6.82	7.50	8.04	8.48	8.85	9.18	9.46	9.72	9.95	10.20	10.40	10.50
	0.99	8.26	10.60	12.20	13.30	14.20	15.00	15.60	16.20	16.70	17.10	17.50	17.90	18.20	18.50
4	0.95	3.93	5.04	5.76	6.29	6.71	7.05	7.35	7.60	7.83	8.03	8.21	8.37	8.52	8.66
	0.99	6.51	8.12	9.17	9.96	10.60	11.10	11.50	11.90	12.30	12.60	12.80	13.10	13.30	13.50
5	0.95	3.64	4.60	5.22	5.67	6.03	6.33	6.58	6.80	6.99	7.17	7.32	7.47	7.60	7.72
	0.99	5.70	6.97	7.80	8.42	8.91	9.32	9.67	9.97	10.20	10.50	10.70	10.90	11.10	11.20
6	0.95	3.46	4.34	4.90	5.31	5.63	5.89	6.12	6.32	6.49	6.65	6.79	6.92	7.03	7.14
	0.99	5.24	6.33	7.03	7.56	7.97	8.32	8.61	8.87	9.10	9.30	9.49	9.65	9.81	9.95
7	0.95	3.34	4.16	4.68	5.06	5.36	5.61	5.82	6.00	6.16	6.30	6.43	6.55	6.66	6.76
	0.99	4.95	5.92	6.54	7.01	7.37	7.68	7.94	8.17	8.37	8.55	8.71	8.86	9.00	9.12
8	0.95	3.26	4.04	4.53	4.89	5.17	5.40	5.60	5.77	5.92	6.05	6.18	6.29	6.39	6.48
	0.99	4.74	5.63	6.20	6.63	6.96	7.24	7.47	7.68	7.87	8.03	8.18	8.31	8.44	8.55
9	0.95	3.20	3.95	4.42	4.76	5.02	5.24	5.43	5.60	5.74	5.87	5.98	6.09	6.19	6.28
	0.99	4.60	5.43	5.96	6.35	6.66	6.91	7.13	7.32	7.49	7.65	7.78	7.91	8.03	8.13
10	0.95	3.15	3.88	4.33	4.65	4.91	5.12	5.30	5.46	5.60	5.72	5.83	5.93	6.03	6.11
	0.99	4.48	5.27	5.77	6.14	6.43	6.67	6.87	7.05	7.21	7.36	7.48	7.60	7.71	7.81
11	0.95	3.11	3.82	4.26	4.57	4.82	5.03	5.20	5.35	5.49	5.61	5.71	5.81	5.90	5.99
	0.99	4.39	5.14	5.62	5.97	6.25	6.48	6.67	6.84	6.99	7.13	7.25	7.36	7.46	7.56
12	0.95	3.08	3.77	4.20	4.51	4.75	4.95	5.12	5.27	5.40	5.51	5.62	5.71	5.80	5.88
	0.99	4.32	5.04	5.50	5.84	6.10	6.32	6.51	6.67	6.81	6.94	7.06	7.17	7.26	7.36
13	0.95	3.0-6	3.73	4.15	4.45	4.69	4.88	5.05	5.19	5.32	5.43	5.53	5.63	5.71	5.79
	0.99	4.26	4.96	5.40	5.73	5.98	6.19	6.37	6.53	6.67	6.79	6.90	7.01	7.10	7.19
14	0.95	3.03	3.70	4.11	4.41	4.64	4.83	4.99	5.13	5.25	5.36	5.46	5.55	5.64	5.72
	0.99	4.21	4.89	5.32	5.63	5.88	6.08	6.26	6.41	6.54	6.66	6.77	6.87	6.96	7.05

Table 6 *(contd)* 0.95 and 0.99 Quantiles of the Studentized Range Distribution for Parameters k, ν

ν		k = 2	3	4	5	6	7	8	9	10	11	12	13	14	15
15	0.95	3.01	3.67	4.08	4.37	4.60	4.78	4.94	5.08	5.20	5.31	5.40	5.49	5.58	5.65
	0.99	4.17	4.83	5.25	5.56	5.80	5.99	6.16	6.31	6.44	6.55	6.66	6.76	6.84	6.93
16	0.95	3.00	3.65	4.05	4.33	4.56	4.74	4.90	5.03	5.15	5.26	5.35	5.44	5.52	5.59
	0.99	4.13	4.78	5.19	5.49	5.72	5.92	6.08	6.22	6.35	6.46	6.56	6.66	6.74	6.82
17	0.95	2.98	3.63	4.02	4.30	4.52	4.71	4.86	4.99	5.11	5.21	5.31	5.39	5.47	5.55
	0.99	4.10	4.74	5.14	5.43	5.66	5.85	6.01	6.15	6.27	6.38	6.48	6.57	6.66	6.73
18	0.95	2.97	3.61	4.00	4.28	4.49	4.67	4.82	4.96	5.07	5.17	5.27	5.35	5.43	5.50
	0.99	4.07	4.70	5.09	5.38	5.60	5.79	5.94	6.08	6.20	6.31	6.41	6.50	6.58	6.65
19	0.95	2.96	3.59	3.98	4.25	4.47	4.65	4.79	4.92	5.04	5.14	5.23	5.32	5.39	5.46
	0.99	4.05	4.67	5.05	5.33	5.55	5.73	5.89	6.02	6.14	6.25	6.34	6.43	6.51	6.58
20	0.95	2.95	3.58	3.96	4.23	4.45	4.62	4.77	4.90	5.01	5.11	5.20	5.28	5.36	5.43
	0.99	4.02	4.64	5.02	5.29	5.51	5.69	5.84	5.97	6.09	6.19	6.29	6.37	6.45	6.52
24	0.95	2.92	3.53	3.90	4.17	4.37	4.54	4.68	4.81	4.92	5.01	5.10	5.18	5.25	5.32
	0.99	3.96	4.54	4.91	5.17	5.37	5.54	5.69	5.81	5.92	6.02	6.11	6.19	6.26	6.33
30	0.95	2.89	3.49	3.84	4.10	4.30	4.46	4.60	4.72	4.83	4.92	5.00	5.08	5.15	5.21
	0.99	3.89	4.45	4.80	5.05	5.24	5.40	5.54	5.65	5.76	5.85	5.93	6.01	6.08	6.14
40	0.95	2.86	3.44	3.79	4.04	4.23	4.39	4.52	4.63	4.74	4.82	4.91	4.98	5.05	5.11
	0.99	3.82	4.37	4.70	4.93	5.11	5.27	5.39	5.50	5.60	5.69	5.77	5.84	5.90	5.96
60	0.95	2.83	3.40	3.74	3.98	4.16	4.31	4.44	4.55	4.65	4.73	4.81	4.88	4.94	5.00
	0.99	3.76	4.28	4.60	4.82	4.99	5.13	5.25	5.36	5.45	5.53	5.60	5.67	5.73	5.79
120	0.95	2.80	3.36	3.69	3.92	4.10	4.24	4.36	4.48	4.56	4.64	4.72	4.78	4.84	4.90
	0.99	3.70	4.20	4.50	4.71	4.87	5.01	5.12	5.21	5.30	5.38	5.44	5.51	5.56	5.61
Infinity	0.95	2.77	3.31	3.63	3.86	4.03	4.17	4.29	4.39	4.47	4.55	4.62	4.68	4.74	4.80
	0.99	3.64	4.12	4.40	4.60	4.76	4.88	4.99	5.08	5.16	5.23	5.29	5.35	5.40	5.45

Table 7 γ-Quantiles of Bechhofer's Statistic D for Parameters k and ν

γ = 0.95

ν \ k	2	3	4	5	6	7	8	9	10
5	2.86	3.45	3.79	4.03	4.21	4.36	4.47	4.58	4.67
6	2.74	3.31	3.62	3.83	4.00	4.13	4.24	4.34	4.41
8	2.63	3.14	3.42	3.61	3.76	3.87	3.97	4.06	4.13
10	2.56	3.04	3.31	3.49	3.62	3.73	3.82	3.90	3.97
12	2.52	2.98	3.24	3.41	3.54	3.65	3.73	3.80	3.87
14	2.39	2.94	3.18	3.34	3.48	3.58	3.66	3.73	3.80
16	2.47	2.91	3.15	3.31	3.44	3.54	3.62	3.69	3.75
18	2.45	2.88	3.13	3.28	3.41	3.51	3.58	3.65	3.71
20	2.43	2.87	3.10	3.25	3.38	3.48	3.55	3.62	3.68
30	2.40	2.81	3.04	3.18	3.30	3.39	3.46	3.54	3.59
60	2.36	2.76	2.97	3.13	3.22	3.32	3.38	3.45	3.51
120	2.35	2.73	2.94	3.08	3.20	3.28	3.35	3.41	3.46
Infinity	2.33	2.71	2.92	3.06	3.15	3.24	3.31	3.37	3.42

γ = 0.99

ν \ k	2	3	4	5	6	7	8	9	10
5	4.77	5.52	5.95	6.26	6.51	6.69	6.86	6.99	7.11
6	4.44	5.11	5.49	5.76	5.95	6.12	6.26	6.38	6.49
8	4.10	4.65	4.96	5.19	5.36	5.49	5.60	5.70	5.78
10	3.90	4.40	4.68	4.88	5.03	5.15	5.25	5.35	5.42
12	3.79	4.26	4.51	4.70	4.84	4.95	5.03	5.12	5.19
14	3.71	4.16	4.40	4.57	4.70	4.81	4.89	4.96	5.03
16	3.65	4.07	4.31	4.48	4.61	4.71	4.79	4.86	4.92
18	3.61	4.02	4.26	4.41	4.54	4.62	4.71	4.78	4.84
20	3.58	3.97	4.20	4.26	4.48	4.57	4.65	4.72	4.78
30	3.48	3.85	4.06	4.20	4.31	4.40	4.47	4.54	4.58
60	3.38	3.73	3.93	4.06	4.16	4.24	4.30	4.36	4.41
120	3.34	3.68	3.86	3.99	4.09	4.16	4.23	4.29	4.33
Infinity	3.29	3.62	3.80	3.92	4.01	4.09	4.14	4.20	4.25

ν = Inf.

γ \ k	2	3	4	5	6	7	8	9	10
0.80	1.1902	1.6524	1.8932	2.0528	2.1709	2.2639	2.3404	2.4049	2.4608
0.90	1.8124	2.2302	2.4516	2.5997	2.7100	2.7872	2.8691	2.9301	2.9829
0.95	2.3262	2.7101	2.9162	3.0552	3.1591	3.2417	3.3099	3.3679	3.4182
0.99	3.2900	3.6173	3.7870	3.9196	4.0121	4.0861	4.1475	4.1999	4.2456
0.995	3.6428	3.9517	4.1224	4.2394	4.3280	4.3989	4.4579	4.5083	4.5523
0.999	4.3703	4.6450	4.7987	4.9949	4.9856	5.0505	5.1047	5.1511	5.1917

$D = \max(\{(Z_i - Z_1)/(W/\nu)^{1/2}, i = 2, \ldots, k\})$, where $Z_1, \ldots, Z_k$ are standard normal, W is Chi-square with n degrees of freedom, and $W, Z_1, \ldots, Z_k$ are independent.

Answers

CHAPTER 1

1.2.1: (a) 60, 15, 244, (4, 0, 8, 4, 12), (1, 1, 1, 0, −1), 244 = 240 + 4. (c) $p(\mathbf{y}|\mathbf{x}_2) = \begin{pmatrix} 0 & 2.5 & \\ 0 & 2.5 & 0 \end{pmatrix}$.

1.3.1: (a) 2 and 3; (b) $V_2 = \mathscr{L}((1, 1, 0, 0), (0, 0, 1, 1))$, $V_3 = \mathscr{L}((1, 1, 0, 0), (0, 0, 1, 1), (1, 0, 1, 0))$; (c) $\mathbf{z} = (1, -1, -1, 1)$ (or any multiple); (d) $\dim(V_4) = 3$, (e) V_3 is the collection of all vectors orthogonal to $\mathbf{z}$. That is, the sum of the first and fourth components is the same as the sum of the second and third components.

1.3.3: (a) $\begin{pmatrix} 4 & 9 & 8 \\ 4 & 9 & \\ 4 & & \end{pmatrix}$, $\begin{pmatrix} 2 & 2 & 0 \\ 0 & -2 & \\ -2 & & \end{pmatrix}$, 290, 274, 16; (b) $\hat{\mathbf{y}} = \sum \bar{y}_i \mathbf{C}_i$.

1.3.4: (a) (3, 9, 5, 3), (−2, 0, 0, 2); (b) $3\mathbf{x}_1$, $(8/7)\mathbf{x}_2$, $3\mathbf{x}_1 + (8/7)\mathbf{x}_2 \neq \mathbf{y}$; (c) $(\mathbf{e}, \mathbf{x}_1) = 0$, $(\mathbf{e}, \mathbf{x}_2) = 0$; (d) 132, 124, 8, 132 = 124 + 8; (e) Example: $\mathbf{v}_1 = (1, 1, 1, 1)$, $\mathbf{v}_2 = \mathbf{x}_2 - 3\mathbf{x}_1 = (1, -2, 0, 1)$, $\mathbf{v}_3 = (1, 0, 0, -1)$, $\mathbf{v}_4 = (1, 1, -3, 1)$; (f) $\hat{\mathbf{y}} = 5\mathbf{v}_1 - 2\mathbf{v}_2$, $\mathbf{y} = \hat{\mathbf{y}} - 2\mathbf{v}_3 - 0\mathbf{v}_4$; (g) $\mathbf{w} = 4\mathbf{v}_1 - 2\mathbf{v}_2 = 4\mathbf{x}_1 - 2(\mathbf{x}_2 - 3\mathbf{x}_1) = 10\mathbf{x}_1 - 2\mathbf{x}_2$, 12 = 8 + 4. $(\mathbf{y} - \hat{\mathbf{y}}) \perp V$, and $(\hat{\mathbf{y}} - \mathbf{w}) \in V$, so that the Pythagorean Theorem applies; (h) true, true; (i) all true.

1.3.5: (a) Let $\mathbf{x}^* = \mathbf{x} - \bar{x}\mathbf{J}$, so $V = \mathscr{L}(\mathbf{J}, \mathbf{x}^*)$; (b) $a_1 = S_{xy}/S_{xx}$, $a_0 = \bar{y}$; (c) $b_1 = a_1 = S_{xy}/S_{xx}$, $b_0 = \bar{y} - b_1\bar{x}$; (d) S_{xy}^2/S_{xx} and $S_{yy} - S_{xy}^2/S_{xx}$; (f) 6, 2, (3, 5, 7, 9), 3, 2, 168, 164, 4.

1.3.6: $\hat{\mathbf{y}} = \begin{bmatrix} 6 & 5 & 1 \\ 10 & 9 & 5 \end{bmatrix}$, $\mathbf{e} = \begin{bmatrix} 1 & 0 & -1 \\ -1 & 0 & 1 \end{bmatrix}$, $\|\mathbf{y}\|^2 = 272$, $\|\hat{\mathbf{y}}\|^2 = 268$, $\|\mathbf{e}\|^2 = 4$.

1.6.3: $(1/11)\begin{bmatrix} 1 & 1 & 3 & 0 \\ 1 & 1 & 3 & 0 \\ 3 & 3 & 9 & 0 \\ 0 & 0 & 0 & 0 \end{bmatrix}$, $(1/2)\begin{bmatrix} 1 & 1 & 0 & 0 \\ 1 & 1 & 0 & 0 \\ 0 & 0 & 2 & 0 \\ 0 & 0 & 0 & 2 \end{bmatrix}$, $(1/22)\begin{bmatrix} 9 & 9 & -6 & 0 \\ 9 & 9 & -6 & 0 \\ -6 & -6 & 4 & 0 \\ 0 & 0 & 0 & 22 \end{bmatrix}$,

$(4, 4, 12, 0)'$, $(-3, -3, 2, 1)'$, $(1, 1, 14, 1)'$.

1.6.4: (a) $(1/2)\begin{bmatrix} 1 & 0 & -1 \\ 0 & 0 & 0 \\ -1 & 0 & 1 \end{bmatrix}$, (b) $(1/2)\begin{bmatrix} 1 & 0 & 1 \\ 0 & 2 & 0 \\ 1 & 0 & 1 \end{bmatrix}$.

1.6.8: $\begin{bmatrix} 8 & 8 & 8 \\ 8 & 8 & 8 \end{bmatrix}, \begin{bmatrix} 2 & 2 & 2 \\ -2 & -2 & -2 \end{bmatrix}, \begin{bmatrix} 3 & -4 & 1 \\ 3 & -4 & 1 \end{bmatrix}, \begin{bmatrix} 13 & 6 & 11 \\ 9 & 2 & 7 \end{bmatrix}, \begin{bmatrix} -1 & 1 & 0 \\ 1 & -1 & 0 \end{bmatrix}$,
$464 = 384 + 24 + 52 + 4$.

1.6.9: (a) $\{(2, 1, -1)\}$.

1.6.10: (b) $43 = (16)(7)/3 + (-2)(1)/2 + (-8)(-5)/6$.

1.7.1: (a) $\lambda_1 = 15, \lambda_2 = 10, \mathbf{v}_1 = (2, -1)'/\sqrt{5}, \mathbf{v}_2 = (1, 2)'/\sqrt{5}$. (b) $\mathbf{U} = (\mathbf{v}_1, \mathbf{v}_2)$, $\Lambda = \text{diag}(\lambda_1, \lambda_2)$. (c) $\mathbf{P}_1 = (1/5)\begin{bmatrix} 4 & -2 \\ -2 & 1 \end{bmatrix}$, $\mathbf{P}_2 = (1/5)\begin{bmatrix} 1 & 2 \\ 2 & 4 \end{bmatrix}$.

1.7.7: The matrix of the quadratic form has diagonal terms 2, 2, 11, above diagonal terms 8, -1, -1. The eigenvalues are 12, 9, -6. The quadratic form is *not* nonnegative definite.

1.7.8: $\mathbf{B} = \begin{bmatrix} 3 & 0 \\ -2 & 4 \end{bmatrix}$.

1.7.9: $\mathbf{G}^{-1} = (1/17)\begin{bmatrix} 12 & -4 & 3 \\ -4 & 7 & -1 \\ 3 & 1 & 5 \end{bmatrix}$.

1.7.10: $\mathbf{P} = \begin{bmatrix} -1/\sqrt{2} & -1/\sqrt{3} \\ 1/\sqrt{2} & -1/\sqrt{3} \\ 0 & -1/\sqrt{3} \end{bmatrix}$, $\mathbf{Q} = (1/\sqrt{2})\begin{bmatrix} -1 & 1 \\ -1 & -1 \end{bmatrix}$, $\mathbf{D} = \begin{bmatrix} 6 & 0 \\ 0 & \sqrt{24} \end{bmatrix}$,
$\mathbf{X}^+ = (1/12)\begin{bmatrix} 2 & 0 & 1 \\ 0 & 2 & 1 \end{bmatrix}$.

CHAPTER 2

2.1.1: (a) The 3×3 zero matrix, (b) $\begin{bmatrix} 8 & 0 & 6 \\ 0 & 15 & 3 \\ 6 & 3 & 13 \end{bmatrix}$, (c) $\begin{bmatrix} -3 & 17 & 9 \\ 1 & 36 & -20 \end{bmatrix}$,
(d) $\begin{bmatrix} 24 \\ 31 \end{bmatrix}$, (e) $\begin{bmatrix} 36 & 25 \\ 25 & 92 \end{bmatrix}$, (f) $\begin{bmatrix} 5/6 & 2/9 & 13/[6\sqrt{5}] \\ 9/[4\sqrt{23}] & 14/[3\sqrt{23}] & 4/\sqrt{115} \end{bmatrix} =$
$\begin{bmatrix} 0.833 & 0.222 & 0.969 \\ 0.469 & 0.973 & 0.373 \end{bmatrix}$,

(g) $\begin{bmatrix} -1/2\sqrt{3} & 17/24 & \sqrt{3}/4 \\ 1/2\sqrt{69} & 9/\sqrt{23} & -5/\sqrt{69} \end{bmatrix} = \begin{bmatrix} -0.289 & 0.708 & 0.433 \\ 0.060 & 0.938 & -0.602 \end{bmatrix}$.

2.1.2: $D[\mathbf{Y}] = (c_{ij})$, where $c_{ij} = \min(i, j)$. $R[\mathbf{Y}] = (\rho_{ij})$, where $\rho_{ij} = \sqrt{\min(i, j)/\max(i, j)}$.

2.1.5: Diagonal terms of $D[\mathbf{Y}]$ are $\sigma_G^2 + \sigma_\varepsilon^2$, off-diagonal terms are σ_G^2. Off-diagonal terms of $R[\mathbf{Y}]$ are $\sigma_G^2/[\sigma_G^2 + \sigma_\varepsilon^2]$. $\text{Var}(Y_1 + Y_2 + Y_3) = 9\sigma_G^2 + 3\sigma_\varepsilon^2$.

2.2.1: 6.

2.2.2: (a) $n(n-1)\sigma^2$, (b) $Q_2(\mathbf{X}) = nQ_1(\mathbf{X})$, (c) $1/[n(n-1)]$.

2.2.3: (a) $\sum \sigma_i^2/n^2$, (b) $1/[n(n-1)]$.

2.3.1: (a) $\hat{\mathbf{Y}} = (4\mathbf{X}_2 + 7\mathbf{X}_3)/7$, (b) 4, 0.4, $\sqrt{0.9}$.

2.3.4: $g(0) = 1$, $g(1) = 5/6$, $g(2) = 0$, $h(X) = 0.7 - (X-1)/2$, $E[g(X)] = E[h(X)] = 0.7$, $\text{Var}(g(X)) = 5/12 - 0.29$, $E[Y - g(X)]^2 = 1/12$, $\text{Var}(\hat{Y}) = 0.10$, $E[Y - \hat{Y}]^2 = 0.11$.

2.3.5: (a) 68.30, 68.09, 1.811, 2.541, 0.460 2, $\hat{Y} = 68.09 + 0.645\,7(X - 68.30)$; (b) 21/22; (c) 13/14.

2.4.5: 0.960 2.

2.5.5: $K = 2/3$, $n_1 = 3$, $n_2 = 2$, $\delta = 77/9$.

CHAPTER 3

3.1.1: (d) $\hat{\mathbf{Y}} = (5, -1, 4, 6)'$, $S^2 = 15$, $S^2(\hat{\boldsymbol{\beta}}) = \begin{bmatrix} 5 & 0 \\ 0 & 5 \end{bmatrix}$.

3.1.2: (b) $\hat{\boldsymbol{\beta}} = \begin{bmatrix} 3 \\ -1 \end{bmatrix}$, $\mathbf{e} = (-3, -4, 3, 1, 1)'$, $S^2 = 12$.

3.1.3: (b) $\hat{\boldsymbol{\beta}} = (100/12)\begin{bmatrix} 5 & -1 & 2 \\ -1 & 5 & 2 \end{bmatrix}\begin{bmatrix} Y_1 \\ Y_2 \\ Y_3 \end{bmatrix}$; (c) $(0.36^2/24)\begin{bmatrix} 5 & -1 \\ -1 & 5 \end{bmatrix}$; (d) $\hat{\boldsymbol{\beta}} = \begin{bmatrix} 2 \\ 3 \end{bmatrix}$, $S^2 = 96 \times 10^{-6}$.

3.1.4: (b) $\hat{\boldsymbol{\beta}} = (100, 1, 2, 3)'$; (c) $\begin{bmatrix} 70 & 70 & 130 & 130 \\ 40 & 40 & 160 & 160 \\ 10 & 10 & 190 & 190 \end{bmatrix}$, SSE $= 616$, $S^2 = 77$,

$S^2(\hat{\boldsymbol{\beta}}) = (1/3{,}600)\begin{bmatrix} 23{,}100 & 0 & 0 & 0 \\ 0 & 77 & 0 & 0 \\ 0 & 0 & 77 & 0 \\ 0 & 0 & 0 & 77 \end{bmatrix}$; (d) $\hat{\mathbf{Y}} = \begin{bmatrix} 66 & 66 & 126 & 126 \\ 39 & 39 & 159 & 159 \\ 15 & 15 & 195 & 195 \end{bmatrix}$,

SSE $= 448$.

3.2.3: (b) (4.84, 25.19) and (−5.37, 35.37).

3.2.4: (52.15, 64.19), (42.63, 49.40), and (1.24, 25.96).

3.2.6: (c) $\hat{\beta}_1 = 1.5$.

3.3.4: For $g(x) \equiv 1$, $\hat{\beta} = 0.684$, $S^2(\hat{\beta}) = 0.000\,325$. For $g(x) = x$, $\hat{\beta} = 0.688$, $S^2(\hat{\beta}) = 0.003\,58$. For $g(x) = x^2$, $\hat{\beta} = 0.692\,5$, $S^2(\hat{\beta}) = 0.031\,2$.

3.5.2: $-2, -3, 4, \begin{bmatrix} 6 & 1 \\ 6 & 9 \\ 22 & 17 \\ 22 & 25 \end{bmatrix}, \begin{bmatrix} 2 & 1 \\ -2 & -1 \\ 1 & -1 \\ -1 & 1 \end{bmatrix}$, 3.5, 0.097 2.

3.6.1: $R^2_{y\cdot x_2} = 0.099\,265$, $R^2_{y\cdot x_2x_3} = 0.485\,294$, $t = -1.224\,745$, $d = 0.75$.

3.6.3: $\text{ESS}_3 = 16$, $\text{EES}_2 = 80$, $\hat{\beta}_3 = 8$, $\|\mathbf{x}_3^{\perp}\|^2 = 1$, $t = 2\sqrt{3}$.

3.7.1: $r_{12} = 0.389\,66$, $r_{13} = 0.480\,38$, $r_{23} = 0.634\,81$, $r_{12.3} = 0.125$.

3.7.3: $R^2_{1.23} = 7/16$.

3.7.5: 95% Confidence Interval on ρ_y: (0.120 9, 0.477 4), 95% C.I. on ρ_x: (0.064 7, 0.350 9).

3.8.1: (a) $\mathbf{A} = \begin{bmatrix} 0 & 0 & 0 & 1 \\ 0 & 1 & -1 & 0 \end{bmatrix}$; (b) (41, −3, −8, 3)′, (0, 32, 40, 8, 20, 20)′, (3, −8)′; (c) $\hat{\mathbf{Y}}_0 = (4, 36, 20, 20, 20, 20)'$, $\mathbf{Y} - \hat{\mathbf{Y}}_0 = (0, 0, 24, -8, -4)'$, $\hat{\mathbf{Y}}_1 = (-4, -4, 20, -12, 0, 0)'$; (d) $\text{SSE}_{\text{FM}} = 224$, $\text{SSE}_{H_0} = 800$, $\|\hat{\mathbf{Y}} - \hat{\mathbf{Y}}_0\| = 576$, $F = 2.571\,4$; (f) $\mathbf{c} = (0, 0, 0, 192)'$, $\mathbf{a}_c = (-4, -4, 20, -12, 0, 0)'$, $t_c = 2.267\,79$, $S^2 = 112$.

3.8.2: $\hat{\mathbf{Y}} = 3\mathbf{J}_1 + \mathbf{J}_2 + 5\mathbf{J}_3 + 2\mathbf{x}$, $\hat{\mathbf{Y}}_0 = 2.085\,7\mathbf{J} + 2.285\,7\mathbf{x}$, $S^2 = 2.666\,7$, $\|\hat{\mathbf{Y}}_1\|^2 = 26.742\,9$, $F = 5.014\,3$, $F_{2,6,0.95} = 5.14$.

3.8.3: $S^2 = 34.8$, $F = 2.155$ for 2 and 5 d.f..

3.8.4: (a) 0.84; (b) $F = 2.0$ for 2 and 16 d.f.; (c) $\mathbf{A} = \begin{bmatrix} 0 & 0 & 1 & 0 \\ 0 & 0 & 0 & 1 \end{bmatrix}$.

3.8.7: (e) (−8.567, 0.567), $t = -2.071\,5$; (f) (−9.147, 1.146 6).

3.8.9: $a = 0.839\,7$, $b = 3.364\,7$, (−10.649, 2.649).

3.10.1: (b) $\delta = [18(\beta_2 - \beta_3)^2 + 114\beta_4^2 - 60\beta_4(\beta_2 - \beta_2)]/\sigma^2$, (c) $\delta = 18$, power = 0.500 4.

3.10.3: (a) 0.3, (b) $n_0 = 10$, (c) $n_0 = 62$.

3.10.4: (a) 0.4, (b) 9.

3.10.5: (b) 0.3, (c) 7.

3.11.1: (a) $\hat{Y} = 0.518\,9 + 0.807x_2$, $S^2 = 0.180\,0$, $r_{yx_2} = 0.575\,9$; (b) For $x_2 = 2.74$ the interval is (2.607, 2.854), (c) For $x_2 = 2.74$, the interval is (1.869, 3.593), (d) $r_{y1.2} = 0.703\,6$.

3.11.2: (b) (10.05, 19.98), (22.35, 37.65); (c) (11.65, 18.35), (23.29, 36.71); (d) (2.18, 3.44).

3.11.4: For $i = 1$ and 2, the 95% confidence intervals are (−6.60, 6.60) and (25.40, 38.60). The 95% prediction intervals are (−7.94, 7.94) and (24.06, 39.94).

3.11.5: (a) The endpoints of the interval are the roots of a quadratic equation in x_0; (b) (7.769, 9.090); (c) (7.487, 9.350).

3.12.1: $\hat{Y} = 62.2206 - 0.3114$ (age), $S^2 = 772.39$, (46.35, 53.18), (38.67, 60.86).

3.12.2: $\hat{Y} = 88.46 - 3.2040x_1 - 0.1504x_2$, $S^2 = 7.168$, SSE $= 200.7158$, $F = 3.35$.

3.12.3: -0.2652, -0.0847.

3.12.4: -0.4209.

3.12.5: 0.1075.

3.12.6: 0.7887.

3.13.1: (a) $\hat{Y} = 3 - 2x_1 + 1.5X_2$; (b) $(-8.45, 4.45)$; (c) $\hat{x}_2 = 2.2 - x_1 + 0.2x_2$; (d) $\sqrt{0.3}$, $r^2 = \hat{\beta}_1(1/\hat{\eta})$; (e) $\hat{Y} = 15.83 + 29.16x_1 + 3.65x_2$, $S(\hat{\beta}_1) = 7.147$, $t_1 = 4.992$.

3.13.2: $\delta = 4.17$, power $\doteq 0.4$.

3.13.3: (a) $\hat{Y} = 61{,}771.7 - 7{,}252.02f + 282.43x_2$, $S(\hat{\beta}_1) = 3{,}685.56$; (b) $-7{,}252.05 \pm 7{,}223.7$; (c) $R = 0.238$.

CHAPTER 4

4.1.1: (8.049, 9.097) and (7.365, 9.782), for the model $Y_i = \beta_0 + \beta_1 \log(x_i) + \varepsilon_i$, ε_i's independent $N(0, \sigma^2)$.

4.1.2: (0.402, 0.482) for the model $Z_i = \log(Y_i) = \beta_0 + \gamma_1 x_i + \varepsilon_i$, ε_i's independent $N(0, \sigma^2)$.

4.1.3: The best two fits seem to be $\hat{y} = 8.724x^{-1.1947}$, $(R^2 = 0.9881)$, and $\hat{y} = -1.4549 + 11.1155/x$, $(R^2 = 0.9845)$.

4.2.1: $(2\beta_2, 3.4\beta_3)'$, $\beta_2(2, -1, -2, -1, 2)' + 1.2\beta_3(-1, 2, 0, -2, 1)'$, $(14\beta_2^2 + 10\beta_3^2)/3$.

4.2.2: $\sigma^2(34/70 + (130/144)x^2) > \sigma^2(1/5 + x^2/10)$.

4.2.3: $C_1 = 148.67$, $C_2 = 34.0$, $C_3 = 2.67$, $C_4 = 6.33$, $C_5 = 6.67$.

4.2.4: (a) 139.07 (with) and 142.97 (without); (b) $\hat{\eta} = (121/202)\beta_2 = 11.96$; (c) $(\beta_2/\sigma) < 1.2513$; (d) 76.01.

4.3.1: $\hat{\mu} = \sum (Y_i/w_i)/\sum (1/w_i)$.

4.3.2: $\mathbf{a}_1 = (3, -3, 5)'/8$, $\mathbf{a}_2 = (3, 5, -3)'/8$, $\begin{bmatrix} 7 & -1 \\ -1 & 7 \end{bmatrix}\sigma^2/16$.

4.3.4: $\hat{\beta}_1 = \sum u_i Z_i/\sum u_i^2$, where $u_i = (x_i - x^*)/\sqrt{k_i}$, $x^* = \sum (x_i/k_i)/[\sum (1/k_i)]$, $Z_i = Y_i/\sqrt{k_i}$. $\text{Var}(\hat{\beta}_1) = \sigma^2(76{,}035/137^2) = 0.2468\sigma^2$, $\text{Var}(\hat{\beta}_{LS}) = 0.3\sigma^2$.

4.3.5: $g(y) = \log(y)$.

4.4.1: (a) $\hat{\boldsymbol{\beta}} = (6, -2)'$, $\hat{\mathbf{Y}} = (12, 10, 8, 6, 4, 2, 0)'$, SSE $= 32$; (b) $\hat{\boldsymbol{\beta}}^* = (82, -36)'/15$, $\hat{\mathbf{Y}}^* = (170, 146, 122, 98, 74, 50, 26)'/15$, SSE* $= 368/15$; (c) $\hat{\boldsymbol{\beta}} = (6.5, -1.5)'$, $\hat{\mathbf{Y}} = (66, 57, 48, 39, 30, 21, 12, 3)'/6$, SSE $= 53$; (d) $\hat{\boldsymbol{\beta}} = (265, -111)'/45$, $\hat{\mathbf{Y}} = (13.29, 10.82, 8.36, 5.89, 3.42, 0.96, -1.51, 15.76, -3.98)$, SSE $= 43.82$.

4.5.1: (a) The consecutive F-ratios are 1/3, 12/7, 150/11, leading to the choice of the model $\mathscr{M}_1$, (b) 0.05, 0.0975, 0.6498, 0.2027, (c) 0.05, 0.16, 0.62, 0.17.

4.5.2: $S(3)$ $\mathbf{A} = \begin{bmatrix} 0.8667 & 1.4000 & -0.2000 & 20.3331 \\ 1.4000 & 0.3429 & 0.2571 & 0.1170 \\ -0.2000 & -0.2571 & 0.0571 & -2.0404 \\ -20.3331 & 0.1170 & 2.0404 & 0.0465 \end{bmatrix}$, (b) $\hat{\boldsymbol{\beta}} =$ $(20.3331, -2.0404)'$, $[-2.112, -1.969]$, $[2.257, 9.843]$, (c) 0.99974.

4.9.1: (a) For $i = 1, 2, 3, 4$ the ith row is $(1, i, i^2, 0, 0)$. For $i = 5, 6, 7, 8$, the ith row is $(1, i, i^2, (i-4)^2, 0)$. For $i = 9, 10$, the ith row is $(1, i, i^2, (i-4)^2, (i-8)^2)$; (b) (19.583, −9.795, 2.932, −3.832, 5.073); (c) (59.412, 61.007); (d) $F = 89.205$, reject.

4.10.1: $\hat{\beta} = 0.7307$, $\hat{\mathbf{y}} = (2.0765, 4.3119, 8.9536)'$.

4.10.2: $\hat{\boldsymbol{\beta}} = (0.462, 1.073)'$, $\hat{\mathbf{y}} = (1.352, 3.956, 11.572)'$.

4.10.3: $\hat{\beta} = 0.5$.

4.10.4: $\hat{\boldsymbol{\beta}}^1 = (4.990, 3.440, -5.939)'$, $Q(\hat{\boldsymbol{\beta}}^0) = 766.15$, $Q(\hat{\boldsymbol{\beta}}^1) = 727.87$, $\hat{\boldsymbol{\beta}} =$ $(1.383, 4.317, -2.924)'$, $Q(\hat{\boldsymbol{\beta}}) = 0$ (to 4 decimals).

4.11.1: (a) 9, 6 + 7(9/14), 1 + 43/4; (b) 5 and 8.

4.11.2: Yes, $\hat{\boldsymbol{\beta}} = (-8, 9)'$.

4.11.5: Huber estimate = 4, least squares estimate = 9, $7 \le y_3 \le 17$.

4.12.1: (a)

m	3	4	5	6	7	8	9
$27P(\bar{X}^* = m)$	3	6	6	7	6	3	1

(b) 2

(c)

m	3	6	9
$27P(\bar{X}^* = m)$	7	13	7

(d) 14/3

4.12.3: (a)

u	−1	0	1	2
$27P(\hat{\beta}^* - \hat{\beta} = u)$	8	12	6	1

(b) (1, 3), (c) 1/3 probability on −1, 0, 1.

CHAPTER 5

5.2.1: (a) (−8.78, 18.78), (−12.78, 14.78), (−15.49, 23.49); (b) (−12.10, 22.10), (−16.10, 18.10), (−20.19, 28.19); (c) $F = 39/15$, $\mathbf{c} = (15, 3)'$, $\mathbf{a}_c = \hat{\mathbf{Y}} =$ $(5, -1, 4, 6)'$; (d) the set $\{\boldsymbol{\beta} \mid (\beta_1 - 5)^2 + (\beta_2 - 1)^2 \le 190\}$ (a circle of radius $\sqrt{190}$ about (5, 1) and its interior).

5.2.2: $A = \{\boldsymbol{\beta} \mid Q(\boldsymbol{\beta}) = 2(\beta_0 - 8)^2 + 7(\beta_1 + 2)^2 + 6(\beta_0 - 8)(\beta_1 + 2) \le 38\}$. No, since $(7, -1) \in A$.

5.2.3: (a) Treatment SSqs. = 42.1, Error SSqs. = 20, $F = 4.21$, do not reject. (b) The Scheffé confidence intervals on $\mu_1 - \mu_2$ and $\mu_1 - \mu_3$ are $(-8.30, 4.30)$ and $(-10.63, 0.63)$. (c) **c** has ith term $(\bar{Y}_i - \bar{Y}_{..})n_i$.

5.2.4: $\hat{g}(x_1) = 47.376 + [(-0.862)(5.327)/1.387](x_1 - 10.586)$, $k_1(x_1) = 2.045h(x_1)^{1/2}$, and $k_2(x_1) = 2.580h(x_1)^{1/2}$, where $h(x_1) = [1/3 + (x_1 - 10.586)^2/57.71$.

5.3.1: 0.951 and 1.036.

5.3.2: $(-2.07, 12.11)$.

5.3.3: $q_{0.95,2,10} = 3.15$, $t_{10,0.975} = 2.228$.

5.3.5: 0.01.

5.3.6: An example: $3\mu_1 - (\mu_2 + \mu_3 + \mu_4)$, relative lengths 0.89.

5.4.1: For $n_1 = 5$, 0.919, 0.965, 0.952. For $n_1 = 10$, 0.919, 0.952, 0.964.

5.5.1: (a) Cable MSq. = 1,923.65, EMSq. = 26.53, $F = 9.06$; (b) $K_B = 1.343$, $K_S = 1.646$, $K_T = 1.299$; (d) Bonferroni; (e) $\mu_9 \geq \mu_0 - 5.086$; (f) Scheffé: $\bar{X}_j - \bar{X}_{j'} \pm 7.626$, Bonferroni: $\bar{X}_j - \bar{X}_{j'} \pm 8.673$.

CHAPTER 6

6.1.1: (b) Training SSqs. = 26.650, Length SSqs. = 15.699, $T \times L$ SSqs. = 4.761, Error SSqs. = 32.038, (Corr.) Total SSqs. = 79.148. (c) On training differences $\alpha_i - \alpha_{i'}$: $\bar{Y}_{i..} - \bar{Y}_{i'..} \pm 0.883$. On length differences $\beta_j - \beta_{j'} \pm 1.125$. (d) For training: $\delta = 30.625$, power $\doteq 0.999$; for length: $\delta = 18.75$, power = 0.951. (e) $\hat{\gamma} = 0.038\,25$, $S^2(\hat{\gamma}) = 88.99 \times 10^{-6}$, $0.038\,25 \pm 0.019\,0$. (f) New Error SSqs. = 29.626, New $S^2 = 0.644$.

6.2.1: (b) Table of $\hat{\mu}_{ij}$ is $\begin{bmatrix} 24 & 16 & 8 \\ 10 & 2 & 6 \end{bmatrix}$, $\hat{\mu} = 11$, $\hat{\alpha}_1 = 5$, $\hat{\beta}_1 = 6$, $\hat{\beta}_2 = -2$, $\widehat{(\alpha\beta)}_{ij}$: $\begin{bmatrix} 2 & 2 & -4 \\ -2 & -2 & 4 \end{bmatrix}$, $S^2 = 8/3$, $F = 1161/56$ for 2 and 6 d.f.; (c) $F = (31{,}618/105)/\ (886/56) = 19.033$ for 1 and 8 d.f.; (d) SSE = 16, SSA = 270, SSB = 201.6, SSAB = 86.4, $F = 16.2$ for 2 and 6 d.f..

6.3.1: (a) MSG = 0.090 0, MSA = 0.923 7, MSResid. = 0.070 9; (b) $\hat{\alpha}_1 = -0.152$, $\hat{\alpha}_2 = 0.658$, $\hat{\alpha}_3 = -0.665$, $\hat{\alpha}_4 = 0.158$, Tukey intervals on $\alpha_j - \alpha_{j'}$ are $\bar{Y}_{.j} - \bar{Y}_{.j'} \pm 0.753\,3$; (c) Use $S_p^2 = 0.028\,5$ for 6 d.f., $F_{GA} = 2.488$, $F_G = 3.16$, $F_A = 32.41$; (c) $y_{34} = 8.927$, $S^2 = 0.064\,5$, MSG = 0.144, MSA = 0.959, $F_G = 1.859$; (e) $(-0.422, 0.502)$ and $(-0.823, 0.209)$.

6.5.1: (a) $\mu = 15$, $\alpha_1 = 5$, $\beta_1 = 8$, $\gamma_1 = 4$, $(\alpha\beta)_{11} = 2$, $(\alpha\gamma)_{11} = 0$, $(\beta\gamma)_{11} = 1$, $(\alpha\beta\gamma)_{111} = 0$; (b) 0.633.

6.5.2: $\hat{\mu} = 15.937\,5$, $\hat{\alpha}_1 = 5.812\,5$, $\hat{\beta}_1 = 7.812\,5$, $\hat{\gamma}_1 = 2.062\,5$, $\widehat{(\alpha\beta)}_{ij} = 1.687\,5$, $\widehat{(\alpha\gamma)}_{ik} = -0.562\,5$, $\widehat{(\beta\gamma)}_{jk} = 1.687\,5$, $\widehat{(\alpha\beta\gamma)}_{ijk} = 1.062\,5$, MSA = 540.56, MSB = 976.56, MSC = 68.06, MSAB = 45.56, MSAC = 5.06, MSBC = 45.56,

MSABC = 18.06, MSE = 29.69; (b) for example, for V_{AB},

$$\mathbf{x} = \begin{bmatrix} 1 & 1 \\ 1 & 1 \\ -1 & -1 \\ -1 & -1 \end{bmatrix} \begin{bmatrix} -1 & -1 \\ -1 & -1 \\ 1 & 1 \\ 1 & 1 \end{bmatrix};$$

(c) 57.95 and 43.15; (d) $\hat{\mu}_{112} = 27.5$, $\text{Var}(\hat{\mu}_{112}) = \sigma^2/4$, (21.22, 33.78); (e) -1.0625.

6.6.1: (a) Analysis of Covariance Table

	SSqs.	MSqs.	γ
A	96.00	96.00	0.333
B	271.06	271.06	0.882
AB	26.18	26.18	-1.000
Error	8.00	8.00	-1.000

$\hat{\mu} = 20$, $\hat{\alpha}_1 = 6$, $\hat{\beta}_1 = 8$, $\widehat{(\alpha\beta)}_{11} = 2$, $\hat{\gamma} = -1$. (c) $3\sigma^2/2$, $\sigma^2/2$, $(-0.512, 12.510)$, (5.636, 18.364). (d) $(-1.866, -0.134)$.

CHAPTER 7

7.1.3: (b) Squares MSq. = 63,034.79, Residual MSq. = 802.88, (Corr.) Total SSqs. = 3,249,266. (c) (10.16, 25.08) and (8,159.79, 20,137.32). (d) Reject for $F > 36.87$, $F = 78.51$, Reject. Power > 0.999. (f) Estimates are Squares MSq./5 = 12,606.9, 2(12,606.9), and 12,606.9/50. (g) For $I = 50$, $J = 5$, $\text{Var}(\bar{Y}_{..}) = 203.6$. For $I = 54$, $J = 2$, $\text{Var}(\bar{Y}_{..}) = 193.5$, for $I = 56$, $J = 1$ (unsuitable) $\text{Var}(\bar{Y}_{..}) = 194.2$.

7.2.1: (b) Machine SSqs. = 45.075, Heads SSqs. = 282.875, Error SSqs. = 642.00, $F_M = 0.598$, $F_H = 1.762$; (c) (0, 0.862) and (0, 9.227).

7.2.2: 0.95.

7.4.3: (a) $\mathbf{a} = (1.114, -1.196, -0.400, 3.557, -2.886, 4.100, -3.500)'$, $Q = 1.521$; (b) $F_A = 19.40$ for 6 and 8 d.f.; (c) for $j = 3$, $\mathbf{A}_2[i, 3] = \begin{cases} 1 - 1/3 & \text{for } i = 2 \\ -1/3 & \text{for } i = 3, 5 \end{cases}$; $\mathbf{Y}_A[i, 3] = \begin{cases} 2.839 & \text{for } i = 2 \\ 0.445 & \text{for } i = 3, \\ -3.284 & \text{for } i = 5 \end{cases}$ $\hat{\mathbf{Y}}_B[i, 3] = 10.433$ for $i = 2, 3, 5$, $\hat{\mathbf{Y}}[i, 3] = \begin{cases} 13.272 & \text{for } 2 \\ 10.888 & \text{for } i = 3; \\ 7.149 & \text{for } i = 5 \end{cases}$ (d) 7/9.

7.4.5: (b) $\mathbf{a} = (1/6)(-66, -11, 8, 16, 31, 22)$, SSA = 520.17, SSB = 20,763.5,

SSE = 77.33, $S^2 = 7.733$; (c) $F_{St} = 13.45$ for 5 and 10 d.f.; (d) confidence intervals have the form $(a_i - a_i') \pm 7.833$; (e) SSRep = 298.5, (SS for blocks within replicates, after removal of storage effects) = 213.4, $F_{Reps} = 3.497$ for 4 and 10 d.f..

CHAPTER 8

8.2.1: (c) 7/64, 0.09, 0.420 2.

8.2.2: $G(A) = 0.00063$. The LeCam upper bound is 0.001 3.

8.2.3: (a) $E(\mathbf{Y}) = \begin{bmatrix} 2 & 3 & 2 \\ 8 & 10 & 2 \end{bmatrix}$, $D[\mathbf{Y}] = \begin{bmatrix} 6.4 & -4.6 & -1.8 \\ -4.6 & 7.1 & -2.5 \\ -1.8 & -2.5 & 4.3 \end{bmatrix}$.

8.2.4: (b) $E(\mathbf{Y}) = \begin{bmatrix} 80 \\ 60 \\ 60 \end{bmatrix}$, $D[\mathbf{Y}] = \begin{bmatrix} 48 & -24 & -24 \\ -24 & 42 & -18 \\ -24 & -18 & 42 \end{bmatrix}$, (c) Exact answer = 0.072 86, bivariate normal approximation 0.070.

8.2.5: 45/42,042, 20/91.

8.2.6: 1/168.

8.2.7: (0.067 6, 0.102 0) and (0.066 8, 0.101 2).

8.2.8: (2.127, 2.50 5).

9.2.9: $g(x) = \sqrt{x}$, and $\arcsin(\sqrt{x})$.

8.2.12: 0.712 6, 0.820 3, $e_T = \lambda T e^{-\lambda T}/[1 - e^{-\lambda T}]$.

8.2.13: $\text{Var}(\hat{h}) \doteq 0.0179$, $P(|\hat{h} - h| \le 0.2) \doteq 0.865$.

8.3.1: (a) 0.088 0, (b) 4,603, (c) 6,636.

8.3.2: (0.046 9, 0.279 6) for $\alpha_1 = \alpha_2 = 0.05$.

8.3.3: (a) (5.432, 18.410), (b) (i) 352.13, 416.57), (ii) (350.81, 415.19), (iii) (350.81, 416.24).

8.3.4: (b) $\hat{\theta}_1 = 3.071$, $\hat{\theta}_2 = 2.574$, $S(\hat{\theta}_1) = 0.341$, $S(\hat{\theta}_2) = 0.117$, 90% confidence interval (−0.096, 1.091), (c) (0.969, 1.459).

8.3.6: (a) (3.182, 5.453), (b) $\hat{\mathbf{p}} = \begin{bmatrix} 7.969 \times 10^{-4} & 4.031 \times 10^{-4} \\ 0.3215 & 0.6773 \end{bmatrix}$, 2.479×10^{-3}, 4.031×10^{-4}, $(7.409. 8.530) \times 10^{-4}$.

8.3.7: (1.172, 5.188).

8.5.1: (a) $\hat{\beta}_1 = \log(y_3)$, $\hat{\beta}_2 = \log[(y_1 + y_2)/(2y_3)]$, (b) $\hat{\boldsymbol{\beta}} = (\log 10, \log 7)$, $\hat{\boldsymbol{\mu}} = \log(17, 17, 10)$, $\hat{\mathbf{m}} = (70, 70, 10)$, $\mathbf{p} = \mathbf{m}/150$.

8.5.2: $\hat{\boldsymbol{\beta}} = (5.3884, -0.01022)$.

8.5.3: $\hat{\boldsymbol{\beta}} = (2.140, 2.016, 1.211, 0.452, 0.724)$, $\hat{\mathbf{m}} = \begin{bmatrix} 8.50 & 7.51 & 3.07 & 1.57 \\ 8.50 & 15.49 & 26.93 & 28.43 \end{bmatrix}$.

8.5.4: (a) $\hat{\mathbf{m}} = \begin{bmatrix} 52.0 & 54.5 & 26.5 & 16.0 & 7.0 \\ 54.5 & 87.0 & 52.0 & 16.0 & 9.5 \\ 26.5 & 52.0 & 83.0 & 45.5 & 33.0 \\ 16.0 & 16.0 & 45.5 & 82.0 & 46.0 \\ 7.0 & 9.5 & 33.0 & 46.0 & 84.0 \end{bmatrix}$,

(b) $\mathbf{m} = \begin{bmatrix} 52.00 & 45.56 & 19.35 & 10.07 & 4.02 \\ 63.44 & 87.00 & 46.25 & 12.48 & 6.83 \\ 33.65 & 57.75 & 83.00 & 40.41 & 27.18 \\ 21.93 & 19.52 & 50.59 & 82.00 & 42.97 \\ 9.98 & 12.17 & 38.82 & 49.03 & 84.00 \end{bmatrix}$.

(b) For (a): $G^2 = 22.53$, $\chi^2 = 22.16$. For (b) $G^2 = 0.593$, $\chi^2 = 0.594$.

8.5.5: (a) $\hat{\theta} = 170/1{,}049$, $\hat{\mathbf{m}} = \hat{\theta}(213, 147, 89, 190, 284, 126)$, (b) $\hat{\boldsymbol{\beta}} = (-5.612, 11.685)$, $\hat{\mathbf{m}} = (18.75, 26.40, 28.24, 21.74, 31.56, 43.31)$.

8.5.6: (a) $\mu_{ij} = \lambda + \lambda_i^1 + \lambda_j^2 + \beta w_{ij}$, where $\mathbf{w} = (1/4)\begin{bmatrix} 3 & 1 & -1 & -3 \\ -3 & -1 & 1 & 3 \end{bmatrix}$.

(b) $\hat{\mathbf{m}} = \begin{bmatrix} 36.83 & 46.20 & 85.11 & 131.86 \\ 62.17 & 52.80 & 65.89 & 69.14 \end{bmatrix}$, $\hat{\beta} = -0.3896$.

(d) $\hat{\mathbf{m}} = \begin{bmatrix} 54 & 54 & 82.36 & 109.64 \\ 45 & 45 & 68.64 & 91.36 \end{bmatrix}$.

8.6.1: $G^2(\mathbf{y}_2, \mathbf{y}_1) = 0.5567$, $G^2(\mathbf{y}_3, \mathbf{y}_2) = 1.9700$, $G^2(\mathbf{y}_3, \mathbf{y}_1) = 2.5267$, $\chi^2(\mathbf{y}_2, \mathbf{y}_1) = 0.5556$, $\chi^2(\mathbf{y}_3, \mathbf{y}_2) = 1.8750$, $\chi^2(\mathbf{y}_3, \mathbf{y}_1) = 2.5000$.

8.6.2: $G^2(\mathbf{y}_2, \mathbf{y}_1) = 0.8517$, $G^2(\mathbf{y}_1, \mathbf{y}_2) = 0.8435$, $\chi^2(\mathbf{y}_2, \mathbf{y}_1) = 0.8602$, $\chi^2(\mathbf{y}_1, \mathbf{y}_2) = 0.8356$, $K(\mathbf{y}_2, \mathbf{y}_1) = K(\mathbf{y}_1, \mathbf{y}_2) = 0.8476$.

8.6.3: (b) $G^2(\hat{\mathbf{m}}, \mathbf{y}) = 0.1757$, $G^2(\hat{\mathbf{m}}^*, \hat{\mathbf{m}}) = 24.80$, $G^2(\hat{\mathbf{m}}^*, \mathbf{y}) = 24.97$.

8.7.1: (a) $\beta_1 = 6.299385$, $\mathbf{m} = (544.237, 244.541, 1211.222)$, (b) $D[\hat{\boldsymbol{\beta}}] = \begin{bmatrix} 2.36 & -4.89 \\ -4.89 & 10.12 \end{bmatrix} \times 10^{-4}$, $D[\hat{\mathbf{m}}] = \begin{bmatrix} 69.99 & 96.53 & -166.53 \\ 96.53 & 113.11 & -229.63 \\ -166.53 & -229.63 & 386{,}16 \end{bmatrix}$, $D[\mathbf{p}] = (1/2{,}000^2)D[\mathbf{m}]$, (c) $\hat{\beta}_2 = 0.8095$, $\hat{\sigma}(\hat{\beta}_2) = 0.001018^{1/2}$, (0.747, 0.872), (d) 0.886, (e) $\text{Var}(\hat{\beta}_1) = 7.36 \times 10^{-4}$, other terms in $D[\hat{\boldsymbol{\beta}}]$ don't change.

8.7.2: (a) $(-0.2359, 0.3650)$, (b) $\hat{\mathbf{m}} = \begin{bmatrix} 197.23 & 302.77 & 502.77 & 797.23 \\ 302.77 & 397.23 & 297.23 & 402.77 \end{bmatrix}$, (c) $(-0.538, 0.060)$.

8.7.3: $(-0.304, 0.127)$.

8.7.4: $\mathbf{m} = \begin{bmatrix} 180 & 120 \\ 100 & 100 \end{bmatrix}$, $\mathbf{m}_2 = \begin{bmatrix} 168 & 132 \\ 112 & 88 \end{bmatrix}$, $\delta = G^2(\mathbf{m}_2, \mathbf{m}) = 4.864$, power $\doteq$ 0.597.

8.7.5: (Scheffé interval length)/(Bonferroni interval length) = 4.585/3.48, so use the Bonferroni method.

8.7.6: (c) $G^2(\hat{\mathbf{m}}, \mathbf{y}) = 7.483$ for 2 d.f., p-value $= \exp(-7.483/2) = 0.023\,7$,

(d) $\mathbf{m} = \begin{bmatrix} 5.32 & 26.19 \\ 13.68 & 31.81 \\ 15.68 & 29.81 \\ 40.32 & 36.19 \end{bmatrix}$, $\boldsymbol{\mu} = \log(\mathbf{m})$, $\lambda = 5.322$, $\lambda_1^1 = -0.385$, $\lambda_1^2 = -0.385$,

$\lambda_1^3 = -0.385$, $\lambda_{11}^{12} = -0.173$, $\lambda_{11}^{13} = -0.173$, $\lambda_{11}^{13} = -0.173$, $\lambda_{111}^{23}, -0.173$, $\lambda_{111}^{123} = 0.038\,5$, (e) $\delta = 5.93$, power $\doteq 0.579$.

8.8.1: (b) $\hat{\boldsymbol{\beta}} = (-3.504,\ 0.294\,3)'$, $\hat{\mathbf{m}} = (13.665,\ 23.357,\ 31.297,\ 41.685)'$, $G^2(\hat{\mathbf{m}}, \mathbf{y}) = 1.73$, $\chi^2 = (\hat{\mathbf{m}}, \mathbf{y}) = 1.707$, (c) $\hat{D}[\hat{\boldsymbol{\beta}}] = \begin{bmatrix} 0.040\,0 & -0.016\,1 \\ -0.016\,1 & 0.008\,4 \end{bmatrix}$, $z = 0.294\,3/0.008\,4^{1/2} = 3.13$, p-value $= 0.001\,7$, $\hat{\mathbf{m}}_2 = (110/2{,}294)\mathbf{n} = (22.44,\ 28.87,\ 29.20,\ 29.49)'$, $G^2(\hat{\mathbf{m}}_2, \hat{\mathbf{m}}) = 9.52$, p-value $= 0.002\,0$.

8.8.2: (a) $\hat{\boldsymbol{\beta}} = (-5.057,\ 3.122)'$, $D[\hat{\boldsymbol{\beta}}] = \begin{bmatrix} 0.218\,9 & -0.121\,0 \\ -0.121\,0 & 0.069\,3 \end{bmatrix}$, (b) $\hat{p}(d) = \exp(\hat{\eta}(d)/[1 + \exp(\hat{\eta}(d)]$, where $\hat{\eta}(d) = -5.057 + 3.122d$, (c) (0.974, 0.993), (d) $\hat{d}_{1/2} = 1.620$, (1.061, 2.179).

8.8.4: $\hat{\boldsymbol{\beta}} = (-5.208,\ 3.219,\ 0.397)'$, $\hat{\mathbf{p}} = (0.120, 0.406, 0.774, 0.169, 0.504, 0.836)$, $\hat{\sigma}(\hat{\beta}_j)$: 0.451, 0.274, 0.198, confidence interval is (0.008, 0.786).

8.8.5: (a) (1.695, 2.397) and (2.081, 2.835), (b) (−0.124, 1.282), (c) (Using 4**x**-vectors) we get the 95% confidence interval on (male slope) − (female slope): (−0.551, 1.174), (d) (3.104, 7.528), (3.321, 8.135).

8.8.6: (c) $\hat{\boldsymbol{\beta}} = (-0.587,\ 0.553,\ 1.141)'$, $\hat{\mathbf{p}}^* = \begin{bmatrix} \text{——} & 0.643 & 0.365 & 0.242 \\ 0.357 & \text{——} & 0.242 & 0.151 \\ 0.635 & 0.758 & \text{——} & 0.357 \\ 0.758 & 0.849 & 0.643 & \text{——} \end{bmatrix}$,

$G^2(\hat{\mathbf{m}}, \mathbf{y}) = 18.63$, (d) $\mathbf{p} = \begin{bmatrix} \text{——} & 0.750 & 0.525 & 0.401 \\ 0.525 & \text{——} & 0.401 & 0.289 \\ 0.750 & 0.832 & \text{——} & 0.525 \\ 0.832 & 0.891 & 0.750 & \text{——} \end{bmatrix}$,

$$D[\hat{\boldsymbol{\beta}}] \doteq \begin{bmatrix} 0.056\,6 & -0.006\,3 & 0.006\,1 & 0.012\,4 \\ -0.006\,3 & 0.160\,0 & 0.068\,8 & 0.066\,0 \\ 0.006\,1 & 0.068\,8 & 0.149\,5 & 0.080\,7 \\ 0.012\,4 & 0.065\,9 & 0.080\,7 & 0.171\,8 \end{bmatrix},$$

(e) $\hat{\boldsymbol{\beta}} = (0.431\,7, -0.611\,6, 0.576\,7, 1.184\,4)'$, $\hat{\mathbf{p}} = \begin{bmatrix} \text{---} & 0.739 & 0.464 & 0.319 \\ 0.455 & \text{---} & 0.319 & 0.203 \\ 0.733 & 0.835 & \text{---} & 0.455 \\ 0.835 & 0.903 & 0.739 & \text{---} \end{bmatrix}$,
$G^2(\hat{\mathbf{m}}, \mathbf{y}) = 15.22$, (f) $G^2(\hat{\mathbf{m}}^*, \hat{\mathbf{m}}) = 3.41$ (or $z = 1.82$), (g) $G^2(\mathbf{m}^*, \mathbf{m}) = 3.46$, power $\doteq 0.229$.

Author Index

Subject Index

THE WILEY SERIES IN PROBABILITY AND STATISTICS

Probability and Statistics

ANDERSON · An Introduction to Multivariate Statistical Analysis, *Second Edition*
*ANDERSON · The Statistical Analysis of Time Series
ARNOLD, BALAKRISHNAN, and NAGARAJA · A First Course in Order Statistics
BACCELLI, COHEN, OLSDER, and QUADRAT · Synchronization and Linearity: An Algebra for Discrete Event Systems
BARNETT · Comparative Statistical Inference, *Second Edition*
BERNARDO and SMITH · Bayesian Statistical Concepts and Theory
BHATTACHARYYA and JOHNSON · Statistical Concepts and Methods
BILLINGSLEY · Convergence of Probability Measures
BILLINGSLEY · Probability and Measure, *Second Edition*
BOROVKOV · Asymptotic Methods in Queuing Theory
BRANDT, FRANKEN, and LISEK · Stationary Stochastic Models
CAINES · Linear Stochastic Systems
CAIROLI and DALANG · Sequential Stochastic Optimization
CHEN · Recursive Estimation and Control for Stochastic Systems
CONSTANTINE · Combinatorial Theory and Statistical Design
COOK and WEISBERG · An Introduction to Regression Graphics
COVER and THOMAS · Elements of Information Theory
*DOOB · Stochastic Processes
DUDEWICZ and MISHRA · Modern Mathematical Statistics
ETHIER and KURTZ · Markov Processes: Characterization and Convergence
FELLER · An Introduction to Probability Theory and Its Applications, Volume I, *Third Edition*, Revised; Volume II, *Second Edition*
FULLER · Introduction to Statistical Time Series
FULLER · Measurement Error Models
GIFI · Nonlinear Multivariate Analysis
GUTTORP · Statistical Inference for Branching Processes
HALD · A History of Probability and Statistics and Their Applications before 1750
HALL · Introduction to the Theory of Coverage Processes
HANNAN and DEISTLER · The Statistical Theory of Linear Systems
HEDAYAT and SINHA · Design and Inference in Finite Population Sampling
HOEL · Introduction to Mathematical Statistics, *Fifth Edition*
HUBER · Robust Statistics
IMAN and CONOVER · A Modern Approach to Statistics
JUREK and MASON · Operator-Limit Distributions in Probability Theory
KAUFMAN and ROUSSEEUW · Finding Groups in Data: An Introduction to Cluster Analysis
LARSON · Introduction to Probability Theory and Statistical Inference, *Third Edition*
LESSLER and KALSBEEK · Nonsampling Error in Surveys
LINDVALL · Lectures on the Coupling Method
MANTON, WOODBURY, and TOLLEY · Statistical Applications Using Fuzzy Sets
MORGENTHALER and TUKEY · Configural Polysampling: A Route to Practical Robustness
MUIRHEAD · Aspects of Multivariate Statistical Theory
OLIVER and SMITH · Influence Diagrams, Belief Nets and Decision Analysis

*Now available in a lower priced paperback edition in the Wiley Classics Library.

Probability and Statistics (Continued)

*PARZEN · Modern Probability Theory and Its Applications
PILZ · Bayesian Estimation and Experimental Design in Linear Regression Models
PRESS · Bayesian Statistics: Principles, Models, and Applications
PUKELSHEIM · Optimal Experimental Design
PURI and SEN · Nonparametric Methods in General Linear Models
PURI, VILAPLANA, and WERTZ · New Perspectives in Theoretical and Applied Statistics
RENCHER · Methods of Multivariate Analysis
RAO · Asymptotic Theory of Statistical Inference
RAO · Linear Statistical Inference and Its Applications, *Second Edition*
ROBERTSON, WRIGHT, and DYKSTRA · Order Restricted Statistical Inference
ROGERS and WILLIAMS · Diffusions, Markov Processes, and Martingales, Volume II: Îto Calculus
ROHATGI · A Introduction to Probability Theory and Mathematical Statistics
ROSS · Stochastic Processes
RUBINSTEIN · Simulation and the Monte Carlo Method
RUZSA and SZEKELY · Algebraic Probability Theory
SCHEFFE · The Analysis of Variance
SEBER · Linear Regression Analysis
SEBER · Multivariate Observations
SEBER and WILD · Nonlinear Regression
SERFLING · Approximation Theorems of Mathematical Statistics
SHORACK and WELLNER · Empirical Processes with Applications to Statistics
SMALL and McLEISH · Hilbert Space Methods in Probability and Statistical Inference
STAPLETON · Linear Statistical Models
STAUDTE and SHEATHER · Robust Estimation and Testing
STOYANOV · Counterexamples in Probability
STYAN · The Collected Papers of T. W. Anderson: 1943–1985
WHITTAKER · Graphical Models in Applied Multivariate Statistics
YANG · The Construction Theory of Denumerable Markov Processes

Applied Probability and Statistics

ABRAHAM and LEDOLTER · Statistical Methods for Forecasting
AGRESTI · Analysis of Ordinal Categorical Data
AGRESTI · Categorical Data Analysis
ANDERSON and LOYNES · The Teaching of Practical Statistics
ANDERSON, AUQUIER, HAUCK, OAKES, VANDAELE, and WEISBERG · Statistical Methods for Comparative Studies
*ARTHANARI and DODGE · Mathematical Programming in Statistics
ASMUSSEN · Applied Probability and Queues
*BAILEY · The Elements of Stochastic Processes with Applications to the Natural Sciences
BARNETT · Interpreting Multivariate Data
BARNETT and LEWIS · Outliers in Statistical Data, *Second Edition*
BARTHOLOMEW, FORBES, and McLEAN · Statistical Techniques for Manpower Planning, *Second Edition*
BATES and WATTS · Nonlinear Regression Analysis and Its Applications
BELSLEY · Conditioning Diagnostics: Collinearity and Weak Data in Regression
BELSLEY, KUH, and WELSCH · Regression Diagnostics: Identifying Influential Data and Sources of Collinearity
BHAT · Elements of Applied Stochastic Processes, *Second Edition*
BHATTACHARYA and WAYMIRE · Stochastic Processes with Applications
BIEMER, GROVES, LYBERG, MATHIOWETZ, and SUDMAN · Measurement Errors in Surveys
BIRKES and DODGE · Alternative Methods of Regression
BLOOMFIELD · Fourier Analysis of Time Series: An Introduction

*Now available in a lower priced paperback edition in the Wiley Classics Library.

Applied Probability and Statistics (Continued)

BOLLEN · Structural Equations with Latent Variables
BOULEAU · Numerical Methods for Stochastic Processes
BOX · R. A. Fisher, the Life of a Scientist
BOX and DRAPER · Empirical Model-Building and Response Surfaces
BOX and DRAPER · Evolutionary Operation: A Statistical Method for Process Improvement
BOX, HUNTER, and HUNTER · Statistics for Experimenters: An Introduction to Design, Data Analysis, and Model Building
BROWN and HOLLANDER · Statistics: A Biomedical Introduction
BUCKLEW · Large Deviation Techniques in Decision, Simulation, and Estimation
BUNKE and BUNKE · Nonlinear Regression, Functional Relations and Robust Methods: Statistical Methods of Model Building
CHATTERJEE and HADI · Sensitivity Analysis in Linear Regression
CHATTERJEE and PRICE · Regression Analysis by Example, *Second Edition*
CLARKE and DISNEY · Probability and Random Processes: A First Course with Applications, *Second Edition*
COCHRAN · Sampling Techniques, *Third Edition*
*COCHRAN and COX · Experimental Designs, *Second Edition*
CONOVER · Practical Nonparametric Statistics, *Second Edition*
CONOVER and IMAN · Introduction to Modern Business Statistics
CORNELL · Experiments with Mixtures, Designs, Models, and the Analysis of Mixture Data, *Second Edition*
COX, BINDER, CHINNAPPA, CHRISTIANSON, COLLEDGE, and KOTT · Business Survey Methods
COX · A Handbook of Introductory Statistical Methods
*COX · Planning of Experiments
CRESSIE · Statistics for Spatial Data, *Revised Edition*
DANIEL · Applications of Statistics to Industrial Experimentation
DANIEL · Biostatistics: A Foundation for Analysis in the Health Sciences, *Sixth Edition*
DAVID · Order Statistics, *Second Edition*
*DEGROOT, FIENBERG, and KADANE · Statistics and the Law
*DEMING · Sample Design in Business Research
DILLON and GOLDSTEIN · Multivariate Analysis: Methods and Applications
DODGE and ROMIG · Sampling Inspection Tables, *Second Edition*
DOWDY and WEARDEN · Statistics for Research, *Second Edition*
DRAPER and SMITH · Applied Regression Analysis, *Second Edition*
DUNN · Basic Statistics: A Primer for the Biomedical Sciences, *Second Edition*
DUNN and CLARK · Applied Statistics: Analysis of Variance and Regression, *Second Edition*
ELANDT-JOHNSON and JOHNSON · Survival Models and Data Analysis
EVANS, PEACOCK, and HASTINGS · Statistical Distributions, *Second Edition*
FISHER and VAN BELLE · Biostatistics: A Methodology for the Health Sciences
FLEISS · The Design and Analysis of Clinical Experiments
FLEISS · Statistical Methods for Rates and Proportions, *Second Edition*
FLEMING and HARRINGTON · Counting Processes and Survival Analysis
FLURY · Common Principal Components and Related Multivariate Models
GALLANT · Nonlinear Statistical Models
GLASSERMAN and YAO · Monotone Structure in Discrete-Event Systems
GROSS and HARRIS · Fundamentals of Queueing Theory, *Second Edition*
GROVES · Survey Errors and Survey Costs
GROVES, BIEMER, LYBERG, MASSEY, NICHOLLS, and WAKSBERG · Telephone Survey Methodology
HAHN and MEEKER · Statistical Intervals: A Guide for Practitioners
HAND · Discrimination and Classification
*HANSEN, HURWITZ, and MADOW · Sample Survey Methods and Theory, Volume I: Methods and Applications

*Now available in a lower priced paperback edition in the Wiley Classics Library.

Applied Probability and Statistics (Continued)

*HANSEN, HURWITZ, and MADOW · Sample Survey Methods and Theory, Volume II: Theory
HEIBERGER · Computation for the Analysis of Designed Experiments
HELLER · MACSYMA for Statisticians
HINKELMAN and KEMPTHORNE: Design and Analysis of Experiments, Volume I: Introduction to Experimental Design
HOAGLIN, MOSTELLER, and TUKEY · Exploratory Approach to Analysis of Variance
HOAGLIN, MOSTELLER, and TUKEY · Exploring Data Tables, Trends and Shapes
HOAGLIN, MOSTELLER, and TUKEY · Understanding Robust and Exploratory Data Analysis
HOCHBERG and TAMHANE · Multiple Comparison Procedures
HOEL · Elementary Statistics, *Fifth Edition*
HOGG and KLUGMAN · Loss Distributions
HOLLANDER and WOLFE · Nonparametric Statistical Methods
HOSMER and LEMESHOW · Applied Logistic Regression
HØYLAND and RAUSAND · System Reliability Theory: Models and Statistical MethodsL
HUBERTY · Applied Discriminant Analysis
IMAN and CONOVER · Modern Business Statistics
JACKSON · A User's Guide to Principle Components
JOHN · Statistical Methods in Engineering and Quality Assurance
JOHNSON · Multivariate Statistical Simulation
JOHNSON and KOTZ · Distributions in Statistics
Continuous Univariate Distributions—2
Continuous Multivariate Distributions
JOHNSON, KOTZ, and BALAKRISHNAN · Continuous Univariate Distributions, Volume 1, *Second Edition*
JOHNSON, KOTZ, and BALAKRISHNAN · Continuous Univariate Distributions, Volume 2, *Second Edition*
JOHNSON, KOTZ, and KEMP · Univariate Discrete Distributions, *Second Edition*
JUDGE, GRIFFITHS, HILL, LÜTKEPOHL, and LEE · The Theory and Practice of Econometrics, *Second Edition*
JUDGE, HILL, GRIFFITHS, LÜTKEPOHL, and LEE · Introduction to the Theory and Practice of Econometrics, *Second Edition*
KALBFLEISCH and PRENTICE · The Statistical Analysis of Failure Time Data
KASPRZYK, DUNCAN, KALTON, and SINGH · Panel Surveys
KISH · Statistical Design for Research
*KISH · Survey Sampling
LANGE, RYAN, BILLARD, BRILLINGER, CONQUEST, and GREENHOUSE · Case Studies in Biometry
LAWLESS · Statistical Models and Methods for Lifetime Data
LEBART, MORINEAU, and WARWICK · Multivariate Descriptive Statistical Analysis: Correspondence Analysis and Related Techniques for Large Matrices
LEE · Statistical Methods for Survival Data Analysis, *Second Edition*
LEPAGE and BILLARD · Exploring the Limits of Bootstrap
LEVY and LEMESHOW · Sampling of Populations: Methods and Applications
LINHART and ZUCCHINI · Model Selection
LITTLE and RUBIN · Statistical Analysis with Missing Data
MAGNUS and NEUDECKER · Matrix Differential Calculus with Applications in Statistics and Econometrics
MAINDONALD · Statistical Computation
MALLOWS · Design, Data, and Analysis by Some Friends of Cuthbert Daniel
MANN, SCHAFER, and SINGPURWALLA · Methods for Statistical Analysis of Reliability and Life Data
MASON, GUNST, and HESS · Statistical Design and Analysis of Experiments with Applications to Engineering and Science
McLACHLAN · Discriminant Analysis and Statistical Pattern Recognition
MILLER · Survival Analysis

Applied Probability and Statistics (Continued)

MONTGOMERY and PECK · Introduction to Linear Regression Analysis, *Second Edition*
NELSON · Accelerated Testing, Statistical Models, Test Plans, and Data Analyses
NELSON · Applied Life Data Analysis
OCHI · Applied Probability and Stochastic Processes in Engineering and Physical Sciences
OKABE, BOOTS, and SUGIHARA · Spatial Tesselations: Concepts and Applications of Voronoi Diagrams
OSBORNE · Finite Algorithms in Optimization and Data Analysis
PANKRATZ · Forecasting with Dynamic Regression Models
PANKRATZ · Forecasting with Univariate Box-Jenkins Models: Concepts and Cases
PORT · Theoretical Probability for Applications
PUTERMAN · Markov Decision Processes: Discrete Stochastic Dynamic Programming
RACHEV · Probability Metrics and the Stability of Stochastic Models
RÉNYI · A Diary on Information Theory
RIPLEY · Spatial Statistics
RIPLEY · Stochastic Simulation
ROSS · Introduction to Probability and Statistics for Engineers and Scientists
ROUSSEEUW and LEROY · Robust Regression and Outlier Detection
RUBIN · Multiple Imputation for Nonresponse in Surveys
RYAN · Statistical Methods for Quality Improvement
SCHUSS · Theory and Applications of Stochastic Differential Equations
SCOTT · Multivariate Density Estimation: Theory, Practice, and Visualization
SEARLE · Linear Models
SEARLE · Linear Models for Unbalanced Data
SEARLE · Matrix Algebra Useful for Statistics
SEARLE, CASELLA, and McCULLOCH · Variance Components
SKINNER, HOLT, and SMITH · Analysis of Complex Surveys
STOYAN · Comparison Methods for Queues and Other Stochastic Models
STOYAN, KENDALL, and MECKE · Stochastic Geometry and Its Applications
STOYAN and STOYAN · Fractals, Random Shapes and Point Fields
THOMPSON · Empirical Model Building
THOMPSON · Sampling
TIERNEY · LISP-STAT: An Object-Oriented Environment for Statistical Computing and Dynamic Graphics
TIJMS · Stochastic Modeling and Analysis: A Computational Approach
TITTERINGTON, SMITH, and MAKOV · Statistical Analysis of Finite Mixture Distributions
UPTON and FINGLETON · Spatial Data Analysis by Example, Volume I: Point Pattern and Quantitative Data
UPTON and FINGLETON · Spatial Data Analysis by Example, Volume II: Categorical and Directional Data
VAN RIJCKEVORSEL and DE LEEUW · Component and Correspondence Analysis
WEISBERG · Applied Linear Regression, *Second Edition*
WESTFALL and YOUNG · Resampling-Based Multiple Testing: Examples and Methods for *p*-Value Adjustment
WHITTLE · Optimization Over Time: Dynamic Programming and Stochastic Control, Volume I and Volume II
WHITTLE · Systems in Stochastic Equilibrium
WONNACOTT and WONNACOTT · Econometrics, *Second Edition*
WONNACOTT and WONNACOTT · Introductory Statistics, *Fifth Edition*
WONNACOTT and WONNACOTT · Introductory Statistics for Business and Economics, *Fourth Edition*
WOODING · Planning Pharmaceutical Clinical Trials: Basic Statistical Principles
WOOLSON · Statistical Methods for the Analysis of Biomedical Data

Tracts on Probability and Statistics

BILLINGSLEY · Convergence of Probability Measures
TOUTENBURG · Prior Information in Linear Models